T0205730

Advances in Intelligent Systems and Computing

Volume 469

Series editor

Janusz Kacprzyk, Polish Academy of Sciences, Warsaw, Poland
e-mail: kacprzyk@ibspan.waw.pl

About this Series

The series "Advances in Intelligent Systems and Computing" contains publications on theory, applications, and design methods of Intelligent Systems and Intelligent Computing. Virtually all disciplines such as engineering, natural sciences, computer and information science, ICT, economics, business, e-commerce, environment, healthcare, life science are covered. The list of topics spans all the areas of modern intelligent systems and computing.

The publications within "Advances in Intelligent Systems and Computing" are primarily textbooks and proceedings of important conferences, symposia and congresses. They cover significant recent developments in the field, both of a foundational and applicable character. An important characteristic feature of the series is the short publication time and world-wide distribution. This permits a rapid and broad dissemination of research results.

More information about this series at http://www.springer.com/series/11156

Suresh Chandra Satapathy · Vikrant Bhateja
Amit Joshi
Editors

Proceedings of the International Conference on Data Engineering and Communication Technology

ICDECT 2016, Volume 2

Editors
Suresh Chandra Satapathy
Department of CSE
Anil Neerukonda Institute of Technology and Sciences
Visakhapatnam, Andhra Pradesh
India

Amit Joshi
Sabar Institute of Technology
Tajpur, Sabarkantha, Gujarat
India

Vikrant Bhateja
Shri Ramswaroop Memorial Group of Professional Colleges (SRMGPC)
Lucknow, Uttar Pradesh
India

ISSN 2194-5357 ISSN 2194-5365 (electronic)
Advances in Intelligent Systems and Computing
ISBN 978-981-10-1677-6 ISBN 978-981-10-1678-3 (eBook)
DOI 10.1007/978-981-10-1678-3

Library of Congress Control Number: 2016944918

Printed on acid-free paper

This Springer imprint is published by Springer Nature
The registered company is Springer Science+Business Media Singapore Pte Ltd.

Preface

The First International Conference on Data Engineering and Communication Technology (ICDECT 2016) was successfully organized by Aspire Research Foundation, Pune during March 10–11, 2016 at Lavasa City, Pune. The conference has technical collaboration with Div-V (Education and Research) of Computer Society of India. The objective of this international conference was to provide opportunities for the researchers, academicians, industry persons, and students to interact and exchange ideas, experience and gain expertise in the current trends and strategies for information and intelligent techniques. Research submissions in various advanced technology areas were received and after a rigorous peer-review process with the help of program committee members and external reviewer, 160 papers in separate two volumes (Vol-I: 80, Vol-II: 80) were accepted. All the papers are published in Springer AISC series. The conference featured seven special sessions on various cutting-edge technologies which were conducted by eminent professors. Many distinguished personalities like Dr. Ashok Deshpande, Founding Chair: Berkeley Initiative in Soft Computing (BISC)—UC Berkeley CA; Guest Faculty: University of California Berkeley; Visiting Professor: New South Wales University, Canberra and Indian Institute of Technology Bombay, Mumbai, India; Dr. Parag Kulkarni, Pune; Prof. Amit Joshi, Sabar Institute, Gujarat; Dr. Swagatam Das, ISI, Kolkata, graced the event and delivered talks on cutting-edge technologies.

Our sincere thanks to all Special Session Chairs (Dr. Vinayak K. Bairagi, Prof. Hardeep Singh, Dr. Divakar Yadav, Dr. V. Suma), Track Manager (Prof. Steven Lawrence Fernandes) and distinguished reviewers for their timely technical support. Thanks are due to ASP and its dynamic team members for organizing the event in a smooth manner. We are indebted to Christ Institute of Management for hosting the conference in their campus. Our entire organizing committee, staff of CIM, student volunteers deserve a big pat for their tireless efforts to make the event a grand success. Special thanks to our Program Chairs for carrying out an immaculate job. We would like to extend our special thanks here to our publication chairs doing a great job in making the conference widely visible.

Lastly, our heartfelt thanks to all authors without whom the conference would never have happened. Their technical contributions to make our proceedings rich are praiseworthy. We sincerely expect readers will find the chapters very useful and interesting.

Visakhapatnam, India — Suresh Chandra Satapathy
Lucknow, India — Vikrant Bhateja
Tajpur, India — Amit Joshi

Organizing Committee

Honorary Chair

Prof. Sanjeevi Kumar Padmanaban

Organizing Committee

Mr. Satish Jawale
Mr. Abhisehek Dhawan
Mr. Ganesh Khedkar
Mayura Kumbhar

Program Committee

Prof. Hemanth Kumbhar
Prof. Suresh Vishnudas Limkar

Publication Chair

Prof. Vikrant Bhateja, SRMGPC, Lucknow

Publication Co-Chair

Mr. Amit Joshi, CSI Udaipur Chapter

Technical Review Committee

Le Hoang Son, Vietnam National University, Hanoi, Vietnam
Nikhil Bhargava, CSI ADM, Ericsson India
Kamble Vaibhav Venkatrao, P.E.S. Polytechnic, India
Arvind Pandey, MMMUT, Gorakhpur (U.P.), India
Dac-Nhuong Le, VNU University, Hanoi, Vietnam
Fernando Bobillo Ortega, University of Zaragoza, Spain
Chirag Arora, KIET, Ghaziabad (U.P.), India
Vimal Mishra, MMMUT, Gorakhpur (U.P.), India
Steven Lawrence Fernandes, Sahyadri College of Engineering and Management
P.B. Mane, Savitribai Phule Pune University, Pune, India
Rashmi Agarwal, Manav Rachna International University, Faridabad, India
Kamal Kumar, University of Petroleum and Energy Studies, Dehradun
Hai V. Pham, Hanoi University of Science and Technology, Vietnam
S.G. Charan, Alcatel-Lucent India Limited, Bangalore
Frede Blaabjerg, Aalborg University, Denmark
Deepika Garg, Amity University, Haryana, India
Bharat Gaikawad, Vivekanand College campus, Aurangabad, India
Parama Bagchi, MCKV Institute of Engineering, Kolkata, India
Rajiv Srivastava, Scholar tech education, India
Vinayak K. Bairagi, AISSMS Institute of Information Technology, Pune, India
Rakesh Kumar Jha, Shri Mata Vaishnodevi University, Katra, India
Sergio Valcarcel, Technical University of Madrid, Spain
Pramod Kumar Jha, Centre for Advanced Systems (CAS), DRDO, India
Chung Le, Duytan University, Da Nang, Vietnam
V. Suma, Dayananda Sagar College of Engineering, Bangalore, India
Usha Batra, ITM University, Gurgaon, India
Sourav De, University Institute of Technology, Burdwan, India
Ankur Singh Bist, KIET, Ghaziabad, India
Agnieszka Boltuc, University of Bialystok, Poland.
Anita Kumari, Lovely Professional University, Jalandhar, India
M.P. Vasudha, Jain University Bangalore, India
Saurabh Maheshwari, Government Women Engineering College, Ajmer, India
Dhruba Ghosh, Amity University, Noida, India
Sumit Soman, C-DAC, Noida, India
Ramakrishna Murthy, GMR Institute of Technology, A.P., India
Ramesh Nayak, Shree Devi Institute of Technology, Mangalore, India

Contents

About the Editors

Dr. Suresh Chandra Satapathy is currently working as Professor and Head, Department of Computer Science and Engineering at Anil Neerukonda Institute of Technology and Sciences (ANITS), Andhra Pradesh, India. He obtained his Ph.D. in Computer Science and Engineering from JNTU Hyderabad and M.Tech. in CSE from NIT, Rourkela, Odisha, India. He has 26 years of teaching experience. His research interests include data mining, machine intelligence, and swarm intelligence. He has acted as program chair of many international conferences and has edited six volumes of proceedings from Springer LNCS and AISC series. He is currently guiding eight scholars for Ph.D. Dr. Satapathy is also a Senior Member of IEEE.

Prof. Vikrant Bhateja is Professor, Department of Electronics and Communication Engineering, Shri Ramswaroop Memorial Group of Professional Colleges (SRMGPC), Lucknow and also the Head (Academics and Quality Control) in the same college. His areas of research include digital image and video processing, computer vision, medical imaging, machine learning, pattern analysis and recognition, neural networks, soft computing, and bio-inspired computing techniques. He has more than 90 quality publications in various international journals and conference proceedings. Professor Bhateja has been on TPC and chaired various sessions from the above domain in international conferences of IEEE and Springer. He has been the track chair and served in the core-technical/editorial teams for the following international conferences: FICTA 2014, CSI 2014 and INDIA 2015 under Springer-ASIC Series and INDIACom-2015, ICACCI-2015 under IEEE. He is associate editor in International Journal of Convergence Computing (IJConvC) and also serving in the editorial board of International Journal of Image Mining (IJIM) under Inderscience Publishers. At present he is guest editor for two special issues received in International Journal of Rough Sets and Data Analysis (IJRSDA) and International Journal of System Dynamics Applications (IJSDA) under IGI Global publications.

Mr. Amit Joshi has an experience of around 6 years in academic and industry in prestigious organizations of Rajasthan and Gujarat. Currently, he is working as Assistant Professor in Department of Information Technology at Sabar Institute in Gujarat. He is an active member of ACM, CSI, AMIE, IEEE, IACSIT-Singapore, IDES, ACEEE, NPA, and many other professional societies. Currently, he is Honorary Secretary of CSI Udaipur Chapter and Honorary Secretary for ACM Udaipur Chapter. He has presented and published more than 40 papers in National and International Journals/Conferences of IEEE, Springer, and ACM. He has also edited three books on diversified subjects including Advances in Open Source Mobile Technologies, ICT for Integrated Rural Development, and ICT for Competitive Strategies. He has also organized more than 25 national and international conferences and workshops including International Conference ETNCC 2011 at Udaipur through IEEE, International Conference ICTCS 2014 at Udaipur through ACM, International Conference ICT4SD 2015 by Springer. He has also served on organizing and program committees of more than 50 conferences/seminars/workshops throughout the world and presented six invited talks in various conferences. For his contribution towards the society, The Institution of Engineers (India), ULC, has given him Appreciation Award on the Celebration of Engineers, 2014 and by SIG-WNs Computer Society of India on ACCE, 2012.

Experimental Analysis on Big Data in IOT-Based Architecture

Anupam Bera, Anirban Kundu, Nivedita Ray De Sarkar and De Mou

Abstract In this paper, we are going to discuss about big data processing for Internet of Things (IOT) based data. Our system extracts information within specified time frame. Data tracker or interface tracks information directly from big data sources. Data tracker transfers data clusters to data controller. Data controller processes each data cluster and makes them smaller after removing possible redundancies. Big data processing is a challenge to maintain data privacy-related protections. Data controller processes big data clusters and sends them through secure and/or hidden channels maintaining data privacy.

Keywords Big data (BD) · Distributed file system (DFS) · Hadoop · Distributed network (DN) · Internet of things (IOT)

1 Introduction

Internet is a great source for information sharing through hypertext documents, applications on World Wide Web (WWW), electronic mail, telephony and peer-to-peer networks. Internet is the framework of cooking massive information. Nowadays, a large number of smart devices are connected to Internet, and con-

A. Bera (✉) · A. Kundu · N.R. De Sarkar · De Mou
Netaji Subhash Engineering College, Kolkata 700152, India
e-mail: drive.abera@gmail.com

A. Kundu
e-mail: anik76in@gmail.com

N.R. De Sarkar
e-mail: nivedita.ray2009@gmail.com

De Mou
e-mail: mou.latu@gmail.com

A. Bera · A. Kundu · N.R. De Sarkar · De Mou
Computer Innovative Research Society, Howrah 711103, West Bengal, India

S.C. Satapathy et al. (eds.), *Proceedings of the International Conference on Data Engineering and Communication Technology*, Advances in Intelligent Systems and Computing 469, DOI 10.1007/978-981-10-1678-3_1

tinuously transmit massive amount of heterogeneous data. It is a challenging job of processing such huge amount of data in real time. Data should be processed in distributed environment in real time. Data could be stored in offline. Size of data is too large and distributed, that it requires use of BD analytical tools for processing [1]. There is number of different application domains widely used for data streaming, storing and processing. These application domains are difficult to maintain due to size of data and nature of network [2, 3]. User could interact with databases to modify, insert, delete, manage information using modern database software or database management system (DBMS) [4]. BD database activities are included in capture, data noise removal, search, sharing, storage, transfer, visualization, and information privacy [5].

2 Related Works

BD is a collection of large, multidimensional, heterogeneous datasets that becomes difficult to manage on typical database management systems. BD refers to data bundles whose volumes are beyond the scope of traditional database software tools [5, 6]. It is a useful tool to use predictive analytics or certain advanced methodology to extract value from data [7]. Typical database management systems failed to provide storage volumes and efficient data processing due to tremendous growth and massive data volume [5]. In 2012, a researcher "Gartner" has described BD as high volume, high velocity and/or high variety information assets which require new forms of processing for enhanced decision-making, insight discovery, and process optimization [8]. In Hadoop [9] mechanism, JobTrackers [10] have processed a block of data (δx) from distributed storage followed by analysis of data blocks to generate final results for storing into a data file. Hadoop has executed MapReduce algorithms for data processing in parallel fashion on distinct CPU nodes. Hadoop works as follows: (i) Place job to Hadoop for required process; (ii) Hadoop job client submits job to JobTracker; (iii) JobTracker executes task as per MapReduce implementation; (iv) JobTracker stores data into data file.

3 Proposed Work

Our proposed model has two segments as follows: (i) one side actively receives all signals from sensors, and forward it to BD database servers; (ii) other side serves request from enterprise personnel for data analysis.

System receives query from end-user and system coverts them to smaller size. After conversion of queries, system interface controller has plugged the information with minimum loaded interface to fetch data from BD server. Fetched records are subdivided into multiple data cubes, and further received at controller side for data analysis. Data analyzer is used to extract all necessary data from data cubes using two levels of data analyses. Finally, data cubes are received at user end through secure data path.

Proposed architecture is shown in Fig. 1. This model has four sections as follows: (i) dynamic query generator; (ii) data retrieval through multiple interface; (iii) two levels of data analyzer; (iv) data transmission through secure channels.

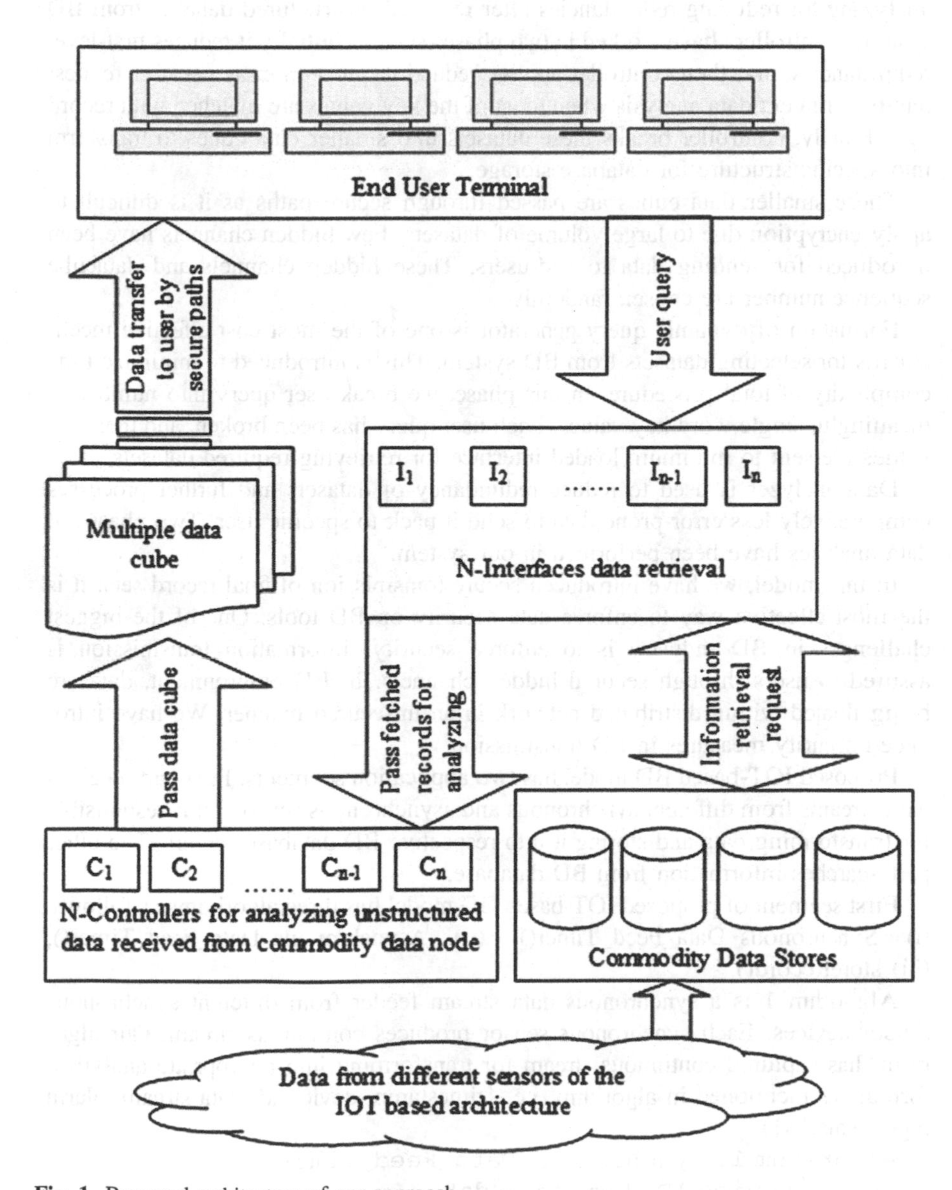

Fig. 1 Proposed architecture of our approach

Dynamic query generator has received user's key string as input. System has splitted the string into multiple single word key values. These single word key values are being passed to minimum loaded interface. Interfaces are linked with backend BD database server, which are large-scale distributed datasets, distributed file system, e.g. NoSQL database. Structured data as well as unstructured data are available in BD system. Data retrieval time using load tracker has been reduced.

Controllers have received large datasets from interface. Controller has started analyzing for reducing redundancies after received unstructured datasets from BD system. Controllers have worked in two phases such as initially it reduces first-level redundancies, and then controller applies reduction procedure, as per user request having stringent data analysis when most of the key values are matched with record sets. Finally, controller breaks these datasets into smaller data cubes to transform into specific structure for database storage.

These smaller data cubes are passed through secure paths as it is difficult to apply encryption due to large volume of datasets. Few hidden channels have been introduced for sending data to end-users. These hidden channels and datacube sequence number are chosen randomly.

Formation of dynamic query generator is one of the most cost effective mechanisms for selecting datasets from BD system. This is introduced to minimize time complexity of total procedure. In this phase, we break user query into number of meaningful single word key values. Each user query has been broken, and then key values are sent to minimum loaded interface for retrieving required datasets.

Data analyzer is used to reduce redundancy of datasets and further processes comparatively less error-prone data to send it back to specific user. Two phases of data analyses have been performed in our system.

In this model, we have introduced secure transmission of final record sets. It is the most effective way to enforce data security on BD tools. One of the biggest challenges in BD industry is to enforce security. Information transmission is assured to users through secured hidden channels. In BD environment, data are being floated within distributed network in an unsecured manner. We have introduced security measures in BD transmission.

Proposed IOT-based BD model has two application segments. First part receives data streams from different synchronous and asynchronous sensors. It is responsible for transforming data and storing it into respective BD database servers. The other part searches information from BD database.

First segment of proposed IOT-based BD model has three algorithms as follows: (i) Synchronous_Data_Feed_Timer(); (ii) Asynchronous_Data_Feed_Timer(); (iii) storeRecord().

Algorithm 1 is a synchronous data stream feeder from different synchronous sensor devices. Each synchronous sensor produces continuous stream. Our algorithm has captured continuous stream for transforming into appropriate analytical formats as mentioned in algorithm (i.e. {timestamp, device id, data stream, alarm type, remarks}).

```
Algorithm 1: Synchronous_Data_Feed_Timer ()
Input: Device ID, Real time data
```

```
  Output: Time Stamp, Device ID, Data, Alarm type
Step 1:Read devID, data
Step 2:l_thr : = call Find_Lower_Thresold (devID)
       /*to get lower threshold of the said device*/
Step 3:h_thr : = call Find_Higher_Thresold (devID)
       /*to get higher threshold of the said device*/
Step 4:if l_thr > data; then alarmType = low
       /*data to be checked for alarm selection*/
Step 5:else if h_thr < data; then alarmType = high
Step 6:else alarmType = withinRange
Step 7:Call storeRecord ({timestamp, devID, data, alarm-
Type, remarks})
Step 8:Stop
```

Algorithm 2 scans and collects data stream from asynchronous devices. It is executed periodically, either when sensor queue is full or after specified time intervals. Finally, it passes all received information for processing by data server.

```
  Algorithm 2: Asynchronous_Data_Feed_Timer ()
  Input: Device ID, data
  Output: Time Stamp, Device ID, Data, Alarm type.
Step 1:if (data < threshold or data > threshold);
       then alarmType: = OFR/*OFR-Out Of Range*/
Step 2:else alarmType: = WIR/*WIR-With In Range*/
Step 3:Call storeRecord ({timestamp, devID, data, alarm-
Type, remarks})
Step 4:Stop
```

Algorithm 3 receives information packs from Algorithm 1 or Algorithm 2, then converts all data items into single data string. This string is stored in a data file. Database writer periodically updates string information to database server.

```
  Algorithm 3: storeRecord ()
  Input: Record set
  Output: Output file
Step 1:For each data: record set
Step 2:Call concate (recordString, dataItem)
Step 3:printfile recordString #file
Step 4:Stop
```

Second segment of our proposed IOT-based BD model has three algorithms as follows: (i) Fetch_record(); (ii) First_Level_Query_Analysis(); (iii) Second_Level_Query_Analysis().

Algorithm 4 acts as dynamic query generator and raw data retriever from BD system. System finds raw data (unstructured data) from distributed datasets in this phase of processing. It receives query string as input and generates output extracting large datasets from BD database.

```
Algorithm 4: Fetch_Record()
Input: Query, level of analysis
Output: Interface id, key value

Step 1:Read query_string, query_level
Step 2:key_values[]: = split_Query_String()
Step 3:For each key_values : key_val
Step 4:interfaceID : = call minimum_Loaded_Interface()
Step 5:record_set[]: = call Find_Record (key_val, interfaceID)
Step 6:End For loop
Step 7:Return {record_set, query_stringm query_level}
Step 8:Stop
```

Algorithm 5 works for first-level data redundancy minimization. It takes large datasets as input from Fetch_Record algorithm. Based on user's query level, either data is transferred for further analysis to BD analyzer, or, transferred for second level reduction.

```
Algorithm 5: First_Level_Query_Analysis ()
Input: Record sets, query string, query level
Output: Less redundant data

Step 1:For each record_set: record
Step 2:if record (common); then delete record
Step 3:End For
Step 4:if query_level : = high then
Step 5:Call Second_Level_Query_Analysis(record_set, query_
string)
Step 6:Call Big_Data_Ananlyzer(record_set)
Step 7:Stop
```

Algorithm 6 exhibits final-level reduction before calling BD analyzer. It receives data from Algorithm 5 for further reduction.

```
Algorithm 6: Second_Level_Query_Analysis ()
Input: record_set, query_string
Output: reduced datasets

Step 1:len: = no_of_keyvalues (query_string)
Step 2:if count (key_val, record_set) < len/2; then delete
record
```

```
Step 3:Call Big_Data_Analyzer (record_set)
Step 4:Stop
```

4 Experimental Results

Following diagrams are exhibited to understand behaviour of our system framework in real time. It also shows that our system has a cost effective and time efficient structure.

In Fig. 2, CPU load has been shown with good performance for executing proposed framework. CPU load is less than 50 % in most of the time. It is an important part to identify how our model works.

Figure 3 represents real-time performance of proposed BD model with an effect on CPU temperature around 45 °C. Most of the time CPU temperature is near to 40 °C.

Figure 4 shows a graph of memory load with respect to time. Primary memory occupancy is quite high due to large raw record sets. It is around 90 % of total memory available. Less temporary space at run time is occupied by our model.

Our system works in two phases, such as data tracker and controller.

In Phase 1, there is '*I*' number of data trackers responsible for data retrieval and each successive request has '*K*' number of key-value pairs. So, we have drawn time equations of data tracker algorithms as follows:

$$T(K) = O(\log_I K), \tag{1}$$

where I = number of data trackers; K = number of key-value pairs;

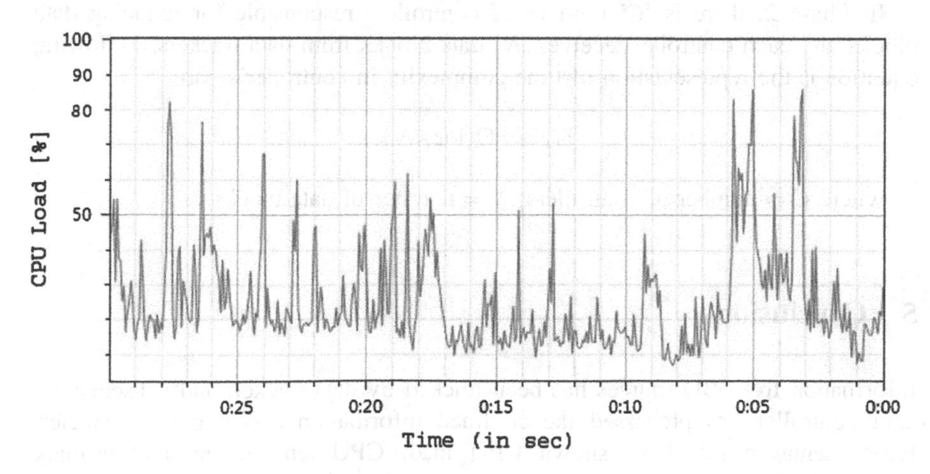

Fig. 2 CPU load distribution in real time

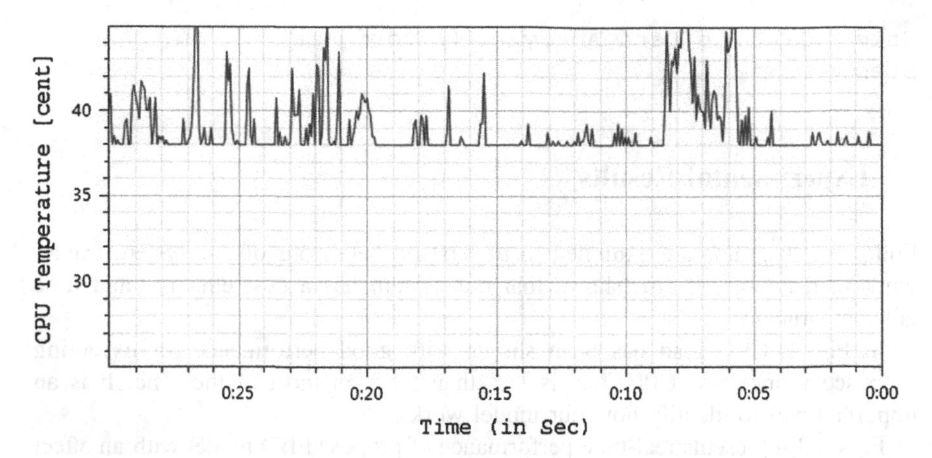

Fig. 3 CPU temperature performance with respect to time

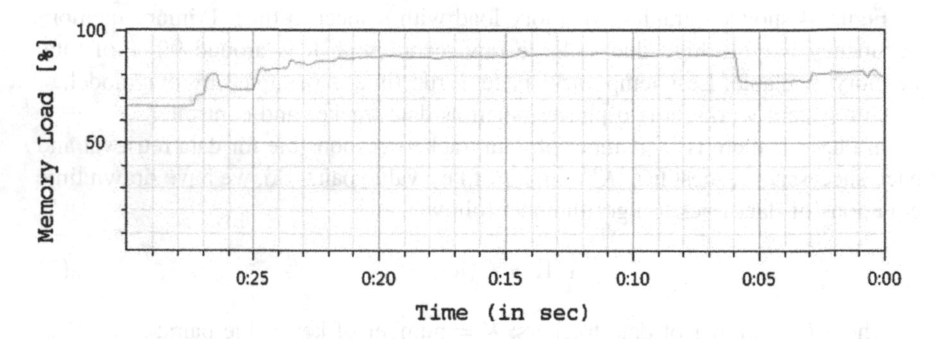

Fig. 4 Primary memory load versus time

In Phase 2, there is '*C*' number of controllers responsible for reducing data blocks and each controller receives '*N*' data blocks from data trackers. Following equation is the representation of time complexity in controller's part:

$$\mathrm{T(N)} = \mathrm{O}(\log_C N), \tag{2}$$

where C = number of controllers; N = number of data blocks;

5 Conclusion

Information from BD sources has been tracked by data tracker, and subsequently data controller has processed the confined information avoiding redundancies. Experimental results have shown CPU load, CPU temperature, and primary memory load in real time to exhibit characteristics of our proposed system. Cost analysis of our system framework is depicted with an aim of achieving better efficiency.

References

1. Charu C. Aggarwal, Naveen Ashish, Amit Sheth, "The Internet of Things: A Survey from the Data-Centric Perspective," Managing and Mining Sensor Data, Springer, 2012, pp. 383–428.
2. D. Nukarapu, B. Tang, L. Wang, S. Lu, "Data replication in data intensive scientific applications with performance guarantee," Parallel and Distributed Systems, IEEE Transactions, 2011, pp. 1299–1306.
3. Chi-Jen Wu, Chin-Fu Ku, Jan-Ming Ho, "A Novel Approach for Efficient Big Data Broadcasting," Knowledge and Data Engineering, IEEE Transactions, 2014, IIS Technical Report-12–006.
4. P. Beynon-Davies, "Database Systems," Palgrave Macmillan, 2004, ISBN 1-4039-1601-2.
5. Sugam Sharma, Udoyara S Tim, Johnny Wong, Shashi Gadia, Subhash Sharma, "A Brief Review on Leading Big Data Models," Data Science Journal, 2014, Vol. 13.
6. M. H. Padgavankar, S. R. Gupta, "Big Data Storage and Challenges," International Journal of Computer Science and Information Technologies, Vol. 5, No. 2, 2014, pp. 2218–2223.
7. Chris Snijders, Uwe Matzat, Ulf-Dietrich Reips, ""Big Data": Big gaps of knowledge in the field of Internet," International Journal of Internet Science, 2012, Vol. 7, No. 1, pp. 1–5.
8. Douglas and Laney, "The importance of 'Big Data': A definition", 2008.
9. "Extract, Transform, and Load Big Data with Apache Hadoop," Intel, Big Data Analytics, White Paper, 2013.
10. "Comparing the Hadoop Distributed File System (HDFS) with the Cassandra File System (CFS)," Datastax Corporation, 2013.

Morphology Based Approach for Number Plate Extraction

Chetan Pardeshi and Priti Rege

Abstract Number Plate Recognition identifies vehicle number without human intervention. It is a computer vision application and it has many important applications. The proposed system consists of two parts: number plate area extraction and character identification. In this paper, morphological operation-based approach is presented for number plate area extraction. Effective segmentation of characters is done after plate area extraction. Histogram-based character segmentation is a simple and efficient technique used for segmentation. Template matching approach is used for character extraction. Number plate with variable character length poses limitation on number identification in earlier reported literature. This is taken care of using histogram-based character segmentation method.

Keywords Histogram • Morphological operations • Number plate extraction • Template matching • Thresholding

1 Introduction

Every country has specific vehicle identification system. These systems are used in the traffic control and surveillance systems, security systems, toll collection at toll plaza and parking assistance system, etc. Human eye can easily recognize these number plates, but designing automated system for this task has many challenges. Blur, unequal illumination, background and foreground color and also many natural phenomena like rain fall, dust in air may create problem in number extraction. Also number plate standards are different for each country, therefore large number of variations are obtained in parameters like, location of number plate, area of number plate and characters, font and size used for numbers and characters (standard font is

Chetan Pardeshi (✉) · Priti Rege
Department of Electronics and Telecommunication, College of Engineering Pune, Pune, India
e-mail: chetan7pardeshi@gmail.com

Priti Rege
e-mail: ppr.extc@coep.ac.in

S.C. Satapathy et al. (eds.), *Proceedings of the International Conference on Data Engineering and Communication Technology*, Advances in Intelligent Systems and Computing 469, DOI 10.1007/978-981-10-1678-3_2

Arial Black), background color (white, yellow or black) and foreground color (black or red), etc., which make the task of number plate extraction difficult.

Number of applications of license plate identification can be listed as parking assistance facility during ticket collection, unmanned toll collection at toll booths, traffic surveillance system, tracking vehicles during signal violation, vehicle's marketing research.

Aim of this paper is to implement an efficient method for number plate extraction. Algorithm proposed in this paper identifies characters present on single line number plate with variable character length.

Rege and Chandrakar [1] has separated text image in document images using run length searing algorithm and boundary perimeter detection. Morphological operation and bounding box analysis are used by Patel et al. [2]. Owamoyo et al. [3] used Sobel filter along with morphological operations. Algorithm presented by Gilly and Raimond [4] stresses on connected component analysis. Bulugu [5] used edge finding method to locate the plate in the scene. Kate [6] proposed morphological operation based on area for searching number plate. Kolour [7] has reviewed a number of license plate detection algorithms and compared their performances in his paper. His experimentation gives a basis for selection of the most appropriate technique for their applications. Parasuraman and Kumar [8] extracted the plate region using edge detection followed by vertical projection. Proposed algorithm has four stages: (i) Acquisition of vehicle image and preprocessing includes conversion of image to gray format, resizing of image, etc. (ii) Marking of area covering the number plate in the vehicle image, (iii) Segmentation of characters from the number plate extracted, and (iv) Recognizing and displaying the segmented characters.

2 Proposed Method for Identification of Letters/Numbers from License Plates

This section elucidates the number plate extraction method for single line number plate. Input to the system is a vehicle image (with clear view of number plate in it) which is captured by digital camera and output is the actual characters present in that vehicle image. Each character present on input number plate image should at least have minimum resolution of 24×42 pixels and distance of number plate from camera should be such that it guarantees clear view of numbers present on the number plate.

Proposed algorithm consists of following steps:

- Image Preprocessing
- Number plate area extraction
- Segmentation of each character area in image
- Image matching for each character
- Output extracted characters in text format

2.1 *Image Preprocessing*

Preprocessing is used to enhance the contrast of the image and for resizing of the image. RGB image is converted to grayscale image which carries intensity information. RGB values are converted to grayscale values by forming a weighted sum of the R, G, and B components:

$$\text{Gray} = 0.2989 * R + 0.587 * G + 0.114 * B$$

2.1.1 Contrast Enhancement

Captured image may have unevenly distributed lighting or darkness. During edge detection fine edge details in dark region of the image are eliminated. Also feature edges in bright regions need to be preserved. Top-hat transformation is used to preserve these edge details as well as prominent ones. The main property of top-hat operator can be applied to contrast enhancement. After applying top-hat filtering, image is converted to binary image from gray scale using Otsu's algorithm. Figure 1a–d shows various stages of image preprocessing.

2.2 *Number Plate Area Detection*

System's speed and accuracy is enhanced using precise number plate area extraction. At this stage, number plate area is extracted from entire preprocessed image. This step reduces processing burden on next stage of identification of numbers from license plate area.

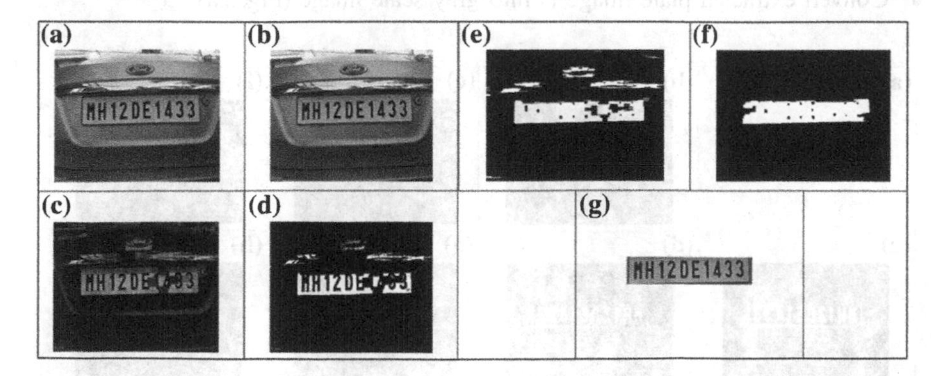

Fig. 1 **a** Original image, **b** Gray image, **c** Top-hat filtered image, **d** Binary image, **e** Dilated image, **f** Selected object, **g** Extracted plate

2.2.1 Plate Area Detection

A morphological operator is employed to the preprocessed image for extracting the plate area. Morphological operator that suits the rectangular shape of number plate is used for this purpose. Binary image is dilated with rectangular box as structuring element.

After dilation operation, horizontal object with Aspect Ratio(R) > 3 is filtered out from dilated image. As number plate generally have larger lengths as compared to its width. Aspect Ratio = width/height and R defines the region of interest.

Upon detection of plate area, top left and bottom right coordinates are extracted and further these coefficients are used for selecting plate area from original image. Figure 1e–g shows various stages involved in plate area extraction. Figure 1e shows dilated image. Filtered region (selected object) from unwanted region is shown in Fig. 1f. Figure 1g shows extracted plate area.

2.3 Character Segmentation

In this step, number plate is segmented to obtain characters present in it.

Morphological operations are used in segmentation process. Dilation of an image I by the structure element H is given by the set operation

$$I \oplus H = \{(p+q) \mid p \in I, q \in H\}$$

Erosion of an image I by the structure element H is given by the set operation

$$I \Theta H = \{(p \in Z^2) \mid (p+q) \in I, \text{for every } q \in H\}$$

Algorithm for segmentation process is as follows:

- Convert extracted plate image is into gray scale image (Fig. 2a).

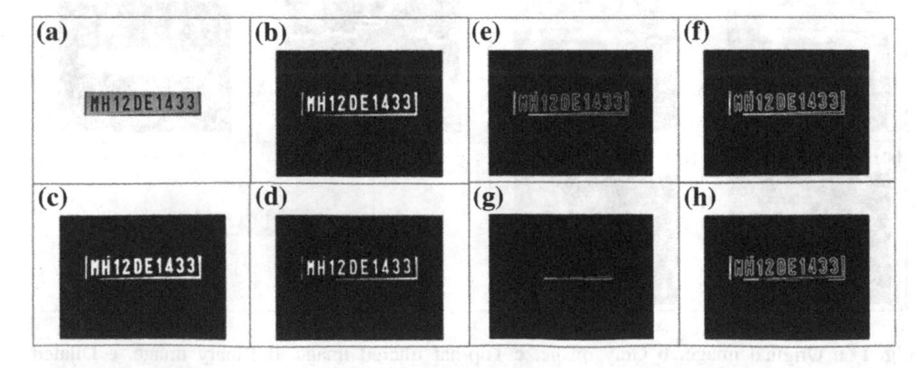

Fig. 2 **a** Gray image, **b** Inverted binary image, **c** Dilated image, **d** Eroded image, **e** Subtracted image, **f** Convolved Image, **g** Erosion with horizontal structuring element, **h** Subtracted after erosion

- Binarized image is using Ostu's algorithm. Invert pixel values of binarized image, i.e., make 0 to 1 and 1 to 0 for further processing. Result of binarization is shown in Fig. 2b.
- Apply morphological gradient for edges enhancement. Figure 2c shows dilated image with disk as structuring element. Figure 2d shows results of erosion on binary image Fig. 2b, e shows result of image subtraction of eroded image from dilated image.
- Convolve subtracted image with [1 1; 1 1] for brightening the edges (Fig. 2f).
- Erode with horizontal line as structuring element to eliminate the possible horizontal lines from the output image after subtraction operation (Fig. 2g).
- Subtract eroded image from convolved image (Fig. 2h).
- Fill all the regions of the image (Fig. 3a).
- Do thinning on the image to ensure character isolation (Fig. 3b).
- Calculated properties of connected components, i.e., objects present in image using 4 and 8 neighborhood labeling algorithms. Following properties of each object are calculated, (i) [x y] which specifies the upper left corner of the object. (ii) [x_width y_width] which specifies the width of the object along each dimension.
- Calculate histogram of all objects based on y-dimensions and y-width. Find intersection of number of objects in histogram based on y-dimensions and histogram based on y-width, to know exact number of characters present in image. Use 4, 8 neighborhood connectivity algorithms to find out bounding box for individual objects in image. Using result of histogram analysis, only those objects which have been identified in intersection of histogram are proposed for further processing. The result of object segmentation is as shown in Fig. 3c.

Results for two more samples are shown in Figs. 4 and 5.

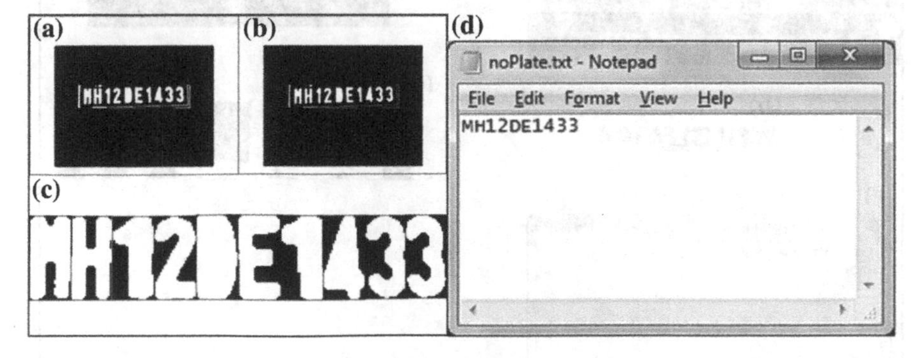

Fig. 3 **a** Holes filled, **b** Thinning, **c** Segmented characters, **d** Extracted characters stored in text file

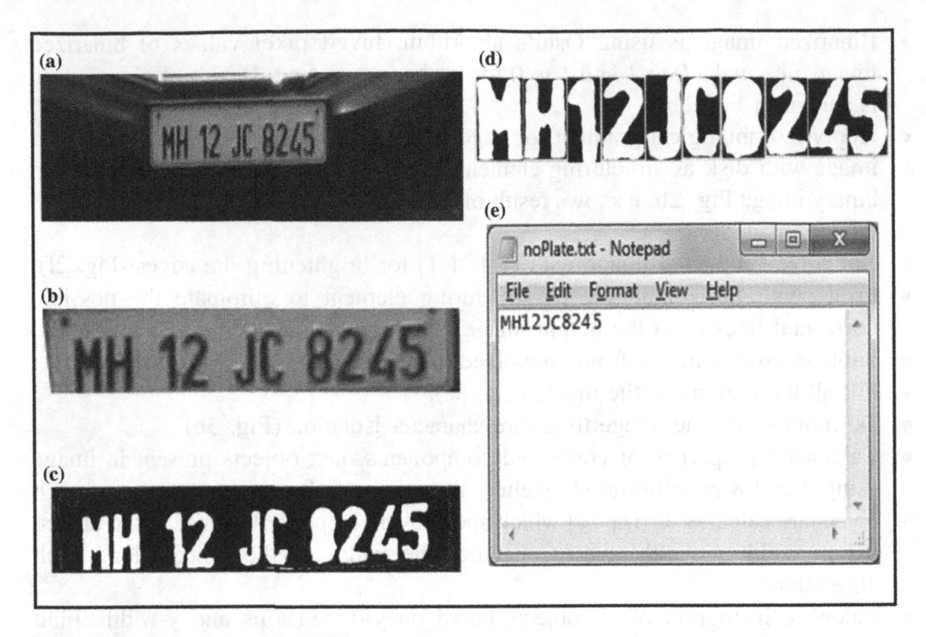

Fig. 4 **a** Sample image Fig. 2, **b** Extracted plate, **c** Processed image, **d** Extracted letters, **e** Extracted characters in text file

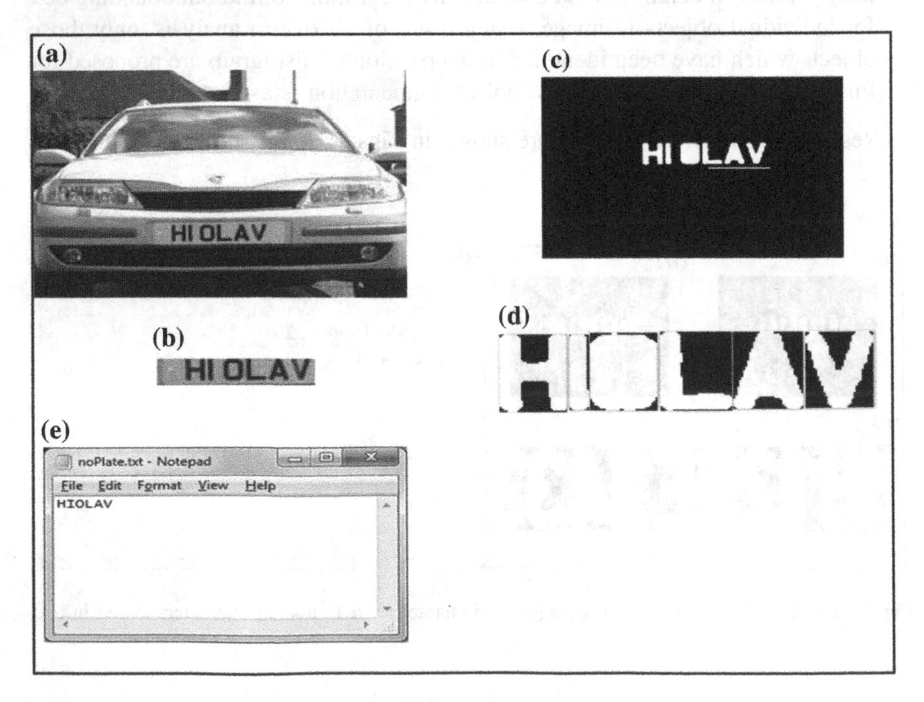

Fig. 5 **a** Sample image Fig. 3, **b** Extracted plate, **c** Processed image, **d** Extracted letters, **e** Extracted characters in text file

2.4 Image Matching

Each segmented character image is then compared against database image (database is set of few samples of character images of different fonts and different style of holes fillings in them) and correlation between test character image and database images is found.

Correlation coefficient between image A and image B can be calculated using

$$r = \frac{\sum_{m}\sum_{n}(A_{mn} - \overline{A})(B_{mn} - \overline{B})}{\sqrt{\left(\sum_{m}\sum_{n}(A_{mn} - \overline{A})^2\right)\left(\sum_{m}\sum_{n}(B_{mn} - \overline{B})^2\right)}},$$

where $\bar{A} = \text{mean}(A)$, and $\bar{B} = \text{mean}(B)$.

The database image for which maximum correlation is obtained is the identified character. This identified character is then stored in text file as shown in Fig. 3(d).

3 Conclusion and Future Works

Algorithm proposed in this paper identifies characters present on single line number plate with variable character length. In future, the extraction of number plate can be integrated with video-based surveillance system for automatic detection of various objects from video sequences.

References

1. Priti P Rege, Chanchal A Chandrakar.: Text-Image Separation in Document Images using Boundary/Perimeter Detection. In: ACEEE International Journal of Signal & Image Processing, 4 (1), pp. 7 (2013)
2. Ronak P Patel, Narendra M Patel and Keyur Brahmbhatt.: Automatic Licenses Plate Recognition. In: International Journal of Computer Science and Mobile Computing (April 2013)
3. Najeem Owamoyo, A. Alaba Fadele and Abimbola Abudu.: Number Plate Recognition for Nigerian Vehicles. In: Academic Research International (2013)
4. Divya Gilly and Dr. Kumudha Raimond.: License Plate Recognition- a Template Matching Method. In: International Journal of Engineering Research and Applications (IJERA) (2013)
5. Isack Bulugu.: Algorithm for License Plate Localization and Recognition for Tanzania Car Plate Numbers. In: International Journal of Science and Research (IJSR) (2013)
6. Rupali Kate.: Number Plate Recognition Using Segmentation. In: International Journal of Engineering Research and Technology (IJERT) (November 2012)

7. Hadi Sharifi Kolour.: An Evaluation of License Plate Recognition Algorithms. In: International Journal of Digital Information and Wireless Communications (IJDIWC) (2011)
8. Kumar Parasuraman and P.Vasantha Kumar.: An Efficient Method for Indian Vehicle License Plate Extraction and Character Segmentation. In: IEEE International Conference on Computational Intelligence and Computing Research (2010)

NeSeDroid—Android Malware Detection Based on Network Traffic and Sensitive Resource Accessing

Nguyen Tan Cam and Nguyen Cam Hong Phuoc

Abstract The Android operating system has a large market share. The number of new malware on Android is increasing much recently. Android malware analysis includes static analysis and dynamic analysis. Limitations of static analysis are the difficulty in analyzing the malware using encryption techniques, to confuse the source, and to change behavior itself. In this paper, we proposed a hybrid analysis method, named NeSeDroid. This method used static analysis to detect the sensitive resource accessing. It also used dynamic analysis to detect sensitive resource leakage, through Internet connection. The method is tested on the list of applications which are downloaded from Android Apps Market, Genome Malware Project dataset and our additional samples in DroidBench dataset. The evaluation results show that the NeSeDroid has the high accuracy and it reduces the rate of fail positive detection.

Keywords Android malware analysis • Network behavior • Network traffic monitoring • Sensitive resource • Hybrid analysis • Mobile device security

1 Introduction

According to the International Data Corporation's statistic [1], the Android operating system always has the highest market share recently. In the second quarter of the year 2015, the Android operating system accounted for 82.8 % market share. According to F-secure's statistic [2], the number of new malware on Android is 99 % when compared with others. Analyzing and detecting malware on Android

N.T. Cam (✉)
Department of Science and Technology, Hoa Sen University, Ho Chi Minh, Vietnam
e-mail: camnguyentan@gmail.com

N.C.H. Phuoc
Department of Information Technology, University of Science, Vietnam National University, Ho Chi Minh, Vietnam
e-mail: nguyencamhongphuoc@gmail.com

S.C. Satapathy et al. (eds.), *Proceedings of the International Conference on Data Engineering and Communication Technology*, Advances in Intelligent Systems and Computing 469, DOI 10.1007/978-981-10-1678-3_3

Table 1 Malicious behavior categories in 2014

Behavior category	2014 (%)	Behavior category	2014 (%)
Steals device data	36	Modifies settings	20
Spies on user	36	Spam	7
Sends premium SMS	16	Elevates privileges	7
Downloader	18	Banking trojan	7
Back door	18	Adware/annoyance	13
Tracks location	9		

operating system become very necessary. There are three methods of analyzing malware: they are static analysis, dynamic analysis, and hybrid analysis. Static analysis is not necessary to run the application and opposite with dynamic analysis. Many researches used the static analysis [3–10]. Dynamic analysis [11–16] will solve the problem of static analysis as encryption, confusing the source. Meanwhile, hybrid analysis [17, 18] combines both static analysis and dynamic analysis. Hybrid analysis technique inherits the advantages of static analysis and dynamic analysis but it needs more cost.

The report of Symantec [19] shows that "steals device data" and "spies on user" are two malicious behaviors that were the most common used behaviors, which mean a study performing on this behaviors is necessary (Table 1).

Based on the network accessing behavior of application to detect malware, dynamic analysis method has many advantages and a lot of research interest [15, 18, 20–22].

Feizollah et al. [20] analyzed malware on Android operating system based on network packet analysis. They used two classification algorithms: K-means and Mini batch K-means to detect anomalies from network packets of the applications. Besides, they compare the performance of each algorithm. In this article, the authors focused subclass of applications compared to the set of known malware, and it had difficulty in analyzing new malware families (zero-day malware). This approach does not care about the semantics of the packet. Needs to expand the content analysis of network connectivity requirements of the application to be able to detect malicious behavior instead of matching network data collected by the application are considered for data collection network from the sample application in practice.

Zaman et al. [15] proposed a method to detect malware based on information of network access requirements from applications. Their system collects the domain which is accessed by application; if the domain name in the list of blacklist domains is defined before, this is malware. Thus, this approach depends on blacklist domains. The same difficulty in detecting malware using the domain name (Command and Control—C&C server) new is not belong to blacklist domains. Consider to find out a method of analyzing the network access behavior of application with this solution to able to detect the malware using new C&C server.

Jun et al. [21] proposed an Android malware analysis method using SVM classifier algorithm on network accessing information. The authors run benign applications for a long time and collect network packet information. Next, they run malware applications to collect network information. The classification is based on two information collected. This approach only detects a comparison with the available malware that have difficulty in detecting the new malware, new behaviors.

Shabtai et al. [22] proposed an approach to analyze malware which updated itself by collecting network information. This approach also allows to distinguish changes in the network access behavior of applications in one of three kinds: changing of user habit, self-update new version, and insert malicious malware from the Internet. This approach only concerned with network access behavior but do not consider the dangerous of stolen information in device, which is then sent to the network via network connection.

Feldman et al. [18] also proposed a solution which allows to detect malware by grouping network information and machine techniques. This method is not only interested in stealing sensitive information but also interested in the behavior of network access, which leads to false alarm applications; normal Internet access application is the malware.

Arzt et al. [9] proposed FlowDroid. FlowDroid is an approach that is used to detect a sensitive flow from sensitive sources to critical sinks. Because FlowDroid uses a static analysis technique, it cannot detect some malicious behaviors that are performed when the application is running. So, an approach based on hybrid analysis technique is necessary to study.

In this paper, we propose a malware analysis method based on network information and behavior of sensitive resources accessing in order to reduce the rate of false positive detection. This mean to reduce case which detects that a benign application is a malware.

The contributions of this study included the following:

- Make a dynamic malware analyzing system based on network information and sensitive resources with low FP.
- Compare with the existing relevant works.
- Add some samples into the existing dataset.

The rest of this paper is organized as follows. Part objective is presented in Sect. 2. Section 3 is a detailed description of malware detection system that is proposed, NeSeDroid. The test results and discussed assessment are presented in Sect. 4, and conclusion in Sect. 5.

2 Objectives

In this section, we present some motivation examples and related research questions. Besides, we outlined the objectives of this study.

```
<uses-permission android:name="android.permission.INTERNET" />
```

Fig. 1 An example of AndroidManifest.xml

The first research question: If an application access to a website on the Internet does not have behavior of leak user's information, can we conclude that it is a malware or not?

Consider the AccessInternet application which wrote to make an experiment in this study. This application just accesses the Internet but does not leak information of devices on the Internet. Applications are the granted permissions that are shown in Fig. 1:

More detail of Java code is shown in Fig. 2.

This application just only access and download the website's content. However, based on the solution of Zaman et al. [15], if the URL is detected in the predefined blacklist, this application is called a malware, even though this site is just a normal website (such as a news website, a education website…). In this case, increase the False Positive of method which is proposed by Zaman et al. and thus need to distinguish between normal access and dangerous access to decrease rate of false detection. An application has the risk of information leakage when this application is accessed to sensitive resources on device, and to the Internet to steal this sensitive information.

Second research question: When analyzing an application's code using static analysis techniques, there are several functions that can access to sensitive resources (e.g., getDeviceId ()) with Internet connection to leak information. Is it a case of malware? Do we have any application that has a malicious signal if it is analyzed by static analysis techniques although it never access the Internet when the application is executed in the real?

We wrote an application, named NeverClick, for sensitive resources access testing using getDeviceId() function, such as strDeviceID = telephonyManager. getDeviceId(); Although source code of this application has functions that are used to send values of strDeviceID to a web server, these functions should never be done because btnNeverClick button is disabled so the onClick event will not be called during running the application. More detail of NeverClick is shown in Fig. 3.

According to FlowDroid [9], this is a malicious application because it has both sensitive source and critical sink. However, the fact is that behavior of AccessInternet cannot happen. Therefore, this is caused False Positive of FlowDroid.

```
URL url = new URL("http://www[redacted]);
HttpURLConnection con = (HttpURLConnection) url.openConnection();
String readStream = readStream(con.getInputStream());
System.out.println(readStream);
```

Fig. 2 Java code of AccessInternet application

```
    TelephonyManager telephonyManager = (TelephonyManager)
            getSystemService(Context.TELEPHONY_SERVICE);
    strDevice = telephonyManager.getDeviceId();
    Button btnNeverClick = (Button) findViewById(R.id.btnNeverClick);
    btnNeverClick.setEnabled(false);
    btnNeverClick.setOnClickListener((view) → { fncPOST(); });
}
public void fncPOST() {
    if (android.os.Build.VERSION.SDK_INT > 9) {
        StrictMode.ThreadPolicy policy = new
                StrictMode.ThreadPolicy.Builder().permitAll().build();
        StrictMode.setThreadPolicy(policy);
    }
    HttpClient httpclient = new DefaultHttpClient();
    HttpPost httppost = new HttpPost("http://[redacted]/default.php");
    List<NameValuePair> sendValue = new ArrayList<~>(1);
    sendValue.add(new BasicNameValuePair("DeviceID:",strDevice));
    try {
        UrlEncodedFormEntity entity = new UrlEncodedFormEntity(sendValue);
        httppost.setEntity(entity);
        ResponseHandler<String> handler = new BasicResponseHandler();
        String content = httpclient.execute(httppost, handler);
```

Fig. 3 Java code of NeverClick application

```
Button btnConnect = (Button) findViewById(R.id.btnConnect);
btnConnect.setEnabled(true);
```

Fig. 4 Java code of SensitiveResource application

We wrote other application, named SensitiveResource. This application is like NeverClick but the difference is that enabled attribution of button btnConnect is True (such as, btnConnect.setEnabled (true)). This application either gets device ID or sends it to a webserver (Fig. 4).

In this case, the sensitive resource access and network access are detected not only in the case of static analysis but also in dynamic analysis. This application is a specific example for the case of information leakage of devices via the Internet.

Through the examples above, this paper sets out specific objectives:

- Reduce the false detection rate in detecting the risk of leaking sensitive information via the Internet by combining static analysis with dynamic analysis.
- Add the sample to the dataset used for the evaluation of malware analysis solutions based on network traffic and sensitive resource accessing.

3 System Design

To achieve the goals set out in Sect. 2, we propose the malware analysis system based on the analysis behavior of sensitive resource accessing and Internet accessing to steal information, named NeSeDroid. NeSeDroid system is operated with three phases:

Phase 1: Create a list of dangerous domains (BlackListDB) and the list of sensitive sources (SensitiveSourceDB). Step (1) transfers the malicious domain information from Virustotal [23] to BlackListDB. Step (2) runs the samples in Genome Dataset [24] to obtain network access information. This information is also included in BlackListDB. Thus, compared to the other, NeSeDroid created backlist from more sources, such as Virustotal and other applications from dataset of Android Malware Genome Project. This makes the list of domains in this blacklist more accurate (Fig. 5).

Step (3) extracts the sensitive sources from source–sink list provided by SuSi [25]. SuSi provide source–sink list maybe included in an Android application. We proceed to extract the sensitive sources which maybe included from this list to make an evidence for detecting sensitive resource and accessing of applications was analyzed.

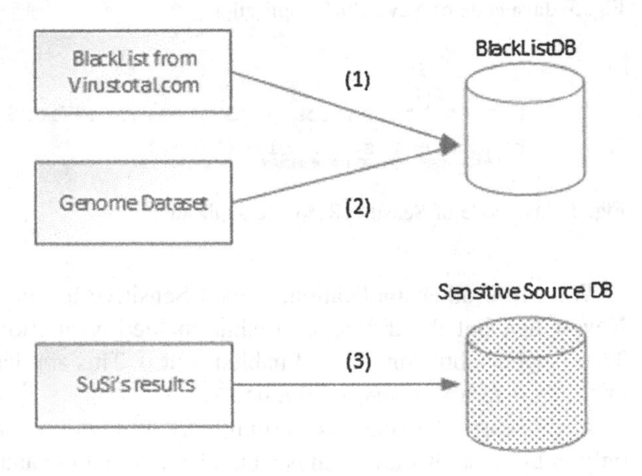

Fig. 5 Three steps in the first phase of NeSeDroid

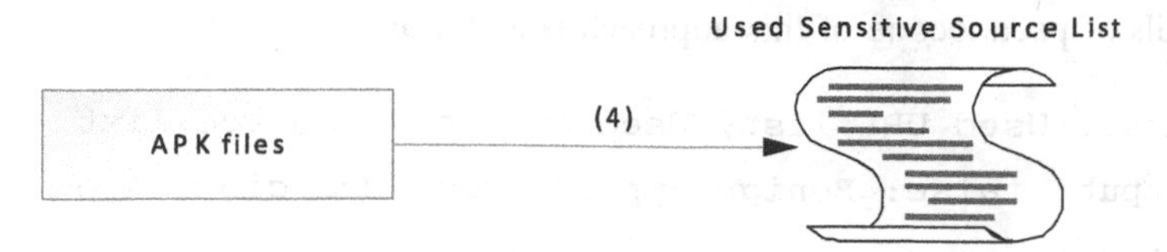

Fig. 6 Extract the used sensitive sources list from apk file

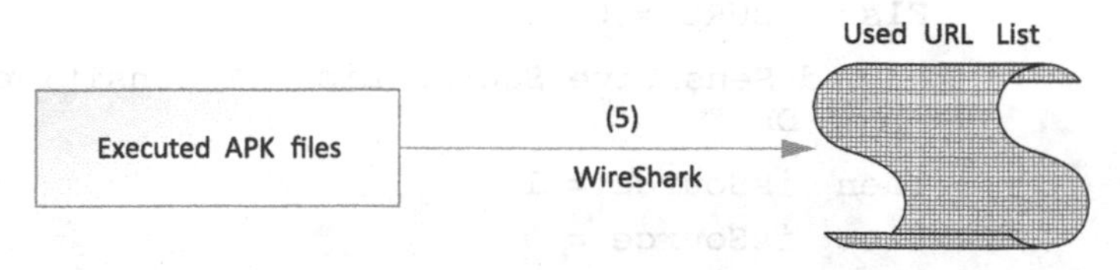

Fig. 7 Capture the used URL list using WireShark

Phase 2: The system extracts the sensitive sources (if any) by analyzing apk file. This information is stored in a list (Used sensitive source list). If there is any sensitive source in the application, turn isSource flat into 1, if not turn isSource into 0. In step (4), we use apkTool [26] and dex2jar [27] to reverse the apk file. Extract the functions related to sensitive resource accession using AndroidManifest.xml file and classes_dex2jar.jar file (Fig. 6).

Phase 3: The system executes a tested application on the Android virtual machine. In this phase, the network accessing information is logged. The gathering of network information was conducted by the Wireshark [28]. Network information will be compared to the black list in Phase 1. If this information is contained in the black list, the system turns isURL flag into 1, and 0 in opposite case (Fig. 7).

If both iSource and isURL are equal to 1, this is a malware. In opposite situation, this is a benign application. Using the information that relate to network accessing and sensitive resource accessing, our system can detect malicious applications with low of false detection rate. However, this system needs more than cost to perform two analysis techniques: static analysis method is used to extract BlackListDB, sensitive source DB, used sensitive source list; and dynamic analysis method is used to create used URL list.

Details of pseudocode of this approach is as follows:

```
Input: Used URL list, Used Sensitive Source list
Output: false: Benign App OR true: Malicious App
Begin
            If Used URL list IN BlacklistDB
              Then  isURL = 1
              Else  isURL = 0
            If Used Sensitive Source List IN Sensitive
            Source DB
              Then  isSource = 1
              Else  isSource = 0
            If isURL==1 AND isSource==1
              Then  isMalware = True
              Else isMalware = False
            Return isMalware
End.
```

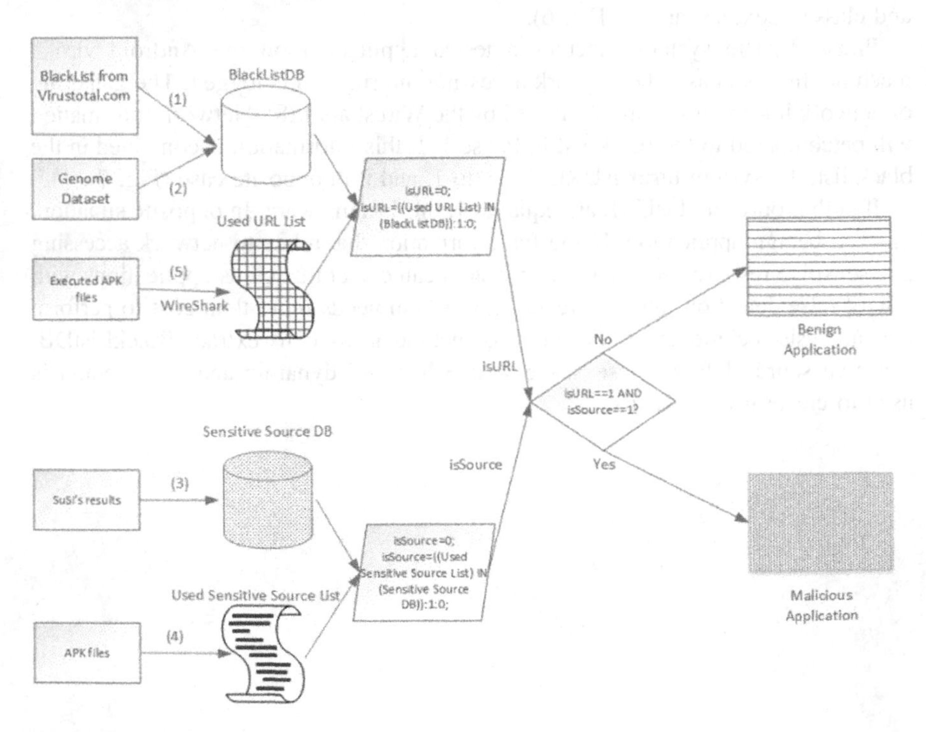

Fig. 8 The NeSeDroid system

In this pseudocode, the return value is isMalware. If isMalware is True, this application is a malicious software, whereas it is a benign software.

The system diagram of NeSeDroid is shown in Fig. 8.

4 Evaluation

NeSeDroid is a hybrid analysis system. NeSeDroid was implemented on the experimental model which is shown in Fig. 9.

In this model, the Android device is a virtual machine running Android 4.4 (KitKat) [29]. This virtual machine connected to the Internet via a gateway machine. Gateway is a computer running Ubuntu 12.04 [30]. Wireshark software is installed on the gateway machine. Internet accessing information from Android Device saved into a file by Wireshark. This file content is used by NeSeDroid in phase 3 of the analysis.

Dataset in this test included the dataset from Android Malware Genome Project [24] and the free applications from Google Play. Moreover, we added some samples (NeverClick, AccessInternet, and SensitiveResource) to the DroidBench dataset.

To compare NeSeDroid with other approaches, we tested two samples which wrote in this study to supplement the DroidBench dataset for malware analysis by analyzing sensitive source accessing and network traffic. AccessInternet is a normal Internet accessing application group without sensitive resources accessing.

NerverClick is an application group having the behavior of access to sensitive sources, but the behavior of Internet access just have in code of application which did not active when this program was executed. This mean the NeverClick application group is not a dangerous application group which is leak information. The last application group is Exploit sensitive resource via Internet. This group includes applications which have behavior of access to sensitive sources and Internet access to steal this sensitive information. Thus, a good malware detecting system needs to detect three of this case with results: the first group is not malware, the second group is not malware, and the third group is malware.

Thus, the experimental results showed that NeSeDroid detected three of application groups accurately when FlowDroid and Zaman et al's solution just have two out of three cases. Table 2 shows that FlowDroid and NeSeDroid detect AccessInternet (a benign application) as a benign application. This means that two methods are successful in this case. Meanwhile, AccessInternet is seen as a malware

Fig. 9 The experimental of NeSeDroid

Table 2 The experimental results

Sample	Type of application	FlowDroid	Zaman et al's solution	NeSeDroid
Access internet	Benign	True	False	True
Never click	Benign	False	True	True
Exploit sensitive resource via internet	Malicious	True	True	True

if we use Zaman et al's solution. In NeverClick case, FlowDroid has the fail detection, because NeverClick is really a benign application while it is seen as a malicious software by FlowDroid. In the last case, SensitiveResource is seen as a malware by all of three methods. In short, the test result shows that the rate of positive false of NeSeDroid is lower than the two remaining studies.

5 Conclusion

In this paper, we propose the Android malware analysis system through the analyzing network information and sensitive resources accessing (NeSeDroid). This static analysis is used to detect sensitive resources accessing and dynamic analysis to detect information leakage via the Internet. The test result shows that this approach has higher accuracy and lower false positive detection. Moreover, we added some samples to the DroidBench dataset that are used to compare difference approaches. Studying a method to shorten time detect malware by this method should be studied in the future, such as the used URL is collected on Android device instead of on the gateway.

Acknowledgments Our thanks to professors of Science and Technology faculty of HoaSen University and professors of Telecommunication and Networking faculty of University of Science (Vietnam National University–Ho Chi Minh City) who have many positive comments and value reviews.

References

1. Corporation, I.D. 2015; Available from: http://www.idc.com/prodserv/smartphone-os-market-share.jsp.
2. F-Secure. 2014; Available from: https://www.f-secure.com/documents/996508/1030743/Mobile_Threat_Report_Q1_2014.pdf.
3. Sanz, B., et al., *MAMA: Manifest Analysis For Malware Detection In Android.* Cybern. Syst., pp. 469–488 (2013).
4. Moonsamy, V., et al., *Contrasting Permission Patterns between Clean and Malicious Android Applications*, in *Security and Privacy in Communication Networks*, T. Zia, et al., Springer International Publishing. pp. 69–85 (2013).

5. Gascon, H., et al., *Structural detection of android malware using embedded call graphs*, in *Proceedings of the 2013 ACM workshop on Artificial intelligence and security*. ACM: Berlin, Germany. pp. 45–54 (2013).
6. Li, L., et al., *IccTA: Detecting Inter-Component Privacy Leaks in Android Apps*, in *The 37th International Conference on Software Engineering (ICSE)*. Firenze, Italy (2015).
7. Li, L., et al. *Automatically Exploiting Potential Component Leaks in Android Applications*. in *Trust, Security and Privacy in Computing and Communications (TrustCom), 2014 IEEE 13th International Conference on*. (2014).
8. Aafer, Y., W. Du, and H. Yin, *DroidAPIMiner: Mining API-Level Features for Robust Malware Detection in Android*, in *Security and Privacy in Communication Networks*, T. Zia, et al., Springer International Publishing. pp. 86–103 (2013).
9. Arzt, S., et al., *FlowDroid: precise context, flow, field, object-sensitive and lifecycle-aware taint analysis for Android apps*, in *Proceedings of the 35th ACM SIGPLAN Conference on Programming Language Design and Implementation*. ACM: Edinburgh, United Kingdom. pp. 259–269 (2014).
10. Bagheri, H., et al., COVERT: Compositional Analysis of Android Inter-App Permission Leakage. Software Engineering, IEEE Transactions on, pp. 1–1 (2015).
11. Dini, G., et al., *MADAM: a multi-level anomaly detector for android malware*, in *Proceedings of the 6th international conference on Mathematical Methods, Models and Architectures for Computer Network Security: computer network security*. Springer-Verlag: St. Petersburg, Russia. pp. 240–253 (2012).
12. Shabtai, A., et al., *"Andromaly": a behavioral malware detection framework for android devices*. J. Intell. Inf. Syst., pp. 161–190 (2012).
13. Zheng, C., et al., *SmartDroid: an automatic system for revealing UI-based trigger conditions in android applications*, in *Proceedings of the second ACM workshop on Security and privacy in smartphones and mobile devices*. ACM: Raleigh, North Carolina, USA. pp. 93–104 (2012).
14. Enck, W., et al., *TaintDroid: an information-flow tracking system for realtime privacy monitoring on smartphones*, in *Proceedings of the 9th USENIX conference on Operating systems design and implementation*. USENIX Association: Vancouver, BC, Canada. pp. 1–6 (2010).
15. Zaman, M., et al. *Malware detection in Android by network traffic analysis*. in *Networking Systems and Security (NSysS)* (2015).
16. Wu, X., et al., *Detect repackaged Android application based on HTTP traffic similarity*. Security and Communication Networks, (2015).
17. Zheng, M., M. Sun, and J.C.S. Lui, *DroidRay: a security evaluation system for customized android firmwares*, in *Proceedings of the 9th ACM symposium on Information, computer and communications security*. ACM: Kyoto, Japan. pp. 471–482 (2014).
18. Feldman, S., D. Stadther, and W. Bing. *Manilyzer: Automated Android Malware Detection through Manifest Analysis*. in *Mobile Ad Hoc and Sensor Systems (MASS), IEEE 11th International Conference on*. (2014).
19. *Mobile security threat report*. 2015 [cited 2015 April 10]; Available from: http://www.sophos.com/en-us/threat-center/mobile-security-threat-report.aspx.
20. Feizollah, A., et al. *Comparative study of k-means and mini batch k-means clustering algorithms in android malware detection using network traffic analysis*. in *Biometrics and Security Technologies (ISBAST), 2014 International Symposium on*. (2014).
21. Jun, L., et al. *Research of android malware detection based on network traffic monitoring*. in *Industrial Electronics and Applications (ICIEA), 2014 IEEE 9th Conference on*. (2014).
22. Shabtai, A., et al., *Mobile malware detection through analysis of deviations in application network behavior*. Computers & Security, pp. 1–18 (2014).
23. *Malware Domain Blocklist*. [cited 2015 July 10]; Available from: http://www.malwaredomains.com/?page_id=23.
24. *Android Malware Genome Project* 2015; Available from: http://www.malgenomeproject.org/.
25. Rasthofer, S., S. Arzt, and E. Bodden, A Machine-learning Approach for Classifying and Categorizing Android Sources and Sinks. (2014).

26. *ApkTool*. 2015; Available from: https://github.com/iBotPeaches/Apktool.
27. *dex2jar: Tools to work with android .dex and java .class files* 2015 [cited 2015 May 20]; Available from: https://github.com/pxb1988/dex2jar.
28. *WireShark*. 2015 [cited 2015 May 20]; Available from: https://www.wireshark.org/.
29. *Android-x86 Project - Run Android on Your PC*. 2015 [cited 2015 May 10]; Available from: http://www.android-x86.org/.
30. *Ubuntu 12.04.5 LTS (Precise Pangolin)*. 2015 [cited 2015 May 2]; Available from: http://releases.ubuntu.com/12.04/.

RiCoBiT—Ring Connected Binary Tree: A Structured and Scalable Architecture for Network-on-Chip Based Systems: an Exclusive Summary

V. Sanju and Niranjan Chiplunkar

Abstract High-performance computing systems consist of many modules of high complexity with billion transistors on a small silicon die. To design these mega functional systems, the common bus architecture poses a serious problem in terms of latency and throughput. To overcome the disadvantages of the common bus architecture, a new paradigm in ASIC design called the Network-on-Chip (NoC) was proposed. Several topologies like 2D mesh, torus, etc., were used to interconnect the different modules of the design using this novel idea. These topologies underperformed when scaled. This paper proposes a new architecture RiCoBiT: Ring Connected Binary Tree. It is a new scalable, structured architecture for Network-on-Chip based systems. An optimal routing algorithm for it has been designed. The paper discusses the different properties and performance parameters like maximum hop count, average hop count, number of wire segments, and wire length used to interconnect the nodes of RiCoBiT. These parameters are compared with that of 2D mesh and torus. The paper also discusses and bounds real-time parameters like latency and throughput.

Keywords Network-on-chip · Topology · Architecture

1 Introduction

The advances in semiconductor physics and successful research outcome in fabrication and lithography processes have enabled integration of more and more logic on a small silicon area. This prompted designs for high-performing mega functional

V. Sanju (✉)
Department of Computer Science & Engineering, Muthoot Institute of Technology and Science (MITS), Varikoli, Ernakulam, India
e-mail: sanjuv@mgits.ac.in

Niranjan Chiplunkar
NMAMIT, Nitte, India
e-mail: niranjanchiplunkar@rediffmail.com

S.C. Satapathy et al. (eds.), *Proceedings of the International Conference on Data Engineering and Communication Technology*, Advances in Intelligent Systems and Computing 469, DOI 10.1007/978-981-10-1678-3_4

modules. Initially, these designs were realized using the common bus architecture. As the system complexity increased, several new techniques like parallel bus architecture, arbitration, and polling were incorporated. As year passed by, these also caused problems in high performance.

To overcome these, a new paradigm in ASIC design called Network-on-Chip (NoC) [1–3] was introduced wherein the concepts of networking were implanted on silicon. The processing modules are placed in different nodes of the direct and indirect network like 2D mesh, torus, tree, octagon, and omega [1, 4] which were proposed to implement this concept. The nodes communicated through packets transferred from one node to the other. These innovations proved advantageous in terms of performance with a small increase in area which allowed the designs to swell further. This continued for sometime. As the designs scaled up, there was a small dip in expected performance because of the latency in packet transmission and reception. This was due to the diameter of the network/length of the path to be traversed by the packet to reach from the source to destination node. The problem of performance dip may be overcome by a design of the solution using a new architecture which has a smaller diameter. This architecture should be structured, scalable, and modular which should not increase the area or wire length substantially [5–10].

This paper proposes a new scalable and structured architecture for Network-on-Chip based systems. An optimal routing algorithm for the same is also presented. Its performance in terms of maximum hop, average hop, number of wire segments, and wire length is studied and compared with that of 2D mesh and torus. During the analysis, it is found that the proposed architecture performs better without any substantial increase in area. Proposed architecture performs better without any substantial increase in area.

2 Proposed Architecture

A new architecture for Network-on-Chip based system called RiCoBit—Ring Connected Binary Tree—is presented in this section.

2.1 RiCoBiT Topology

The proposed RiCoBiT topology resembles the spider's cob web as shown in Fig. 1. The topology consists of growing interconnected concentric rings. The rings are numbered from 1 to K, where K is the number of rings in the configuration. The number of nodes in ring L is 2^L numbered from 0 to 2^L–1. Therefore, the number of nodes Nr in the configuration with K rings is

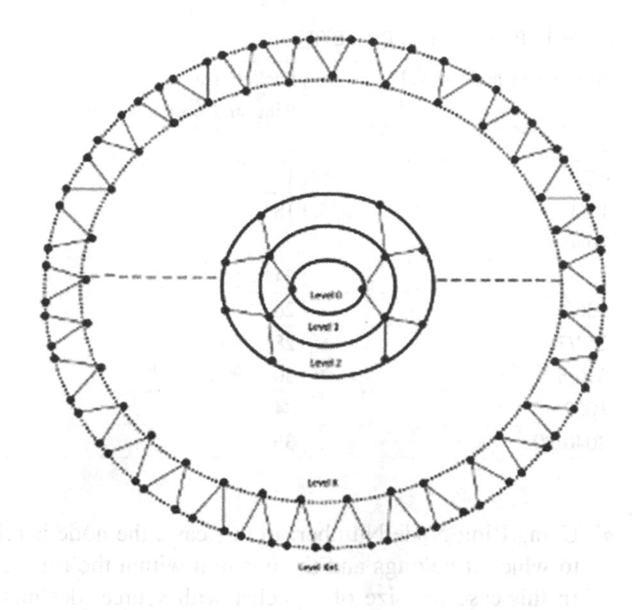

Fig. 1 Proposed architecture with $K + 1$ rings

$$Nr = \sum_{L=1}^{k} 2^L$$

The node n in ring L is connected to its neighboring nodes $2n$ and $2n + 1$ in ring $L + 1$. The number of wire segments to connect the nodes in ring L to those in ring $L + 1$ is therefore 2^{L+1}. Similarly, the nodes within each ring are also interconnected by wire segments. The number of such wire segments in ring L is 2^L.

The proposed topology has several advantageous properties. The topology is symmetric and regular in nature. It is structured, modular, and scalable limited only by the fabrication technology. The scalability property is well complemented with high performance with marginal increase in chip area. The performance study is presented in section three.

2.2 RiCoBiT Addressing

The processing nodes are placed and addressed relative to the ring to which they belong and the position within the ring. The rings are numbered starting from one and the nodes within the ring are numbered starting from zero.

It is observed that node addressing scheme plays an important role in performance of the system. The transmission time of a packet between the interfaces depends on the number of bits the packet contains. This implies that if the transmission time would increase as the packet size grows thereby decreasing the performance.

Let us consider two addressing schemes for RiCoBiT.

Table 1 Packet size in RiCoBiT

Number of nodes/level	Packet size	
	Ring and node number	Only node number
2/1	10	10
6/2	14	14
14/3	18	16
30/4	20	18
62/5	24	20
126/6	26	22
254/7	28	24
510/8	30	26
1022/9	34	28
2046/10	36	30

- Using Ring Node Number: In this case, the node is referred by the ring number to which it belongs and the position within the ring. This is as shown in figure. In this case the size of a packet with source, destination addresses, and K bits data can be expressed as $2*(\log_2 R + R) + K$, where R is the maximum number of rings in the topology.
- Using Only Node Number: In this case, the nodes are addressed using continuous numbers starting from the inner ring as shown in figure. In this case the size of a packet with source, destination addresses, and K bits data can be expressed as $2*(\log_2 N) + K$, where N is the number of nodes in the topology.

It is observed that the packet size is reduced when using only node numbers. Table 1 illustrates the same for a byte data. This trend increases as the topology grows.

2.3 *RiCoBiT Routing Algorithm*

This section presents an optimal routing algorithm for RiCoBiT. The proposed algorithm to route the packet from the source to the destination is presented below.

Step 0: Check destination address. Initialize current_src as node address and current_dest as destination address.

Case 1: current_src and current_dest in same ring.

1. Compute the minimum hop count between current_src and current_dest. If it is greater than 3, set current_src and current_dest as their parent nodes, respectively, in the adjacent ring below.
2. Repeat Step 1 till shorter hop count is greater than 3.
3. Move to the next node from current_src toward current_dest along the minimum hop count path. Set the current_src as the next node.

4. Repeat Step 3 until current_src is equal to current_dest.
5. Consider a complete binary tree with top right node of the current_src as root. If destination node lies in the tree, mark the right node as set current_src else mark the sibling of the left node as current_src. Move to the current_src.
6. Repeat Step 5 until current_src is equal to destination address.

Case 2: When destination node is in a ring above that of the source node

1. Move one step at a time from the current_dest to the ring below until ring of the current_src is reached. Set current_dest as the node obtained in the ring of the source.
2. Repeat the steps as in Case 1.
3. Repeat Step 5 and 6 of Case 1.

Case 3: when destination node is in a ring below that of the source node

1. Move one step at a time from current_src to the ring below until the ring of current_dest is reached. Set current_src as the node obtained.
2. Repeat the procedure as in case 1.

A packet when received by a node is checked by the routing logic of the interface. The logic is built around three cases based on the ring to which the current source node and the destination of the packet, respectively, belong. If the current source node and the destination of the packet are in the same ring, the packet is routed to the parent node on the adjacent ring until the node difference is greater than three. Then the node traverses the lateral path along the shortest route to current destination. After reaching the current destination, the packet travels up the ring toward the destination.

Whenever the destination is above the current node, the packet is initially routed along the same ring to the parent node of the destination. Then the packet moves up toward the destination.

Whenever the destination node is below the current node, the packet moves toward the ring of the destination along the parent node of current node. Then the routing logic routes the packet toward the destination.

From the above, it is evident that the packet uses binary tree traversal which is known for its optimal routing and a lateral traversal through the shortest path to reach the destination. This proves that the packet is routed through the shortest path. Therefore, the proposed algorithm is optimal.

Statement: The proposed algorithm is optimal, i.e., the routing algorithm would route the packet from the source to the destination for any configuration with K rings through the shortest path.

The statement is proved by mathematical induction.

Case 1: $K = 1$

Let us consider the algorithm for the smallest case, i.e., for ring $K = 1$ with $N = \sum_{L=1}^{k} 2^L = 2$ nodes. The source and destination being adjacent to each other, there are two alternate paths each consisting of single hop. Note that this is the shortest possible.

Case 2: $K = 2$

1. If source and destination are both in ring 1, then it is same as case 1.
2. If the source and destination are both in ring two, then the shortest path consists of one hop in case source and destination are adjacent to each other. Otherwise, the shortest path consists of two hops.
3. If the source and destination are in two different rings, then the shortest path consists of one hop in case they are adjacent. Otherwise, the shortest path consists of two hops. It is noted that such a path has an intermediate node in ring one which is adjacent to source and destination both. Therefore, it is of length two hops and is shortest. When the intermediate node is in ring one, it is concluded that if a configuration with a single ring others shortest path between adjacent nodes in ring one then a configuration with two rings also others shortest path for any source destination pair in two different rings.

From this we infer that there is a shortest path between any source destination pair when the configuration has two rings.

Case 3: $K = 3$

1. If source and destination are confined to ring one and ring two, then it is same as Case 2.
2. If the source and destination both are in ring three, then the shortest path consists of one hop in case these are adjacent to each other. Otherwise, there is a shortest path along the ring three consisting of two hops or three hops or four hops if there are one or two or three intermediate nodes between source and destination pair, respectively. In case of three intermediate nodes, there are alternate shortest paths of four hops each which pass through ring two.
3. If one of the nodes is in ring three and the other one in ring two, then the shortest path is of one hop when the source and destination are adjacent to each other. Else there exists an intermediate node in ring two which is adjacent to the node in ring three.

According to the Case 2, it is known that there exists a shortest path for any source destination pair for configuration of two rings. Hence, there exists a shortest path between a node in ring three and any node either in ring one or two. From this we infer that there exists a shortest path between any source destination pair when the configuration has three rings.

Case 4: $K = 4$

1. If source and destination are confined to ring one, ring two, and ring three, then it is same as Case 3.
2. If the source and destination both are in ring four, then the shortest path consists of one hop in case these are adjacent to each other. Otherwise, there is a shortest path along ring four, consisting of two hops or three hops or four hops if there

are one or two or three intermediate nodes between source and destination pair, respectively. In case of three intermediate nodes, there are alternate shortest paths of four hops each which pass through ring three. When the source and destination are separated by more than four hops in ring four, then the shortest path passes through ring three to the destination at ring four. The number of hops in such shortest path varies from four to six on case to case basis.

3. If one of the nodes is in ring four and the other node in ring three, then the shortest path is of one hop when the source and destination are adjacent to each other. Else there exists an intermediate node in ring three which is adjacent to the node in ring four. According to the case of ring three, there is a shortest path for any source destination pair for configuration of three rings. Hence, there is a shortest path between a node in ring four and any node in other rings.

From this we infer that there exists a shortest path between any source destination pair when the configuration has four rings.

Case $K - 1$:

Assume that there are shortest paths between any source destination pair in a configuration consisting of $K - 1$ rings.

Case K:

1. If the source and destination both are in ring K, then the shortest path consists of one hop in case these are adjacent to each other. Otherwise, there is a shortest path along the same ring consisting of two hops or three hops or four hops if there are one or two or three intermediate nodes between source and destination pair, respectively. When the source and destination are separated by more than four hops in ring K, then the shortest path passes through ring $K - 1$ to the destination at ring K.
2. If one of the nodes is in ring K and the other node in ring $K - 1$, then the shortest path is of one hop when the source and destination are adjacent to each other. Else there exists an intermediate node in ring $K - 1$ which is adjacent to the node in ring K. In view of the assumption for the Case $K - 1$ above, there is a shortest path between any source destination pair for configuration of K rings passing through the adjacent intermediate node between ring K and $K - 1$.

From the above discussion it is evident that the routing algorithm routes the packets from any source to any destination through the shortest route. Thus the proof.

3 Performance Parameters and Comparison

In this section, the performance of RiCoBiT is studied and compared with two of the most popularly used architectures in Network-on-Chip based systems namely 2D mesh and torus. In this comparison, some important parameters like the maximum hop for a given order (size/no of nodes), the number of wire segments, and the wire length used for the interconnection are considered. These are the major factors governing the system performance in terms of latency, throughput, and area of a given design (which in turn are decisive factors for power consumption and dissipation).

3.1 Maximum Hop Count

Consider the set of shortest path(s) between all possible source–destination pair of nodes for a given configuration. The largest element in the set is referred to as maximum hop count. As the number of hops traversed by a packet from the source to the destination increases, the latency of the system also increases and the performance dips. Hence, this is one of the most important parameters which determine the performance and throughput of a system.

In case of RiCoBiT, the packet travels from the source to the destination with a maximum hop count

$$\text{Max Hop} = 2\log_2(N_r + 2) - 4\,;\ N_r \text{ is the no of nodes}$$

Proof In a complete binary tree of *Nr* nodes, the maximum distance from the root to the leaf node is expressed as

$$\text{Max Distance} = \log_2(N_r + 1) - 1$$

Therefore, the distance from the extreme left to the extreme right would be twice of the above expression

$$\text{Max Distance} = 2 \times \log_2(N_r + 1) - 2$$

By construction, complete the figure to form a balanced complete binary tree with $N_r + 1$ nodes as

$$\text{Max Distance} = \log_2(N_r + 2) - 1$$

Therefore, the distance from the extreme left or the extreme right would be twice of the above expression:

$$\begin{aligned}\text{Max Distance} &= 2 \times (\log_2(N_r + 2) - 1)\\ &= 2 \times \log_2(N_r + 2) - 2\end{aligned}$$

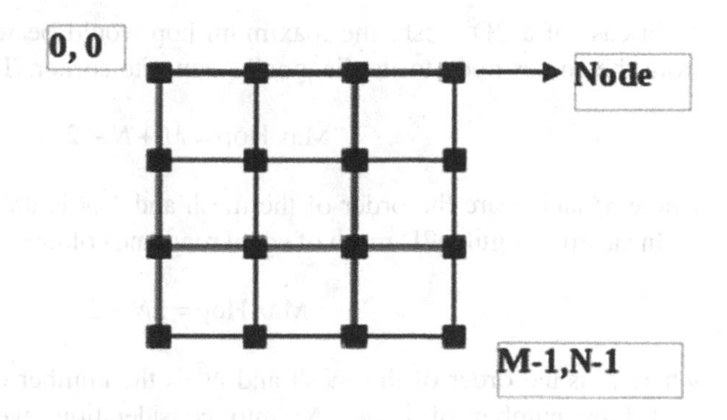

Fig. 2 A 2D mesh of M rows and N columns showing the modules at the intersections with the addressing

Now removing the two extra edges from the construction, we get

$$\begin{aligned}\text{Max Distance} &= 2 \times \log_2(N_r + 2) - 2 - 2\\ &= 2 \times \log_2(N_r + 2) - 4\end{aligned}$$

Thus the proof.

Now let us consider a 2D mesh. It is one of the most popular, simplest, and widely used topologies in Network-on-Chip based systems. In this topology, the processing modules are placed at the intersection of the row and column as shown in Fig. 2.

In the comparison of the parameter maximum hop, we consider the proposed architecture of L rings with $N_r = \sum_{i=1}^{L} 2^i$ nodes. This is compared with a 2D mesh with M nodes along its row and N nodes along its column having total MN nodes. The equation below shows the relation between the number of nodes of 2D mesh and RiCoBiT:

$$N_r = MN - 2$$

where N_r is the number of nodes in RiCoBiT and MN is the number of nodes in 2D mesh. Whenever $MN \gg 2$, N_r tends to MN or $N_r \approx MN$. For example, in 2D mesh of 32 rows and 32 columns we have $MN = 1024$ nodes, which is compared to $Nr = MN - 2 = 1024 - 2 = 1022$ nodes of RiCoBiT.

Extending the above equation for a square mesh, the equation reduces to

$$\begin{aligned}N_r &= N^2 - 2\\ &\approx N^2\\ N &= \sqrt{Nr}\end{aligned}$$

In case of a 2D mesh, the maximum hop would be when a packet can travels from the corner node to its diagonally opposite corner. This is expressed as

$$\text{Max Hop} = M + N - 2$$

where M and N are the order of the mesh and MN is the number of nodes.

In case of a regular 2D mesh of equal rows and columns, the expression reduces to

$$\text{Max Hop} = 2N - 2$$

where N is the order of the mesh and N^2 is the number of nodes.

Taking number of nodes N_r into consideration, we can rewrite the above equation as

$$\begin{aligned} \text{Max Hop} &= 2\sqrt{N_r} - 2 \\ &\approx 2\sqrt{N_r} \end{aligned}$$

Torus is another popular, simple, and widely used topology in Network-on-Chip based systems. The topology is evolved from 2D mesh by adding an extra edge between the end points of the mesh topology. Like 2D mesh, the processing modules are placed at the intersection of the row and column as shown in Fig. 3 below.

The discussion on the number of nodes of torus and its comparison with RiCoBiT remains the same as in the previous case. The maximum hop parameter in the case of torus is when the packet travels from the corner node to the middle of the system and is given by

$$\text{Max Hop} = \lfloor M/2 \rfloor + \lfloor N/2 \rfloor$$

where M and N are the order of the mesh and MN is the number of nodes.

Reducing the equation for same dimension of rows and columns,

$$\text{Max Hop} = 2 \text{X} \lfloor N/2 \rfloor$$

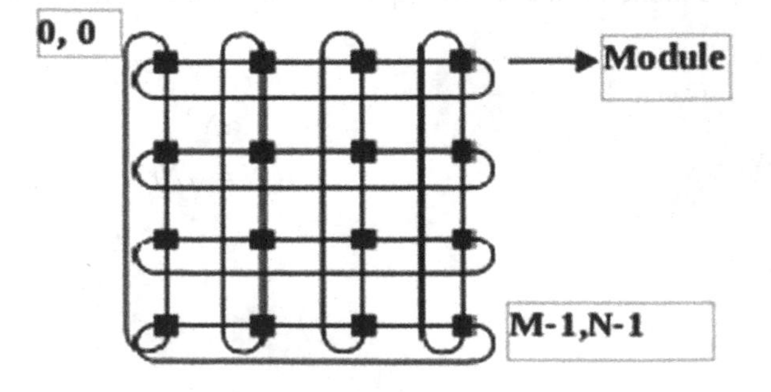

Fig. 3 A 2D torus of M rows and N columns showing the modules at the intersections with the addressing

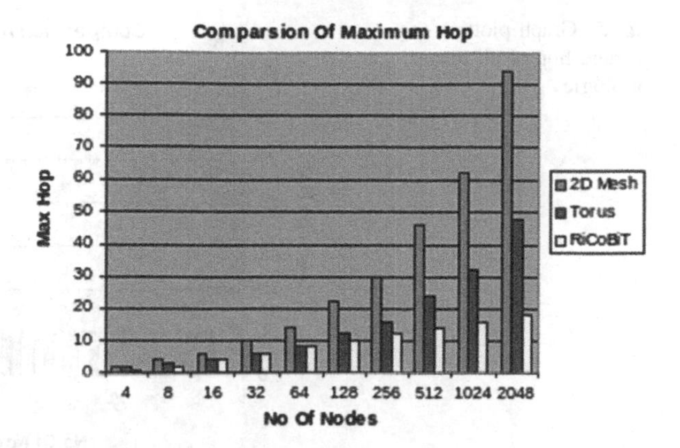

Fig. 4 Graph plotting the maximum hop of all the topologies

where N are the order of the mesh and N^2 is the number of nodes.

Taking number of nodes N_r into consideration, we can rewrite the above equation as

$$\text{Max Hop} \approx \sqrt{N_r}\ ;\ \text{Where } N_r \text{ is the number of nodes}$$

The graph (Fig. 4) below compares the maximum hop of the mesh and torus with RiCoBiT. From the graph and equations, it is quite evident that the maximum hop is highest in mesh. The parameter is half for torus as compared to mesh and the maximum hop is lowest in the proposed architecture. The maximum hop is almost the same when the number of modules is less. As the number of modules scales up, the parameter rises substantially in both mesh and torus but the growth is very slow in case of our proposed architecture. This implies that the performance is not compromised when design is scaled. All the other parameters are compared in the same manner.

3.2 Maximum Hop (Average Case)

As a part of the comparison we also compute an average value of the maximum hop. This parameter helps us to compare the overall performance of the packet transfer in the system. This parameter is calculated by taking the arithmetic mean of the maximum hop for all possible node pairs. From the graph (Fig. 5) below, it is evident that the average hop is substantially less when compared to the other architectures. This also proves that the proposed architecture performs well when compared to other architectures.

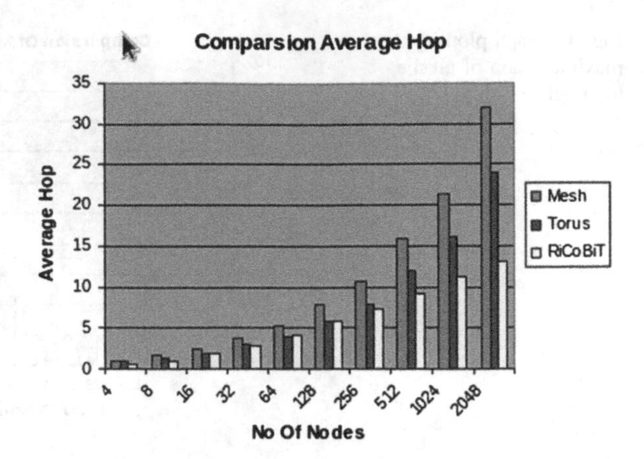

Fig. 5 Graph plotting the average hop of all the topologies

3.3 *Number of Wire Segments*

This is another important parameter which would govern the area and routing complexity of the design.

If the design has more wire segments, it would increase the area, power consumption, and also pose difficulty for CAD tool to place and route the wire segments. The parameter expressed in terms of number of nodes Nr for the proposed architecture is

$$\text{No. of Segments } W = 2 * N_r - 2$$
$$\approx 2 * Nr$$

Proof Let us prove the above equation by method of induction.

Consider the equation for the base case with 2 nodes at ring 1:

$$\text{No Of segments} = W = 2 * 2 - 2 = 2$$

The statement is true for $N = 2$.

Now let us express the number of nodes N in terms of ring L:

$$N = \sum_{i=1}^{L} 2^i$$

Let the equation for some ring K be true:

$$W = 2 * \sum_{i=1}^{k} 2^i - 2$$

Now let us prove for ring $K + 1$

$$\mathrm{W} = 2 * \sum_{i=1}^{k} 2^i - 2 + 2^{k+1} + 2 \sqcap 2^k$$

where the term 2^{K+1} represents the number of links to interconnect nodes at $K + 1$ ring and the term $2 * 2^K$ is for the number of links to interconnect the nodes at ring K and $K + 1$. Reducing the above equation,

$$\begin{aligned}
\mathrm{W} &= 2 * \sum_{i=1}^{k} 2^i - 2 + 2^{k+1} + 2^{k+1} \\
&= 2 * \sum_{i=1}^{k} 2^i + 2 \sqcap 2^{k+1} - 2 \\
&= 2 * \left\{ \sum_{i=1}^{k} 2^i + 2^{k+1} \right\} - 2 \\
&= 2 * \left\{ \sum_{i=1}^{k+1} 2^i \right\} - 2 \\
&= 2 * \mathrm{N} - 2
\end{aligned}$$

where the number of nodes at ring $K + 1 = \sum_{i=1}^{k+1} 2^i$ Thus the proof.

The number of wire segments for an $M \times N$ order 2D mesh is given as

$$W = 2M \times N - M - N$$

where M and N are the order of the mesh and MN is the number of nodes.

Reducing the equations for equal rows and columns,

$$\begin{aligned}
W &= 2 * N^2 - 2N \\
&= 2N(N - 1)
\end{aligned}$$

where N is the order of the mesh and N^2 is the number of nodes.

Considering the number of nodes N_r, we can express the relation as

$$\text{No. of Segments } W \approx 2N_r$$

In the case of torus, the number of wire segments would increase due to the addition of extra wire segments to join the end points. This would give rise to an expression as given below:

$$W = 2M \times N$$

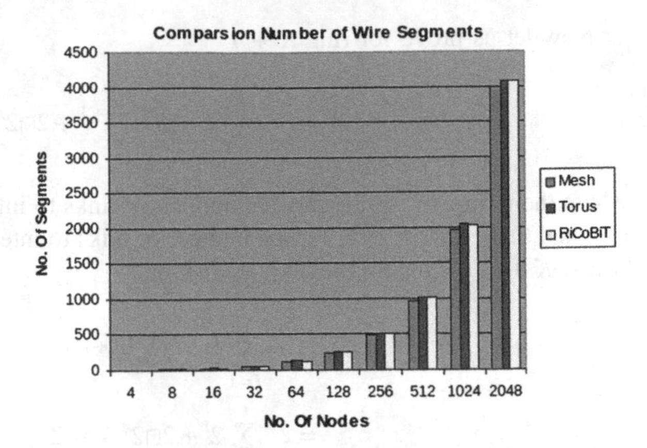

Fig. 6 Graph plotting the number of segments for all the topologies

Reducing the equations for equal rows and columns,

$$W = 2N^2$$

Expressing the above expression in terms of number of nodes *Nr*,

$$\text{No. of Segments } W \approx 2N_r$$

From the graph (Fig. 6) below, we could observe that the number of segments is almost the same as other architectures. This implies that the area requirement of RiCoBiT for a given number of nodes would remain the same with advantage of performance over the others.

3.4 Wire Length

This is also an important parameter affecting the performance and area of the design. The same effects of number of wire segments hold good even in this parameter. For considering the wire length of the proposed architecture, we consider unit length for interconnects between the modules with an additional wire of length equal to the number of nodes to connect the ends. The parameter for the proposed architecture is calculated as

$$\begin{aligned} WL &= 3N_r - 4 \\ &\approx 3N_r \end{aligned}$$

Proof Let us prove the above equation by method of induction.

Consider the equation for the base case with 2 nodes at ring 1:

$$\text{Wirelength} = WL = 3 * 2 - 4 = 2$$

The statement is true for $N = 2$.
Now let us express the number of nodes N in terms of ring L:

$$\mathrm{N} = \sum_{i=1}^{L} 2^i$$

Let the equation for some ring K be true:

$$WL = 3 * \sum_{i=1}^{K} 2^i - 4$$

Now let us prove for ring K + 1:

$$WL = 3 * \sum_{i=1}^{K} 2^i - 4 + \{2^{K+1} - 1\} + 2 * 2^K + \{(2^{K+1} - 1) + 2\}$$

where the term $2^{K+1} - 1$ represents the number of links to interconnect nodes at $K + 1$ ring and the term $2 * 2^K$ is for the number of links to interconnect the nodes at ring K and $K + 1$. Considering a unit length for each link, the number of links would be equal to the wire length. Also, we would require $(2^{K+1} - 1) + 2$ units for interconnecting the two end points.

Reducing the above equation,

$$\begin{aligned} WL &= 3 * \sum_{i=1}^{K} 2^i - 4 + 2^{K+1} + 2^{K+1} + 2^{K+1} \\ &= 3 * \sum_{i=1}^{K} 2^i + 3 * 2^{K+1} - 4 \\ &= 3 * \{\sum_{i=1}^{K} 2^i + 2^{K+1}\} - 4 \\ &= 3 * \sum_{i=1}^{K+1} 2^i - 4 \\ &= 3 * N - 4 \end{aligned}$$

where the number of nodes at ring is $K + 1 = \sum_{i=1}^{K+1} 2^i$. Thus the proof.

Similarly, considering a unit and equal length for all the interconnects, the wire length for 2D mesh of order $M \times N$ can be computed as

$$WL = 2M \times N - M - N$$

Reducing the equations for equal rows and columns,

$$\begin{aligned} WL &= 2N^2 - 2N \\ &= 2N(N-1) \end{aligned}$$

Expressing the above in terms of number of nodes N_r,

$$\begin{aligned} WL &\approx 2N_r - 2\sqrt{N_r} \\ &\approx 2N_r \end{aligned}$$

In case of torus, considering a unit and equal length for all the interconnects between internal nodes and the length of the end points wires to be approximately equal to the order of the torus, the wire length can be computed as

$$WL = 4M \times N - 2M - 2N$$

Reducing the equations for equal rows and columns,

$$\begin{aligned} WL &= 4N^2 - 4N \\ &= 4N(N-1) \end{aligned}$$

Expressing the above in terms of number of nodes Nr,

$$\begin{aligned} WL &\approx 4N_r - 4\sqrt{N_r} \\ &\approx 4N_r \end{aligned}$$

It may be noted from the graph (Fig. 7) that the wire length of the proposed architecture is more than mesh but is lesser than torus. We could take this marginal hike in the wire length for the performance that it would deliver.

Summarizing the discussion above, the proposed architecture is better in terms of maximum hop and average hop which is a parameter indicating performance. It is also comparable in terms of number of wire segments and wire length which represents area. From the above two parameters it is evident that the proposed architecture fairs better than the existing popular ones. The section below discusses performance analysis of the proposed architecture deducing bounds for parameters like latency and throughput.

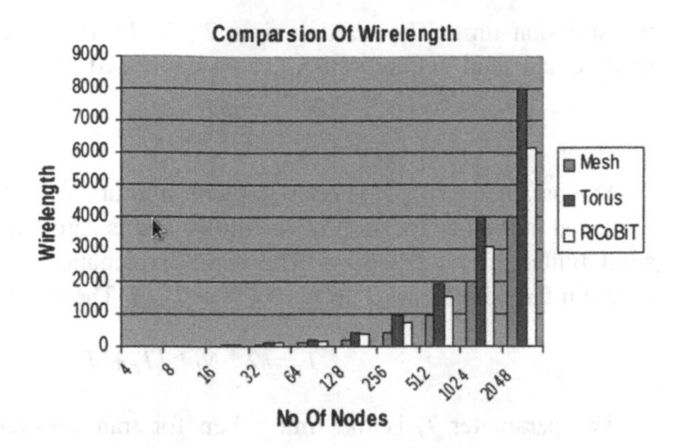

Fig. 7 Graph plotting the wire length for all the topologies

4 Performance Analysis

Various performance analyses were carried out on the above-discussed topologies using analytical deductions and verified using simulation. The topologies during simulation were subjected to various operating conditions and the performance parameters were recorded. The operating condition includes loading the topology with packets of different sizes under varying buffer size and varying buffer load to study different parameters like overall packet delivery time, packet latency, throughput, etc.

Let us first understand the assumption and context under which these analyses are carried out. A node communicates with each other through their interface using simple request acknowledgment protocol. The communication between the nodes starts with rising of the request signal when there is a packet of size p bits present in the send buffer of N_b size and waits T_{ack} time for acknowledgment. The adjacent node issues an acknowledgment only if its N_b size received buffer is not full after which the data is serially transferred. The packet in the receive register is processed to find the next transit node by the routing logic in T_p amount of time. Now the packet is placed in the send buffer of the interface along the destination if there is a space in it else its own receive buffer.

Consider a system realized using the above topologies for store and forward strategy with n nodes. The packet (p bits) contains source, destination addresses, and the data bits. Assuming the number of data bits and number of buffers per node to be equal, the number of cycles (*T*) required to transfer a packet between any two nodes can be modeled as follows:

$$T = T_p + T_w + T_{ack} + T_t$$

where T_p is the constant processing time, T_w is the waiting time in the buffers, T_{ack} is the time for receiving the acknowledgment after the request signal, and T_t is the transmission time. The waiting time T_w is the time spent by the packet in both receive and send buffers which can be expressed as

$$T_w = T_{\text{receive}} + T_{\text{send}}$$

The parameter T_{receive} is bounded within limits N_b * $(T_t + T_{ack})$ when the packet is placed in the last location of the buffer and is zero when it placed directly in the send buffer after processing. The same explanation holds good for send buffer wherein the bounds are 1 to N_b * $(T_t + T_{ack})$. Therefore,

$$T_w = 2 * N_b * (T_t + T_{ack})$$

The parameter T_t is the time taken for transmission of the packet of p bits including the request/acknowledgment which is expressed as $T_t = 2 + p$. Now let us estimate the time T_{ack}. The best case is when there is a free space and the acknowledgement is given without any delay, i.e., $T_{ack} = 0$. In the worst case, in a deadlock free scenario the parameter is bounded by $T_{ack} = (n-1) * T_t$. This can be explained as follows.

Consider a source node S1 sends a request to its adjacent node *T*1. Since the receive register of *T*1 is not free it cannot generate an acknowledgment. To generate an acknowledgement the packet in the receive register of *T1* should be placed in one of the send buffers which in turn is full. The send buffer of the link for which the packet in the receive register is meant for should create a space by sending a request signal and the cycle continues over other nodes. This can be modeled as a graph problem where the path traced by the congested bu_ers now forms a Hamilton path which can be of length n −1 for n nodes at the worst case.

Summing the above set of equations

$$\begin{aligned} T &= T_p + T_w + Ta_{ck} \\ &= 2 * N_b * (T_t + (n-1) * T_t) + (n-1) * T_t + T_t \\ &= T_p + 2 * N_b * T_t * n + (n-1) * T_t + T_t \\ &= T_p + T_t * n(2 * N_b + 1) \end{aligned}$$

Applying bounds on the parameter T,

$$T_p + T_t \leq T \leq T_p + T_t * n(2 * N_b + 1)$$

4.1 *Packet Consumption Rate*

This parameter refers to the time taken by the system to consume all the generated packets. Consider a system with n nodes and the longest path between any nodes in the system. The time taken for a packet to reach from source to destination can be expressed as

$$T_{\text{long}} = T * \text{Maxhop}$$

Now consider a node sends one packet to all other nodes. This implies that the node would generate n–1 packet. The time taken for all the packets to be consumed by the systems can be expressed as

$$\begin{aligned} T_{1;All} &= (n-1) * T + T * \text{Maxhop} \\ &= T * (\text{Maxhop} + n - 1) \end{aligned}$$

Similarly, consider a system where X nodes generate one packet each to other nodes. By keeping the parameters same, the expression for time taken for all the packets to be consumed by the system can be expressed as

$$T_{All} = T * (\text{Maxhop} * (n - X))$$

We have already proved that the maximum hop is very less in RiCoBiT when compared to mesh and torus. Also, the time taken for transfer of packet from one node to another node is equal in all the cases. Hence the above equations, it is quite evident that the system using RiCoBiT topology absorbs the entire packet in a time which is much lower than when compared to mesh and torus.

4.2 Throughput

Considering the construction of the topology, it could be observed that the node in RiCoBiT is connected to five other nodes as compared four that of mesh and torus. This means that node of RiCoBiT can send or receive more packets, and hence process more data. Also from the above equations, it is clear that the time taken for packet to reach the destination is also less to the other compared topologies. This implies that the more packets can be processed per unit time using the proposed topologies thereby providing increased throughput. The throughput of a system depends on the amount of output it can generate per unit time. This implies that the node interface in topology should be able to get more packets to generate more results which are a function of the waiting time of the packets. So for a given number of nodes, the throughput is minimum when the delay is high and worst case is when the delay at the node is minimum. Considering the above equations derived for time bounds, we can bound the throughput as given below:

$$\frac{1}{Tp + Tt \sqcap n(2 \sqcap Nb + 1)} \leq Th \leq \frac{1}{Tp + Tt}$$

From the delay estimation it is proved that the proposed architecture performs well for all the cases. Hence, the throughput is all better for the proposed architecture.

The above-mentioned topologies are subjected to experimental setup using simulation. All the above-discussed cases were simulated with varying buffer size, packet size, and traffic loads. It was observed that proposed topology performs better than mesh and torus.

5 Conclusion

A new architecture for Network-on-Chip based systems namely RiCoBiT—Ring Connected binary Tree—is proposed in this paper. The proposed architecture is scalable and structured and is limited only by the fabrication technology. The paper also proposes with proof, an optimal routing algorithm which routes the packet from the source to the destination though the shortest path. The comparative study of the performance parameters namely maximum hop count, average hop count, number of wire segments, and wire length is also presented. It also discusses and bounds latency and throughput parameters. An expression for these parameters is analytically deduced and the validity is computationally verified. These performance parameters are compared with that in 2D mesh and torus. It is observed that the length of the shortest path is much lower when compared to 2D mesh and torus of the same order. This enables the packets to reach the destination much faster thereby improving the performance and throughput. Also, the requirement in terms of area is almost the same as others. This implies that the proposed architecture performs better without increase in the area. These parameters were also verified using simulation experimental setup.

The proposed architecture is being implemented using a FPGA. This is to enable the study of performance and area factors under real-time conditions.

References

1. William J. Dally, Brian Towles, Route Packets Not Wires: on chip interconnection network, DAC 2001 June 2001.
2. Ahmed Hemani, Axel Jantsch, Shashi Kumar, Adam Postula, Johnny berg, Mikael Millberg, Dan Lindqvist, Network on a Chip: An architecture for billion transistor era, DAC 2001.
3. J. Nurmi: Network on Chip: A New Paradigm for System on Chip Design. Proceedings 2005 International Symposium on System on Chip, 15 17 November 2005.
4. Ville Rantala, Teijo Lehtonen, Juha Plosila, Network On Chip Routing Algorithms, TUCS Technical Report No 779, August 2006.
5. Tobias Bjerregaard And Shankar Mahadevan, A Survey of Research and Practices of Network on Chip, ACM Computing Survey March 2006.
6. Radu Marculescu, IEEE, Natalie Enright Jerger, Yatin Hoskote, Umit Y. Ogras and Li Shiuan Peh, Outstanding Research Problems in NoC Design: System, Micro architecture, and Circuit

Perspectives, IEEE Transactions on computer aided design of integrated circuits and systems, vol. 28, no. 1, January 2009.
7. L. Benini, G. De Micheli. Networks on chips: A new SoC paradigm. IEEE Computer. 35(1), 2002.
8. S. Kumar, A. Jantsch, J. Soininen, M. Forsell, M. Millberg, J. Oberg, K. Tiensyrja, and A. Hemani, A Network on Chip Architecture and Design Methodology, Proceedings International Symposium VLSI (ISVLSI), pp. 117–124, 2002.
9. Dan Marconett, A Survey of Architectural Design and Implementation Tradeoffs in Network on Chip Systems, University of California, Davis.
10. Nilanjan Banerjee, Praveen Vellanki, Karam S. Chatha, “A Power and Performance Model for Network-on-Chip Architectures”, Proceedings of the Design, Automation, and Test in Europe Conference and Exhibition (DATE), p. 21250, 2004. Transactions on Computers, vol. 54, no. 8, pp. 1025–1040, August, 2005.

Application of Compressed Sensing (CS) for ECG Signal Compression: A Review

Yuvraj V. Parkale and Sanjay L. Nalbalwar

Abstract Compressed Sensing (CS) is a fast growing signal processing technique that compresses the signal while sensing and enables exact reconstruction of the signal if the signal is sparse with a few numbers of measurements only. This scheme results in reduction of storage requirement and low power consumption of system compared to Nyquist sampling theorem, where the sampling frequency must be at least double the maximum frequency present in the signal for the exact reconstruction of the signal. This paper presents an in-depth study on recent trends in CS focused on ECG compression. Compression Ratio (CR), % Root-mean-squared Difference (% PRD), Signal-to-Noise Ratio (SNR), Root-Mean Square Error (RMSE), Sparsity and power consumption are used as the performance evaluation parameters. Finally, we have presented the conclusions based on the literature review and discussed the major challenges in CS ECG implementation.

Keywords Compressed sensing (CS) · ECG compression · SPIHT · Sparsity

1 Introduction

Electrocardiogram (ECG) is an important and advanced diagnostic tool used for various heart diseases diagnosis, such as arrhythmia, myocardial ischemia, and cardiac infarction. The detail interpretation provides information about patient's health. The different heart activities are represented by different waves. A normal ECG includes a P wave followed with QRS complex wave and lasts with T wave.

Y.V. Parkale (✉) · S.L. Nalbalwar
Department of Electronics and Telecommunication Engineering,
Dr. Babasaheb Ambedkar Technological University (DBATU),
Lonere (Established & Funded by Government of Maharashtra), Lonere, India
e-mail: yuvrajparkale@gmail.com

S.L. Nalbalwar
e-mail: slnalbalwar@gmail.com

S.C. Satapathy et al. (eds.), *Proceedings of the International Conference on Data Engineering and Communication Technology*, Advances in Intelligent Systems and Computing 469, DOI 10.1007/978-981-10-1678-3_5

Figure 1 shows the normal ECG waveform. The P wave is caused produced by the depolarization of the atria muscles and associated with their contraction, while the QRS complex wave is initiated by the depolarization of ventricles before to their contraction. The detection of the QRS complex is an important step in automated ECG analysis. Because of their distinctive shape, the QRS complex used as the reference tip for automated heart rate monitoring and as the initial point for additional evaluation.

Therefore, the long-term ECG data records become an important aspect to perceive information from these heart signals, resulting in increase in memory size of data. Hence, the ECG data compression becomes an essential for decreasing the data storage and the transmission times. ECG compression methods are categorized into two main groups: (i) Direct compression methods (ii) Transform compression methods. In the transform methods, the discrete wavelet transform-based techniques are simple to implement and provides good localization in both time and frequency scale.

The further best work in this area is described by embedded zero tree wavelet (EZW) [1] and a set partitioning in hierarchical tree (SPIHT) [2, 3] protocols, which work on the self similarity basis of the wavelet transform.

Compared to traditional ECG compression scheme, CS [4, 5, 6] transfers the computational load from the encoder side to the decoder side, and thus provides simple encoder hardware implementations. Also, there is no need to encode locations of the significant wavelet coefficients. This paper presents the detailed review of different aspects of CS-based ECG compression. The paper is outlined as: Sect. 2 describes the CS acquisition and recovery model. Section 3 presents different performance measures of signal. In Sect. 4, we have presented the detailed literature reviews on current CS-based ECG compression followed by conclusions in Sect. 5.

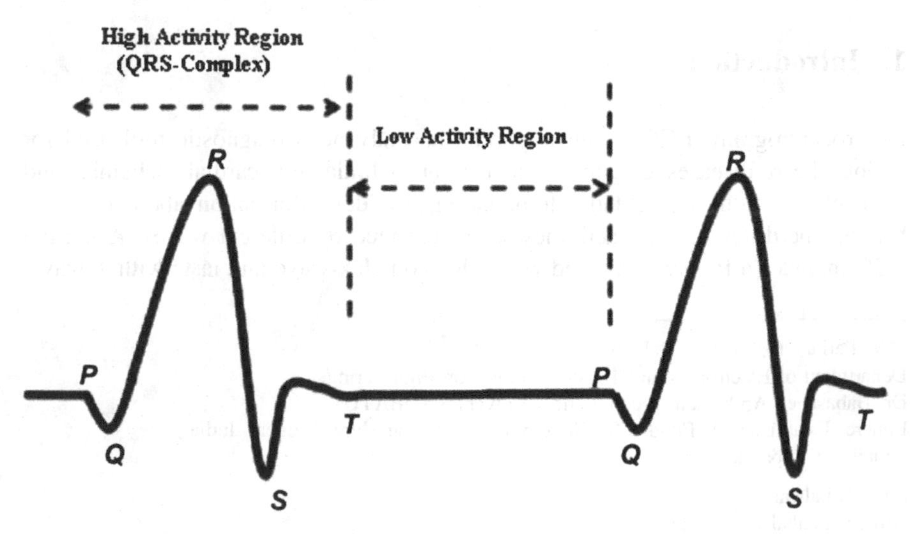

Fig. 1 Normal ECG waveform

2 Compressed Sensing (CS) Framework

2.1 Background

CS is a fast developing signal compression technique that acquires the signal and exactly recovers with few numbers of samples than Shannon–Nyquist sampling. CS takes advantage of signal sparsity property in some domain like wavelet transform. Nyquist sampling depends on the greatest amount of rate of alteration of a signal, whereas CS depends on the greatest amount of rate of knowledge in a signal. The CS executes two main operations: compression and recovery of signal.

2.2 CS Sensing Model

Compressed sensing scheme is represented as follows:

$$y = \Phi f \tag{1}$$

where, f is the original input signal of length $N \times 1$, y is the compressed output signal of length $M \times 1$, and Φ is the $M \times N$ sensing matrix. Here, as Φ is always constant, CS is nonadaptive scheme.

The input signal x is further defined as

$$f = \Psi_x \tag{2}$$

where, x is the nonsparse input signal with length N and Ψ is the $N \times N$ sparsifying basis. Combined form of Eqs. (1) and (2) as follows:

$$y = \Theta_x \tag{3}$$

where, $\Theta = \Phi\Psi$ is commonly referred to as the sensing matrix. CS acquires $M << N$ observations from N samples utilizing random linear estimate. In order to implement a CS algorithm correctly, there are three key requirements:

(a) The input signal f must be a sparse in some domain.
(b) The Ψ and Φ must be incoherent.
(c) The Φ should satisfy the Restricted Isometric Property (RIP) [7] and defined as

$$(1 - \delta_s)\, \|f\|_2^2 \le \|\Theta f\|_2^2 \le (1 + \delta_s)\, \|f\|_2^2 \tag{4}$$

For exact and stable compression, and signal recovery, Candes et al. [8, 9] recommends a minimum number of compressed measurements (m) based on the sparsity and coherence constraints outlined.

$$m \geq c.\mu^2(\Psi, \Phi)s.\log(N) \tag{5}$$

where, s is number of nonzero elements and c is a small fixed value. The sensing matrix can be designed with random distribution [10] values from Bernoulli, Gaussian, and uniform probability density functions.

2.3 CS Signal Reconstruction Model

Because Φ is not a square matrix, this CS problem becomes under determined problem with many possible solutions. These CS recovery algorithms require information of a representation basis where the signal is compressible or sparse for approximate or accurate recovery of signal. Reconstruction algorithms in CS exploit the sparse solution by minimizing L0, L1, L2 norm over solution space. L0-norm minimization will accurately reconstruct the original signal under the sparse condition, with slow speed and is NP (Non-Polynomial) hard. The L2-norm minimization is fast but it does not find the sparse solution resulting in error. The L1-norm gives exact sparse solution with efficient reconstruction speed. Hence, L1-norm is the good alternative to L0-norm and L2-norm minimization to find the accurately sparse solution. Finally, original signal can be reconstructed by calculating x using L1 norm minimization as given by equation below:

$$\min \|\hat{x}\|_1 \text{ Subject to } y = \Theta\hat{x} = \Psi\Phi\hat{x} \tag{6}$$

The sparsity and incoherence conditions guarantee the high probability of sparse exact solution using Eq. (6).There are different CS reconstruction algorithms available based on convex optimization method for e.g. Basis Pursuit (BP) [11], BP denoising (BPDN) [11], M-BPDN [12], LASSO [13] and greedy methods like OMP [14, 15], CoSaMP [16].

3 Performance Evaluation Parameters

There are different distortion measures used for performance evaluation of signal like % root-mean squared difference (PRD), Compression ratio (CR), Root-mean square error (RMS), Signal-to-noise ratio (SNR), and sparsity.

3.1 Percentage Root-Mean Squared Difference (PRD)

PRD is the measure of the difference between the input signal and recovered signal and given as:

$$PRD(\%) = \frac{\sum_{n=1}^{N} (f(n) - \hat{f}(n))^2}{\sum_{n=1}^{N} f^2(n)} \tag{7}$$

Measurement of PRD without the DC level in the input signal is given as

$$PRD(\%) = \frac{\sum_{n=1}^{N} (f(n) - \hat{f}(n))^2}{\sum_{n=1}^{N} (f(n) - \bar{f})^2} \times 100 \tag{8}$$

Here, $f(n)$ is the input signal, $\hat{f}(n)$ is the recovered signal, $\bar{f}$ is the mean of the signal, and N is the length of signal.

3.2 Compression Ratio (CR)

CR is used to measure the reduction in the dimensionality of the signal and given as:

$$CR = \frac{M}{N} \tag{9}$$

where the input signal is of the length N and M *is* the number of measurements taken from sensing matrix.

3.3 Root-Mean Square Error (RMSE)

RMSE is given as:

$$RMS = \sqrt{\frac{\sum_{n=1}^{N} (f(n) - \hat{f}(n))^2}{N}} \tag{10}$$

where, $f(n)$ is the original signal and $\hat{f}(n)$ is the recovered output signal and N is the length of input signal.

3.4 Signal-to-Noise Ratio (SNR)

SNR is given as

$$SNR = 10 \times \log\left(\frac{\sum_{n=1}^{N} (f(n) - \bar{f})^2}{\sum_{n=1}^{N} (f(n) - \hat{f}(n))^2}\right) \tag{11}$$

where $f(n)$ is the original input signal, $\hat{f}(n)$ is the recovered output signal, $\bar{f}$ is the average of the signal, and N is the length of input signal.

3.5 Sparsity

When a given signal contains only few no. of nonzero (K) coefficients, it is called as sparse signal and given as

$$\text{Sparsity}(\%) = \frac{(N-K)}{N} \times 100 \tag{12}$$

where, N is the signal length, K is the number of nonzero coefficients of the signal and *(N-K)* is the number of discarded coefficients of the signal.

4 Application of Compressed Sensing (CS) for ECG Signal Compression

An extensive literature survey have been performed on CS-based ECG compression papers. Table 1 shows the comparative summary of literature papers for CS-based ECG signal compression.

Pooyan et al. [2] tested wavelet transform on the ECG signal and SPIHT technique is used to encode the coefficients. SPIHT achieves CR = 21.4, PRD = 3. 1 with a very good reconstruction quality and outperforms all others compression algorithm. SPIHT has a low computational complexity and easy to implement. Polania et al. [17] proposed a 1-lead compression method. In this, the ECG signal's quasi periodic nature is exploited in between adjacent beats of samples. The author utilized the distributed compressed sensing to explore the common support between samples of adjacent beats. The drawback of the scheme is increase in the computational complexity at encoder. Experimentation is performed using the MIT-BIH Arrhythmia Database. The proposed CS-based scheme accomplishes a good CR for low PRDs and also out performs SPIHT. Polania et al. [18] incorporate two properties of signal structure; in the first property the wavelet scale dependencies are included into the recovery methods and second used the great common support

Table 1 Comparative summary of literature for CS-based ECG signal compression

Author	Sparsity basis, sensing matrix, reconstruction algorithm used	PRD (%)	CR	Strengths/Findings
Pooyan [3]	Set partitioning in hierarchical trees (SPIHT) algorithm	3.1	21.4	High efficiency, easy to implement, computationally simple
Polania [17]	Daubechies wavelets (db4), random gaussian matrix, SOMP	2.57	7.23	Achieves a good compression ratio, outperforms SPIHT as the PRD increases, offers a low complexity encoder
Polania [18]	Wavelet transform-based tree-structure model, bernoulli matrices, MMB-CoSaMP and MMB-IHT (Record no.100)	IHT = 3.65, CoSaMP = 3.86	6.4	Simple hardware design, less data size, and small computational need
Daniel [22]	Wavelet bases, random bernoulli matrix, BP	9	2.5	Simple encoder design
Ansari-Ram [23]	Biorothogonal4.4, nonuniform binary matrix, convex optimization	8.58	5	Increases the overall PRD
Akanks-ha Mishra [19–21]	29 different wavelet families like coiflets, daubichies, symlets, biorthogonal, reverse biorthogonal, random gaussian matrix and KLT sensing matrix, BP	0.01 0.01	2 2	rbio3.9 shows best sparsity. KLT sensing matrix superior than gaussian matrix
Anna Dixon [24, 25]	Random gaussian matrix, convex optimization, OMP, CoSaMP, ROLS, NIHT	–	–	CoSaMP and L1-norm convex gives best accuracy, CoSaMP preferred in noisy conditions, OMP is best for low computational complexity
Mamaghanian [33, 36]	Wavelet basis, random gaussian matrix, BP	<9	3.44	PRD = 0–9 % is “good or very good” ECG signal quality

of wavelet domain coefficients. The two model-based algorithms namely modified model-based MMB-CoSaMP and MMB–IHT are evaluated. For compression ratio CR = 4, the PRD of MMB–IHT is 3.65 and for MMB-CoSaMP is 3.86. For nearly all the ECG records, MMB-IHT shows superior execution than MMB-CoSaMP.

- *Selection of Best Sparsity Basis for CS ECG Compression*

One of the primary research areas for ECG compression is the choice of sparsity basis. Mishra et al. [19, 20, 21] evaluated a best wavelet basis for ECG Signal compression using Compressed Sensing approach by comparing several wavelet families like Haar, Daubechies, Reverse Biorthogonal, and biorthogonal etc. These families were evaluated using the performance metrics such as MSE, PSNR, PRD, and Correlation Coefficient (CoC). Here, L1 optimization is used as the signal reconstruction method. The result analysis is performed for five different compression ratios, like 2:1, 4:1, 6:1, 8:1, 10:1 and from each CR the best fit wavelet family was identified. From reverse biorthogonal wavelet family, rbio3.7 and rbio3.9 shows the best results. Finally, the best Daubechies wavelet (db) is selected for specific CR. For CR 2:1, db4 is chosen. Similarly, for CR 4:1, db8 is selected. For CR 6:1, db4 is most suitable Daubechies wavelet.

- *Different Sensing Matrices for ECG Signal Compression*

Polania et al. [17] used Random Gaussian Matrix with zero mean and 1/m variance achieves a good compression ratio. Polania et al. [18] and Chae et al. [22] tested Bernoulli matrices results in small data storage, less computational complexity, and simple encoder hardware design. Mishra et al. [19, 20, 21] compared the random Gaussian matrix and KLT sensing matrix performance for different compression ratios: for Random Gaussian matrix with CR = 2, PSNR = 57.57, PRD (%) = 0.01, MSE = 1.57, COC = 0.9994 where for KLT sensing matrix with CR = 2, PSNR = 59.92, PRD (%) = 0.01, MSE = 1.20, COC = 0.9996. This result shows that KLT sensing matrix shows better performance than random Gaussian matrix. Ansari-Ram and Hosseini-Khayat [23] proposed a nonuniform measurement matrix. This matrix acquires the QRS complex, i.e., region of interest and increases the total PRD value. This method has a weakness since it is also required to transmit sensing matrix to decoder end.

- *CS Reconstruction Algorithms for ECG Compression*

Dixon et al. [24, 25] evaluated comparative analysis on state-of-the-art CS recovery algorithms: BP, convex optimization, OMP, CoSaMP, ROLS and NIHT based on metrics like computational time, accuracy and noise tolerance. When accuracy is needed CoSaMP and L1 based convex are the natural choices. In noisy conditions CoSaMP outperforms L1-norm convex. OMP is preferable where computational complexity is important like real-time implementation with low power consumption.

- *Performance of CS ECG with Different Sparsity and Noisy Conditions*

Chae et al. [22] evaluated the CS ECG compression performance under the noisy ECG signal acquisition and with varied heartbeat rate because of body movements. From results, we can conclude that the CS with noise is quite difficult to minimize because of nonlinear nature of the recovered noise. The CS performance is compared to TH-DWT for ECG compression, where TH-DWT outperforms CS in the sense of CR. Care should be taken while applying CS for ECG compression as the CS is very susceptible to noise and sparseness of the signal. The TH-DWT method attains a PRD of 9 % at a CR = 5 whereas for the same PRD the CS attains a CR = 1.67. With noisy signals, the TH-DWT shows a flat SNR up to a CR = 2.5 after which have a slight fall in SNR, while the CS performance sharply decreased from a CR = 1.25.

- *Application of Dictionary Approach for ECG Compression*

Polania and Barner [26] proposed the multiscale dictionary learning approach for the recovery of ECG signals and evaluated the results with wavelet and single scale dictionary approach. Results show that these dictionary learning-based schemes provide better performance than the CS scheme using standard wavelet dictionaries. A multiscale dictionary in CS schemes improves the quality of the recovered ECG signal when compared to single scale methods. The proposed method utilizes the different wavelet subbands information at various scales to efficiently learn sparse and redundant dictionaries for representation of the ECG. Pant and Sridhar Krishnan [27] proposed the dictionary learning algorithms which produce a dictionary which can be used with L_p^{2d}–RLS method. This approach significantly improves signal recovery performance of L_p^{2d}–RLS method for ECG signals.

Fira et al. [28, 29] obtained the best results for the CPCS method using optimized projection matrix and patient-specific dictionary. The optimized dictionaries show the excellent results for all the records. The great extent of quality score achieved, for a 15:1 CR, 0.97 PRD and 15.46 QS is 50 % with CPCS technique with patient-specific heart beats (PRD = 0.51, QS = 29.13) without preprocessing. Singh and Dandapat [30] proposed distributed CS (DCS) for multichannel ECG signals with sparse learned dictionary, which are suitable for sparsity control of a signal. Pathology-specific and normal overcomplete dictionaries are used as the sparsifying basis and learnt using K-SVD algorithm. This improves the efficiency of the conventional CS in views of data size and reconstruction time.

- *Bayesian Learning Approach for ECG Compression*

Zhang et al. [31] proposed the Bayesian learning approach for the reconstruction of ECG signal which is sparser than existing compressed sensing solutions and it is also faster due to the improved sparsity. However, Bayesian approach has a drawback such as lack of theoretical foundation between the Bayesian approach, RIP, and incoherence. Some of the approaches fulfill the condition of RIP and incoherence, but in case of the full rank Fourier matrix, the BCS is failed. Zhang et al. [32] successfully applied the block sparse Bayesian learning (BSBL) structure to compress as well as recover non-sparse FECG signal and improved the performance using the correlation structure of signals. Experiments are performed on the DaISy Dataset and the OSET Dataset. Same sensing matrix is used with every experiment for all the CS methods used to compress FECG recordings. Two categories CS algorithms are tested. First category of recovery methods do not utilize block structure of signals includes CoSaMP, BP, SL0, Elastic-Net, and EM-GM-AMP which are from the class of greedy algorithms. The Basis Pursuit algorithm is used to reconstruct adult ECG recordings. The second group of algorithms utilized the structure of signals which includes, Block Basis Pursuit, Block-OMP, StructOMP, BM-MAP-OMP, and CluSS-MCMC. Block Basis Pursuit and Block-OMP requires a priori information of the block division. The result of comparison shows satisfactory quality with BSBL-BO algorithm. The BSBL-BO is evaluated for various factors such as impact of, the block partition effects, effect of compression ratio, and the impact of number of nonzero column entries of the sensing matrix. The proposed framework can be employed to many other telemedicine applications, such as wireless electroencephalogram, and electromyography.

- *On Quality of the Recovered ECG in the Views of Clinical usage*

Mamaghanian et al. [33] evaluated the PRD values for NIHT, EIHT, and FLIHT using the wavelet tree model. Among all, EIHT shows the great degrees of performance for S-Sparse signals. For model-based reconstruction methods MB-NIHT accomplish 95 % and more successful recoveries for M = 160 number of measurements. The same is achieved for M = 224 with EIHT and M = 192 with MB-FLIHT. Hence, NIHT and FLIHT results in a improved signal recovery performance in view of the probability of signal reconstruction and the output recovered PRD. This research shows that PRD values from 0–9 % are considered as "good or very good" ECG signal quality for clinical diagnosis. Zigel et al. [34] suggested a novel distortion metric named weighted diagnostic distortion (WDD) for ECG signal compression. The result shows that the suggested WDD metric is most appropriate for evaluating ECG recovered signals compared to the PRD metric. Drawback of WDD is that it is expensive to calculate compared to inexpensive measure of PRD.

- *Real-Time CS Hardware Implementation for ECG Signal Compression*

In the recent years significant research focus and efforts are made on the design, development and implementation of real-time CS hardware for different applications. Duarte et al. [35] proposed the one pixel imaging using CS. Body area network (BAN) is one of the recent application which is explored by many publications. Mamaghanian et al. [36] evaluated the performance of CS for wireless BAN on shimmer embedded platform which outperform DWT based lossy compression technique. Mishali et al. [37] designed "analog-domain CS system—a modulated wideband converter (MWC)". Chen et al. [38, 39] presented digital based CS for ECG and EEG signals. Dixon et al. [25] evaluated the performance of bio-medical signal sensors for BAN application on real time hardware.

5 Conclusion

In this review paper, we have presented a complete survey of the CS area for 1-dimensional biomedical application focused on CS-based ECG signal compression. We have investigated the basic of CS technique and discussed theoretical and mathematical basis of the important concepts. We have presented the reviews on some of important areas of CS-based ECG compression like choice of most excellent wavelet basis function for maximum sparsity of ECG signal, different sensing matrices for ECG signal compression, evaluation of CS reconstruction algorithms for ECG signal recovery with good accuracy and less reconstruction time, performance of CS ECG signal in different sparseness and noisy conditions, application of dictionary approach for ECG compression, Bayesian learning approach for ECG signal compression or recovery and real-time CS hardware implementation for ECG signal compression. Research on CS has demonstrated that CS is suitable alternative compared to the state of the art ECG compression techniques like SPIHT. From the review summary we can conclude that biorthogonal wavelet family rbio3.7 and rbio3.9 gives best sparsity, Bernoulli's matrices results in simple encoder design, OMP is the best choice for real time implementation. Dictionary learning approach will improve the CS reconstruction performance, while recently emerged Bayesian learning approach will even outperform CS-based approaches. Low power consumption and reconstruction quality of signal are the major challenges faced by real-time CS hardware implementation.

References

1. M.L. Hilton: Wavelet and Wavelet Packet Compression of Electrocardiograms. IEEE T-BME, vol. 44, no. 5, pp. 394–402, (1997).
2. Lu Z., Kim D. Y., Pearlman W. A.: Wavelet Compression of ECG Signals by the Set Partitioning in Hierarchical Trees Algorithm. IEEE T-BME, vol. 47, no. 7, pp. 849–856 (2000).
3. M. Pooyan, Ali Taheri, Morteza Moazami-Goudarzi, Iman Saboori: Wavelet Compression of ECG Signals Using SPIHT Algorithm. International Journal of Information and Communication Engineering, vol. 1 (2005).
4. D. Donoho: Compressed Sensing. IEEE Trans. Inf. Theory, vol. 52, no. 4, pp. 1289–1306 (2006).
5. araniuk R. G.: Compressive Sensing. IEEE Signal Processing Magazine, 24, 118–121 (2007).
6. Emmanuel J. Candes, Michael B. Wakin: An Introduction to Compressive Sampling. IEEE Signal Processing Magazine, pp. 21–30 (2008).
7. Emmanuel J. Candes: The Restricted Isometry Property and its Implications for Compressed Sensing. *Comptes Rendus Mathematique*, 346, 589–592. (2008).
8. E. Candes, J. Romberg, T. Tao: Stable signal recovery from incomplete and inaccurate measurements. Commun. Pure Applied Math., vol. 59, no. 8, pp. 1207–1223 (2006).
9. Emmanuel J. Candes, Romberg, T. Tao: Robust Uncertainty Principles: Exact Signal Reconstruction from Highly Incomplete Frequency Information. IEEE Trans. Inf. Theory, 52 (2), 489–509 (2006).

10. E. Candes, T. Tao: Near-optimal signal recovery from random projections: Universal encoding strategies. IEEE Trans. Inf. Theory, vol. 52, no. 12, pp. 5406–5425 (2006).
11. S. S. Chen, D. L. Donoho, M. A. Saunders: Atomic decomposition by Basis Pursuit. Journal on Scientific Computing, vol. 20, pp. 33–61 (1999).
12. W. Lu, N. Vaswani: Modified basis pursuit denoising (modified BPDN) for noisy compressive sensing with partially known support. ICASSP, pp. 3926–3929 (2010).
13. R. Tibshirani: Regression shrinkage and selection via the lasso. J. Royal Statistical Society. Series B (Methodological), pp. 267–288 (1996).
14. J. Tropp, A. Gilbert: Signal recovery from random measurements via orthogonal matching pursuit. IEEE Trans. Inf. Theory, vol. 53, no. 12, p. 4655 (2007).
15. D. Donoho, Y. Tsaig, I. Drori, and J. Starck: Sparse solution of underdetermined linear equations by stagewise orthogonal matching pursuit (2006).
16. D. Needell, J. Tropp: CoSaMP: Iterative signal recovery from incomplete and inaccurate samples. Applied Comp. Harmonic Analysis, vol. 26, no. 3, pp. 301–321 (2009).
17. Luisa F. Polania, Rafael E. Carrillo, Manuel Blanco-Velasco, Kenneth E. Barner: Compressed Sensing Based Method for ECG Compression. ICASSP, Page no.761–764 (2011).
18. Luisa F. Polania, Rafael E. Carrillo, Manuel Blanco-Velasco, Kenneth E. Barner: Exploiting Prior Knowledge in Compressed Sensing Wireless ECG Systems. IEEE Jour. of Biomedical and Health Informatics (2014).
19. A. Mishra, Falgun Thakkar, Chintan Modi, Rahul Kher: ECG signal compression using Compressive Sensing and wavelet transform. Ann. International Conference of the IEEE Engineering in Medicine and Biology Society (EMBC), pp. 3404–3407, 2012.
20. A. Mishra, F. N. Thakkar, C. Modi, R. Kher: Selecting the Most Favorable Wavelet for Compressing ECG Signals Using Compressive Sensing Approach. International Conf. on Comm. Systems and Network Technologies (CSNT), pp. 128–132 (2012).
21. A. Mishra, Falgun Thakkar, Chintan Modi, Rahul Kher: Comparative Analysis of Wavelet Basis Functions for ECG Signal Compression through Compressive Sensing. International Jour. of Comp. Science and Telecommunications, vol. 3, p. 9(2012).
22. D. H. Chae, Y. F. Alem, S. Durrani, R. A. Kennedy: Performance study of compressive sampling for ECG signal compression in noisy and varying sparsity acquisition. In ICASSP, pp. 1306–1309 (2013).
23. F. Ansari-Ram and S. Hosseini-Khayat: ECG signal compression using compressed sensing with nonuniform binary matrices. In 16th CSI Inter. Symposium on, Artificial Intelligence and Signal Processing (AISP), pp. 305–309, (2012).
24. A. M. R. Dixon, E. G. Allstot, A. Y. Chen, D. Gangopadhyay, D. J. Allstot: Compressed Sensing reconstruction: Comparative study with applications to ECG bio-signals. In IEEE International Symposium on Circuits and Systems (ISCAS), pp. 805–808 (2011).
25. A. M. R. Dixon, E. G. Allstot, D. Gangopadhyay, D. J. Allstot: Compressed Sensing System Considerations for ECG and EMG Wireless Biosensors. IEEE Transactions on Biomedical Circuits and Systems, vol. 6, pp. 156–166 (2012).
26. Luisa F. Polania, Kenneth E. Barner: Multi-Scale Dictionary Learning For Compressive Sensing ECG. IEEE DSPSPE (2013).
27. Jeevan K. Pant, Sridhar Krishnan: Compressive Sensing of Electrocardiogram Signals by Promoting Sparsity on the Second-Order Difference and by Using Dictionary Learning. IEEE Transactions on Biomedical Circuits and Systems, Vol. 8, No. 2 (April 2014).
28. Catalina Monica Fira, Liviu Goras, Constantin Barabasa, Nicolae Cleju: ECG Compressed Sensing Based on Classification in Compressed Space and Specified Dictionaries. 19th European Signal Processing Conference (EUSIPCO 2011), Barcelona, Spain (2011).
29. Monica Fira, Liviu Goras: On Projection Matrices and Dictionaries in ECG Compressive Sensing - a Comparative Study. NEURAL, Serbia (2014).
30. Anurag Singh, S. Dandapat: Distributed Compressive Sensing for Multichannel ECG Signals over Learned Dictionaries. Annual IEEE India Conference (INDICON) (2014).

31. Zhang Hong-xin, Wang Hai-qing, Li Xiao-ming, Lu Ying-hua, Zhang Li-kun: Implementation of compressive sensing in ECG and EEG signal processing. The Journal of China Universities of Posts and Telecommunications, ScienceDirect, pp. 122–126 (2010).
32. Zhilin Zhang, Tzyy-Ping Jung, Scott Makeig, Bhaskar D. Rao: Compressed Sensing for Energy-Efficient Wireless Telemonitoring of Noninvasive Fetal ECG via Block Sparse Bayesian Learning. IEEE Transactions on Biomedical Engineering, Vol. 60, No. 2, pp. 300–309 (2013).
33. H. Mamaghanian, N. Khaled, D. Atienza and P. Vandergheynst: Structured sparsity models for compressively sensed electrocardiogram signals: A comparative study. In IEEE Biomedical Circuits and Systems Conference (BioCAS), pp. 125–128 (2011).
34. Y. Zigel, Arnon Cohen, A. Katz: The weighted diagnostic distortion (WDD) measure for ECG signals compression. IEEE Transactions on Biomedical Engineering, vol. 47, pp. 1422–1430 (2000).
35. M. F. Duarte, M. A. Davenport, D. Takhar, J. N. Laska, Sun Ting, K. F. Kelly, R. G. Baraniuk: Single-Pixel Imaging via Compressive Sampling. IEEE Signal Processing Magazine, vol. 25, pp. 83–91 (2008).
36. Hossein Mamaghanian, Nadia Khaled, David Atienza, Pierre Vandergheynst: Compressed Sensing for Real-Time Energy-Efficient ECG Compression on Wireless Body Sensor Nodes. IEEE Transactions on Biomedical Engineering, Vol. 58, No. 9, (2011).
37. M. Mishali, Y. C. Eldar, O. Dounaevsky, E. Shoshan: Xampling: Analog to digital at Sub-Nyquist rates. IET Circuits, Devices Syst.,vol. 5, pp. 8–20, (2011).
38. F. Chen, A. P. Chandrakasan, V. Stojanovic: A signal-agnostic compressed sensing acquisition system for wireless and implantable sensors. In Proc. IEEE Custom Integrated Circuits Conf., pp. 1–4 (2010).
39. F. Chen, A. P. Chandrakasan, V. Stojanovic: Design and analysis of a hardware-efficient compressed sensing architecture for data compression in wireless sensors. IEEE J. Solid-State Circuits, vol. 47, pp. 744–756 (2012).

Tracking Pointer Based Approach for Iceberg Query Evaluation

Kale Sarika Prakash and P.M.J. Pratap

Abstract Extracting small set of data from large or huge database is a challenge in front of data warehouse system. The queries executed on data warehouse are of the nature aggregation function followed by having clause. This type of query is called as iceberg query. Present database system executes it just like normal query so it takes more time to execute. To increase execution speed of iceberg query on large database is the challenge in front of researchers. Previous research uses tuple scan approach and bitmap index pruning strategy to execute query which is time-consuming and it faces the problem of fruitless bitwise AND-XOR operation. They focus on only COUNT and SUM aggregate functions. To address these problems and improve efficiency of iceberg query the proposed research makes use of tracking pointer concept. It avoids fruitless bitwise AND-XOR operations and also it minimizes the futile queue pushing problem that occurs in previous research. Along with COUNT and SUM function this study creates framework for MIN, MAX, COUNT, and SUM aggregate functions. This proposed work uniquely distinguishes the MIN, MAX, and SUM operations which is not found in existing systems.

Keywords Data warehouse (DW) · Iceberg queries (IBQ) · Aggregate functions—COUNT · SUM · MIN · MAX · Bitwise AND-XOR operation

K.S. Prakash (✉) · P.M.J. Pratap
Department of Computer Science and Engineering,
St. Peters Institute of Higher Education and Research,
St. Peters University, Chennai, India
e-mail: kalesarikaprakash@gmail.com

P.M.J. Pratap
e-mail: drjoeprathap@gmail.com

S.C. Satapathy et al. (eds.), *Proceedings of the International Conference on Data Engineering and Communication Technology*, Advances in Intelligent Systems and Computing 469, DOI 10.1007/978-981-10-1678-3_6

1 Introduction

The users of the data warehouse are the decision makers, business analyst and knowledge workers of respective organization. They make use of data from data warehouse to predict some issues about their business. Accordingly they take decisions about business. Once they have taken decision they concentrate on implementation of the same. As they make use of data for planning means they do not use data warehouse frequently. The information which they collect from data warehouse is small information from huge data set. To extract such a type of data the query executed is of the nature aggregation of some value on some specified condition or threshold. This type of query is called as iceberg query. Iceberg queries were first studied by scientist named Fang et al. [1]. According to him, iceberg query is defined as the type of query which performs aggregation function on some set of attributes followed by having clause on some condition or on threshold value. The output of iceberg query is small set of data from huge data set as input. Because of this type of nature it is called as iceberg query. Syntax of an iceberg queries on a relational table T (Col1, Col2… Coln) is stated below:

SELECT Col1, Col2, …, Coln AGG (*) FROM T GROUP BY Col1, Col2…, Coln, HAVING AGG (*) > = Threshold, where Col1, Col2,…, Coln represents a subset of attributes in R which are aggregate attributes. AGG is aggregate functions used in iceberg query like MIN, MAX, SUM, AVG, and COUNT. The greater than equal to (>=), less than or equal to (<=), or is equal to (= =) are special symbols used as a comparison predicate in having clause.

Due to threshold condition iceberg query returns very small distinct data set as output which looks like the tip of an iceberg. The output of iceberg query is very small that is on the tip of the iceberg compared to the input so such a type of query must answer quickly even it executes on huge data set. But current database systems do not take advantage of this feature of iceberg query. The relational database systems like Oracle, SQL server, Sybase, MySQL, and PostgreSQL are all using general aggregation algorithms [2–4] to execute iceberg query. They did not consider it as special query so the time required to execute such queries on these databases are more. Existing optimization techniques for processing iceberg queries [1, 5] can be categorized as the tuple-scan-based method, which requires minimum one table scan to read data from disk. They focus on reducing the number of passes when the data size is large. They did not make use of iceberg query property for efficient query processing. Due to this a tuple-scan-based method generally takes a long time to answer iceberg queries on large database. Besides these tuple-scan-based approaches, Ferro et al. [6] designed a two-level bitmap index which can be used for processing iceberg queries. However, it suffers from the massive empty bitwise AND results problem.

The research proposed in [7] provides dynamic pruning and vector alignment strategy tries to overcome empty bitwise AND operation problem. But empty bitwise XOR operation issue is occurred in this research. This empty bitwise XOR operation problem is overcome by [8, 9] but it faces the problem of futile queue pushing which degrades the performance of query. Proposed research concentrates on preprocessing of bitmap index. Due to this futile queue pushing cost is reduced. If the number of attributes in bitmap vector is increased the maintenance of number of queues is also increased. The preprocessing of bitmap vector reduces the queue maintenance cost. To achieve this proposed research makes use of tracking pointer concept along with making array of 1 bit position in each vector prior to perform followed operations. Existing research [7–11] only concentrate on COUNT aggregate function which is not sufficient to execute generalized iceberg query so to address this problem in our proposed research along with COUNT function MIN, MAX, and SUM operation worked out.

The remaining sections of this paper are structured as follows: Sect. 2 discusses related work. Section 3 discusses proposed tracking pointer strategy and in Sect. 4 comparative study of different iceberg query evaluation algorithm is discussed followed by conclusion in Sect. 5.

2 Related Work

Due to involvement of Internet there is rapid increase in input sources to form a data warehouse. So to build a data warehouse for large data set has become easier but extracting knowledge from it is a challenging task for researchers in the area of data mining, decision support system, and OLAP systems. The queries to be executed on the data warehouse are of the nature of iceberg query. In this section, we are highlighting the iceberg query execution strategies developed by different researchers and application of bitmap index to execute iceberg query.

The iceberg query processing is first described by Fang et al. [1] in 1998. In this study author proposed Hybrid and Multi-bucket algorithms by extending probabilistic technique used in [12]. In this research sampling and multi-hash functions are combined to improve the performance of iceberg query and reduce memory requirement. But these algorithms do not scale to large data sets.

To overcome this problem [1] suggests algorithms which consist of sampling and bucket counting mechanism. This mechanism generates false positive values which are candidates for the final result but it does not exceed the threshold limit. It also generates false negative values which are in the final result but it is not in the candidate list. The focus of this research is to minimize false positive. Different optimization methods are used which consist of hashing, multiple hashing, combination of multiple hash functions and multiple scan over relation. These methods reduce number of false positive values but it takes more time to execute query as it requires multiple scan of relation.

Iceberg query processing is also proposed by [5], focus of this study is to reduce number of table scans so that time required to execute the query will get reduced. It introduces methods to select candidate values using partitioning and POP (Postpone Partitioning) algorithms. This overcomes the problem of multiple scan over relation occurs in sampling and bucket counting mechanism [1]. The result of this study shows that performance of these algorithms degraded due to data order and memory size. If database is sorted then performance is excellent without regard to memory size.

A comparison was presented for Collective Iceberg Query Evaluation (CIQE) [13] using three standard methods like Sort Merge Aggregate (SMA), Hybrid Hash Aggregate (HHA), and ORACLE. CIQE indicates that performance of SMA is better on data sets with low to moderate number of targets than moderate to high skew. HHA performance was not robust and quite bad when the number of targets was high. There was a considerable performance gap between the online algorithms and ORACLE, indicating a scope for designing better iceberg query processing algorithms.

Above-mentioned approaches come under the group of tuple scan based method, which requires one physical table scan to read data from disk. These algorithms focus on reducing number of tuple scan but no one of them uses property of iceberg query.

Bitmap index is usually a better choice for querying the massive, high-dimensional scientific data sets. It supports fastest data accesses and reduced the query response time on both high-and low-cardinality values with a number of techniques [14]. Generating the bit map index of attribute will not affect on the performance of query because generated bitmap by database system is in compressed mode [15]. Therefore, use of bitmap index to execute iceberg query avoids the complete table scan.

However, [6] tries to make use of this property of iceberg query and uses bitmap index but it suffers from empty bitwise AND results problem. This problem is minimized by [7] using dynamic pruning and vector alignment approaches. This work leverages the antimonotone property of iceberg query and develops dynamic pruning algorithm using bitmap indexing. However, they notice that there is problem of massively empty bitwise AND results. To overcome this challenge they develop vector alignment algorithm which uses priority queue handling. The problem with this technique is that all vectors may not have 1 bit at same position and if it is not at same position then all AND as well as XOR operations are fruitless and time-consuming.

In this way both the above approaches suffer from fruitless AND as well as XOR operations. Research [8] try to handle empty XOR operation problem but did not able to solve fruitless bitwise AND operation problem. Both the research [7, 8] use priority queue concept for all vectors and faces the problem of futile queue pushing. In proposed research by making use of tracking pointer we are trying to minimize the number of fruitless bitwise XOR and AND operation. It will help to minimize futile queue pushing problem.

None of the above research [7–11] works on other aggregate functions like MIN, MAX, SUM. In proposed research we are developing framework for aggregate functions like MIN, MAX, SUM, and COUNT, only our strategy is not applicable for AVERAGE function.

3 Details About Proposed Tracking Pointer Strategy

This section highlights the methodology used in tracking pointer to improve the efficiency of iceberg query. The block diagram and pseudocode of proposed methods are included in this section.

3.1 Description of Tracking Pointer-Based IBQ Evaluation Method

Figure 1 shows the working flow of tracking pointer-based IBQ evaluation method. This model is used to interface with any data warehouse/large database system. Iceberg query is executed through this model. As per query attributes bitmap index is prepared. The preprocessing of bitmap index is important part. In this part the analysis of bitmap vector as per query attribute will be done. With this analysis we can directly prune the vector which reduces fruitless bitwise AND and XOR operation. Bitwise AND operation is performed to find probability of tuples to be subset of final answer and XOR operation is performed to generate new vector. In this strategy tracking pointer keeps track on POP out element list from priority

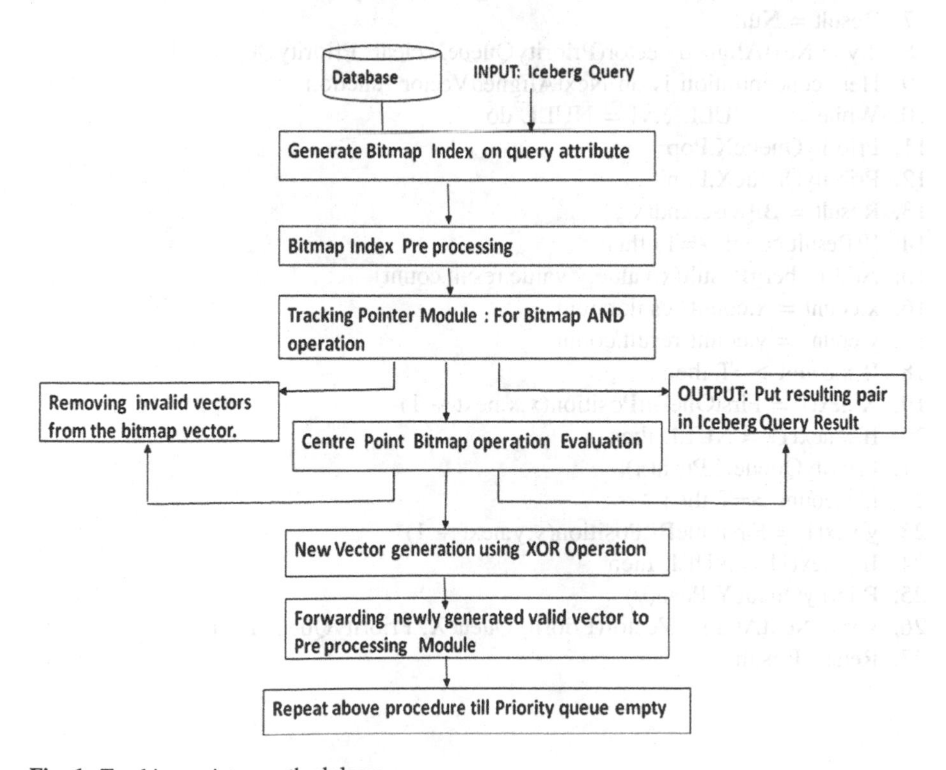

Fig. 1 Tracking pointer methodology

queue. Then check for longest first 1 bit of both the list has to be done. Tracking pointer points to first 1 bit of vectors and compares it with second vectors bits. This will repeat until both vectors have 1 bit on same position. As shown in Fig. 1 center point bitmap operation evaluation logic declares to prune vector and to perform complete operation to generate iceberg result.

3.2 Pseudocode for Iceberg Query Evaluation Using Tracking Pointer Concept

Input: Iceberg Query
Iceberg Result(attribute X, attribute Y, threshold T)
Output: iceberg results
Algorithm

1. PriorityQueueX.clear, PriorityQueueY.clear
2. for each vector x of attribute X do
3. if x.count > = T then
4. x.next1 = FirstOneBitPosition(x,0)
5. PriorityQueueX.Push(x)
6. Same logic is applied to push the highest priority vector in PriorityQueueY
7. Result = Null
8. x,y = NextAlignedVector(PriorityQueueX.clear, PriorityQueueY,T)
9. Here concentration is on NextAlignedVector function.
10. While x! = NULL &y! = NULL do
11. PriorityQueueX.Pop
12. PriorityQueueX.Pop
13. Result = BitwiseAnd(x,y)
14. If(Result.count >=T) then
15. Add IcebergResult(x.value,y.value,result.count)
16. x.count = x.count-result.count
17. y.count = y.count-result.count
18. If x.count >=T then
19. X.next1 = FirstOneBitPosition(x,x.next + 1)
20. If x.next1! = NULL then
21. PriorityQueueX.Push(x)
22. If y.count >=T then
23. y.next1 = FirstOneBitPosition(y,y.next + 1)
24. If y.next1! = NULL then
25. PriorityQueueY.Push(y)
26. x,y = NextAlignedVector(PriorityQueueX, PriorityQueueY,T)
27. Return Result

Same logic is applicable whenever new vector is generated after XOR operation. The fruitless XOR operations are reduced by assuming A1 AND B1 = R1 then New A1 = A1-R1 but before performing this XOR operation count number of 1 bits of A1 and R1. If A1-R1 or R1-A1 is greater than threshold then only perform New A1 = A1 XOR R1. After getting new vector apply above NextAlignedVector (PriorityQueueX, PriorityQueueY,T) function and repeat the procedure till both the queue will be empty.

4 Comparative Study Between Different Algorithms

In this section, we are showing the analysis of solving some iceberg query on given tables using different iceberg query processing strategies. The examples considered here are from [7, 8] and number of bitwise AND operation, XOR operations and number of iterations required to solve query are summarized.

Example: Select Month, category, COUNT (*) from R group by Month, category having COUNT (*) > 2.

Month	Category
July	Milk
June	Veg
July	Fruit
July	Milk
June	Veg
July	Fruit
July	Milk
July	Fruit
June	Veg
July	Milk
Aug	Fruit
Aug	Fruit

Table R

June	July	Aug	Fruit	Milk	Veg
0	1	0	0	1	0
1	0	0	0	0	1
0	1	0	1	0	0
0	1	0	0	1	0
1	0	0	0	0	1
0	1	0	1	0	0
0	1	0	0	1	0
0	1	0	1	0	0
1	0	0	0	0	1
0	1	0	0	1	0
0	0	1	1	0	0
0	0	1	1	0	0

Bitmap Indices for Month, category

This work solves above query using Bitmap Indexing, Dynamic Pruning, Vector Alignment, and Tracking pointer strategy. From this we collected the count of number of AND and XOR operations required to solve query. Similarly, number of iterations required for query execution is calculated through implemented module of proposed strategy. Table 1 shows the summary of results obtained from above-mentioned methods and implemented module. Experimental result shows that number of AND operation, XOR operation, and iteration count get reduced in case of tracking pointer strategy. This increases the execution speed of query and reduces I/O access.

Table 1 Comparative analysis

Iceberg query evaluation strategy	Bitwise AND operation	Bitwise XOR operations	Number of iterations required
Query1:			
Bitmap indexing	9	0	108
Dynamic pruning strategy	5	6	60
Vector alignment strategy	5	6	66
Tracking pointer strategy	3	1	51
Query2:			
Bitmap indexing	9	0	98
Dynamic pruning strategy	5	2	56
Vector alignment strategy	3	2	47
Tracking pointer strategy	3	0	39

5 Conclusion

Generally, knowledge discovery systems (KDS) and decision support systems (DSS) compute aggregate values of interesting attributes by processing it on large databases. In particular, iceberg query is a special type of aggregation query that computes aggregate values above threshold values provided in query. Normally, only a small number of results will satisfy the threshold constraint. The results are on the tip of iceberg means we are extracting small information from large database. Proposed research makes use of the antimontone property to speed up iceberg query execution by tracking pointer concept. With our analysis and experimental results we found that tracking pointer requires less number of AND operation, XOR operations, and number of iteration. Along with COUNT function proposed research is also concentrating on Aggregate functions like MIN, MAX, and SUM. To support AVERAGE function proposed strategy is not appropriate. AVERAGE function does not support antimontone property. Our research makes use of antimontone property of iceberg query which supports only COUNT, MIN, MAX, and SUM aggregate function. The challenge in front of current research is to develop framework for AVERAGE aggregate function.

References

1. M. Fang, N. Shivakumar, H. Garcia-Molina, R. Motwani, and J.D. Ullman, “Computing Iceberg Queries Efficiently,” Proc. Int’l Conf. Very Large Data Bases (VLDB), pp. 299–310, 1998.
2. G. Graefe, “Query Evaluation Techniques for Large Databases,” ACM Computing Surveys, vol. 25, no. 2, pp. 73–170, 1993.
3. W.P. Yan and P.A. Larson, “Data Reduction through Early Grouping,” Proc. Conf. Centre for Advanced Studies on Collaborative Research (CASCON), p. 74, 1994.

4. P.A. Larson, "Grouping and Duplicate Elimination: Benefits of Early Aggregation," Technical Report MSR-TR-97-36, Microsoft Research, 1997.
5. J. Bae and S. Lee, "Partitioning Algorithms for the Computation of Average Iceberg Queries," Proc. Second Int'l Conf. Data Warehousing and Knowledge Discovery (DaWaK), 2000.
6. A. Ferro, R. Giugno, P.L. Puglisi, and A. Pulvirenti, "BitCube: A Bottom-Up Cubing Engineering," Proc. Int'l Conf. Data Warehousing and Knowledge Discovery (DaWaK), pp. 189–203, 2009.
7. Bin He, Hui-I Hsiao, Ziyang Liu, Yu Huang and Yi Chen, "Efficient Iceberg Query Evaluation Using Compressed Bitmap Index", IEEE Transactions On Knowledge and Data Engineering, vol 24, issue 9, sept 2011, pp. 1570–1589.
8. C.V. Guru Rao, V. Shankar, "Efficient Iceberg Query Evaluation Using Compressed Bitmap Index by Deferring Bitwise- XOR Operations" 978-1-4673-4529-3/12/$31.00c 2012 IEEE.
9. C.V. Guru Rao, V. Shankar, "Computing Iceberg Queries Efficiently Using Bitmap Index Positions" DOI: 10.1190/ICHCI-IEEE.2013.6887811 Publication Year: 2013,Page(s): 1 – 6.
10. Vuppu.Shankar, Dr. C.V. Guru Rao, "Cache Based Evaluation of Iceberg Queries", International conference on Computer and Communications Technologies (ICCCT), 2014, DOI: 10.1109/ICCCT2.2014.7066694,Publication Year: 2014, Page(s): 1–5.
11. Rao, V.C.S.; Sammulal, P., "Efficient iceberg query evaluation using set representation", India Conference (INDICON), 2014 Annual IEEE DOI: 10.1109/INDICON.2014.7030537. Publication Year: 2014, Page(s): 1–5.
12. K.-Y. Whang, B.T.V. Zanden, and H.M. Taylor, "A Linear-Time Probabilistic Counting Algorithm for Database Applications," ACM Trans. Database Systems, vol. 15, no. 2, pp. 208–229, 1990.
13. K.P. Leela, P.M. Tolani, and J.R. Haritsa, "On Incorporating Iceberg Queries in Query Processors" Proc. Int'l Conf. Database Systems for Advances Applications (DASFAA), pp. 431–442, 2004.
14. Ying Mei, Kaifan Ji*, Feng Wang, "A Survey on Bitmap Index Technologies for Large-scale Data Retrieval" 978-1-4799-2808-8/13 $26.00 © 2013 IEEE.
15. F. Delie`ge and T.B. Pedersen, "Position List Word Aligned Hybrid: Optimizing Space and Performance for Compressed Bitmaps," Proc. Int'l Conf. Extending Database Technology (EDBT), pp. 228–239, 2010.

Performance Evaluation of Shortest Path Routing Algorithms in Real Road Networks

Nishtha Kesswani

Abstract Dijkstra's algorithm is one of the established algorithms that is used widely for shortest path calculation. The algorithm finds the shortest path from the source node to every other node in the network. Several variants of the Dijkstra's algorithm have been proposed by the researchers. This paper focusses on Multi-parameter Dijkstra's algorithm that uses multiple parameters for shortest path calculation in real road networks. The major contributions of this paper include (1) Comparison of the Dijkstra's algorithm to Multi-parameter Dijkstra's algorithm with special focus on real road networks, (2) Performance evaluation of Multi-parameter Dijkstra's algorithm, (3) Time complexity analysis of Different modifications of Dijkstra's algorithm.

Keywords Shortest path • Fastest path • Multi-parameter Dijkstra's algorithm

1 Introduction

Due to its capability of solving single-source shortest path problem, Dijkstra's algorithm is widely used. For instance, given a set of cities, representing the vertices V in a graph and the edges in the graph are represented by E, weight of the edges w representing the distance between the cities, Dijsktra's algorithm can be used for calculating the shortest route between any two cities. The graph representing the network can be represented as follows:

$$G = \langle V, E \rangle$$

Dijkstra's algorithm relaxes the vertices continuously until the shortest path is obtained. Though in real road networks, using the weight w or the distance between the edges may not be sufficient. Multi-parameter Dijkstra's algorithm proposed in [1]

Nishtha Kesswani (✉)
Central University of Rajasthan, BandarSindri, Ajmer, Rajasthan, India
e-mail: nishtha@curaj.ac.in

S.C. Satapathy et al. (eds.), *Proceedings of the International Conference on Data Engineering and Communication Technology*, Advances in Intelligent Systems and Computing 469, DOI 10.1007/978-981-10-1678-3_7

uses multiple parameters like real-time congestion and time taken to travel from source to destination into consideration. This paper compares Dijkstra's algorithm to Multi-parameter Dijkstra's (MPD) algorithm [1] and validates its use in real road networks.

Section 2 provides an overview of the original Dijkstra's algorithm and various other shortest path algorithms. Section 3 provides a comparative analysis of Dijkstra's algorithm and Multi-parameter Dijkstra's algorithm. Section 4 analyzes the time complexity of various algorithms. Conclusions and Future directions are given in Sect. 5.

2 Shortest Path Algorithms

The popularity of shortest path algorithms is due to their application in varied scenarios. The shortest path algorithms are also used for routing problems. Many algorithms have been suggested to compute the single-source shortest path or all pairs shortest path problems [2, 3]. While Dijkstra's algorithm solves the single-source shortest path problem, Bellman–Ford algorithm solves the single-source shortest path problem with negative edge weights. Algorithms such as Floyd–Warshall and Johnson's algorithm solve the all-pairs shortest path problem.

There are several approaches proposed in the literature. Some of them include privacy preserving shortest path calculation [4] using Private Information Retrieval in which the authors have proposed the involvement of third party in shortest path calculation and thus ensuring location privacy.

A framework for road networks has been suggested in [5]. The authors have devised highway-based labeling and an algorithm for the road networks.

Several modifications of the Dijkstra's algorithm have also been suggested in [3]. These include Dijkstra's algorithm with Double buckets (DKD), Dijkstra's algorithm with overflow bag implementation (DKM), and Dijkstra's algorithm with approximate buckets implementation (DKA). A comparison of various implementations of Dijkstra's algorithm has been given in [1].

Dijkstra's algorithm has also been extended to solve bi-objective shortest path problem in [6]. The authors have proposed an algorithm that runs in $O(N(m+n\log n))$ time to solve one-to-one and one-to-all bi-objective shortest path problems.

Next section compares Dijkstra's algorithm to its modified version, Multi-parameter Dijkstra's algorithm [1].

3 Multi-parameter Dijkstra's Algorithm Versus Dijkstra's Algorithm

Given a set of Edges E and a set of vertices V, the cities can be represented as Vertices and edges represent the distance between any two cities. The technique that has been suggested by Dijkstra's algorithm is to update the distances as soon as a shorter path is found between any two nodes. The existing path is updated with the newly discovered shorter path. This process is continued until all the neighboring nodes are visited. In this manner the shortest distance between any two nodes is calculated.

A shortest path algorithm can be used to find the fastest way to get from a node *A* to node *B* or find the least expensive route from *A* to *B* or it may be used to find the shortest possible distance from *A* to *B* [7]. Though Dijkstra's algorithm is able to calculate the shortest possible distance, but this might not necessarily be the fastest route or the least expensive one. In real road networks, users may want to calculate the route that fulfills other criteria as well.

Though distance between two nodes is an important factor while computing the shortest path between any two nodes, there are several other factors that can be taken into consideration while calculating the shortest path for real road networks. Apart from distance, Multi-parameter Dijkstra's algorithm [1] takes other factors such as the time taken to travel form one node to another and real-time congestion factor into consideration. As in real road networks, it would be difficult if only distance is taken into consideration. Since the time taken to travel from one city to another may be affected by other factors like congestion and blockage or ongoing construction on the road. It makes sense if the users are provided a large number of choices and according to the preference, the algorithm adapts. In Multi-parameter Dijkstra's algorithm, if the user gives the preference of distance as a basis of shortest path calculation then this algorithm functions as the original Dijkstra's algorithm. But, if the user chooses time or congestion as a deciding factor then the distance is weighted as per the time factor or congestion factor respectively. The time factor and congestion factor are weighted on a scale of 1–10, with higher value indicating that the route takes more time or is more congested. If w indicates the distance between any two nodes $v1$ and $v2$, the adjusted weight w can be given as follows:

$$w^{'} = w * \varphi$$

where φ indicates the congestion factor for the path. Also, the time taken to travel a path may increase or decrease irrespective of the distance. As other factors like some blockage on the road or construction work going on the road may affect the time taken in real life scenarios. Sensors may be used to capture the real-time information of the route before the user can actually decide on which route can be adopted. As per the real-time information received, the congestion and time factor may be adjusted. The revised weight or distance between any two nodes using time factor may be calculated as:

$$w^{'} = w * \delta$$

where δ indicates the time factor with which the weight needs to be adjusted. The revised weight w can now be used for shortest path calculation. The corresponding weight update can be calculated as per Algorithm 1.

Algorithm 1 Weight Update using MPD

---

1. If user preference = distance
   $w' = w$
2. Else if user preference = time then
   $w' = w * \delta$
3. Else if user preference = congestion then
   $w' = w * \phi$
4. Return w'

For evaluation of the Multi-parameter Dijkstra's algorithm, real-time information about 500 nodes in Jaipur City was collected. This information included the distance between the nodes, congestion factor and the time taken to travel between any two nodes or cities. This information was collected at different time stamps as in real-time scenarios the values are highly dependent on the time at which they are collected. The Multi-parameter Dijkstra provides higher adaptability and calculates the shortest path as per the preference of the user. For instance, let us consider the graph illustrated in Fig. 1.

As shown in Fig. 1, the original Dijkstra's algorithm gives {1, 2, 5, 6} as the shortest path. In real road networks, other factors such as congestion and the time taken to travel a path may be taken into consideration for better calculation of shortest path. For example, let us consider the congestion factor during real-time analysis is as shown in Table 1.

As shown in Fig. 2, congestion factor is independent of the distance or weight between two nodes. In real-time scenarios the value of congestion factor may vary from one time stamp to another.

The congestion factor may also vary as per the time at which the data is collected. Figure 3 shows the congestion factor for two nodes at different moments of time.

Such information was collected for 500 nodes across Jaipur City. The weight/congestion collected for the sample graph of Fig. 1 is shown in Fig. 4.

Taking the above real-time data into consideration, the shortest path can now be recalculated using Multi-parameter Dijkstra's algorithm. Figure 5 indicates the revised shortest path after taking Congestion into consideration.

Fig. 1 Shortest path calculation using Dijkstra's algorithm {1, 2, 5, 6}

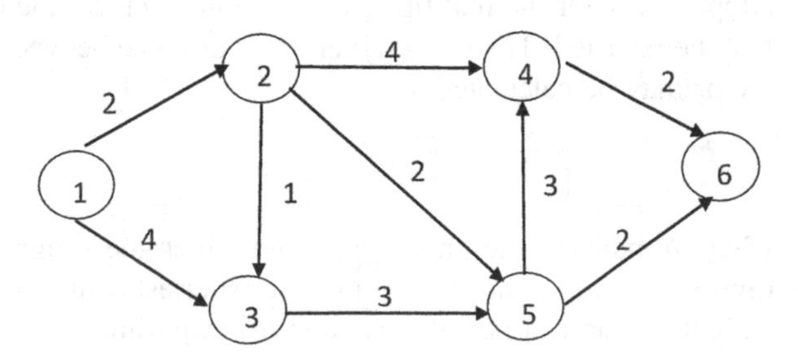

Table 1 Revised weight with congestion factor

Source node	Destination node	Weight (w)	Congestion factor (φ)	Revised weight (w')
5	6	2	2	4
1	2	2	3	6
2	3	1	7	7
3	5	3	3	9
2	5	2	8	16
4	6	2	8	16
5	4	3	6	18
2	4	4	5	20
1	3	4	9	36

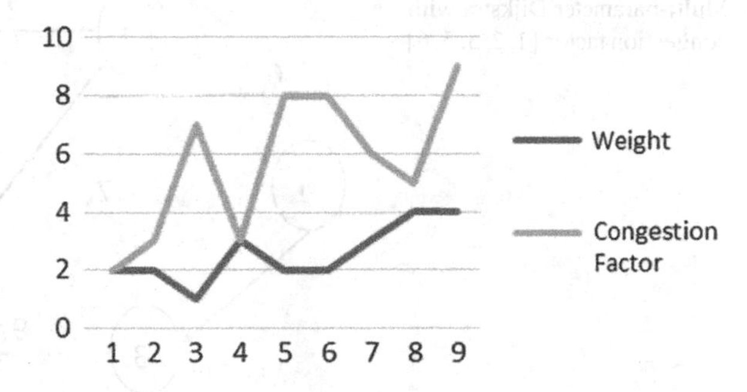

Fig. 2 Weight and congestion factor for different nodes 1–9

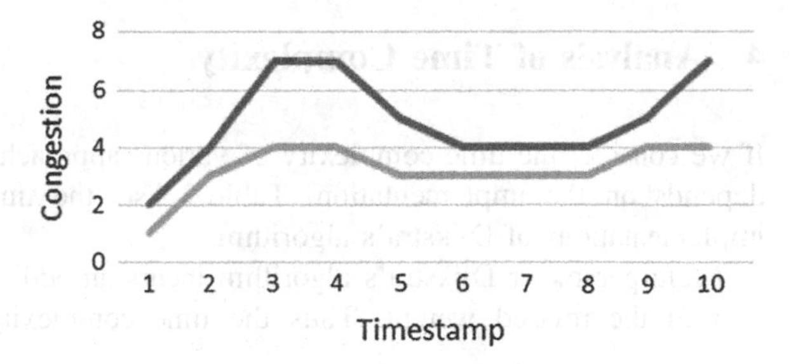

Fig. 3 Congestion factor at different time stamps for n = 2

Similarly it can be shown that time factor is also an important criterion for deciding the shortest route. Thus in real road networks, the shortest path may vary depending on the real-time information about the traffic and road conditions.

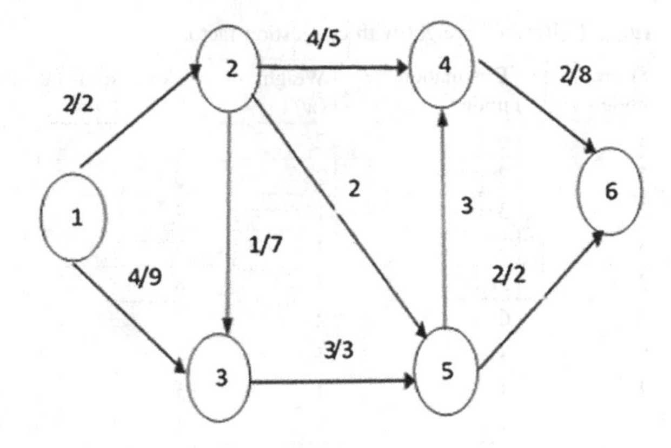

Fig. 4 Weight/congestion for the sample graph

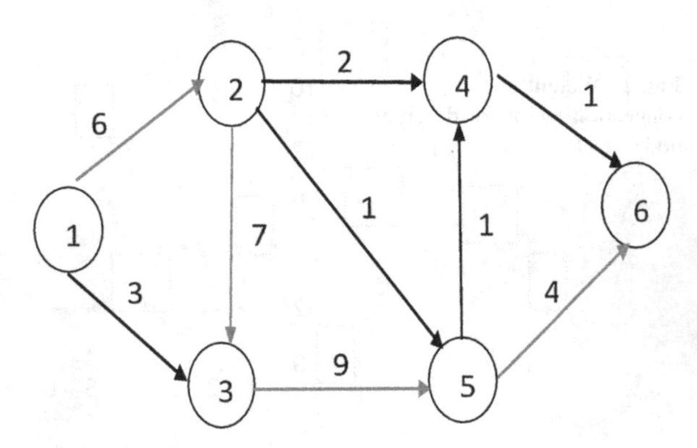

Fig. 5 Shortest path using Multi-parameter Dijkstra with congestion factor {1, 2, 3, 5, 6}

4 Analysis of Time Complexity

If we consider the time complexity of various approaches, the complexity largely depends on the implementation. Table 2 lists the time complexity of various implementations of Dijkstra's algorithm.

Multi-parameter Dijkstra's algorithm incurs an additional overhead of calculation of the revised weight. Thus the time complexity of the Multi-parameter

Table 2 Time complexity of different algorithms

Algorithm	Time complexity
Dijkstra's algorithm with list	$O(V^2)$
Dijkstra's algorithm with modified binary heap	$O((E + V) \log V)$
Dijkstra's algorithm with Fibonacci heap	$O(E + V \log V)$
Johnson-Dijkstra	$O(EV + V^2 \log V)$
Multi-parameter Dijkstra	$O(E \log V) + E$

Dijkstra's algorithm is $O(E \log V) + E$ where the factor E is added in order to compute the revised weight for all the edges. But in turn a well-updated information about shortest path can be a boon in real road networks. Also it can provide the fastest route between any two set of nodes.

5 Conclusion

Various approaches to shortest path calculation have been discussed in this paper. Multi-parameter Dijkstra's algorithm can be used to incorporate the state-of-the-art information in the real road networks. Factors such as congestion and time taken to reach the destination may affect the shortest path. Though computing the revised weight has an additional overhead but it can contribute towards saving time of travel in real road networks. And thus aide in finding the shortest and fastest route in the real road networks.

Acknowledgments This paper provides Performance Evaluation of our previous work on Multi-parameter Dijsktra's Algorithm published in International Journal of Advances in Engineering Research [1].

References

1. Nishtha Kesswani, Dinesh Gopalani: Design and implementation of Multi-Parameter Dijkstra; s algorithm: A Shortest path algorithm for real road netwroks, International J. of Advances in Engineering Research, Vol.2, Issue 3 (2011).
2. Anany Levitin, Introduction to the design & analysis of algorithms, Pearson Education (2009).
3. F. Benjamin Zhan, "Three Fastest shortest path algorithms on Real-road networks", Journal of Geographic information and decision analysis, Vol. 1, No.1, pp. 69–82 (2010).
4. Yong Xi, Loren Schwiebert, "Privacy preserving shortest path routing with an application to navigation, Pervasive and mobile computing", Elsevier, Volume 13, pp 142–149 (2014).
5. Akiba, Takuya, et al. "Fast Shortest-path Distance Queries on Road Networks by Pruned Highway Labeling." ALENEX. 2014, pp 147–154 (2014).
6. Antonio Sedeno-Noda, Andrea Raith, "A Dijkstra-like method computing all extreme supported non-dominated solutions of the biobjective shortest path problem", Computers & Operations Research, Elsevier, Volume 57 (2015).
7. A Goswami, DD Chakraborty, Optimization 2015 http://textofvideo.nptel.iitm.ac.in/111105039/lec19.pdf.

An Outlook in Some Aspects of Hybrid Decision Tree Classification Approach: A Survey

Archana Panhalkar and Dharmpal Doye

Abstract Decision tree one of the complex but useful approach for supervised classification is portrayed in this review. Today's research is deemed toward the use of hybridized decision tree for the need of various applications. The recent approaches of decision tree techniques come with hybrid decision tree. This survey, it has been elaborating the various approaches of converting decision tree to hybridized decision tree. For classification of data SVMs and other classifier in decision tree are generally used at the decision node to improve accuracy of decision tree classifier. Then the more penetration is given to some aspects which less likely used by researchers which gives more scope. The ideas of various hybridized approaches of decision tree are given like use of clustering, naïve Bayes, and AVL tree, fuzzy and genetic algorithm.

Keywords Data mining · AVL · Fuzzy clustering · Naïve Bayes · SVM · Cuckoo optimization

1 Introduction

This paper deals with advanced approaches used in decision tree classifier and the approaches on which researchers not focused. This paper summarizes the advancement done and various approaches used in decision tree. Like mining the gold from hefty area of coal, mining useful data from large database is intricate task. The Latest era in data mining is classification. Classification involves predicating

Archana Panhalkar (✉)
Amrutvahini College of Engineering, Sangamner, Maharashtra, India
e-mail: archana10bhosale@rediffmail.com

Dharmpal Doye
Shri Guru Gobind Singhji Institute of Engineering and Technology,
Nanded, Maharashtra, India
e-mail: dddoye@yahoo.com

S.C. Satapathy et al. (eds.), *Proceedings of the International Conference on Data Engineering and Communication Technology*, Advances in Intelligent Systems and Computing 469, DOI 10.1007/978-981-10-1678-3_8

the approved class of the given sample data. Classification comprises various methods but decision tree classifier is one of the scorching research areas.

When the data grows in size, decision tree is one of the preferred selections of many researchers for classification. Besides complexity of implementation, it is having significant accuracy as compared to other classification methods. It creates supervised model to show the relationship between instances and attributes [1]. It takes polynomial time to create the model. Decision tree is advantageous for various applications like speech recognition, medical research, energy modeling, intrusion detection and many other real time applications. Decision tree approach is hybridized by many researchers for increasing accuracy of classification and reducing training and testing times.

1.1 Decision Tree Construction and Decisive Factors

Decision tree is the visual representation of data which consist of nodes and edges where each node examine attribute, branches represents outcome of test and leaf node represents predicted class. A tree constructs using top-down recursive and divide and conquer approach [1]. Decision tree is generated in two phases

(a) Tree Construction: At start keep all training records at the root node and then partition the records recursively based on the selected attribute.
(b) Tree Pruning: Identify the branches which reflect the noise or outlier and then remove these branches.

Decision tree construction is more emphasized by the strategy used for selecting attributes. Sufficient approaches are provided by researchers to select the attribute which best split the training records. Ben-Bassat [2] categorized these attribute selection measures as use of information theory, distance measure and dependence measure. Some of the measures are enlisted below.

(a) **Measure of Information Theory**

The basic criteria used by Quinlan [3] are information gain and gain ratio for selecting best attribute which increases the accuracy of decision tree learning. Information Gain measure for selecting attribute is used by ID3 decision tree construction algorithm while Gain ratio is used by C4.5 algorithm which are choices of majority researchers. Jun et al. [4] modified entropy by considering the base of the logarithm as the number of successors to the node. This strategy can handle huge amount of data efficiently.

(b) **Measure of Distance**

Gini index is the measure based on distance used by CART (Classification and Regression Tree) for selection of best attribute [5]. It uses separability, deviation and discrimination between classes [6]. Brieman et al. [6] proved that if the dataset contains large number of classes then Gini index fails to give performance. Murthy et al. [7] proposed binary criteria towing index which divides multiple classes in two superclasses and chooses splitting criteria based on these two super classes.

Rounds [8] used approach of Kolmogorov–Smirnov distance as an attribute selection measure based on distance whose feature is to produce an optimal classification decision at each node. Utgoff and Clouse [9] used the same measure which modified measure to deal with multiclass attributes and also its contribution is in missing value attributes. Martin [10] has analyzed the difference of splitting criteria like distance, orthogonality, a Beta function and two chi-squared tests.

Chandra et al. [11] presented new splitting measure called as Distinct Class Splitting Measure to produce purer partitions for the dataset with distinct classes.

1.2 Decision Tree Learners

The field of decision tree learning is crafted by researchers from last four decades and done immeasurable improvements in methods of decision tree creation. Some basic algorithms are narrated below.

ID3: ID3 (Iterative Dichotomiser 3) [12] decision tree is a simple decision tree algorithm which is simple and base for all the researchers. ID3 uses information gain for selecting best attribute for partitioning the dataset. Information gain gives poor performance when the feature takes multiple values. No pruning method is used by ID3. It fails to handle numeric attributes, missing and noisy data.

C4.5: The C4.5 algorithm [3] the first choice of researchers is an extension of ID3 algorithm which uses information Gain as splitting criteria. It can handle categorical as well as numerical data. It removes all the drawbacks of ID3 algorithm. Error-based pruning is used by the C4.5. New version of C4.5 is proposed called as the C5.0. The changes in C5.0 encompass new capabilities as well as much-improved efficiency. It is also capable to easily handle missing values.

CART: Classification and regression tree (CART) developed by Breiman et al. [6] creates two ways tree called as binary tree. CART uses Gini index for selecting attribute which divides the data in two child left and right. It can handle missing values and can be applied for any type of data. To reduce size of tree it uses cost-complexity pruning.

With these basic algorithms researchers modified and hybridized these basic techniques to improve performance and accuracy. According to the need of latest era any researchers applied for large datasets with modified approaches. Next section is emphasized on the latest hybridized approaches used in decision tree learning changed according to the application and requirement of area.

2 Hybridized Approaches of Decision Tree

In the present era, majority researchers are contributing toward the use of hybridized approaches to construct and evaluate decision trees. Because of these methods the performance and accuracy of decision tree is increased. These hybridized methods deals with major areas. Following are some of the methods used by researchers to hybridize the learning of the decision tree.

2.1 *Clustering-Based Decision Tree Induction*

Clustering is unsupervised learning method which is applied for large data set without class values. To deal with large data set many researchers combined unsupervised learning with the supervised learning.

Jayanta and Krishnapuram [13] proposed interpretable hierarchical clustering method to construct unsupervised decision tree. In this method the leaf node represents the clusters. The rule is constructed from root to leaf. Interpretable hierarchical clustering demonstrates the performance of four attribute selection criteria. This method chooses attribute from unlabeled data in such a way that after performing clustering the in homogeneity get reduced. The results of unsupervised decision trees are outperforming than the supervised decision trees. As the data streams are unbounded data so requires only on scan so Qian and Lin [14] developed efficient approach which produces cluster decision tree. Its main aim to solve the problem of concept drifting produced in data streams. It clusters the data which is not classified by the Very Fast Decision Trees (VFDT). With these new clusters it updates the tree. Due to these strategies the classification accuracy is increased for unlabeled data and works for noisy data.

Esfandiary and Moghadam [15] used *K* means clustering to create decision trees. In this algorithm, data is clustered using *K* means clustering and then sorted data in the cluster. Each cluster creates one layer of decision tree. It uses depth growth constraint to control the size of decision tree. The author argued that this clustering decision tree uses very simple strategy and provides better accuracy, processing time and complexity. Horng et al. [16] used fuzzy strategy to create decision tree. The hierarchical fuzzy clustering decision tree (HFCDT) proposed which is designed for the dataset consisting with large number of classes. It performs better for continuous valued attribute. It uses division degree based hierarchical clustering which successively creates reduced fuzzy rule set. The strategy reduces computational complexity because of reduced size of fuzzy rule set and increases classification rate due to finer fuzz partition. Cheng et al. used the method on real time system that is ion implantation problem where the method performs outstanding.

Recently many researches [17–21] used clustering to construct decision tree for large dataset by considering different parameters of construction and evaluation. These algorithms aimed to create decision tree which is accurate, comprehensible and can be applicable for real time data by considering its size and different factors. In today's era many researchers developing hybridized decision tree approaches by combining clustering. From literature it is observed that *K* means clustering is the choice of many researchers because of its better performance and simplicity to understand.

2.2 Height Balanced Decision Tree Induction

In 80s decade, to reduce size of decision trees height was considered as a major factor. Laber and Nogueira [22] proved that decision tree with minimum height can completely identify the given objects with the given tests.

Gerth et al. [23] had given the solution to the problem of minimum cache. The novel method called dynamic binary tree with small height $\log n + O(1)$. It embeds this tree in the static binary tree which get embedded in an array in cache unaware manner. The new improved height balanced approach is provided [24] which had used AVL trees for classification. This is one of the research areas where researchers are failed to concentrate. Due to the use of AVL trees quality and stability of decision tree is increased without affecting performance and accuracy of the decision trees. Due to successive recursive partitioning the data subset becomes small, so partitioning such data set is not possible. To solve this problem exception threshold is taken after which data subset is not partitioned. It uses merging of nodes if child stores same class as parent. After inserting node in decision tree, height of left and right sub tree is checked. If any imbalance is occurred then the basic rotations are performed to balance tree. This method improves quality of decision tree.

Larsen [25] had given the proof of complexity for the relaxed AVL tree. He suggested whenever use the complex structure, there is a need to know in advance how it is efficient. To find out rebalancing, parallelism and efficiency there is need to give the proof of complexity otherwise it will create inefficient research.

2.3 Decision Tree with Naïve Bayes Classifier

Naive Bayes (NBC) is one of the straightforward techniques used to create classifier models that assign class labels to data instances. These data instances are represented as vectors of feature values in which the class labels are drawn from some finite set. All naive Bayes classifiers presume that value of every feature is different for the given the class variable. For example, to identify fruit as an apple considers values as it should be red and maximum diameter is 10 mm. A naive Bayes classifier thinks each of these features to contribute independently to the probability that this fruit is an apple, regardless of any possible correlations between the color, roundness and diameter features.

Many researchers [26–33] hybridized decision tree by taking facilities of features of naïve Bayes classifiers. This is one of the aspects of decision tree where the many researchers concentrated. Here we are giving brief literature about it. Ratanamahatana and Gunopulos [26] had given new aspect of decision tree classifier. C4.5 decision tree is used for feature classification and then naive Bayes classifier is applied. According to author the method reduced the size of training and testing data as well as the running time get minimized. The different classifiers with different characteristics given improved performance.

Kohavi [27] proposed the NBTree which hybrid approach combines decision tree with NBC. NBTree constructs using univariate splits at decision nodes and its leaves contains NBC. It will produce accurate hybridized trees which outperforms for many datasets especially for large datasets. Lewis [29] proposed self adaptive NBTree which combines decision tree with NBC. This self adaptive NBTree outperforms for continuous valued attributes. NBC is used for collection of attributes and finds proper ranges. Use of NBC node solves the problem of overgeneralization and overspecialization. So the method avoids negative effect of information loss.

Farid et al. [33] proposed two independent hybrid classifiers that is hybrid DT and hybrid NB classifier. These methods applied for multi-class classification problem. In hybrid DT algorithm, noisy instances are removed using NB classifier and then decision tree is induced. In hybrid NB classifier, decision tree is used to select subset of attributes and then produced naïve assumptions for class conditional independence.

By the literature it is observed that the when decision tree approach for feature selection and NB classifier for classification gives more efficient, accurate classification rate.

2.4 *Decision Tree with Support Vector Machine Classifier*

Support Vector Machine (SVM) is supervised classifier performs non linear classification using kernel strategy. It uses the hyperplanes to classify the given data. The hyperplane which maximizes the classification is selected. The decision tree with SVM is contributed from last three decades, from which few [34–46] are analyzed in this review.

In [35], decision tree based multiclass support vector machine is proposed. Top node uses hyperplanes to classify the classes. If hypotheses create plural classes then apply hyperplane until single class is obtained. At the top node most separable classes are separated which solves the problem of generalization. It produces highly generalized classifier. Multiclass problem is also tackled in [36–38] which create SVM based decision tree. The method uses modified Self Organizing Map called Kernel SOM. Decisions in binary tree are made by SVM which contribute to solve the problems of multiclass in the database.

Oblique decision trees [41] are the decision trees which performs the multivariate search with the advantage that it performs polygonal partitions of the attribute space. Internal nodes uses SVM generated hyperplanes which maximizes the boundary between different classes so no need of pruning which is requirement of any general tree. SVMs achieve best accuracy and capable to operate on mixed data [44]. SVM based hierarchical classifier gives fruitful results. SVM-based decision classifier provides high testing accuracies, low computational and storage cost, interpretable structures used for classification of images [44].

Kumar and Gopal [46] given the approach of hybrid SVM-based decision tree which performs both unit variant and multivariate decisions. Instead of reducing number of support vectors, it aims to reduce number of test data points which requires SVM's decision in classification. Fuzzy SVM-based decision tree [42–44] was interest of last decade.

From the study of literature, we can conclude that SVMs in decision tree are generally used at the decision node to improve accuracy of decision tree classifier.

2.5 Fuzzy and Evolutionary Approach to Decision Tree

In today's hybridized era, decision capabilities of decision tree are increased by countless methods. Fuzzy and evolutionary are the approach is and was the significantly used by the researchers in combination with basic algorithms. Fuzzy decision tree are the decision trees in which the data partition among children is done using fuzzy membership function. Fuzzy decision tree approach is very old. In this review we are going to give some aspects [47–57] of these fuzzy and genetic algorithm in induction of decision trees.

The fuzzy decision tree is similar to the standard decision tree methods characterized by using recursive binary tree. At each node during the construction process of a fuzzy decision tree, the most stable splitting region is selected and the boundary uncertainty is estimated based on an iterative resampling algorithm [48]. The boundary uncertainty estimate is used within the region's fuzzy membership function to direct new samples to each resulting partition with a quantified confidence [48]. The fuzzy membership function recovers those samples that lie within the uncertainty of the splitting regions. One of the best example of combining fuzzy and genetic algorithm based decision tree is proposed in [48, 49]. Fuzzy clustering is carried out using genetic algorithm which creates the nodes of the trees. When tree grows the nodes split into clusters of lower in homogeneity. Effect of these methods highly accurate decision tree gets produced.

Hanmandlu and Gupta [51] applied fuzzy decision tree for dynamic data. It creates fuzzy decision tree in top down manner using new discrimination measure which is used to calculate entropy of fuzzy events. This method best performs time changing data called temporal data. The size of the decision tree is reduced by one of the approach presented in [55] which assigns trapezoidal function of membership to assign fuzzy membership value to the attributes. By fixing the value of attributes the size of fuzzy decision tree is reduced.

Wang et al. [56] proposed fuzzy rule based decision tree which uses multiple attributes for testing at internal nodes. Due to this strategy the size of the decision tree get reduced. To produce the leaves with most purity, the fuzzy rules are used.

3 Scope of Research

From the literature survey, today's research is more centered to produce hybrid decision trees which combines basic decision tree with multiple approaches like clustering, fuzzy, genetic algorithm and many more methods. But as the sky is limit so to improve the accuracy and performance of decision tree we are suggesting to concentrate on the AVL height balanced decision tree approach where there is more scope to do the improvement. There are new optimizations algorithms like cuckoo optimization algorithm [58, 59] which is not tried in literature for the decision trees. There are many aspects of decision tree where the rays are not reached. Like reducing complexity of decision tree by converting dataset into simplified manner and compressing the data without affecting the accuracy.

4 Conclusion

In this paper we are penetrating some important aspects of decision tree. This literature survey given the basic decision tree aspects along with hybridized decision tree which is today's hot research topic. We have tried to provide some important methods which are used by many researchers to make the hybridized tree. This literature survey will be definitely used for new coming decision trees researchers to combine any soft computing approach to the basic methods. Lastly, we have given the suggestion of new interesting algorithm for optimization which may produce efficient decision tree.

References

1. Jiawei Han, Micheline Kamber: Data mining: concepts and techniques, Morgan Kaufmann Publishers Inc., San Francisco, CA, (2000).
2. Moshe Ben-Bassat (1987). Use of distance measure, Information measures and error bounds on feature evaluation. In Sreerama Murthy (1), pp. 9–11.
3. J. R. Quinlan: C4.5: Programming for Machine Learning. San Francisco, CA: Morgan Kaufman (1993).
4. Byung Hwan Jun, Chang Soo Kim, Hong-Yeop Song and Jaihie Kim:A new criterion in selection and discretization of attributes for the generation of decision trees. *IEEE Transactions on Pattern Analysis and Machine Intelligence*, Vol. 19, No. 12, pp. 1371–1375 (1997).

5. Mark Last and Oded Maimon: A compact and accurate model for classification. *IEEE Transactions on Knowledge and Data Engineering*, Vol. 16, No. 2, pp. 203–215 (2004).
6. Leo Breiman, Jerome H. Friedman, Richard A. Olshen, and Charles J. Stone: *Classification and Regression Trees.* Wadsworth International Group, Belmont, California (1984).
7. S. K. Murthy, Simon Kasif and Steven Salzberg: A system for induction of oblique decision trees *Journal of Artificial Intelligence Research 2*, pp. 1–33. (1994).
8. E. Rounds (1980). A combined nonparametric approach to feature selection and binary decision tree design. *Pattern Recognition,* Vol. 12, pp. 313–317 (1980).
9. P. E. Utgoff and J. A. Clouse: A Kolmogorov-Smirnoff metric for decision tree induction. Tech. Rep. No. 96–3, Dept. Comp. Science, University Massachusetts, Amherst (1996).
10. J. K. Martin: An exact probability metric for decision tree splitting and stopping. *Machine Learning,* Vol. 28, No. 2-3, pp. 257–29 (1997).
11. B. Chandra, R. Kothari, P. Paul.: A new node splitting measure for decision tree construction Pattern Recognition Vol. 43, Elsevier Publishers, pp. 2725–2731 (2010).
12. Quinlan, J. Ross. "Induction of decision trees" Machine learning 1.1 (1986): 81–106.
13. R. Basak Jayanta, and Raghu Krishnapuram. "Interpretable hierarchical clustering by constructing an unsupervised decision tree." Knowledge and Data Engineering, IEEE Transactions on 17.1 (2005): 121–132.
14. Qian, Lin, and Liang-xi Qin. "A framework of cluster decision tree in data stream classification." Intelligent Human-Machine Systems and Cybernetics (IHMSC), 2012 4th International Conference on. Vol. 1. IEEE, 2012.
15. Esfandiary, Nura, and Amir-Masoud Eftekhari Moghadam. "LDT: Layered decision tree based on data clustering." Fuzzy Systems (IFSC), 2013 13th Iranian Conference on. IEEE, 2013.
16. Horng, Shih-Cheng, Feng-Yi Yang, and Shieh-Shing Lin. "Hierarchical fuzzy clustering decision tree for classifying recipes of ion implanter." Expert Systems with Applications 38.1 (2011): 933–940.
17. Barros, Rodrigo C., et al. "A clustering-based decision tree induction algorithm." Intelligent Systems Design and Applications (ISDA), 2011 11th International Conference on. IEEE, 2011.
18. Lin, C., Chen, W., Qiu, C., Wu, Y., Krishnan, S., & Zou, Q. (2014). LibD3C: ensemble classifiers with a clustering and dynamic selection strategy. Neurocomputing, 123, 424–435.
19. Dror, Moshe, et al. "OCCT: A one-class clustering tree for implementing one-to-many data linkage." Knowledge and Data Engineering, IEEE Transactions on 26.3 (2014): 682–697.
20. Hu, Yakun, Dapeng Wu, and Antonio Nucci. "Fuzzy-clustering-based decision tree approach for large population speaker identification." Audio, Speech, and Language Processing, IEEE Transactions on 21.4 (2013): 762–774.
21. Fraiman, Ricardo, Badih Ghattas, and Marcela Svarc. "Interpretable clustering using unsupervised binary trees." Advances in Data Analysis and Classification 7.2 (2013): 125–145.
22. Laber, Eduardo S., and Loana Tito Nogueira. "On the hardness of the minimum height decision tree problem." Discrete Applied Mathematics 144.1 (2004): 209–212.
23. Brodal, Gerth Stølting, Rolf Fagerberg, and Riko Jacob. "Cache oblivious search trees via binary trees of small height." Proceedings of the thirteenth annual ACM-SIAM symposium on Discrete algorithms. Society for Industrial and Applied Mathematics, 2002.
24. Ali, Mohd Mahmood, and Lakshmi Rajamani. "Decision Tree Induction: Data Classification using Height-Balanced Tree." In IKE, pp. 743–749. 2009.
25. Larsen, Kim S. "AVL trees with relaxed balance." In Parallel Processing Symposium, 1994. Proceedings., Eighth International, pp. 888–893. IEEE, 1994.
26. Ratanamahatana, Chotirat ann, and Dimitrios Gunopulos. "Feature selection for the naive bayesian classifier using decision trees." Applied artificial intelligence 17.5–6 (2003): 475–487.
27. Kohavi, Ron. "Scaling Up the Accuracy of Naive-Bayes Classifiers: A Decision-Tree Hybrid." In KDD, pp. 202–207. 1996.

28. Zadrozny, Bianca, and Charles Elkan. "Obtaining calibrated probability estimates from decision trees and naive Bayesian classifiers." In ICML, vol. 1, pp. 609–616. 2001.
29. Lewis, David D. "Naive (Bayes) at forty: The independence assumption in information retrieval." In Machine learning: ECML-98, pp. 4–15. Springer Berlin Heidelberg, 1998.
30. Wang, Li-Min, Xiao-Lin Li, Chun-Hong Cao, and Sen-Miao Yuan. "Combining decision tree and Naive Bayes for classification." Knowledge-Based Systems 19, no. 7 (2006): 511–515.
31. Farid, Dewan Md, Nouria Harbi, and Mohammad Zahidur Rahman. "Combining naive bayes and decision tree for adaptive intrusion detection." arXiv preprint arXiv:1005.4496 (2010).
32. Meretakis, Dimitris, and Beat Wüthrich. "Extending naive bayes classifiers using long itemsets." In Proceedings of the fifth ACM SIGKDD international conference on Knowledge discovery and data mining, pp. 165–174. ACM, 1999.
33. Farid, Dewan Md, Li Zhang, Chowdhury Mofizur Rahman, M. A. Hossain, and Rebecca Strachan. "Hybrid decision tree and naive Bayes classifiers for multi-class classification tasks." Expert Systems with Applications 41, no. 4 (2014): 1937–1946.
34. Bennett, Kristin P., and Jennifer A. Blue. "A support vector machine approach to decision trees." In Neural Networks Proceedings, 1998. IEEE World Congress on Computational Intelligence. The 1998 IEEE International Joint Conference on, vol. 3, pp. 2396–2401. IEEE, 1998.
35. Takahashi, Fumitake, and Shigeo Abe. "Decision-tree-based multiclass support vector machines." In Neural Information Processing, 2002. ICONIP'02. Proceedings of the 9th International Conference on, vol. 3, pp. 1418–1422. IEEE, 2002.
36. Cheong, Sungmoon, Sang Hoon Oh, and Soo-Young Lee. "Support vector machines with binary tree architecture for multi-class classification." Neural Information Processing-Letters and Reviews 2, no. 3 (2004): 47–51.
37. Mao, Yong, Xiaobo Zhou, Daoying Pi, Youxian Sun, and Stephen TC Wong. "Multiclass cancer classification by using fuzzy support vector machine and binary decision tree with gene selection." BioMed Research International 2005, no. 2 (2005): 160–171.
38. Oh, Juhee, Taehyub Kim, and Hyunki Hong. "Using binary decision tree and multiclass svm for human gesture recognition." In Information Science and Applications (ICISA), 2013 International Conference on, pp. 1–4. IEEE, 2013.
39. Heumann, Benjamin W. "An object-based classification of mangroves using a hybrid decision tree—Support vector machine approach." Remote Sensing 3, no. 11 (2011): 2440–2460.
40. Saimurugan, M., K. I. Ramachandran, V. Sugumaran, and N. R. Sakthivel. "Multi component fault diagnosis of rotational mechanical system based on decision tree and support vector machine." Expert Systems with Applications 38, no. 4 (2011): 3819–3826.
41. Barros, Rodrigo C., Ricardo Cerri, Pablo Jaskowiak, and André CPLF De Carvalho. "A bottom-up oblique decision tree induction algorithm." In Intelligent Systems Design and Applications (ISDA), 2011 11th International Conference on, pp. 450–456. IEEE, 2011.
42. Sahin, Y., and E. Duman. "Detecting credit card fraud by decision trees and support vector machines." In International MultiConference of Engineers and Computer Scientists, vol. 1. 2011.
43. Moustakidis, Serafeim, Giorgos Mallinis, Nikos Koutsias, John B. Theocharis, and Vasilios Petridis. "SVM-based fuzzy decision trees for classification of high spatial resolution remote sensing images." Geoscience and Remote Sensing, IEEE Transactions on 50, no. 1 (2012): 149–169.
44. Madzarov, Gjorgji, and Dejan Gjorgjevikj. "Multi-class classification using support vector machines in decision tree architecture." In EUROCON 2009, EUROCON'09. IEEE, pp. 288–295. IEEE, 2009.
45. Moustakidis, S. P., J. B. Theocharis, and G. Giakas. "A fuzzy decision tree-based SVM classifier for assessing osteoarthritis severity using ground reaction force measurements." Medical engineering & physics 32, no. 10 (2010): 1145–1160.
46. Kumar, M. Arun, and Madan Gopal. "A hybrid SVM based decision tree." Pattern Recognition 43, no. 12 (2010): 3977–3987.

47. Chang, Pei-Chann, Chin-Yuan Fan, and Wei-Yuan Dzan. "A CBR-based fuzzy decision tree approach for database classification." Expert Systems with Applications 37, no. 1 (2010): 214–225.
48. Fan, Chin-Yuan, Pei-Chann Chang, Jyun-Jie Lin, and J. C. Hsieh. "A hybrid model combining case-based reasoning and fuzzy decision tree for medical data classification." Applied Soft Computing 11, no. 1 (2011): 632–644.
49. Shukla, Sanjay Kumar, and Manoj Kumar Tiwari. "GA guided cluster based fuzzy decision tree for reactive ion etching modeling: a data mining approach." Semiconductor Manufacturing, IEEE Transactions on 25, no. 1 (2012): 45–56.
50. Costa, Herbert R. do N., and Alessandro La Neve. "Fuzzy Decision Tree applied to defects classification of glass manufacturing using data from a glass furnace model." In Fuzzy Information Processing Society (NAFIPS), 2012 Annual Meeting of the North American, pp. 1–6. IEEE, 2012.
51. Kumar, Amioy, Madasu Hanmandlu, and H. M. Gupta. "Fuzzy binary decision tree for biometric based personal authentication." Neurocomputing 99 (2013): 87–97.
52. Marsala, Christophe. "Fuzzy decision trees for dynamic data." In Evolving and Adaptive Intelligent Systems (EAIS), 2013 IEEE Conference on, pp. 17–24. IEEE, 2013.
53. Popescu, Adrian, Bogdan Popescu, Marius Brezovan, and Eugen Ganea. "Image semantic annotation using fuzzy decision trees." In Computer Science and Information Systems (FedCSIS), 2013 Federated Conference on, pp. 597–601. IEEE, 2013.
54. Dai, Jianhua, Haowei Tian, Wentao Wang, and Liang Liu. "Decision rule mining using classification consistency rate." Knowledge-Based Systems 43 (2013): 95–102.
55. Bajaj, Shalini Bhaskar, and Akshaya Kubba. "FHSM: Fuzzy Heterogeneous Split Measure algorithm for decision trees." In Advance Computing Conference (IACC), 2014 IEEE International, pp. 574–578. IEEE, 2014.
56. Wang, Xianchang, Xiaodong Liu, Witold Pedrycz, and Lishi Zhang. "Fuzzy rule based decision trees." Pattern Recognition 48, no. 1 (2015): 50–59.
57. Al-Obeidat, Feras, Ahmad T. Al-Taani, Nabil Belacel, Leo Feltrin, and Neil Banerjee. "A Fuzzy Decision Tree for Processing Satellite Images and Landsat Data." Procedia Computer Science 52 (2015): 1192–1197.
58. García, Salvador, et al. "Evolutionary selection of hyperrectangles in nested generalized exemplar learning." Applied Soft Computing 11.3 (2011): 3032–3045.
59. Rodrigues, Durval, et al. "BCS: A binary cuckoo search algorithm for feature selection." Circuits and Systems (ISCAS), 2013 IEEE International Symposium on. IEEE, 2013.

Content Search Quaternary Look-Up Table Architecture

D.P. Borkute, P.K. Dakhole and Nayan Kumar Nawre

Abstract Scaling of devices due to reduction in technology has reached to a limit. Such nanoscale devices below certain limit are difficult to fabricate for reasons of physical characteristics, dimensions and cost involved in it. Use of multiple value logic (MVL) is solution to this in terms of cost, area. With use of multi-value logic less number of bits is required to store any information as compared to binary number system. In this work look-up table memory structure is chosen to implement. Radix selected for that is 4, called quaternary logic designs. Look-up table begin faster as compared to conventional memory structure which overcome issue of speed which is a limitation of MVL design. Circuit design for quaternary logic is done using available binary circuits using single power supply. In this design instead of Static RAM a Content based search method is used for design which facilitates searching, matching of data and initialize the next stage of circuit if required. Proposed circuit tries reduction of interconnections in binary systems, without power consumption overhead. This works particularly implements Look-up table memories using content-based search.

Keywords Quaternary system · Quaternary SRAM · Content search based memory · Power dissipation

---

D.P. Borkute (✉) · N.K. Nawre
Department of Electronics and Telecommunication, Pimpri Chinchwad College of Engineering, University of Pune, Pune 411044, India
e-mail: deepti_pachu@yahoo.com

N.K. Nawre
e-mail: nayan.naware22@gmail.com

P.K. Dakhole
Department of Electronics and Communication, Yeshwantrao Chavhan College of Engineering, Nagpur, Maharashtra, India
e-mail: pravin_dakhole@yahoo.com

S.C. Satapathy et al. (eds.), *Proceedings of the International Conference on Data Engineering and Communication Technology*, Advances in Intelligent Systems and Computing 469, DOI 10.1007/978-981-10-1678-3_9

1 Introduction

Multiple value logic is based on quaternary number system with radix 4. This system can represent data in four different stable and discrete levels. Such designs affect important parameters like power, delay, interconnect and speed [1]. These circuits are designed using single power supply method ensuring logic optimization stated by Multi-Valued Logic [2]. Look-up Tables (LUT) is the kind of memories that uses RAM structures. Look-up tables are structures implement an arbitrary binary logic function. Data to be stored is to converted proper format required for quaternary system. Look-up tables are implemented using CAM-based architecture having facility of storing, searching data and observed how interconnect and power can be saved.

2 Inception of Quaternary Number System in Look-Up Tables

Look-up tables (LUT) are digital blocks which can store data depending on function implemented. The desired result or data are programmed in SRAM and attached to the LUTs. A n-bit look-up table is designed using data selector with select lines as inputs to the LUT [3]. The capacity of an LUT |C| is given by [3, 4].

$$C = n \times b^k \tag{1}$$

n: outputs, *k*: inputs and *b*: logic values for a function. The total functions implemented in an LUT with *k* input are given by [3, 4].

$$F = b^{|C|} \tag{2}$$

Comparing these numbers for binary look-up tables (BLUT) and quaternary look-up tables (QLUT), equivalent logic complexities can be found. These numbers are compared in Table 1.

Table 1 Comparison of logic complexity of BLUT and QLUT

No. of outputs	No. of inputs	Quaternary system radix	Binary system radix	Capacity		Possible functions implemented	
N	*K*	Base (*b*)		BLUT	QLUT	BLUT	QLUT
1	2	4	2	2	4	2^4	4^4
	4			16	64	2^{16}	4^{64}
	8			256	65536	2^{256}	4^{65536}

3 Quaternary Look-Up Table (LUT)

The basic element of look-up table is multiplexer. Number of stages required to implement given functionality define the switching of data. The multiplexers to be used depend upon the number of input processed in LUT and radix of number system. The proposed design quaternary look-up table (QLUT) has 16 inputs and the basic multiplexer used is 4 × 1 MUX. It have 4 quaternary input, a delta literal circuit are connected as select input to multiplexer [3, 4].

3.1 Design of 4 × 1 Data Selector Using Quaternary Transmission Gates

The design of data selector depends upon radix of number system. In quaternary logic design radix is 4 so basic design circuit is 4 × 1 multiplexer as in Fig. 1. To ensure optimum complexity transmission gates are used in design. These circuits use variable threshold voltages transistors and operate with four stable and discreet logic levels, corresponding to ground that is 0 V and power supply of 1/3Vdd, 2/3Vdd and Vdd. The quaternary signal is divided in 3 binary signals by Down Literal Circuits [4–6] for controlling the operation. These control signals given to the pass transistor. It consist of 3 multiplexing stages of transmission gates whose

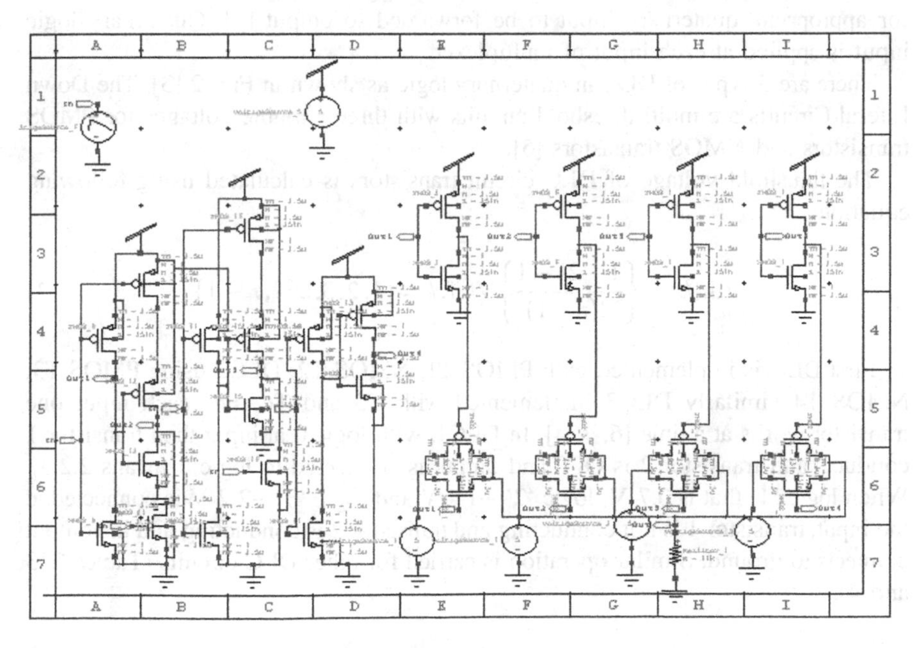

Fig. 1 Design of quaternary transmission gate

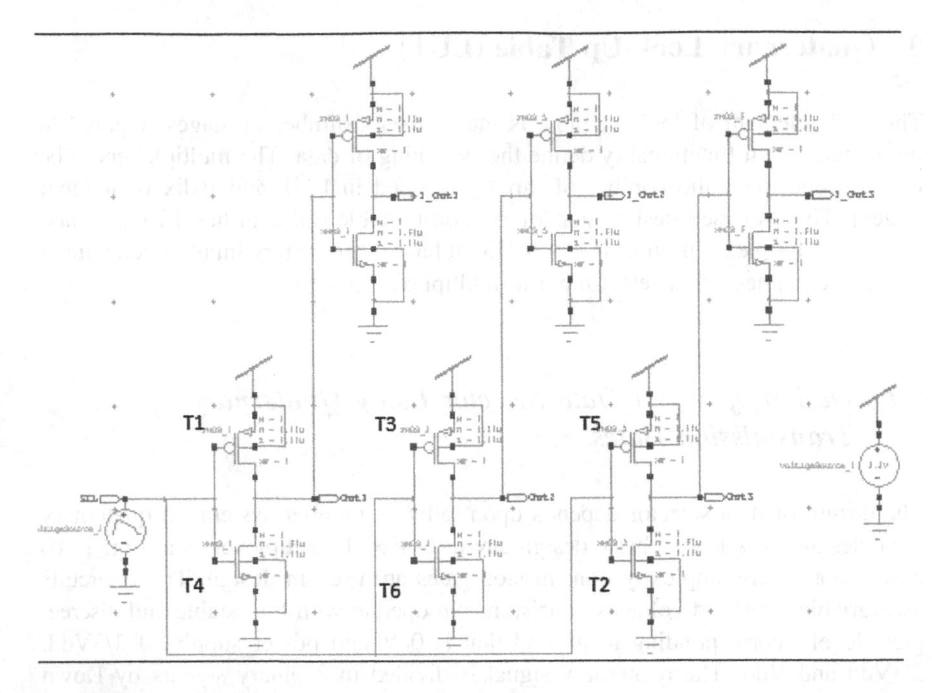

Fig. 2 Down literal circuit

select signal is provided to DLC circuit which triggers the correct transmission gate for appropriate quaternary input to be forwarded to output [4]. Quaternary logic input is applied at each input of multiplexer.

There are 3 types of DLC in quaternary logic as shown in Fig. 2 [3]. The Down Literal Circuits are multi threshold circuits with three variable voltages for PMOS transistors and NMOS transistors [5].

The threshold voltage of DLC circuit transistors is calculated using following equation:

$$\mathrm{Vth}(k) = \left\{\frac{1 - 2k - 1}{2(n-1)}\right\}\mathrm{Vref},\, k = \{1, 2, 3, \ldots, n-1\} \tag{3}$$

First DLC is implemented with PMOS T1, NMOS T2. DLC2 using PMOS T3, NMOS T4 similarly DLC3 implemented with T5 and T6. For each input one transistor works at a time [6, 3, 4]. In DLC1, with logic 0 at input then transistor 1 conducts and transistor 2 is OFF and output is connected to logic 3 means 2.2 V. When logic ‘1’ that is 0.7 V, logic ‘2’ −1.4 V and logic ‘3’ −2.2 V is connected at the input, transistor 1 is not conducting and transistor 2 is conducting and the output connects to ground. Similar operation is carried for other DLC circuits (Tables 2, 3 and 4).

Table 2 Calculations of capacitance for write operation

Cin		87.29 fF
Cov	(Cov)n = CGD0 × W × 11 = 5.45 fF (Cov)p = CGD0 × W × 9 = 3.29 fF	Avg cov = 4.37 fF
Cjb	(Cjb)n = 0.001078 F, (Cjb)p = 0.001003 fF	Avg Cjb = 0.0010405 F
Cjsw	(Cjsw)n = 2.874×10^{-10}, (Cjsw)p = 2.9×10^{-10}	Avg Cjsw = 0.000287 fF

CL total = Cin + Cov + Cjb + Cjsw = 91.66 fF, $P_{Dynamic}$ = 131.99 μW

Table 3 Calculations of capacitance for read, search operation, match entries

Cin		87.29 fF
Cov	(Cov)n = CGD0 × W × 11 = 4.905 fF (Cov)p = CGD0 × W × 9 = 3.29 fF	Avg Cov = 4.097 fF
Cjb	(Cjb)n = 0.001078 F, (Cjb)p = 0.001003 fF	Avg Cjb = 0.0010405 F
Cjsw	(Cjsw)n = 2.874×10^{-10}, (Cjsw)p = 2.9×10^{-10}	Avg Cjsw = 0.000287 fF

CL total = Cin + Cov + Cjb + Cjsw = 89.125 Ff, $P_{Dynamic}$ = 128.34 μW

Table 4 Calculations of capacitance for read operation, search operation, mismatch

Cin		94.192 fF
Cov	(Cov)n = CGD0 × W × 11 = 5.995 fF (Cov)p = CGD0 × W × 9 = 3.29 fF	Avg Cov = 4.645 fF
Cjb	(Cjb)n = 0.001078 F, (Cjb)p = 0.001003 fF	Avg Cjb = 0.0010405 F
Cjsw	(Cjsw)n = 2.874×10^{-10}, (Cjsw)p = 2.9×10^{-10}	Avg Cjsw = 0.000287 fF

CL total = Cin + Cov + Cjb + Cjsw = 98.837 fF, $P_{Dynamic}$ = 142.326 μW

3.2 Design of Quaternary SRAM

Static RAM is designed to store quaternary data bits. SRAM uses two quaternary inverters connected back to back. The data to be stored in the quaternary SRAM is interpreted as a logic 0, 1, 2, 3. Quaternary SRAM is designed with Quaternary inverters [1, 7]. Figure 3 shows the schematic of the quaternary SRAM with two quaternary inverters and access transistors.

Simulation results of QSRAM cell as shown in Fig. 4. When word line is equals to '1' data is written into the SRAM which is at available bit line through access transistor, and when write line is equals to '0' is available for reading.

3.3 Design of Quaternary LUT

The quaternary multiplexer 16 × 1 is designed using 5 quaternary 4:1 MUX. It contains 6 DLCs, 6 binary inverters and 84 transistors as in Fig. 5. This circuit has 16 inputs decoded as quaternary signals and one output and two control signals that

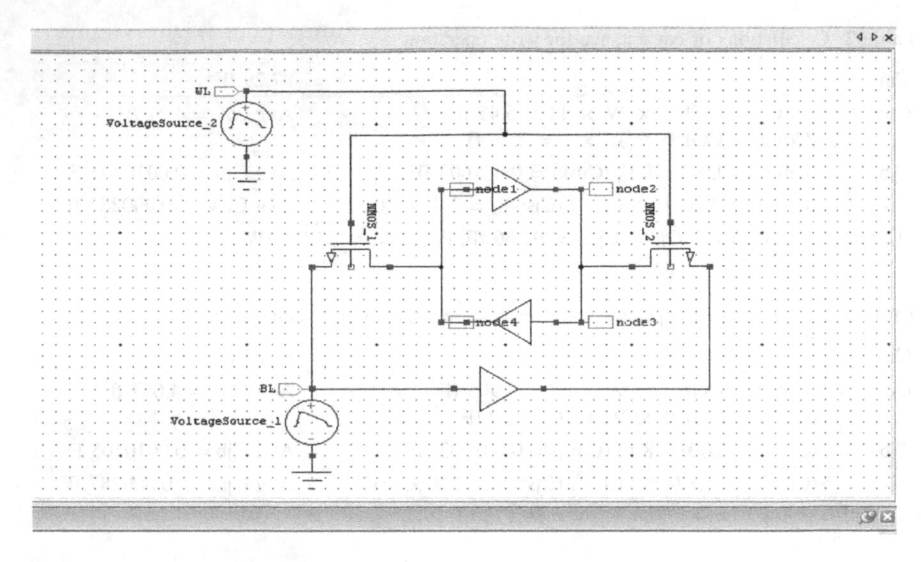

Fig. 3 Quaternary SRAM

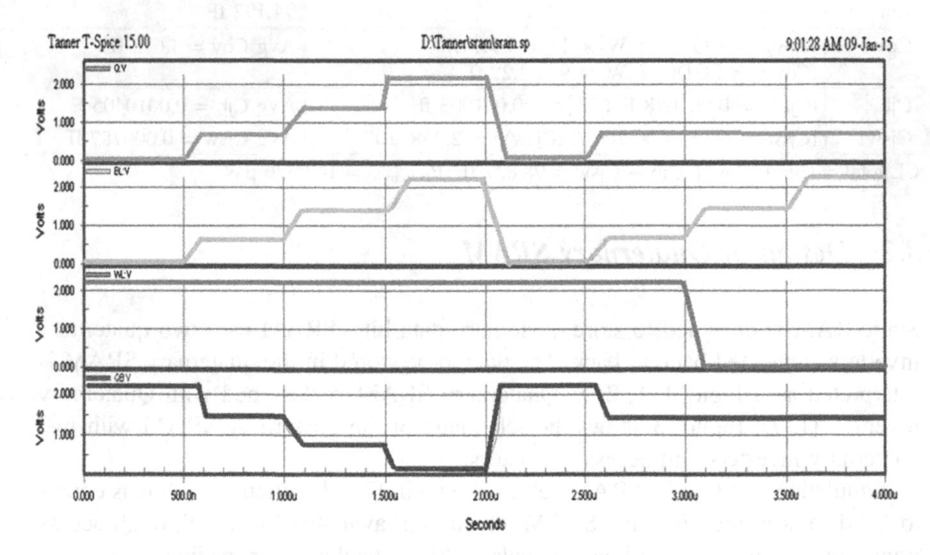

Fig. 4 Simulation results of quaternary SRAM

change the output according to control inputs. To each input quaternary content addressable memory cell is provided to store the data that is to be searched in look-up table.

That Fig. 6 shows For Bit Line (BL0) = 0, Select Line (SL) = 1 and word Line (W)L = 1 it can be seen that Match Line (ML) = 0, i.e. the no match state. For such all possibilities the structure is simulated for correctness.

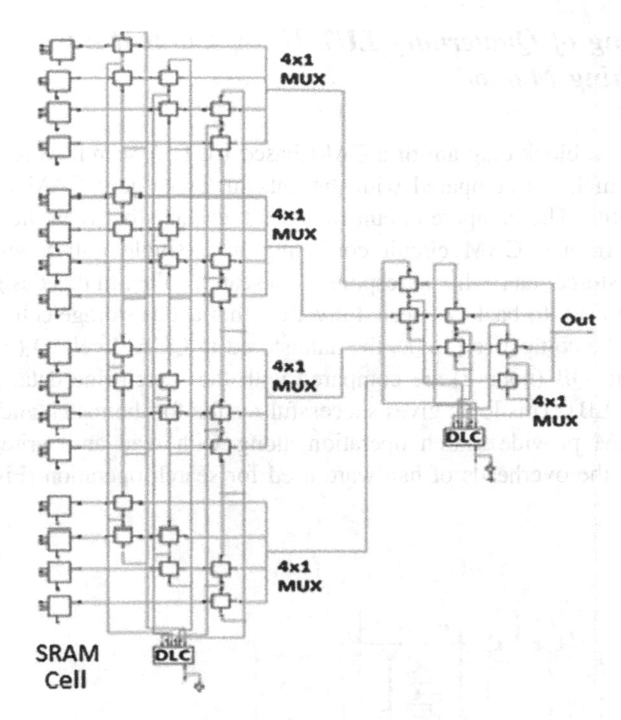

Fig. 5 Quaternary look-up table using SRAM

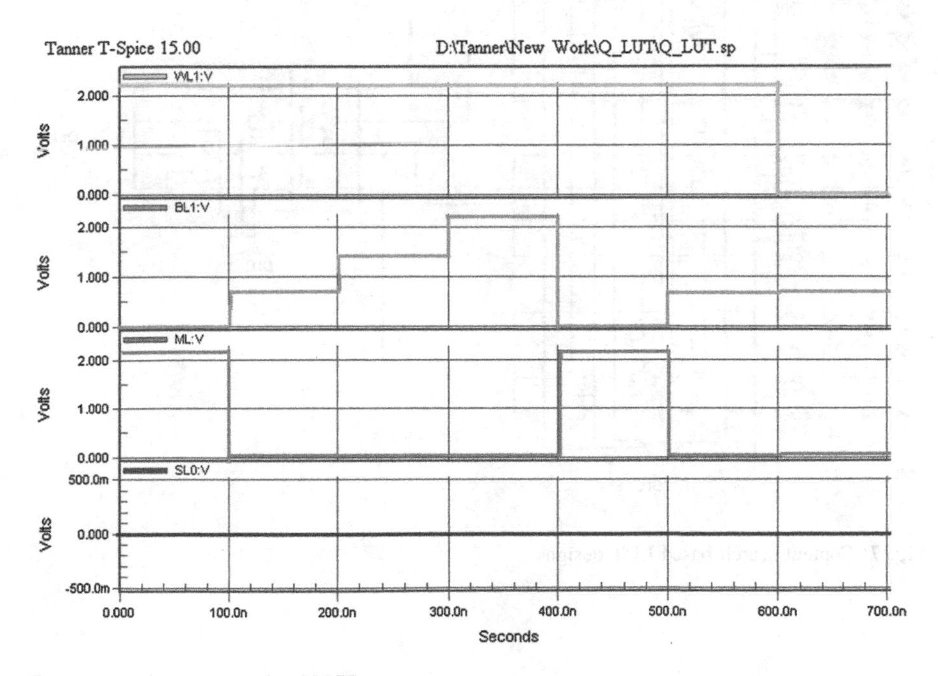

Fig. 6 Simulation result for QLUT

3.4 Working of Quaternary LUT Using Content Based Searching Method

Figure 7 shows a block diagram of a CAM-based LUT. A search word is given to search lines which are compared with the data in Quaternary CAM Cell having compares circuits. The compare circuit includes two paths between the match line and ground terminal. CAM circuit consisting of a single pull-down transistor coupled to the stored data value in response to a search value. In this design the data is stored in the back-to-back connected inverters that act as storage cell, same as in QSRAM. For the comparison logic, the data stored in QCAM cell, Q (node 1) and its complement, QB (node 2) are compared with the search line data, SL and its complement (SLB). This logic gives successful results for the match and mismatch condition. CAM provide search operation along with read and write operation which reduces the overheads of hardware used for search operation (Fig. 8).

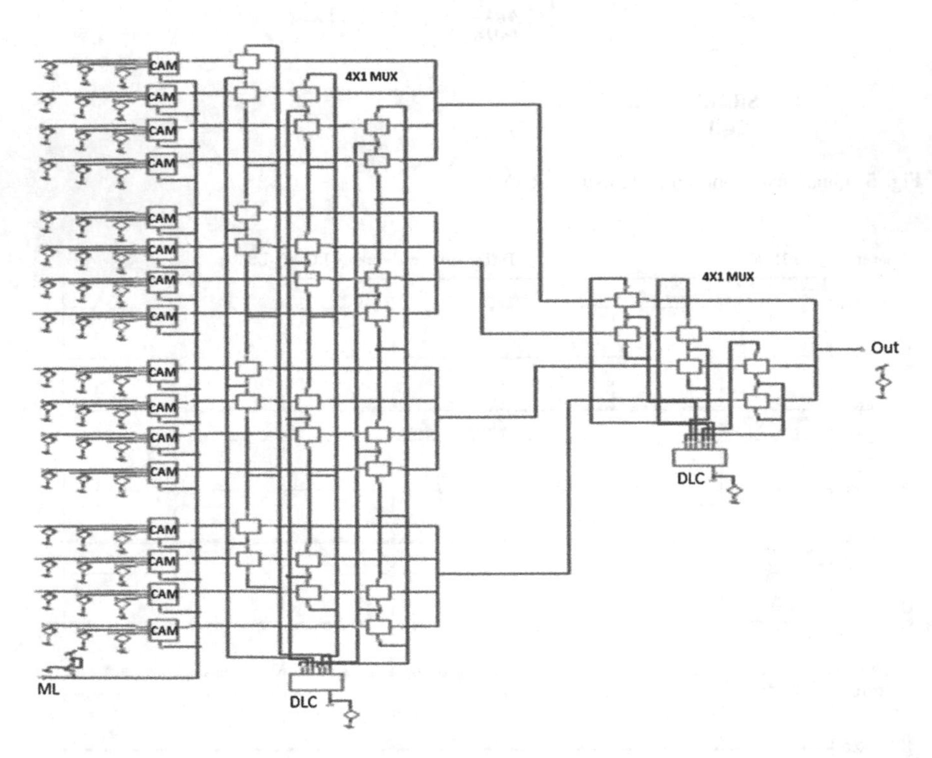

Fig. 7 Content search-based LUT design

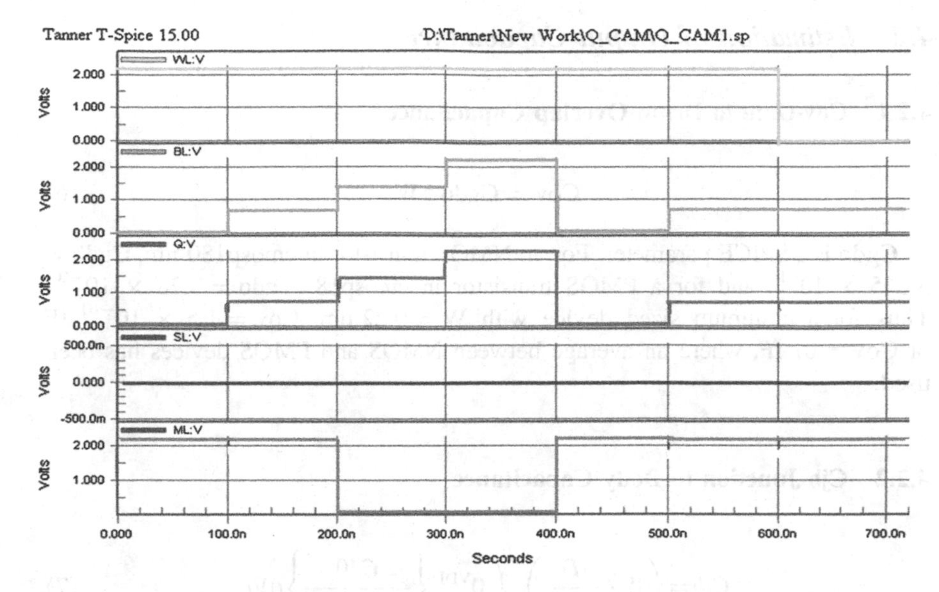

Fig. 8 Simulation result of QLUT using content search method

4 Capacitance and Power Calculations

4.1 Estimation of Input Capacitance

The input capacitance of QLUT is estimated by summing the gate capacitances connected to input under consideration [8],

$$\text{Cin} = \sum_{i=0}^{n} \text{Cgate}, i \quad (4)$$

The Cgate for a transistors can be calculated as below,

$$\text{Cgate} = \text{Cox} \sum_{i=0}^{n} (\text{WL})i \quad (5)$$

Cox is capacitance per unit of square area (Cox = εox/tox, where tox is the oxide thickness). For cmos process 180 nm technology, tox is around 4.1 nm. Cox = εox/tox = 3.85 * 8.85 × 10^{-12} Fm / 4.1 × 10^{-9} m = 8.31 fF/μm^2 Cgate = 8:31 fF/μm^2 * 12:0672 μm^2 = 10:08 fF.

4.2 Estimation of Output Capacitance

4.2.1 Cov-Gate to Drain Overlap Capacitance

$$\mathrm{Cov} = \mathrm{Cgdo} * W \tag{6}$$

Cgdo is a SPICE parameter. For an NMOS transistor in cmosp180 nm, Cgd0 = 3.665 × 10^{-10} and for a PMOS transistor in cmosp18, Cgdo = 3.28 × 10^{-10}. Thus for a minimum sized device with W = 0.22 μm, Cov = 1.5 × 10^{-10} fF or Cov = 07 fF, where an average between NMOS and PMOS devices has been used.

4.2.2 Cjb Junction to Body Capacitance

$$Cjb = \left(W \cdot \frac{D}{\mathrm{VDD}}\right) \int 0^{\mathrm{VDD}} \left\{ \frac{CJ0}{\left[1 + \frac{Vj}{Vb}\right]^{mj}} \right\} \mathrm{d}Vj \tag{7}$$

Vj is drain/source to body junction voltage, *Vb* is the built-in voltage of any junction, *mj* is grading coefficient [8, 4, 9]. *D* is given as length of the drain/source contact. Here VDD = 2.2 V and taking an average between PMOS and NMOS devices, the integral evaluates to 5.2 × 10^{-4} F/m^2 * *W* * *D*. For a minimum sized device of *D* = 0.48 μm and W = 0.22 μm, Cjb = 1.56 fF.

4.2.3 Drain Sidewall Capacitance Cjsw

$$Cjsw = ((W \cdot 2D + W)/\mathrm{VDD}) \int 0^{\mathrm{VDD}} \left\{ \frac{\mathrm{CJsw0}}{\left[1 + \frac{Vj}{Vb}\right]^{\mathrm{mjsw}}} \right\} \mathrm{d}Vj. \tag{8}$$

Cjsw0 = 2.3 × 10^{-10} F/m^2, VDD = 2.2 V, the integral in Eq. 8 evaluates to Cjsw = 1.2 × 10^{-10} fF/m^2 (2*D* + *W*). Cjsw = 1. 092 Ff.

Summing all capacitances it comes out to be

Load Capacitance is CL = Cin + Cov + Cjb + Cjsw = 0.7 pF.

Dynamic Power at 1 MHz frequency is Pdyn = CL * Vdd^2 * Freq.

Pdyn = 47 uW.

4.3 *Read/Write Ratio of QCAM Cell*

a. For analysis it is found that Power dissipated = 128.34 μW for read operation and match found entries and D_{ynamic} = 131.99 μW for write operation. The ratio for the same is = 0.97.
b. So, read operation for QCAM cell consumes approximately 9 % less power than write when match is found during search operation.
c. For read mismatch the Power dissipated D_{ynamic} = 142.326 μ, Ratio = 1.07.
d. During first mismatch Power dissipated is high by 10.7 % but further search line operations are disabled to power consumption of later stages are saved.

5 Conclusion

CMOS quaternary logic circuits are designed using multi-threshold circuit and quaternary power lines. 16 × 1 QLUT circuit have been simulated using Tanner tool at 0.18 μm technology. QCAM works properly on 4 stable states defined for Quaternary Logic under storing, searching operation. Cross-coupled stages properly defines mismatch and match operations. Propagation delay of QSRAM is lessened by 75 % and static power dissipation is reduced by 40 % and propagation delay of QLUT is reduced up to 29.71 %. On comparing propagation delay of BLUT and QLUT by connecting capacitive load of 100, 400 and 500 fF. It is found that delay of QLUT is reduced by 14.33 %, 22.08 % and 23.53 % respectively. For 1000 fF load capacitance, delay of QLUT is reduced by 29.71 %. Such circuits can be used for modern routers which can store routing tables and need very fast lookups than conventional SRAM based routers.

References

1. Vasundara Patel, Gurumurthy\Applications in Multi Valued Logic" International Journal of VLSI Design & Communication Systems(IJVCS)Vol1, No.1, March 2010.
2. Kostas Pagiamtzis, Ali Sheikholeslami, "Content-Addressable Memory (CAM) Circuits and Architectures: A Tutorial and Survey "IEEE JOURNAL OF SOLID-STATE CIRCUITS, VOL. 41, NO. 3, MARCH 2006.
3. R.C.G daSilva, C. Lazzari, H. Boudinovc, L. Carro, "CMOS voltage-mode quaternary look-up tables for multi-valued FPGAs" Microelectronics Journal 40(2009) 14661470, 2007.
4. Deepti P. Borkute, Dr. Dakhole P.K. "Impact of Quaternary Logic on Performance of Look-Up Table" IOSR Journal of VLSI and Signal Processing (IOSR-JVSP) Volume 5, Issue 3, Ver. II (May–Jun. 2015), pp. 36–48 e-ISSN: 2319-4200.
5. Diogo Brito, Taimur G. Rabuske, Jorge R. Fernandes, Paulo Flores and Jos Monteiro, \Quaternary Logic Lookup Table in Standard CMOS " IEEE Transactions On Very Large Scale Integration (Vlsi) Systems 1063–8210 2014 IEEE.

6. Cristiano Lazzari, Paulo Flores, Jose Monteiro, Luigi Carro, "A New Quaternary FPGA Based on a Voltage-mode Multi-valued Circuit" 978-3-9810801-6-2/DATE10 2010 EDAA.
7. Deepti P. Borkute, Pratibha Patel, "Delay performance and Implementation Of QuaternaryLogic Circuits," ICCUBEA International Conference 2015, PCCOE, Pune.
8. David J. Grant and Xiuling Wang "4-bit CMOS Transmission Gate Adder Module" Department of Electrical & Computer Engineering, University of Waterloo April 14, 2003April 14, 2003.
9. Deepti P. Borkute, Pratibha Patel, \ Power, Delay and Noise Margin Comparison of Binary and Quaternary SRAM, "IEEE Devices, Circuits and Systems (ICDCS), 2014 2nd International Conference on, 6-8 March 2014, Coimbatore.

Exhaust Gas Emission Analysis of Automotive Vehicles Using FPGA

Ratan R. Tatikonda and Vinayak B. Kulkarni

Abstract This paper presents the exhaust gas emissions analysis of automotive vehicles which is a quick and accurate way to determine the running conditions of an engine. The main aim of the paper is intended to read the data from Sensors such as Oxygen and Carbon monoxide (CO) interfaced to FPGA board to get emissions gas information and to carry out its analysis. The Xilinx ISE timing simulation along with MATLAB waveform results are presented. The final observed concentration for CO gas detection is found to be 287 ppm. The basic principles regarding why and how exhaust analysis carried out is briefly discussed for better understanding along with different methods of gas analysis. This is a very effective reason for designing such a safety system.

Keywords Sensors · Oxygen · Carbon monoxide (CO) · FPGA

1 Introduction

Analysis is one of the important factor in present day life for all aspects whether it be automobiles or any kind of applications, without which the results are not confirmed and as such needed more work and research. The problem of air quality has reaching to an increasing level if taken into account developing country like India. Due to this, proper monitoring and analysis is to be carried out for the excess gas emissions [1]. The main aim of the paper is carry out analysis using FPGA tool and sensors.

R.R. Tatikonda (✉)
Department of Electronics, MIT Academy of Engineering, Alandi(D), University of Pune, Pune, India
e-mail: ratan.tatikonda@gmail.com

V.B. Kulkarni
Department of Electronics & Telecommunication, MIT Academy of Engineering, Alandi(D), University of Pune, Pune, India
e-mail: vbkulkarni@entc.maepune.ac.in

S.C. Satapathy et al. (eds.), *Proceedings of the International Conference on Data Engineering and Communication Technology*, Advances in Intelligent Systems and Computing 469, DOI 10.1007/978-981-10-1678-3_10

Three main reasons for exhaust gas analysis are: First is to identify mechanical problems and engine performance and second is to test the running efficiency of the engine and the third one is exhaust emissions test for against federal standards and state. Analysis can be done by using external rugged portable analyzer or by using different sensors available in industry for measuring the gas contents in the exhaust pipe. The different methods of measurement required to done for analysis are by Exhaust Gas constituents, Gas Analysis, Conductivity Methods of Analysis and Combustion Methods of Analysis. Exhaust gas is emitted from the exhaust pipe of the vehicle due to the combustion (burning) of fuels such as petrol, natural gas and diesel. FPGA is being used as an intermediate for gas analysis with use of different sensors like Oxygen sensor and Carbon Monoxide (CO) sensor. The use of other previously used portable gas analyzers is not taken into account in this paper because the world is moving to green environment and in future days there will be existing electronic devices for measuring, monitoring and analysing the different application factors [2, 3].

2 Methods of Exhaust Gas Analysis

There are various methods of measurement of exhaust gas analysis of which each is described below.

2.1 Constituents of Exhaust Gas

The exhaust gases composition from the burning of fuels in an internal ignition engine is quite modifiable, which depends to some expanse upon the operation characteristics, but to a greater expanse upon the quantity of air provided per unit of burned fuel. When a large surplus of air is supplied, the fuel hydrocarbons are completely converted to water (H_2O) and carbon dioxide (CO_2), and some oxygen emerge in the analysis. The nitrogen (N_2) present in the air used for burning is passed through the action of combustion which is not changed and so emerges in the final products. When the air to fuel proportion is minimized, the Carbon combustion is incomplete and carbon monoxide (CO) emerges. Similarly, the hydrogen (H) combustion to water is not complete and free hydrogen is present in the products of exhaust. Since, water made gas analyses, the water vapour is generated by the burning of H_2 which is not appeared in the analysis. In general, a full gas analysis will result the constituents such as CH_4, N_2, O_2, CO_2, CO, H_2 [4].

2.2 *Conductivity and Combustion Method of Analysis*

The mixture of gaseous analysis by virtue of contrast in the thermal conductivity of the constituents has presented a great attention. Not all gaseous mixtures confer themselves to this type of analysis. Where high thermal conductance gases such as helium or hydrogen are present, however, even in small quantity, there outcomes an appreciable contrast in the conductivity as compared to air or overall thermal conductivity [4].

On the rich side of a theoretically perfect mixture, considerable quantity of combustible gases is present. The quantity of such gases (CO, CH_4, H_2) increases as the mixture becomes richer and this factor has been utilized as a basis for instruments that determine the liberated heat and burn the residual combustible. The combustion is attained by disclosing air along with the exhaust gases and afterwards burning the combustible catalytically upon the surface of a warm platinum wire or a mass catalyst. In either case, the temperature increase gives the quantitative indication [4, 5].

3 Emission Standards

The emission standards are instituted by the Indian Government to control the output of air pollutants from internal combustion engine equipment, including motor vehicles. The exhaust emission limits must be known to keep the vehicle under normal conditions. Below is the table which shows the CO concentration in g/km for 2 and 4 wheeler gasoline vehicles with respect to year's progress (Tables 1 and 2).

In above table Stage BS means Bharat Stage emission standards [6]. As per the results discussed below in next topic it is shown that concentration of CO is measured in ppm (parts per million) and the emission standards are in g/km. Below Eq. (1) is the conversion from ppm to g/km [7].

$$\mathrm{CO(g/km)} = 9.66 \times 10^{-3} \times \mathrm{CO\,(ppm)} \tag{1}$$

The CO concentration for 287 ppm in g/km is 2.77 by calculating using above equation which is excess than the limits mentioned in table above as per the present year to be taken in consideration.

Table 1 Emission standard for 2-wheeler gasoline vehicles (g/km)

Year	Stage	CO
2000	BS I	2.0
2005.04	BS II	1.5
2010.04	BS III	1.0

Table 2 Emission standard for 4-wheeler gasoline vehicles (g/km)

Year	Stage	CO
2000	BS I	2.72–6.90
2005	BS II	2.2–5.0
2010	BS III	2.3–4.17
2010	BS IV	1.81–2.27

4 Sensors

The different sensors used for exhaust gas emission measurement are Oxygen (O_2) sensor and Carbon monoxide (CO) sensor. The description of these sensors is taken into account below.

Oxygen (O_2) sensor is one of the important sensors in this system which is mounted in the exhaust manifold to check unburned oxygen in the exhaust which tells the system computer if the fuel mixture is burning rich referred as (less oxygen) or burning lean referred a (more oxygen). The O_2 sensor used in vehicles is a voltage generating sensor [8].

CO sensor: When the vehicle with the doors closed is entrapped in a traffic signal or in a traffic jam, the emission of CO from the exhaust of other nearby vehicles will be easily hauled into the vehicle cabin which creates the major tragedy to the persons inside the cabin. The system has a sensor which identifies the presence of CO gas inside the vehicle cabin. A CO sensor is a device that detects the presence of the CO gas in order to stop the poisoning from CO. MQ-7 CO gas sensor is mounted on the vehicle, which will be giving automatic update of the status of the vehicle continuously. So, we can easily detect the excess gas emission status of the vehicle and prevent the engine and canny the user. Sensitive material of MQ-7 gas sensor is tin Dioxide SnO_2, which with lower conductivity in clean air. It make detection by method of cycle high and low temperature, and detect CO when low temperature (heated by 1.5 V). The architecture configuration of sensor and is shown below [9].

Structure and configuration of MQ-7 CO gas sensor is shown in Fig. 1 which is made by micro AL_2O_3 ceramic tube. The necessary work conditions provided by heater for work of sensitive components. The enveloped MQ-7 have 6 pin, 4 are used to take ferry signals, and other 2 are used for generating heating current.

The above Fig. 2 is basic test circuit of the sensor with two voltages, heater voltage (V_H) and test voltage (V_C). V_H is used to supply working temperature to the sensor, while V_C used to perceive voltage (V_{RL}) on load resistance (R_L) which is in series with sensor. The sensor has light polarity, so DC power is needed for V_C. For better performance of the sensor, suitable R_L value is needed. The equations for Power of Sensitivity body (P_S) and Resistance of sensor (R_S) are:

$$P_S = V_C^2 \times R_S / (R_S + R_L)^2 \quad (2)$$

$$R_S = (V_C / V_{RL} - 1) \times R_L \quad (3)$$

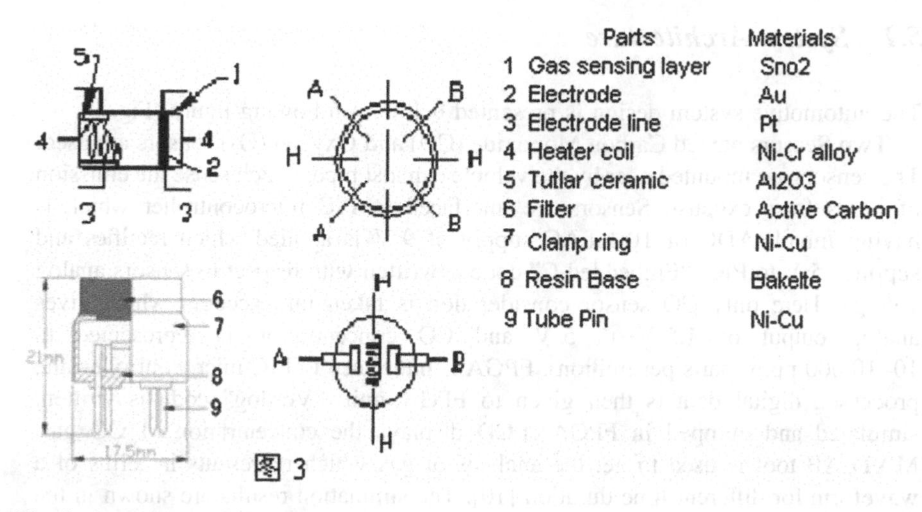

Parts
Materials
1 Gas sensing layer
Sno2
2 Electrode
Au
3 Electrode line
Pt
4 Heater coil
Ni-Cr alloy
5 Tutlar ceramic
Al2O3
6 Filter
Active Carbon
7 Clamp ring
Ni-Cu
8 Resin Base
Bakelte
9 Tube Pin
Ni-Cu
图 3

Fig. 1 Architecture of MQ-7 CO gas sensor [11]

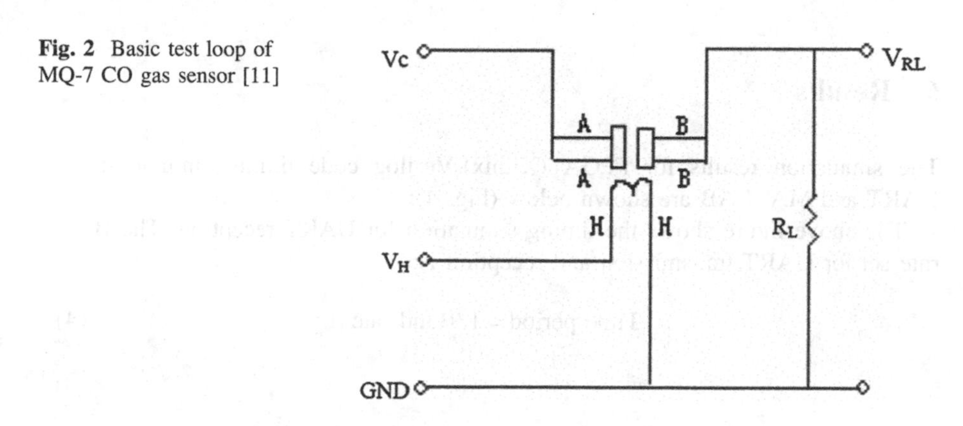

Fig. 2 Basic test loop of MQ-7 CO gas sensor [11]

5 FPGA-Based Exhaust System

The FPGA used for exhaust analysis is device XC3S500E of Spartan 3E family because it meets the high volume and cost sensitive consumer electronic applications. The two reasons for selecting these FPGA is firstly because of its features such as low cost, high performance, embedded processor cores and multi-level memory architecture and secondly because PIC node is a end device and FPGA is Master node to which many end devices can be connected. PIC 16F877A is used as a A-to D converter purpose only because start–stop bit coding in verilog is to be written once for every sensor.

5.1 System Architecture

The automotive system design is presented below in following figure (Fig. 3).

Two Sensors named Carbon Monoxide (CO) and Oxygen (O_2) sensors are used. The sensors are mounted near by the vehicle exhaust pipe, which sense the emission of gases from exhaust. Sensors are interfaced to PIC microcontroller which is having inbuilt ADC of 10 bit.AC supply of 9 V is applied which rectifies and supplies 5 V to PIC. "Embedded C" code is written with respect to sensors analog voltage. Here only CO sensor consideration is taken into account which gives analog output of 1.5 V to 5 V and CO concentration is represented in 10–10,000 ppm (parts per million). FPGA is interfaced to PIC microcontroller, the processed digital data is then given to FPGA unit. "Verilog" code is written, simulated and dumped in FPGA. LCD displays the concentration of CO gas. MATLAB tool is used to get the analysis of gas which represents in terms of a waveform for different time duration [10]. The simulation results are shown in the next topic.

6 Results

The simulation results for FPGA (Xilinx)-Verilog code timing simulation for UART and MATLAB are shown below (Fig. 4):

The above figure shows the timing simulation for UART reception. The Baud rate set for UART transmission and reception is:

$$\text{Time period} = 1/\text{Baud rate} \tag{4}$$

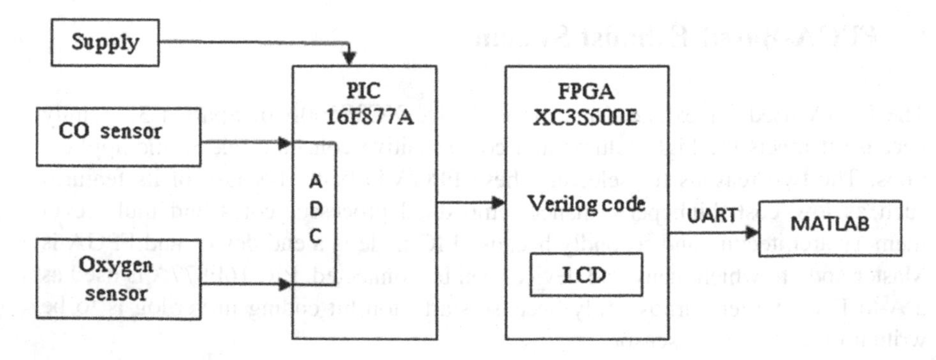

Fig. 3 Block diagram of automotive system using FPGA

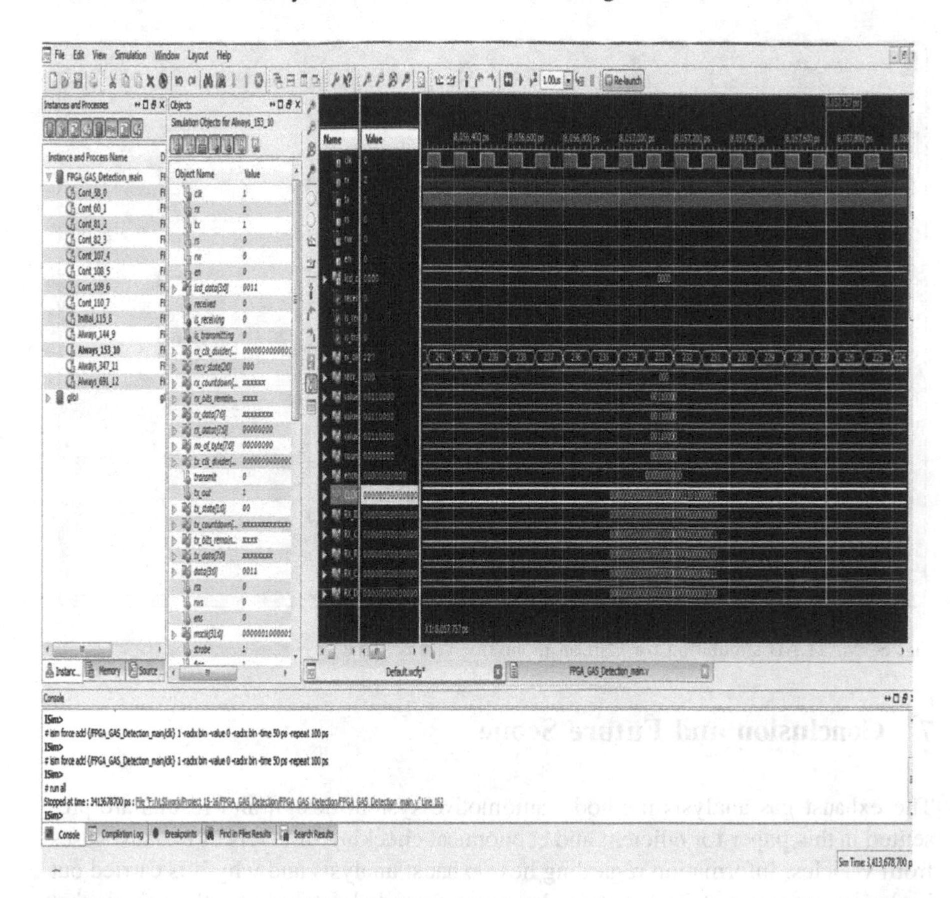

Fig. 4 Xilinx ISE timing simulation for UART reception

Equation 2 shows the time period required according to the baud clock. Hence Time period for 1 bit = 104 μs, since 8 data bit is required for transmission, 1 start bit and 1 stop bit. Therefore total time period for 8 bit to transfer is 104*8 = 832 μs. Input is firstly set to 0 so as to start reception then it is lock till the last bit is received. Sampling time is the important part of UART reception which is set for baud clock and then data is received and is made for better reception to get the required data. The RX buffer helps in reception of data with the help of reception enable signal (Fig. 5).

The MATLAB simulation shows the different levels of CO gas detection with respect to certain time. Y-axis in above figure represents Carbon monoxide in parts per million and X–axis represents the corresponding time duration. 287–CO which displayed in the figure means the last recorded result of CO in parts per million.

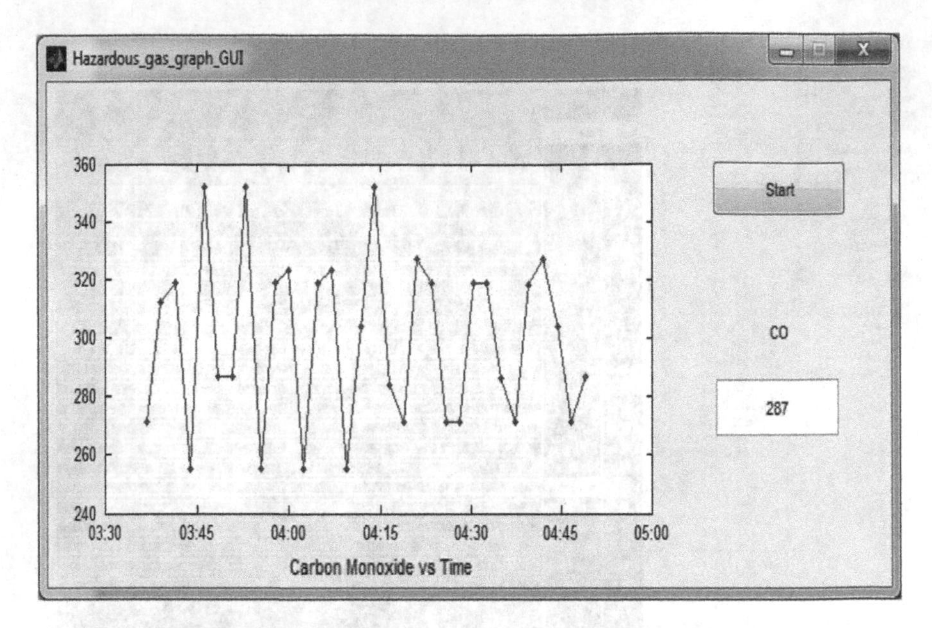

Fig. 5 MATLAB simulation for Carbon monoxide versus Time

7 Conclusion and Future Scope

The exhaust gas analysis methods, automotive system design and results are presented in this paper for efficient and economical checking of levels of exhaust gases from vehicles. Information regarding how exhaust analysis and why it is carried out is also been presented. The system also gives upgraded database for the vehicle that causes excess emission and this database can be used further for applications like vehicle speed control, Vehicle fitness and Pollution control. The emission standards are discussed briefly for excess emission limits. Verilog is used as the design language to achieve the module of UART.

In Future years, the exhaust gas emissions from automated vehicle should be controlled by regular basis checking and by vehicle maintenance analysis to be done per week using the above system design FPGA method and sensors which is to be embedded in a vehicle and from the outcomes the necessary immediate action will be taken.

References

1. S. B., V. R., H. S. K., and G. S.: Analysis of exhaust emissions from gasoline powered vehicles in a sub-urban indian town, International Research Journal of Environment Sciences, vol. 2, no. 1, pp. 37–42, January (2013)

2. Konstantas, G., Stamatelos, A.: Quality assurance of exhaust emissions test data, vol. 218, pp. 901–913 (2014)
3. Singh, R., Maji, S.: Performance and exhaust gas emissions analysis of direct injection cng-diesel dual fuel engine, International Journal of Engineering Science and Technology (IJEST), vol. 4, no. 03, pp. 833–846, March (2012)
4. GRAF, S.H., GLEESON, G.W., PAUL, W.H.: Interpretation of Exhaust Gas Analyses, no. 04
5. Bhandari, K.: Performance and emissions of natural gas fueled internal combustion engine: A review, JSIR vol. 64, pp. 333–338 (2005)
6. Emission Standards India, https://www.dieselnet.com/standards/in/2wheel.php
7. E. Muzenda, T.J. Pilusa, M.M. Mollagee.: Reduction of vehicle exhaust emissions from diesel engines using the whale concept filter, Aerosol and Air Quality Research, pp. 994–1006, 2012
8. William Allen, http://www.wellsve.com/sft503/counterp_v2_i3_1998.pdf
9. Deepa, T.: Fpga based pollution control system for vehicles using special sensors, International Journal of Engineering Sciences & Research Technology, vol. 4, no. 3, April (2014)
10. Bollampalli.: Fpga interface for an analog input module, International Journal of Application or innovation in Engineering & Management, vol. 2, no. 4, April (2013)
11. Carbon Monoxide gas sensor MQ-7, https://www.pololu.com/file/0J313/MQ7.pdf

A Graph-Based Active Learning Approach Using Forest Classifier for Image Retrieval

Shrikant Dhawale, Bela Joglekar and Parag Kulkarni

Abstract Content-Based Image Retrieval System is comprised of large image collections which find their use in applications such as statistical analysis, medical diagnosis, photograph archiving, crime prevention, face detection, etc. This poses a challenge for pattern recognition techniques used in image retrieval, which require being both efficient and effective. These techniques involve high computational burden in training phase, to separate samples from distinct classes, for active learning. In active learning paradigm, system first returns a small image set. Inference is drawn from this image set based on relevance as per user perception. Hence image retrieval based on context suffers in terms of precision and recall, as retraining and interactive time response is involved. A classifier known as Optimum-Path Forest (OPF) reduces this computational overhead by transforming the problem of classification as a graph obtained from dataset samples in feature space. It involves fast computation of trees built through the forest classifier in a graph resulting from training dataset samples. An optimum path is conquered from least cost approximation of a shortest path through the current prototype sample to the image under consideration.

Keywords Forest classifier · Active learning · Minimum spanning tree · Image retrieval

---

Shrikant Dhawale (✉) · Bela Joglekar
Department of Information Technology, Maharashtra Institute of Technology, Pune, India
e-mail: shrikantdhwl7@gmail.com

Bela Joglekar
e-mail: bela.joglekar@mitpune.edu.in

Bela Joglekar · Parag Kulkarni
Bharti Vidyapeeth University, Pune, India
e-mail: paragindia@gmail.com

Parag Kulkarni
Anomaly Solutions Pvt. Ltd, Pune, India

S.C. Satapathy et al. (eds.), *Proceedings of the International Conference on Data Engineering and Communication Technology*, Advances in Intelligent Systems and Computing 469, DOI 10.1007/978-981-10-1678-3_11

1 Introduction

Image Retrieval based on context attempts to retrieve most relevant images in database. Nature of problem is dynamic depending on the relevance inferred. User's perception differs for the similar query image. Such systems usually depend on active learning. In active learning paradigm, system first returns a small set of images and user specifies about the relevance. As large image collections are available for content-based image retrieval system which requires feedback to infer the perception of user, retraining and interactive time response is involved. Existing classifiers such as Support Vector Machine, Artificial Neuronal Network and *k*-Nearest Neighbor take more computation time for big datasets especially in training phase [1]. The problem of computational burden for training is there for above-mentioned classifiers. Although the Support Vector Machine classifier has attained high recognition rate in numerous applications, when training sample set becomes large, its learning becomes improbable due to burden of huge training size over classifier. Other classifiers such as Artificial Neuronal Network, Radial Basis Functions, Self-Organizing Maps and Bayesian classifiers have same problem [2].

A novel approach for pattern classifier based on fast computation of a forest resulting from training samples is addressed in order to overcome such challenges [1, 3]. Optimum-Path Forest can achieve similar effectiveness to above-mentioned machine learning techniques and can be faster in training phase [4]. Content-Based Image Retrieval (CBIR) consists of image database characterized by feature vectors which encodes features of an image such as color, texture, and/or shape. The similarity between images can be a measure of difference between their respective features vectors often called as visual descriptors. For given query image, CBIR System ranks more similar images by comparing their distance from query image. But, a semantic gap exists in retrieved result due to inability of low-level features from an image to meet user's expectations. Existing algorithms fail to precisely relate the high-level concept, or the semantic aspect of image, to lower level content. To minimize the semantic gap, feedback-based learning approach is significant. In typical image retrieval system based on context, user first gives query image and based on similarity measure the small set of images is returned.

The organization of paper is, Sect. 2 provides survey done in similar area, Sect. 3 will give overall idea about the system framework and structural design of project, Sect. 4 presents mathematical modeling, Sect. 5 provides experimental result obtained and Sect. 6 is conclusion.

2 Related Work

The goal of Context-Based Image Retrieval (CBIR) can be divided into two main categories, first one being efficiency and the second one, effectiveness. Efficiency deals with indexing structure involved in CBIR framework where more enhancement is possible and there is room for improvement in scalability due to the huge collection of image data set. This can be achieved by techniques such as Principal Component Analysis (PCA) and Clustering and Single Value Decomposition (CSVD) by reducing dimension scale in search space [5]. But to address effectiveness, we need to bridge semantic gap problem. Recently focus is on feedback-based learning, synonymously used for active learning, to figure out composite descriptor which improves ranking of the most relevant images to develop a pattern classifier, responsible for retrieving the most relevant candidates from image database and presenting them to the user [5]. The image retrieval system based on relevance feedback has two approaches, namely, greedy and planned approach [2]. In this paper, greedy and planned image retrieval system based on OPF is discussed. The OPF classification algorithm and two optimization techniques to improve the retrieval process by adding multiple distance spaces called as Multi-Scale Parameter Search (MSPS) and Genetic Programming (GP) algorithms are presented in [5]. Further, many optimizations of OPF have been proposed to speed up classification process [6]. Optimum-Path Forest (OPF) training phase consists of two steps, the first one being Minimum Spanning Tree (MST) generation for detection of prototype nodes, and second is OPF formation. The actual proposed forest classifier optimization method performs both above-mentioned steps simultaneously, that is computation of MST and OPF at the same time, which will further save computational efforts.

3 System Framework

The OPF models the classification problem as a graph obtained from feature space as shown in Fig. 1.

System Architecture, shown in Fig. 2, depicts active learning through Optimum-Path Forest Classifier using feedback technique, which is divided into three modules. In the user interface module, user can give query image for searching as an input. Thereafter, the user will be able to retrieve relevant results and give appropriate feedback. Second module is learning module where feature

Fig. 1 Image descriptor (v, d)

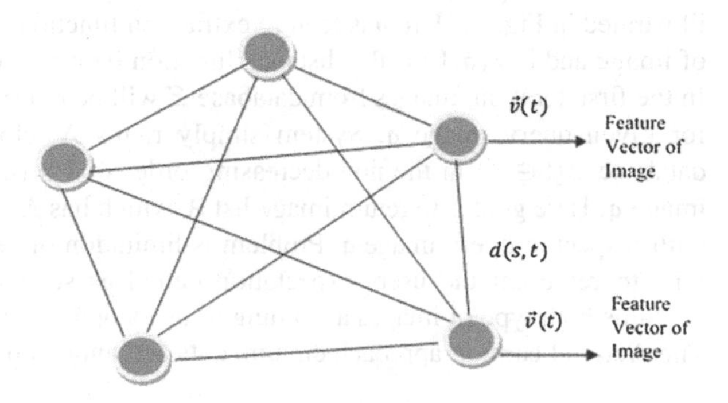

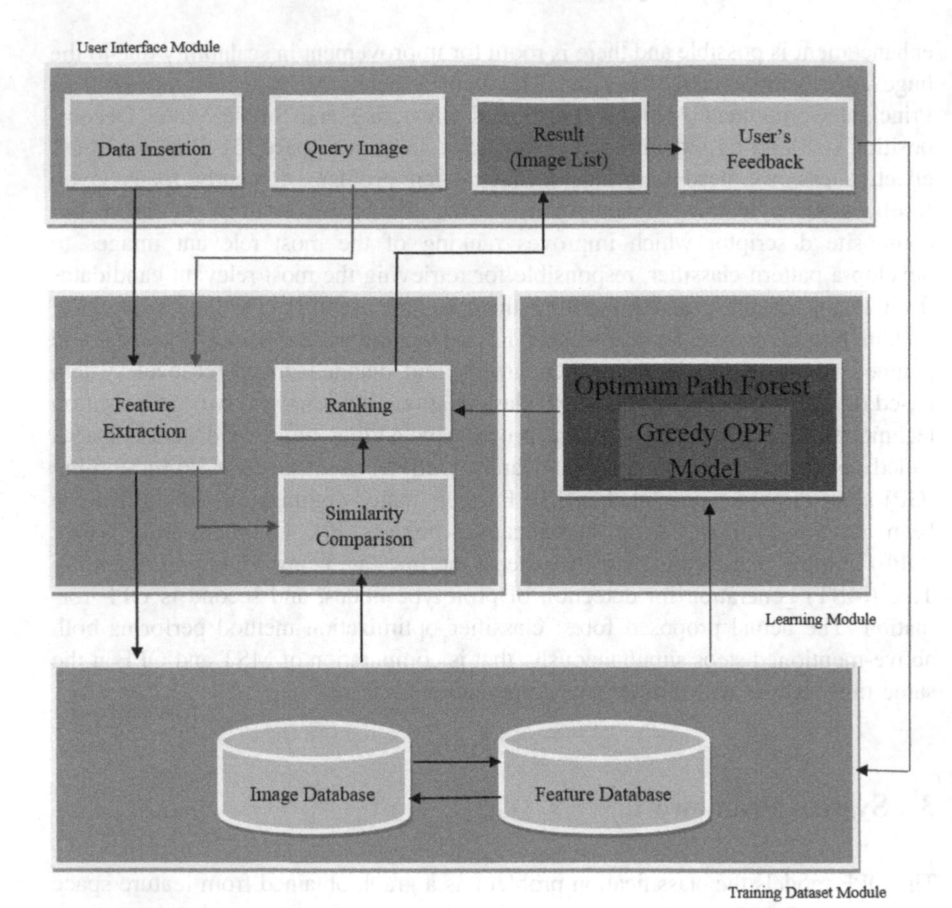

Fig. 2 System architecture

extraction, similarity comparison based on extracted features, image ranking, retraining and learning by OPF classifier takes place. Third module is a database module where all the features extracted from image dataset are saved. The simple descriptor $\mathcal{D}=(\mathfrak{v},\mathfrak{d})$ shows how nodes are spread in feature space, which is illustrated in Fig. 1. Here $\mathfrak{v}$ is feature extraction function which extracts the features of image and $\mathfrak{d}=(\mathfrak{a},\mathfrak{b})$ is the distance function between two image representations. In the first iteration, images from database $\mathcal{Z}$ will be returned according to similarity for given query image $\mathfrak{q}$. System simply ranks $\mathcal{N}$ closest image ($\mathcal{Z}$) set from database $\mathcal{Z}(\mathfrak{t}\in\mathcal{Z})$ in the non-decreasing order of $\mathfrak{d}=(\mathfrak{a},\mathfrak{b})$ with respect to query image $\mathfrak{q}$. Here goal is to return image list $\mathcal{X}$ which has $\mathcal{N}$ most relevant images in $\mathcal{Z}$ with respect to query image $\mathfrak{q}$. Problem is limitation of descriptor $\mathcal{D}=(\mathfrak{v},\mathfrak{d})$ which fails to represent the users expectation called as semantic gap, such that list $\mathcal{X}$ contains both type of images according to users opinion, i.e. relevant and irrelevant. The Active Learning approach circumvents semantic gap problem. User's feedback for the relevancy in returned image set, for a small number of iterations is taken. Here, user marks which images are relevant in $\mathcal{X}$, thereby forming a labeled training dataset $\mathcal{T}$ which increases with new images at each iterations so $\mathcal{T} \leftarrow \mathcal{T} \cup \mathcal{X}$ [5]. We model our OPF classifier on this training set. The process consists of estimating prototypes (adjacent node with different class labels) first and generation of forest as a second step. The training process interprets samples of a training set as a complete graph, which has nodes all from sample image elements of training set $\mathcal{T}$ and arcs between the sample nodes is given by $\mathfrak{d} = (\mathfrak{a}, \mathfrak{b})$. Thereafter, a minimum spanning tree (MST) over underlying graph is formed to obtain prototype nodes [7, 8]. All paths in the graph have a certain cost defined by $f_{\max}$ and minimum cost paths are computed from prototypes to all image nodes of training set $(\mathfrak{t} \in \mathcal{T})$, such that classifier is optimum-path forest rooted in prototype [2].

Thereafter, the classifier is used to evaluate images in the database from prototype to all image nodes $(\mathfrak{t} \in Z/\mathcal{T})$ in incremental manner.

4 Mathematical Modeling

Working of OPF Classifier is transforming the problem of classification in the form of a graph, partitioned in a certain feature space. The nodes of a graph are denoted by feature vector and edges connect all pairs of nodes, thereby forming a fully connected graph. Competition process then partitions the graph, between prototype samples which offer optimum path to left over nodes of the graph [9]. The OPF is a generalization made in Dijkstra's algorithm, to calculate optimum paths from source to remaining nodes [3]. The only difference is that OPF uses a set of source nodes (prototypes) instead of single source node with a more smoothly defined path cost function.

$$\text{Given } Z_1 \cup Z_2;$$

where Z_1 = training set,
Z_2 = test set,
Assume B = Prototype Samples.

An optimum path π with sequence of image samples of terminus $\mathfrak{j}$ can be represented as $\pi_{\mathfrak{j}}$. For each optimum path π, consisting of a sequence of samples, we assign a cost function $f(\pi)$. $f(\pi_{\mathfrak{j}})$ will compute the cost of optimum path for image sequence till terminus $\mathfrak{j}$, such that $f(\pi_{\mathfrak{i}}) \leq f(\pi_{\mathfrak{j}})$ [4, 8, 10]. Smooth path cost function, f_{max}, which is defined as

$$f_{\max}(\langle i \rangle) = \begin{cases} 0 & \text{if } \ i \in B, \\ +\infty & \text{otherwise,} \end{cases} \quad [2,3,4,5]$$

$$f_{\max}(\prod_{i}\cdot\langle i,j\rangle) = \max\{f_{\max}\left(\prod_{i}\right), d(i,j)\}\ [2,3,4,5]$$

The maximum cost of the path will correspond to higher distance value between sample i and j in path, which is a set of adjacent sample of images. OPF consist of two different phases, namely, Training Phase and Classification Phase, further classification consist of fit and predict steps [10], shown in Fig. 3.

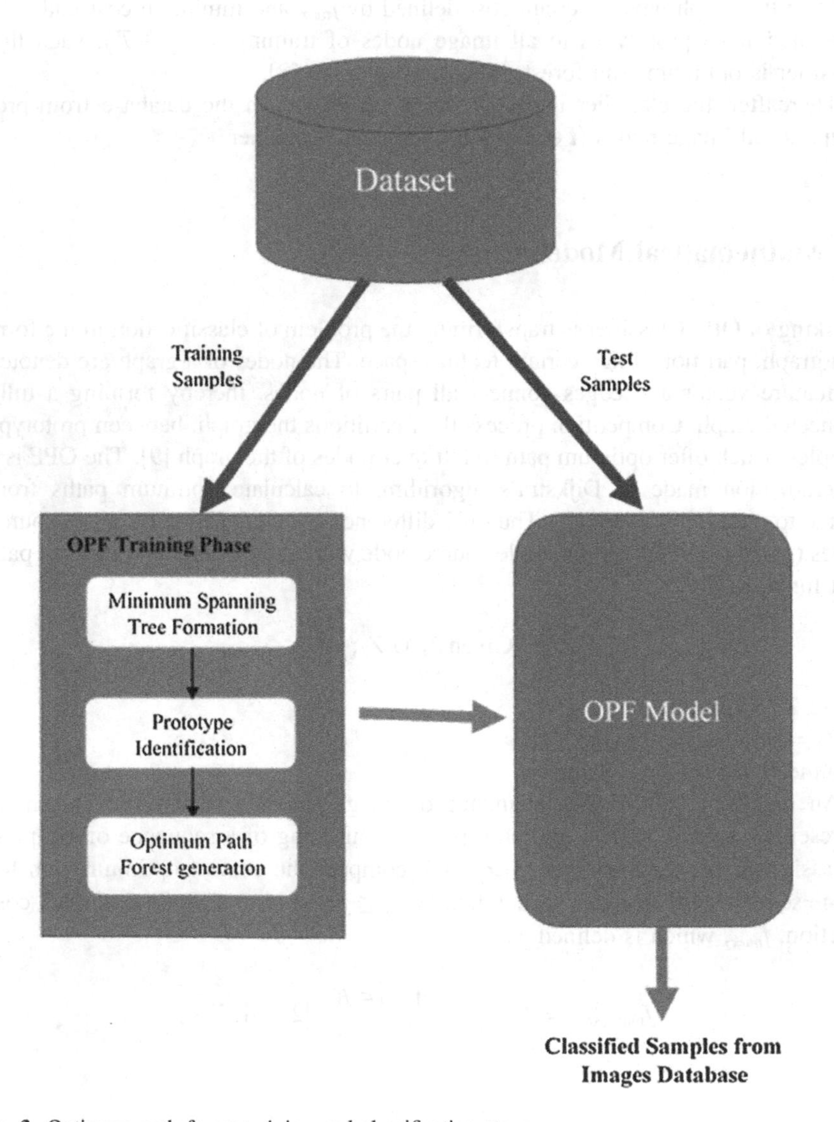

Fig. 3 Optimum-path forest training and classification steps

4.1 Optimum-Path Forest Algorithm

The input to OPF algorithm is training set T. In our case there are only two classes, $S_R \subset T$ is the set of relevant prototypes which is a subset of training set T and $S_I \subset T$ is the set of irrelevant prototypes and, an image descriptor (v, d).

The outcome of this algorithm is a set (P, C, R, T'), optimum-path forest (P) which is Predecessor Map, Path Cost Map (C), Root Map (R) which will give information about root of image and ordered list of training node (say T').

Priority queue and cost variable (cst), are used as supplementary requisite.

1. *Initialization of image nodes which belongs to training set*
   a. For each image belongs to training set but not prototype ($\mathfrak{i} \in T \backslash S_R \cup S_I$)
   b. Set cost of image $C(\mathfrak{i}) \leftarrow$ infinity
2. *Initialization of image nodes which belongs to prototype set*
   a. For each image belong to prototype set i.e. $\{\mathfrak{i} \in S_R \cup S_I\}$
   b. Set cost of image node $\mathfrak{i}$, $C(\mathfrak{i}) \leftarrow 0$
   c. Set predecessor of image node $\mathfrak{i}$, $P(\mathfrak{i}) \leftarrow$ Nil
   d. Set root of image node $\mathfrak{i}$, $R(\mathfrak{i}) \leftarrow \mathfrak{i}$ (image node itself)
   e. Insert this image node $\mathfrak{i}$ into queue Q
3. *Now until queue does not become empty do*
   Remove image node ($\mathfrak{i}$) from queue with minimum cost $C(\mathfrak{i})$ and insert image node ($\mathfrak{i}$) in ordered list of training nodes T'
   A. For each node which belongs to training node set $\mathfrak{t} \in$ T such that cost of node $C(\mathfrak{j})$ is greater than cost of node $C(\mathfrak{i})$ i.e. $\{C(\mathfrak{j}) > C(\mathfrak{i})\}$, compute cst $\leftarrow$ maximum of Cost of image node $\mathfrak{i}$, i.e. $C(\mathfrak{i})$, and distance between node $\mathfrak{i}$ and $\mathfrak{j}$, i.e. cst $\leftarrow \max\{C(\mathfrak{i}), d(\mathfrak{i}, \mathfrak{j})\}$
   B. End
4. End

In classification step, for every image sample that has to be classified from the image dataset but not training set $(\mathfrak{j} \in \mathcal{Z}/T)$, the optimum path that has terminus node $\mathfrak{j}$ can be obtained by the discovery of which training image node $\mathfrak{i}^*$ gives less value in following equation [9],

$$C(\mathfrak{j}) = \min_{(\forall \mathfrak{i} \in T)} \{\mathrm{Max}\{C(\mathfrak{i}), d(\mathfrak{i}, \mathfrak{j})\}\}$$

Image node $\mathfrak{i}^*$ is the predecessor $P(\mathfrak{j})$ in the optimum path with terminus image node $\mathfrak{j}$, hence image node $\mathfrak{j}$ is classified as label of its predecessor node, i.e. label function $\lambda(R(\mathfrak{i}^*))$ [1, 4].

4.2 Complexity Analysis

Consider $Z1$ and $Z2$ as training and test sets respectively having number of samples $|Z1|$ and $|Z2|$ respectively. OPF works in two phases, training and classification. Now consider $|n|$ is the number of sample (considered as vertex) present in the image training set $Z1$ and $|e|$ is the number of edges (arc between vertices given by distance function) in the image training set, represented by graph model of samples [7]. Minimum spanning tree is obtained with the help of Prim's algorithm which requires $O(|e| \log |n|)$ complexity. For finding prototype sets, it requires $O(|n|)$ complexity. OPF runs in $O(|n^2|)$ because of the complete graph obtaining from training samples. Total training phase complexity will be $O(|n| \log |e|) + O(|n|) + O(|n^2|)$, in which $O(|n^2|)$ is dominating factor [7].

The second phase is classification, which evaluates the samples in test set and assigns label to samples. It requires $O(|Z_2| \cdot |n|)$ as $|Z_2|$ is the test set size on which evaluation of OPF will be carried out. An image sample $t \in Z_2$ to which label gets assigned for the classification purpose is connected to all the training image samples (|n|). Therefore, the expected OPF computational complexity is $O(|n^2|) + O|Z_2| \cdot |n|)$.

5 Experiment and Results

Experiment is carried out on Dataset used which is Corel heterogeneous image collection. The average number of marked images per query is N. The experiments were carried out on randomly generated training set $Z1$ and test set $Z2$ for OPF classifier and for different training set sizes. The system is implemented through following screenshots and results obtained are shown below.

As shown in Fig. 4, minimum spanning tree is obtained from underlying graph of training samples through pruning unnecessary nodes so as to reduce the overall complexity involved.

Figure 5, describes the OPF generated through competition process by offering optimum distance. Each prototype conquers the samples most strongly connected to it.

The experiment evaluates the accuracy on dataset. In the experiment, the dataset is divided into two part training set and test set. Figure 6 shows the result obtained by OPF Classifier on increasing number of dataset. The experiments were performed on *Intel® core* I_5 processor with 8 GB RAM. The result proves that OPF is far more accurate in terms of recognition rate and faster in terms of execution time in training phase over huge datasets. The OPF proves to be superior to the sequential random forest in terms of shorter navigation time and reduction in losses. Sequential random forest considers adaptive weight distribution updated in stage-wise growing tree, where the losses are integrated at each level. On the contrary, OPF reduces the navigation time through shortest weighted path. Hence, it

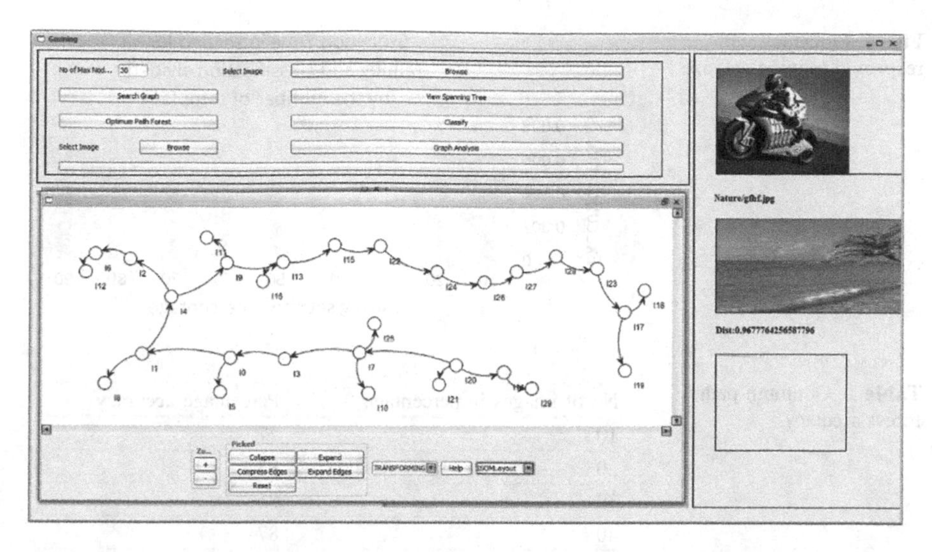

Fig. 4 Minimum spanning tree of graph

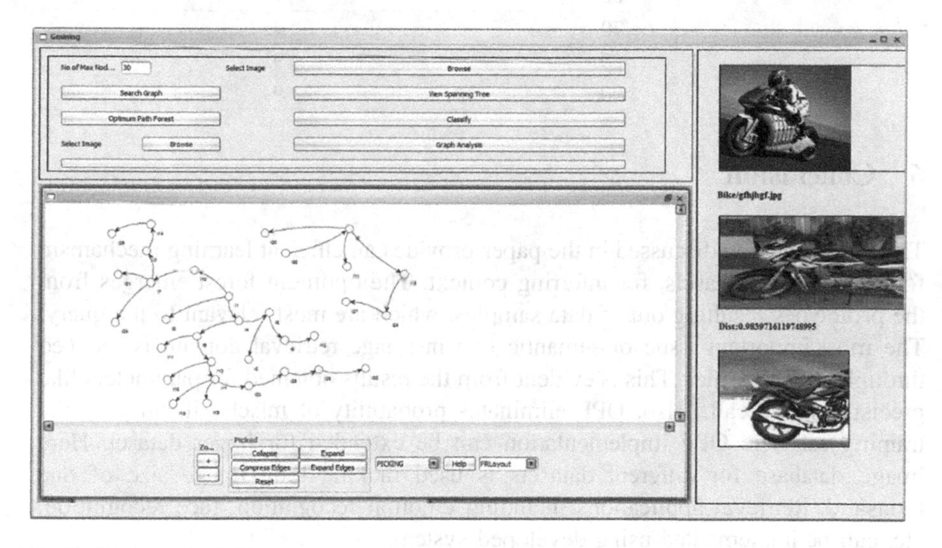

Fig. 5 Optimum-path forest

is clear that OPF can attain greater precision and efficiency for much bigger training set, highly reducing the probability of misclassification. Random Sampling method is not suitable for the said system as it is simple probability sampling technique, used in situation where not much information about the population is available and for a small data size. Here OPF works on huge sample size, hence it is used as classifier for active learning (Table 1).

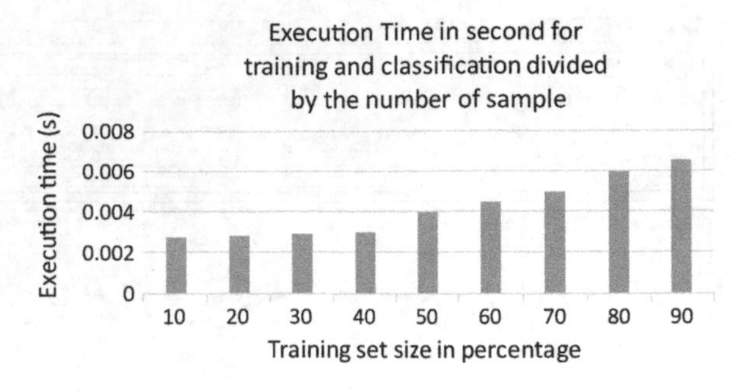

Fig. 6 Efficiency with respect to training set size

Table 1 Optimum-path forest accuracy

No of images in percentage	Percentage accuracy
10	77
20	83
30	85
40	87
50	88
60	90
70	91
80	92
90	94

6 Conclusion

The OPF approach discussed in the paper provides an efficient learning mechanism for large image datasets, for inferring context. The optimum forest emerges from the prototypes resulting out of data samples, which are most relevant to the query. The most important issue of semantic gap in image retrieval domain is reduced through OPF classifier. This is evident from the results obtained for parameters like precision and recall. Also, OPF eliminates probability of misclassification in the training set. The OPF implementation can be extended for larger dataset. Here image database for different datasets is used making total image size of one thousand. Retrieval applications including emotion recognition, face recognition, etc. can be implemented using developed system.

References

1. J.P. Papa, A.X. Falcão, V.H.C. Albuquerque, J.M.R.S. Tavares, Efficient supervised optimum path forest classification for large datasets, Pattern Recognition, Elsevier, 45 (2012) 512–520.
2. A. T. Silva, A. X. Falcão, L. P. Magalhães, Active learning paradigms for CBIR systems based on Optimum path Forest classification, Pattern Recognition, Elsevier, 44 (2011) 2971–2978.

3. J.P. Papa, A.X. Falcão, C.T.N. Suzuki, Supervised pattern classification based on optimum path forest, Int. J. Imaging Syst. Technol. 19 (2) (2009) 120–131.
4. J. P. Papa, A. X. Falcao, A. M. Levada, D. Correa, D. Salvadeo, N. D. A. Mascarenhas, Fast and accurate holistic face recognition through optimum path forest, in: Proceedings of the 16th International Conference on Digital Signal Processing, Santorini, Greece, 2009, pp. 1–6.
5. André Tavares da Silva, Jefersson Alex dos Santos, Alexandre Xavier Falcão, Ricardo da S. Torres, Léo Pini Magalhães, Incorporating multiple distance spaces in optimum-path forest classification to improve feedback-based learning, Computer Vision and Image Understanding, Elsevier, 116 (2012) 510–523.
6. A.S. Iwashita, J.P. Papa, A.N. Souza, A.X. Falcão, R.A. Lotufo, V.M. Oliveira,Victor Hugo C. de Albuquerque, João Manuel R.S. Tavares, A path- and label-cost propagation approach to speed up the training of the optimum-path forest classifier, Pattern Recognition Letters, Elsevier, 40 (2014) 121–127.
7. Greice M. Freitas, Ana M. H. Avila, J. P. Papa, A. X. Falcao, Optimum Path Forest Based Rainfall Estimation, 2009, IEEE.
8. J. P. Papa, A. X. Falcao, Fabio A. M. Cappabianco, Optimising Optimum Path Forest Classification for Huge Datasets, International conference on pattern recognition, IEEE, 2010.
9. Luis C. S. Afonso, J. P. Papa, Apaarecido N. Marana, Ahmad poursaberi and Svetlana N. Yanushkevich, A fast scale iris database classification with optimum path forest technique: A Case Study, WCCI, IEEE, 2012.
10. Roberto Souza, Roberto Lotufo, Leticia Rittner, A Comparison between Optimum Path Forest and k-Nearest Neighbors Classifiers, Conference on Graphics, Patterns and Images, IEEE, 2012.

Comparative Analysis of Android Malware Detection Techniques

Nishant Painter and Bintu Kadhiwala

Abstract In recent years, the widespread adoption of smartphones has led to a new age of information exchange. Among smartphones, Android devices have gained huge popularity due to the open architecture of Android and advanced programmable software framework to develop mobile applications. However, the pervasive adoption of Android is coupled with progressively uncontrollable malware threats. This paper gives an insight of existing work in Android malware detection. Additionally, this paper highlights the parametric comparison of existing Android malware detection techniques. Thus, this paper aims to study various Android malware detection techniques and to identify plausible research direction.

Keywords Android · Malware · Signature-based · Machine-learning-based

1 Introduction

Android smartphone operating system has covered 80.7 % of market share worldwide by 2014, retiring its competitor operating systems iOS at 15.4 %, Windows at 2.8 %, BlackBerry at 0.6 % and other OS at 0.5 % [1]. Android adoration has inspired developers to proffer ingenious applications commonly called *apps*. Official Android market Google Play hosts the third-party developer apps for nominal fee [2]. However, enormous acceptance of Android has boosted various Android malwares performing malicious activities [3]. Android malware is a malicious application that transmits sensitive user information to a remote location from the mobile device [4], abuses the telephony service to extract monetary benefits [5] and exploits platform vulnerabilities to gain smartphone control [6].

Nishant Painter (✉) · Bintu Kadhiwala
Computer Engineering Department, Sarvajanik College of Engineering and Technology, Surat, Gujarat, India
e-mail: nishant.painter@gmail.com

Bintu Kadhiwala
e-mail: bintu.kadhiwala@scet.ac.in

S.C. Satapathy et al. (eds.), *Proceedings of the International Conference on Data Engineering and Communication Technology*, Advances in Intelligent Systems and Computing 469, DOI 10.1007/978-981-10-1678-3_12

Android malware includes SMS Trojans, backdoor, worm, botnet, spyware, aggressive adware and ransomware [7–9].

As malicious activities are committed in the background, user is unaware of these activities [10]. To address this issue, numerous schemes have been proposed by Android security researchers from both academia and industry [11, 12]. Android malware detection techniques can be classified into signature-based detection approach and machine-learning-based detection approach [3]. Signature-based detection approach detects malware by signature matching [13]. Commercial anti-malware solutions use signature-based detection approach for the sake of simplicity and implementation efficiency [14]. Machine-learning-based detection approach trains machine-learning algorithms with the help of noted malware behaviour and identifies unknown and novel malware [13].

This paper aims to make survey of existing Android malware detection techniques. Moreover, we highlight the parametric evaluation of these Android malware detection techniques.

The rest of the paper is structured as follows. Section 2 discusses different Android malware detection techniques. The parametric evaluation of Android malware detection techniques is presented in Sect. 3. Finally, Sect. 4 concludes with a discussion of future research directions in this area.

2 State-of-the-Art

This section gives a detailed description of various Android malware detection techniques based on different approaches. Android malware detection approaches can be classified into signature-based detection approach and machine-learning-based detection approach [3].

2.1 Signature-Based Detection Approach

Signature-based detection approach detects malware by regular expression. Regular expression consists of rules and policies. Various techniques using this approach to detect Android malware are as follows.

2.1.1 Isohara et al.

Isohara et al. [15] implement a system consisting of a log collector in the Linux layer and a log analysis application to effectively detect malicious behaviours of unknown applications. Log collector notes system calls from process management activity and file I/O activity. Log analyzer selects process tree of interested application's activity to remove log data of uninterested process. Signature is described by regular expression and 16 patterns of signature are generated for evaluation of threats including information leakage, jailbreak and abuse of root. The discussed prototype system is evaluated on 230 applications in total to effectively detect malicious behaviours of the unknown applications.

2.1.2 DroidAnalytics

Zheng et al. [16] propose DroidAnalytics that automates the process of malware collection, signature generation, information retrieval and malware association. DroidAnalytics implements application crawler to perform regular application downloads. Dynamic payload detector determines malicious trigger code and traces downloaded file behaviour. DroidAnalytics uses *three-level* signature generation scheme and generates signature at method, classes and application level. DroidAnalytics can detect zero-day repackaged malware. Efficacy of DroidAnalytics is demonstrated using 1,50,368 Android applications and successfully determining 2,475 Android malwares from 102 different families.

2.1.3 APK Auditor

In [17], Talha et al. develop a static analysis system APK Auditor that uses permission feature for classification of benign and malicious applications. APK Auditor consists of signature database, Android client and central server. APK Auditor signature database is a relational database that stores the results of analyzed Android applications. APK Auditor client is an application installed on the system that communicates with the central server through web service. APK Auditor central server accesses signature database and manages malware analysis process. To test system performance, 1,853 benign applications and 6,909 malicious applications, 8,762 applications in total, were collected and analyzed.

2.2 *Machine-Learning-Based Detection Approach*

Machine-learning-based detection approach trains machine-learning algorithms based on noted malware behaviour. This approach detects anomalies, which is essential to discover new malware. Various techniques using this approach to detect Android malware are as follows.

2.2.1 DroidMat

Wu et al. [13] present static feature-based mechanism DroidMat for detecting the Android malware. DroidMat extracts information including permission, deployment of components, Intent messages passing and API calls for characterizing the Android applications behaviour. It applies K-means algorithm that uses *Singular Value Decomposition* (*SVD*) method on the low rank approximation to decide the number of clusters. Finally, it uses kNN algorithm with $k = 1$ to classify the application as benign or malicious. Total 1,738 applications, comprising of 238 malware applications and 1,500 benign applications, were collected to evaluate the system.

2.2.2 Crowdroid

In [18], Burguera et al. propose Crowdroid for dynamic analysis of application behaviour and detection of Android malware. Crowdroid is a lightweight client that monitors system call and sends it to remote server. Remote server parses the data and creates a system call vector. Finally, K-means partitional clustering algorithm is used to cluster dataset with known value of $k = 2$ as input parameter. This system is analyzed using artificial malware created for test purposes and real malware found in the wild.

2.2.3 Yerima et al.

Yerima et al. [19] present an effective Android malware detection approach using Bayesian classification method. Java-based Android package profiling tool is implemented for automatic reverse engineering of the APK files. After reverse engineering an APK, a set of detectors including *API call detectors, Command detectors* and *Permission detectors* are applied to check properties and to map them into feature vectors for the Bayesian classifier. Feature ranking and selection reduces feature while the training function calculates the probability of feature occurrence in malicious and benign applications. The set of 2,000 applications of different categories, comprising of 1,000 malware applications and 1,000 benign applications, were made up in order to cover a wide variety of application types.

2.2.4 Samra et al.

In [20], Samra et al. apply *K*-means clustering algorithm for Android malware detection and use machine-learning algorithm for automatic detection of malware applications. Proposed system extracts the permission requested by the application. ARFF file is prepared from extracted features and WEKA (Waikato Environment for Knowledge Analysis) tool is used to train *K*-means clustering algorithm. While training, the clustering algorithm identifies two clusters and assigns an application to a cluster having the minimum distance from the centroid of clusters. To test system performance, 4,612 business applications and 13,535 tools applications, 18,174 Android applications in total, were evaluated.

2.2.5 Android Application Analyzer (AAA)

Almin and Chatterjee [21] propose a system for analyzing and removing harmful applications using clustering and classification algorithms. For classification purpose, the analysis of the applications is done based on permissions requested by the applications. In this system, *K*-means clustering algorithm is used to create malicious clusters based on some known families of malicious applications. Then, Naïve Bayesian classification algorithm is used to classify applications as benign and malicious. The DogWar, ICalender and SuperSolo samples of malicious applications are used for testing.

3 Parametric Evaluation

In case of Android application, features consist of elements such as permission, java code, certification, system call, API call, etc. should be considered for the purpose of application classification. These features can be divided into static, dynamic and hybrid. Based upon the selected features, the technique can be further classified into static analysis technique and dynamic analysis technique. Furthermore, feature selection leads to more accurate results and removes noisy and irrelevant data. Parametric evaluation of the discussed detection techniques is described in Table 1.

Table 1 Parametric evaluation of android malware detection techniques

Technique	Detection approach	Analysis	Features	Clustering algorithm	Classification algorithm	Type of applications in dataset	Number of applications in dataset
Isohara et al. [15]	Signature-based approach	Dynamic	System calls	–	–	Benign application and malware application	230
DroidAnalytics [16]	Signature-based approach	Static	API calls	–	–	Benign application and malware application	150368
APK auditor [17]	Signature-based approach	Static	Permission	–	–	Benign application and malware application	8762
DroidMat [13]	Machine-learning-based approach	Static	Permission, deployment of component, intent, API calls	*K*-means algorithm, EM algorithm	kNN algorithm, naïve Bayesian algorithm	Benign application and malware application	1738
Crowdroid [18]	Machine-learning-based approach	Dynamic	System calls	*K*-means algorithm	–	Self-written malware and malware application	–
Yerima et al. [19]	Machine-learning-based approach	Static	API calls, command, permission	–	Bayesian algorithm	Benign application and malware application	2000

(continued)

Table 1 (continued)

Technique	Detection approach	Analysis	Features	Clustering algorithm	Classification algorithm	Type of applications in dataset	Number of applications in dataset
Samra et al. [20]	Machine-learning-based approach	Static	Permission, XML-based features	*K*-means algorithm	–	Business application and tools application	18174
Android application analyzer [21]	Machine-learning-based approach	Static	Permission	*K*-means algorithm	Naïve Bayesian algorithm	Malware application	–

4 Conclusion and Future Work

As discussed, Android malware detection techniques can be classified into Signature-based detection approach and Machine-learning-based detection approach. Techniques based on Signature-based detection approach require frequent updates to detect new malware. Small modification in a malware generates varying malware that can easily bypass the existing signature database matching process. More time is required by Signature-based detection approach for larger signature database. Finally, it is only capable of detecting malwares which are included in signature database. However, Machine-learning-based detection approach overcomes the limitations of signature-based detection approach by quickly detecting novel or unknown malware from the behaviour of known malwares without frequent updates.

Future work may include devising a more efficient technique based on Machine-learning-based detection approach that considers more features. Moreover, various feature selection methods can be applied to reduce feature set and to consider the most efficient features. Finally, various clustering and classification algorithms can be used to improve the modelling capability.

References

1. G. Inc., Gartner Says Smartphone Sales Surpassed One Billion Units in 2014, http://www.gartner.com/newsroom/id/2996817
2. Parvez Faruki, Ammar Bharmal, Vijay Laxmi, Vijay Ganmoor, Manoj Singh Gaur, Mauro Conti, Muttukrishnan Rajarajan.: Android Security: A Survey of Issues, Malware Penetration, and Defenses. In: IEEE communications surveys & tutorials, vol. 7, no. 2, pp. 998–1022 (2015)
3. Ali Feizollah, Nor Badrul Anuar, Rosli Salleh, Ainuddin Wahid Abdul Wahab: A review on feature selection in mobile malware detection. In: Digital Investigation, vol. 13, pp. 22–37 (2015)
4. Android.Bgserv, http://www.symantec.com/security_response/writeup.jsp?docid=2011031005-2918-99
5. Android Hipposms, http://www.csc.ncsu.edu/faculty/jiang/HippoSMS/
6. RageAgainstTheCage, https://github.com/bibanon/android-development-codex/blob/master/General/Rooting/rageagainstthecage.md
7. C. A. Castillo.: Android malware past, present, future. In: Mobile Working Security Group McAfee, Santa Clara, CA, USA, Tech. Rep. (2012)
8. W. Zhou, Y. Zhou, X. Jiang, P. Ning.: Detecting repackaged smartphone applications in third-party android marketplaces. In: Proc. 2nd ACM CODASPY, New York, NY, USA, pp. 317–326 (2012)
9. Android Malware Genome Project, http://www.malgenomeproject.org/
10. M. Eslahi, R. Salleh, N. B. Anuar.: Mobots: a new generation of botnets on mobile devices and networks. In: IEEE Symposium on Computer Applications and Industrial Electronics (ISCAIE), pp. 262–266 (2012)

11. Sufatrio, Darell J. J. Tan, Tong-Wei Chua, Vrizlynn L. L. Thing.:Securing Android: A Survey, Taxonomy, and Challenges. In: ACM Computing Surveys, vol. 47, no. 4, article 58, pp. 58–103 (2015)
12. Zheran Fang, Weili Han, Yingjiu Li.: "Permission based Android security: Issues and countermeasures". In: Computers & Security, pp. 1–14 (2014)
13. Dong-Jie Wu, Ching-Hao Mao, Te-En Wei, Hahn-Ming, Kuo-Ping Wu: DroidMat: Android Malware Detection through Manifest and API Calls Tracing. In: Seventh Asia Joint Conference on Information Security, IEEE, pp. 62–69 (2012)
14. E. Fernandes, B. Crispo, M. Conti.:FM 99.9, radio virus: Exploiting FM radio broadcasts for malware deployment. In: IEEE Trans. Inf. Forensics Security, vol. 8, no. 6, pp. 1027–1037 (2013)
15. Takamasa Isohara, Keisuke Takemori, Ayumu Kubota: Kernel-based Behavior Analysis for Android Malware Detection. In: Seventh International Conference on Computational Intelligence and Security, IEEE, pp. 1011–1015 (2011)
16. Min Zheng, Mingshen Sun, John C.S. Lui: DroidAnalytics: A Signature Based Analytic System to Collect, Extract, Analyze and Associate Android Malware. In: 12th IEEE International Conference on Trust, Security and Privacy in Computing and Communications, IEEE, pp. 163–171 (2013)
17. Kabakus Abdullah Talha, Dogru Ibrahim Alper, Cetin Aydin: APK Auditor: Permission-based Android malware detection system. In: Digital Investigation, vol. 13, pp. 1–14 (2015)
18. Iker Burguera, Urko Zurutuza, Simin Nadjm-Tehrani: Crowdroid: Behavior-Based Malware Detection System for Android. In: SPSM, ACM, pp. 15–26 (2011)
19. Suleiman Y. Yerima, Sakir Sezer, Gavin McWilliams, Igor Muttik: A New Android Malware Detection Approach Using Bayesian Classification. In: IEEE 27th International Conference on Advanced Information Networking and Applications, IEEE, pp. 121–128 (2013)
20. Aiman A. Abu Samra, Kangbin Yim, Osama A. Ghanem: Analysis of Clustering Technique in Android Malware Detection. In: Seventh International Conference on Innovative Mobile and Internet Services in Ubiquitous Computing, IEEE, pp. 729–733 (2013)
21. Shaikh Bushra Almin, Madhumita Chatterjee: "A Novel Approach to Detect Android Malware", in International Conf. on Advanced Computing Technologies and Applications (ICACTA). In: Procedia Computer Science, vol. 45, pp. 407–417 (2015)

Developing Secure Cloud Storage System Using Access Control Models

S.A. Ubale, S.S. Apte and J.D. Bokefode

Abstract Access control model-based security plays very crucial role in security. Cloud computing is one of the emerging and challenging fields. In the implementation explained in this paper, issues related with cloud security are tried to resolve to some extent using access control model security. Encryption, also being part of access control-based model, also incorporated for better security. Role-Based Access Control (RBAC) model along with AES–RSA encryption algorithm is implemented to achieve efficient security. The main aim of this work is to design a framework which uses the cryptography concepts to store data in cloud storage and allowing access to that data using role perspective with the smallest amount of time and cost for encryption and decryption processes.

Keywords Access control model · RBAC · AES · RSA · Cloud

1 Introduction

Cloud computing is amongst the emerging, required and promising field in the area Information Technology. It provides services to a customer over a network as per their requirements. Third party owns the infrastructure and provides the cloud computing services. It offers the organizations to adopt IT services with minimum cost. Cloud computing has a number of benefits but the most organizations are worried for accepting it due to security issues and challenges that are with cloud. With the third party cloud storage, we may lose control over physical security. In a public cloud, an organization does not have information about where the resources

S.A. Ubale (✉) · S.S. Apte
Walchand Institute of Technology, Solapur, Maharashtra, India
e-mail: swapnaja.b.more@gmail.com

S.S. Apte
e-mail: aptesulabha@gmail.com

J.D. Bokefode
SVERI College of Engineering, Gopalpur, Pandharpur, Maharashtra, India

S.C. Satapathy et al. (eds.), *Proceedings of the International Conference on Data Engineering and Communication Technology*, Advances in Intelligent Systems and Computing 469, DOI 10.1007/978-981-10-1678-3_13

are stored, run and who manages them. There are many organizational and distributed aspects that need to be solved at enterprise level. One of the security requirements is to design such models that solve such aspects for the information security at the enterprise level. Such models need to present the security policies intended to protect information against unauthorized access and modification stored in cloud. This work describes a logical approach to modeling the security requirements according to the job functions and work performed in an organization. In this work, we combined two algorithms RSA and AES for securing data, and role-based access control model is used to provide access according to role perspective [1].

The cloud computing provides three main services, Information as a Service (IaaS), Platform as a Service (PaaS), Software as a Service (Saas). It reduces the cost of hardware required to store data that could have been used at user end. Instead of purchasing the infrastructure that is required to store data and run the processes we can lease the assets according to our requirements. The cloud computing provides the number of advantages over the traditional computing and it include: quickness, lower cost, scalability, device independency and location independency. Security in cloud computing is one of the most critical aspects, owing to importance and sensitivity of data on the cloud. But many major issues as data security, user access control, data integrity, trust and performances issues exists for the cloud. In order to solve these problems, many schemes are proposed under different systems and security models [2–6].

Whenever security of cloud computing is concerned, there are various security issues related to the cloud. Some of these security problems and their solutions are described here. Due to sharing, computing resources with another company physical security is lost. User does not have knowledge and control of where the resources run and stored. It can be insured using secure data transfer. Second, maintaining the consistency or integrity of the data. It can be insured by providing secure software interfaces. Third, privacy rights may be violated by cloud service providers and hackers. It can be ensured using cryptographic technique. Forth, when cryptographic technique was used, then who will control the encryption/decryption keys? It can be ensured by giving rights to the users/customers. So implementing security in cloud computing is a must, which will break the difficulty of accepting the cloud by the organizations. There are varieties of security algorithms which can be implemented to the cloud. There are two types of algorithms; symmetric key and asymmetric key. DES, Triple-DES, AES and Blowfish are some symmetric algorithms that can be used to implement cloud security. RSA algorithm is an asymmetric algorithm that can be used to generate encryption and decryption key for symmetric algorithms. In cloud computing, both symmetric-key and asymmetric-key algorithms are used for encryption and decryption. In the presented work, RSA algorithm is used to generate encryption and decryption keys for AES symmetric algorithm.

Another major issue is how to manage user access to cloud storage system. For that, different access control mechanism can be enforced for cloud users. Access Control is nothing but about specifying rights to the users to deal with any of the specific system resources or applications or also any system. There are three access control models, such as MAC (Mandatory access control model), DAC (Discretionary access control model) and RBAC Role-based access control models [7]. These access control models specify the set rules or criteria to access the system and its resources. In MAC, the administrator has all the privileges to assign the user's roles according to his wishes. And end users do not have the authority to change the access policies specified by the administrator; therefore it is very restrictive and a less-used access control model. It can be used in a very sensitive environment. For example, military and research centers. In DAC, the end users have authority to change the access policy for its own objects. In RBAC, first different roles or jobs can be specified and then these roles can be assigned to cloud user so these users can get access according to their jobs requirement. It is effectively and mostly used access control model within an organization because access to particular data and resources can be given according to the roles [8, 9]. In the presented work, we used central control of admin to create role and assign user to corresponding roles as in MAC. Each role can access the objects for which they have the access and it cannot be changed by other user, as in DAC. All rules and access are governed and controlled by roles of the system as in RBAC model [10]. Along with basic access control models as MAC, DAC, RBAC, encryption algorithms used as AES and RSA plays crucial role. Why AES and RSA?—is explained below.

2 Methodology

In symmetric-key algorithm, for encrypting and decrypting data, it uses same secret key. In contrast to asymmetric-key cryptography algorithm, symmetric-key cryptography algorithm like AES (Advanced Encryption Standard) is high speed and it requires low RAM requirements, but because of the same secret key used for both encryption and decryption, it faces big problem of key transport from sender side to receiver side. But in asymmetric-key algorithm, it needs two different keys for encryption and decryption, one of which is private key and one of which is public key. The public key can be used to encrypt plaintext; whereas the private key can be used to decrypt cipher text. Compared to symmetric-key algorithm, asymmetric-key algorithm does not have problem while key exchanging and transporting key, but it is mathematically costly [11, 12].

To solve the problem of key transport and get better performance, these two algorithms can be combined together. In this data, receiver generates the key pairs using asymmetric-key algorithm, and distributes the public key to sender. Sender uses one of the symmetric-key algorithms to encrypt data, and then sender uses asymmetric-key algorithm to encrypt the secret key generated by the symmetric-**key algorithms with the help of receiver's public**. Then the receiver uses its private key to decrypt the secret key, and then decrypt data with the secret key. In this paper, asymmetric-key algorithm is used only for encrypting the symmetric key, and it requires negligible computational cost. It similarly works like SSL. For encrypting the files or file data, AES (Advanced Encryption Standard) algorithm is used, and RSA (Rivest, Shamir and Adleman) is used to encrypt AES key [11, 12]. These encrypted files can be uploaded according to role perspective. In proposed system, role-based access control is used for authenticating the users to access files uploaded or given rights for the specific roles and to maintain the data privacy and integrity AES and RSA algorithms are used.

General description of AES and RSA algorithms is given below.

AES algorithm starts with an Add round key stage, followed by nine rounds of four stages and a tenth round of three stages. The four stages are Substitute bytes, Shift rows, Mix Columns and Add Round Key. The tenth round does not perform the Mix Columns stage. These stages also apply for decryption. The first nine rounds of the decryption algorithm consist of Inverse Shift rows, Inverse Substitute bytes, Inverse Add Round Key and Inverse Mix Columns. Again, the tenth round does not perform the Inverse Mix Columns stage. [11]. RSA makes use of measured exponential for encoding and decoding symmetric key that is secret key generated by the AES algorithm. Let us consider that S is secret key and C is cipher key, then at encryption $C = S \bmod n$ and at decryption side $S = C \bmod n$. n is very large number which is created during key generation process [11, 12].

In the implemented scheme, the administrator of the system defines different job functionalities required in an organization; then according to the needs of the organization he adds users or employees. After that, owner of the data encrypts the data in such a way that only the users with appropriate roles as specified by a RBAC policy can decrypt and view this data. The role manager assigns roles to users who are appropriate for that role and he can also remove the users from assigned role. The cloud provider (who owns the cloud infrastructure) is not able to see the contents of the data. A role manager is able to assign a role for particular user after the owner has encrypted the data or file for that role. A user assigned to particular role can be revoked at any time in which case, the revoked user will not have access rights to data or file uploaded for this role [13]. Revocation of user from role will not affect other users and roles in the system. This approach achieves an efficient encryption and decryption at the client side.

Our system uses the AES for the function of encrypting and decrypting the data. AES key length used is 128 bits. The main purpose of using this algorithm is for providing more security for data which will be uploaded on the cloud. The system is developed in asp.net. For the public cloud we have taken instance from Microsoft azure and private cloud is created using the Windows Server with i5 processor and 8 GB RAM. The use of latest processor will reduce the response time for uploading and delivering of the data to the owner and user, respectively. The size of the decryption key is another important factor in cloud storage system. The decryption key must be portable as users may use the storage service from different clients [14] (Figs. 1 and 2).

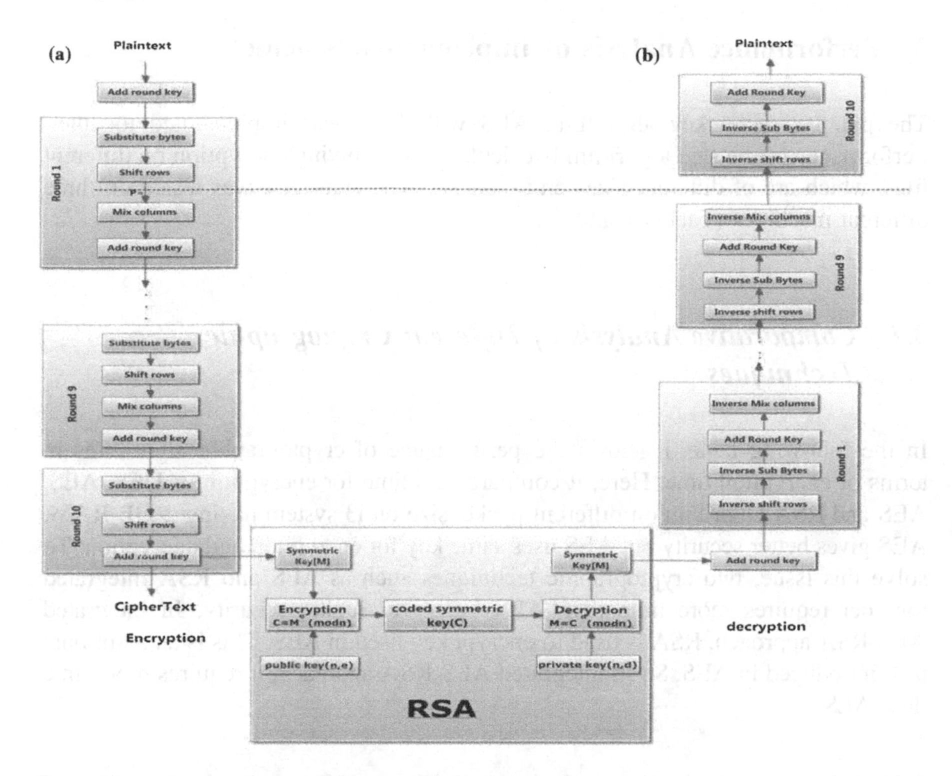

Fig. 1 Flow of encryption algorithm used in implementation

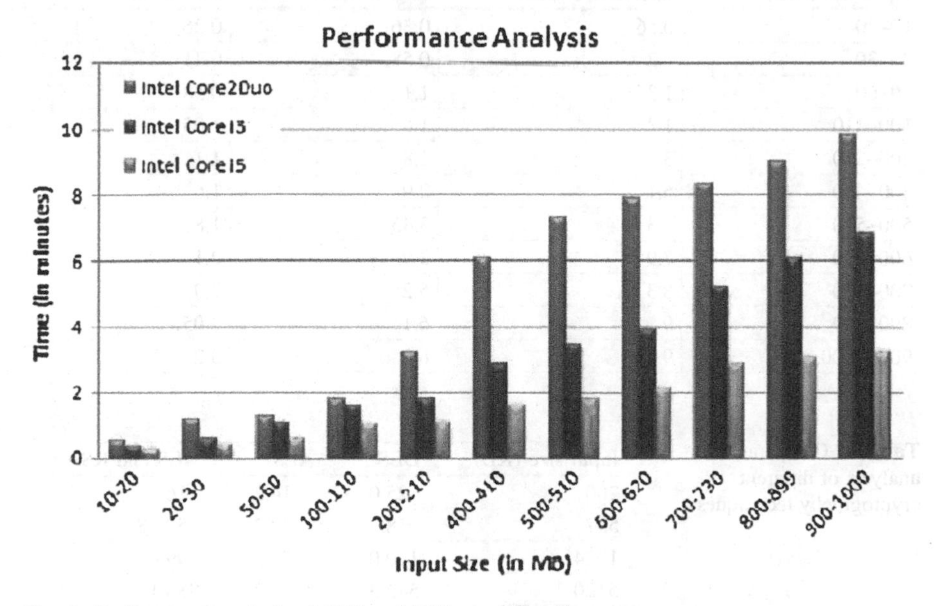

Fig. 2 Performance analysis of AES and RSA on different processor

3 Performance Analysis of Implemented Scheme

The popular secret key algorithm, AES with RSA was implemented together. Performance of the said algorithm is calculated by applying encryption on different files, which are of different sizes and contents. The algorithm was tested on three different machines (Tables 1 and 2).

3.1 Comparative Analysis of Different Cryptographic Techniques

In the following table, it shows the performance of cryptographic algorithms in terms of encryption time. Here, it compares the time for encryption of DES, AES, AES and RSA algorithm on different packet size on i3 system having 2 GB RAM. AES gives better security but AES uses same key for encryption and decryption. To solve this issue, two cryptographic techniques such as AES and RSA integrated together requires more time than AES but gives higher security. In integrated AES-RSA approach, RSA is used to encrypt key used in AES. This is an additional task introduced in AES. So in integrated AES-RSA approach, it requires more time than AES.

Table 1 Testing result of implemented scheme on different platforms

Input size (MB)	Intel Core2 Duo	Intel Core i3	Intel Core i5
10–20	0.56	0.36	0.26
20–30	1.2	0.58	0.41
50–60	1.3	1.1	0.58
100–110	1.8	1.6	1.05
200–210	3.2	1.8	1.1
400–410	6.1	2.9	1.6
500–510	7.3	3.45	1.8
600–620	7.9	3.9	2.1
700–730	8.3	5.2	2.9
800–899	9	6.1	3.05
900–1000	9.8	6.8	3.2

Table 2 Comparative analysis of different cryptography techniques

Input size (KB)	DES	AES	AES and RSA
200	25.0	14.2	16
557	58.2	38.2	52.38
1024	110.0	72.2	99.0
5120	542.3	1362.2	488.1

4 Conclusion

The main aim of this work is to design a framework, which uses the cryptography concepts to store and access data in cloud storage and allowing access to that data using role perspective. Security approach used in this work is based purely on access control-based models of security.

References

1. Zhidong Shen, Li Li, Fei Yan, Xiaoping Wu. Cloud Computing System Based on Trusted Computing Platform. In Proc.International Conference on Intelligent Computation Technology and Automation, Volume 1, May 2010, pp. 942–945.
2. Pearson S., Benameur A. Privacy, Security and Trust Issues Arises from Cloud Computing, In Proc. IEEE Second International Conference on Cloud Computing Technology and Science (CloudCom). 2010, pp. 693–702.
3. Rohit Bhadauria and Sugata Sanyal, A Survey on Security Issues in Cloud Computing and Associated Mitigation Techniques. International Journal of Computer Applications, Volume 47- Number 18, June 2012, 47–66.
4. Mohammed E.M., Ambelkadar H.S, Enhanced Data Security Model on Cloud Computing, In Proc. 8th International Conference on IEEE publication 2012, pp. 12–17.
5. Sang Ho. Na, Jun-Young Park, Eui-Nam Huh, Personal Cloud Computing Security Framework, In Proc. Service Computing Conference (APSSC) IEEE publication, Dec 2010, pp. 671–675.
6. Wang, J.K.; Xinpei Jia, Data Security and Authentication in hybrid cloud computing model, Global High Tech Congress on Electronics (GHTCE) on IEEE publication, 2012, 117–120.
7. R. Sandhu. The next generation of access control models: Do we need them and what should they be? In SACMAT–01, May 2001, p. 53.
8. Prof. S.A. Ubale and Dr. S.S. Apte, Study and Implementation of Code Access Security with . Net Framework for Windows Operating System, International Journal of Computer Engineering & Technology (IJCET), Volume 3, Issue 3, 2012, pp. 426–434.
9. Prof. S. A. Ubale, Dr. S. S. Apte, Comparison of ACL Based Security Models for securing resources for Windows operating system, IJSHRE Volume 2 Issue 6, p. 63.
10. Bokefode J.D, Ubale S. A, Apte Sulabha S, Modani D. G, Analysis of DAC MAC RBAC Access Control based Models for Security, International Journal of Computer Applications, Volume 104–No. 5, October 2014.
11. Daemen, J., and Rijmen, V., Rijndael: The Advanced Encryption Standard. Dr. Dobb 's Journal, March 2001.
12. R.L. Rivest, A. Shamir, and L. Adleman, A Method for Obtaining Digital Signatures and Public–Key Cryptosystems, Communication of the ACM, Volume 21 No. 2, Feb. 1978.
13. H. L. F. Ravi S. Sandhu, Edward J. Coyne and C. E. Youman. Role-based access control models. IEEE Computer, February 1996, pp. 38–47.
14. W. Stallings, Cryptography and Network Security Principles and Practices Fourth Edition, Pearson Education, Prentice Hall, 2009.

ETLR—Effective DWH Design Paradigm

Sharma Sachin and Kumar Kamal

Abstract Prior to start of any data warehouse project, developers/architects have to finalize architecture to be followed during project life cycle. Two possibilities are (a) Use of commercial ETL tool or (b) the development of in-house ETL program. Both options are having merits and demerits. The scope of this article is to optimize the ETL process with retrieval by making use of technology mix and keep in consideration all the factors which are not been considered by ETL tools.

Keywords ETL · ETLR · DWH · Optimization · Big data

1 Introduction

The tremendous digitization of information in recent past, coupled with the growth of processes and transactions, has resulted in a data flooding. Majority of data of any organization is in unstructured form which needs data cleansing while extraction or transformation. The process of extraction of data from source, cleansing and transformation followed by Loading is ETL (Extraction Transformation and Loading) and is integral part of every Data warehouse. Organizations need to study structured and unstructured data for decision-making and future forecasting. The beauty of ETL in data warehouse is (a) extract data from any source like CSV, mainframe data input, excel data, flat files or any other type of

Sharma Sachin (✉)
M. M. University, Mullana, Ambala, Haryana, India
e-mail: er.sachinsharma@gmail.com

Kumar Kamal
University of Petroleum and Energy Studies, Dehradun, Uttrakhand, India
e-mail: kkumar@ddn.upes.ac.in

S.C. Satapathy et al. (eds.), *Proceedings of the International Conference on Data Engineering and Communication Technology*, Advances in Intelligent Systems and Computing 469, DOI 10.1007/978-981-10-1678-3_14

data (b) clean and transform the data to the required format and (c) Load data to any database/flat files or any other destination in the required format. The tools responsible for this task are called ETL tools.

Available research has only concentrated on designing and managing ETL workflows. Research approaches have focus on standalone problems and problems related to web data only. When it comes to huge data volumes on daily basis which needs to be loaded after cleansing and transformation, it is always a big challenge. When we refer to "huge data volumes", it signifies data in Terabytes/Petabytes which was not common in recent past. Transformation of every industry in terms of digitization motivates organizations to maintain every aspect of data in data warehouse which eventually leads to emerging field of "Big Data Analytics". Not only ETL is a big challenge, there are many bottleneck in data retrieval from destination data source because there needs meta-data processes and various data mining techniques to fetch relevant information out of data warehouse.

Research area in ETL optimization should also include optimization of operating system parameters, storage hardware and its optimization in terms of storage allocation techniques available and storage configuration, optimization of SQL queries during extraction and transformation, use of parallelism, optimizing DBMS parameters, knowledge of staging data source format for best performance, optimization of ETL tools parameters and their configuration, optimizing memory usage during full load of ETL processes, purging policy optimization, network-level optimization, source data optimization, filter required information at source level and optimization of meta-data information. We should focus on optimizing various area of ETL in which ETL tools are not able to handle and optimize data retrieval techniques.

The scope of this research is to optimize ETL process with retrieval and to tune and optimize every sub-area at micro-level, such that data arriving in huge volumes on daily basis can be extracted, transformed, and loaded in time bound manner without the use of ETL tools. This includes optimization of ETL process in all areas where ETL tools were never been worked upon. The research will result in the complete solution for ETL problem with the retrieval in data warehouse which results in reporting of the intended data from data warehouse in highly efficient way. The beauty of the research includes unique combination of techniques of OS and DBMS utilities.

2 Related Work

A data warehouse is "subject–oriented, integrated, time-varying, non–volatile collection of data in support of the management's decision making process" [10]. We make use of customized ETL process or ETL tools to build a data warehouse. Majority of commercial ETL tools work on application layer and make use of DBMS application programming interface and libraries [20]. Existing tools facilitate programmers to interact with various data sources and play graphically with data for analytics [4].

Research has been done in ETL field including (a) ARKTOS II [1], (b) AJAX [17], and Potter's wheel [18]. Research has been carried out in the area of real-time data warehouse for business intelligence [5]. There are challenges in the field of data warehouse as architecture is different for various domains including clinical data [6]. Semi-automated tools have also been proposed for ETL processes [7]. Oracle contributed by providing many features like partitioning, data capture, query rewrite for data warehouse [8], and SQL loader-like utilities [16]; Oracle also contributed by performance tuning guidelines [14] and IBM contributed in many ways including IBM data warehouse manager [11] which aimed at decision-making using data analysis [9]. Research has been carried out for ETL using incremental loading techniques and break point transmission [15].

Research on workflows revolves around three factors: (a) modeling [12], where the authors concentrate on providing a meta-model for workflows; (b) correctness issues [3], where criteria are established to determine whether a workflow is well formed, and (c) workflow transformations [19]. The research was oriented towards fetching huge volumes of data from source and its transformation and pushes more and more data from staging/dimensional area to fact/destination area to form data mart followed by data warehouse [1]. Schafer, Becker, and Jarke have worked on case study of data warehouse of Deutche Bank. The authors describe the management of metadata for huge data warehouse of terabytes in size keeping in view of the business intelligence [2] goals and application-oriented architecture which involves broad spectrum of the business methodologies so as to provide precise reporting in a time-bound manner [13]. The study give emphasis on querying data warehouse as reporting from the data warehouse is the ultimate goal behind establishment of data warehouse.

3 Problem Definition

We can load data in tables much faster without having indexes and other dependencies, as indexes and data loading is inversely proportional to each other. On huge volume databases, it is a more challenging job as if we speed up the data loading speed, we drop indexes and we recreate/rebuild indexes after data loading which again takes a very long time and our motive will not be achieved, as overall time to give usable data set will remain the same without any optimization. The conventional data load speed using different techniques is shown in Fig. 1.

The problem is to extract data from source which may come as ongoing multiple files (which may be thousands per hour) from middleware or any other source system each having fixed size or fixed characters or variable size, read each file content, perform the transformation, perform data cleansing, put it in staging area, and then feed data into production database without compromising retrieval

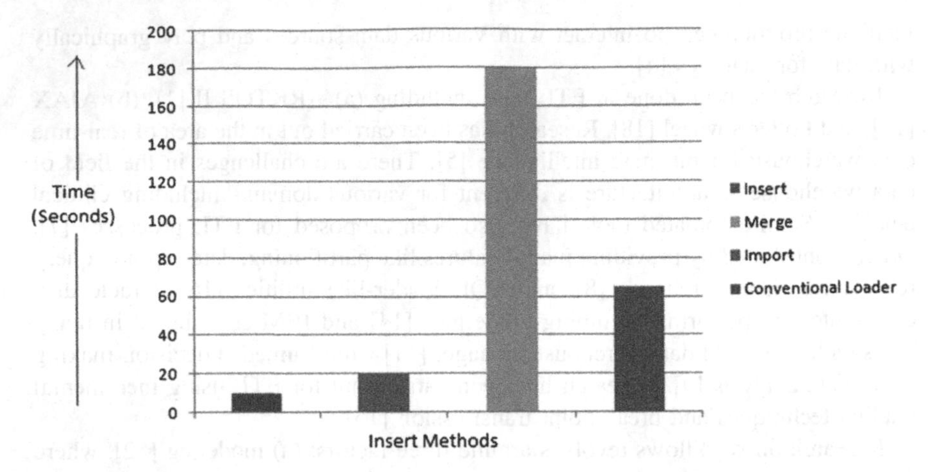

Fig. 1 Conventional data load speed [18]

process. The major problem is to maintain data in production database with all the dependencies and without dropping indexes on huge tables.

4 ETLR—Effective DWH Design Paradigm

Extraction, Transformation, and Load should be further elaborated as Extraction, Transformation, Load, and Retrieval, i.e., **ETLR**. ETLR is the new way of designing data warehouse such that we design and optimize data warehouse by keeping in mind the data retrieval requirements and optimization at various levels as shown in Fig. 2.

For better understanding—we take into consideration an example of an organization which wants to build a data warehouse having very huge data volume on daily basis. Data comes from mediation system in the form of raw files which continues to receive persistently all time.

We propose a different approach to address the issue and build a DWH. Going one step further at source system, tune source system and perform filtration process at source system itself and segregate data at source system. Multiple paths at DWH file server has been identified such that data can be dumped into respective destination paths. Data from source system/mediation is set in such a way that it is being generated in parallel mode in multithreading and parallel data streams are generating which dumps data persistently into DWH server. One file of source system splits into multiple files that contains data of respective filter condition and dumped into predefined destination path at DWH server. By making use of this step, we reduced processing time at DWH server.

After many ETL mechanisms adopted, we propose to make use of external table concept on the raw data files, keeping files in bunch and provide it to external table

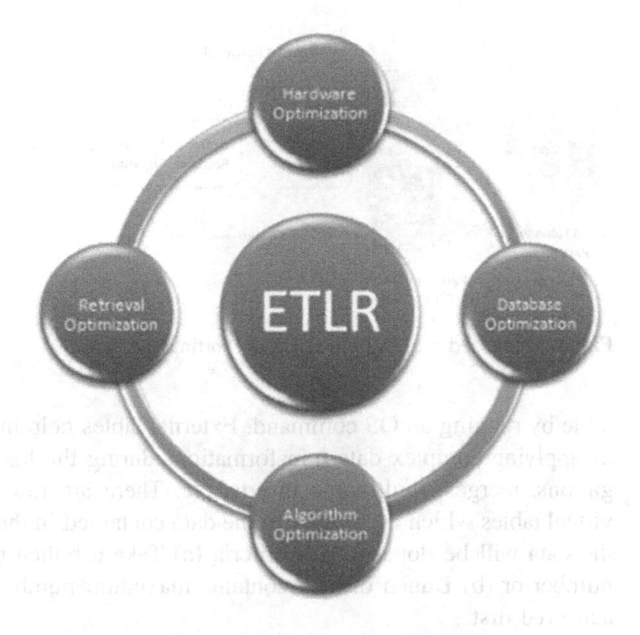

Fig. 2 ETL and retrieval optimization for overall performance

engine. External table engine of the DBMS limits the size of the files to be shown together and show it in the database in tabular form on applying SQL scripting within the database. We are still making use of the OS memory and not DBMS memory to show data in the delimited files. Upon getting data visible in tabular form in the DBMS, view functionality in combination with copy utility of the DBMS can be used to put data from staging area to production area in the required format. This process maintains the integrity of the constraints and indexes while loading with minimum impact on performance. The switching of files happens and parallel execution of different modules incorporated so that desired data loading can be achieved in best minimum time. Error mechanism and its rectification have been incorporated without making use of any ETL tool. Data loading will be done in multiple destination database objects of respective zones and further per zone multiple database objects to take the benefit of modularity among database objects. Retrieval process has been designed in such a way that it calculates the desired statistics before referring to the destination data object and in parallelism mode.

5 Implementation

In DWH server, there are multiple locations where data from mediation system is pushed containing domain-wise data in each location. We will make use of external table concept to load data in which we can access data much before its loading into the database and can perform transform operations without making use of database memory; as it uses the OS memory only. We can change the data in an external

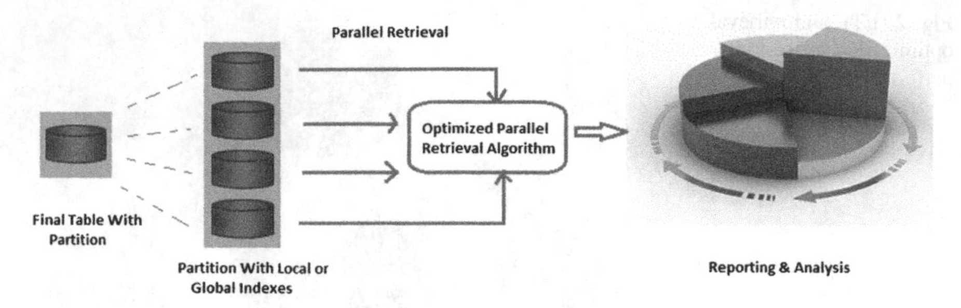

Fig. 3 Optimized retrieval process for reporting and analysis

table by running an OS command. External tables help in efficient loads, by means of applying complex data transformations during the load process includes aggregations, merges, multi-table inserts, etc. There are raw table definitions of individual tables which are similar to the data contained in the raw files. The loading of the data will be done on two criteria (a) Take a bunch of files having predefined number or (b) Bunch of files contains maximum number of records whichever is achieved first.

Further, due to large volume of the data, every database object is range partitioned such that we have multiple database objects each having range partition. Every database objects needs fast retrieval, so we have identified fields on which indexing will be there for fast retrieval. Now we have range portioned database objects with indexes on predefined fields of each table. For transformation operation on raw tables, we make use of predefined views on all raw tables. The raw data will be automatically converted into production format as and when it reflects into the raw objects. To achieve this, we make use of "COPY" functionality of database which takes reference of the data available in raw datasets after processing from different views and then it feeds the data into production database in parallel. The data in production tables can be used for reporting and analysis as shown in Fig. 3.

The data will be loaded into range partitions of the destination objects and then partition swap mechanism has to be used such that the loaded data can be just sandwiched with the table. The partition swap takes only seconds thus ensuring the data availability at the earliest.

6 Result and Discussion

In this paper, we have come up with a different theoretical framework along with its implementation for an effective data warehouse. On combining various tools and techniques and optimization of already available methodologies of ETL, we have a data warehouse of large size with very less maintenance activities associated with it. The techniques like external table, partition swap, and copy command functionality of database combined together with the help of operating system programming/utilities gives us a unique way of ETL process to build a data warehouse which has significantly large volume of data. The main objective of the research work is to maintain effective ETL with retrieval dependencies.

A traditional data warehouse process can only optimize a portion of ETL process, i.e., either ETL tool or in-house ETL program. A bird's eye view is missing in existing ETL process. Not only swift ETL process, but hassle free and quick availability of data is must. The parallelism and layered optimization with partition exchange and scalability is shown in Fig. 4.

To ensure quick ETL process with data availability in time-bound environment, the ETLR architecture with the use of parallel architecture at each layer with optimization gives us the performance we can just imagine of. The scalability is not an issue with ETLR paradigm; just add partitions in tables with optimized hardware and database configuration for storage and scale the data warehouse to the extent you want.

The ETLR objective of the research work has been achieved as constraints and indexes available on the destination data sets are intact during whole process which ensures the data availability all the time. The speed of overall process is at least two times faster when compared with the existing methodologies of ETL process. Now, the further course of action is to write effective extraction programs with knowledge of internal structure of the data warehouse. The reporting of the required data from data warehouse can be done with the hints functionality of the databases and the queries can be optimized on the logic of partitioned tables. For future, we can maintain the same architecture as mentioned above; in building distributed data warehouse and also it is scalable for the cloud networks.

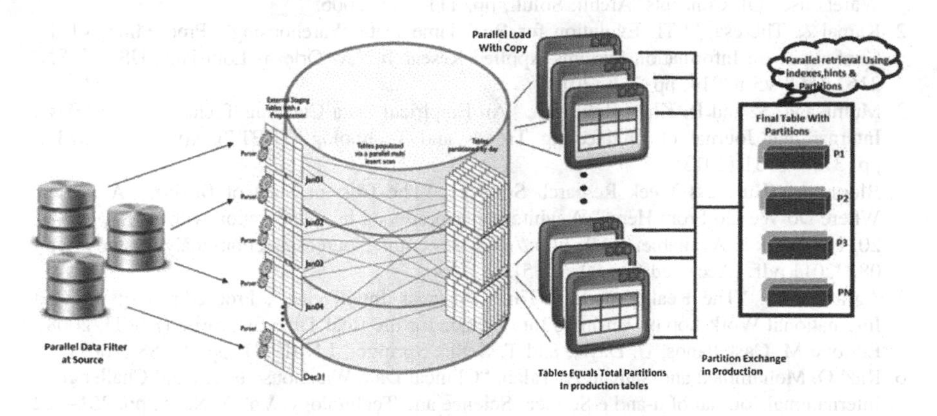

Fig. 4 ETLR—parallelism and layered optimization

7 Conclusion and Future Work

In this paper, we have presented state-of-the art issues in existing ETL technology. Practically a designer uses commercial ETL tool or develop customized product in-house for establishment of data warehouse. We have concentrated on the fact that ETL cannot be optimized by only making use of ETL tool or optimizing only workflows but it is a many-fold optimization which should be applied at each layer of data warehouse. We have introduced one more layer for optimization along with ETL, i.e., retrieval. Not only a new layer has been introduced, but we have discussed every aspect of individual layer along with optimization techniques used at each level. Additionally, we detailed the major issues and problems during ETL process and we have come up with a new paradigm for developing a data warehouse addressing retrieval part which was totally unexplored.

Finally, we have pointed out that developing a data warehouse system is not only to choose a good ETL tool or write custom code for ETL, but it is about all phases of life cycle of data warehouse considering optimization of every aspect throughout phases which give us not only efficient ETL process, but also ensures quick availability of data. Further, we have indicated list of open research issues: conventional ETL and parallel ETL with retrieval and we have provided several future directions.

References

1. Simitisis, P. Vassiliadis, S. Skiadopoulos, and T. Sellis, "Data warehouse refreshment," Data Warehouses Ol. Concepts, Archit. Solut., pp. 111–134, 2006.
2. Kamal & Theresa, "ETL Evolution for Real-Time Data Warehousing", Proceedings of the Conference on Information Systems Applied Research New Orleans Louisiana, USA; ISSN: 2167-1508 v5 n2214, pp. 7–8 (2012).
3. Mummana, S. and R. Kiran Rompella, "An Empirical Data Cleaning Technique for CFDs", International Journal of Engineering Trends and Technology (IJETT). 4(9). 3730–3735., pp. 3730–3731 (2013).
4. Bloomberg Business Week Research Services, "The Current State of Business Analytics: Where Do We Go From Here?" A white paper produced in collaboration with SAS, pp. 5–8, 2011. [Online]. Available: SAS, http://www.sas.com/resources/asset/busanalyticsstudy_wp_08232011.pdf. [Accessed Sep 09, 2015].
5. Agrawal, D., "The Reality of Real-Time Business Intelligence", Proceedings of the 2nd International Workshop on Business Intelligence for the Real Time Enterprise (BIRTE 2008), Editors: M. Castellanos, U. Dayal, and T. Sellis, Springer, LNBIP 27, pp. 75–88 (2009).
6. Razi O. Mohammed and Samani A. Talab, "Clinical Data Warehouse Issues and Challenges", International Journal of u-and e-Service, Science and Technology, Vol. 7, No. 5, pp. 251–262 (2014).
7. Bergamaschi, S., Guerra et al., "A Semantic Approach to ETL Technologies, Data & Knowledge Engineering", 70(8), pp. 717–731. (2011).
8. P Lane, "Oracle Database Data Warehousing Guide 11 g", Oracle Corporation USA, Release 2 (11.2) E25554–02, (2013).

9. Kushaoor & JNTUA.: ETL Process Modeling In DWH Using Enhanced Quality Techniques, International Journal of Database Theory and Application. Vol. 6. No. 4, pp. 181–182 (2013).
10. W.H. Inmon, "Building the Data Warehouse",(third ed.), John Wiley and Sons, USA, (2002).
11. IBM, "IBM Data Warehouse Manager", 2003 [Online], Available: http://www3.ibm.com/software/data/db2/datawarehouse, [Accessed Sep 20, 2015].
12. Simitsis, A., "Modeling and optimization of extraction-transformation-loading(ETL) processes in data warehouse environments", Doctoral Thesis, NTU Athens, Greece (2004).
13. Muhammad Arif,Ghulam Mujtaba, "A Survey: Data Warehouse Architecture", International Journal of Hybrid Information Technology, Vol. 8, No. 5, pp. 349–356, (2015).
14. Immanuel Chan, Lance Ashdown, "Oracle Database Performance Tuning Guide, 11 g", Oracle Corporation USA, Release 2 (11.2) E41573–04 (2014).
15. Qin Hanlin; Jin Xianzhen; Zhang Xianrong, "Research on Extract, Transform and Load (ETL) in land and Resources Star Schema data Warehouses" Computational Intelligence and design (ISCID), 2012 fifth International IEEE Symposium on (Volume: 1), pp. 120–123, ISBN-978-1-4673-2646-9 (2012).
16. Burleson Consulting, "Hypercharge SQL*Loader load speed performance" [Online], Available: http://www.dba-oracle.com/art_orafaq_data_load.htm (2008).
17. H. Galhardas, "Achieving Data Quality With AJAX", [Online], Available: https://fenix.tecnico.ulisboa.pt/downloadFile/3779571376272/ajax.pdf, 2006–07, Accessed: Sep 04, 2015.
18. V. Raman and J. M. Hellerstein, "Potter's Wheel: An Interactive Data Cleaning System," Data Base, vol. 01, pp. 381–390, 2001.
19. Shaker H. Ali El-Sappagha, Abdeltawab M. Ahmed Hendawib, Ali Hamed El Bastawissyb, "A proposed model for data warehouse ETL processes", Journal of King Saud University - Computer and Information Sciences.
20. P. Vassiliadis et al., "A generic and customizable framework for the design of ETL scenarios", Information Systems, vol. 30, no. 7, pp. 492–525, Elsevier Science Ltd.

Prediction of Reactor Performance in CATSOL-Based Sulphur Recovery Unit by ANN

Gunjan Chhabra, Aparna Narayanan, Ninni Singh and Kamal Preet Singh

Abstract For the prediction Artificial Neural Network has been used, which is inspired from human biological neural network. Artificial Neural Network tool has very vast application in almost every field. ANN is a very powerful tool in the field of AI and spreading over in every field of Computer Science, IT, Refinery, Medical, etc. We have used MATLAB software for the use of built-in neural network toolbox. MATLAB platform is very user friendly and includes all the important transfer functions, training functions which helps in comparison of results with each other and allow us to predict the best result using ANN. Further, this describes application of ANN in CATSOL process. After complete study about ANN and CATSOL process, our experimental results of using different transfer functions and training functions are shown which results in the overall best network for the CATSOL process.

Keywords Artificial neural network · CATSOL process · Biological neural network

1 Introduction

Fuels are any materials that store potential energy in forms that can be practicably released and used as heat energy [1]. This energy can be used further for the smooth functioning of various machines. One such fuel is natural gas. Natural gas is

Gunjan Chhabra (✉) · Aparna Narayanan · Ninni Singh · K.P. Singh
Department of Centre of Information Technology, University of Petroleum and Energy Studies, Dehradun, Uttrakhand, India
e-mail: g_chhabra@yahoo.com

Aparna Narayanan
e-mail: appunara@gmail.com

Ninni Singh
e-mail: ninnisingh1991@gmail.com

K.P. Singh
e-mail: kamalpreet1010@gmail.com

S.C. Satapathy et al. (eds.), *Proceedings of the International Conference on Data Engineering and Communication Technology*, Advances in Intelligent Systems and Computing 469, DOI 10.1007/978-981-10-1678-3_15

generally regarded as a natural mixture of hydrocarbons found issuing from the ground, or obtained from specially driven wells. Its constituent components can be used as fuels for various purposes [2]. Raw natural gas primarily comes from one of three types of wells, either crude oil well or gas well or condensate well. The major contaminants present in natural gas are: 1. Heavier gaseous hydrocarbons (ethane, propane, butane), 2. Acid gases (carbon dioxide, hydrogen sulphide and mercaptons), 3. Other gases (nitrogen, helium), 4. Water. The raw natural gas needs to be purified before it can be converted to useful proportions [3]. Water vapour from raw natural gas is removed first and sent to amine treating unit for the removal of acid gases. The resultant gas obtained is known as sweetened natural gas [4]. The acid gases obtained after the gas sweetening process cannot be directly released into the atmosphere due to various environmental concerns. Maximum limit of hydrogen sulphide gas released to the atmosphere is 10 ppmv [5]. In refineries, the acid gas generated from amine treating unit contains more than 50 % H_2S [6]. The acid gas is sent to a Sulphur Recovery Unit (SRU) based on CLAUS process [7]. One-third of H_2S is burnt into SO_2 and forms sulphur by reaction with SO_2 and remaining in a combustion chamber. In case of Gas Processing Complex, the acid gas generated from gas sweetening unit contains less than 5 % H_2S [8]. This type of acid gas cannot be handled in CLAUS type SRU due to unstable flame inside the burner of combustion chamber. SRU based on Liquid Redox Process is used for efficient removal of H_2S from such type of acid gas [9–11]. EIL [11] and GNFC [12] developed an indigenous Liquid Redox Process called CATSOL Process for the conversion of H_2S into elemental sulphur in the presence of metal-based catalyst. Catalyst is regenerated by air. Both the absorption and regeneration is taking place inside a single reactor called as auto circulation reactor [13, 14]. Elemental sulphur is recovered from the acid gas settles down and is removed regularly as part of slurry. Sulphur is separated from slurry using a filter system and sent to sulphur yard for storage [15]. Various chemicals used in the process for maintaining the pH, settling of sulphur particles, avoiding foaming inside reactor, avoiding biological growth inside reactor, etc. The chemicals required during the operation of the SRU are: Catalyst, Chelate, KOH solution, Surfactant, Defoamer and Biocide [16–21]. This process is working based on a number of parameters like pressure of the feed acid gas, temperature of the feed acid gas, concentration of H_2S in feed acid gas, pressure of regeneration air, temperature of regeneration air, pH of the solution, concentration of catalyst and chelate dosing of other chemicals, etc. [22–25]. A mathematical model is required for predicting the performance of the reactor according to all the above-mentioned parameters. Artificial Neural network (ANN) [26] is used for modelling complex relationships between inputs and outputs or to find patterns in data by connectionist approach. ANN is a very good tool for predicting the performance of the reactor which depends on the number of process parameters mentioned above [27–29]. CATSOL Process is a complex process in which a number of process parameters are interlinked non-linearly. ANN is a good tool for finding out the connections within all process parameters.

2 Process Description

The CATSOL process uses iron chelate solution as catalytic reagent for conversion of H_2S in acid gas to elementary sulphur in a reactor [30, 31]. During conversion of H_2S, the iron catalyst is reduced from ferric ion to ferrous ion. Subsequently, ferrous ion is oxidized by air in the same reactor. The whole process is shown in Fig. 1. The prominent features of this technology are

1. Overall reaction is carried out at low temperature. 45–50 °C.
2. Technology treats acid gas with very low H_2S concentration and can also handle acid gases with variation in H_2S concentration.
3. Technology shows high H_2S removal efficiency and ensures H_2S less than 10 ppmv in the gas discharge to the atmosphere.
4. Use of filter minimizes the catalyst loss and less maintenance in the plant. The reactor of SRU consists of two main sections—Absorption Section (draft tube) and Regenerator Section (annular). In the absorption section, H_2S is absorbed in

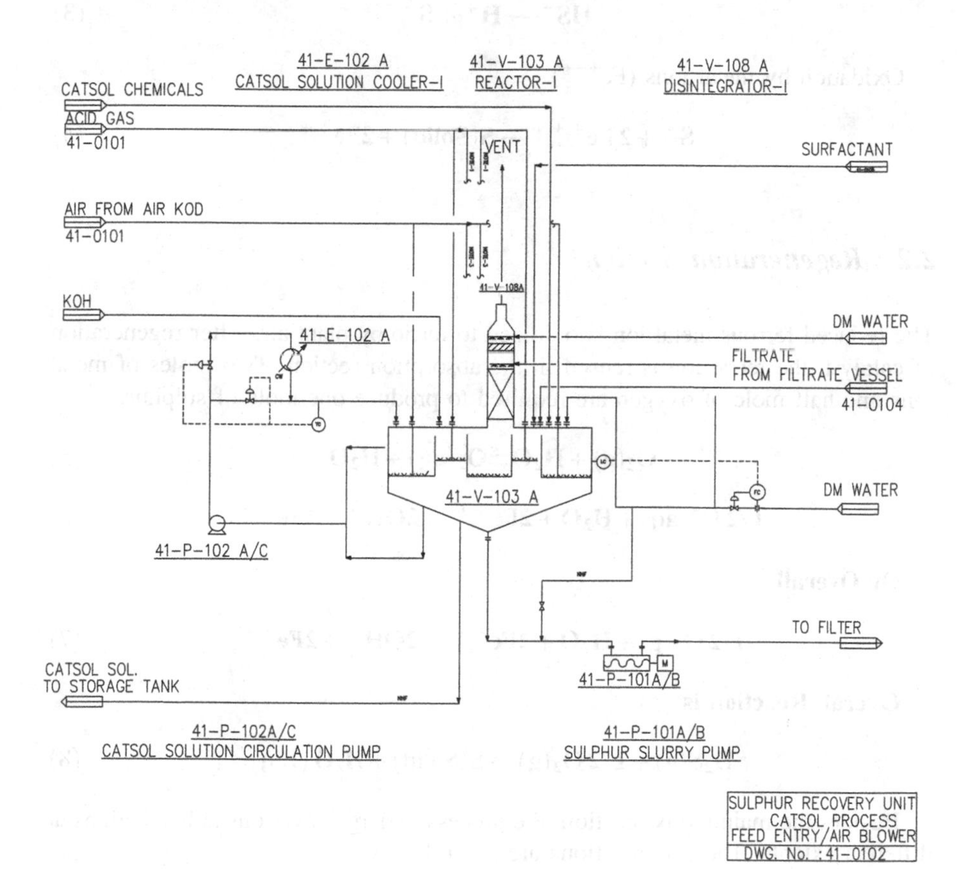

Fig. 1 SRU for CATSOL process

the alkaline catalyst solution as S_2 ions. Subsequently, sulphide ion is converted to sulphur in the presence of ferric ion and ferric ion is reduced to ferrous ion [32–35]. The reaction in absorption section is given below.

2.1 Absorption Section

$$H_2S(g) + H_2O \rightleftarrows H_2S\,(aq) + H_2O \quad (1)$$

First ionization:

$$H_2S(aq) \rightarrow H^+ + HS^- \quad (2)$$

Second ionization:

$$HS^- \rightarrow H^+ + S^- \quad (3)$$

Oxidation by metal ions (Fe^{+++})

$$S^- + 2\,Fe^{+++} \rightarrow S(Solid) + 2Fe^{++} \quad (4)$$

2.2 Regeneration Section

The reduced ferrous metal ion is oxidized to ferric by plant air. After regeneration of catalyst, the ferric ion is reused in the absorption section. Two moles of metal ions and half mole of oxygen are required to produce one mole of sulphur.

$$O_2(g) + H_2O \rightleftarrows O_2(aq) + H_2O \quad (5)$$

$$1/2\,O2(aq) + H_2O + 2Fe^{++} \rightarrow 2OH^- + 2Fe^{+++} \quad (6)$$

Or Overall

$$1/2\,O_2(g) + H_2O + 2Fe^{++} \rightarrow 2OH^- + 2Fe^{++} \quad (7)$$

Overall Reaction is

$$H_2S(g) + 1/2\,O_2(g) \rightarrow S(Solid) + H_2O\,(Liq) \quad (8)$$

Besides the main redox reaction, the process undergoes various side reactions at different extents. The side reactions are given below

(a) **Reaction** 1

$$2HS + 2O_2(g) \rightarrow S_2O_3^{--} + H_2O \tag{9}$$

Combining reaction (2) and (9)

$$2H_2S + 2O_2(g) \rightarrow 2H^+ + S_2O_3^{--} + H_2O \tag{10}$$

(b) **Reaction** 2

$$H_2O + CO_2(g) \rightarrow H_2CO_3 \tag{11}$$

Although overall reaction (Reaction 8) shows no H^+ ion formation, H^+ ions generated in the solution due to side reaction leads to unfavourable situation. To prevent the lowering of pH of the solution, KOH solution is added either intermittently or continuously and pH of the solution is maintained good enough to get high degree of H_2S absorption in the absorption section [36]. Operation at high pH causes formation of thiosulfate. As concentration of thiosulfate increases in the solution, density of the solution increases and it results in more purging of the solution. Generally, pH range of 7.8 and 8.1 is favoured for better performance [37].

Slurry filter is used for separating sulphur from the catalyst solution. After separation, the filtrate is returned to the reactor. Filter system ensures minimum loss of catalyst and chemicals with solid sulphur cake.

It is required to analyse the catalyst solution intermittently for checking of the metal and chelate contents in the solution [38, 39]. The catalyst and chelate solutions are added to the reactor to maintain the right concentration and addition is done based on the analysis results [40].

2.3 Parameter Selection

There are variable process parameters which affect the CATSOL process and can be elemental in the amount of sulphur recovered and concentration of H_2S released into the atmosphere. Lists of these parameters are: Air Flow Rate, Concentration of H_2S in the acid gas, Acid gas flow rate, pH, ORP, Temperature, Pressure of Air, KOH solution added, Concentration of the chelate added, Catalyst, Vent release and Absorber level.

The first 10 parameters listed above are input parameters and the last two are output parameters. From the set of input parameters, only the first five are most important ones. The other input parameters are not of that much importance as the rest are maintained at almost a constant level due to which there is no non-linear change in their presence to make them useful while modelling the process. For modelling of the CATSOL process the parameters selected are given below.

Input Parameters:

1. **Air flow rate**: Air is used for the regeneration of the chelate and also helps in the circulation of the aqueous solution.
2. **Concentration of H_2S in the acid gas**: Concentration of H_2S in acid gas is in the range of 0–5 %v. The acidic gas is sent to the SRU to reduce the amount of H_2S gas released into the atmosphere.
3. **Acid gas flow rate**: Input parameter that takes into account the flow of the acid gas from gas sweetening unit into the CATSOL Unit.
4. **pH**: The pH of the solution needs to be constant within the range of 7–8.3. Due to unwanted reactions that take place, there is a change in pH, KOH is added continuously or intermittently to maintain it
5. **ORP**: The oxidation reduction potential of the solution gives information whether catalyst is regenerated properly or not.

Output Parameters:

1. **Vent release**: The amount of H_2S released into the atmosphere
2. **Absorber level:** The level of catalyst solution in the reactor

3 Artificial Neural Network

Now, based on the selected set of parameters the data with respect to them need to be collected from various sources or recorded from various measurements of the CATSOL process. The data that is collected should contain values for both input and output parameters. This set of data would be used for training, validation and testing of the network. Multilayer networks can be trained to generalize well within the range of inputs for which they have been trained. However, they do not have the ability to accurately extrapolate beyond this range, so it is important that the training data span the full range of the input space.

The data collected from the CATSOL process has been divided into three subsets viz. Training data, Validation data and testing data. Each of these three different sets would be used for training the network, validating the network and testing the trained network.

The training of the network with a set of sample data, the number of neurons in hidden layer was varied and the results were recorded. The following result is shown in the Fig. 2. It can be inferred easily from the graph that a hidden layer containing 11 neurons gives us best result. The graph clearly depicts that till 11

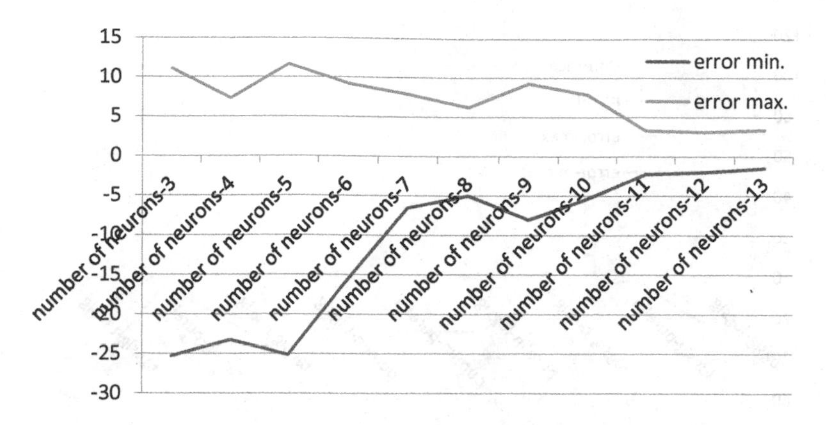

Fig. 2 Hidden layer

neurons there is a lot of change in the performance of the trained network. From 11 neurons onwards it is clearly visible that the graph smoothens up and there is not much change leading to the belief that 11 neurons in the hidden layer would provide the best trained network.

The various network training function were used and the best of them all was chosen. From the Fig. 3 it can be inferred that trainbr and trainlm give the best result with regards to training the network. But even though trainlm has a higher

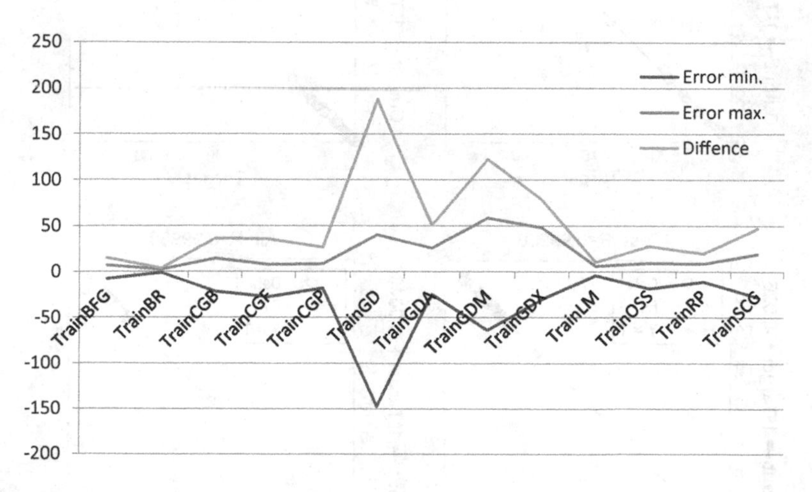

Fig. 3 Training function

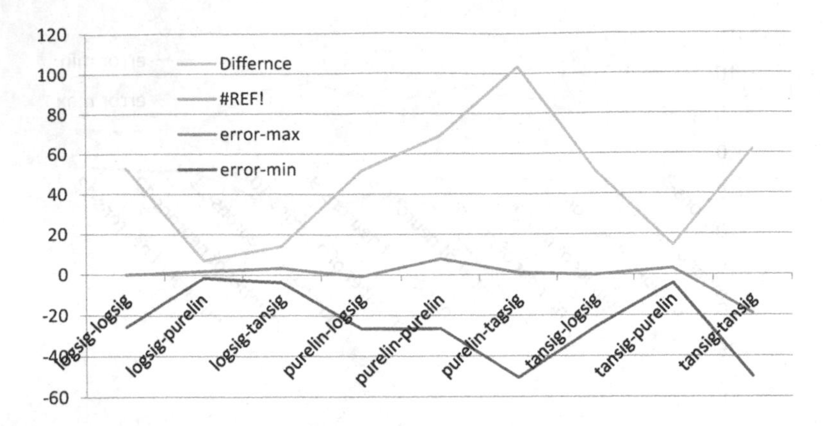

Fig. 4 Transfer function

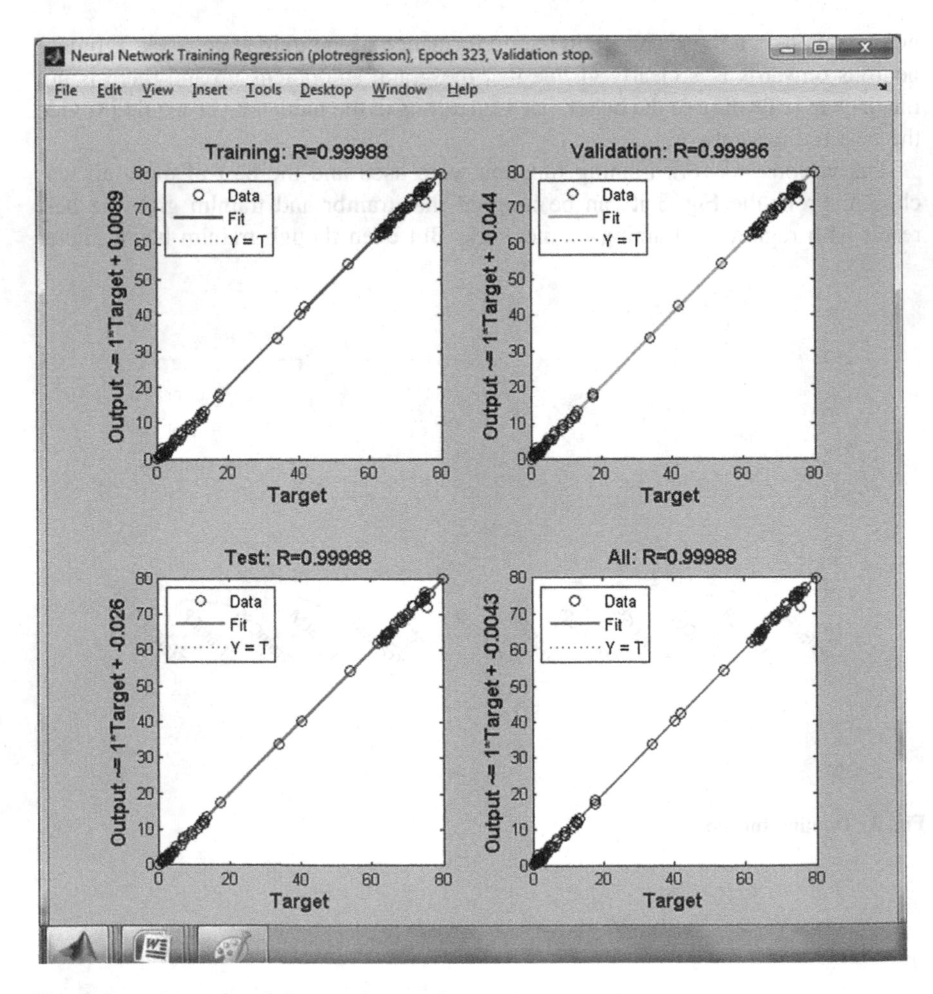

Fig. 5 Regression plot of the neural network

error range than trainbr, it validates the network which the latter neglects to do. This results in the belief that the training function trainlm produced the best result.

For the training of the network, there is a need to specify the transfer functions with respect to each non-input layer which is what makes artificial neural network non-linear in nature. There are various transfer functions which can be used. There are two main types of transfer function viz. Linear and non-linear. The linear transfer functions are used in the final layer of multilayer networks as function approximates. The non-linear transfer functions can be used in the hidden layers of the network as they are differentiable in nature. A graph depicting the result of the use of various transfer functions is given in Fig. 4. From this it can be inferred that purelin transfer function for the output layer and logsig function for the hidden layer provides the best result (Fig. 5).

4 Strength of Proposed Technique

In the industry it is not an easy process to take corrective action only after the entire reaction process is complete as it may lead to adverse situations. The proposed system would help in predicting the reactor's performance on varying the inputs and observing what it results in without actually having to let the reaction take it course. This would also help in producing the desirable outcome when the reaction reaches it completion along with reducing the wastage and any kind of contamination of pollution. And with 97 % accuracy we can say that the amount of gas release into the atmosphere would be within the permissible limits. The presence of equipment in the reactor for measurement of the inputs and outputs of the reactor adversely affects the reaction that takes place within the reactor. The use of the proposed system would reduce this risk also (Table 1).

Table 1 Results of the neural network

Sr no.	Acid gas flow rate (knm^3/h)	H_2S conc. in acid gas (ppmv)	Air flow rate (knm^3/h)	pH	ORP (mV)	Absorber level (%)	Vent (ppmv)
1	5.82	4679	9.15	7.66	−22	68.3201	2.3645
2	5.8	4702	9.48	7.62	−22	73.3791	2.0173
3	5.97	4722	9.41	7.6	−25	69.7314	2.8937

5 Conclusion

The Neural Network which has been trained, validated and tested against data collected from the CATSOL process. It is now ready to be used for prediction of the output parameter values given the input parameters. This model would be very helpful in gas processing units as it is not an easy task to repeatedly test the various output parameters and take action accordingly. The modelled network can instead be used to predict the values that the output parameters would give and use it as a basis to perform further actions as required in the reactor with regards to the CATSOL process.

The Network currently simulated is only with regards to the data obtained from the CATSOL process. But the same can be extended for various other processes which are a part of the gas processing/refining process. This would make it a lot easier to predict what kind of result could be obtained for a particular set of inputs.

The neural network is 97 % accurate in providing results.

6 Future Scope

In this scope, the study was based on only the five most important factors that affect the reaction. But further studies can be conducted to find effect of the other factors on the reaction. Also the CATSOL process is just a small part of the refinery unit and the input to it is controlled by another process. A study on the purification of natural gas and its relation to the CATSOL process needs to be determined.

As both are two separate process and output of one forms the input to the other an uncertainty may arise, which can be reduced by the use of Fuzzy Logic.

The same predictive analysis can be performed with other techniques and can be compared with the results obtained from the use of ANN for better results. And in case a single technique is not sufficient for this then hybrid AI techniques can be utilized for better results.

References

1. Inverse function - Wikipedia, the free encyclopedia.
2. Ayub, Shahanaz, and J. P. Saini. "Abnormality detection in ECG using artificial neural networks." *Electronic Journal of Biomedicine, Spain, ISSN* (2010): 47–52.
3. Feed-Forward Neural Network.
4. Train neural network - MATLAB train - MathWorks India.
5. Resilient back propagation - MATLAB trainrp - MathWorks India.
6. Scaled conjugate gradient back propagation - MATLAB trainscg - MathWorks India.
7. Bayesian regulation back propagation - MATLAB trainbr - MathWorks India.
8. Levenberg-Marquardt back propagation - MATLAB trainlm - MathWorks India.
9. Neural Network Toolbox Documentation - MathWorks India.

10. Inverse transfer function - MATLAB netinv - MathWorks India.
11. Inverse transfer function - MATLAB netinv - MathWorks India.
12. ann application in oil refinery - Google Search.
13. Artificial neuron - Wikipedia, the free encyclopedia.
14. Neuron Model - MATLAB & Simulink - MathWorks India.
15. customizing transfer function (neural net domian) - Newsreader - MATLAB Central.
16. Neuron Model - MATLAB & Simulink - MathWorks India.
17. Neural Network Toolbox Documentation - MathWorks India.
18. matlab transfer function neural network toolbox - Google Search.
19. Create transfer function model, convert to transfer function model - MATLAB tf MathWorks India.
20. Merichem I Case Studies.
21. http://www.merichem.com/resources/case_studies/LO-CAT/Desulfurization.pdf.
22. Cascade-forward neural network - MATLAB cascadeforwardnet - MathWorks India. Feedforward neural network - MATLAB feedforwardnet - MathWorks India.
23. Function fitting neural network - MATLAB fitnet - MathWorks India.
24. Radial basis function network - Wikipedia, the free encyclopedia.
25. Probabilistic neural network - Wikipedia, the free encyclopedia.
26. Linear layer - MATLAB linearlayer - MathWorks India.
27. Hopfield network - Wikipedia, the free encyclopedia.
28. Design generalized regression neural network - MATLAB newgrnn - MathWorks India.
29. Competitive layer - MATLAB competlayer - MathWorks India.
30. Resilient backpropagation - MATLAB trainrp - MathWorks India.
31. Levenberg-Marquardt backpropagation - MATLAB trainlm - MathWorks India.
32. Gradient descent with adaptive learning rate backpropagation - MATLAB traingda - MathWorks India.
33. Cascade-forward neural network - MATLAB cascadeforwardnet - MathWorks India.
34. Layer recurrent neural network - MATLAB layrecnet - MathWorks India.
35. Competitive Learning - MATLAB & Simulink - MathWorks India.
36. Yonaba, H., F. Anctil, and V. Fortin. "Comparing sigmoid transfer functions for neural network multistep ahead streamflow forecasting." *Journal of Hydrologic Engineering* 15.4 (2010): 275–283. ftp://icsi.berkeley.edu/pub/ai/jagota/vol2_6.pdf.
37. Transfer Functions in Artificial Neural Networks - A Simulation-Based Tutorial — Brains, Minds & Media.
38. Neural Network Toolbox - MATLAB - MathWorks India.
39. KV, Narayana Saibaba, et al. "APPLICATION OF ARTIFICIAL NEURAL NETWORKS AND STATISTICAL METHODS IN COCONUT OIL PROCESSING." International Journal of Advanced Computer and Mathematical Sciences , Vol 3, Issue 2, 2012, pp. 209–214.
40. http://www.ogj.com/articles/print/volume-97/issue-20/in-this-issue/refining/gulf-coast.

A Multilevel Clustering Using Multi-hop and Multihead in VANET

G. Shanmugasundaram, P. Thiyagarajan, S. Tharani
and R. Rajavandhini

Abstract Clustering is the process of grouping similar objects, to provide secure and stable communication in wireless network. Clustering plays a vital role in wireless networks, as it assists on some important concepts such as reusing of resources and increasing system capacity. In current research scenario, among different types of clustering, multilevel clustering is considered to the hot research topic as it is not exploited to its full potential. Multilevel clustering is the process of partitioning the nodes within the cluster in hierarchy manner. This multilevel method results in very fast clustering. Among the multilevel clustering techniques mentioned in the literature, only few of them forms the cluster with equal size. The major problem in the multilevel clustering techniques is the farthest nodes from head are not able to be communicated within the short span of time. To overcome the above-mentioned problems, this paper proposes an algorithm which satisfies the multilevel approach by combine the multi-hop and multihead concept. The proposed method of cluster is justified and implemented by using OMNET++ simulator.

Keywords Multilevel clustering · VANET clustering · Location · Distance · Cluster stability · Life time · OMNET++

G. Shanmugasundaram (✉) · S. Tharani · R. Rajavandhini
Department of Information Technology, Sri Manakula Vinayagar
Engineering College, Pondicherry, India
e-mail: anitguy2006@gmail.com

S. Tharani
e-mail: dharshinisegar@gmail.com

R. Rajavandhini
e-mail: vandhini93@gmail.com

P. Thiyagarajan
Department of Computer Science and Engineering, Mahindra Ecole Centrale,
Hyderabad 500043, India
e-mail: thiyagu.phd@gmail.com

S.C. Satapathy et al. (eds.), *Proceedings of the International Conference on Data Engineering and Communication Technology*, Advances in Intelligent Systems and Computing 469, DOI 10.1007/978-981-10-1678-3_16

1 Introduction

Vehicular Ad hoc Networks (VANET) are a type of network which is emerging in peak among various wireless networks. VANET [1] is a form of Mobile Ad hoc Network (MANET) which is basically a vehicle to vehicle and vehicle to roadside communication. Now it has recently drawn significant research attention since it provides the infrastructure that enhances driver's safety, less accidents, and avoid congestion. It also helps drivers to acquire real-time information about road conditions in prior so that they react properly on time. This prior information about road conditions and traffic helps the driver to take new routes by avoiding the vehicles to enter crash dangerous zone with the help of Vehicle to Vehicle (V2V) communication. VANET is characterized by high vehicle mobility and thus the changes in topology occur. Many control mechanisms and topology management has been proposed [1]. Clustering permits the formation of dynamic virtual backbone and it is a great challenge of clustering in highly dynamic environments. Most of the clustering algorithms for VANET are drawn from MANET clustering concepts. These algorithms lack to capture the mobility characteristics and other important drawback is formation of cluster that is based only on the position and direction of vehicles in geographic locations regardless of relative speed. And thus clustering is concentrated more than another. To enhance the stability, the clustering models are redefined based on the elements such as location, speed difference, and direction.

As discussed by Yvonne et al. [2], clustering techniques can be classified into many types. One such clustering technique is multilevel clustering which uses hierarchical structure for clustering. Multilevel clustering is gaining its importance over other clustering technique as it is faster than other clustering techniques when cluster formed are equal in size.

The proposed algorithm involves four main steps in formation of clusters, they are initiating clustering process, head selection, slave selection, and merging the clusters. There are several advantages in the proposed method as multihead approach is used, these merits are validated in simulation environment. The rest of the paper is structured as: Sect. 2 contains existing works in clustering techniques. Section 3 explains the proposed approach and algorithm for multilevel clustering with multihead and multi-hop concepts. Section 4 deals with simulation of the proposed algorithm and its details. And finally Sect. 5 concludes the paper.

2 Existing Work

Literature survey has been made on both clustering and multilevel clustering algorithms in which the multilevel clustering has to be subjugated more. Various algorithms have been implemented for multilevel clustering by considering network parameters. The existing algorithms are surveyed and those are summarized below.

In [3], cluster-based location routing (CBLR) was proposed, in which the nodes uses HELLO messages to share out their state. When a new node enters the system, it stays in the undecided state and announces itself as the cluster head to all other nearby nodes by receiving the HELLO message. If it does not receive the HELLO message for a period of time from other nodes then it register it as the cluster member. For the infrastructure changes, nodes maintain the table which contains the list of neighboring nodes to exchange the messages. The main objective of the protocol is to improve the routing effectiveness in VANET. The nodes are assigned to know their position and the position of their destination and therefore, the packets are directly forwarded toward the destination. But the straight forwarding of the packets to the destination is completely dependent on the table. If any problem in updating the table occurs then the complete method is not achieved which is the drawback of this technique.

In [4], authors proposed a heuristic clustering approach that is equivalent to the computation of minimum dominating sets (MDS) used in graph theory for cluster head elections. This technique is called position-based prioritized clustering (PPC). To build the cluster structure priorities associated with the vehicles traffic information and geographic position of nodes are used. To attain clustering, each node must transmit small amount of message to itself and also to its neighbors. The message contains five tuples (cluster head ID, node location, node priority, node ID, and ID of the next node). A node which has highest priority becomes a cluster head. Based on the node ID, eligibility function and current time the priority of the node is calculated.

Hung hai Bui et al. [5] proposed another algorithm for multilevel clustering "Bayesian non parametric multilevel clustering with group—level contexts" they presents Bayesian nonparametric framework for multilevel clustering which utilizes group level context information to discover the low dimensional structure of the group contents and partitions the groups into clusters. They use Dirichlet process as the building block; Dirichlet process is a set of probability distributions which use Bayesian inference to describe the knowledge about the distribution of random variables. The Dirichlet process is specified by a concentration parameter (base distribution H and positive real number α). This work concentrates on multilevel clustering problem in multilevel analysis, the theme is to join two clusters based upon the content data and group level context information.

Meirav galum et al. [6], introduced a scheme for clustering formation in which various aspects such as cluster noise, dimensionality, and size are considered. A large variety of clustering algorithm has been proposed but they focus on the spherical shape, average linkage, and spectral methods. To conquer this problem they introduced fast multilevel clustering for different shape of cluster. In this scheme they apply recursive coarsening process, resulting in a pyramid of graphs. The pyramid provides a hierarchical decomposition of the data into clusters in all resolutions. This algorithm is easily adaptable to handle different types of data sets and different clustering shapes.

Yan Jine et al. [7], proposed an energy efficient multilevel clustering protocol for heterogeneous environment. This algorithm is proposed for minimum energy consumption in heterogeneous wireless sensor network. There are three common type of resources heterogeneity: computation, energy, and link. Among above three, the most important heterogeneity is energy because computational and link depends on energy resources. In energy heterogeneity the nodes is line powered and its battery is replaceable. If some heterogeneity nodes are placed in the sensor network, then the average energy consumption will be lower in forwarding packet from nodes to the sink. Thus the network lifetime will increase.

The existing clustering algorithms suffer from unequal sized clusters [8, 9]. All the existing algorithms reported in the literature does not consider about the important parameters such as location, relative speed and distance while forming clusters. To overcome the above-mentioned problems, this paper proposes an algorithm which uses both the multihead and multi-hop concept. With the help of cluster head the members of the clusters can be communicated in short span of time. It also paves way for formation of equal-sized clusters. The proposed algorithm also generates an alert message that can be easily passed to others clusters without any congestion.

3 Proposed Work

In existing multilevel clustering the coarsening of the process are carried out in which the proposed system overcomes many presented drawback. In planned system according to Dedicated Short Range Communication (DSRC) terms, data link layer provide transmission range of up to 1,000 m for a channel. By using control channel, the cluster head can be able to communicate with neighboring cluster heads and vehicles can gather information about neighboring vehicles. In the proposed work, there are four processes to form effective clusters, they are:

1. Initiating clustering process
2. HEAD selection
3. Slave selection
4. Merging the clusters.

3.1 Initiating Clustering Process

In the algorithm first process is to initialize the cluster. When the node enters into the Road Side Unit (RSU) range the node sends the initialize message to the RSU. After the initialize the message the new node sends the beacon message to all the neighbor nodes to form a node list. The beacon message contains the information about the node ID and this message share to all the nodes within the particular range. The pseudocode for initiating clustering process is given below in Fig. 1.

Fig. 1 Pseudocode for initiating clustering process

```
If node enters the range of RSU
{
    Send message ("initialize")
} // End if
```

3.2 Head Selection

In the second step, once the node list is maintained with the help of that head selection will be done. The nodes which have the maximum list will be elected as Head (i.e., master). In the cluster formation Head plays the major role because with the help of the Head all the information (road condition, traffic and accidents) from the RSU is passed to all the neighbor nodes within the transmission range. The pseudocode for head selection process is given below in Fig. 2.

3.3 Slave Selection

In the third step, slave selection will be done. To reduce the cluster overhead we introduce the slave method. The Head act as master and the other nodes act as slave.

After the Head selection, the head check the distance of all the nodes by sending the beacon message. The nodes which are approximately at 200 distance is elected as slave because each node (car) will have the transmission range of 200 m, whereas remaining nodes beyond the 200 m distances will act as slave members. The pseudocode for slave selection process is given below in Fig. 3.

Fig. 2 Pseudocode for head selection process

```
On receiving ("initialize") message
For 1... Size (nodelist)
{
    Check suitability ()
    If suitability == true then
    {
        Send message ("head") to the node
    } // End if
} // End for
```

Fig. 3 Pseudocode for slave selection process

```
For 1... Size (nodelist)
{
    getposition()
    if (Check distance ~ 200)
    {
        Send message ("slave")
    } // End if
} // End for
```

Fig. 4 Pseudocode for merging of clusters

```
For j=i... Size (head list)
{
   For j=i+1... Size (head list)
   {
      getposition( i )
      getposition( j )
      if Check distance (i, j) ~ 200
      {
          Send message (i, "head")
          Send message (j, "slave")
      } // End if
   } // End for j } // End for i
```

3.4 *Merging of Clusters*

In the fourth step, merging of the clusters will be done to reduce the cluster congestion. When two neighboring clusters merge within its transmission range, the role of Cluster Head (CH) will be loosed by the head which has less number of members then it joins other cluster and become a cluster member. Here cluster restructuring will be take places, if the losing node has cluster members, then it became member of new cluster. While restructuring the cluster two processes take place either the members join any nearby clusters or form a new cluster if they could not find a cluster to join. The pseudocode for merging of clusters is given below in Fig. 4.

4 Simulation and Performance Evaluation

To evaluate the performances of our proposed work, OMNET++ simulation tool is used by considering different road traffics and different network parameters such as transmission speed of the message, congestion, and cluster overhead [10]. In OMNET++ existing sample GOOGLE EARTH demo is used to simulate a wireless vehicular ad hoc network. The simulation contain cars as mobile nodes that move randomly over a 2 km-by-2 km area in which nodes have identical radios, with transmission range of 500 m. When nodes (cars) come within same transmission range then they communicate with each other by ad hoc formation to share the information like accidents, road condition, and traffic. Once the nodes are assigned start the simulation in fast or express mode. Then we are in the platform where the browser supports. We can run this simulation in Internet with help of Google earth plugin. For the simulation purpose, we have different network parameters. The data rate is set in Mbps and the periodic messages are sent, the size of the message contains the mobility information in the form of bytes. The data rate is supported by the dedicated short range communication (DSRC) standard.

The parameters which we consider majorly in our proposed system and algorithm are position, direction of vehicle (car), speed and velocity of the vehicle, cluster overhead, and cluster maintenance.

4.1 Cluster Overhead

To reduce the cluster overhead we should form a stable cluster by considering the number of nodes in each cluster. The first process is to form a cluster in equal size in which one cluster acts as head and others as slave. So to form a stable cluster we consider certain limit for numbers of nodes. When the nodes reach the limit cluster formation take places with the help of that we can reduce the cluster overhead.

4.2 Cluster Maintenance

For the topology change there are three types of events in VANET which can be defined as follows: a mobile node joins the network, a node leaves the cluster formation, and two cluster heads come into direct transmission range. When the new node joins the network and has non-clustered members, then these nodes will form a new cluster according to the rules used by each clustering method. And if it has the cluster head as a neighbor then it tries to join by sending the message to CH. And finally, this cluster joining overhead is same for both weight-based (WB) and position-based (PB) methods. When two neighboring clusters merge within its transmission range, the role of CH will be loosed by the head which has less number of members then it joins other cluster and become a cluster member. If the losing node has cluster members, then the members are processed for cluster restructuring. The members either join any nearby clusters or form a new cluster if they could not find a cluster to join.

4.3 Cluster Stability

To increase cluster stability direction of the vehicles play a major role. Transmission speed, reclustering, communicating with RSU (to share the information about traffic and road condition) will be difficult when vehicles moves in different directions. Therefore, in multilevel clustering unidirectional vehicles must be considered for effective cluster formation. The number of cluster changes is dependent on following conditions, they are (i) A node leaves its cluster and forms a new one (ii) A node leaves its cluster and joins a nearby cluster (iii) A CH merges with a nearby cluster.

5 Conclusion

Clustering methods should be designed to adapt to the VANET atmosphere. These methods should take into account all vehicle dynamics. In this paper, we proposed a novel VANET multilevel cluster formation algorithm that tends to group vehicles showing similar mobility pattern in one cluster. This algorithm explains the speed difference among vehicles as well as the position and the direction during the cluster formation process. After the simulation experiment, it is observed that our technique increases the cluster lifetime and reduces the transmission time of messages between the nodes. And the cluster head to head communication is made possible and the proposed algorithm significantly increases the stability of the clusters.

References

1. Saif Al-Sultan, MoathM. Al-Doori, AliH. Al-Bayatti, HussienZedan, "A Comprehensive Survey on Vehicular Adhoc network ", Journal of Network and Computer Applications, No 37, 380–392 (2014).
2. G Yvonne, W Bernhard, G Hans Peter, Medium access concept for VANETs based on clustering, in Proceedings of the 66th IEEE Vehicular Technology Conference (VTC), vol. 30, 2189–2193, Baltimore (2007).
3. RA Santos, RM Edwards, NL Seed, Inter vehicular data exchange between fast moving road traffic using ad-hoc cluster based location algorithm and 802.11b direct sequence spread spectrum radio, Post-Graduate Networking Conference, (2003).
4. W Zhiagang, L Lichuan, Z MengChu, A Nirwan, A position-based clustering technique for ad hoc intervehicle communication. IEEE Trans Man Cybern. 38(2), 201–208 (2008).
5. Vu Nguyen Dinh Phung XuanLong Nguyen Svetha Venkatesh Hung Hai Bui "Bayesian Nonparametric Multilevel Clustering with Group-Level Contexts" International Conference on Machine Learning, vol 32, (2014).
6. Ya'ara Goldschmidt, Meirav Galun, Eitan Sharon, Ronen Basri, Achi Brandt," Fast Multilevel Clustering" www.wisdom.weizmann.acil.
7. Yan Jine, Ling Wang, Yoohwan Kim, Xiaozong Yang, " An Energy Efficient Multilevel Clustering Algorithm for large scale wireless sensor Networks", Journal of Computer Networks vol 52, No 542–562 22 February (2008).
8. Yun-Wei Lin, Yuh-Shyan Chen, Sing-Ling Lee, "Routing Protocols In Vehicular Ad Hoc Networks: A Survey And Future Perspectives", Journal Of Information Science And Engineering vol 26, 913–932, (2010).
9. Rasmeet SBalia, NeerajKumar, Joel J.P.C. Rodrigues," Clustering in vehicular ad hoc networks: Taxonomy, challenges and solutions", Journal of Vehicular Communications, 134–152, (2014).
10. Pengfei Guo, Bo Xu, Huan Zhou "QoS Evaluation of VANET Routing Protocols ", Journal Of Networks, vol. 8, no 1, (2013).

Patient-Specific Cardiac Computational Modeling Based on Left Ventricle Segmentation from Magnetic Resonance Images

Anupama Bhan, Disha Bathla and Ayush Goyal

Abstract This paper presents three-dimensional computational modeling of the heart's left ventricle extracted and segmented from cardiac MRI. This work basically deals with the fusion of segmented left ventricle, which is segmented using region growing method, with the generic deformable template. The multi-frame cardiac MRI image data of heart patients is taken into account. The region-based segmentation is performed in ITK-SNAP. The left ventricle is segmented in all slices in multi-frame MRI data of the whole cardiac cycle for each patient. Various parameters like myocardial muscle thickness can be calculated, which are useful for assessing cardiac function and health of a patient's heart by medical practitioners. With the left ventricle cavity and myocardium segmented, measurement of the average distance from the endocardium to the epicardium can be used to measure myocardial muscle thickness.

Keywords Patient-specific · Cardiac model · Computational modeling · Left ventricle · Segmentation · Magnetic resonance images

1 Introduction

Cardiac disease is the number one leading cause of deaths worldwide. Hence, cardiological research is underway to build patient-specific computational models of patient cardiac anatomy in order to aid the cardiologist's diagnosis and evaluation of cardiac function of a patient's heart. In cardiac modeling, the left ventricle is of particular significance. The heart consists of four chambers, namely the left ventricle, right ventricle, left atrium, and right atrium. A ventricle is one of two large chambers that gathers and ejects blood received from an atrium towards the peripheral beds within the body and lungs. The anatomy of the heart indicating the left ventricle is shown in Fig. 1. The pumping in the heart is primed by the atrium (an adjacent/upper

---

Anupama Bhan (✉) · Disha Bathla · Ayush Goyal
Amity University Uttar Pradesh, Noida 201313, Uttar Pradesh, India
e-mail: abhan@amity.edu

S.C. Satapathy et al. (eds.), *Proceedings of the International Conference on Data Engineering and Communication Technology*, Advances in Intelligent Systems and Computing 469, DOI 10.1007/978-981-10-1678-3_17

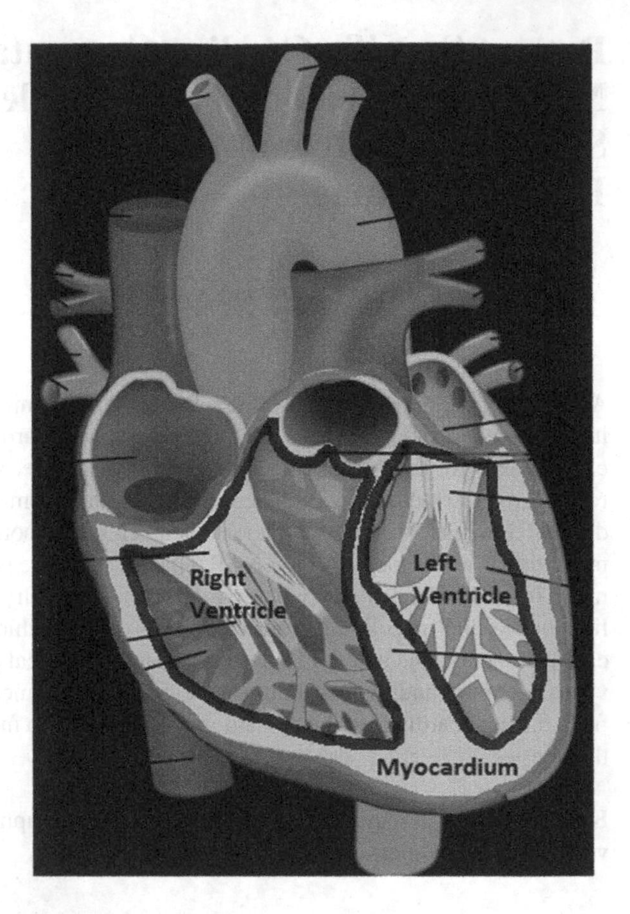

Fig. 1 Anatomy of the heart

heart chamber that is smaller than a ventricle). Extracting and reconstructing the left ventricle from cardiac MRI images is an important aspect of creating a computational model of a patient's heart.

Segmentation of the left ventricle function is essential for diagnosing and planning therapy for the cardiac diseases. For this purpose, there are various cardiac imaging modalities such as echocardiography, X-ray angiography, multispice spiral computed tomography (CT), and magnetic resonance imaging (MRI). Out of all these, MRI is most commonly used to image and quantify the function of the left ventricle due to its advantages of being noninvasive, radiation-free, and high quality resolution. The most important advantages of MRI are that it is nonintrusive and radiation-free. Second, it has high quality resolution as compared to other methods. Third, soft tissue can be easily distinguished in MRI images. A cardiac MRI image labeled with the left ventricle (LV), right ventricle (RV), and myocardial muscle (MYOCARDIUM) is shown in Fig. 2.

A number of methods are commonly used for LV segmentation in MRI images [1–4] of the heart such as edge detection, region-based segmentation, and pixel

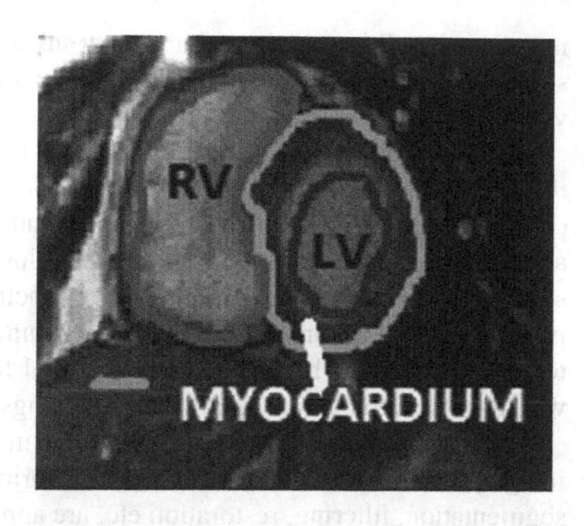

Fig. 2 Labeled MRI image showing LV, RV, and myocardium

classification. The method based on tracing edges is edge based detection [1]. The region-based method is based on pixel similarity of image regions [2]. There are various pixel-based classification methods like k-means, fuzzy c-means clustering, and graph cuts methods. These methods divide the pixels into groups based on pixel similarity measures such as intensity, texture, color or gray level [5–7]. In this research, the segmentation is done with the region-based method in the ITK-SNAP tool. The segmented binary image can then be used for deforming a generic template of the cardiac anatomy for generating a patient-specific computational model.

In our body, the four-chambered muscular organ is called the human heart. The human heart is situated behind the breastbone in the chest usually at par with the level of thoracic vertebrae. The pericardial sac is a double membrane that encloses the heart wall. Three layers of tissue comprise the heart wall. The outer one is designated as the epicardium, the one in the middle is known as the myocardium, and one inside is referred to as endocardium. The heart is divided into four chambers where we have upper two atria and lower two ventricles. The atria can be termed as receiving cavities whereas ventricles are described as discharging cavities. The function of two atria is to receive blood flowing through the veins. The function of two ventricles is to pump blood to the rest of the body. The differences in wall thickness of cardiac chambers are because of the variation in myocardium. Two kinds of blood flows in the body one is deoxygenated and other is termed as oxygenated. The deoxygenated blood is received by the right atria and oxygenated blood is received by left atria. The oxygenated blood goes to pulmonary veins as a medium for flowing.

In any pump we need valves so that fluid flows in unidirection. Similarly our heart has two types of valves, known as cuspid valves and semilunar valves. The valves which lie between the atria and ventricles are called cuspid or atrioventricular valves. The function of the cuspid walls is not to allow blood flowing back into the atrium. The other set is semilunar valves which are located at the lower edges of the large vessels quitting the ventricles and have a function of blocking blood flowing back into ventricles when ventricles tighten. The atrioventricular valve on the right is known as the tricuspid valve. The atrioventricular valve at the left is known as metrial valve. Between the pulmonary trunk and right ventricle lies the pulmonary semilunar valve. Between the aorta and the left ventricle lies the aortic semilunar valve.

The blood flow is conveniently described at right side then at the left side. Other thing which is to be noted is that both atria and ventricles have identical timing of contraction and expansion. To understand the function of heart we assume it as two pumps, with both the pumps operating in opposite directions but working simultaneously. First, we understand the function of first pump which functions to pump the blood streamed from the right atrium into the right ventricle, to the lungs to be oxygenated. The second pump is used for systemic circulation of the blood which streams to the left atrium from the lungs and then to the left ventricle.

The heart has a complex structure and in order to study its anatomy various image processing techniques are applied. Various stages of imaging system such as segmentation, filtering, restoration etc. are applied on anatomical structures mostly heart and brain. Apart from this various mathematical transforms are also used.

To comprehend the motion of the left ventricle and to model and conceptualize the elaborate three-dimensional motion and shape of the LV, segmentation of patient cardiac MRI is necessary [8–11]. The methodology section describes the approach used in this work.

2 Methodology

2.1 *Image Segmentation Algorithm*

Region-growing is a simple image segmentation method. It comes under pixel-based image segmentation method and this method requires initialization that is a pre seed point. Basic approach consists of starting with the seed point and the regions are grown by applying to neighbouring pixels having same properties such as specific ranges of gray levels. These properties may be defined under predefined criteria. When the predefined criteria is not given the procedure is to compute at every pixel same set of properties that will be used to assign pixels to regions during growing process. However if we use description alone it leads to misleading criteria such as connectivity and adjacency. The method actually constitutes two parts first determining the seed point and second determine criteria for growing the region. The basic formulation is:

1. $\cup_{i=1}^{n} R_i = R$
2. R_i is a connected region for all i=1,2,...,n
3. $R_i \cap R_j$ = NULL for all i=1,2,...,n
4. $P(R_i)$ = TRUE for all i=1,2,...,n
5. $P(R_i \cup R_j)$ = FALSE for any adjacent region R_i and R_j

Step 1 means that each and every pixel should be taken into account while doing segmentation. Step 2 means that the connectivity in the pixels should be there else it will lead to misleading results. Step 3 indicates that regions which are taken into account should not overlap. Step 4 indicates the properties which are predefined for the pixels. For example, $P(R_i)$ = True if all pixels in R_i have similar intensity values. Step 5 indicates that P, the regions R_i and R_j are non identical.

2.2 Model Reconstruction Procedure

The entire MRI image dataset was written into VTK format acceptable to the ITK-Snap tool which allows segmentation. ITK-SNAP is a user friendly software platform. The three-dimensional medical images can be visualized using this software. We can perform the delineation to obtain the object outline, and segment images automatically. The software also uses snakes for delineating anatomical structures in an image. A snake is also referred to as an energy minimizing deformable spline. ITK-SNAP is a user friendly software and is mostly used with magnetic resonance imaging (MRI) and computed tomography (CT) data sets. The software is an interactive software used on various platforms.

2.3 Image Navigation

Using the image navigation component,the orthogonal planes can be visualized in 3-D plane. The cut planes are linked by common cursor, such that moving the cursor in one plane updates other plane too. The cut planes are observed by dragging a mouse. Another high point of this software is an option of linked cursor which works over several windows sessions, making it possible to direct various artifacts.

2.4 Manual Segmentation

ITK-SNAP allows for segmentation of physical objects in images. Under manual segmentation we study the anatomy of the heart wall and manually delineate the object of interest. We can segregate the region using various tools.

2.5 Automatic Segmentation

Using the level set method ITK-SNAP provides automatic functionality segmentation. The level set functions make use of numerical methods to be applied on the surfaces. Even the deformed surfaces can be taken into account using this method.

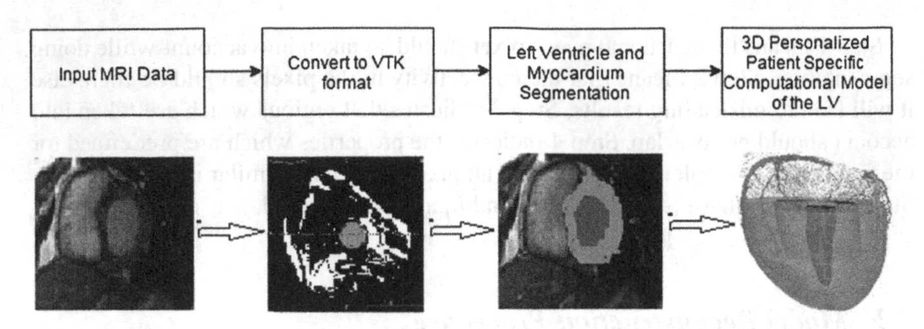

Fig. 3 Left ventricle and myocardium extraction, segmentation, and reconstruction flowchart diagram

ITK-SNAP is software which is written in C++. It can be downloaded free of cost at various platforms. It is provided with documentation and support and is user friendly. ITK-SNAP can work with many medical image formats, including DICOM, NIfTI, VTK, etc. It is a great help for researchers in field of medical imaging. The MRI images for patients are loaded in ITK-SNAP after first writing them in the VTK format in MATLAB. The images are then segmented and boundary is traced once they are read into ITK-SNAP in the VTK format. The flowchart of the work carried out is shown in Fig. 3.

2.6 Computational Model Generation

To accurately display clinical cardiac ventricular information, 3-D models can be used. The technique for creating a personalized patient-specific computational cardiac model is fusing a generic template mesh to the binary segmentation of the left ventricle. The least square circular fitting technique is used to deform the generic template of the left ventricle to the binary segmentation of the left ventricle. The flowchart of the work is shown in Fig. 4.

3 Results

Figure 5 shows that the left ventricle cavity has been segmented and myocardial muscle has been extracted. This procedure is rapid because it does not require any iterations which require more processing time and initialization from user intervention. Hence, the method presented in this paper can be performed on several MRI image frames for fast fully multi-frame segmentation.

The segmentation of the left ventricle is next saved as a binary 3-D image. This creates the patient's binary segmentation 3-D image of the left ventricle. The generic

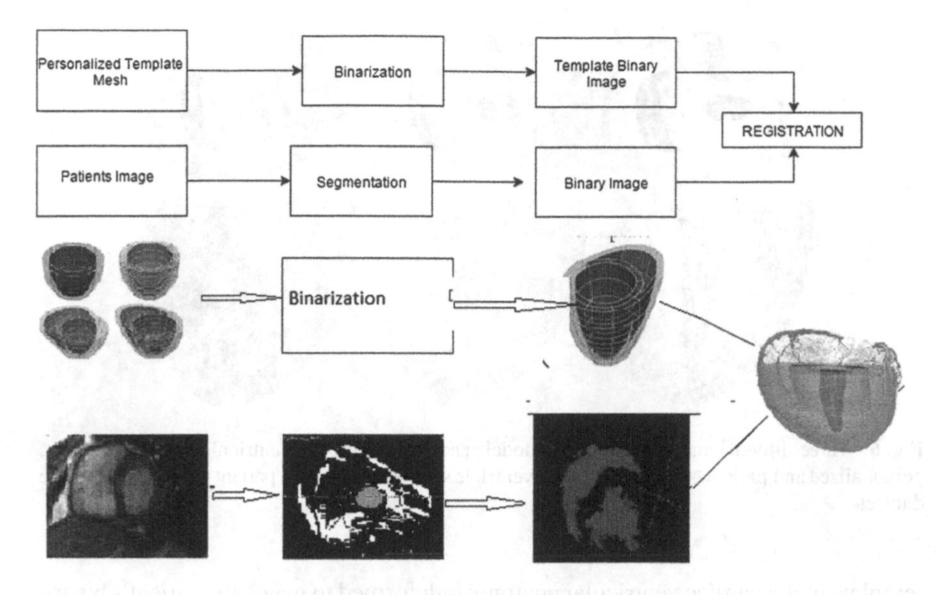

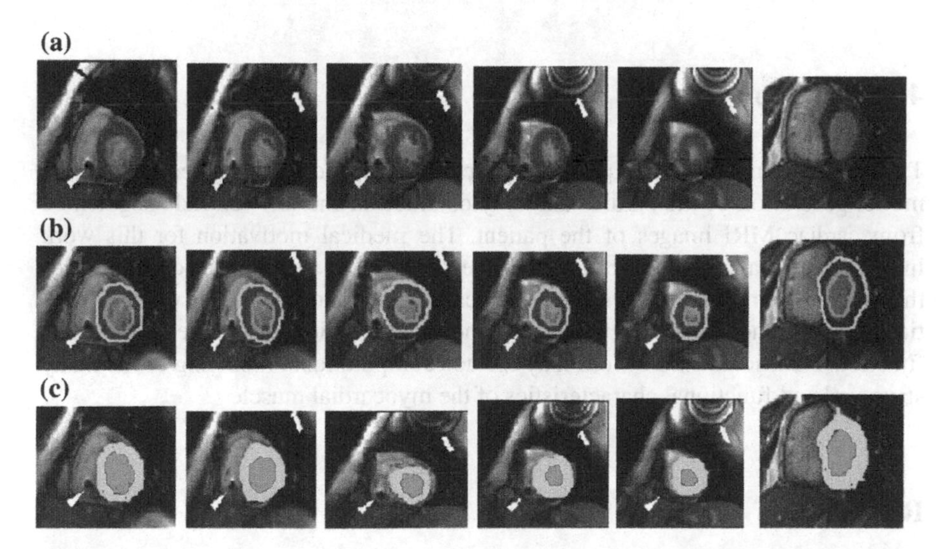

Fig. 4 Computational cardiac ventricular model reconstruction flowchart diagram

Fig. 5 Left ventricle and myocardium extraction: **a** *first row* shows the original MRI Images, **b** *second row* shows the endo-cardial and epi-cardial boundaries of the left ventricle, and **c** *third row* shows the colored segmented left ventricle and myocardium regions

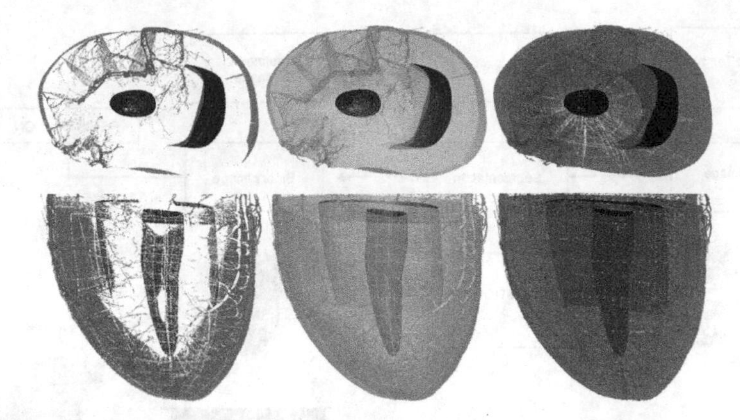

Fig. 6 Three-dimensional computational model generation of the left ventricular cardiac anatomy, personalized and patient-specific to the left ventricle segmentation from patient cardiac MRI image dataset

template of the cardiac ventricular anatomy is deformed to match the patient's binary segmentation 3-D image of the left ventricle. The results of the personalized patient-specific computational model of the cardiac anatomy are shown in Fig. 6 below.

4 Discussion

The research in this paper presents patient-specific personalized computational model generation of the cardiac anatomy obtained of the left ventricle segmented from cardiac MRI images of the patient. The medical motivation for this work includes personalized visualization of the heart muscle of a patient and calculation of the myocardium muscle thickness for clinical evaluation of the patient cardiac function. This parameter is very important for medical practitioners for assessing cardiac function. Additional future work may include computational modeling of the many structural and functional characteristics of the myocardial muscle.

References

1. Duncan, J.S., Smeulders, A., Lee, F., and Zaret, B.: Measurement of End Diastolic Shape Deformity Using Bending Energy. Computers in Cardiology, pp. 277–280 (1988)
2. von Schutthess, G.K.: The Effects of Motion and Flow on Magnetic Resonance Imaging. Morphology and Function in MRI, Ch. 3, pp. 43–62 (1989)
3. Goshtasby, A., Turner, D.A.: Segmentation Of Cardiac Cine MR Images For Extraction Of Right And Left Ventricular Chambers. IEEE Transactions on Medical Imaging, vol. 14, no. 1, pp. 56–64 (1995)

4. Vandenberg, B., Rath, L., Stuhlmuller, P., Melton, H., Skorton, D.: Estimation Of Left Ventricular Cavity Area With An On-Line, Semiautomated Echocardiographic Edge Detection System. Circulation, vol. 86, no. 1, pp. 159–166 (1992)
5. Mhlenbruch, G., Das, M., Hohl, C., Wildberger, J., Rinck, D., Flohr, T., Koos, R., Knackstedt, C., Gnther, R., Mahnken, A.: Global Left Ventricular Function In Cardiac CT. Evaluation Of An Automated 3D Region-Growing Segmentation Algorithm. Eur Radiol, vol. 16, no. 5, pp. 1117–1123 (2005)
6. Paragios, N.: A Level Set Approach For Shape-Driven Segmentation And Tracking Of The Left Ventricle. IEEE Transactions on Medical Imaging, vol. 22, no. 6, pp. 773–776 (2003)
7. Chen, C., Luo, J., Parker, K.: Image Segmentation Via Adaptive K-Mean Clustering And Knowledge-Based Morphological Operations With Biomedical Applications. IEEE Transactions on Image Processing, vol. 7, no. 12, pp. 1673–1683 (1998)
8. Cetitjean, C., Dacher, J.: A Review Of Segmentation Methods In Short Axis Cardiac MR Images. Medical Image Analysis, vol. 15, no. 2, pp. 169–184 (2011)
9. Lynch, M., Ghita, O., Whelan, P.: Automatic Segmentation Of The Left Ventricle Cavity And Myocardium In MRI Data. Computers in Biology and Medicine, vol. 36, no. 4, pp. 389–407 (2006)
10. Kaus, M., Berg, J., Weese, J., Niessen, W., Pekar, V.: Automated Segmentation Of The Left Ventricle In Cardiac MRI. Medical Image Analysis, vol. 8, no. 3, pp. 245–254 (2004)
11. Suri, J.: Computer Vision, Pattern Recognition and Image Processing in Left Ventricle Segmentation: The Last 50 Years. Pattern Analysis and Applications, vol. 3, no. 3, pp. 209–242 (2000)

A Cryptographic Key Generation on a 2D Graphics Using RGB Pixel Shuffling and Transposition

Londhe Swapnali, Jagtap Megha, Shinde Ranjeet, P.P. Belsare and Gavali B. Ashwini

Abstract Now a day with incredible change in social media network like mobile communication and computer, all type of a data, such as audio, video, images are used for the communication. Privacy for that data is an important issue. Cryptography is one of the techniques used for stopping unauthorized access and increasing integrity of that data. In research area encryption and decryption scheme is used based on image pixel shuffling and transposition. For security purpose, we can use cipher algorithm for generating key using RGB values of the pixel instead of using only pixel values. For that purpose in our proposed system we are using *m**n* size image on which different operations can be performed. Our proposed system is more secure as compare to existing system.

Keywords Cryptography · Encryption · Decryption · Cipher text · Pixel · 2D graphics image · Integrity · Key network

Londhe Swapnali (✉) · Jagtap Megha · Shinde Ranjeet · P.P. Belsare · G.B. Ashwini
Department of Computer Engineering, S. B. Patil College of Engineering,
Savitribai Phule Pune University, Pune, India
e-mail: tejswapn@gmail.com

Jagtap Megha
e-mail: jagtapmd95@gmail.com

Shinde Ranjeet
e-mail: ranjeetshinde307@gmail.com

P.P. Belsare
e-mail: pritambelsare@gmail.com

G.B. Ashwini
e-mail: dnyane.ash@gmail.com

S.C. Satapathy et al. (eds.), *Proceedings of the International Conference on Data Engineering and Communication Technology*, Advances in Intelligent Systems and Computing 469, DOI 10.1007/978-981-10-1678-3_18

1 Introduction

We cannot think about the world without communication. Nowadays hiding of data from unauthorized person is an important task. There are many online services present in which communication takes place through social media network. In that case there is high probability of hacking [1]. Integrity of a data is not maintained well.

Different techniques are used to maintain a security in all social networks communication. The techniques such as cryptography and stenography. Cryptography is the best technique to increase the security between communications. It is applied on a different type of data such as text, image, video, etc. In cryptography two different processes are there, first is encryption in which specific key is used to encrypt the data. It is a process of converting plain text into cipher text and second is decryption in which cipher text is converting into plain text. Two things are important for performing encryption and decryption, first is algorithm and second is key. Plain text is combined with the algorithm to form a key. This generated key is then used for the encryption process. This cipher text is applied to the algorithm for generating the plain text. Symmetric and asymmetric are the two approaches of encryption. In symmetric same cryptographic key is used for encryption and decryption but in decryption different key's are used for processing. In this approach keys are identical or combination of different keys.

Image is made up of different color pixels. Every pixel having different colors at a single point, so pixel is a single color dot. Digital approximation can be resulted from that color dot using this values reconstruction of that image take place. Digital image having two types first is color image made up of color pixels. Color pixel holds primary colors such as red, blue and green. This values proportion used for creating secondary colors. If each primary contain the 256 levels then four byte memory required for each. Bi-level image having single bit to represent each pixel. It's having only two states with color black and white [2].

In this paper we focused on key part. In image-based cryptographic technique, cipher algorithm of $m*n$ size image used for fetching RGB pixel values. Encryption and decryption of an $m*n$ size image is based on the RGB pixel values. Property of 2D graphics image is only viewing and listing image dimensions is sometimes impossible for generating the image.

Rest of the paper is organized as, Sect. 2 is related to related work. Sect. 3 gives information related to proposed system, Sect. 4 represent the advantages, Sect. 5 proposes experimental work with some mathematical part. Section 6 provides the overall summary of our paper with conclusion of the paper.

2 Related Work

Xiukun Li, Xianggian Wu*, Ning Qi, Kuanguan Wang has proposed a Novel Cryptographic algorithm containing Iris features nothing but the iris textural features. Its algorithm having Add/minus operational and read—Solomon error

Table 1 Analysis of existing system

Year of publication	Paper title	Drawback	Solution
2012	A novel cryptographic key generation method using image features	Create a novel algorithm for secure communication with image features	Image features and properties are used to key generation
2012	Cryptographic algorithm based on matrix and a shared secrete key	Image key generation using pixel	using pixel values key generation Techniques take place
2014	Color image encryption decryption using pixel shuffling with Henon chaotic system.	Generate a key for 2d graphic image	Scrambling the position of a pixel depend on 2D Henon chaotic system
2014	A new color-oriented cryptographic algorithm based on unicode and RGB color model	Encryption of a data using unicode data	Unicode uses RGB values for key generation

correcting. He defines the Region of Interest which used to differentiate different images [3]. Using another technique for a novel cryptographic some extension is given by a shujiangXu.yinglong Wang, Yucui Geo and Cong Wang. He uses a nonlinear chaotic map and means of XOR operations to encrypt a novel image. In which two rounds are proposed for encryptions [4] (Table 1).

Krishan, G.S, and Loganathan proposed a new scheme based on a visual cryptography in which binary images is used as the key input. The secret communicated images than decomposed into three monochromatic images based on YCbCr color space and this monochromatic images converted into binary images were encrypted using key called share—1 [5]. This technique is extended by christy and seenivasgam. He uses Back Propagation Network (BPN) to produce the two shares. Using this technique images has been produced having same features like original [6].

B. santhi, K.S. Ravichandran, A.P. Arun and L. Chakkarapani explains using images Features. Its uses Gray Level co-occurrence matrix of an image to extract the properties of an image [7]. Kester, QA proposed a cryptographic algorithm, which is based on a matrix and shared secrete key. He extends his research and use RGB Pixel to encrypt and decrypt the images [8].

RuisongYe and Wei Zhou proposed a chaos-based scheme for image encryption. In which chaotic orbits is constructed by using a 3d skew tent map with three control parameters. This generation is used to scramble the pixel positions. This approach having same good qualities such as sensitivity to initial control parameters pseudorandomness [9]. Extension to this approach for increasing the security purpose, AsiaMahdi Naseralzubaidi propose a encryption technique using pixel shuffling with Henon chaotic. He divides scrambled image into 64 blocks rotate each one in clockwise direction with go angle. For making distortion, two dimension Arnold Cat Mapping is applied and original position of pixel is reordered back to its original position [10].

AmneshGoel and Nidhi proposed a contrastive method. This method uses RGB values of pixels for rearranging the pixel within image than encryption take place [11].

Panchami V, Varghese Paul, Amithab Wahi proposed a techniques in which massage encoded using Unicode and encrypted using the RGB color. Hexadecimal values present for cipher color its distributes into two equal parts and 1st parts act as a message and 2nd is a key.

3 Proposed System

In previous existing system key generation is take place using the pixel values of the 2D graphics image. In which proposed algorithm used for extracting the pixel values as well as creating the key from pixel [1]. For increasing the security of image during the communication and transmission we propose new techniques. In last module we use RSA algorithm for encryption and same key is used for decryption process of 2D graphics image [2].

In this paper user having freedom to generate any type of pattern signature, etc. having fixed size of design pattern. Any image is made up of using different colors pixels. Each pixel having three components ass RGB. This RGB values first of all extracted and after shuffling we get cipher image. It is done only by using a RGB values.

In this method, we not use the bit values of pixel as well as pixel expansion at the end of the encryption and decryption process. Pixel having numeric values which interchanged or the RGB values displaced from their original position to create a cipher text. When we add all values in between image no change take place in original size and shape in image. All the features of an image remain unchanged during the process of encryption and decryption [2].

3.1 Architecture of Proposed System

Above Fig. 1 describes the architecture of encryption and decryption. We provides user interface for designing the pattern. According to cipher algorithm we fetch the all RGB values and manipulation take place using reshuffling, transposition techniques for generating key. Image decomposed into three components which act upon the cipher algorithm. That components form the all features and characteristics of original image. Within image boundaries all the RGB values are shuffled and interchanged, created array is different for all components.

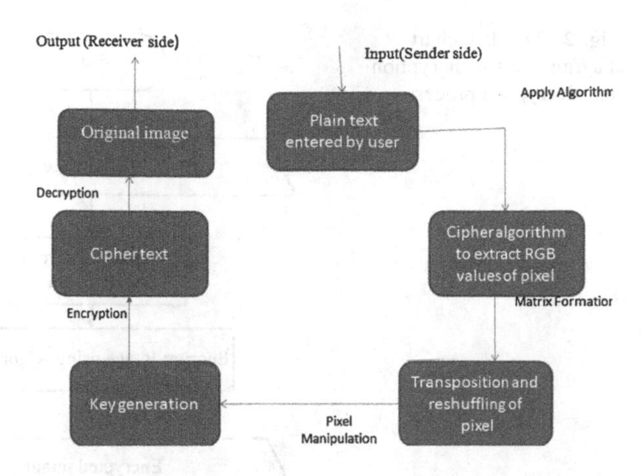

Fig. 1 Block diagram of proposed system

3.2 Flow of Proposed System

Figure 2 describes the flow of encryption and decryption process. In which first of all we accept the image, fetch all the RGB components using cipher algorithm and convert plain text into cipher text. We use RSA algorithm for the encryption process. In which key is combined with cipher text for performing encryption process.

Decryption process take place by importing all data and using reveres process of encryption. This process is done for generating original image with its shape and size. We can extract all pixel values only when image is given as an input. We can apply cipher algorithm having following steps.

1. Start
2. Import data and form image by interpreting each element.
3. Extract all r, g, b component from image.
4. Reshape all r, g, b component into one-dimensional array for each.
5. Let, $t = [y; 1; p]$ which is a column matrix.
6. Transpose 't'.
7. Reshape 't' into one dimensional array.
8. Let, $n =$ Total number of array.
9. Let, $r =$ (1st part of n): (1/3rd part of n) as one-dimensional array.
10. Let, $b =$ (1/3rd part of n): (2/3rd part of n) as one-dimensional array.
11. Let, $g =$ (2/3rd part of n): (nth) as one-dimensional array.
12. Transform it is with its original dimensions.
13. Finally all data will convert into an image.

For decrypt the image from cipher text to plain text inverse of algorithm is used.

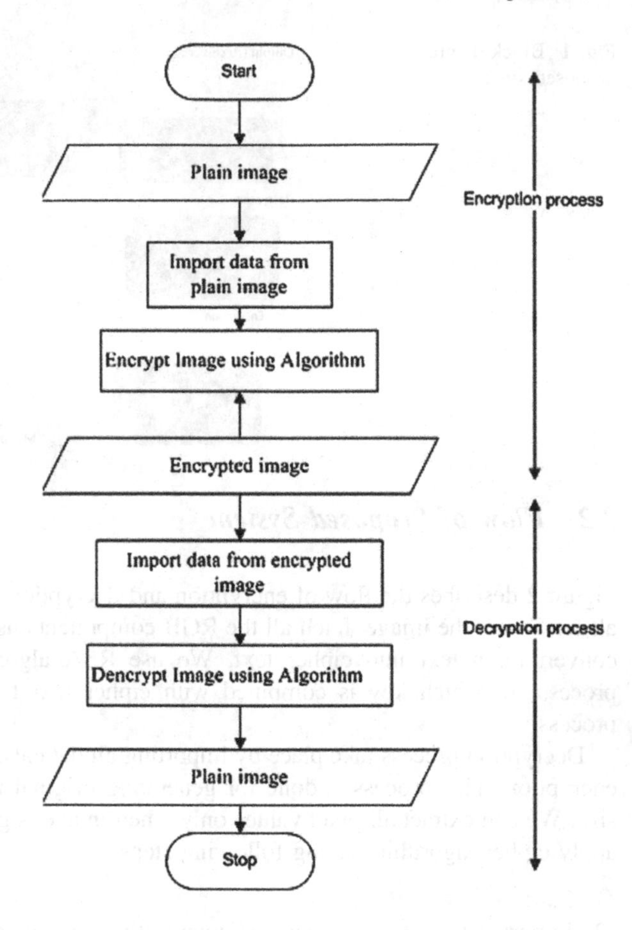

Fig. 2 The flow chart diagram for the encryption and decryption process

4 Advantages

In terms of security analysis transposition and reshuffling of the RGB values of 2D graphics is really effective. Security of image against all the attacks is increased due to extra swapping of component. Using RGB values we extract all the exact features of the image and it is quite easy to generate exact shape and size of the original image.

5 Future Scope

In existing system security is less as compared to our proposed system because we are using RGB pixel values as an input from the given image but existing system uses only pixel values of the given image. Similarly, our proposed system gives accurate and same image on receiver's side as that of a sender's side but in existing system there may be few modifications on receiver's side (Table 2)

Table 2 Comparative analysis/study of proposed system with existing system

	Input	Security	Image features
Existing system	Pixel values	Less	On receivers side there may be some changes in image after sending
Proposed system	RGB values	More	On receivers side there will not be any changes in image after sending

6 Conclusion

In this paper, we have proposed cryptographic key generation on a 2D graphics using RGB pixel shuffling and transposition. For our proposed system we use any size of image as an input. Small change in image creates a number of keys. In proposed system at each stage cipher algorithm generates different keys using the RGB pixel values on which different operations are performed. Existing system uses pixel values only but our proposed system extracts RGB pixel values of given image. This will be helpful for increasing security of given 2D graphics image. In future, we can combines different RGB values of pixel with each other and generate a new pattern for key generation management.

References

1. Pratik Shrivastava, Reteshjain, K.S. RaghuWanshi, A modified. Approach of key manipulation in cryptography using 2D graphics Image, Published in Proceeding ICESC '14 Proceedings of the 2014 International Conference on Electronic Systems, Signal Processing and Computing Technologies Pages 194–197, IEEE Computer Society Washington, DC, USA ©2014.
2. Xiukum Li, Xianggian Wu*, Ning Qi, KuanquanWang. A novel cryptographic Algorithm based on iris feature, 2008.
3. ShujiangXu, YinglongWang, YucuiGuo, Cong Wang, "A Novel Image Encryption Scheme based on a Nonlinear Chaotic Map", IJIGSP, vol. 2, no. 1, pp. 61–68, 2010.
4. Krishnan, G.S.; Loganathan, D.;, "Color image cryptography scheme based on visual cryptography," Signal Processing, Communication, Computing and Networking Technologies (ICSCCN), 2011 International Conference on, vol., no., pp. 404–407, 21–22 July 2011.
5. Christy, J.I.; Seenivasagam, V., "Construction of color Extended Visual Cryptographic scheme using Back Propagation Network for color images," Computing, Electronics and Electrical Technologies (ICCEET), 2012 International Conference on, vol., no., pp. 1101–1108, 21–22 March 2011.
6. B. Santhi, K.S. Ravichandran, A.P. Arun and L. Chakkarapani Novel Cryptographic Key Generation Method Using Image Features, 2012.
7. Kester, Quist-Aphetsi;, "A public-key exchange cryptographic technique using matrix," Adaptive Science & Technology (ICAST), 2012 IEEE 4th International Conference on, vol., no., pp. 78–81, 25–27 Oct. 2012.
8. RuisongYe, WeiZhou, "A Chaos-based Image Encryption Scheme Using 3D Skew Tent Map and Coupled Map Lattice", IJCNIS, vol. 4, no. 1, pp. 38–44, 2012.
9. Asia mahdiNaserAlzubaid, Cotor Image Encryption & decryption using pixel shuffling with Henon chaotic syste, 2014.

10. AmneshGoel, NidhiChandra, "A Technique for Image Encryption with Combination of Pixel Rearrangement Scheme Based On Sorting Group Wise Of RGB Values and Explosive Inter-Pixel Displacement", IJIGSP, vol. 4, no. 2, pp. 16–22, 2012.
11. Panchami V, Varghes Amithab Wahi, A New Color ORIENTED Cryptographic Algorithm Based on Unicode And RGB Color Model, 2014.

Methods for Individual and Group Decision Making Using Interval-Valued Fuzzy Preference Relations

B.K. Tripathy, Viraj Sahai and Neha Kaushik

Abstract Although a lot of research has been done in studying and exploiting, both, interval valued fuzzy sets and intuitionistic fuzzy sets, we still believe a lot more focus is still required on interval-valued fuzzy preference relations (IVFPR). In this paper, our focus shall be IVFPRs as we put forward two algorithms, based on the additive and multiplicative consistencies, to select the best alternative from a certain set of alternatives. Fuzzy preference relations have few advantages in preciseness and consistency as they give the freedom, to the experts, to evaluate the alternatives relatively. Further, we also extend our algorithms to group decision making and demonstrate their efficacy through numerical illustrations.

Keywords Interval-valued fuzzy preference relations · Group decision making · Individual decision making · Additive consistency · Multiplicative consistency

1 Introduction

Decision making is a process that we go through every day. No matter how different each decision-making scenario seems, there will always be a set of alternatives to choose from. A decision maker's ultimate realization of goal occurs with a successful choice of the best alternative. However, this is not as easy in reality. When Zadeh and Klaua first introduced their concept of fuzzy sets [1] in 1965, it was lacking in certain aspects- as the inability to express uncertainty of a decision maker's decision about an alternative. Zadeh's fuzzy sets used one fundamental

B.K. Tripathy (✉) · Viraj Sahai · Neha Kaushik
School of Computing Science and Engineering, VIT University, Vellore, Tamil Nadu, India
e-mail: tripathybk@vit.ac.in

Viraj Sahai
e-mail: virajsahai32@hotmail.com

Neha Kaushik
e-mail: nkaushik098@gmail.com

S.C. Satapathy et al. (eds.), *Proceedings of the International Conference on Data Engineering and Communication Technology*, Advances in Intelligent Systems and Computing 469, DOI 10.1007/978-981-10-1678-3_19

value, μ, as the membership function and 1 − μ was defined as the nonmembership function. As observed, there was no function to consider decision maker's abstinence in the process. In 1975, Zadeh extended his work to define interval valued fuzzy sets [2]. These sets defined the membership functions of fuzzy sets as a range of values having range of values with r^- and r^+ as the lower and upper range. They took into consideration the hesitancy factor for decision makers by using ranges to define the extent of certainty. Later in 1986, Atanassov proposed the concept of intuitionistic fuzzy sets [3] that also took into consideration the concept of hesitancy. However, these sets can still be imprecise as they ask the decision makers to rate alternatives in a more absolute sense. This can be overcome by utilizing fuzzy preference relations, intuitionistic, or interval. Here, we shall be focusing on IVFPRs and shall propose algorithms for decision making.

Section 2 establishes the basic definitions for building the concerned models. In Sects. 3 and 4, we propose the decision-making models for additive and multiplicative consistent IVFPR, respectively. Section 5 extends the concept to group decision-making problems. Section 6 will demonstrate the methods using a numerical. Section 7 marks the conclusion of the paper.

2 Preliminaries

Interval value fuzzy sets are defined as [2]:

$$A = \{(x_i),\ r^-(x_i),\ r^+(x_i) | x_i \in X\} \tag{1}$$

where

$$\begin{aligned} r^- &: X \to [0, 1], \\ r^+ &: X \to [0, 1], \\ r^- &\le r^+ \end{aligned}$$

Definition 2.1 [2]: An IVFPR, R on X is defined as a matrix $R = (r_{ij})_{n \times n}$ $X \times X$ where $r_{ij} = [r_{ij}^-, r_{ij}^+]$ for all i, $j = 1, 2, 3, \ldots, n$. Here, r_{ij} represents the interval-valued preference degree of x_i over x_j, such that:

$$\begin{aligned} & r_{ij}^+ \ge r_{ij}^- \ge 0, \\ & r_{ij}^- + r_{ij}^+ = r_{ij}^+ + r_{ij}^- = 1, \end{aligned}$$

$$\begin{aligned} & r_{ij}^- = r_{ij}^+ = 0.5 \\ & \text{for all } i, j = 1, 2, 3, \ldots, n. \end{aligned} \tag{2}$$

Definition 2.2 [1]: A fuzzy interval preference relation $R = (r_{ij})_{n\times n}$ where $r_{ij} = [r_{ij}^{-}, r_{ij}^{+}]$ is additive consistent if a normalized priority vector exists, $w = [w_1, w_2, w_3, \ldots, w_n]$ so that,

$$r_{ij}^{-} \le 0.5(w_i - w_j + 1) \le r_{ij}^{+},\ \text{for all } i = 1, 2, 3, \ldots, n-1;\ j = i+1, \ldots, n,$$
$$w_i \ge 0,\ i = 1, 2, 3, \ldots, n$$
$$\sum_{i=1}^{n} w_i = 1 \quad (3)$$

Definition 2.3 [1]: A fuzzy interval preference relation $R = (r_{ij})_{n\times n}$ is multiplicative consistent if a normalized priority vector exists, $w = [w_1, w_2, w_3, \ldots, w_n]$ so that,

$$r_{ij}^{-} \le \frac{w_i}{w_i + w_j} r_{ij}^{+},\ \text{for all } i = 1, 2, 3, \ldots, n-1;\ j = i+1, \ldots, n,$$
$$w_i \ge 0,\ i = 1, 2, 3, \ldots, n$$
$$\sum_{i=1}^{n} w_i = 1 \quad (4)$$

3 Priority Vector from Additive Consistent IVFPR

Let there be an interval-valued fuzzy preference relation $R = (r_{ij})_{n\times n}$ where $r_{ij} = [r_{ij}^{-}, r_{ij}^{+}]$. If additive consistency holds for $R = (r_{ij})_{n\times n}$ there will be some priority vector $w = [w_1, w_2, w_3, \ldots, w_n]^T$ that satisfy Eq. (3). However, this might not always be the case as the opinions of decision makers might be fairly subjective [1]. Thus, as suggested by Xu [4] we relax the Eq. (3) by introducing two non-negative real deviation variables d_{ij}^{-} and d_{ij}^{+}:

$$r_{ij}^{-} - d_{ij}^{-} \le 0.5(w_i - w_j + 1) r_{ij}^{+} + d_{ij}^{+},\ \text{for all } i = 1, 2, 3, \ldots, n-1;\ j = i+1, \ldots, n,$$
$$w_i \ge 0,\ i = 1, 2, 3, \ldots, n$$
$$\sum_{i=1}^{n} w_i = 1 \quad (5)$$

The smaller the values d_{ij}^{-} and d_{ij}^{+} the closer the interval-valued fuzzy preference R gets closer to becoming additive consistent.

Basing on his initial efforts of Xu [5] for intuitionistic FPR we establish a model for IVFPRs as:

$$\delta = \text{Min} \sum_{i=1}^{n-1} \sum_{j=i+1}^{n} \left(d_{ij}^{-} + d_{ij}^{+}\right)$$

such that,

$$
\begin{aligned}
&0.5(w_i - w_j + 1) + d_{ij}^{-} \geq r_{ij}^{-}\\
&0.5(w_i - w_j + 1) - d_{ij}^{+} \leq r_{ij}^{+}\\
&w_i \geq 0, = 1, 2, 3 \ldots, n\\
&\textstyle\sum_{i=1}^{n} w_i = 1\\
&d_{ij}^{-}, d_{ij}^{+} \geq 0\\
&\text{for all } i = 1, 2, 3, \ldots, n-1; j = i+1, \ldots, \qquad (\text{MOD}-1)
\end{aligned}
$$

For arriving at the optimal deviation values, d_{ij}^{-} and d_{ij}^{+}, we need to solve model (MOD-1). If $\delta = 0$ then $d_{ij}^{-} = d_{ij}^{+} = 0$, for all $i = 1, 2, 3, \ldots, n-1; j = i+1, \ldots, n$ and R is an additive consistent interval-valued fuzzy preference relation. If not, then we need to improve its consistency by using the nonzero deviation values:

$$
\begin{aligned}
\dot{R} &= (\dot{r}_{ij}),\\
\dot{r}_{ij} &= (\dot{r}_{ij}^{-}, \dot{r}_{ij}^{+}),\\
\dot{r}_{ij}^{-} &= \dot{r}_{ij}^{-} - d_{ij}^{-},\\
\dot{r}_{ij}^{+} &= \dot{r}_{ij}^{+} + d_{ij}^{+}
\end{aligned} \tag{6}
$$

Considering the above improved preference relation, we use the following optimization models to calculate the priority vector:

$$w_i^{-} = \min w_i$$

such that,

$$
\begin{aligned}
&0.5(w_i - w_j + 1) \geq \dot{r}_{ij}^{-}\\
&0.5(w_i - w_j + 1) \leq \dot{r}_{ij}^{+}\\
&w_i \geq 0,\ i = 1, 2, 3, \ldots, n\\
&\textstyle\sum_{i=1}^{n} w_i = 1\\
&\text{for all } i = 1, 2, 3, \ldots, n-1; j = i+1, \ldots, n \qquad (\text{MOD}-2)
\end{aligned}
$$

$$w_i^{+} = \max w_i$$

such that,

$$
\begin{aligned}
&0.5(w_i - w_j + 1) \geq \dot{r}_{ij}^{-}\\
&0.5(w_i - w_j + 1) \leq \dot{r}_{ij}^{+}\\
&w_i \geq 0,\ i = 1, 2, 3, \ldots, n\\
&\textstyle\sum_{i=1}^{n} w_i = 1\\
&\text{for all } i = 1, 2, 3, \ldots, n-1; j = i+1, \ldots, n \qquad (\text{MOD}-3)
\end{aligned}
$$

Solving the above two models, namely (MOD-2) and (MOD-3), we can calculate the priority vector intervals $[w_i^-, w_i^+]$. If both happen to be equal, one unique weight vector can be calculated and defined as $w = [w_1, w_2, w_3, \ldots, w_n]^T$.

4 Priority Vector from Multiplicative Consistent IVFPR

Similarly, using already established equations and relaxing the definition for multiplicative consistency we get the following as before:

$$r_{ij}^- - d_{ij}^- \le \frac{w_i}{w_i + w_j} \le r_{ij}^+ + d_{ij}^+, \text{ for all } i = 1, 2, 3, \ldots, n-1;\ j = i+1, \ldots, n,$$
$$w_i \ge 0, i = 1, 2, 3, \ldots, n$$
$$\sum_{i=1}^n w_i = 1 \tag{7}$$

$$\delta^* = \text{Min} \sum_{i=1}^{n-1} \sum_{j=i+1}^{n} \left(d_{ij}^- + d_{ij}^+\right)$$

such that,

$$\begin{aligned} &\frac{w_i}{w_i + w_j} + d_{ij}^- \ge r_{ij}^- \\ &\frac{w_i}{w_i + w_j} - d_{ij}^+ \le r_{ij}^+ \\ &w_i \ge 0,\ i = 1, 2, 3, \ldots, n \\ &\sum_{i=1}^n w_i = 1 \\ &d_{ij}^-, d_{ij}^+ \ge 0 \\ &\text{for all } i = 1, 2, 3, \ldots, n-1; j = 1 + 1, \ldots, n \end{aligned} \tag{MOD – 4}$$

$$\begin{aligned} \dot{R} &= (\dot{r}_{ij}), \\ \dot{r}_{ij} &= (\dot{r}_{ij}^-, \dot{r}_{ij}^+), \\ \dot{r}_{ij}^- &= r_{ij}^- - d_{ij}^-, \\ \dot{r}_{ij}^+ &= r_{ij}^+ + d_{ij}^+ \end{aligned} \tag{8}$$

$$w_i^- = \min w_i$$

such that,

$$\frac{w_i}{w_i + w_j} \geq \dot{r}_{ij}^{-}$$
$$\frac{w_i}{w_i + w_j} \leq \dot{r}_{ij}^{+}$$
$$w_i \geq 0,\ i = 1,\ 2,\ 3,\ \ldots,\ n$$
$$\sum_{i=1}^{n} w_i = 1$$
$$\text{for all } i = 1,\ 2,\ 3,\ \ldots, n-1;\ j = i + 1,\ \ldots,\ n \qquad \text{(MOD} - 5\text{)}$$

$$w_i^{+} = \max w_i$$

such that,

$$\frac{w_i}{w_i + w_j} \geq \dot{r}_{ij}^{-}$$
$$\frac{w_i}{w_i + w_j} \leq \dot{r}_{ij}^{+}$$
$$w_i \geq 0,\ i = 1,\ 2,\ 3,\ \ldots,\ n$$
$$\sum_{i=1}^{n} w_i = 1$$
$$\text{for all } i = 1,\ 2,\ 3,\ \ldots, n-1; j = i + 1,\ \ldots,\ n \qquad \text{(MOD} - 6\text{)}$$

5 Extension to Group Decision Making

If we have the individual IVFPRs for, say, m experts and if we know the weight vector for these experts as $\lambda_k = [\lambda_1,\ \lambda_2,\ \lambda_3,\ \ldots,\ \lambda_m]^T$ we can easily extend the models by aggregating the IVFPRs as:

$$\begin{aligned} \bar{r}_{ij}^{-} &= \sum_{k=1}^{m} \lambda_k \left(r_{ij}^{-}\right)^{(k)} \\ \bar{r}_{ij}^{+} &= \sum_{k=1}^{m} \lambda_k \left(r_{ij}^{+}\right)^{(k)} \\ \bar{r}_{ij} &= \left[\bar{r}_{ij}^{-}, \bar{r}_{ij}^{+}\right] \\ \bar{R} &= \left(\bar{r}_{ij}\right) n \times n \end{aligned} \tag{9}$$

Using the above obtained equations, the set of preference relations of all the experts can be aggregated into one interval-valued fuzzy preference matrix. Now all the three models (MOD-1), (MOD-2), and (MOD-3) can be applied, to get:

$$\delta = \text{Min} \sum_{i=1}^{n-1} \sum_{j=i+1}^{n} \left(d_{ij}^{-} + d_{ij}^{+}\right)$$

such that,

$$\begin{aligned}
&0.5\left(w_i - w_j + 1\right) + d_{ij}^{-} \geq \bar{r}_{ij}^{-} \\
&0.5\left(w_i - w_j + 1\right) - d_{ij}^{+} \leq \bar{r}_{ij}^{+} \\
&w_i \geq 0,\ i = 1, 2, 3, \ldots, n \\
&\textstyle\sum_{i=1}^{n} w_i = 1 \\
&d_{ij}^{-},\ d_{ij}^{+} \geq 0 \\
&\text{for all } i = 1, 2, 3, \ldots, n-1;\ j = i+1, \ldots, n
\end{aligned} \qquad (\text{MOD} - 7)$$

For determining the optimal deviation values for d_{ij}^{-} and d_{ij}^{+}, we need to solve model (MOD-7). If $\delta = 0$ then $d_{ij}^{-} = d_{ij}^{+} = 0$ and $\bar{R}$ is an additive consistent interval-valued fuzzy preference relation. If not, then we need to improve its consistency as before:

$$\begin{aligned}
\dot{R} &= \left(\dot{r}_{ij}\right), \\
\dot{r}_{ij} &= \left(\dot{r}_{ij}^{-},\ \dot{r}_{ij}^{+}\right), \\
\dot{r}_{ij}^{-} &= \bar{r}_{ij}^{-} - d_{ij}^{-}, \\
\dot{r}_{ij}^{+} &= \bar{r}_{ij}^{+} + d_{ij}^{+}
\end{aligned} \qquad (10)$$

Use the following optimization models to calculate the priority vector:

$$\bar{w}_i^{-} = \min w_i$$

such that,

$$\begin{aligned}
&0.5\left(w_i - w_j + 1\right) \geq \dot{r}_{ij}^{-} \\
&0.5\left(w_i - w_j + 1\right) \leq \dot{r}_{ij}^{+} \\
&w_i \geq 0,\ i = 1, 2, 3, \ldots, n \\
&\textstyle\sum_{i=1}^{n} w_i = 1 \\
&\text{for all } i = 1, 2, 3, \ldots, n-1;\ j = i + 1, \ldots, n
\end{aligned} \qquad (\text{MOD} - 8)$$

$$\bar{w}_i^{+} = \max w_i$$

such that,

$$\begin{aligned}&0.5(w_i - w_j + 1) \le \dot{r}_{ij}^{\;-}\\&0.5(w_i - w_j + 1) \le \dot{r}_{ij}^{\;+}\\&w_i \ge 0,\ i = 1, 2, 3, \ldots, n\\&\textstyle\sum_{i=1}^{n} w_i = 1\\&\text{for all } i = 1, 2, 3, \ldots, n-1;\ j = i+1, \ldots, n\end{aligned} \qquad \text{(MOD – 9)}$$

The solution from (MOD-8) and (MOD-9) determines the priority weight vector as $\bar{w} = [\bar{w}_1, \bar{w}_2, \bar{w}_3, \ldots, \bar{w}_n]^T$ for the alternatives.

Similarly for multiplicative consistent IVFPR, following the same approach, we get:

$$\begin{aligned}&\frac{w_i}{w_i + w_j} + d_{ij}^{\;-} \ge \bar{r}_{ij}^{\;-}\\&\frac{w_i}{w_i + w_j} - d_{ij}^{\;+} \le \bar{r}_{ij}^{\;+}\\&w_i \ge 0,\ i = 1, 2, 3, \ldots, n\\&\textstyle\sum_{i=1}^{n} w_i = 1\\&d_{ij}^{\;-}, d_{ij}^{\;+} \ge 0\\&\text{for all } i = 1, 2, 3, \ldots, n-1;\ j = i+1, \ldots, n\end{aligned} \qquad (11)$$

$$\delta = \text{Min} \sum_{i=1}^{n-1} \sum_{j=i+1}^{n} \left(d_{ij}^{\;-} + d_{ij}^{\;+}\right)$$

such that,

$$\begin{aligned}&\frac{w_i}{w_i + w_j} + d_{ij}^{\;-} \ge \bar{r}_{ij}^{\;-}\\&\frac{w_i}{w_i + w_j} - d_{ij}^{\;+} \le \bar{r}_{ij}^{\;+}\\&w_i \ge 0,\ i = 1, 2, 3, \ldots, n\\&\textstyle\sum_{i=1}^{n} w_i = 1\\&d_{ij}^{-}, d_{ij}^{+} \ge 0\\&\text{for all } i = 1, 2, 3, \ldots, n-1; j = i+1, \ldots, n\end{aligned} \qquad \text{(MOD – 10)}$$

$$\begin{aligned}\dot{R} &= (\dot{r}_{ij}),\\\dot{r}_{ij} &= (\dot{r}_{ij}^{\;-}, \dot{r}_{ij}^{\;+}),\\\dot{r}_{ij}^{\;-} &= \dot{r}_{ij}^{\;-} - d_{ij}^{\;-},\\\dot{r}_{ij}^{\;+} &= \dot{r}_{ij}^{\;+} + d_{ij}^{\;+}\end{aligned} \qquad (12)$$

$$\bar{w}_i^{\;-} = \min w_i$$

such that,

$$\begin{aligned}
&\frac{w_i}{w_i + w_j} \geq \dot{r}_{ij}^{-}\\
&\frac{w_i}{w_i + w_j} \leq \dot{r}_{ij}^{+}\\
&w_i \geq 0,\ i = 1, 2, 3, \ldots, n\\
&\textstyle\sum_{i=1}^{n} w_i = 1\\
&\text{for all } i = 1, 2, 3, \ldots, n-1;\ j = i+1, \ldots, n
\end{aligned} \qquad \text{(MOD} - 11\text{)}$$

$$\bar{w}_i^{+} = \max w_i$$

such that,

$$\begin{aligned}
&\frac{w_i}{w_i + w_j} \geq \dot{r}_{ij}^{-}\\
&\frac{w_i}{w_i + w_j} \leq \dot{r}_{ij}^{+}\\
&w_i \geq 0,\ i = 1, 2, 3, \ldots, n\\
&\textstyle\sum_{i=1}^{n} w_i = 1\\
&\text{for all } i = 1, 2, 3, \ldots, n-1;\ j = i+1, \ldots, n
\end{aligned} \qquad \text{(MOD} - 12\text{)}$$

The solution from (MOD-11) and (MOD-12) determines the priority weight vector as $\bar{w} = [\bar{w}_1, \bar{w}_2, \bar{w}_3, \ldots, \bar{w}_n]^T$ for the alternatives.

6 Numerical Illustration

Let us take one example for an IVFPR and try to solve it using the proposed models (MOD-1)-(MOD-3).

Example There is a multiple criteria decision-making problem having five criteria of measure namely x_i (i = 1, 2, 3, 4, 5). The IVFPR matrix is as follows:

$$R = \left(r_{ij}\right)_{n \times n} = \begin{bmatrix} [0.5, 0.5] & [0.6, 0.7] & [0.3, 0.7] & [0.7, 0.8] & [0.4, 0.5] \\ [0.35, 0.4] & [0.5, 0.5] & [0.5, 0.7] & [0.6, 0.9] & [0.3, 0.4] \\ [0.2, 0.5] & [0.3, 0.5] & [0.5, 0.5] & [0.6, 0.8] & [0.4, 0.5] \\ [0.2, 0.3] & [0.1, 0.4] & [0.2, 0.4] & [0.5, 0.5] & [0.35, 0.4] \\ [0.5, 0.65] & [0.6, 0.7] & [0.5, 0.6] & [0.6, 0.7] & [0.5, 0.5] \end{bmatrix}$$

After solving the above with (MOD-1), we get, δ = 0.075. Since δ is not equal to 0, the interval valued fuzzy preference relation is additive inconsistent and the optimum deviation values are as follows:

$$d_{14}^{-} = 0.025,\ d_{24}^{-} = 0.025,\ d_{34}^{-} = 0.025$$

Apply (MOD-2) and (MOD-3) and get the following:

$$w_1^{-} = w_1^{+} = 0.30,\ w_2^{-} = w_2^{+} = 0.25,\ w_3^{-} = w_3^{+} = 0.20,$$
$$w_4^{-} = w_4^{+} = 0.25,\ w_5^{-} = w_5^{+} = 0$$

Therefore, we have found a unique priority vector $w = [[0.30, 0.30], [0.25, 0.25], [0.20, 0.20], [0.25, 0.25], [0, 0]]^T$. Arranging these priorities in decreasing order we get the rank of the alternative as $x_1 \succ x_2 \sim x_4 \succ x_3 \succ x_5$.

7 Conclusion

There is still scope for developing algorithms using transitivity property of IVFPRs. Also, the algorithm can be improved for missing information case. As from our end, we have tried to propose a linear model, as well as, a nonlinear model for decision making IVFPRs and extended it to encompass group decision making. These may be used for individual and group decision making with a wide area of application in the supply chain management, risk assessment and prioritizing, natural disaster prediction system or personnel evaluation system.

References

1. Z.S. Xu, Consistency of interval fuzzy preference relations in group decision making, Appl. Soft Comput. 11 (2011) 3898–3909.
2. H. Bustince, J. Montero, M. Pagola, E. Barrenechea, D. Gómez, A survey of IVFS, in: W. Pedrycz (Ed.), Handbook of Granular Computing, Wiley, New Jersey, 2008.
3. K. Atanassov, Intuitionistic fuzzy sets, Fuzzy Sets Syst. 20 (1986) 87–96.
4. Bustince H, Burillo P (1995) Correlation of interval-valued intuitionistic fuzzy sets. Fuzzy Sets and Systems 74: 237–244.
5. Z.S. Xu, A method for estimating criteria weights from intuitionistic preference relations, Fuzzy Inform. Eng. (ICFIE), ASC (2007) 503–512.

A New Approach to Determine Tie-Line Frequency Bias (B) in Interconnected Power System with Integral Control AGC Scheme

Charudatta Bangal

Abstract This work is of investigative type to suggest a new approach for determining appropriate value of tie-line frequency bias '*B*' for an interconnected power system. The present discussion is not for challenging the conventional trends of selecting values of '*B*', however, simulations of single and two area power systems with integral control AGC scheme have been carried out using Matlab and Simulink. Exhaustive investigation has been carried out so as to correlate the parameter '*B*' with governor regulation '*R*,' power system constant '*Kp*' and integral controller gain '*K*.' These investigations are limited to (i) making certain remarks on *B* and *R*, (ii) determining relation between *B* and *Kp* under wide range of operating conditions of power system and (iii) suggesting a possible way in which the AGC scheme can be made to determine appropriate value of '*B*' on its own in real time as per prevailing load situations so as to give satisfactory overall performance.

Keywords Automatic generation control (AGC) • Area frequency response characteristic (AFRC) • Interconnected power systems • Tie-line frequency bias parameter

AFRC Area Frequency Response Characteristic
AGC Automatic Generation Control
GRC Generation Rate Constraint
Δf_1 Deviation in frequency of area 1
Δf_2 Deviation in frequency of area 2
P_D Prevailing load on power system
P_r Rated capacity of power system
ΔP_{D1} Load perturbation of area 1
ΔP_{D2} Load perturbation of area 2
ΔP_{tie} Perturbation in tie-line power

Charudatta Bangal (✉)
Sinhgad Technical Education Society, Pune, India
e-mail: charudatta_bangal@yahoo.com

S.C. Satapathy et al. (eds.), *Proceedings of the International Conference on Data Engineering and Communication Technology*, Advances in Intelligent Systems and Computing 469, DOI 10.1007/978-981-10-1678-3_20

1 Introduction

In AGC studies of interconnected (multi-area) power system, the question of selecting most appropriate value of tie-line frequency bias parameter '*B*' is very important and the same has been much hotly and controversially discussed in all the past years [1, 2]. From most of the literature related to AGC of interconnected power system, it is evident that the value of '*B*' is conventionally taken equal to the area frequency response characteristic 'β' for certain reasons [1–9].

i.e.,

$$B = \beta = \frac{1}{Kp} + \frac{1}{R}$$

Where '*Kp*' is the power system constant and '*R*' is the governor regulation (droop). Further, in majority of the related literature and published papers, the analysis of interconnected system is done with the help of models which almost always assume $Kp = 120$ and $Tp = 20$ [8–14].

By definition, $Kp = \frac{1}{D}$, where, $D = \frac{\partial P_D}{\partial f}$. '*D*' is the rate of change of existing (prevailing) load with change in prevailing value of frequency. For example, if at a certain instant the prevailing load on a power system is 50 % of its rated capacity (i.e., when $P_D = 0.5P_r$) with a normal frequency of 60 Hz, then any addition or removal of load on the system (ΔP_D) at this instant will cause $D = \frac{0.5}{60} = 0.008333$ pu MW/Hz and value of *Kp* at this instant will be $K_P = \frac{1}{D} = 120$ Hz/pu MW. Also, by definition, the power system time constant is given by; $Tp = \frac{2H}{fD}$, where $H =$ inertia constant, which is usually taken as 5 s [5, 7]. For the present case, $Tp = 20$ s. Thus, it is evident that, the values $Kp = 120$ and $Tp = 20$ correspond to a specific condition, i.e., when the power system is operating at 50 % of its rated capacity.

However, in practice, the load perturbation can occur at any random operating condition of power system. Hence, the values of Kp and Tp solely depend on amount of prevailing load and frequency at the time of perturbation. Thus, if the prevailing load is assumed to be at any value in the range of *Pr* to 0.1 *Pr*, the

Table 1 Some values of *Kp* and *Tp* in entire operating range

Prevailing load on power system (MW)	Kp (Hz/pu MW)	Tp (second)
1.0 Pr	60	10
0.9 Pr	66.6666	11.1111
0.8 Pr	75	12.5
0.7 Pr	85.7143	14.2857
0.6 Pr	100	16.6666
0.5 Pr	120	20
0.4 Pr	150	25
0.3 Pr	200	33.3333
0.2 Pr	300	50
0.1 Pr	600	100

corresponding values of *Kp* will be in the range of 60–600 Hz/pu MW and corresponding values of *Tp* will be in the range of 10–100 s. as given in "Table 1."

It is therefore necessary to consider appropriate values of *Kp* and *Tp* while studying the behavior of interconnected power system under AGC scheme. In present discussion, the term 'satisfactory overall performance' is limited only to initial overshoot in dynamic responses of Δf_1 and Δf_2, final steady-state values of Δf_1 and Δf_2 and final steady-state value of tie-line power ΔP_{tie}. The exhaustive simulations were carried out using Matlab and Simulink.

2 Procedure

As a first step, a single area power system using non-reheat turbine along with integral control AGC scheme was simulated. The simulation model is shown in "Fig. 1" [5, 7].

The Regulation '*R*' was adjusted to 4 % pu, i.e., 2.4 Hz/pu MW. The load perturbation 'ΔP_D' was assumed to be 1 % of Pr, i.e., 0.01 pu. Various prevailing load conditions were assumed as shown in "Table 1." By setting values of *Kp* and *Tp*, every time the integral controller gain '*K*' was adjusted to critical level i.e., to a value such that the dynamic response of Δf just avoided positive overshoot. These values of *K* were recorded and averaged to get a single value of *K*. In this case, i.e., for 4 % regulation, the average value of *K* was found to be 0.2025.

After determining value of *K* as above, an interconnected system comprising of two identical areas was simulated. The simulation model is shown in "Fig. 2."

The integral controller gains for both areas, i.e., K_1 and K_2 were set to 0.2025. The regulations for both areas, i.e., R_1 and R_2 were kept at same value, i.e., 2.4 Hz/pu MW. The load perturbations for two areas were assumed as $\Delta P_{D1} = 0.01$ pu and $\Delta P_{D2} = 0$.

For different operating states as shown in "Table 1," the values of *Kp* (ranging from 60 to 600) and corresponding *Tp* (ranging from 10 to 100) were set and every time the value of tie-line frequency bias parameter '*B*' was so adjusted that final

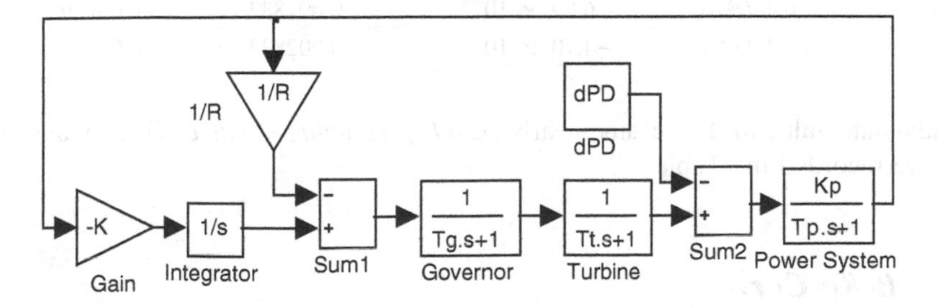

Fig. 1 Single area power system simulated in Matlab and Simulink

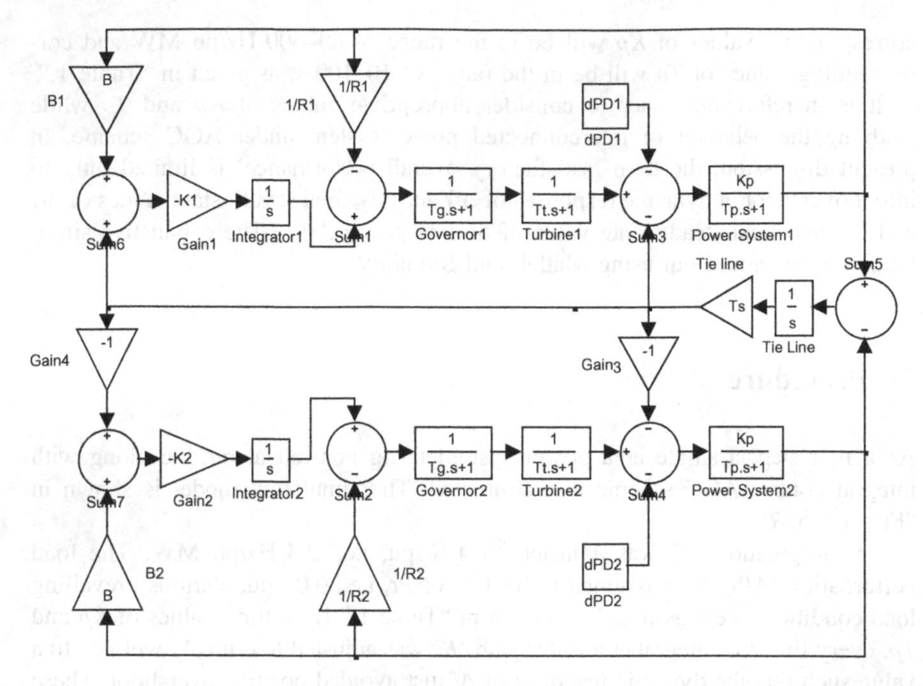

Fig. 2 Two area power system simulated in Matlab and Simulink

Table 2 Values obtained for '*B*' for various values of *Kp* and *Tp* for R = 2.4 Hz/pu MW

Kp	B	Δf_1 (p.u.)	Δf_2 (p.u.)	ΔP_{tie} (p.u.)
60	0.424477	-1.18×10^{-9}	−0.002831	−0.001211
66.6666	0.423530	-2.64×10^{-9}	−0.002831	−0.001209
75	0.422585	-5.20×10^{-9}	−0.002832	−0.001208
85.7143	0.421643	-4.31×10^{-9}	−0.002832	−0.001206
100	0.420704	-9.15×10^{-10}	−0.002833	−0.001204
120	0.419767	-2.17×10^{-9}	−0.002833	−0.001202
150	0.418833	-3.31×10^{-9}	−0.002833	−0.001200
200	0.417903	-1.16×10^{-9}	−0.002833	−0.001198
300	0.416976	-6.09×10^{-9}	−0.002833	−0.001196
600	0.416052	-4.70×10^{-9}	−0.002833	−0.001194

steady-state value of Δf_1 became nearly zero *before going positive*. These values of *B* were recorded in "Table 2".

2.1 B-Kp Curve

The values of *B* when plotted against the respective values of *Kp*, gave a smooth hyperbolic curve as shown in "Fig. 3"

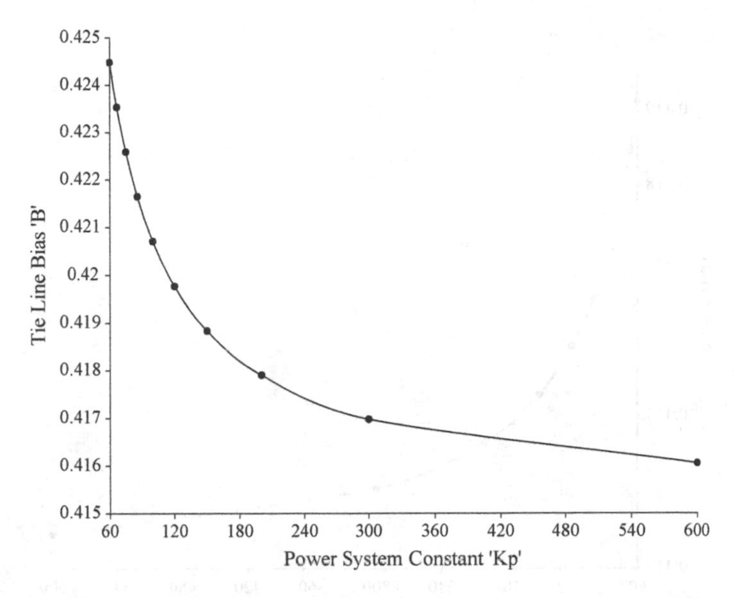

Fig. 3 Variation of tie-line frequency bias '*B*' with power system constant '*Kp*' at governor regulation value of 2.4 Hz/pu MW and controller gain of 0.2025

2.2 B-Kp Curve for Various Other Values of Regulation

The entire procedure as mentioned above was repeated for regulation of 3 % p.u. and 5 % p.u., i.e., 1.8 Hz/pu MW and 3 Hz/pu MW, respectively. The average integral controller gains were calculated as, 0.2383 and 0.1685, respectively. In both cases, the variation of *B* with *Kp* was again observed to be smooth hyperbolic curves. These curves are shown in "Figs. 4 and 5," respectively. "Figure 6" shows the curves for all the three cases on a common scale.

2.3 Equations for 'B'

Using curve fitting techniques, the equations for hyperbolic functions were determined for the above three cases. The hyperbolic function was assumed to be of a form $B = \frac{C_1}{Kp} + C_2$, where C_1 and C_2 are constants. The equations obtained for the three cases as above are shown in "Table 3." It should be noted that, the values of C_1 and C_2 are calculated approximately, however, they can be calculated more accurately using appropriate mathematical techniques.

The two area model was then investigated with conventional values of *B* (i.e. with $B = \beta$) and comparison was made between the performances obtained in two cases.

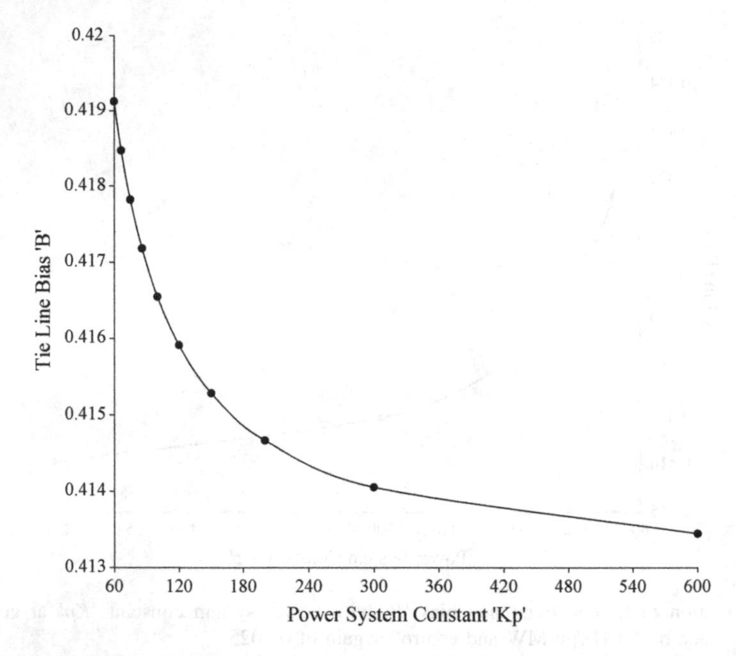

Fig. 4 Variation of tie-line frequency bias '*B*' with power system constant '*Kp*' at a governor regulation value of 1.8 Hz/pu MW and controller gain of 0.2383

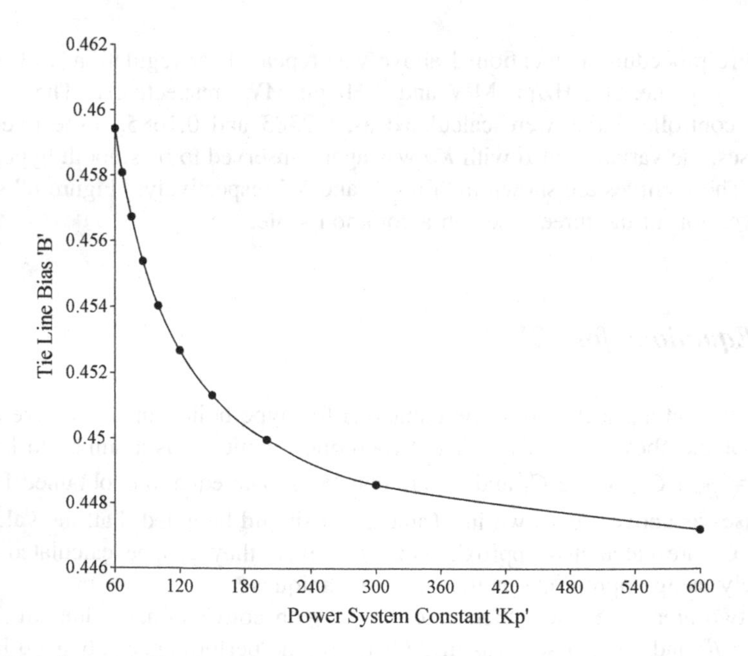

Fig. 5 Variation of tie-line frequency bias '*B*' with power system constant '*Kp*' at a governor regulation value of 3.0 Hz/pu MW and controller gain of 0.1685

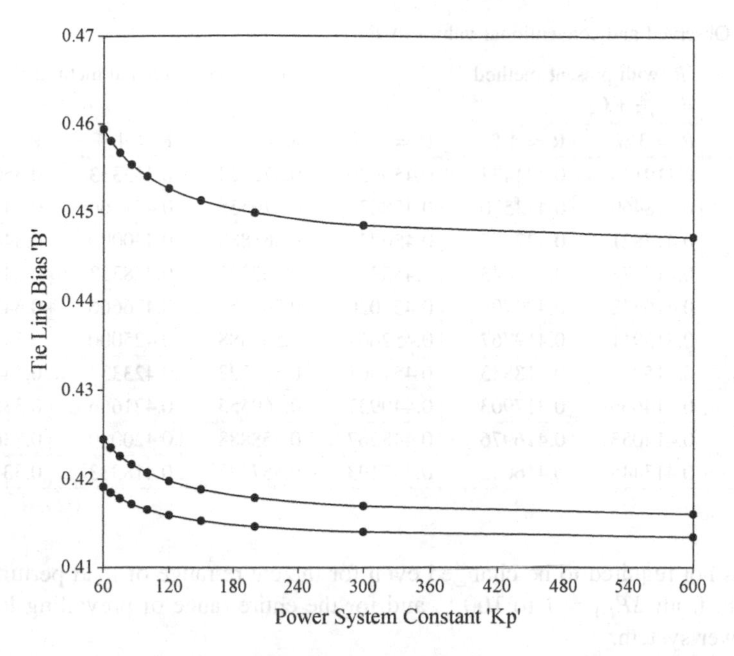

Fig. 6 Variation of 'B' with 'Kp' for R = 1.8, 2.4, and 3.0 on common scale. The lower curve corresponds to R = 1.8, middle to R = 2.4 and the upper corresponds to R = 3.0 Hz/pu MW

Table 3 Equations for 'B' at various values of regulation

Regulation	Equation for B
3 % p.u. (1.8 Hz/pu MW)	$B = \frac{0.37853}{Kp} + 0.41281$
4 % p.u. (2.4 Hz/pu MW)	$B = \frac{0.56166}{Kp} + 0.41511$
5 % p.u. (3.0 Hz/pu MW)	$B = \frac{0.81513}{Kp} + 0.44583$

The values of B determined with this method and the conventional values of B are tabulated in "Table 4".

3 Observations and Concluding Remarks

(1) With conventional value of B (i.e., when $B = \beta$), every time when ΔP_{D1} was changed, the integral controller gain was required to be changed for getting comparable results for Δf_1, Δf_2, and ΔP_{tie}, which could be difficult to achieve in practical systems, whereas, with the values of B as determined in present method, the integral controller gain, once set to a value as calculated before,

Table 4 Observed and conventional values of *B*

Kp	'B' with present method $B = \frac{C_1}{Kp} + C_2$			'B' with conventional method $B = \beta = \frac{1}{Kp} + \frac{1}{R}$		
	R = 3 %	R = 4 %	R = 5 %	R = 3 %	R = 4 %	R = 5 %
60	0.419123	0.424477	0.459420	0.572222	0.433333	0.350000
66.6666	0.418469	0.423530	0.458074	0.570555	0.431666	0.348334
75	0.417821	0.422585	0.456725	0.568888	0.430000	0.346666
85.7143	0.417178	0.421643	0.455374	0.567222	0.428333	0.344999
100	0.416542	0.420704	0.454020	0.565555	0.426666	0.343333
120	0.415911	0.419767	0.452662	0.563888	0.425000	0.341666
150	0.415286	0.418833	0.451301	0.562222	0.423333	0.340000
200	0.414666	0.417903	0.449937	0.560555	0.421666	0.338333
300	0.414053	0.416976	0.448567	0.558888	0.420000	0.336666
600	0.413445	0.416052	0.447193	0.557222	0.418333	0.334999

was not required to be changed even for the entire range of load perturbation right from $\Delta P_{D1} = 1$ to 100 % and for the entire range of prevailing load on power system.

(2) Since C_1 and C_2 are constants, once they are determined for a particular value of regulation, the corresponding equation for *B* becomes standardized for that value of regulation, as shown in "Table 3."

(3) With this method, if the power system is enabled to sense the prevailing load in real time so as to determine the current value of *Kp* at the time of load perturbation, the value of *B* can be determined and adjusted accordingly almost instantaneously.

(4) The methodology adopted here to determine equation for *B* would be equally applicable for any other model, including nonlinearities and for load disturbances in both areas.

(5) The main benefit of present method of determining the value of *B* is that, the four parameters namely *B*, *Kp*, *R*, and *K* are correlated with each other in such a way that, the value of integral controller gain '*K*,' once set for a particular value of regulation, need not be changed for entire range of operating conditions and for any amount of load perturbation [15, 16].

Parameters values assumed:

Tg Governor time constant = 0.08 s
Tt Turbine time constant = 0.4 s
Ts Tie-line synchronizing coefficient = 0.0707
H Inertia constant = 5 s
f Normal operating frequency = 60 Hz

References

1. T. Kennedy, S. M. Hoyt, and C. F. Abell: Variable Non-linear Tie Line Frequency Bias for Interconnected Systems Control. IEEE TPS, 1244–1253 (1988).
2. Louis S VanSlyck, Nasser Jaleeli, and W. Robert Kelley: Implications of Frequency Control Bias Settings on Interconnected System Operation and Inadvertent Energy Accounting. IEEE Transactions on Power Systems, Vol. 4, No. 2, (1989).
3. N. Cohn: Some Aspects of Tie Line Bias Control on Interconnected Power System. Amer. Inst. Elect. Eng. Trans., Vol 75, 1415–1436 (1957).
4. B. Oni, H. Graham, and L. Walker: Investigation of Non-linear Tie Line Bias Control of Interconnected Power Systems. IEEE Transactions on Power Apparatus and Systems., Vol. PAS-100, No. 5, 2350–2356 (1981).
5. O. I. Elgerd: Electric Energy Systems Theory. Mc-Graw Hill, Newyork (1983).
6. Nasser Jaleeli, Louis S. VanSlyck, Donald N. Ewart, and Lester H. Fink: Understanding Automatic Generation Control. A report of the AGC task force of the IEEE/PES/PSE/system Control Sub-committee, Transaction on Power Systems, Vol. 7, No. 3 (1992).
7. D. P. Kothari and I. J. Nagrath: Modern Power System Analysis. 3rd edition, Mc-Graw Hill (2003).
8. Le-Ren Chang-Chien, Naeb-Boon Hoonchareon, Chee-Mun Ong, and Robert A. Kramer: Estimation of β for Adaptive Frequency Bias Setting in Load Frequency Control. IEEE Transactions on Power Systems, Vol. 18, No. 2 (2003).
9. Charudatta B. Bangal "Integral Control AGC of Interconnected Power Systems Using Area Control Errors Based On Tie Line Power Biasing" International Journal of Innovative Research in Electrical, Electronics, Instrumentation and Control Engineering". Vol. 2, Issue 4, April 2014.
10. IEEE Standard Definitions of Terms for Automatic Generation Control on Electric Power Systems. IEEE Transactions on Power Apparatus and Systems, Vol PAS-89, No. 6 (1970).
11. E. C. Tacker, T. W. Reddoch, O. T. Tan, and T. D. Linton: Automatic Generation Control of Electric Energy Systems-A Simulation Study. Work supported in part by US Air Force under contract F44620-68-C-0021 (1973).
12. S. C. Tripathi, G. S. Hope, and O. P. Malik: Optimization of Load Frequency Control Parameters for Power Systems with Reheat Steam Turbines and Governor Dead Band Non-linearity. IEE Proc-Part C 129, 10–16 (1982).
13. Robert P. Schulte: An Automatic Generation Control Modification for Present Demands on Interconnected Power Systems. IEEE Transactions on Power Systems, Vol 11, No. 3 (1996).
14. J. Nanda, M. parida, and A. Kalam, "Automatic Generation Control of Multi-area Power System with Conventional Integral Controllers". Proceedings of AUPEC 2006, Melbourne, Australia TS13 - Load and Frequency Control 2, December 10–19, 2006.
15. A. J. Connor, F. I. Denny, J. R. Huff, T. Kennedy, and C. J. Frank: Current Operating Problems Associated With Automatic Generation Control. IEEE Transactions on Power Apparatus and Systems, Vol. PAS-98, No. 1 (1979).
16. Ibraheem, Prabhat Kumar and Dwarka P. Kothari, "Recent Philosophies of Automatic Generation Control Strategies in Power Systems" IEEE Transactions on Power Systems, Vol. 20, Issue: 1, pages 346–357, Feb. 2005.

Significance of Frequency Band Selection of MFCC for Text-Independent Speaker Identification

S.B. Dhonde and S.M. Jagade

Abstract This paper presents significance of Mel-frequency Cepstral Coefficients (MFCC) Frequency band selection for text-independent speaker identification. Recent studies have been focused on speaker specific information that may extends beyond telephonic passband. The selection of the frequency band is an important factor to effectively capture the speaker specific information present in the speech signal for speaker recognition. This paper focuses on development of a speaker identification system based on MFCC features which are modeled using vector quantization. Here, the frequency band is varied up to 7.75 kHz. Speaker identification experiments evaluated on TIMIT database consisting of 630 speaker shows that the average recognition rate achieved is 97.37 % in frequency band 0–4.85 kHz for 20 MFCC filters.

Keywords Speaker recognition • Feature extraction • Mel scale • Vector quantization

1 Introduction

Speaker recognition is nothing but to recognize the person from known set of voices. Speaker recognition is classified into speaker identification and speaker verification. Speaker identification is nothing but to identify a person from the known set of voices. It is a task of identifying who is talking from known set of voice samples. While, speaker verification is to verify claimed identity of a speaker,

S.B. Dhonde (✉)
Department of Electronics Engineering, All India Shri Shivaji
Memorial Society's Institute of Information Technology, Pune 411001, India
e-mail: dhondesomnath@gmail.com

S.M. Jagade
Department of Electronics and Telecommunication Engineering,
TPCT COE, Osmanabad 413501, India
e-mail: smjagade@yahoo.co.in

S.C. Satapathy et al. (eds.), *Proceedings of the International Conference on Data Engineering and Communication Technology*, Advances in Intelligent Systems and Computing 469, DOI 10.1007/978-981-10-1678-3_21

i.e., Yes or No decision. Speaker identification is further classified into text-dependent identification and text-independent identification. Text-dependent speaker identification requires same utterance in training and testing phase. Whereas, in text-independent speaker identification training and testing utterances are different. Speaker identification system consists of two distinct phases, a training phase and testing phase. In training phase, the features computed from voice of speaker are modeled and stored in the database. In testing phase, the features extracted from utterance of unknown speaker are compared with the speaker models stored in database to identify the unknown person.

Feature extraction step in speaker identification transforms the raw speech signal into a set of feature vectors. The raw speech signal is represented in compact, less redundant feature vectors [1]. Features emphasizing on speaker specific properties are used to train speaker model. As feature extraction is the first step in speaker identification, the quality of the speaker modeling and classification depends on it [2].

In the computation of MFCCs, the spectrum is estimated from windowed speech frames. The spectrum is then multiplied by triangular Mel filter bank to perform auditory spectra analysis. Next step is the logarithm of windowed signal followed by discrete cosine transform. An important step in the computation of MFCC is the Mel filter bank [1, 3]. The MFCC technique computes speech parameters based on how human hears and perceives sound [2]. However, MFCC does not consider the contribution of piriform fossa, which results in high frequency components [4].

The auditory filter created by the cochlea inside human ear has frequency bandwidth termed as critical band. The existence of auditory filter is experimented by Harvey Fletcher [5]. The auditory filters are responsible for frequency selectivity inside the cochlea which helps the listener for discrimination between different sounds. These critical band filters are designed using frequency scales, i.e., the Mel scale and the Bark scale [5]. The MFCCs are widely used in speaker recognition system [2, 6–8]. In previous work, many researchers have demonstrated the dominant performance of MFCCs and contributed to enhance the robustness of MFCC features as well as speaker recognition system. Such efforts are [2, 7–15]. The importance of speaker specific information present in the wideband speech in demonstrated in [16].

This paper presents the importance of frequency band selection. The speaker specific information extends beyond telephonic pass band [16]. The performance of MFCC scheme in different frequency bands is demonstrated in this paper. The organization of this paper is as follows. Section 2 discusses frequency warping scale Mel scale and MFCC computation process. Experimental set-up is discussed in the Sect. 3. Section 4 discusses the results followed by conclusion in Sect. 5.

2 Mel Scale and MFCC

Nerves in human ear perception system responds differently to various frequencies in a listened sound. For example, sound of 1 kHz triggers nerves while sound of other frequencies will keep quite. This scale is roughly nonlinear in nature. It is like a band-pass filter that looks like triangular in shape. This was observed for how human ear perceives Melody sound. Mel scale is based on pitch perception [5]. Mel scale uses triangular-shaped filters and is roughly linear below 1 kHz and logarithmically nonlinear above 1 kHz. The relationship between Mel scale frequencies and linear frequencies is given as per the following equation,

$$F_{\text{mel}} = 2595^{*} \log_{10}\left(1 + \frac{F_{\text{Linear}}}{700}\right) \quad (1)$$

Figure 1 shows Mel scale filter bank. MFCC procedure starts with pre-emphasis which boosts the higher frequencies. The high-pass filter given by transfer function, $H(z) = 1 - az^{-1}$ where, $0.9 \le a \le 1$ is generally used for pre-emphasis. The pre-emphasized signal is divided into frames of duration 10–30 ms with 25–50 % overlap to avoid loss of information. Over this short duration, speech signal is assumed to remain stationary. Then, each frame is multiplied with Hamming window in order to smooth the speech signal. After windowing step, fast Fourier transform is used to estimate the frequency content present in speech signal. Next, the windowed spectrum is integrated with Mel filter bank which is based on Mel scale as given in Eq. (1). The vocal tract response is separated from excitation signal using logarithm of windowed spectrum integrated with Mel filter bank followed by discrete cosine transform.

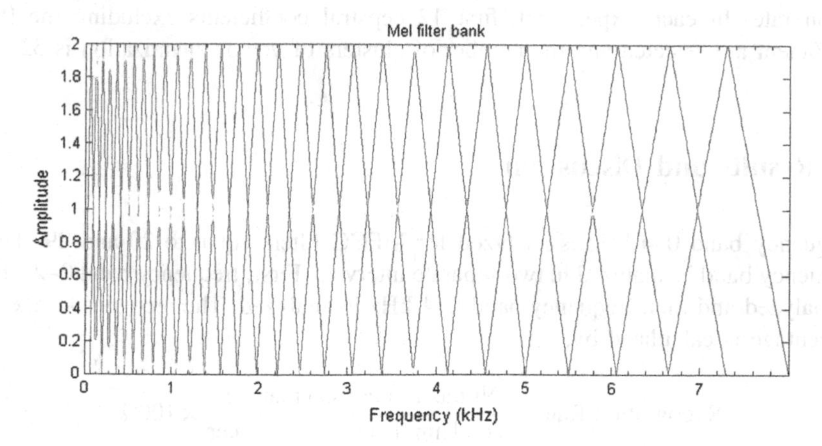

Fig. 1 Mel filter bank

3 Experimental Set-up

In this paper, the performance of Mel-frequency cepstral coefficients frequency band selection for text-independent speaker identification system is evaluated on TIMIT [17] database. TIMIT database consists of a total number of recordings of 630 speakers among which 438 are male speakers and 192 are female speakers. There are ten different sentences of each speaker of sampling frequency 16 kHz which makes a total of 6300 sentences recorded from 8 dialect region of the United States. For training of speaker model, eight sentences, five SX and three SI (approximately 24 s) were used. For testing purpose, two remaining SA sentences (sentences of 3 s each) were used. All the experiments have been performed using HP Pavilion g6 laptop with CPU speed of 2.50 GHz, 4 GB RAM, and MATLAB 8.1 signal processing tool.

The speech signal has been pre-emphasized with the first-order high pass filter given by equation $H(z) = 1–0.95z^{-1}$. The signal is divided into 256 samples per frame with 50 % overlap followed by the Hamming window. The spectrum of the windowed signal is calculated by fast Fourier transform (FFT). The spectrum is multiplied by Mel-filter bank followed by logarithm and discrete cosine transform (DCT) to obtain MFCCs. Speaker model is generated for each speaker from the MFCCs using vector quantization (LBG algorithm). This speaker model is stored in the database. In testing phase, MFCC features of an unknown speaker are extracted. Next, Euclidean distance between MFCC features and speaker model stored in the database is calculated. The speaker is recognized on the basis of minimum Euclidean distance computed between MFCC features in testing phase and speaker model stored in database. The experiments are carried out for different number of MFCC filters, i.e., 20 and 29 in the frequency band 0–4 kHz. Next, frequency is varied up to 7.75 kHz with 20 MFCC filters. The number of MFCC filters is varied as 13, 20, and 29 in the significant frequency band to observe the average recognition rate. In each experiment, first 12 cepstral coefficients excluding the 0th coefficient are selected and the number of clusters of vector quantization is 32.

4 Results and Discussion

Frequency band 0–4 kHz is analyzed for MFCC filters equal to 20 and 29. This frequency band is analyzed in two separate intervals. First, frequency band 0–2 kHz is analyzed and then frequency band 0–4 kHz is analyzed. The recognition rate in percentage is calculated by,

$$\text{Recognition Rate} = \frac{\text{Number of correct matches}}{\text{Total number of test speaker}} \times 100\%$$

Table 1 Recognition rate for frequency range 0–4 kHz

Sr. no.	Frequency band (kHz)	Sampling frequency (kHz)	No. of filters	Average recognition rate
1	0–2	0–4	20	81.18
2	0–2	0–4	29	83.41
3	0–4	0–8	20	95.95
4	0–4	0–8	29	95.80

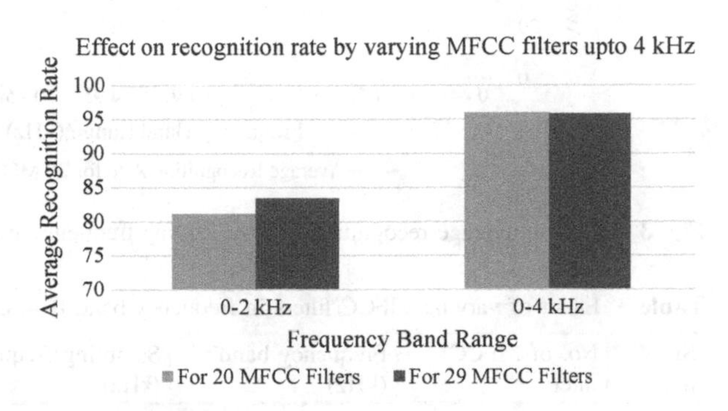

Fig. 2 Effect on recognition rate by varying MFCC filters up to 8 kHz

The following Table 1 and Fig. 2 shows the average recognition rate observed in these bands.

It is observed that the frequency band 0–4 kHz has provided a good resolution as compared to frequency band 0–2 kHz. This is because average recognition rate of 95.95 % is observed for 20 MFCC filters in frequency band 0–4 kHz. This indicates that speaker specific information is present up to 4 kHz. Also, varying the number of filters in these bands has less effect on recognition rate as compared to variation in frequency band. In addition to number of filters, it is also important to select a frequency band which is having good resolution for speaker identification.

In next subsequent experiments 20 MFCC filters are chosen and frequency band is varied up to 7.75 kHz. Table 2 and Fig. 3 shows the effect on recognition rate by varying frequency band.

Table 2 Effect on average recognition rate by varying frequency band up to 7.75 kHz for 20 MFCC filters

Sr. no.	No. of filters	Frequency band (kHz)	Sampling frequency (kHz)	Average recognition rate
1	20	0–4	0–8	95.95
2	20	0–4.75	0–9.5	96.66
3	20	0–4.85	0–9.7	97.37
4	20	0–4.9	0–9.8	96.90
5	20	0–4.95	0–9.9	96.58
6	20	0–5	0–10	96.66
7	20	0–6	0–12	81.58
8	20	0–7.75	0–15.5	61.83

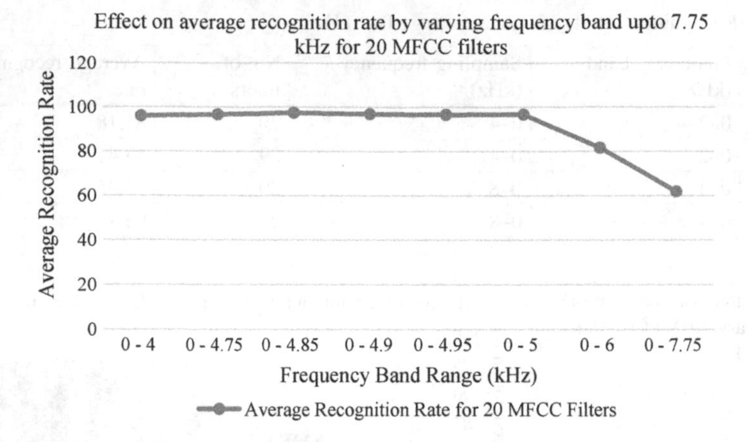

Fig. 3 Effect on average recognition rate by varying frequency band for 20 MFCC filters

Table 3 Effect of varying MFCC filters in frequency band 0–4.85 kHz

Sr. no.	No. of MFCC filters	Frequency band (kHz)	Sampling frequency (kHz)	Average recognition rate
1	13	0–4.85	0–9.7	95.32
2	20	0–4.85	0–9.7	97.37
3	29	0–4.85	0–9.7	97.30

From Table 2, it is observed that frequency band 0–4.85 kHz is the significant frequency band. This is because the maximum average recognition rate achieved is 97.37 % in this frequency band for 20 MFCC filters. Thereafter, average recognition rate is decreasing as shown in Table 2. It is observed that speaker specific information extends beyond 4 kHz, and therefore, it is important to select frequency band. Next, in the significant frequency band, i.e., 0–4.85 kHz, MFCC filters are varied and effect on average recognition rate is observed. Following table shows the effect of varying MFCC filters in frequency band 0–4.85 kHz.

From Table 3 and Fig. 4, it is observed that there is no much more improvement in average recognition rate by varying number of filters in the significant frequency band.

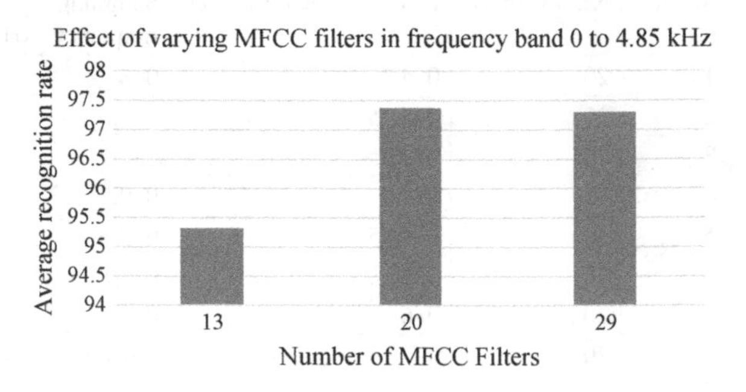

Fig. 4 Effect of varying MFCC filters in frequency band 0–4.85 kHz

5 Conclusion

In this paper, the significance of selection of Mel-frequency Cepstral Coefficients (MFCC) frequency band for speaker identification is proposed. First, frequency band 0–2 kHz is selected and MFCC filters are varied in this frequency band. Next, frequency band is varied 0–4 kHz and MFCC filters are varied in this band. It is found that speaker specific information is present in the frequency band 0–4 kHz is much more as compared to 0–2 kHz. Further, frequency band is varied up to 7.75 kHz. It is observed that the average recognition rate achieved is 97.37 % in the frequency band 0–4.85 kHz for 20 MFCC filters. This indicates that speaker specific information is present up to 4.85 kHz. Thereafter, recognition rate is decreasing. In the significant frequency band 0–4.85 kHz, MFCC filters are varied as 13, 20, and 29 and it is observed that there is no much more improvement in the average recognition rate.

References

1. Frédéric Bimbot, Jean-François Bonastre, Corinne Fredouille, Guillaume Gravier, Ivan Magrin-Chagnolleau, Sylvain Meignier, Teva Merlin, Javier Ortega-García, Dijana Petrovska-Delacrétaz, Douglas A. Reynolds: A tutorial on text-independent speaker verification, EURASIP Journal on Applied Signal Processing 2004, Hindawi, pp. 430–451 (2004).
2. Md Jahangir Alam, Tomi Kinnunen, Patrick Kenny, Pierre Ouellet, Douglas O'Shaughnessy: Multitaper MFCC and PLP features for speaker verification using i-vectors, Journal on Speech Communication, Elsevier, vol. 55, no. 2, pp. 237–251 (2013).
3. Claude Turner, Anthony Joseph, Murat Aksu, Heather Langdond: The Wavelet and Fourier Transforms in Feature Extraction for Text-Dependent, Filterbank-Based Speaker Recognition, Journal onProcedia Computer Science, Elsevier, vol. 6, pp. 124–129 (2011).
4. Mangesh S. Deshpande, Raghunath S. Holambe: New Filter Structure based Admissible Wavelet Packet Transform for Text-Independent Speaker Identification, International Journal of Recent Trends in Engineering, vol. 2, no. 5, pp. 121–125 (2009).
5. Dr. Shaila D. Apte: Speech Processing Applications, in Speech and Audio Processing, Section 1, Section 2 and Section 3, pp. 1–6, 67, 91–92, 105–107, 129–132, Wiley India Edition.
6. Tomi Kinnunen, Haizhou Li: An overview of text-independent speaker recognition: From features to supervectors, Journal onSpeech Communication, Elsevier, vol. 52, no. 1, pp. 12–40 (2010).
7. Tomi Kinnunen, Rahim Saeidi, FilipSedlák, Kong Aik Lee, Johan Sandberg, Maria Hansson-Sandsten, Haizhou Li: Low-Variance Multitaper MFCC Features: A Case Study in Robust Speaker Verification, IEEE Transactions Audio, Speech and Language Processing, vol.20, no.7, pp. 1990–2001 (2012).
8. Pawan K. Ajmera, Dattatray V. Jadhav, Ragunath S. Holambe: Text-independent speaker identification using Radon and discrete cosine transforms based features from speech spectrogram, Journal on Pattern Recognition, Elsevier, vol. 44, no. 10–11, pp. 2749–2759 (2011).
9. WU Zunjing, CAO Zhigang: Improved MFCC-Based Feature for Robust Speaker Identification, TUP Journals & Magazines, vol.10, no 2, pp. 158–161 (2005).

10. Jian-Da Wu, Bing-Fu Lin: Speaker identification using discrete wavelet packet transform technique with irregular decomposition, Journal on Expert Systems with Applications, Elsevier, vol. 36, no. 2, pp. 3136–3143 (2009).
11. R. Shantha Selva Kumari, S. Selva Nidhyananthan, Anand.G: Fused Mel Feature sets based Text-Independent Speaker Identification using Gaussian Mixture Model, International Conference on Communication Technology and System Design 2011, Journal on Procedia Engineering, Elsevier, vol. 30, pp. 319–326 (2012).
12. Seiichi Nakagawa, Longbiao Wang, and Shinji Ohtsuka: Speaker Identification and Verification by Combining MFCC and Phase Information, IEEE Transactions Audio, Speech and Language Processing, vol.20, no.4, pp. 1085–1095 (2012).
13. Sumithra Manimegalai Govindan, Prakash Duraisamy, Xiaohui Yuan: Adaptive wavelet shrinkage for noise robust speaker recognition, Journal on Digital Signal Processing, Elsevier, vol. 33, pp. 180–190 (2014).
14. Noor Almaadeed, Amar Aggoun, Abbes Amira: Speaker identification using multimodal neural networks and wavelet analysis, IET Journals and Magazines, vol. 4, no. 1, pp. 18–28 (2015).
15. Khaled Daqrouq, Tarek A. Tutunji: Speaker identification using vowels features through a combinedmethod of formants, wavelets, and neural network classifiers, Journal on Applied Soft Computing, Elsevier, vol. 27, pp. 231–239 (2015).
16. Pradhan, G.; Prasanna, S.: Significance of speaker information in wideband speech, in Communications (NCC), 2011 National Conference on, pp. 1–5, (2011).
17. J.S. Garofolo, L.F. Lamel, W.M. Fisher, J.G. Fiscus, D.S. Pallett, N.L. Dahlgren, V. Zue, TIMIT acoustic-phonetic continuous speech corpus, http://catalog.ldc.upenn.edu/ldc93s1, 1993.

Ensuring Performance of Graphics Processing Units: A Programmer's Perspective

Mayank Varshney, Shashidhar G. Koolagudi, Sudhakar Velusamy and Pravin B. Ramteke

Abstract This paper mainly focuses on the usage of automation system for ensuring the performance of graphics driver created at Intel Corporation. This automation tool takes into account a client-server structural planning which can be utilized by the developers or the validation engineers so as to guarantee whether the graphics drivers are programmed and modified accurately or not. The tool additionally actualizes some of the Driver Private APIs (it allows any application to talk directly with the driver) which will guarantee the properties of the features which are not bolstered by the Operating System (OS).

Keywords Graphics Processing Unit (GPU) · Operating System (OS) · Windows Display Driver Model (WDDM) · Driver Private APIs

1 Introduction

The integrated graphics component, specifically called the Graphics Processing Unit (GPU) resides on the same chip die as the Central Processing Unit (CPU), and communicates with the CPU via on-chip bus, with internal memory and with output device(s). As Intel GPUs have evolved, it occupies a significant percentage of space on the chip, and provides high performance and low-power graphics processing by eliminating the need to purchase a separate video card. A driver enables OS and

Mayank Varshney (✉) · S.G. Koolagudi · P.B. Ramteke
National Institute of Technology Karnataka, Surathkal, India
e-mail: mvarshney.technologist@gmail.com

S.G. Koolagudi
e-mail: koolagudi@nitk.edu.in

P.B. Ramteke
e-mail: ramteke0001@gmail.com

Sudhakar Velusamy
Intel Technology India Pvt. Ltd., Bengaluru, India
e-mail: sudhakar.velusami@intel.com

S.C. Satapathy et al. (eds.), *Proceedings of the International Conference on Data Engineering and Communication Technology*, Advances in Intelligent Systems and Computing 469, DOI 10.1007/978-981-10-1678-3_22

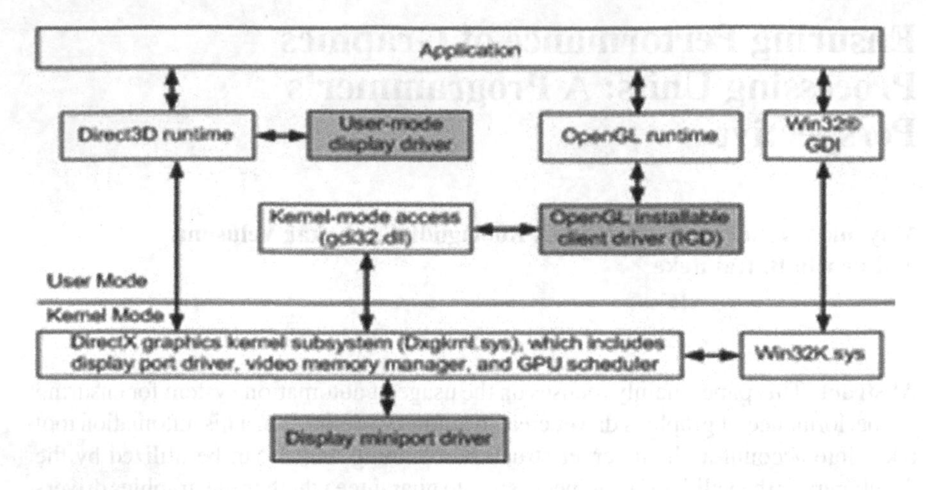

Fig. 1 Architecture required to support WDDM

other computer programs to access hardware functions without the knowledge of the hardware being used.

The display drivers at Intel Corporation are written according to the Windows Display Driver Model (WDDM) [1] as shown in Fig. 1. The display driver model architecture for WDDM, available starting with Windows Vista, is composed of user-mode and kernel-mode parts. WDDM drivers contribute towards greater operating system (OS) stability and security because of small code execution in kernel-mode where it can access system address space and possibly cause crashes. At whatever point an engineer implements any new feature/fixes any bug for the GPU device driver and/or device drivers expected to be redesigned much of the time keeping in mind the end goal to allow better use of the GPU, one needs to verify that the GPU is working and customized effectively for the particular device driver including different display's setup having distinctive variations of the displays.

This tool explores the development of an automation system for totally ensuring the performance of the device drivers for the cutting edge GPUs. This tool ensures that GPU is customized and works accurately for the particular device driver by setting distinctive display setups and features including different variations of the displays. The hardware may bolster different properties of the features which may not be supported by the Operating System. Keeping in mind the end goal to guarantee those properties this tool will implement several new Driver Private APIs that will let the developer guarantee those properties by bypassing the operating system. Each graphics driver created at Intel Corporation is initially validated by the developers themselves and afterward validated by the validation team before releasing, yet the graphics driver created may have some bugs. Since there is a two-level approval mechanism for every display driver or driver redesign, developers/validation engineers dependably attempt to catch as many bugs as possible in the drivers before releasing them to the customers, yet at the same time the bugs may be there due to missing some corner case validation or a percentage of the bugs may be sporadic bugs which happens at unpredictable interim or just with a very unusual kind of scenario which may have been missed by the developers or the validation engineers.

Automation [2] system helps in expanding the effectiveness, adequacy and scope of validation of the item before it is released to the customers. The principle point of interest of test automation is to keep away from the human-inclined bugs while executing the tests. For each release of the graphics driver it must be validated on all supported operating systems, platforms and with all supported display configurations. Physically testing these systems is costly and tedious. Utilizing an automation tool, these validation techniques can be rehashed various times, with no extra cost and they are much speedier than the validation done by the developers and/or validation engineers. Automation system can decrease the time to run monotonous approval strategies from days to hours.

2 Motivation

The performance of a GPU device driver is a measure that how effectively the GPU is functioning and modified. Ensuring the performance of the device driver for a GPUs is an extremely complex assignment since as it includes manual choice of distinctive displays setups including display variations like DP (Display Port), EDP (Embedded Display Port), HDMI (High Definition Multimedia Interface), VGA (Video Graphics Array), and diverse display related features like, modeset, scaling, deep color, display switching, wide gamut, Gamma, and many others. These determinations frame a situation and for better performance check display driver ought to be confirmed over all such possible situations. This is driving and prompting development of automation system where these situations can be produced in type of experiments and GPU can be checked over all these experiments.

The test improvement is a real exertion and regardless of the fact that an expansive arrangement of tests is composed, a developer or validation engineer cannot humanly expect every single possible thing that can happen and create tests appropriately. A validation group comprises of number of specialists, every in charge of an outline unit. A validation engineer will outline a unit base on his instinct, suspicions and his comprehension of how the modules to which his module will convey, carries on. There are risks that a validation engineer comprehension of outside environment is inadequate and thus risks of bugs. Here again, the validation engineer cannot in any way, shape or form foresee all the issues that can come up. So it is an extremely monotonous and the shots of accurately confirming the quality are likewise exceptionally thin.

3 Literature Survey

Few benchmark tools like HP Quick Test Professional (QTP) and IBM Rational Functional Tester (RFT) have been clarified in a nutshell.

3.1 HP QuickTest Professional [3]:

HP QuickTest Professional gives practical and relapse test automation for programming applications and situations. Some piece of the HP Quality Center tool suite, HP QuickTest Professional can be utilized for big business quality confirmation. HP QuickTest Professional backings essential word and scripting interfaces and features a graphical client interface. It utilizes Visual Basic Scripting Edition (VB Script) scripting dialect to indicate a test method, and to control the articles and the application under approval. It can be utilized for taking following operations:

- Verification: Checkpoints confirm that throughout approval system, the genuine application conduct/state is steady with the normal application conduct/state.
- Exception Handling: HP QuickTest Professional oversees special case taking care of utilizing recuperation situations; the objective is to keep running approval system if a startling disappointment happens
- Data driven acceptance: HP QuickTest Professional backings information driven approval. For instance, information can be yield to an information table for reuse somewhere else.

3.2 IBM Rational Functional Tester [4]:

IBM Rational Functional Tester is programming test computerization device utilized by quality affirmation groups to perform robotized relapse acceptance. Execution designers make scripts by utilizing a test recorder which catches a client's activities against their application under approval. The recording component makes an approval script from the activities. The approval script is delivered as either a Java or Visual Basic. net application is spoken to as a progression of screenshots that frame a visual storyboard. Execution specialist can alter the script utilizing standard charges and sentence structure of these dialects or by acting against the screen shots in the storyboard. Acceptance scripts can then be executed by Rational Functional Tester to accept application usefulness. To utilize the IBM Rational Functional Tester execution specialist ought to have learning of Java or Visual Basic.net application.

3.3 Test Automation for Web Applications:

Numerous, maybe most, programming applications today are composed as online applications to be keep running in an Internet browser. The viability of testing these applications fluctuates generally among organizations and associations. Selenium [5] is a set of different software tools each with a different approach to supporting test automation specially for the web applications. The entire suite of tools results in a rich set of testing functions specifically geared to the needs of testing web applications of all types.

The study [6] has been done on what are all the different features which must be tried keeping in mind the end goal to ensure the performance of the Graphics Driver created at Intel Corporation. It incorporates distinguishing different display design and display variations which are conceivable with every era of Intel Graphics Processing Chipset.

4 Challenges and Need

Key components needed for a Test Automation [7] to be effective include Committed Management, Budgeted Cost, Process, Dedicated Resources, and Realistic Expectation. Test computerization system advancement is a multi-stage process. Every stage includes various difficulties to be tended to that are defined beneath

1 Clear Vision: Clear vision of what needs to be attained out of this mechanization must be characterized and reported. To figure the reasonable vision, the accompanying needs to be recognized (refer Fig. 2):

   a Testing Model—To be adjusted (Enterprise, Product, Project)
   b Types of testing under the extension in light of the testing model
   c Prioritizing the mechanization action taking into account application/ module (for instance: Conducting practical testing and afterward execution testing)
   d Identify the zones that need to be mechanized (for instance: Registration, Order preparing)
   e Challenging acceptances that need to be considered (for instance: Communicating from windows to Linux machine and executing the tests)
   f Critical and required functionalities

2 Tool Identification And Recommendation: Tool recognizable proof methodology is a critical one, as it includes discriminating components to be considered. The various components are:

   a Creating a standard tool assessment agenda which needs to be made by considering sorts of testing, groups included, authorizing expense of the

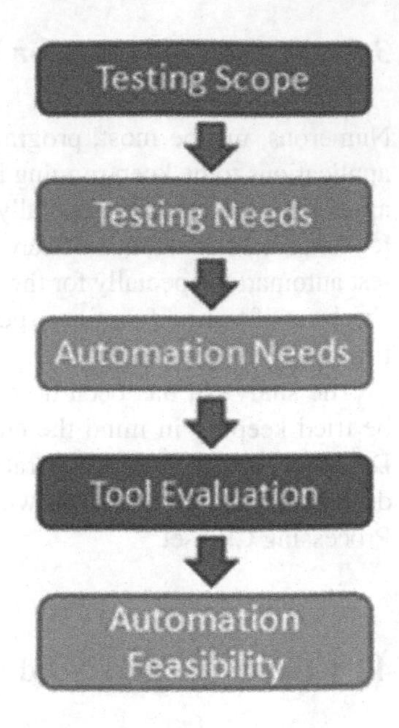

Fig. 2 Testing automation feasibility

instrument, upkeep cost, preparing and backing, device's extensibility, instrument's execution & dependability and so forth.

b Testing necessities which may incorporate sorts of testing, for example, Functional, Performance, and Web Service and so forth.

c The need of various instruments to perform distinctive sorts of testing on the item/venture.

3 Framework Design: System configuration includes distinguishing prerequisites from different ranges. At an abnormal state, this incorporates (not restricted to):

a Identification of important utility/parts identified with application functionalities.

b Types of information store to be conveyed for information stream.

c Communication between the utilities/segments (for instance: information registration parts imparting to the lumberjack).

d Communication between the frameworks and utility/segment advancement identified with the same.

e Tool developing abilities—Developing utilities/segments for the approvals not bolstered by the distinguished test mechanization tool, if any.

5 The Tool

The thought is to give a bound together standard stage to accept the nature of Intel Graphics Driver by the architects on any obliged stage, i.e., distinctive graphics cards and diverse display devices. The tool depends on the current display configuration and the register qualities being sent to the display port. The inward graphics pipeline stages are disregarded. This is in light of the way that when the same image is shown at the same display configuration, the register values for the yield port will likewise be the same. In order to guarantee the features which are upheld by the hardware, however, not by the working OS, new Driver Private API (Driver Private APIs is the particular case that let any application to talk with the driver) will be executed to sidestep the operating system and makes fake calls which should be finished by the operating system.

5.1 Implementation:

This tool gives a client-server model [8] where the developers/validation engineer can calendar tests remotely from their PCs and get the report of tests conveyed once the test is over. Hence, the user no longer has to worry about setting up the system to run the test and whatever other needed (abstraction of tests from user). The general approach followed for the design of tool is shown in Fig. 3.

It will be implemented in three parallel tracks:

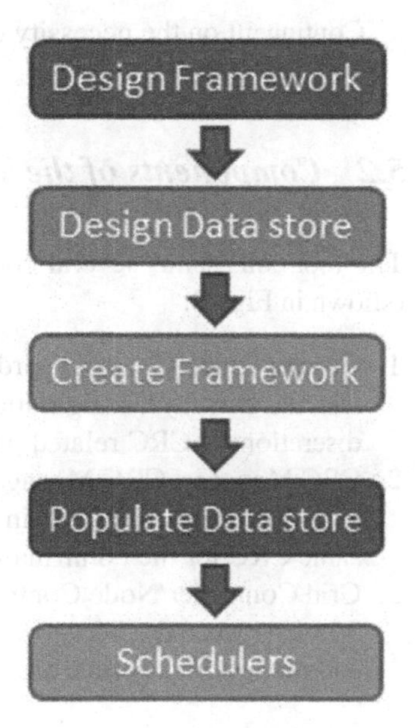

Fig. 3 Design methodology

1. Track 1: Implementation of Client Server Architecture [9] so as to dispatch the test remotely from any area and implementing web services in place guarantee the CRC (Cyclic Redundancy Check) with the end goal of accepting the yield of the test for different graphics related features including different displays variations (HDMI, DP, EDP, and VGA) and diverse era of processors too.
2. Track 2: Implementation of mechanizing different graphics related features, catching CRC with the end goal of approving the further tests.
3. Track 3: Integration of new Driver Private APIs which are to be utilized to bypass the OS. Whatever should be finished by the OS, these driver APIs will do those fake capacity calls and afterward we can guarantee the nature of those features which are not upheld by OS.

The specimen command line which is to be utilized to ensure Mode set feature on the Tri Extended display design: TestSuite.exe EDP 1 DP 1 HDMI 1 FEATURE MODESET TESTLEVEL 1 DISPCONFIG TE

The tool will supports three distinctive level of test execution regarding the multifaceted nature of the test:

1. Test level 1 (Basic): Used to guarantee the predefined feature with fundamental test situations like for ModeSet with just two resolutions
2. Test level 2 (Intermediate): Used to guarantee the predefined feature with some more unpredictable test situations like for ModeSet with some more distinctive resolutions
3. Test level 3 (Extensive): Used to guarantee the predefined feature with all the conceivable different features which may be influenced by any progressions to the predetermined feature.

Contingent on the necessity of the engineers, they can utilize the test as need be.

5.2 Components of the Tool:

The tool consists of several components which provide related set of services are (shown in Fig. 4):

1. Control Panel: Control board is utilized to choose the different parameters for the test like Display Configuration, feature to be tried, show variations, test level and discretionary CRC related parameters (Fig. 5).
2. CRC Manager: CRC Manager is in charge of giving the CRC related administrations like capturing CRC in the CRC database and afterward later utilizing the same CRC for the confirmation reason.
3. Grid Controller/Node Controller Services: This is utilized to give the correspondence between the Server which dispatches the test and in charge of the CRC stuff and the test hubs which are only the genuine equipment.

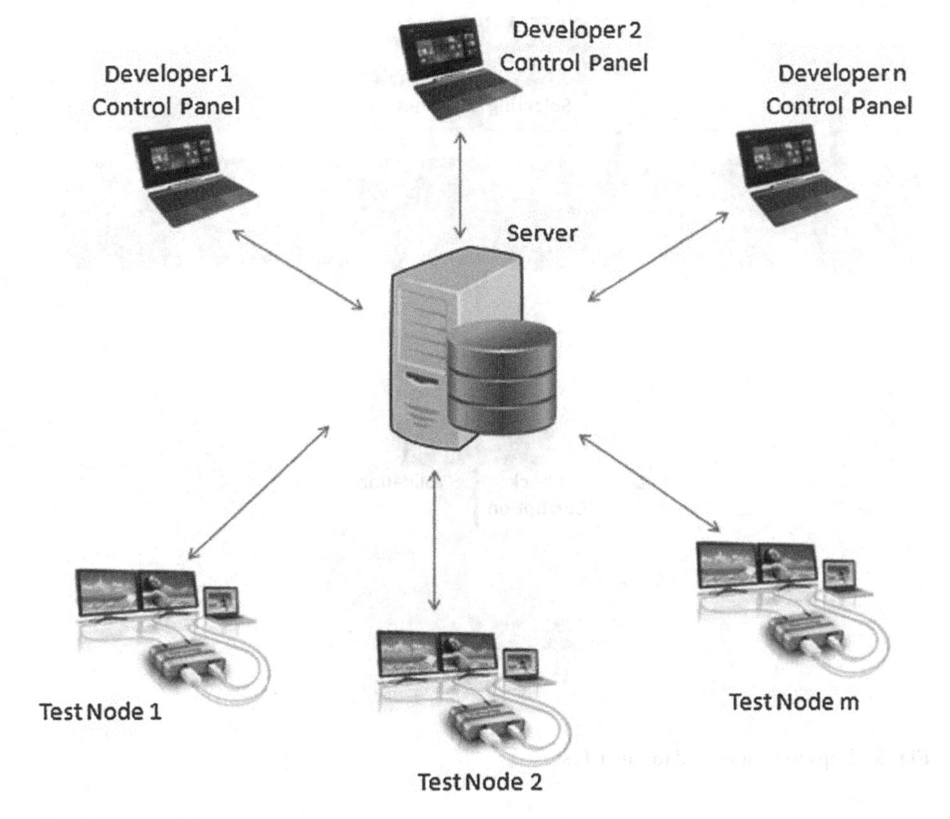

Fig. 4 Tool architecture

4. Test Nodes: The real equipment machine where one of the test is launched and executed.
5. Log Manager: Responsible for creating a wide range of logs and including doing the register dump moreover.
6. Interface Manager: gives the from now on correspondence between the equipment, illustrations CUI and the test.
7. Test Suite: Collection of every last one of tests for the distinctive feature and for each of the three unique levels.
8. Feature Suite: Consists of the computerization of every last one of design related feature which can be tried.

6 Results

This task brought about a computerization tool which is three times quicker than different devices and lessened the work power and the time needed to approve the display driver created at Intel Corporation. This tool helps in expanding the accuracy

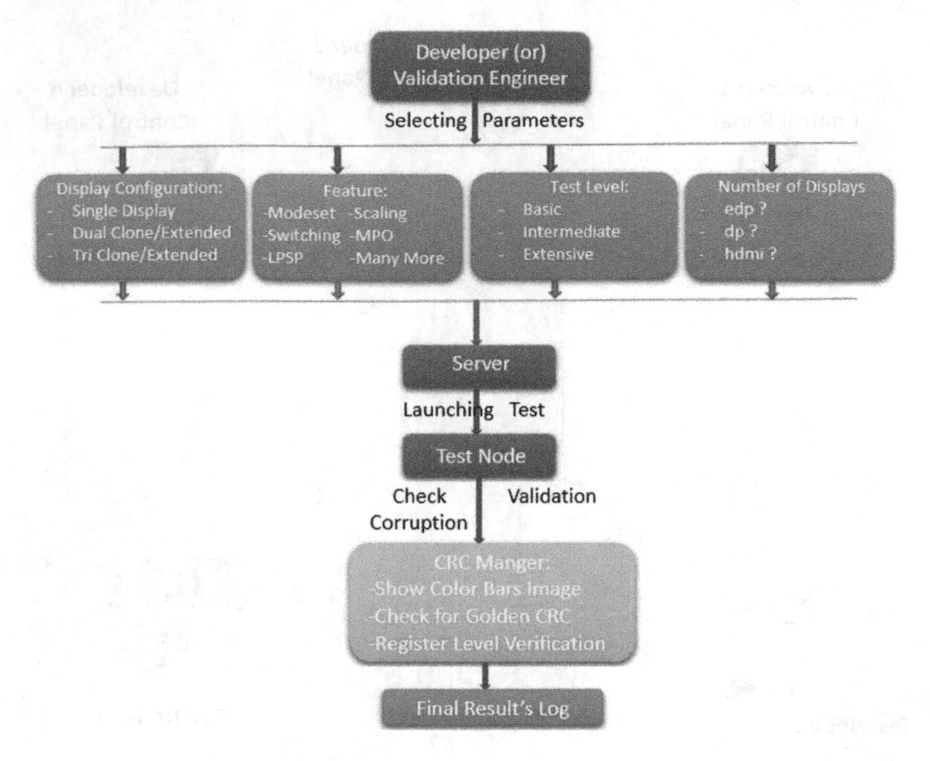

Fig. 5 Steps to be followed to run a Test

of the graphics driver and in the meantime spares the time required for ensuring the performance of the graphics driver. Presently, with the assistance of this tool the developer/validation engineer require not to invested their important energy for performing Unit Level Testing, it will be taken into consideration by the tool and in the instance of failure the developer/validation engineer will be getting the detailed logs which can be utilized to examine the failure that happens.

7 Conclusion

Considering the end goal to automate [10], a great deal of time ought to be spent in examination about the styles and methods of automation and discover a building design that fits nature. In Testing GPUs the client needs to set up distinctive setups to run every test which is a pointless overhead. Besides this is additionally brings about underutilization of assets as the setup needed for one test may be obliged to run some other test for testing an alternate GPU.

This automation system takes out the need of equipment setup for every individual necessity, as test hubs will be pre-configured. Any client can plan tests on these hubs without having to stress over where these tests will run physically. Regardless of the fact that no hubs are accessible right now to launch the test, the test will get queued. The server will run the test once a free hub with coordinating equipment setup is accessible. This system expands asset usage likewise as the same test hubs can be utilized by any number of clients so taking out the need of an alternate setup for every group. Likewise, the tests can be requested specifically from the developers/validation engineer's PCs through the Control Panel which speaks with the server remotely, hence extraordinarily expanding the accommodation of the client.

This undertaking introduces a savvy method for automation of approval procedures for the graphics driver developed by Intel Corporation. This tool has a superior execution, when contrasted with alternate devices accessible in the business sector, regarding the speed of execution and the circle space utilized. It is easy to use and simple to learn. It automates the procedure of acceptance effectively with little assistance from the developer and the individual who automates the application needs to take after basic Test Descriptive Language linguistic structure while composing the approval explanations.

References

1. Windows Display Driver Model (WDDM) Architecture https://msdn.microsoft.com/enus/library/windows/hardware/ff570589(v=vs.85).aspx.
2. Automated Software Testing. Addison Wesley, 2008.
3. RadhikaSrikantha (radhu1068@gmail.com), Prof. Suthikshn Kumara, Sai Arun Junjarampalli.Cisco Smart Automation Tool,Internet (AH-ICI), 2011 Second Asian Himalayas International Conference on Internet.
4. HP Unified Functional Testing software Documentation.
5. Automation tool comparison by Adel Shehadeh.
6. IBM Rational Functional Tester 7.0 Documentation.
7. Anish Srivastava, Automation Framework Architecture for Enterprise Products: Design and Development Strategy http://www.oracle.com/technetwork/articles/entarch/shrivastava-automated-frameworks-1692936.html.
8. Google WebDriver and Selenium Documentation 2010.
9. Intel platform and component validation- A commitment to quality, reliability and compatibility White paper published by Intel Corp. Pvt ltd, http://download.intel.com/design/chipsets/labtour/PVPTWhitePaper.pdf, (23 f eb2014).
10. Kaner, Cem, Architectures of Test Automation, August 2000 http://www.kaner.com/pdfs/testarch.pdf.

Analytical Study of Miniaturization of Microstrip Antenna for Bluetooth/WiMax

Pranjali Jumle and Prasanna Zade

Abstract In this paper, we propose a different size reduction approach of MSA for Bluetooth, WiMAX applications. Prime emphasize of this work is to study the reactance of MSA, i.e., effect of inductive and capacitive reactance and it is validated by Smith chart. A novel technique of insertion of two parallel plates in the middle of patch and in close vicinity to the feed position to achieve the inductive effect for different applications is presented with 82 % size reduction. By single shorting plate along the edge of MSA and near to the feed position separately 2.4 and 3.5 GHz band with 75, 110 MHz BW and 3.4, 3.8 dBi gain, respectively, and 73 % size reduction is obtained. Capacitive effect by varying and selecting optimum feed position without any complexity for dual band WiMAX and WLAN 115, 150 MHz BW with gain 3.8, 4.3 dBi reported and 36 % of size reduction is reported.

Keywords Miniaturized patch antenna • Single/two parallel shorting plate • Smith chart • Capacitive and inductive reactance

1 Introduction

Compact designing of microstrip patch antennas has received much interest due to the increasing demand of small antennas for private communications systems such as portable and handheld wireless devices.

Pranjali Jumle (✉)
Department of Electronics and Tele-Communication, Rajiv Gandhi College of Engineering and Research, Wanadongri, India
e-mail: jumle.pranjali03@gmail.com

Prasanna Zade
Department of Electronics and Tele-Communication, Yeshwantrao Chavan of Engineering, Wanadongri, India
e-mail: zadepl@yahoo.com

S.C. Satapathy et al. (eds.), *Proceedings of the International Conference on Data Engineering and Communication Technology*, Advances in Intelligent Systems and Computing 469, DOI 10.1007/978-981-10-1678-3_23

Microstrip antennas having the advantages of low profile, light weight, low fabrication cost, and compatibility with MMIC [1, 2], are half-wavelength structures which are bulky in personal wireless devices. Many techniques have been adopted to reduce the size of the patch antennas, such as employing high dielectric substrate [3, 4], introducing fractal geometries [5], introducing shorting post [3, 6], loading chip-resistor and chip-capacitor [7, 8], loading of complementary split-ring resonator [9], loading reactive components [10], lengthening electrical path of the current by meandering [3, 11], introducing slots and slits into the patch [12–14], loading suspended plate [15], also many papers have quoted the metamaterial structures to reduce the size of patch antennas [16].

Also, loading distributed reactive components yield in miniaturization. In [17], paper presents two by two multiple-input multiple-output patch antenna systems with complementary split-ring resonator loading on its ground plane for antenna miniaturization. In [18], a loaded shorting elements is introduced which are parasitic to the driven patch, but are shorted to the ground plane which results in size reduction as well as a good radiation pattern. In [19], a via-patch is introduced under the radiating element to create a capacitive coupling effect and lowering the resonating frequency results in reduction of antenna size by almost 50 %. Also, a slight change in the height and location of via can regulate the resonant frequency of the antenna and is also presented. In [20], a method for size reduction of patch antennas is presented by introducing an irregular ground structure to provide capacitive and inductive loading to the patch.

In this paper, a size reduction of microstrip patch antenna and study of impedance characteristics from smith chart are presented. Paper is organized into three parts; first the effect of feed position and shorting plate is studied by varying distance between their locations and shifting of resonant frequency is studied with Smith chart. By means we study and explain the effect of dominant inductive reactance. Second, a very simple approach to obtain dual band characteristics with simply optimizing feed position and study of capacitive reactance with the help of Smith chart is validated. Finally, a novel concept of insertion of two parallel plates in the middle of patch and in the vicinity of feed position is presented, size reduction of 82 % is reported.

2 Design

To design a MSA, we use FR4 Substrate with dielectric constant 4.4 and loss tangent 0.0019 is used throughout this study. Simulation and design are performed in IE3D software.

2.1 Case I: Study of Inductive Effect by Adding Shorting Plate

As shown in Fig. 1a, the RMSA, rectangular microstrip patch antenna is analyzed at the operating frequency of 3.5 GHz (see Fig. 3a). Feed point located along length, the fundamental resonating mode obtained is TM_{10} can be calculated from Eq. (1).

$$f_o = \frac{c}{2\sqrt{\varepsilon_r}} \sqrt{\left[\left(\frac{m}{L}\right)^2 + \left(\frac{n}{W}\right)^2\right]} \quad (1)$$

It is important to note that a 3.5 GHz is showing slightly capacitive in the Smith chart shown in Fig. 2a. Figure 1b shows the RMSA with shorting plate along edge (see Fig. 1c), which introduces the inductive effect, i.e., 2.4 GHz is obtained from 3.5 GHz by adding single shorting plate (see Fig. 3b).

3.5 GHz is modified to reduce antenna physical size by adding single plate in the vicinity of feeding probe. It shows better impedance matching if the distance between plate and feed is about 1.45 mm. If distance between plate and feed is reduced to less than 1.45 mm, i.e., if feed point is close to plate, then frequency shifts to higher side.

Smith chart in Fig. 2b indicates that by adding plate along the edge, 3.5 GHz of basic patch shift to inductive region, i.e., by adding shorting plate, inductive effect

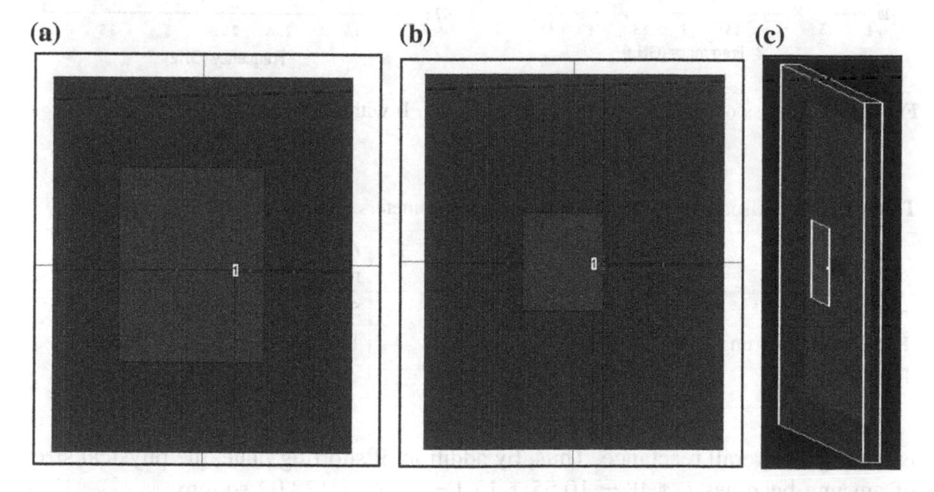

Fig. 1 RMSA in IE3D, Zeland Software, **a** without shorted-plate, **b** with shorted-plate, **c** 3D View of part **b**

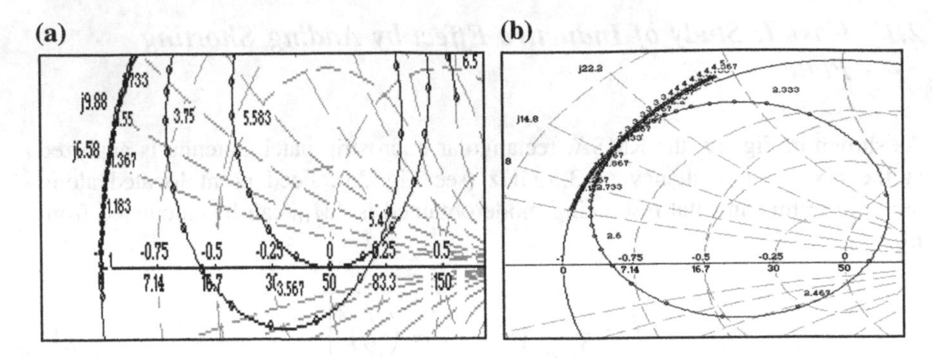

Fig. 2 Smith chart of RMSA, **a** without shorted-plate, **b** with shorted-plate

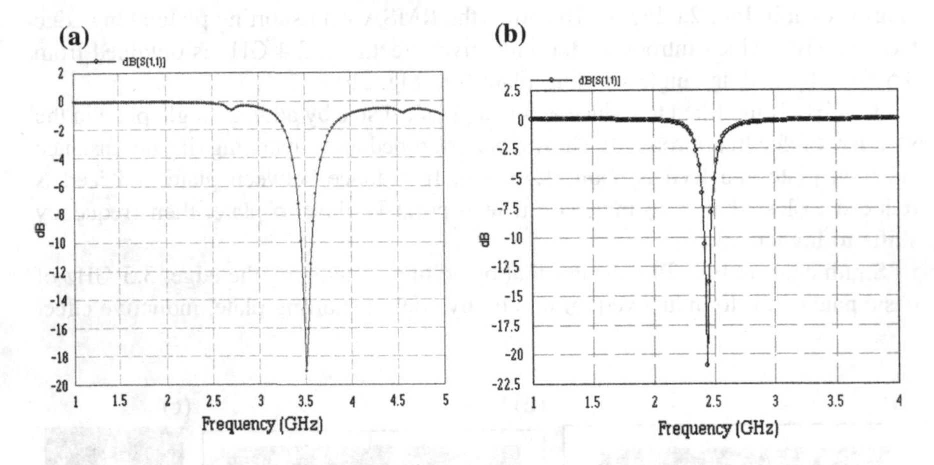

Fig. 3 Return loss of RMSA, **a** without shorted-plate, **b** with shorted-plate

Table 1 Comparison between RMSA dimension parameters without and with shorting plate

	Length (mm)	Width (mm)	Area (sq. mm)	% of size reduction
RMSA	19.97	26.08	509.6	72.91 %
RMSA with shorting plate	10.55	13.1	138.02	

is leading in overall reactance. Thus, by addition of shorting plate, the physical size of antenna becomes $L * W = 10.55 * 13.1 =$ Area $= 138.02$ sq.mm.

Percentage size reduction $= \frac{509.6 - 138.02}{509.6} * 100 = 72.91\%$ as compared to original patch (see Table 1).

2.2 Case II: Study of Capacitive Effect by Changing the Feed Position for Dual Band

Changing the original feed location from 4.28 to 4.60, i.e., (0.32 mm) from TM_{10} mode, next higher order mode can be excited. By optimizing feed point and width we get 5.5 GHz. Smith chart in Fig. 4b shows shifting of 3.5 GHz to capacitive half portion as compared to basic 3.5 GHz (see smith chart in Fig. 4a), where 3.5 GHz is along resistive line.

In this case, there is no electrical as well as physical size reduction but we make the dual band antenna (see Fig. 5) by only well optimizing feed point location.

$$\text{Size reduction in frequency} = \frac{5.5 - 3.5}{5.5} * 100 = 36.36\,\%$$

2.3 Case III: Study of Inductive Effect by Adding Shorting Parallel Strips

Figure 6a shows the layout of RMSA having calculated parameters are $L = 12.16$ mm, $W = 16.6$ mm in IE3D, Zeland Software operated at 5.5 GHz. Smith chart in Fig. 6b shows 5.5 GHz location less capacitive.

Adding shorting parallel strips in the vicinity of feed point, a lower order resonating mode for the antenna having same parameter for 5.5 GHz can be achieved. The same structure is now resonating at 2.4 GHz by observing Smith chart in

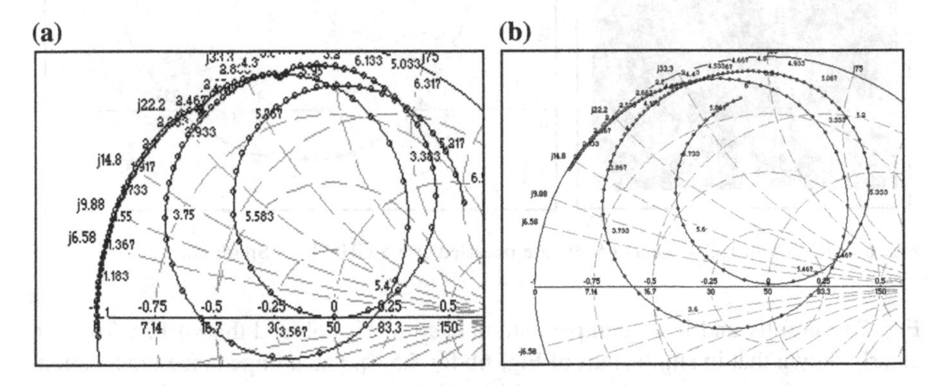

Fig. 4 Smith chart of RMSA, **a** 3.5 GHz along resistive line, **b** 3.5 GHz toward capacitive region

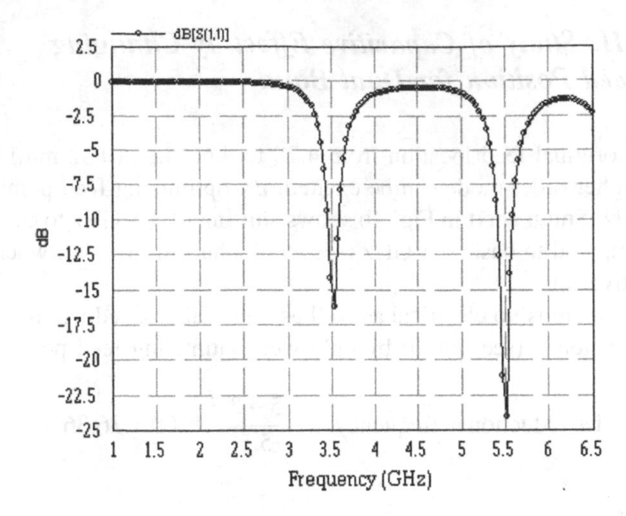

Fig. 5 Return loss of RMSA incorporating 5.5 GHz along with 3.5 GHz, i.e., higher order mode excited

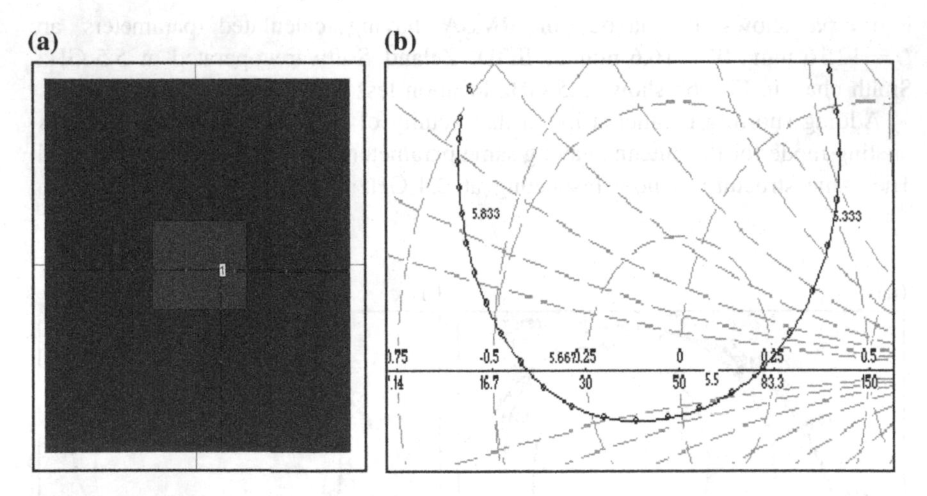

Fig. 6 **a** RMSA in IE3D, Zeland Software operated at 5.5 GHz, **b** its Smith chart

Fig. 7b carefully and if we compare smith chart of Fig. 6b and that of Fig. 7b, it is clearly shown that in smith chart of Fig. 6b for 5.5 sq. patch impedance the location is near to resistive line with less capacitive effect, i.e., overall reactance is capacitive but after addition of two parallel plate in the vicinity of feed point, the lower order mode is excited. It is clearly evident from Smith chart in Fig. 7b where it is shown the shift of 5.5 GHz location to more upper part of inductive region.

Physical size reduction (see Table 2) $= \frac{1119.9 - 206.3}{1119.9} * 100 = 81.60\,\%$

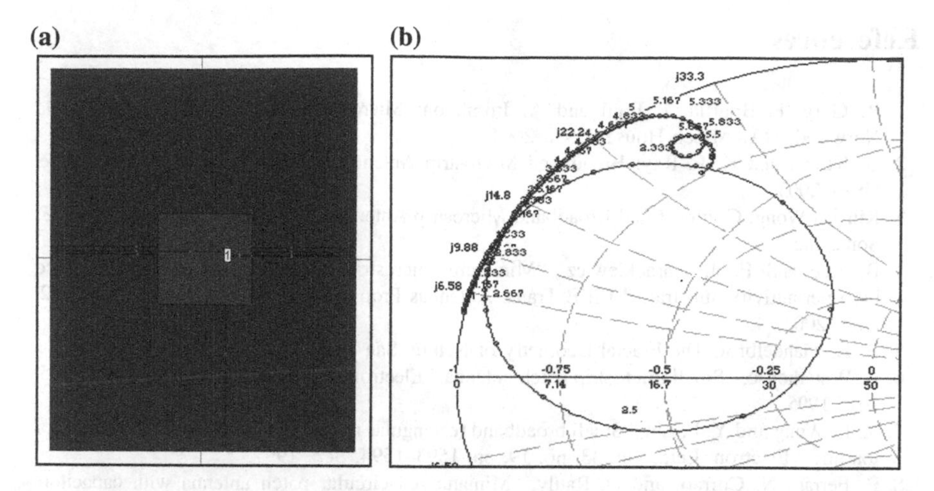

Fig. 7 **a** RMSA in IE3D, Zeland Software operated at both 2.4 and 5.5 GHz, **b** its Smith chart

Table 2 Comparison between 5.5 and 2.4 GHz dimensional parameters

Resonating frequency	W (mm)	ereff	Leff (mm)	2ΔL (mm)	L (mm)	Area (mm^2)
5.5 GHz	16.59	3.85	13.89	1.45	12.44	206.3
2.4 GHz	38.04	4.08	30.92	1.48	29.44	1119.9

3 Conclusion

In this paper, a microstrip antenna is presented which is compacted by 82 % after loaded with two parallel shorted plates. By studying the above cases, it is concluded that loading the inductance in overall reactance causes shifting of resonance frequency toward lower end while loading the capacitance in overall reactance causes excitation of higher order modes. The experimentation is carried out for Bluetooth/WiMax applications (Table 3).

Table 3 Study of gain and BW for Bluetooth/WiMax/WLAN

	Bluetooth	WiMax	WLAN
Frequency	2.4 GHz	3.5 MHz	5.5 GHz
Gain (dBi)	3.4	3.8	4.3
Bandwidth (MHz)	70	115	150

References

1. R. Garg, P. Bhartia, I. Bahl and A. Ittipiboon: Microstrip Antenna Design Handbook, Norwood, MA, Artech House, USA, 2001.
2. G. Kumar and K. P. Ray: Broadband Microstrip Antennas, Norwood, MA, Artech House, USA, 2003.
3. Kin-Lu Wong: Compact and Broadband Microstrip Antennas. Copyright 2002 John Wiley & Sons, Inc.
4. B. Lee and F. J. Harackiewicz, "Miniature microstrip antenna with a partially filled high-permittivity substrate," IEEE Trans. Antennas Propag., vol. 50, no. 8, pp. 1160–1162, Aug. 2002.
5. B. B. Mandelbrot, The Fractal Geometry of Nature. San Francisco, CA: Freeman, 1983.
6. R. Waterhouse, "Small microstrip patch antenna," Electron. Lett., vol. 31, no. 8, pp. 604–605, Apr. 1995.
7. K. L. Wong and Y. F. Lin, "Small broadband rectangular microstrip antenna with chip-resistor loading," Electron. Lett., vol. 33, no. 19, pp. 1593–1594, Sep. 1997.
8. P. Ferrari, N. Corrao, and D. Rauly, "Miniaturized circular patch antenna with capacitors loading," in Proc. IEEE MTT-S Int. Microw. And Optoelectron. Conf., 2007, pp. 86–89.
9. J. H. Lu, C. L. Tang, and K. L. Wong, "Slot-coupled compact broadband circular microstrip antenna with chip-resistor and chip-capacitor loading," Microw. Opt. Technol. Lett., vol. 18, no. 5, pp. 345–349, Aug. 1998.
10. Z. N. Chen and M. Y. W. Chia, "Broadband suspended plate antenna fed by double L-shaped strips," IEEE Trans. Antennas Propag., vol. 52, no. 9, pp. 2496–2500, Sep. 2004.
11. K. L.Wong, C. L. Tang, and H. T. Chen, "A compact meandered circular microstrip antenna with a shorting pin," Microw. Opt. Technol. Lett., vol. 15, no. 3, pp. 147–149, Jun. 1997.
12. Z. N. Chen, "Suspended plate antennas with shorting strips and slots," IEEE Trans. Antennas Propag., vol. 52, no. 10, pp. 2525–2531, Oct. 2004.
13. K.M. Luk, R. Chair, and K. F. Lee, "Small rectangular patch antenna," Electron. Lett., vol. 34, no. 25, pp. 2366–2367, Dec. 1998.
14. Nasimuddin, Z. N. Chen, and X.M. Qing, "A compact circularly polarized cross-shaped slotted microstrip antenna," IEEE Trans. Antennas Propag., vol. 60, no. 3, pp. 1584–1588, Mar. 2012.
15. Z. N. Chen, "Broadband suspended plate antenna with concaved center portion," IEEE Trans. Antennas Propag., vol. 53, no. 4, pp. 1550–1551, Apr. 2005.
16. Y. D. Dong, H. Toyao, and T. Itoh, "Compact circularly-polarized patch antenna loaded with metamaterial structures," IEEE Trans. Antennas Propag., vol. 59, no. 11, pp. 4329–4333, Nov. 2011.
17. M. S. Sharawi, M. U. Khan, A. B. Numan, and D. N. Aloi, "A CSRR loaded MIMO antenna system for ISM band operation," IEEE Trans. Antennas Propag., vol. 61, no. 8, pp. 4265–4274, Aug. 2013.
18. H. Wong, K. K. So, K. B. Ng, K. M. Luk, C. H. Chan, and Q. Xue, "Virtually shorted patch antenna for circular polarization," IEEE Antennas Wirel. Propag. Lett., vol. 9, pp. 1213–1216, 2010.
19. C. Y. Chiu, K. M. Shum, and C. H. Chan, "A tunable via-patch loaded PIFA with size reduction," IEEE Trans. Antennas Propag., vol. 55, no. 1, pp. 65–71, Jan. 2007.
20. D. Wang, H. Wong, and C. H. Chan, "Small patch antennas incorporated with a substrate integrated irregular ground," IEEE Trans. Antennas Propag., vol. 60, no. 7, pp. 3096–3103, Jul. 2012.

Novel All-Optical Encoding and Decoding Scheme for Code Preservation

Varinder Kumar Verma, Ashu Verma, Abhishek Sharma and Sanjeev Verma

Abstract This paper presents an all-optical code preserving encoder and decoder. The proposed encoder and decoder serve to preserve the code in between the communication channel for enhanced security of binary data. It is designed using the XGM effect in SOAs. The communication includes encoder and decoder simulations and results are evaluated in terms of logic received and Extinction ratio of systems. The encoder and decoder are verified to preserve the code appropriately.

Keywords All-optical · Encoder · Decoder · XGM · SOA

1 Introduction

The interminable need for superior processing speed for high bandwidth and information processing application is placing stern pressure on researchers to design techniques that can remarkably promise massively parallel data processing. More and more novel techniques and methods are being explored with increase in bit rates, number of optical channels. Study in the arena of new modulation formats is in a great pace for past few years. Alteration from simple line coding to DPSK-modulated system is gaining speed as it has made possible the upliftment of operational bit rate from 10 to 40 Gbps [1]. DPSK justifies interest in using it by

V.K. Verma (✉) · Ashu Verma
Luz Labs, Patiala, India
e-mail: er.varinderkumarverma@gmail.com

Ashu Verma
e-mail: ashuverma0000@gmail.com

Abhishek Sharma
BGIET, Sangrur, India
e-mail: engineers.abhishek@gmail.com

Sanjeev Verma
AS College, Khanna, India
e-mail: verma05sanjeev@gmail.com

S.C. Satapathy et al. (eds.), *Proceedings of the International Conference on Data Engineering and Communication Technology*, Advances in Intelligent Systems and Computing 469, DOI 10.1007/978-981-10-1678-3_24

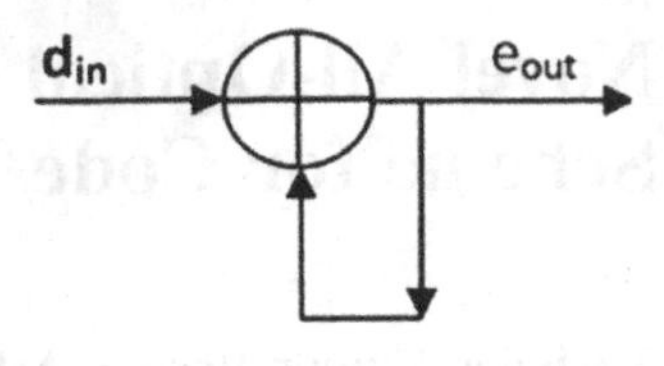

Fig. 1 Encoder schematic

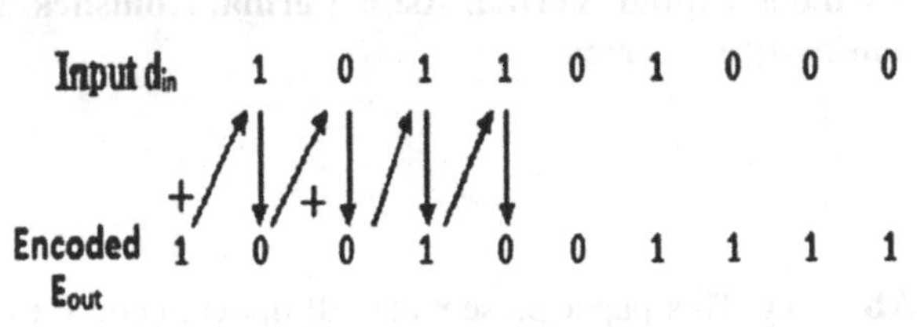

Fig. 2 Encoding operation illustration

presenting forbearance to nonlinear impairments occurring in optical communication links [2]. In this communication, we present an all-optical code preserving encoder and decoder using cross phase modulation in semiconductor optical amplifiers. The simulation shows encoding, phase reversal, and decoding process.

Digital bit streams passing by numerous communication circuits and channels are unintentionally but usually inverted [2]. This is known as phase ambiguity. Thus comes the use of differential encoder to shield against the possibility. Differential encoding is one of the uncomplicated forms of encoding done on a baseband sequence prior to external modulation to safe guard it from the above discussed possibility.

Let us consider $\mathbf{d_{in}}$ is the input data sequence and $\mathbf{e_{out}}$ be the output of differential encoder. Encoding is based on modulo 2 adder operation as shown in Fig. 1. For illustration, we take a data sequence $\mathbf{d_{in}} = 101101000$. A reference bit (0 or 1) is taken. The incoming data stream is added to this reference bit and thus generates the second bit of differentially encoded sequence [3]. This bit is then added to next incoming bit to prolong the process as shown in Fig. 2. The approach to this encoding circuit replication in all-optical domain is made by utilizing Cross Gain Modulation Properties of SOA [4].

2 System Description and Operation

The setup used for encoding comprises of two SOAs exploiting the effect of XGM as shown in Fig. 3. Both the SOAs share a common pump laser source at 1540 nm and 0.25 mw input power. The SOA in upper arm along with pump power is fed with input data sequence at 1550 nm and 0.316 mw. The lower arm SOA is fed with pump laser and output of encoder after filtration from a 20 GHz 1540 nm Gaussian optical filter. For this purpose an initializer is used in simulation. XGM effect is exploited in SOA. The exact course occurring in SOA, the mathematical equations under consideration are taken from [5] and [6]. Assuming that temporal

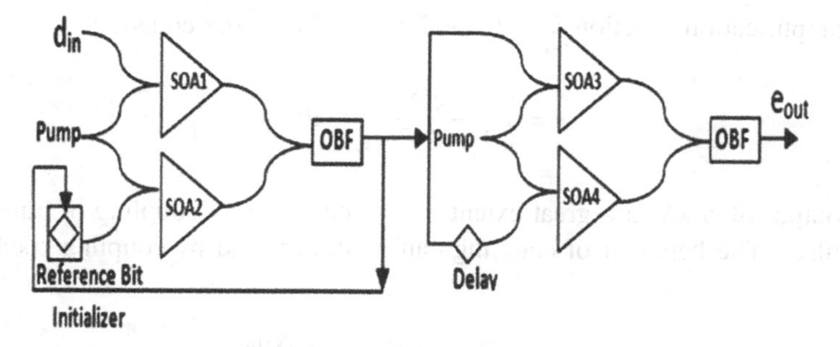

Fig. 3 Schematic diagram of simulation setup

width of input pump and probe pulses in same and pulses are overlapping perfectly, the powers are set as above mentioned values.

When two co-polarized pulses from pump and probe propagate into SOA and interference between them occurs the pump pulse at w_1 center frequency and probe at w_0, provokes a bit of carrier density pulsation at detuning frequency of $\Omega = w_1 - w_0$.

This leads to creation of a pulse at frequency $w_2 = w_0 - \Omega = 2\,w_0 - w_1$. The newly generated pulse is a negated copy of probe pulse and can be taken out from all signals using an optical filter timed at optimum frequency [5].

Here, $A_j\,(z, t), j = 0, 1, 2$ corresponds to slowly changing pump envelopes probe and negated pulses respectively. Taking into account following equation:

$$A_0(L,t) = A_0(0,t)e^{(1/2)(1-i\alpha)h} \tag{1}$$

where $A_0\,(0, t)$ describes amplitude of input pump pulse at initial end of SOA and $A_0\,(L, t)$ is amplitude of input pump pulsed at length L of SOA at time t [6].

$$\begin{aligned} A_1(L,t) = {} & A_1(0,t)e^{1/2[(1-i\alpha)h-\eta 10|Ao(0,t)|^2(e^h-1)]} \\ & \times \cosh\left[\left(\frac{1}{2}\right)\sqrt{\eta 02\tilde{\eta}01}|Ao(0,t)|^2(e^h-1)\right] \end{aligned} \tag{2}$$

and

$$\begin{aligned} A_2(L,t) = {} & \frac{-\hat{A}\,1(L,t)Ao(L,t)}{\acute{\hat{A}}\,o(L,t)}\sqrt{\eta 01/\tilde{\eta}02} \\ & \times \sinh\left[\left(\frac{1}{2}\right)\sqrt{\eta 01\tilde{\eta}02|}\,A0(0,t)|^2\,e^{-h}(e^h-1)\right] \end{aligned} \tag{3}$$

The amplification function h is defined by the following equation:

$$h = \int_0^L \frac{g(z,t)}{1 + \in Po(z,t)} dz \tag{4}$$

The output of SOA to a great extent also depends upon coupling of pump and probe pulses. The behavior of coupling can be understood by coupling coefficient *nij* [7].

$$\eta 01 = \eta 01^{\mathrm{CD}} + \eta 01^{\mathrm{CH}} + \eta 01^{\mathrm{SHB}} \tag{5}$$

where

CD is carrier depletion,
CH is carrier heating, and
SBH is spectral hole-burning

and

$$\eta_{01}^{\mathrm{CD}} = \frac{\in_{\mathrm{CD}}(1 - i\alpha)}{(1 + i\Omega\tau_s)(1 + i\Omega\tau_1)} \tag{6}$$

$$\eta_{01}^{\mathrm{CH}} = \frac{\in_{\tau}(1 - i\alpha_T)}{(1 + i\Omega\tau_h)(1 + i\Omega\tau_1)} \tag{7}$$

$$\eta_{01}^{\mathrm{SHB}} = \frac{\in_{\mathrm{SHB}}(1 - i\alpha_{\mathrm{SHB}})}{(1 + i\Omega\tau_1)} \tag{8}$$

Where ϵ_{CD} is the coefficient allied to carrier depletion, ϵ_{T} and ϵSHB are non-linear gain compression factors satisfying the following equation:

$$\in = \in_{T} + \in_{\mathrm{SHB}} \tag{9}$$

Also α_T and α_{SHB} are line width enhancement factors related to carrier heating and spectral hole-burning [7].

The signal thus received from filter is encoded sequence. For decoding the bit sequence to original, the above process is reversed. The incoming sequence and one bit delayed version of it are added together to recreate the original sequence as (Fig. 4):

For this purpose same architecture as encoder is used. Now, as the problem definition, the polarity reversal may occur in between the communication circuitry [8]. For replicating this reversal into simulation, we use an all-optical NOT gate. This gate is also a SOA-based architecture and reverses the whole sequence and makes it as demonstrated in Fig. 5.

So, in either case the decoding circuit is capable to get the original bit sequence.

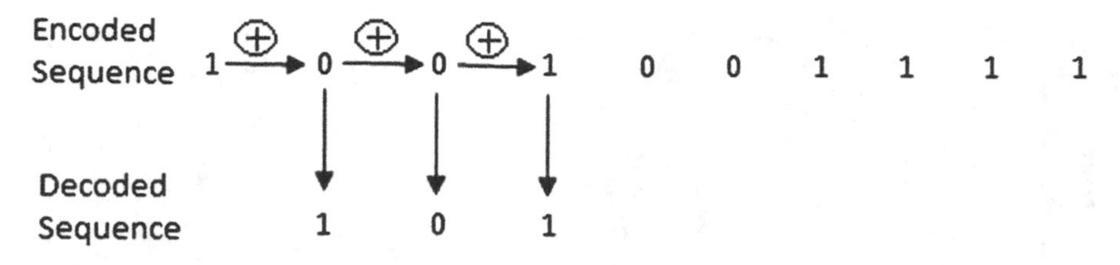

Fig. 4 Decoding operation illustration

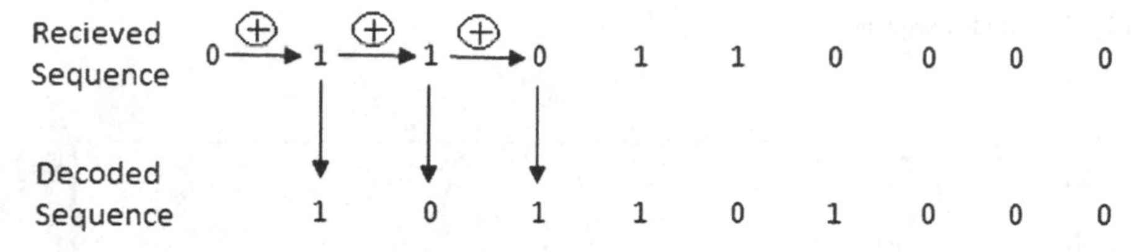

Fig. 5 Decoding operation of polarity reversed sequence

3 Results and Discussion

The simulation is performed at 10 Gbps operational speed. The complete sequence input, encoding, reversal, and decoded pulses in time domain are shown in figure. The input sequence 101101000 is fed to encoder and after the effect of XGM the output of encoder comes out as 100100111 (Figs. 6 and 7).

In case of polarity reversal the sequence becomes 011011000. This case is simulated by an all-optical NOT gate using SOA (Fig. 8).

Even after the reversal of pulse the concluding output comes to be 101101000, i.e., same as the original sequence. The system extinction ratio is found to be 11.02 dB (Fig. 9).

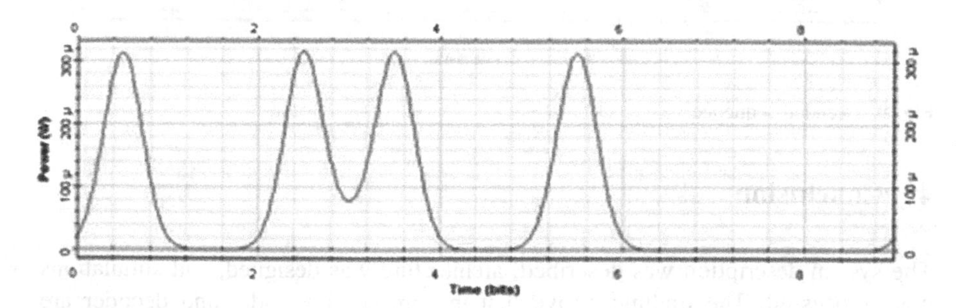

Fig. 6 Input sequence

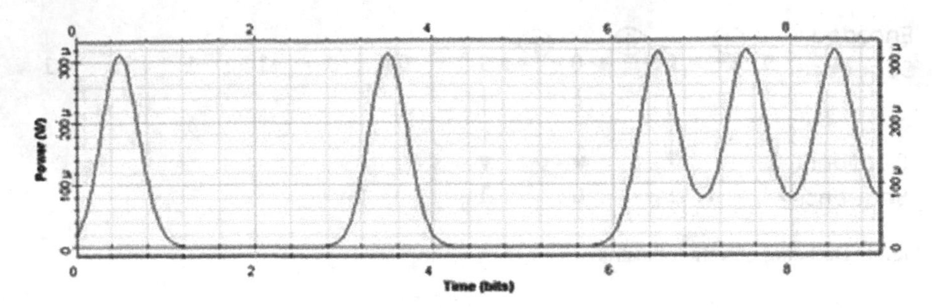

Fig. 7 Encoded sequence

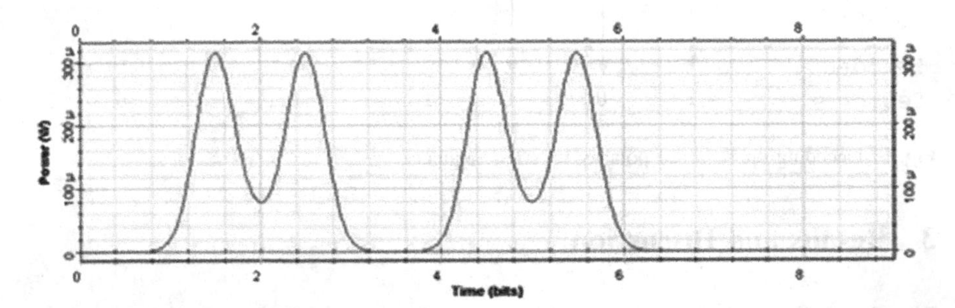

Fig. 8 Sequence after polarity reversal

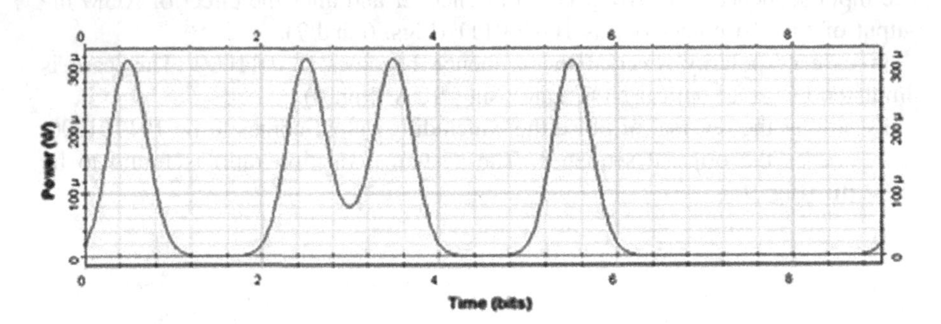

Fig. 9 Decoded sequence

4 Conclusion

The system description was described, architecture was designed, and simulations were processed. The findings prove that the proposed encoder and decoder are observed and as well found valid for all input sequences. The results obtained from simulation match with their logical counterparts. The system extinction ratio is found to be satisfactory. The model is made to operate at 10 Gbps. More operational speed can be achieved with a trade off with extinction ratio. The proposed setup can be used for duobinary encoding in advanced modulation formats and generation of DPSK, DEBPSK-modulated signals.

References

1. W. Kaiser, T. Wuth, M. Wichers and W. Rosenkranz, "A simple system upgrade from binary to duobinary" National Fiber Optic Engineers Conference Technical proceedings, 2001, pp 1043–1050.
2. Vorreau, P. Marculescu, A.; Wang, J.; Bottger, G.; Sartorius, B.; Bornholdt, C.; Slovak, J.; Schlak, M.; Schmidt, C.; Tsadka, S.; Freude, W.; Leuthold, J. IEEE Photonics Technology Letters 10/2006; 18(18):1970–1972. DOI:10.1109/LPT.2006.880714.
3. Differential Encoding tutorial, Signal Processing & Simulation Newsletter,Contemporary report, 2008.
4. W. Kaiser, T. Wuth, M. Wichers, W. Rosenkranz, IEEE Photonics Technology Letters, vol. 13, no. 8, pp. 884–886, 2001.
5. Vikrant K. Srivastava, Devendra Chack, Vishnu Priye, Chakresh Kumar, "All-Optical XOR Gate Based on XGM Properties of SOA", *CICN*, 2010, International Conference on, Computational Intelligence and Communication Networks, International Conference on 2010, pp. 544–547, doi:10.1109/CICN.2010.107.
6. S Junqlang et al., "Analytical solution of Four wave mixing between picosecond optical pulses in semiconductor optical amplifiers with cross gain modulation and probe depletion", Microwave and Optical technology Letters, Vol. 28, No. 1, pp 78–82, 2001.
7. Hari Shankar, "Duobinary modulation for optical systems", IEEE Journal of Light wave Technology, vol. 23, no. 6, 2006.
8. Rosenkranz, W.: "High Capacity Optical Communication Networks - Approaches for Efficient Utilization of Fiber Bandwidth", First Joint Symposium on Opto- & Microelectronic Devices and Circuits (SODC 2000), 10.-15.04.2000, Nanjing, China, pp. 106–107.

XPM-Based Bandwidth Efficient WDM-to-OTDM Conversion Using HNLF

Abhishek Sharma and Sushil Kakkar

Abstract With the growing requirement for advanced speeds and better capacity brought about by speedy data expansion on the Internet, Optical time division multiplexing attracted much attention for its high-speed operations and ability to overcome the electronic bottleneck problems. In this article, 4 × 10 Gb/s WDM/OTDM conversion using supercontinuum (SC) generation at 100 GHz is proposed for 12, 24, and 36 km using SPWRM (Symmetrical pulse width reduction module). HNLF-based multiplexing eliminate the requirement of extra pump sources and provide cost-effective solution. FWM (four wave mixing) demultiplexing using SOA (semiconductor optical amplifiers) is achieved for all the 4 channels with acceptable limits of BER = 10^{-9} at 24 km.

Keywords Pulse width reduction · Four wave mixing · Supercontinuum · Semiconductor optical amplifier · Wavelength division multiplexing

1 Introduction

Wavelength division multiplexing (WDM) and optical time division multiplexing are the promising technologies characterized by capacity of high transmission in optical communication systems possess their own benefits [1]. In current scenario, it is quite obvious to expand the reach of optical systems and to enhance all-optical processing in order to overcome bottleneck problems, which replace conventional E/O conversion with electronic switches in high-speed OTDM networks. WDM–to-OTDM systems to provide all-optical conversion, such solution becomes a significant move for providing system operations at photonic gateways [2]. Basics of WDM/OTDM system is to aggregate or transmultiplex lower data

---

Abhishek Sharma (✉) · Sushil Kakkar
Department of ECE, BGIET, Sangrur, India
e-mail: engineers.abhishek@gmail.com

Sushil Kakkar
e-mail: kakkar778@gmail.com

S.C. Satapathy et al. (eds.), *Proceedings of the International Conference on Data Engineering and Communication Technology*, Advances in Intelligent Systems and Computing 469, DOI 10.1007/978-981-10-1678-3_25

rate tributaries operated at dissimilar frequencies for the realization of high-speed core networks [3]. Till now, lot of work has been reported for WDM/OTDM conversion using different nonlinear mediums such as electro-absorption modulator (EAM), semiconductor optical amplifiers(SOA), and MZI-SOA [4]. However, the limitation of modulators and amplifiers is the restricted frequency response. Highly nonlinear fibers provide efficient solution and attracted much attention for realization of optical signal processing. Another approach for OTDM/WDM conversion using four wave mixing and XPM (cross phase modulation) [5]. However, additional pump sources are required to realize conversion based on Raman compression [6], nonlinear optical loop mirror (NOLM). Supercontinuum generation using HNLF has been reported experimentally for 4 × 10 Gb/s WDM/OTDM conversion at 200 GHz spacing [7]. Advantage of using SC generation is a cost-effective solution to eliminate the requirement of extra pump sources.

In this article, we investigated a simulation setup of WDM/OTDM conversion using SC generation at 100 GHz channel spacing for long distance transmission employing symmetrical pulse reduction technique. Conversion of 4 × 10 Gb/s WDM to 40 Gb/s OTDM has been realized at less channel spacing with time interleaving provided by delay blocks and isolation of different wavelengths achieved with 4 × 4 array waveguide grating in demultiplexed mode. FWM demultiplexing is used for reception of each channel incorporating semiconductor optical amplifier (SOA). Therefore, a bandwidth efficient all-optical WDM/OTDM system with the cost-effective solution measuring BER $<10^{-9}$ for 24 km has been proposed.

2 Functioning Theory

M wavelengths $\lambda 1$, $\lambda 2$ …λ_M are generated using MZM (Mach–Zehnder modulator) and followed by EAM (electro-absorption modulator) for generating RZ line coding converted with a drive from sine generator synchronized to NRZ in order to realize ultra short pulse. Transmitted data from *M* channels with *D* rate is calculated using $B = M \times D$ for total data rate. AWG is operated in demultiplexing mode for routing of each wavelength to different output port followed by time delay module. Each channel delayed by t for time interleaving and fed to HNLF for SPM (Self phase modulation). SPM is calculated as [8]

$$\Delta\omega = \frac{w}{c} n2z \frac{Ic}{\tau} \tag{1}$$

SPM cause spectrum broadening in HNLF and overlapped spectrum achieved with common wavelength is referred as λc. Optical filter is used to select the λc from the broader and overlapped spectrum. Symmetrical pulse reduction technique is used with *L*/2 km length of SMF before and after DCF of length *L* Km in order to mitigate the effects of pulse broadening.

3 Supercontinuum Generation

Highly nonlinear fibers (HNLF) are used to generate phase shift and spectrum broadening referred as supercontinuum generation due to lower effective area (A_{eff}) [8]. SC generation is very popular and attracted much attention because of its application in WDM systems. At higher launched powers, response of HNLF becomes prominent and introduce frequency shift to larger extent. Major advantage of using HNLF as a channel for SC generation is that its nonlinear response is ultra fast and eliminates the requirement of additional pumps to introduce spectral broadening.

4 Simulation Setup

Figure 1a. Represents the Transmitter setup for 40 Gb/s WDM/OTDM system at 100 GHz channel spacing employing symmetrical pulse width reduction technique. A continuous wave laser array is used to transmit four optical reference signals starting from 1557.35 to 1559.79 nm with 0.8 nm channel spacing followed by a Mach–Zehnder modulator driven by NRZ line coder from PRBS of order 2^7–1 biased at 10 Gb/s. EAM converts NRZ to RZ for generating a ultra short pulse which synchronized with 10 GHz sine pulse generator. 4 × 4 AWG is used in demux mode for routing of different wavelengths to different output ports. Each separated wavelength temporarily interleaved or shifted by 0, 0.025, 0.05, 0.075 ns and amplified by EDFA then fed to HNLF for SC generation as shown in Fig. 1b. Specification of HNLF are given in Table 1. Overlapped spectrum wavelength λc at 1558.88 nm is filtered out using optical Bessel filter of bandwidth 0.3 nm. Power booster EDFA with 10 dB gain incorporated after optical filter and converted signal transmitted over symmetrical pulse width reduction module consisting of single-mode fiber (SMF), EDFA and DCF. Specification of PWR module is given in Table 2. Figure 1c shows the demultiplexing of each channel at receiver part using SOA of 0.3 confinement factor. BER visualize evaluates the system performance in terms of eye opening, power penalty, eye closer penalty, and OSNR.

5 Results and Discussion

Figure 1a represents the four different wavelengths each biased with 10 Gb/s data signal after MZM. Signal broadening spectrum of all the four channels after SC generation due to SPM shown in Fig. 1b. Overlapped spectra achieved after HNLF and λc is filtered out using optical Bessel filter as 40 Gb/s OTDM signal represented in Fig. 2c.

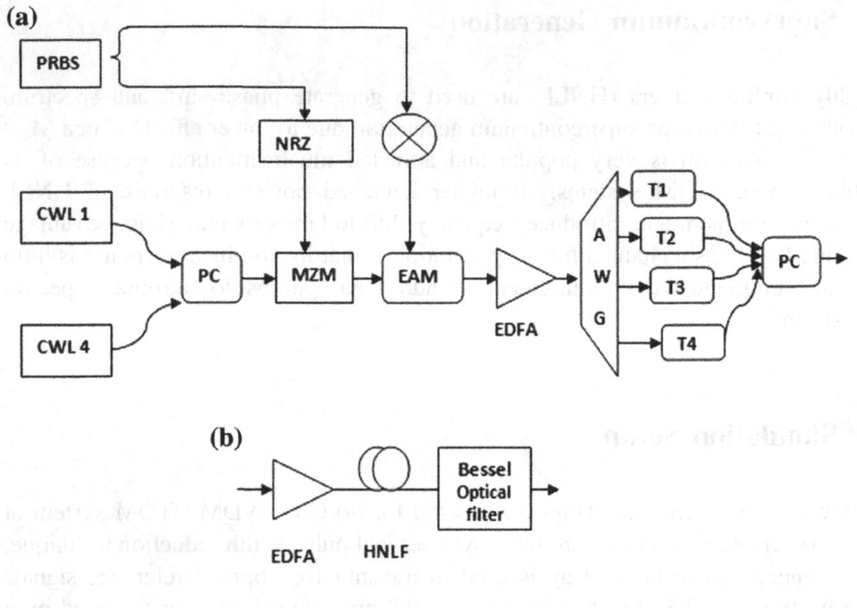

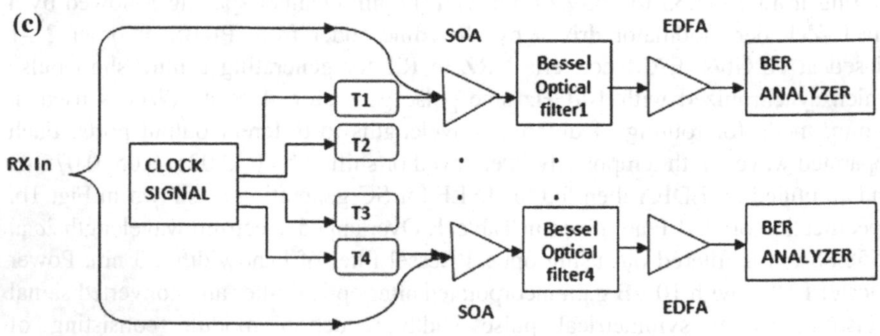

Fig. 1 **a** Transmitter setup for 40 Gb/s WDM/OTDM system at 100 GHz. **b** Super continuum Generation. **c** FWM based demultiplexing using SOA

Table 1 Specification of HNLF

Quantity	Values
HNLF length	2 km
Second-order dispersion	−2.2 ps/nm/Km
Third-order dispersion	0.032 ps/nm^2/Km
A_{eff}	11 um^2
Attenuation	0.55 dB/Km

Table 2 SPWR module specifications

Parameters	Values
SMF	10 km/each loop
DCF	2 km/each loop
Attenuation $_{SMF}$	0.2 dB/Km
Attenuation $_{DCF}$	0.5 dB/Km
Aeff $_{SMF}$	72 um^2
Aeff $_{DCF}$	22 um^2
Dispersion $_{SMF}$	17 ps/nm/Km
Dispersion $_{DCF}$	−85 ps/nm/Km

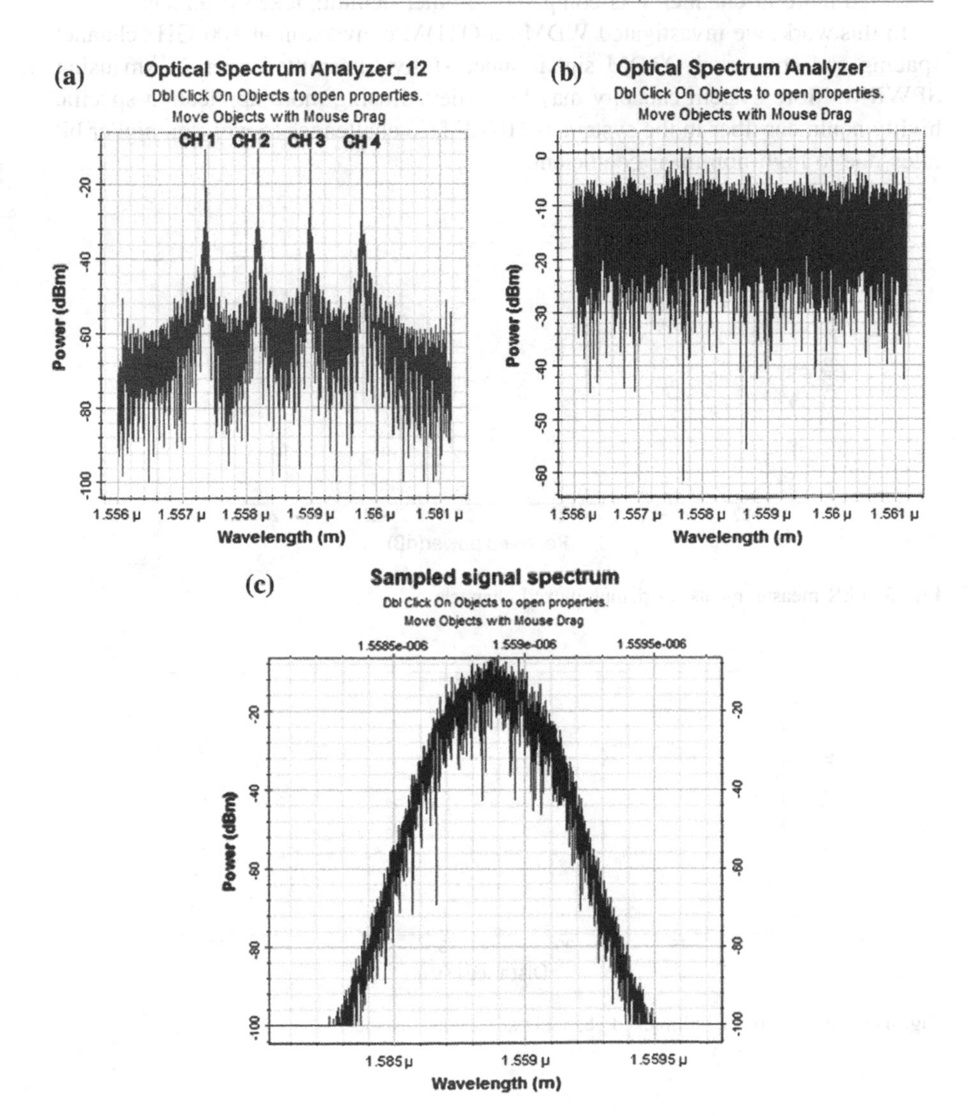

Fig. 2 Optical power spectrum of **a** 4 × 10 Gb/s WDM channels. **b** After SC generation. **c** Filtered λc

OTDM signal transmitted over optical fiber transmission module consisting of SMF of 10 km and DCF of 2 km followed by EDFA with 5 dB gain after each SMF and DCF in order to compensate attenuation effects. Symmetrical pulse width reduction arrangements are used and system performance is investigated for 12, 24, and 36 km. Figure 3 shows the graphical representation of all demultiplexed channels at different distance in terms of Q-factor. Also BER performance of all the channels with respect to received power has been evaluated as shown in Fig. 4. System works successfully for 12 km of link distance with BER value 10^{-11} and maximum achievable distance evaluated at BER 10^{-9} for 24 km. Power penalty is observed more at channel 4 as compared to other demultiplexed channels.

In this work, we investigated WDM–to-OTDM conversion at 100 GHz channel spacing and converted OTDM signal successfully transmitted over 24 km using SPWRM. More system capacity may be achieved using more application specific highly nonlinear fiber in the system as HNLF face challenges to work on higher bit rates due to high nonlinear coefficient.

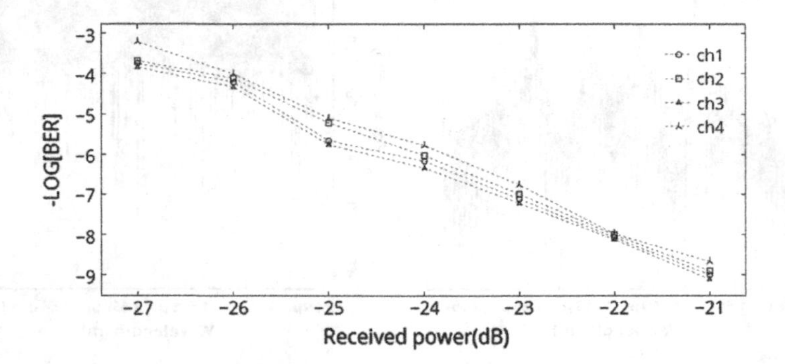

Fig. 3 BER measurements for demultiplexed channels

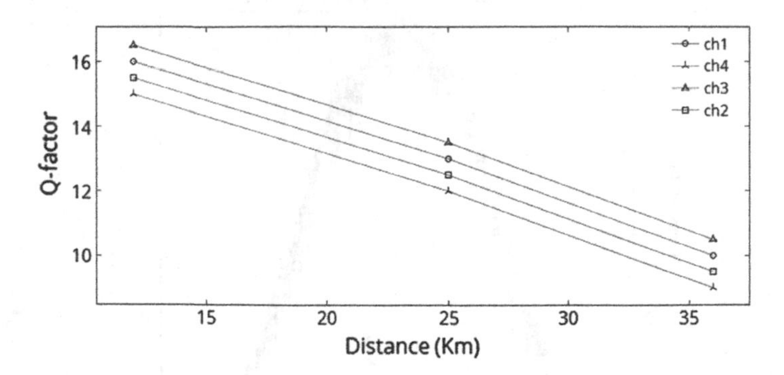

Fig. 4 Evaluation of demultiplexed channels at varying distance

6 Conclusion

All optical 4 × 10 Gb/s WDM-to-OTDM conversion at 100 GHz has been successfully investigated in this work using super continuum generation based on SC multiplexing by incorporating HNLF. Further 40 Gb/s OTDM signal transmitted over 24 km link of fiber using SPWRM with acceptable BER of 10^{-9}. Demultiplexing of four different channels realized with semiconductor optical amplifiers (SOAs) using four wave mixing (FWM). Advantage of using SC generation in the system is, elimination of additional pumps required for the WDM-to-OTDM conversion.

References

1. T. Ohara et al., "Over-1000-channel ultradense WDM transmission with supercontinuum multicarrier source," J. Lightw. Technol., vol. 24, no. 6, pp. 2311–2317, Jun. 2006.
2. M. Nakazawa, T. Yamamoto, and K. R. Tamura, "1.28 Tb/s-70 km OTDM transmission using third-and fourth-order simultaneous dispersion compensation with a phase modulator," Electron. Lett., vol. 36, no. 24, pp. 2027–2029, Nov. 2000.
3. H. Sotobayashi, W. Chujo, and K.-I. Kitayama, "Photonic gateway: TDM-to-WDM-to TDM conversion and reconversion at 40 Gbit/s(4 channels × 10 Gbits/s)," J. Opt. Soc. Amer. B, vol. 19, no.11, pp. 2810–2816, Nov. 2002.
4. V. Polo, J. Prat, J. J. Olmos, I. T. Monroy, and A. M. Koonen, "All optical FSK-WDM to intensity modulation-OTDM transmultiplexing for access passive optical networks," J. Opt. Netw., vol. 5, no. 10, pp. 739–746, Oct. 2006.
5. B.-E. Olsson and D. J. Blumenthal, "WDM to OTDM multiplexing using an ultrafast all-optical wavelength converter," IEEE Photon. Technol. Lett., vol. 13, no. 9, pp. 1005–1007, Sep. 2001.
6. Q. Nguyen, M. Matsuura, and N. Kishi, "All-optical WDM-to-OTDM Conversion using a multiwavelength picoseconds pulse generation in Raman compression," IEEE Photon. Technol. Lett., vol. 24, no. 24, pp. 2235–2238, Dec. 15, 2012.
7. Q. Nguyen, M. Matsuura, and N. Kishi, "WDM-to-OTDM Conversion Using Supercontinuum Generation in a Highly Nonlinear Fiber," IEEE Photon. Technol. Lett., vol. 26, no. 18, pp. 1882–1885, Sep.15, 2014.
8. G. P. Agrawal, Nonlinear Fiber Optics. New York, NY, USA: Academic, 1995.

Analysis of a Sporting Event on a Social Network: True Popularity & Popularity Bond

Anand Gupta, Nitish Mittal and Neeraj Kohli

Abstract Events in a social network and their popularity are described by the quantitative participation of its users. A special occasion in an event is an activity that may hamper or strengthen its popularity or popularity of its entities. Such a study helps the researchers to analyze the trends to know how they change with time during an occasion. Till now popularity is computed by considering the number of tweets. To the best of our knowledge, no study has been done on computing the number of tweets considering the population. Here in this paper, we coin the following terms (a) true popularity, which is the number of tweets normalized with the population. Through this we compare intra-group popularity of entities, and (b) popularity bond, so as to study concentration of tweets for pairs of entities. Through this we compare the inter-group popularity of entities. Experiments are carried out on the content posted by users on Twitter during the Cricket World Cup 2015. Experimental study indicates the effectiveness of the coined terms in providing better insights.

Keywords Data analysis · Data mining · Social network · Sports · Event · Popularity

1 Introduction

The social networking services facilitate to connect people who wish to share interests and activities across political, economic, and geographic borders. It leads to a massive growth in terms of research on the enormous data so generated and its impact. Such social networks provide people a platform to create, evaluate, express,

---

Anand Gupta (✉) · Nitish Mittal · Neeraj Kohli
Netaji Subhas Institute of Technology, University of Delhi, Delhi, India
e-mail: omaranand@gmail.com

Nitish Mittal
e-mail: nitishmittal94@gmail.com

Neeraj Kohli
e-mail: nks1977@gmail.com

S.C. Satapathy et al. (eds.), *Proceedings of the International Conference on Data Engineering and Communication Technology*, Advances in Intelligent Systems and Computing 469, DOI 10.1007/978-981-10-1678-3_26

and convey information, preferences, and opinion. As described in [1], Twitter, a social network service, is considered to be superior over its other competitors because of its massive following, large use of hashtags, and real-time connection through conversations that are 140 characters long. Hence it attracts mass attention of millions of its users.

An event in real world is a planned activity that happens or takes place. In a social network, a trending event is a real-world occurrence associated with a time period, a time order sequence of documents about the occurrence and published during the time period having mass following. The users express their opinions on the event as a whole or one of its entities. A special occasion in an event is an activity that hampers or strengthens the popularity of an event or its entity on a social network. The studies that have been done so far have focused only on the number of tweets for popularity. Here we feel that if we incorporate the population of the country to which an entity belongs along with the number of tweets for estimating its popularity, we are likely to see a different picture altogether. Further, the relationship between the different pairs of entities of an event on the basis of popularity is likely to give informative insights. We try to explain these concepts through the following example.

Consider an international sporting event of game X that is taking place from January 1 to January 31. The teams are divided into two groups—Group Alpha and Group Beta. For initial stages, the teams only play against the teams in their groups. In this example, we shall consider three teams that are a part of this event. One of them is team Challengers. This team belongs to a country Qwerty and is in group Alpha. Second team we shall consider is team Hitters. This team belongs to a country Trew and is also in group Alpha. Third team under consideration is team Attackers. This team belongs to a country Uiop and is a part of group Beta.

Most of the international sporting events are a great source of remuneration for their organizers. Also such events turn out to be very beneficial for the advertisers who are able to promote their product amongst the mass. However, the organizers are not firm about how to arrange the teams in the groups so that there can be matches which can lead to massive earnings. Similarly an advertising company is always in a dilemma of choosing the right matches so that they can obtain maximum promotion of their product.

In this example, we discuss how the popularity of event and its entities is of importance for the organizers of the event and the advertisers. Consider a company Z interested in advertising in this sporting event but it has limited number of funds. So it is on a lookout for the matches that will be high on the viewership for its advertising.

The tweets obtained can be broadly broken down under three categories. There can be General tweets with content about the event. There can be tweets talking about a particular team like the performance of team Attackers. There can also be tweets concerning a particular pair of teams like a match between team Challengers and team Hitters.

Consider team Challengers. On the day of their match, there will be a large number of tweets for them as compared to their rest days. Suppose there is a huge craze for the sport X in Qwerty, there will be large number of tweets coming with geolocation as Qwerty. This is the best indication for the company Z to invest in the advertisements on the matches related to Challengers in Qwerty. If Qwerty has a large population, they also stand a big chance of being the most popular team in the sporting event. Similar trends will exist for the other teams.

Now considering the team Hitters. They belong to the same group as that of Challengers. The intra-group analysis can be made on the popularity in terms of number of tweets about the teams amongst their group. This popularity can be further analyzed after normalizing it with the population of the countries these teams belong to. We propose to define this through a term "true popularity" subsequently in this paper.

Now let us take into account the third team under consideration that is team Attackers. Suppose Qwerty and Uiop are arch rivals. The match between Challengers and Attackers will be highly anticipated and there will be a large inflow of tweets concerning this match. Hence this is used for studying inter-group relations among the teams in different groups. We propose to define this through a term "popularity bond" subsequently in this paper. The company Z must eye this match for their advertisement.

For the organizers it is very important to tactfully arrange the teams to get maximum viewership, both on and off the stands. Number of viewers for terminal matches is always greater than the initial ones. They have to make sure that they get large viewership for these matches too. As Challengers and Attackers are a part of different groups, there will be no match between them at initial stages. Hence the total number of Challengers versus Attackers matches in the World Cup will be very few. As the popularity of the few matches held between these teams at later stages in the event is high, this is a clear indicator for organizers to keep these teams in the same group for the next edition of this sporting event. This will lead to more viewership for the intra-group initial stage, more number of advertisers for these matches and hence leading to more monetary profit.

We have now coined two new terms related to popularity through the above example. Before we define these terms, it is important to have a look at the existing research on popularity trends in social networks which we describe in the subsequent subsection.

1.1 Related Work

There are many event based studies on which researchers have worked. Becker [1] has given definitions of an event and a trending event for different mediums. The study of trending events during peak activity periods has been studied by Kairam et al. [2]. Persistence and decay of trending events on Twitter has been studied by Asur et al. [3]. Becker et al. [4] and Kim et al. [5] have explored approaches for identifying real-world event related Twitter posts and studied the diffusion patterns of such events.

Content on the web like news articles, blog posts, and posts in social networks about such events has a temporal activity and popularity. Arora et el. [6] have studied this temporal activity for a inter-state sporting event Indian Premier League 2013 on social network. They have also determined that geo-tagged data showed major activity from metropolitan suburbs. Clusters in tweets for the same event held in 2011 have been identified by Kewalramani [7] on the basis of the teams that were a part of a particular league. Mazumdar et al. [8] have introduced the notion of Time-aware News Concept Graph to depict the temporal dynamics for news articles. Credibility of content on Twitter for crisis situations has been studied by Gupta and kumaraguru [9].

Temporal popularity on online platforms mentioned earlier has a different meaning for different people. Sun and Ng [10] have proposed the comment arrival model to measure popularity. They have considered four parameters, viz, total number of comments, average comment frequency, peak comment frequency, and highest rising rates of comments.

1.2 Limitations and Motivation

First there has been research on basic popularity trends [6] for a national level cricketing event but no research on a common method for international events. Second clustering of tweets [7] has been identified but the relationship among different clusters or amongst the same cluster has not been observed. Third temporal popularity of entities has been calculated through different techniques in [10] but the significance of population has not been taken into account. These limitations motivate us to overcome them in our study. Further the study by Mazumdar et al. [8] is a motivation to analyze the time-aware trends of popularity of the entities. Research by Asur et al. [3] also motivates us to study the growth and decay of the event under consideration.

2 Notations and Definitions

Some of the essential notations that are further used in the definitions below are formally introduced.

An event is denoted by E. An occasion in an event is denoted by O. There are n entities taking part in an event denoted by set $\rho = \{e_1, e_2, e_3, ..., e_n\}$. G_1 is a set of entities in group-1 and G_2 is a set of entities in group-2, such that $G_1 \cup G_2 = \rho$. We shall now consider the sets of tweets. D denotes the complete set of tweets. g represents a set of general tweets, t a set of team specific tweets and p a set of pair specific tweets such that $g \cup t \cup p = T$.

We first formally define the constituents of our dataset in terms of a Document Stream.

Definition 1 (*Document Stream* [1]) A document stream is a time-ordered sequence of documents; each document represents a set of features, or terms. In this paper, a document is a tweet whereas the document stream is the set of tweets D. There exists a specific time period for which the document stream is large quantitatively.

Definition 2 (*Trending Time Period* [1]) A trending time period for a feature over a document stream is a time period where the document frequency of the feature in the document stream is substantially higher than expected. In our example, the trending time period for the international sporting event is the complete month of January. When we consider a document stream over such a time period for an event, it gives rise to what is known as a trending event.

Definition 3 (*Trending Event* [1]) A trending event is a real-world occurrence with (1) an associated time period T, (2) a stream of documents D about the occurrence and published during time T, and (3) one or more features that describe the occurrence and for which T is a trending time period over document stream D. For a trending event, the popularity of the entities is of great concern. In our example, the international sporting event is a trending event and the popularity of its entities like the Team Challengers is of interest for the advertisers. However, just the tweet count for the determination of popularity may lead to wrong analysis. For this purpose, we take into account the population of the countries to which these teams belong and define True Popularity.

Definition 4 (*True Popularity (TP)*) The true popularity of an entity e_i in a trending event E is the number of occurrence of e_i in t normalized by its population, that is n_{e_i}/X is the true popularity of e_i where n_{e_i} is the number of occurrences of e_i in g and X is the population of the country to which e_i belongs.

Definition 5 (*Popularity Bond*) A popularity bond between e_i and e_j is the quantitative measure of the occurrences of (e_i, e_j) in p. The strength of a popularity bond is the relative measure of the popularity bonds among different entities $\in \rho$. There may exist a strong or a weak popularity bond depending on whether the occurrences of the pair is high or low respectively. For inter-group popularity bonds, we shall consider the occurrences $\forall(e_i, e_j) \in G_1 \times G_2$ in p. In our example, team Challengers and team Attackers are from two different groups and thus have an inter-group bond. Their bond is of strong strength due to rivalry between the countries to which these two teams belong. For intra-group popularity bonds, we shall consider the occurrences $\forall(e_i, e_j) \in G_k \times G_k$ in p where $i \neq j$. Team Challengers and team Hitters share an intra-group popularity bond.

3 Dataset

3.1 Data Extraction

The tweets for the Cricket World Cup 2015 have been collected using Twitter REST API. Further we made use of reverse geocoding APIs for studying the geolocation of the users.

Event Extraction The posts made by users of a social network on the network is a large dataset. Such a big dataset is filtered according to the keywords or hashtags particular to a sporting event. We obtained official hashtag used for World Cup tweets that is #CWC15. We then used this hashtag for querying the Twitter API for fetching these tweets. The results returned by the API is a collection of JSON objects containing the tweet content, user information and the geolocation of the user, if mentioned. There exists a limit of 180 queries per 15 min window in Twitter API. Hence we used multiple instances of our crawler to pipeline the requests and obtain efficiency in terms of quantity of tweets per hour.

Parsing of Geolocation provided by Twitter API The Twitter API provides the longitude and latitude coordinates of the current location of a user, if he has allowed Twitter to post with information about his location. We converted these coordinates returned by the Twitter API to convert it into the country code to which it belongs. For this conversion we made use of reverse geocoding APIs provided by Bing, Google and Data Science Toolkit.

Entity Extraction We developed an entity classifier which filtered out the tweets depending upon the team (s) it is referring to. For this we built regular expressions to tackle the possible combinations for a team name. All the combinations are checked as separate words, usernames (@[expression]) and hashtags (#[expression]). All the tweets are checked for all the teams to obtain the type of tweets and the teams involved in each of them.

3.2 Data Description

The cricket world cup was held from February 14, 2015 till March 29, 2015. However, the data has been collected from February 9 till April 3 to analyse the pre and post match tweets as well. Our data set consists of 3,162,124 (~3 million) tweets collected through Twitter API. Out of the total number of tweets in the data set, 1,334,964 (~42 %) of the tweets are geo-tagged.

The tweets obtained can be broadly divided into three categories—General, Team Specific, and Match Specific as mentioned in Table 1. This shows us the interests of the tweeters related to the World Cup, whether it is general or attachment with a particular team or concerning matches. On analyzing the tweets it is found that 51.1 % of the total tweets (~1.6 million) are talking about a particular match. The team specific tweets are the least, indicating more interest in matches rather than a team.

Table 1 Category of tweets

General world cup tweets	51.1 %
Team specific tweets	32.5 %
Match specific tweets	16.4 %

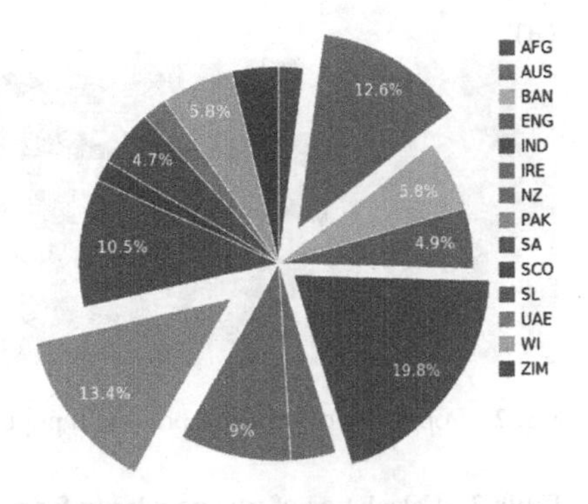

Fig. 1 Popularity of teams

There were 14 teams that took part in the event. They are India, Australia, New Zealand, Ireland, Kenya, Sri Lanka, Bangladesh, South Africa, Zimbabwe, England, UAE, Pakistan, Scotland, Afghanistan. Figure 1 gives the domination of a team in the World Cup in terms of the percentage of the number of tweets. The three offset tweets in the plot indicate the top three popular teams. India is the most popular team in the cricket World Cup 2015 having 19.8 % tweets trending about them followed by Pakistan with 13.4 % and Australia with 12.6 %.

4 Experiments and Discussion

4.1 Intra-Pool Analysis

In order to determine the popularity of a team we use Tree Maps. A tree map is used to depict hierarchical data as a set of nested rectangles of varying sizes. Here we use it to measure the true popularity of a team. The most accurate calculation of popularity is made in terms of the number of tweets viz-a-viz the population size. Figure 2 represents Tree Maps for teams in the two pools. The size of the rectangle is relative to the popularity of the team in the world cup while the color is relative to the population of the country to which the team belongs.

From Fig. 2a and Table 2, it can be observed that the number of tweets obtained for Australia is greater that New Zealand in Pool A but the true popularity shows an opposite trend. Bangladesh has the third highest number of tweets but lies at the last spot in terms of the true popularity. In Fig. 2b and Table 3 it can be seen that Ireland having a small number of tweets is the most popular in terms of true popularity while India having the largest number of tweets is at the least position in Pool B. It can be inferred that cricket world cup is of major interest in Ireland with mass attraction.

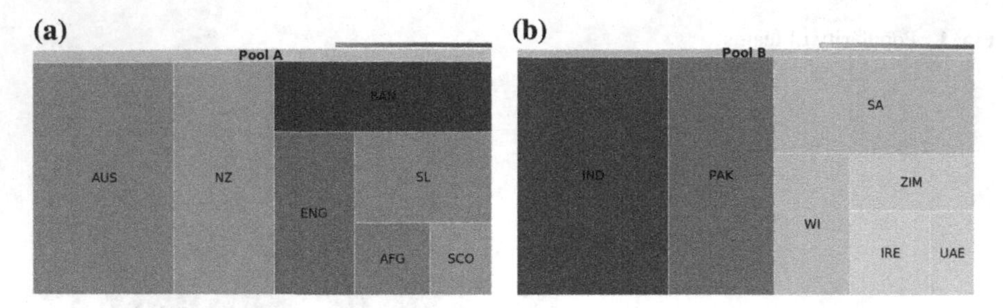

Fig. 2 Popularity of teams in pool with population of associated country. **a** Pool A. **b** Pool B

Table 2 Calculation of true popularity for pool A (population in millions)

Team	Tweets:Population	TP
NZ	351837:5	70367
AUS	493991:23	21477.87
SCO	67117:5	13423.4
SL	185435:20	9271.75
ENG	193013:53	3641.75
AFG	80939:30	2697.97
BAN	228070:156	1461.99

Table 3 Calculation of true popularity for pool B (population in millions)

Team	Tweets:Population	TP
IRE	145762:5	29152
ZIM	151657:14	10832.64
UAE	79801:9	8866.78
SA	409112:53	7719.09
WI	225109:39	5772.03
PAK	525278:182	2886.14
IND	776248:1252	620

4.2 Inter-Pool Analysis

For the purpose of analysis the inter-pool trends, we have used a Sankey diagram. Sankey diagrams are primarily used to show the flow of energy or matter through a system. These have been also used by Ogawa et al. [11] in similar studies to depict dynamic interaction between entities. In the Sankey diagram in Fig. 3, the nodes are the teams and the weightage of the links determines the strength of relationship among the teams of the two different pools. It can be observed that Afghanistan, England and Scotland from Pool A and UAE, Ireland and Zimbabwe from Pool B

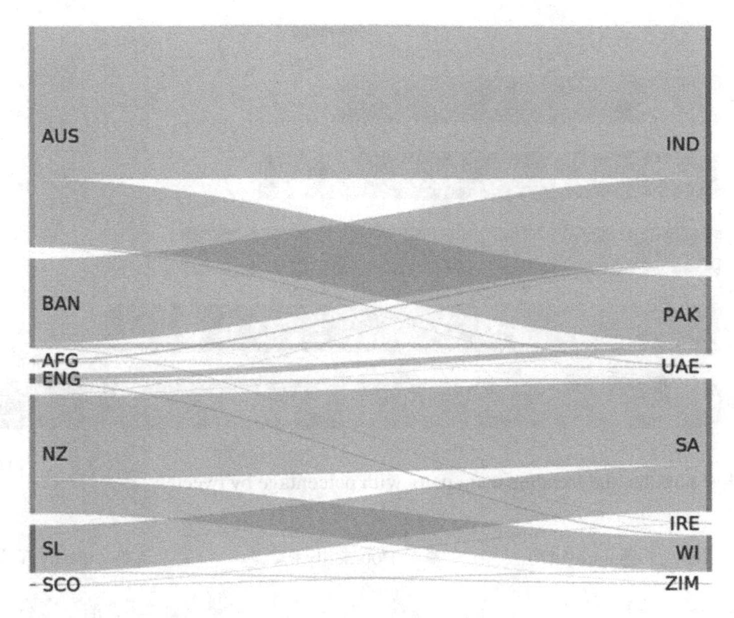

Fig. 3 Relationship between teams from pool A and pool B

are the least popular in terms of inter-pool relationship. They share a weak popularity bond.

On further analysis it is observed that the team pair of Australia and India is the most talked about on Twitter. Hence these two teams share the strongest popularity bond among the other inter-pool pairs. This pair is followed by Bangladesh and India, New Zealand and South Africa, Australia and Pakistan in order of decreasing bond strength.

4.3 Team-Oriented Analysis

We have till now discussed about the teams as a group in their respective pools. Now we will study the behavior of popularity pattern in terms of team-specific analysis. We first depict the time-aware popularity trends of the three most popular teams of the cricket world cup 2015—India, Pakistan, and Australia (Fig. 4).

Figure 5 represents time-aware popularity line charts which show the day-wise popularity trend of the top three trending teams in terms of the percentage of total tweets obtained for the team. We analyze some idiosyncratic peaks in the charts. There is a massive peak of over 50 % popularity of India on March 26, 2015. The day is the semi-final match between India and Australia. There is one major peak of over 45 % popularity of Pakistan. The day is the quarter-final match between Australia and Pakistan. The special interest can be related to the fact that if Pakistan would have

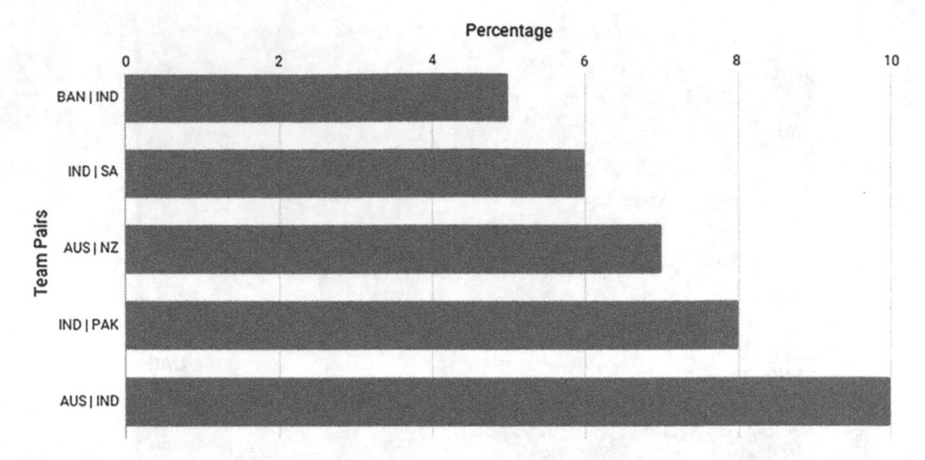

Fig. 4 Most popular and trending team pairs with percentage by tweets

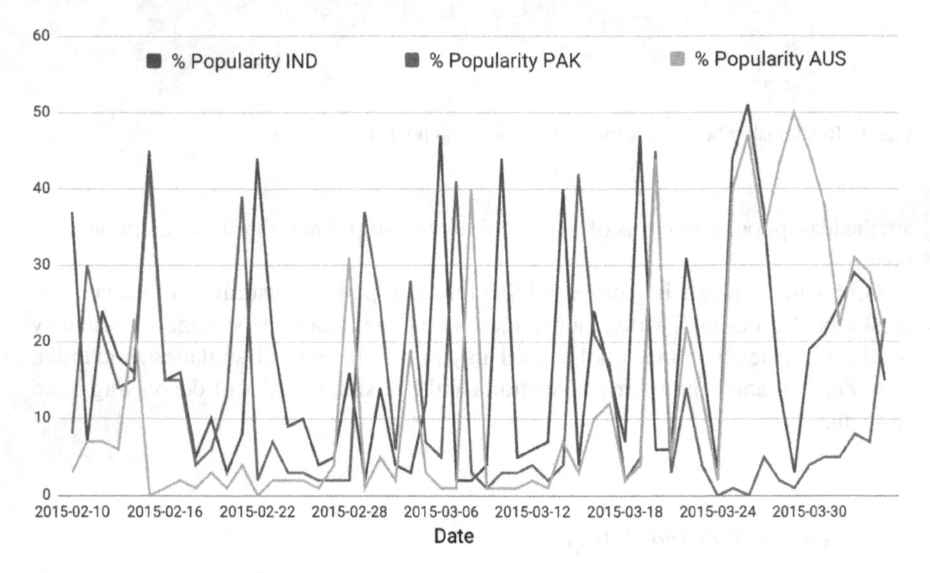

Fig. 5 Time-aware popularity of top 3 most popular teams

won the match, the next match would have been against its arch rival India. Next we observe an exception in the usual popularity trend of Australia for obtaining about 50 % of the total tweets on March 29, 2015. The day is the final match of the world cup. The final win by Australia further justifies the special popularity it has earned that day.

Next we will talk about the popularity of team pairs among the complete cross product of the entities. Figure 4 represents a Bar Chart to show the popularity of team pairs throughout the World Cup on the basis of the percentage tweets domination. The chart represents the top five trending pairs. It can be observed that Australia and India are the most popular and trending pair in this World Cup with 10 % of the tweets. It is followed by the arch rivals, India and Pakistan having a domination by 8 % and Australia and New Zealand having a domination by 7 %.

4.4 Geo-Analysis

We analyze the geo-tagged tweets obtained in our data set to better understand the geographic distribution of World Cup activity on social media. About 42.1 % of the total tweets obtained are geo-tagged as mentioned in Sect. 3.

Table 4 represents the domination by the number of tweets obtained from the country geographically associated with each team. It can be seen that India accounted for more than a quarter of the total geotagged tweets. However significant interest is also seen from Australia, New Zealand, South Africa, and Pakistan.

Figure 6 represents a geo chart which shows the popularity of the World Cup by tweets all over the world. It can be observed that World cup is the most trending event in India followed by Pakistan, USA, and Indonesia. Significant interest can also be observed from UAE, South Africa, Australia, and New Zealand. Brazil, South America and major part of Europe show negligible interest in the World Cup.

Table 4 Team-wise statistics of geo-tagged dataset (top 5)

Team	Geo-Tagged tweets	Percentage
India	363,036	27.19
Australia	27,974	2.09
New Zealand	40,319	3.02
South Africa	32,200	2.41
Pakistan	117,186	8.77

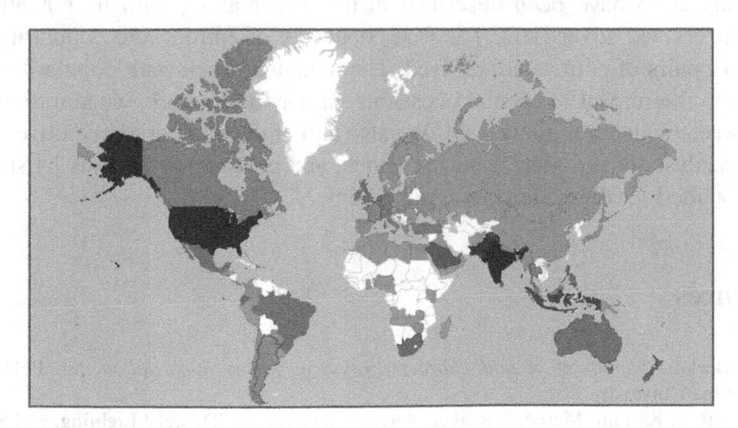

Fig. 6 Stream of world cup tweets from different parts of the world

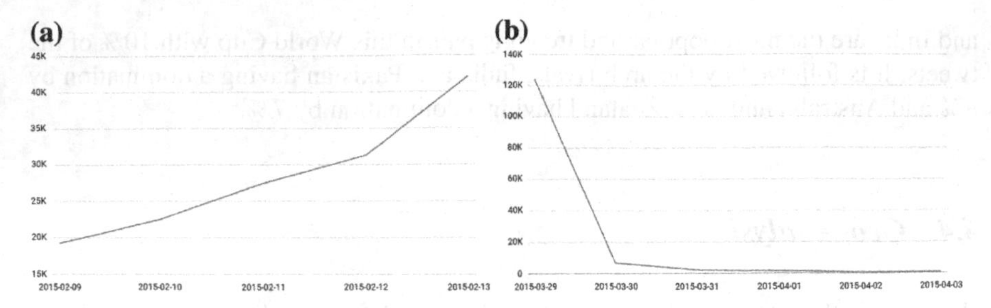

Fig. 7 Popularity trend of world cup. **a** Growth in popularity before start. **b** Decay in popularity after end

4.5 Growth and Decay of Popularity

Cricket World Cup is a scheduled event about which the people already knew. Such events become popular gradually and the popularity of the event finally attains maximum popularity closer to the event date. In Fig. 7a, the number of tweets related to world cup rise from 20 thousand to about 45 thousand, 3–4 days before the world cup which is a gradual growth.

The impact of an event can be judged by observing its decay. Short lived events have a steep decay whereas events with a longer presence on social networks have a gradual fall [11]. In Fig. 7b, the number decreases from about 120 thousand to a few hundred after the world cup. It shows how rapidly popularity for the cricket world cup diminishes after its end.

5 Conclusion and Future Work

Geographical, social and temporal characteristics of the popularity trend of Cricket World Cup 2015 have been described in this paper along with its potential use for organizers and advertisers. The true popularity of entities and popularity bond strength of pairs of entities are derived. The obtained time-aware popularity trends correlate to the impact of special occasions. In our future work, we aim to work to on the tweets which are not textual. We also aim at developing a predictive system through analysis of pre-match, during match, and post-match analysis by studying tweets obtained for these short intervals of time.

References

1. Hila Becker. *Identification and characterization of events in social media*. PhD thesis, Columbia University, 2011.
2. Sanjay Ram Kairam, Meredith Ringel Morris, Jaime Teevan, Daniel J Liebling, and Susan T Dumais. Towards supporting search over trending events with social media. In *ICWSM*, 2013.

3. Sitaram Asur, Bernardo A Huberman, Gabor Szabo, and Chunyan Wang. Trends in social media: Persistence and decay. *Available at SSRN 1755748*, 2011.
4. Hila Becker, Mor Naaman, and Luis Gravano. Beyond trending topics: Real-world event identification on twitter. *ICWSM*, 11:438–441, 2011.
5. Minkyoung Kim, Lexing Xie, and Peter Christen. Event diffusion patterns in social media. In *ICWSM*, 2012.
6. Megha Arora, Raghav Gupta, and Ponnurangam Kumaraguru. Indian premier league (ipl), cricket, online social media. arXiv preprint arXiv:1405.5009, 05 2014.
7. Mohit Naresh Kewalramani. *Community detection in Twitter*. University of Maryland, Baltimore County, 2011.
8. Sahisnu Mazumder, Bazir Bishnoi, and Dhaval Patel. News headlines: What they can tell us? In *Proceedings of the 6th IBM Collaborative Academia Research Exchange Conference (I-CARE)*, I-CARE 2014, pages 4:1–4:4. ACM, 2014.
9. Aditi Gupta and Ponnurangam Kumaraguru. Twitter explodes with activity in mumbai blasts! a lifeline or an unmonitored daemon in the lurking? *IIIT, Delhi, Technical report, IIITD-TR-2011-005*, 2011.
10. Beiming Sun and Vincent Ty Ng. Lifespan and popularity measurement of online content on social networks. In *Intelligence and Security Informatics (ISI), 2011 IEEE International Conference on*, pages 379–383. IEEE, 2011.
11. Michael Ogawa, Kwan-Liu Ma, Christian Bird, Premkumar Devanbu, and Alex Gourley. Visualizing social interaction in open source software projects. In *Visualization, 2007. APVIS'07. 2007 6th International Asia-Pacific Symposium on*, pages 25–32. IEEE, 2007.

Performance Analysis of LPC and MFCC Features in Voice Conversion Using Artificial Neural Networks

Shashidhar G. Koolagudi, B. Kavya Vishwanath, M. Akshatha and Yarlagadda V. S. Murthy

Abstract Voice Conversion is a technique in which source speakers voice is morphed to a target speakers voice by learning source–target relationship from a number of utterances from source and the target. There are many applications which may benefit from this sort of technology for example dubbing movies, TV-shows, TTS systems and so on. In this paper, analysis on the performance of ANN-based Voice Conversion system is done using linear predictive coding (LPC) and mel-frequency cepstral coefficients (MFCCs). Experimental results show that Voice Conversion system based on LPC features is better than the ones based on MFCC features.

Keywords Voice conversion · Morphing · Mel-frequency cepstral coefficients · Linear predictive coding and neural networks

1 Introduction

Speech signal is a convolution of excitation source signal from vocal cords and transfer function of vocal tract. Excitation features contain information about the pitch while transfer function has information about voice quality. Frequency domain transformation of speech signal is used to separate these functions. Cepstral knowledge thus obtained can then be used to find mel-frequency cepstral coefficients (MFCCs) of the speech signal which are helpful characterise to every individual. Formants are

S.G. Koolagudi (✉) · B.K. Vishwanath · M. Akshatha · Y.V.S. Murthy
National Institute of Technology Karnataka, Surathkal Karnataka 575 025, India
e-mail: koolagudi@yahoo.com
URL: http://cse.nitk.ac.in

B.K. Vishwanath
e-mail: kavyabvishwanath25@gmail.com

M. Akshatha
e-mail: akshatham@gmail.com

Y.V.S. Murthy
e-mail: urvishnu@gmail.com

S.C. Satapathy et al. (eds.), *Proceedings of the International Conference on Data Engineering and Communication Technology*, Advances in Intelligent Systems and Computing 469, DOI 10.1007/978-981-10-1678-3_27

the spectral peaks in the sound spectrum where high energy is concentrated. These formants are unique for every phoneme. Feature extraction using linear predictive coding (LPC) mechanism uses this information, where speech signal is analysed after removing the effect of these formants, intensity and frequency of the remaining buzz is analysed this information together with the formants and residue can be used to construct the speech back. LPC and MFCCs are widely used features for auditory modelling. LPC predicts the future values based on the previous samples whereas MFCC considers the nature of the speech while it extracts the features.

Voice Conversion is a technique to modify source speakers speech utterance to sound as if it is spoken by a target speaker. There are two important things to be considered in developing an efficient voice morphing system: One is optimal selection of features and then a model to evaluate a transform function from source speech signal to that of target signal.

Several approaches have been followed in the development of a voice conversion system. A framework for voice conversion using pitch-shifting algorithm by time-stretching with P-SOLA and re-sampling is proposed in [1]. The authors had done testing the performance of the proposed algorithm on a set of Arabic vowels spoken by male and females. A complete voice morphing system and the enhancements needed for dealing with the various artifacts, including a novel method for synthesising natural phase dispersion is described in [2]. Morphing of certain Telugu-Hindi voiced speech and vowels and the conversion between male, female and child speech is tested in [3]. A neural network-based voice conversion system is proposed in [4] where system is trained using MFCC features from the source and target voices. The authors have compared the performance of GMM and ANN. The experiments show that the performance of ANN system is better than GMM. They also address the issue of dependency of voice conversion techniques on parallel data between the source and the target speakers. In this paper, we compare the performance of LPCs and MFCCs features using ANN voice conversion system.

The paper is organised as follows. Section 2 overviews the motivation of work. In Sect. 3 dataset information is discussed. Section 4 describes the experiment simulation. Section 6 describes the result analysis followed by Sect. 7 concludes the work.

2 Motivation

Voice conversion or voice morphing has extensive applications in industrial and commercial sector. For example, it is used in security-related applications to hide the identity of the user. Vocal pathology, voice restoration, games are few other systems where voice conversion is extensively used. One very important step to be considered while building this system is selection of features. There has been a lot of work in building voice conversion system each of them have their own set of features to train the system. In this paper, we considered to evaluate the performance of two such systems considering linear predictive coding (LPC) and mel-frequency cepstral coefficients (MFCCs) as features and artificial neural network (ANN) to train the system. We have analysed performance of this system for four different scenarios considering voice conversion from male to female, female to male, male to male and female to female. Under each of these scenarios, we have analysed the performance of ANN-based conversion system. ANN in this system acts as transform function to convert the voice from source to that of target.

3 Corpus

The bench mark dataset of CMU Arctic database is used for experimentation. The data set is recorded in English language with USA male and female speakers. Six parallel utterances are spoken by them at sampling frequency of 16000 Hz. The silence portion of the speech samples are removed automatically using silence removal algorithm and used them to convert into target speakers' voice.

4 Feature Extraction Process

As described in the previous sections, our paper concentrates on the performance analysis of two important features of speech signal LPC and MFCCs. In this section, we briefly explain these features.

4.1 Linear Predictive Codig (LPC)

LPC methods are the most commonly used in speech coding, speech recognition, speech synthesis, speaker recognition, speaker verification and for speech storage. They are used to represent spectral envelop of a speech in a compressed form using linear predictive model. Basic steps in LPC calculation:

- *Preempahsis:* The speech signal $s(n)$ which is digitised, is sent through a low order digital system, to spectrally flatten the signal this makes it less susceptible to finite precision effects later in the signal processing.
- *Segmentation:* Find $s(n)$ and $R(0)$ through segmentation which will be required to find further $R(i)$ using autocorrelation mechanisms.
- Apply Levinson–Durbins Algorithm to find α coefficients which converts each frame of $p + 1$ autocorrelations into LPC parameter.
- Transmit the output of previous step to a decoder [5].

4.2 Mel-Frequency Cepstral Coefficients (MFCCs)

MFCC features are coefficients that put together make up Mel-frequency Cepstrum. These coefficients represent short-term power spectrum of a sound, based on a linear cosine transform of a log power spectrum on a nonlinear mel scale of frequency. Steps in MFCC calculation are given below:

- Take Fourier transform (FFT) of speech signal.
- Using triangular overlapping windows like hamming window map the powers of spectrum obtained in the previous step onto the mel scale.
- At each of the mel frequencies take logs of the power. Then take the discrete cosine transform on the list of mel log powers.
- The amplitudes of the resulting spectrum are MFCCs.

LPCs or MFCCs thus obtained from the above processes are fed to ANN. In next section, we analyse performance of these features in our voice recognition system.

5 Experiment Simulation

A set of 13 MFCC features were obtained from each frame of size 25 ms with an overlap of 10 ms which is used as an input to ANN for mapping source to target features. Experiments are conducted for four different cases, male to female, female to male, male to male and female to female. Performance of system under each of these cases was analysed.

With varying number of layers in ANN, it is observed that four-layered model with two hidden layers consisting of 50 neurons [13N 50N 50N 13N] performed better than a three-layered or five-layered architecture and hence all our simulations are based on a four layered ANN model. In order to train our system, we have used speech samples consisting of around 20 words and analysed performance under all the above mentioned cases.

6 Result Analysis

In this section, we explain our results and conclusions drawn about the performance of MFCC and LPC features for voice conversion system using ANN training model. The system was trained using voice samples from CMU_ARTIC Database which consists of around 1000 parallel utterances of same sentences from four speakers which include two male and two female speakers. LPC coefficients as feature set for training Voice Conversion system is observed to outperform the system compared when compared with MFCC as training set. The error histogram plots further explain the conclusions drawn from the experiments are shown in Fig. 1.

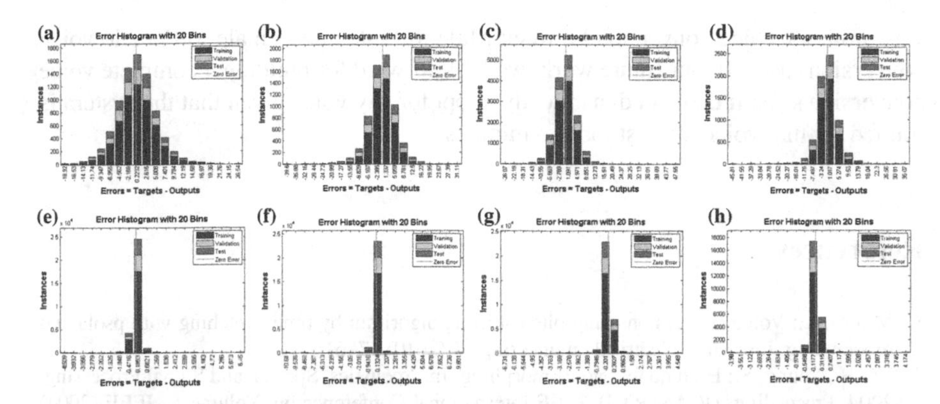

Fig. 1 Error histograms of different combinations of gendres for MFCCs and LPCs. **a** M-F conversion using MFCCs. **b** F-M conversion using MFCCs. **c** M-M conversion using MFCCs. **d** F-F conversion using MFCCs. **e** M-F conversion using LPCs. **f** F-M conversion using LPCs. **g** M-M conversion using LPCs. **h** F-F conversion using LPCs

From the plots, it is seen that the voice conversion system using LPC coefficients as feature vectors have less error compared with that of MFCCs. This could be because LPC coefficients use linear prediction mechanisms to predict expected co-efficients, whereas MFCC uses mapping of features as a mechanism and hence LPC can predict and learn better using ANN compared to that of MFCC.

Figure 1b and f explains the behaviour for voice conversion from Female to Male even in this case it is observed that LPC outperforms MFCC further the output error in the case of MFCC was observed to be more than the previous case.

The behaviour of Voice Conversion system from Male to Male is plotted in Fig. 1c and g. As before LPC coefficients feature vectors were found to be better than MFCC. One important observation made is that the error in this case was observed to be quite higher than the previous cases because here the voice conversion is between Male to Male where the difference between features is less and hence difficult to map the feature vectors and train the system.

Figure 1d and h describes the observations made for Female to Female voice conversion system. As clear from the results LPC features are better to be used as feature vectors when compared to that of MFCC and even in this case the error rate was observed to be more than first two cases as the difference between features are less compared to the first two cases (Male to Female and Female to Male).

7 Conclusion and Future Work

In this paper, we have analysed the performance of MFCC and LPC feature vectors to train Voice Conversion System and in most of the cases it is found that the system using LPC features performed better than those using MFCC further the error rate was more for voice conversion between Male to Male and Female to Female voice conversion cases. In our future work, we look forward for building a complete voice conversion system that can dynamically adopt for any voice given that the system is trained for that voice at-least for two minutes.

References

1. Mousa, A.: Voice conversion using pitch shifting algorithm by time stretching with psola and re-sampling. Journal of electrical engineering **61** (2010) 57–61
2. Ye, H., Young, S.: High quality voice morphing. In: Acoustics, Speech, and Signal Processing, 2004. Proceedings.(ICASSP'04). IEEE International Conference on. Volume 1., IEEE (2004) I–9
3. Shafee, S., Anuradha, B.: Voice conversion using different pitch shifting approach over td-psola algorithm. International Journal of Advanced Research in Computer and Communication Engineering (2013)
4. Desai, S., Black, A.W., Yegnanarayana, B., Prahallad, K.: Spectral mapping using artificial neural networks for voice conversion. Audio, Speech, and Language Processing, IEEE Transactions on **18** (2010) 954–964
5. Mohammed, E.M., Sayed, M.S., Moselhy, A.M., Abdelnaiem, A.A.: Lpc and mfcc performance evaluation with artificial neural network for spoken language identification. (2013)

Person Detection and Tracking Using Sparse Matrix Measurement for Visual Surveillance

Moiz Hussain and Govind Kharat

Abstract Research in the domain of human processor interaction is associated to target segmentation and tracking. The significant analysis in this sector is encouraged as of a view that the multiple field of applications, consisting of observation, computer–human interaction, will benefit from a vigorous and efficient result. In this paper, we propose a framework for the object tracking using sparse matrix and AdaBoost classifier. The technique includes three steps: at the first stage, we use extracted image features. Later, frame features are represented as a sparse matrix. At the final stage, AdaBoost classifier is used to classify correctly the sparse matrix values based on which the tracking task is performed. Experimental results presented in the paper shows that the framework gives improved performance in comparison with other techniques of object tracking.

Keywords Adaboost classifier • Video surveillance • Sparse matrix

1 Introduction

A basic problem for most of the involuntary visual surveillance framework is to identify objects of concern in a known visual scene. A regularly used method for segmentation and detection of moving targets is background differencing [1–5]. Although, the great amount of research is done in the area of the surveillance system, yet it is a tough job to design efficient and robust models for object segmentation and tracking owing to issues such as shape variation, complete and mutual occlusion, lighting change, and motion blur. Today's tracking models frequently update the framework with patches from observations in current frames.

Moiz Hussain (✉)
Anuradha Engineering College, Chikhli, India
e-mail: mymoiz2004@yahoo.co.in

Govind Kharat
Sharadchandra Pawar College of Engineering, Otur, India
e-mail: gukharat@gmail.com

S.C. Satapathy et al. (eds.), *Proceedings of the International Conference on Data Engineering and Communication Technology*, Advances in Intelligent Systems and Computing 469, DOI 10.1007/978-981-10-1678-3_28

Still, many issues are remaining to be addressed; in spite of a great deal of success has been achieved in the image processing applications.

First, most tracking models frequently come across the change in pixel problems [6, 7]. Second, as all these frameworks are data dependent, at the beginning there is a not sufficient amount of information to learn the model. The system performance can be degraded by the fact of over learning; misclassified samples are added to the model. A classic surveillance application includes 3 blocks [8–11]: "Motion" detection, "Object" tracking and "Behaviour" analysis. The job of most of the surveillance system applications is to identify, track and categorize the targets.

The available modules of video surveillance for object segmentation and tracking can be categorized into three key classes: silhouette-based frameworks [12, 13], region-based frameworks [14–16] and attribute point-based frameworks [17–19]. Moreover, the popular automatic surveillance system uses affine flow/optical flow technique for object segmentation and tracking. This is based on the theory that flow [20–22] cannot be estimated regionally, since simply one sovereign dimension is available from the frame string at a point, while the optical flow rate has 2 elements a second restraint is crucial. Intelligent programmed tracking frameworks must be capable to predict the alteration generated by an approaching new object, whereas nonstationary background regions, such as smoke, curtains waving in the air or drizzle must be predicted as a part of the background image. Optical flow; calculates a separate approximation of motion at every pixel; this generally engages decreasing the brightness between successive pixels added over the frame. It is considered that these changes are due to motion variation and because of other consequence, for example, illumination variation. The elucidation can change in an inside and an exterior frame sequence [23, 24]. The elucidation in room sequence can alter as the luminosity changes least 60 times per second by reason of within room illumination [25]. Therefore, elucidation has an effect on the camera captured frame [26, 27]. The elucidation in an open-air also changes due to varieties of causes [28], for instance, the day radiance change, and these changes are enormously sudden in comparison to the interior sequence.

The framework described in this paper is a key to the difficulty of successfully tracking the moving individuals with the easy model and low cost of computation. One of the major merits of the model, compared to the former techniques is that it decreases the incorrect classification, as feature matrix is generated using representation of sparse and the pixels are precisely categorized using AdaBoost network.

The structure of the proposed framework is illustrated in Fig. 1. The Figure shows various levels involved in the model. At first, we extract the frames from video, after that, 'multiscale' frame attributes are estimated using matrix (sparse). Extracted frame characteristic vectors are then given as an input to AdaBoost network for the tracking job. The rest of this paper is structured as follows: Sect. 2 describes literature review. Section 3 represents the system framework. Section 4 presents the database selection for testing work. In Sect. 5 experimental results are shown and finally, Sect. 6 concludes the paper.

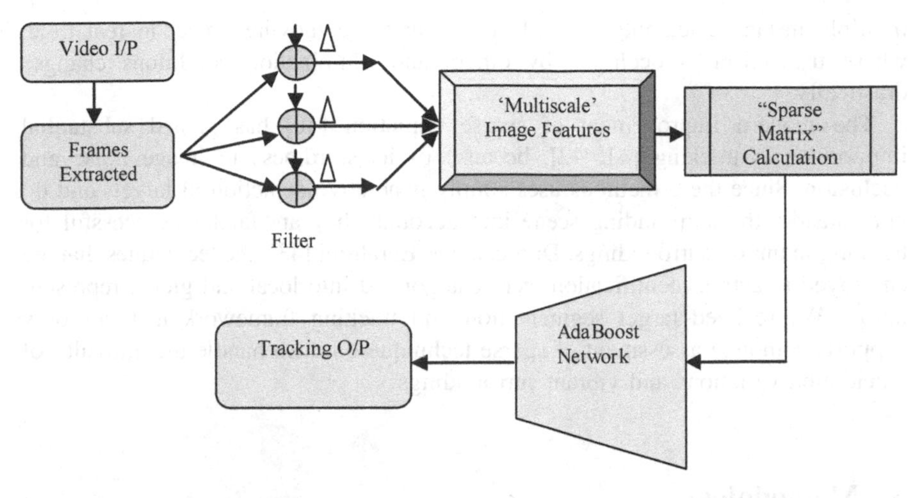

Fig. 1 Tracking system model stages

2 Related Work

Enormous quantity of research is done in the field of surveillance for human/object motion segmentation and tracking. In our [29] latest work, we have presented a threshold initialization model to the segment target from an image frame based on the hypothesis that the mobile targets are visually removed with no partly covered regions. Avidan [30] uses SVM neural n/w classifier and persist it inside the optical flow structure for tracking of an object. An irregularity ratio of interested target and surrounding classes to fix on discriminative features for object tracking is used by Collins et al. [31].

Ravi et al. [32] and Bao and Intille [33], have presented systems to guess actions like standing, walking, sit-ups, running and others using characteristics from accelerometer information and a multiple number of training techniques. On the contrary, they have not applied this information for an interior location. Since this technique focuses on intra image illumination discrepancies and spatial compared to sequential, inter image elucidation changes, it is a divergent collection of illumination compensation. In a study by Cwojek et al. [34], described a technique for some human action recognition in the workplace location with direct tracking. In this model, image and sound attributes are employed to use a multilevel Hidden Markov Model (HMM) technique. A procedure developed to describe a moving individual in indoor surroundings was developed by Pfinder [35]. It detects and tracks solo non-overlapped human/objects in composite scenes. Cai [36] developed a mixture model of 3 parameters with an EM method to model appearance variations during tracking. The technique graph cut, was also designed for motion tracking by estimating a lighting invariant flow field Musa et al. [37]. Ross et al. [38], proposed a technique that shows sturdiness to large variations in pose, size and elucidation using PCA. The Babenko [39], presented algorithm for online multiple instances learning that effectively tracks a moving object in real time, where the object is occluded by others and illumination conditions changes drastically.

The modern improvement of sparse depiction [40] has gained substantial importance in tracking [41, 42], because of its sturdiness to image noise and occlusion. Since these methods uses simply generative depiction of targets and do not consider the surrounding scene into account, they are far less successful for tracking in untidy surroundings. Durucan and Ebrahimi [43], the techniques that are employed in action identification were categorized into local and global representation. We realized target segmentation and tracking framework that not only supports complete assessment of sparse technique but also handle the difficulty of elucidation variations and vibrant surroundings.

3 Methodology

Relatively effortless, yet a competent and effectual tracking model based on attributes extracted [44] from frame space with data basis is presented. The proposed algorithm utilizes a non-adaptive arbitrary projection which protects the formation of the image quality space of an object. A sparse matrix is developed to capably take out the characteristics for the appearance model. Pattern images are compacted of the target and the surroundings due to the similar sparse matrix. The tracking job is devised as a binary categorization via an AdaBoost network with an update in the condensed area. A broad to narrow search strategy is implemented to further decrease the estimation difficulty in the recognition procedure.

The presented tracking framework runs at real-time speed and performs positively against the other methods on different challenging video clips in terms of exact tracking and sturdiness.

3.1 Image Representation

At first, from video sequence all the image frames are grabbed. The 'multiscale' image depiction is developed by convolving the input frame with a Gaussian filter of dissimilar spatial variances. For every input frames its 'multiscale' depiction is developed by convolving the image with a bank of filters at 'multiscale' values $\{I_{1,1} \ldots I_{w,h}\}$ denoted by

$$I_{w,h}(x,y) = \frac{1}{wh} \times \begin{cases} 1, & 1 \leq x \leq w, 1 \leq y \leq h \\ 0, & \text{Otherwise} \end{cases}$$

Here, w is width and h is the height of a rectangle filter. Every filtered image is then symbolized as a vector in column R^{wh} and combine these vectors as an extremely large-dimensional image attribute vector $v = (v\_1, \ldots v\_n)^T \in R^n$ where $n = (wh)^2$. The value n is normally of the order $10E^6$–$10E^{10}$.

3.2 Integral Image

Rectangle image attributes can be estimated exceptionally quickly by means of a centre illustration for the frame which is known as the integral image. The image position x, y has the summation of the pixels over and to the left of x, y inclusive:

$$jj(x, y) = \sum_{x' \leq x, y' \leq y} j(x', y')$$

$jj(x, y)$ is an integral image and $j(x, y)$ is the unique image. With subsequent pair equations:

$$c(x, y) = c(x, y - 1) + i(x, y)$$
$$jj(x, y) = jj(x - 1, y) + c(x, y)$$

where $c(x, y)$ is a collective row sum, $c(x, -1) = zero$, and $jj(x - 1, y) = zero$ integral frame can be estimated in 1 pass over the unique image.

3.3 Image Features

To reduce the inter class variations of the pixel and to increase the outer class variability in comparison to the raw data, the features are used as a substitute of pixel value [45, 46]. This technique makes the classification process easier. Features normally encode information concerning to the area, which is complex to be trained from the finite and raw set of data input. General and an especially big bank of simple Haar features in combination with attribute collection, consequently can improve the capability of the learning model. The feature evaluation pace is also an essential characteristic as approximately every object segmentation and tracking frameworks slide a specific size pane at each scales above the input frame.

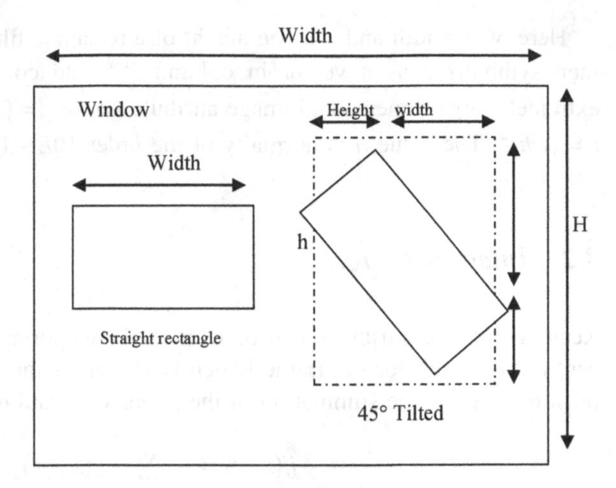

Fig. 2 45° titled and the straight rectangle

The proposed attribute selection method is encouraged with the comprehensive set of Haar-like characteristics initiated through Lienhart and Maydt in [47]. Let us presume the essential component for checking for the occurrence of a target is a window of *Width* × *Height* pixels. Furthermore, presume that we have an extremely quick means of calculating the summation of pixels of any vertical and 45° tilted rectangle within the window. A rectangle is described by a region with $reg = (x, y, w, h, \beta)$ through $zero \leq x, x + w \leq W,$ $zero \leq y, y + h \leq H,$ $x, y \geq zero,$ $w, h \geq zero,$ $\beta \in \{\text{zero}^\circ, 45^\circ\}$ and its pixel summation is indicated by $RSum(reg)$. Two cases of such rectangles are shown in Fig. 2. The presented attribute set is subsequently the collection of every probable attributes of the form

$$fe_1 = \sum_{i \in I = \{1, \ldots N\}} w_i \cdot RSum(reg_i)$$

where network weights $w_i \in R,$ the rectangles reg_i and N are randomly selected.

3.4 Fast Feature Calculation

Every single feature for the object detection can be calculated extremely fast and in steady time for every dimension using two supplementary images. For vertical "rectangles" the supplementary image is an Area of Summed Table, i.e. $AT = (x, y)$. $AT(x, y)$ this is described as addition of pixels of the vertical rectangle starting with the apex left corner at (0, 0) to base right corner (x, y).

$$AT(x, y) = \sum_{x' \leq x, y' \leq y} I(x', y')$$

3.5 *Classifier Design*

Let us consider that all requisites in v are alone dispersed an formulate them with an AdaBoost training model [48]. Known training set and attribute set of frames, any quantity of system training ways might be used to train a categorization function. In this framework, a modification of AdaBoost network is utilized equally to choose a selected set of attributes and to learn the classifier. In its unique form, the AdaBoost training technique is applied to enhance the categorization result of a simple (occasionally called weak) training algorithm. AdaBoost learning model provides the number of formal guarantees.

For every attribute, the weak learner chooses the most favourable threshold classification constraint, due to which the least numbers of samples are wrong classified. AdaBoost classifier $Bj(x)$ thus includes a feature fj, a polarity pj and a threshold θj representing the direction of the dissimilarity sign:

$$Bj(x) = \begin{cases} \text{One} & pjfj(x) < pj\theta j \\ \text{Zero} & \text{else} \end{cases}$$

In terms of calculation, AdaBoost network is possibly faster than the most other training mechanism. Unfortunately, the simple method to enhance the tracking result, adding attributes to the network, directly enlarges the calculation period.

4 Database Collections

In our study we used two datasets. First "Weizmann" dataset which is accessible for research purpose publically and second is "self" developed dataset. This dataset includes 86 video sequences consisting of four and five classes of human actions that includes Jumping, Running, Walking, Side walking, Handwaving and Jogging in that order executed by 19 different subjects in two dissimilar environmental conditions e1 and e2. e1: Indoor Environment e2: Outdoor Environment (Playground + elucidation variations) (Figs. 3 and 4).

Fig. 3 "Self" dataset images. Indoor sample images: **a** Jumping. **b** Sidewalk. **c** Running. **d** Walking. Outdoor (playing field + illumination variations): **a** Jogging. **b** Jumping. **c** Walking. **d** Handwaving

Fig. 4 "Weizmann" dataset images. **a** Bending. **b** Jumping. **c** Running. **d** Skipping

5 Experimental Result

The presented algorithm for the video surveillance system using motion analysis is evaluated with the general motion technique for the state of tracking accuracy and computational time. The settings for experimentation work are briefed in Table 1.

The sparse technique uses raw an attribute collection of an object. To increase the absolute detection of the moving target AdaBoost network is trained with extracted 'multiscale' attributes and the nonstationary target is precisely segmented and tracked. The technique gives the best tracking output as depicted in Fig. 5 as compared to the other motion segmentation method in terms of preciseness.

Table 1 Experimental conditions

Sequence	1280 × 720 pixels, 24 bit colour, 30 fps
CPU	Corei3, 2.13 GHz
RAM	4 GB
OS	Windows 7.0
MATLAB	Version 7.10.0.499 (R2010a)

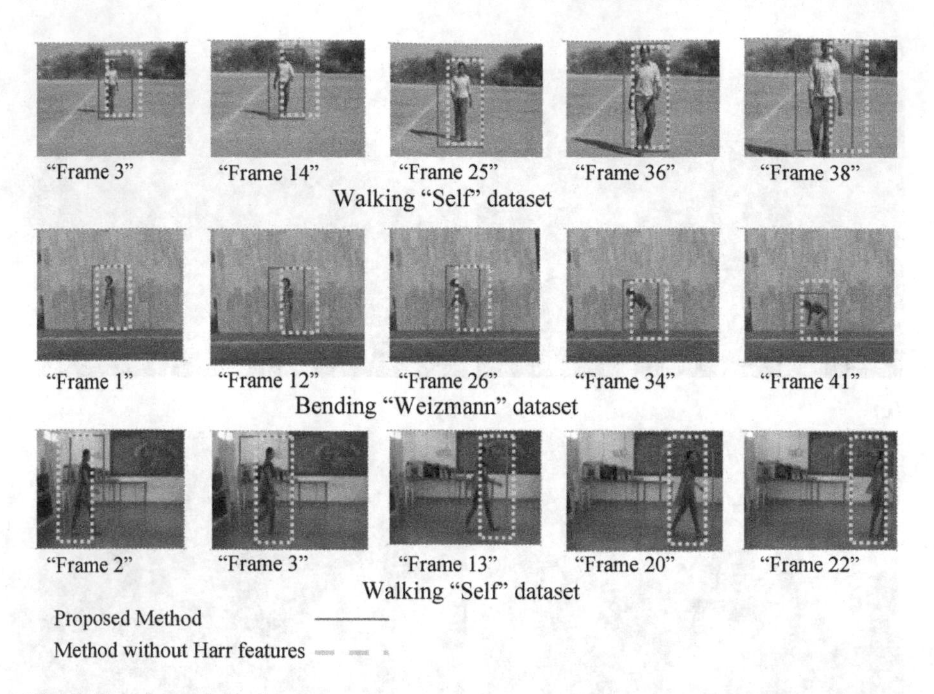

Fig. 5 Experimental results

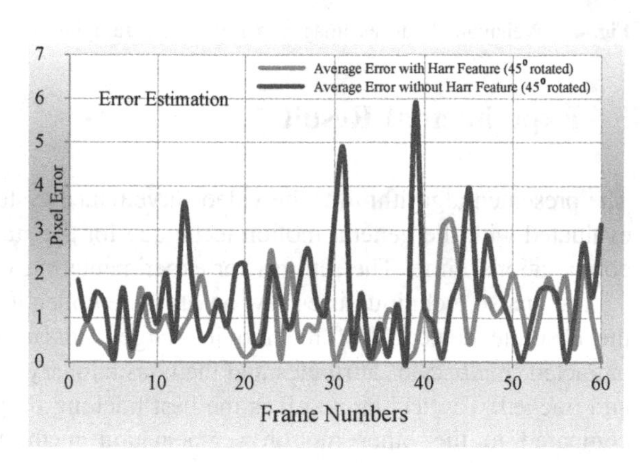

Fig. 6 Error of the tracking window for different frames

Figure 6 shows the error between the actual target and the tracking window position in terms of pixel value distance for a different number of frames. The highest error is restricted in 3 pixels for the classifier considering the 45° rotated Harr features and is nearly six pixels (Double Error) for the classifier without 45° rotated Harr features, which shows that the proposed algorithm provides a reliable prediction performance.

6 Conclusion

In this paper, the authors have presented a vigorous and proficient 'multiscale' sparse matrix attributes based on a object tracking model that uses the consistent sparse nature of image particle representation using a dictionary of object and background templates. The authors have modeled the visual surveillance tracking system as a classification problem that is regularized with the AdaBoost classifier at the stage of elements, and described a competent solution. The authors have systematically analyzed the result of the tracking framework against a challenging method on different demanding video datasets. Quantitative and qualitative experimental results show that the described detection and tracking algorithms outperform other methods, especially in the occurrence of pose variations and illumination changes.

References

1. Lee B., Hedley M: Background Estimation for video Surveillance: IVCNZ 2002, vol. 1, pp. 315–350, 2002.
2. R. T. Collins, A. J. Lipton, and T. Kanade: Introduction to the special section on video surveillance: IEEE Trans. Pattern Anal. Machine Intell., vol. 22, pp. 745–746, Aug. 2000.
3. N. Otsu: A Threshold Selection Method from Gray-Level Histograms: IEEE Transactions on Systems, Man, and Cybernetics, vol. 9, pp. 62–66, 1979.
4. Jun Yang, Tusheng Lin, Bi Li: Dual Frame Differences Based Background Extraction Algorithm: ICCP Proceedings, 2011.
5. Y. Dhome, N. Tronson: A Benchmark for Background Subtraction Algorithms in Monocular Vision: a Comparative Study: In International Conference on Image Processing Theory, Tools and Applications (IPTA 2010). 7–10 July 2010.
6. S. Herrero & J. Bes. Background subtraction techniques: Systematic evaluation & comparative analysis. Advanced Concepts Int. Vision Systems, 33–42. Springer, 2009.
7. D. Parks and S. Fels: Evaluation of background subtraction algorithms with post-processing: In Proc. of IEEE Int. Conf. on Advanced Video and Signal Based Surveillance, pages 192–199, 2008.
8. J. Steffens, E. Elagin, and H. Neven: Person spotter-fast and robust system for human detection, tracking and recognition: In Proc. IEEE Int. Conf. Automatic Face and Gesture Recognition, 1998, pp. 516–521.
9. R. T. Collins, A. J. Lipton, T. Kanade, H. Fujiyoshi, D. Duggins, Y. Tsin, D. Tolliver, N. Enomoto, O. Hasegawa, P. Burt, and L. Wixson: A system for video surveillance and monitoring: Carnegie Mellon Univ., Pittsburgh, PA, Tech. Rep., CMU-RI-TR-00–12, 2000.
10. I. Haritaoglu, D. Harwood, and L. S. Davis: W: Real-time surveillance of people and their activities: IEEE Trans. Pattern Anal. Machine Intell., vol. 22, pp. 809–830, Aug. 2000.
11. R. Poppe: Vision-based human motion analysis: An overview: Comput. Vis. Image Understanding, vol. 108, pp. 4–18, Oct. 2007.
12. D. Serby, E. K. Meier, and L. V. Gool: Probabilistic Object Tracking Using Multiple Features: IEEE Proc. of International Conf on Pattern Recognition Intelligent Transportation Systems, Vol. 6, pp. 43–53, March 2004.
13. P. Viola, M. Jones and D. Snow: Detecting pedestrians using patterns of motion and appearance: ICCV, vol. 02, pp. 734–741, 2003.

14. L. Li, S. Ranganath, H. Weimin, and K. Sengupta: Framework for Realtime Behavior Interpretation Form Traffic Video: IEEE Tran. on Intelligen Transportation Systems, March 2005, Vol. 6, No. 1, pp. 43–53.
15. N. A. Ogale: A survey of techniques for human detection from video: http://www.cs.umd.edu/scholarlypapers/neetiPaper.pdf, Department of Computer Science, University of Maryland, College Park.
16. J. Fernyhough, A. G. Cohn, and D. C. Hogg: Constructing qualitative event models automatically from video input: Image Vis. Comput., vol. 18, no. 9, pp. 81–103, 2000.
17. Z. Zivkovi: Improving the selection of feature points for tracking. In pattern Analysis and Applicarions, vol. 7, no. 2, Copyright Springer Verlag London Limited, 2004.
18. J. Lou, T. Tan, W. Hu, H. Yang, and S. H. Maybank: 3D Model-based Vehicle Tracking: IEEE Trans. on Image Processing, Vol. 14, pp. 1561–1569, October2005.
19. Rachel Kleinbauer: Kalman Filtering Implementation with MatLab: University Stuttgart Institute of Geodesy, Helsinki Nov, 2004.
20. I. Austvoll: A Study of the Yosemite Sequence Used as a Test Sequence for Estimation of Optical Flow: pages 659–668. Lecture Notes in Computer Science. Springer, Berlin 2005.
21. L. Alvarez, R. Deriche, T. Papadopoulo, and J. S´anchez: Symmetrical dense optical flow estimation with occlusions detection: In Proc. ECCV, pages 721–735, 2002.
22. B. K. P. Horn. Robot Vision. (McGraw-Hill, New York, NY, U.S., 1986), 185–216, 1986.
23. M. J. Hossain, J. Lee, and O. Chae: An Adaptive Video Surveillance Approach for Dynamic Environment. Proc. Int. Symposium on Intelligent Signal Processing and.
24. Y. Wang; T. Tan, and K. Loe: A probabilistic method for foreground and shadow segmentation. Proc. Int. Conference on Image Processing: Vol. 3. 937–940, 2003.
25. Jin, H., Favaro, P., Soatto, S.: Real-Time Feature Tracking and Outlier Rejection with Changes in Illumination. In: International Conference on Computer Vision, Los Alamitos, IEEE Computer Society (2001) 684–689.
26. A. F. Bobick: Movement, activity and action: The role of knowledge in the perception of motion: Philos. Trans. Roy. Soc. B, Biol. Sci., vol. 352, pp. 1257–1265, 1997.
27. Li Ying-hong & Li Zheng-xi: An intelligent tracking technology based on kalman and mean shift algorithm: IEEE Second Intl. Conf. on Computer Modelling & Simulation 2010.
28. Zalili Musa: Multi-camera Tracking System in Large Area Case: In IEEE at Korea 2009.
29. Moiz A. Hussain, Dr. G.U.kharat: Video Based Human Detection System Using Movement Analysis: IEEE, International conference on convergence of technology, I2CT-2014.
30. S. Avidan: Support vector tracking: In CVPR, pages 184–191, 2001.
31. R. T. Collins, A. J. Lipton, T. Kanade: Introduction to the special section on video surveillance: IEEE Trans. on Pattern Analysis & Machine Intel. 22 (8) (2000) 745–746.
32. N. Ravi, N. Dandekar, P. Mysore, and M. L. Littman: Activity recognition from accelerometer data: In Proc. 17th Conf. Innovat. Appl. Artif. Intell., 2005, pp. 1541–1546.
33. S. S. Bao and L. Intille: Activity Recognition from User-Annotated Acceleration Data: New York: Springer-Verlag, 2004, pp. 1–17.X.
34. Christian Wojek, Kai Nickel, Rainer Stiefelhagen: Activity Recognition and Room Level Tracking: IEEE International Conference on Multi sensor Fusion and Integration for Intelligent Systems, Heidelberg, Germany; pp 1–6, 2006.
35. C. R. Wren, A. Azarbayejani and A. P. Pentland: Pfinder: real-time tracking of the human body: IEEE Trans. Pattern Analysis and Machine Intelligence, vol. 19, pp. 780–785, 1997.
36. J. Cai, M. Shehata & M. Pervez: An Algorithm to Compensate for Road Illumination Changes for AID Systems. Proc. IEEE Intelligent Systems Conference: 980–985, 2007.
37. A. D. Jepson, D. J. Fleet, and T. F. El-Maraghi: Robust online appearance models for visual tracking: PAMI, 25:1296–1311, 2003.
38. D. A. Ross, J. Lim & M. Yang: Incremental learning for robust visual tracking: IJCV, 77 (1–3):125–141, 2008.
39. B. Babenko: Visual tracking with online multiple instance learning. CVPR, 2009.
40. J. Wright, A. Y. Yang, A. Ganesh, S. S. Sastry, and Y. Ma: Robust face recognition via sparse representation: PAMI, 31(2):210–227, 2009.

41. X. Mei and H. Ling: Robust visual tracking using l1 minimization: In ICCV, 2009.
42. B. Liu, L. Yang, J. Huang, P. Meer, L. Gong, and C. Kulikowski: Robust and fast collaborative tracking with two stage sparse optimization: In ECCV, pages 624–637, 2010.
43. E. Durucan and T. Ebrahimi: Moving object detection between multiple and color images: In IEEE Conf. Advanced Video and Signal Based Surveillance, pages 243–251, 2003.
44. D. Serby, E. K. Meier, and L. V. Gool: Probabilistic Object Tracking Using Multiple Features: IEEE Proc. of International Conf on Pattern Recognition Intelligent Transportation Systems, vol. 6, pp. 43–53, March 2004.
45. R. Fablet and M.J. Black: Automatic detection and tracking of human motion with a view-based representation: In ECCV, 2002, vol. 1, pp. 476–491.
46. H. Sidenbladh: Detecting human motion with support vector machines: Iin ICPR, 2004, vol. 2, pp. 188–191.
47. Rainer Lienhart and Jochen Maydt: An Extended Set of Haar-like Features for Rapid Object Detection: Intel Labs, Intel Corporation, Santa Clara, CA 95052, USA.
48. P. Viola and M. Jones: Rapid object detection using adboosted cascade of simple features: In Proceedings of the Conference on Computer Vision and Pattern Recognition, Kauai, Hawaii, USA, volume 1, pages 511–518, 2003.

Improvisation in Frequent Pattern Mining Technique

Sagar Gajera and Manmay Badheka

Abstract At present, so many techniques are available which can be applicable to wide range of datasets. They provide an effective way to mine frequent pattern from the datasets. Most of them use different kind s of data structures for the processing which provide variations in requirement of time and space. Generally, traditional techniques are restricted to the narrow area or provide effective results only in the specific environment. So, it requires continuous optimization and updation. Dynamic data structure and mapping shows more effectiveness compared to the traditional techniques in terms of time and space requirement for processing.

Keywords Association rule mining · Frequent pattern mining · Frequent itemset · A priori algorithm

1 Introduction

Data Mining is a process of deriving knowledge from a database or data warehouse. It is also known as Data Dredging [1].

The main task of the data mining activity is to mine precious nuggets of knowledge from a data set and converts it into a transparent structure which can be used for additional analysis.

Apart from the analysis step, it contains various tasks like data preprocessing, decision-making, pattern recognition, consideration of complexity, pictorial representation, updating knowledge, schema generation, and online analytical processing. Frequent pattern mining is a technique of finding hidden pattern from a data which occurs frequently [1]. It is emerged as an interesting area of research in recent years. It is presented in the form of Association Rule (R $\Rightarrow$ S) [1].

Sagar Gajera (✉) · Manmay Badheka
L J Institute of Engineering and Technology, Ahmedabad, India
e-mail: sdgajera14@gmail.com

Manmay Badheka
e-mail: manmaybadheka@gmail.com

S.C. Satapathy et al. (eds.), *Proceedings of the International Conference on Data Engineering and Communication Technology*, Advances in Intelligent Systems and Computing 469, DOI 10.1007/978-981-10-1678-3_29

It is also known as Association Rule Mining. This technique is useful in many sectors such as banking, bioinformatics, telecommunication, financial corporation, etc. Basic approach is to find an association and correlation among the data in a database but it should be interesting. The interesting measures are shown below [1].

$$\text{Support}\,(R \Rightarrow S) = P\,(R\,U\,S) \tag{1}$$

$$\text{Confidence}\,(R \Rightarrow S) = P\,(R|S) \tag{2}$$

Items which have values of interesting measures above the predetermined threshold are said to be the frequent item set.

Dynamic programming is one of the techniques to design an efficient algorithm [2]. This technique is used for those problems where optimization of task is needed. This technique stores the previous solutions [3]. So, whenever it reappears, it can be directly accessed from those pre-calculated values without creating more overhead. It has the important feature named memorization [3]. The word memorization comes from memo means recording of solutions, which relates to storage of the solutions of sub problems.

The structure of paper is as follows. Section 2 represents existing work and theories. Section 3 presents designed technique. Section 4 gives a complete analysis proposed algorithm. Section 5 concludes a paper.

2 Related Work

The various algorithms for frequent pattern mining are described below.

2.1 *Association Rule Mining*

Suppose that Q = {q1, q2, …, q*m*} is the set of data items, while P = {p1, p2, …, p*n*} is the collection which has the relevant data, T is the subset of Q. The association rule is the implication in the form of R ⇒ S, where R, S ⊂ I and R ∩ S = ϕ [1]. Association rules are judged by using interesting measures. The purpose of the association rule mining is to discover correlation among itemsets which is used for decision-making [4].

2.2 A Priori Algorithm

The A priori algorithm is a basic algorithm for frequent pattern mining and association rule generation. It usually works with transactional datasets which contains transaction id and set of items called itemset.

A priori is using level-wise searching procedure for exploration of association rules. There are mainly two steps for generating frequent itemsets in this algorithm [1, 5].

- Join Step: The candidate itemset C_k is acquired by combining previous frequent itemset L_{k-1} with itself.
- Prune Step: Look the dataset to acquire the occurrence count of each candidate itemset in C_k.

This procedure is recursively done until no more frequent itemsets can be produced from that dataset [6]. It has one important property called anti-monotonicity [1]. This property states that the subset of any frequent itemsets is also a frequent itemsets. At each step, the possible itemsets are getting multiplied. This makes more use of memory and time [6].

2.3 A Priori Algorithm with Dynamic Programming

This is a variation of A priori algorithm which stores the support count for every itemset using dynamic programming approach. It overcomes the main disadvantage of A priori algorithm, i.e., A priori needs a new pass for each itemsets generation.

It uses the special data structure named count table which is used to store the occurrences of the itemsets [5]. Using count table, only one dataset scan is needed to discover 1-itemsets and 2-itemsets. In this algorithm, only one scan is performed on dataset. During scan, one by one transaction is scanned and processed in terms of deriving combinations of 1-itemsets and 2-itemsets. All the entries of count table which relates to these combinations are incremented by one.

2.4 Pattern Indexing

Pattern Indexing algorithm works in two steps. In the first step, unique pattern indexing takes place. Initially, the pattern will be encoded to a numerical value. This numerical value is unique and stored in a HashMap [7]. While encoding, the pattern is searched into HashMap. If it is not present then new entry is made. Next, the pattern is replaced by its index [7]. In second step, unique pattern is searched and counted by searching and counting its unique index into the input data. This pattern indexing makes task easier of searching any pattern.

2.5 AIS Algorithm

AIS is the oldest algorithm for frequent pattern mining. It generates and counts candidate itemsets at run time [8]. It makes multiple passes of database for generation of candidate itemsets. It uses frontier set which undergoes extension during the scan. In each pass, the supports of certain itemsets are counted which is known as candidate itemsets and also determine whether the itemsets is to be added for the further pass. When this frontier set becomes null, the algorithm ends up with the final result [8].

2.6 Direct Hashing and Pruning

Direct Hashing and Pruning algorithm uses new data structure called Hash Bucket for generation of candidate itemsets [8]. In the first step, it generates hash table for 2-itemsets from 1-itemsets. In the second step, set of candidate itemsets is hashed into hash entry which is frequent. In third step, the pruning procedure is carried out.

2.7 Partition Algorithm

Partition algorithm is the variation of A priori Algorithm. It generally overcomes the problems of multiple passes which exists in A priori and Direct Hashing & Pruning. It consists of two passes on the dataset. It logically partitions database into n partitions and scans the whole database only two times [8]. In the first pass, it scans the entire database as a group of partitions and searching is made for any pattern. It gives the local frequent patterns for a specific part of database. In the second pass, it finally derives the globally frequent pattern which occurs frequently in the database.

2.8 Quantitative Association Rule Generation for Weather Forecast

This is a specific constructed algorithm for the analysis of local weather. It works mainly in three steps and uses real-time weather database [4]. In first step, Database Transformation is performed as a change of each attributes according to local situations. In second step, Quantitative Association Rules are generated. In third step, Dynamic Interdimension Rules are generated from which weather is forecasted.

3 Improved Technique

Using Dynamic data structure and Mapping technique, the performance is improved for generation of candidate itemsets. As it uses Dynamic data structure, the approach stores the occurrence counts for the itemsets. Here, the data structure used is dynamic so it uses the least memory space. It only stores occurrence counts, not its corresponding itemsets. Using Mapping equations, it converts item numbers of itemsets to the unique index in the data structure. Thus, it is efficient and reduces the overhead of storing the itemsets with their occurrence counts. Here, the improved technique is explained for maximum 4-itemset generation. This technique can be extended up to *k*-itemset by extending mapping technique on the data structure.

Input

Transactional Dataset D with No. of Transactions T and No. of Items N.
Minimum Support and Minimum Confidence.

Output

Association Rules generated from Frequent 2, 3 and 4-itemsets.

Procedure

```
For each transaction, item list X holds the items in the transactions.
for items in X
        make possible combination of length 1, 2, 3 and 4.
        for each combination of items
                find unique index n in data structure using Mapping
                increment value at index n by 1
        end for
end for
```

Special Results

1-itemset: a = 0, b = 0 and c = d
2-itemset: a = 0, b = 0 and c & d are distinct (c < d)
3-itemset: a & b are distinct (a < b) and c = d
4-itemset: a, b, c and d are distinct such that a < b < c < d

Mapping

From Special Results, a, b, c and d are initialized such that a < b < c ≤ d. (Exception: a = b when a = 0 and b = 0)

$t = \frac{(x-b)(x-b+1)}{2}$ where, x is total number of items.

$m = \frac{(d-b-1)(d-b)}{2} + 1$ where, m points first element for block d.

$n = m + (c - b) - 1$ where, n is index of stored support.

m and n ≤ t

For example, Table 1 shows transactional dataset with items I = {I1, I2, I3, I4} and five transactions. Using the Procedure and Mapping equations, the data structure will be created as Fig. 1a. For itemset (*a*, *b*, *c*, *d*), the data structure is

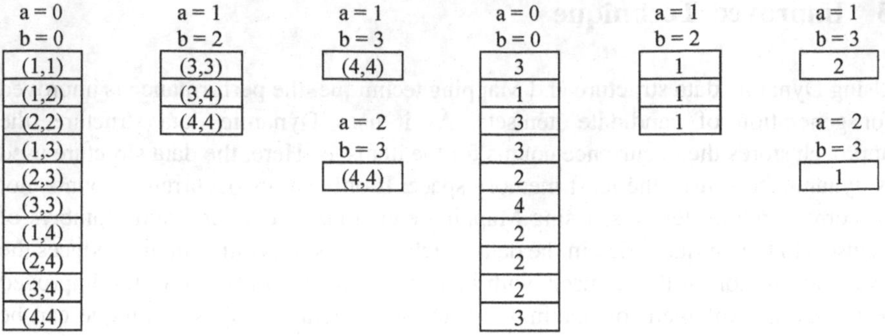

a = 0, b = 0: (1,1) (1,2) (2,2) (1,3) (2,3) (3,3) (1,4) (2,4) (3,4) (4,4)
a = 1, b = 2: (3,3) (3,4) (4,4)
a = 1, b = 3: (4,4)
a = 2, b = 3: (4,4)

a = 0, b = 0: 3 1 3 3 2 4 2 2 2 3
a = 1, b = 2: 1 1 1
a = 1, b = 3: 2
a = 2, b = 3: 1

Fig. 1 Dynamic data structure (**a**, **b**) with values (**c**, **d**) and values of example

identified with values of a and b while index of occurrence count is determined by values c and d.

For Table 1, the values of occurrence counts in the data structure can be shown as Fig. 1b. Here, the occurrence count of any itemset means support count.

Example If the occurrence count of itemset (I1, I2, I4) is to be stored or retrieved from the data structure, the unique index of that itemset is to be calculated.

From Special Results, $a = 1, b = 2, c = 4$ and $d = 4$.
Next, $x =$ total number of items = 4.
$t = \frac{(4-2)(4-2+1)}{2} = 3$. So, data structure with $a = 1$ and $b = 2$ contain three values.

$$m = \frac{(4-2-1)(4-2)}{2} + 1 = 2.$$

$n = 2 + (4 - 2) - 1 = 3$. Where, $m = 2 \le 3 = t$ and $n = 3 \le 3 = t$.

Thus, the itemset (I1, I2, I4) has 3rd index in the data structure $a = 1$ & $b = 2$. Here, value at above index is 1. So, the occurrence count or support count of itemset is 1.

Table 1 Transactional dataset

TID	Itemset
T1	I1, I3
T2	I2, I4
T3	I1, I2, I3, I4
T4	I2, I3
T5	I1, I3, I4

4 Experiments and Results

To exhibit the effectiveness of the research work, we performed several experiments on a Supermarket and Synthetic dataset. Supermarket dataset contains 4627 transactions, 213 items. Synthetic dataset is generated using the random function which provides uniform distribution over the given ranges. A system with core i3 processor and 4 GB RAM is used for performing experiment and analyzing results.

Experiment 1: (Measures of Quality)
This experiment compares the total number of combinations with the actually generated total number of combinations. Figure 2 concludes both the combinations and compares their values on the base of the total items. From the figure, it is clear that the total combinations are increased as the total numbers of items are increased.

As shown in figure, this method makes the less number of combinations than the total number of combinations which can be generated from the items. Thus, this approach stores less number of combinations and utilizes the space.

Experiment 2: (Measures of Performance)
This experiment shows the effect on the execution time by varying the size of the data (number of records) and the total number of items.

Figure 3 shows the effect when the numbers of records are increased. From figure, the execution time varies linearly with increasing the number of records. Here, the total numbers of records varies from 1000 to 5000.

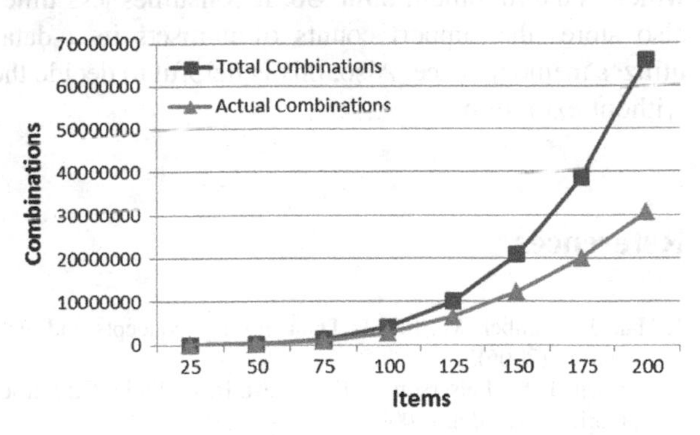

Fig. 2 Comparison of total combinations and actual combinations

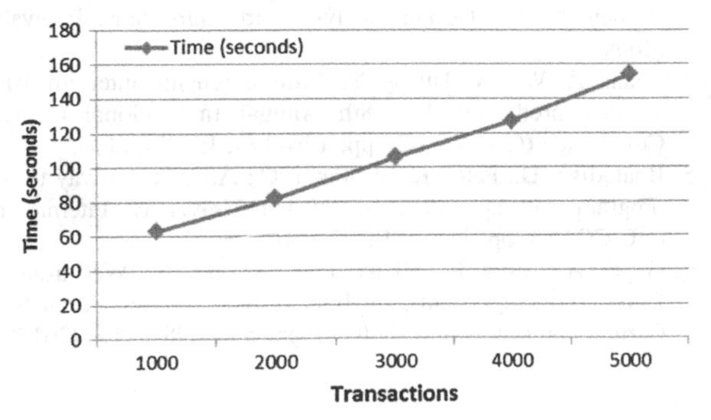

Fig. 3 Execution time by varying number of records

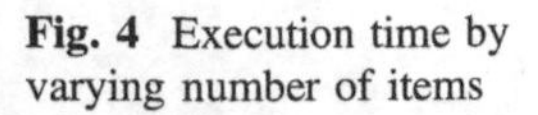

Fig. 4 Execution time by varying number of items

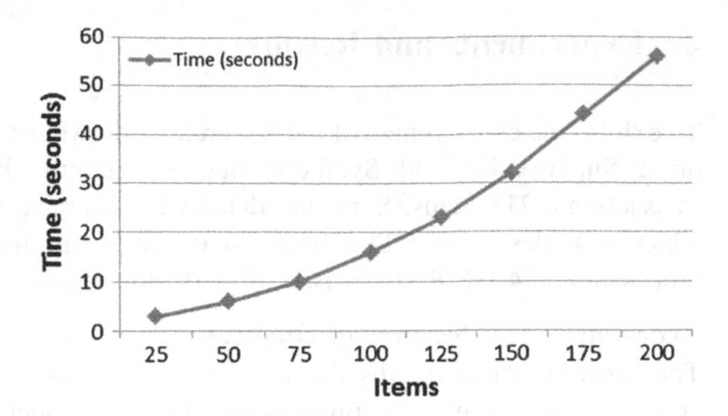

Figure 4 shows the effect on the execution time when the total numbers of items are increased. From figure, the execution time varies linearly with the number of items. Here, the total numbers of items varies from 25 to 200.

5 Conclusion

From the generated results, it is concluded that this method removes some limitations of other techniques of frequent pattern mining. It uses only one dataset scan which is the minimum limit. So, it consumes less time for k-itemsets generation. It also stores the support counts of itemsets in a data structure efficiently which utilizes memory space. Also, this is helpful to decide the strength of association rule without extra effort.

References

1. Han, J., Kamber, M., Pei, J.: Data mining: concepts and techniques. Morgan Kaufmann, San Francisco (2006).
2. Cormen, T. H., Leiserson, C. E., Rivest, R. L., Stein, C.: Introduction to Algorithms. MIT Press, Cambridge, London (2009).
3. Pandey, H. M.: Design Analysis and Algorithms. Firewall Media Publication, New Delhi (2008).
4. Zhang, Z., Wu, W., Huang, Y.: Mining dynamic interdimension association rules for local-scale weather prediction. In: 28th Annual International Computer Software and Applications Conference (COMPSAC), pp. 146–149. IEEE (2004).
5. Bhalodiya, D., Patel, K. M., Patel, C.: An efficient way to find frequent pattern with dynamic programming approach. In: Nirma University International Conference on Engineering (NUiCONE), pp. 1–5. IEEE (2013).
6. Gupta, A., Arora, R., Sikarwar, R., Saxena, N.: Web usage mining using improved Frequent Pattern Tree algorithms. In: International Conference on Issues and Challenges in Intelligent Computing Techniques (ICICT), pp. 573–578. IEEE (2014).

7. Mutakabbir, K. M., Mahin, S. S., Hasan, M. A.: Mining frequent pattern within a genetic sequence using unique pattern indexing and mapping techniques. In: 3rd International Conference on Informatics, Electronics & Vision, pp. 1–5. IEEE (2014).
8. Aggarwal, S., Kaur, R.: Comparative Study of Various Improved Versions of Apriori Algorithm. In: International Journal of Engineering Trends and Technology, Vol. 4, (4) (2013) 687–690.

Design and Simulation of Hybrid SETMOS Operator Using Multiple Value Logic at 120 nm Technology

Raut Vaishali and P.K. Dakhole

Abstract Motivation behind this paper is to bring new method for multi-feature compact designs with low power requirement which is beyond the capacity of binary logic as well as use of only CMOS technology. Simplicity, compactness, and low power requirement can be achieved by hybridization of single electron transistor and CMOS with quaternary logic. The proposed implementation overcomes several limitations found in the previous quaternary implementation such as MIN, MAX, XOR gate and Full Adder which is simulated in 120 nm technology requires less number of transistors as well as with very low power dissipation.

Keywords Single electron transistor · CMOS · Quaternary · Multivalue

1 Introduction

An important aspect of multi-valued logic systems is of choosing the radix value and the choice of radix is available in actual or conceptual domains. Actual and conceptual domains may be different. The choice mainly depends on the ease of manipulations of variables in conceptual domain and manipulations of voltage, current, charge, etc., in the actual circuits. In this thesis, logic levels are represented as voltage levels and are 0, 1, 2, and 3. The number of bits required for binary representation of an n-digit base-r value is n[$\log_2 r$]. When the radix is not a power

Raut Vaishali (✉)
Department of Electronics & Telecommunication, G.H.Raisoni College of Engineering & Management, Pune, India
e-mail: vaishraut02@gmail.com

P.K. Dakhole
Department of Electronics, Y.C.C.E, Nagpur, India
e-mail: pravin_dakhole1@yahoo.com

S.C. Satapathy et al. (eds.), *Proceedings of the International Conference on Data Engineering and Communication Technology*, Advances in Intelligent Systems and Computing 469, DOI 10.1007/978-981-10-1678-3_30

of 2, more number of bits are required than minimum and this leads to an inefficient representation. Hence, when $r = 2^k$ exact k bits are required. Among the radix values that are powers of 2, the system based on 4 shows considerable promise. The values 8 and 16 are potentially useful. Quaternary logic is quite feasible since the implementations can be designed using available circuitry, and no additional special components are required. Hence, quaternary radix is selected to realize the logic circuits [1–4].

2 Single Electron Transistor

The SET is a four-terminal device. It has a source and drain along with two separate gates. One gate resembles the functionality similar to CMOS transistor gate and second gate is specifically used as back gate for controlling purpose. A single electron transistor consists of a small conducting island coupled with source and drain, leads by tunnel junctions, and capacitively coupled to one or more gates [5]. Tunnel junction is a combination of capacitor and resistor. Figure 1 shows symbol of single electron transistor and PMOS. Single electron transistor which is shown in Fig. 1 consists of two tunnel junction formed by R1C1 and R2C2 whose values are given in Table 1. For hybrid design SET [6] and BSIM4.1.0 PMOS model used.

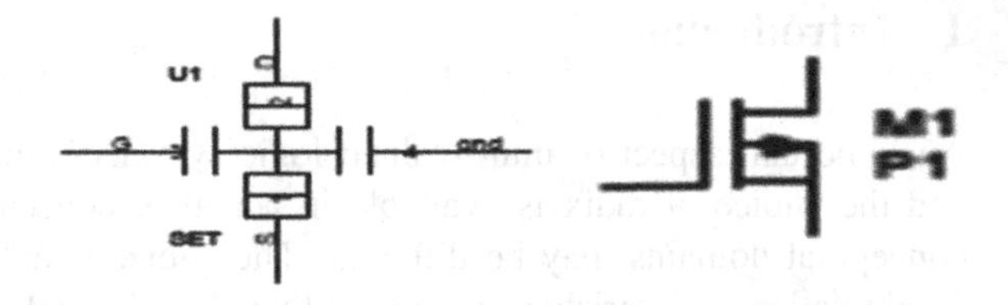

Fig. 1 Symbol of single electron transistor and PMOS

Table 1 Description of single electron transistor and P MOSFET

SET	PMOS [BSIM 4.1.0]
SET	**PMOS**
C1 = 1E-18	Vth = −0.42 V
C2 = 1E-18	W/L = 600 nm/120 nm
R1 = R2 = 1E5	
Cg1 = 1E-18	
Cg2 = 0	

3 Quaternary Single Electron Transistor Logical and Arithmetic Operators

An assortment of logical and arithmetic operators is requisite for countless applications in the processing units. These various operators are available in binary logic in which interconnection increases as the number of components increases. Hence use of binary logic for complex circuits is not worthwhile. Similarly, CMOS-based circuits consume more power which is also not suitable in today's world of low power VLSI. In this section, single electron transistor technology and hybrid CMOS with SET in quaternary logic-based MIN, MAX, XOR gates are designed and simulated which can be used as logical operators for future applications.

3.1 Quaternary SETMOS MAX Gate

The MAX gate is designed by using only two N-type single electron transistors and one P-type MOSFET. The power dissipation of MAX gate is almost zero. As compared with various previous quaternary MAX gate the implemented quaternary MAX gate is much more efficient as well as optimized (Fig. 2 and Table 2).

$$\mathrm{MAX(A, B)} = \mathrm{A} + \mathrm{B} \quad \begin{array}{l} \mathrm{A\ when\ A > B} \\ \mathrm{B\ when\ B > A} \end{array}$$

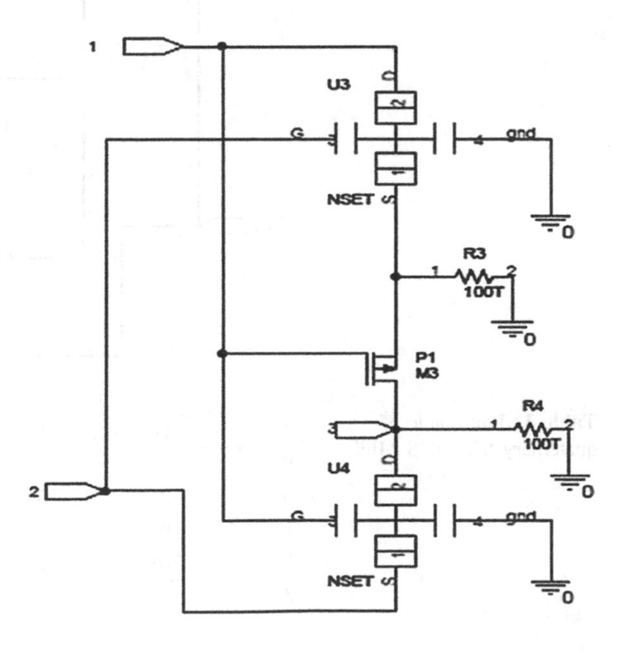

Fig. 2 SETMOS quaternary MAX gate

Table 2 Truth table of quaternary SETMOS MAX gate

A	0	1	2	3
B				
0	0	1	2	3
1	1	1	2	3
2	2	2	2	3
3	3	3	3	3

3.2 *Quaternary SETMOS MIN Gate*

The MIN gate is designed by using only two N-type single electron transistors and one P-type MOSFET. The power dissipation of MIN gate is almost zero. As compared with various previous quaternary MIN gate the implemented quaternary MIN gate is much more efficient as well as optimized (Fig. 3 and Table 3).

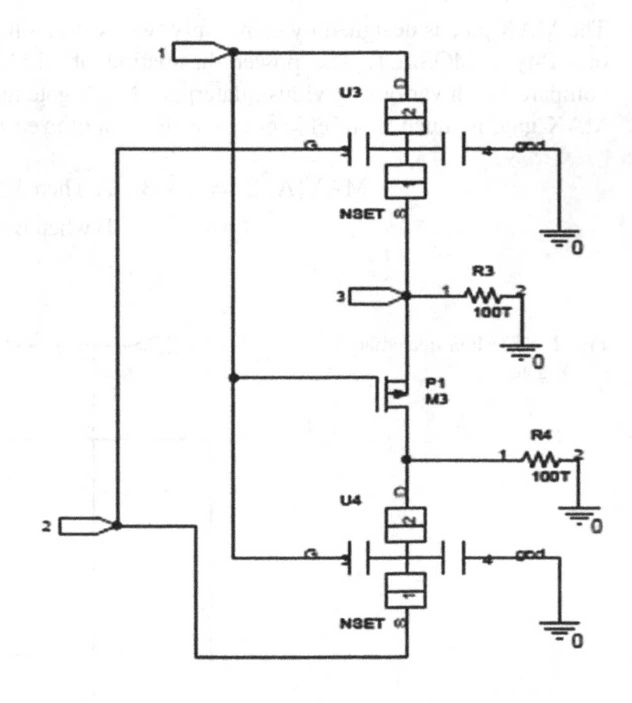

Fig. 3 SETMOS Quaternary MIN gate

Table 3 Truth table of quaternary SETMOS MIN gate

A	0	1	2	3
B				
0	0	0	0	0
1	0	1	1	1
2	0	1	2	3
3	0	1	2	3

$$\mathrm{MIN}(A, B) = A \cdot B \quad \begin{matrix} A \text{ when } A < B \\ B \text{ when } B < A \end{matrix}$$

3.3 Quaternary SETMOS XOR Gate

A new Quaternary SETMOS XOR gate is designed which is shown in Fig. 4 (Table 4). For the design of Quaternary SETMOS XOR gate one MAX, two MIN, one NAND and one AND gate are used. Comparative analysis of Quaternary SETMOS XOR with different XOR gate such as binary three types of XOR gate which is designed with only using single electron transistor and XOR gate which is designed with combination of single electron transistor and MOSFET [7] in terms of power dissipation is given in Table 5.

The graphical representation of power dissipation of XOR gate is shown in Fig. 5 which shows that power dissipation of Quaternary SETMOS XOR gate is less.

3.4 Quaternary SETMOS FULL ADDER

For the design of Full adder various operators are used such as Level 3, Level 0, TSUM, TDIFF [8, 9]. The operators which are introduced in [8, 9] are designed with combination of depletion and enhancement type of MOSFETs as well as with some change of W/L ratios. While in the proposed design of full adder all the four operators along with MIN and MAX gates are used but in that use of N-type single electron and P-type MOSFETs are used. Due to which there is no need of using

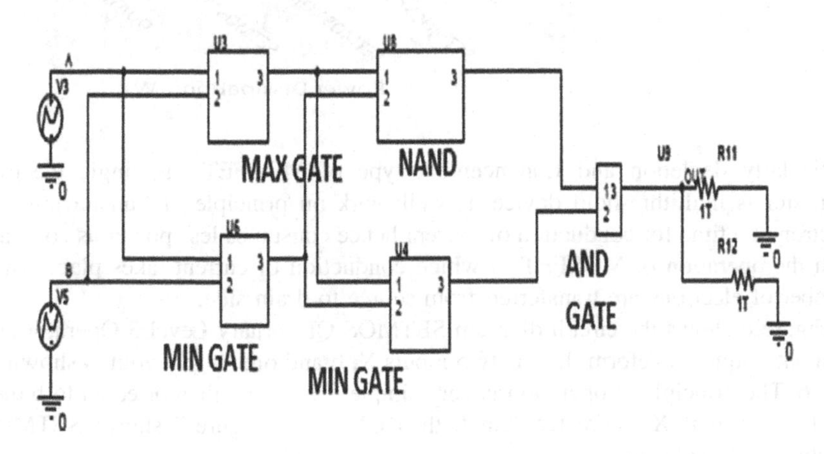

Fig. 4 SETMOS quaternary XOR gate

Table 4 Truth table of quaternary SETMOS XOR gate

A	0	1	2	3
B				
0	0	1	2	3
1	1	0	3	2
2	2	2	0	1
3	3	3	1	0

Table 5 Comparison of power dissipation Of XOR gate with quaternary SETMOS XOR

Sr. no.	Gate type	Power dissipation (μW)
1	Single electron transistor xor (set-xor)	4.2
2	Hybrid single electron transistor-MOSFET xor (SETMOS xor-type i)	10.87
3	Hybrid single electron transistor-MOSFET xor (SETMOS xor-type ii)	6.49
4	Hybrid single electron transistor-MOSFET xor (SETMOS xor-type iii)	5.87
5	Quaternary hybrid single electron transistor-MOSFET xor (quaternary SETMOS xor)	0.2

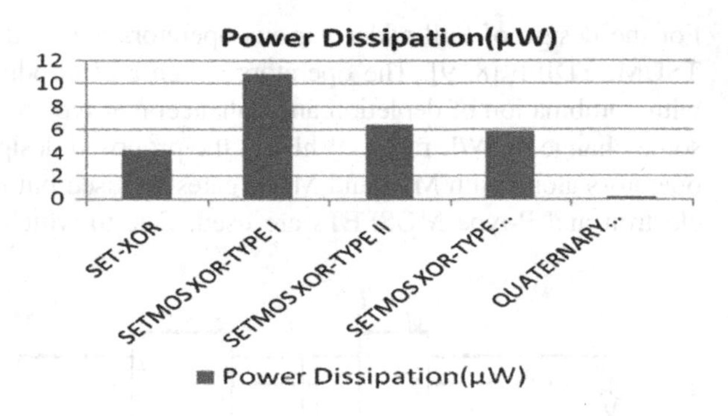

Fig. 5 Comparison of various XOR gate with SETMOS quaternary XOR gate

particularly depletion and Enhancement type of MOSFETs as single electron transistor is multithreshold device as well work on principle of transferring one electron at a time for conduction of current hence consumes less power as compare with the operation of MOSFET in which conduction of current takes place when number of electrons are transferred from source to drain side.

Figure 6 shows the circuit diagram SETMOS Quaternary Level 3 Operator and its input output waveform. In this two inputs X, b and one output vout is shown in Fig. 6. The principle of operation is very simple if X is less than or equal to b then OUT = 3 and if X is greater than b then OUT = *b*. Figure 7 shows SETMOS Quaternary Level 0 operator.

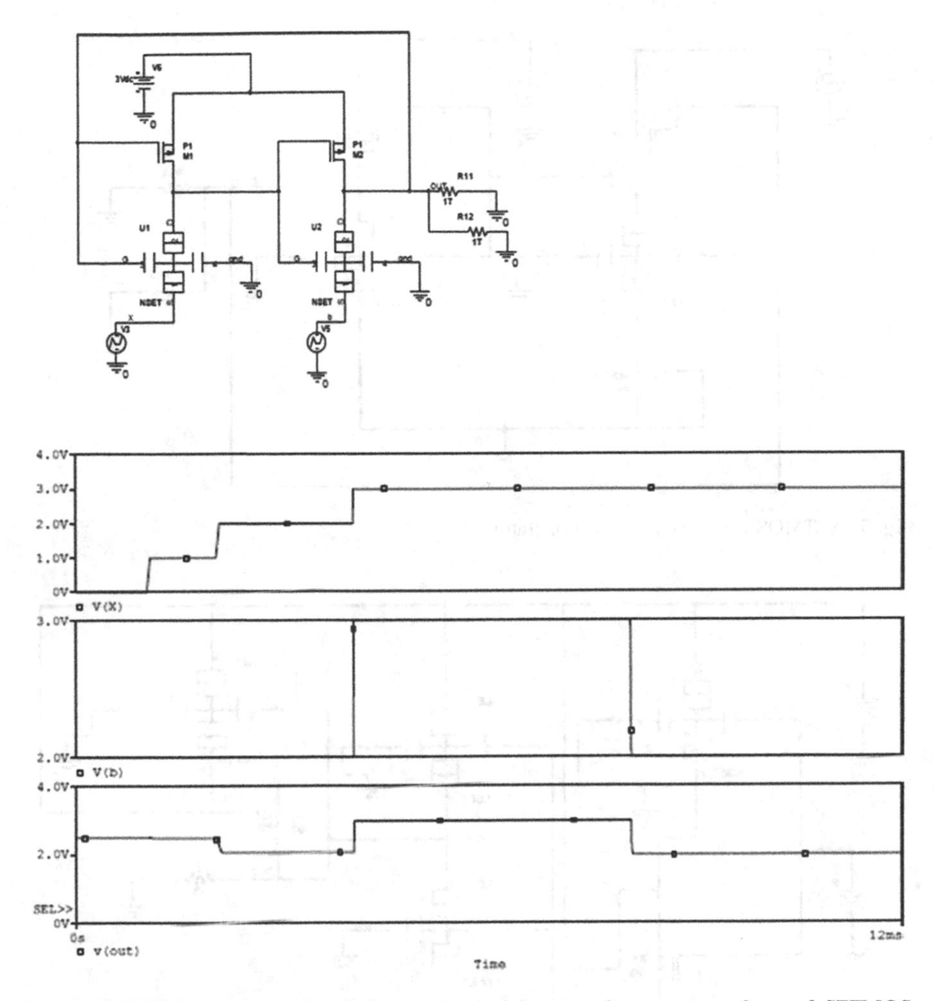

Fig. 6 SETMOS quaternary Level 3 operator and input and output waveform of SETMOS quaternary Level 3 operator

In that when $x = b = 0$ allowing the level of 0v to appear at output. Figures 8 and 9 shows the TSUM operator circuit diagram and input/output waveform. In this, at the output level of input voltage is incremented by one which is shown in Fig. 9. Similarly, Fig. 10 shows TDIFF operator in which input will be decremented by one at output. Table 6 shows the comparison between implemented operator and proposed operator in terms of TOX values and worst case delays. With the use of all operators including MIN and MAX initially a half adder is created. By using this half adder a full adder is designed and whose symbol is shown in Fig. 11.

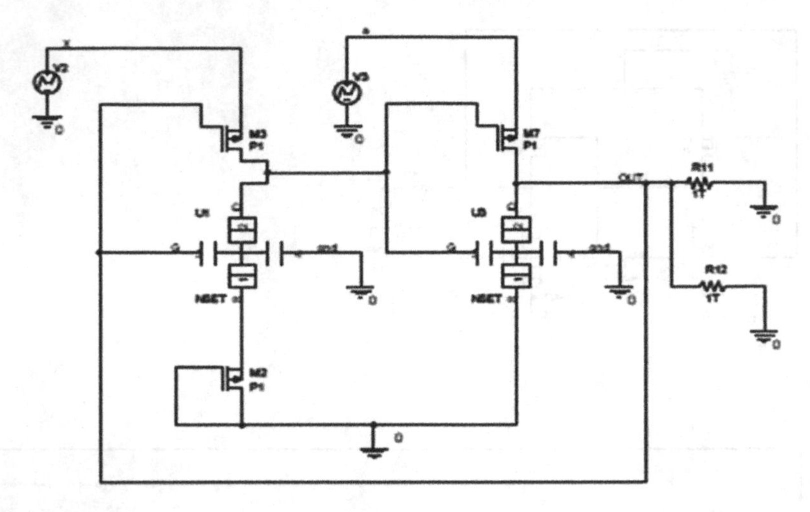

Fig. 7 SETMOS quaternary Level 0 operator

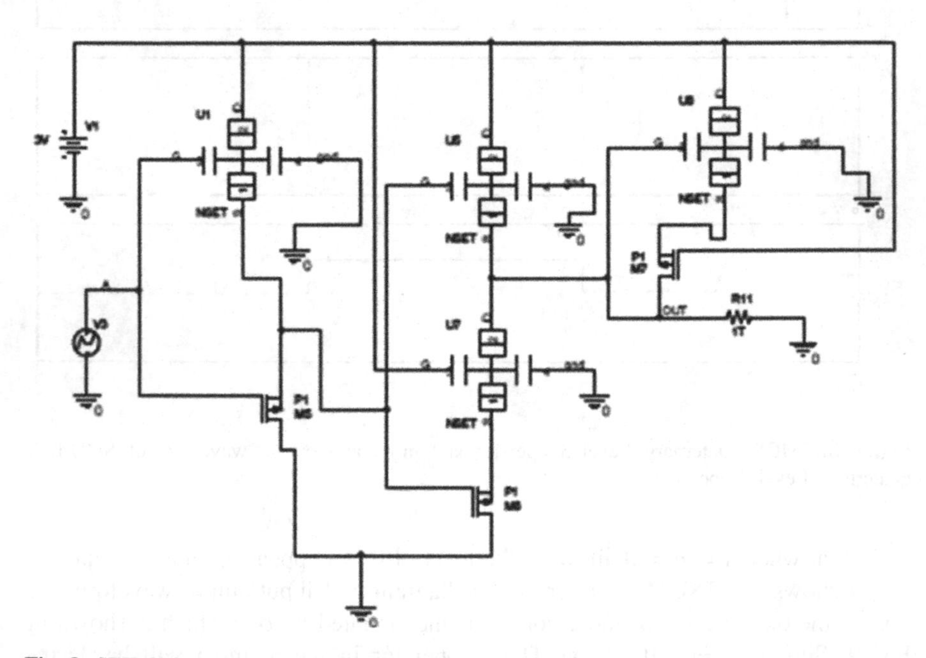

Fig. 8 SETMOS quaternary TSUM operator

Figure 12 and Table 7 shows the comparative analysis full adder in terms of power dissipation. For comparison full adder designed using single electron transistor and combination of single electron transistor with MOSFETs are used in that three types of hybrid full adder are designed with different method.

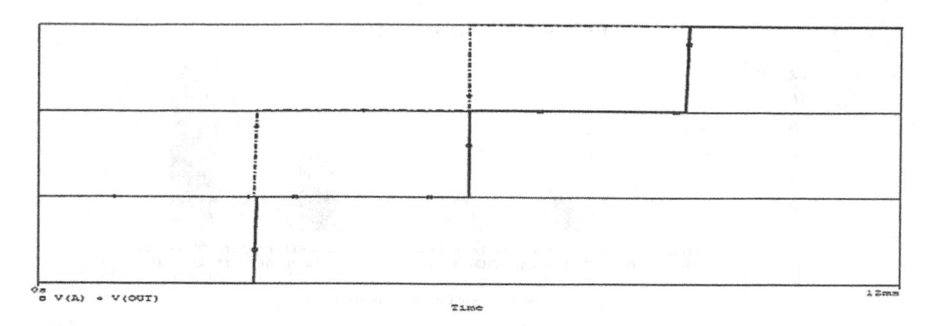

Fig. 9 Input and output waveform of SETMOS quaternary TSUM operator

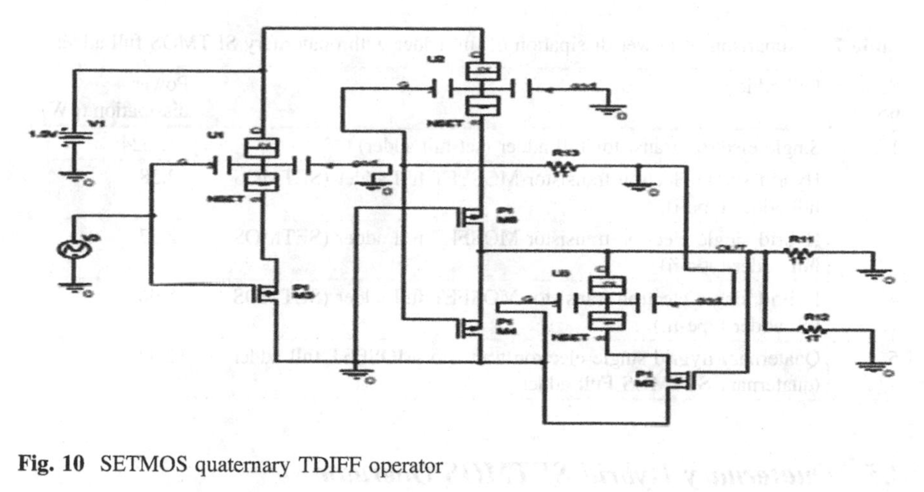

Fig. 10 SETMOS quaternary TDIFF operator

Table 6 Comparisons between Implemented operator and proposed operators

Implemented operator	Operation's worst delay (ns)						
	TOX (Å)	MAX	MIN	TSUM	TDIFF	Level 3	Level 0
Watanabe et al. [8]	400	<1	<1	~300	~300	~400	~400
Thoidis et al. [9]	250	<1	<1	6	9	22	17
Proposed	100	<1	<1	6	10	14	12

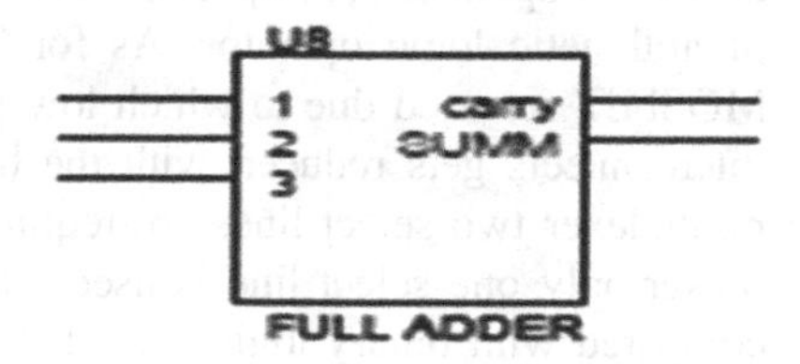

Fig. 11 SETMOS quaternary full adder symbol

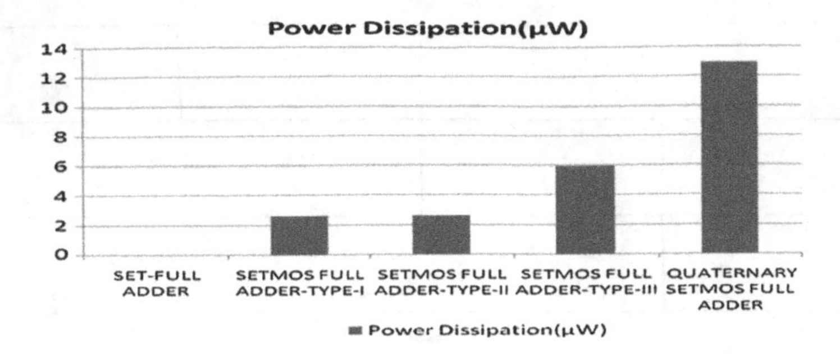

Fig. 12 Comparison of power dissipation of full adder with SETMOS quaternary full adder

Table 7 Comparison of power dissipation of full adder with quaternary SETMOS full adder

Sr. no.	Full adder	Power dissipation (μW)
1	Single electron transistor full adder (set-full adder)	1. 624
2	Hybrid single electron transistor-MOSFET full adder (SETMOS full adder-type-i)	2.58
3	Hybrid single electron transistor-MOSFET full adder (SETMOS full adder-type-ii)	2.57
4	Hybrid single electron transistor-MOSFET full adder (SETMOS full adder-type-iii)	5.96
5	Quaternary hybrid single electron transistor-MOSFET full adder (quaternary SETMOS Full adder)	12.92

3.5 *Quaternary Hybrid SETMOS Operator*

Figure 13 shows the block diagram of Hybrid SETMOS quaternary operator. In that selection of operator is implemented with quaternary 4:1 multiplexer. When select line is equal to zero then MAX operation will be executed. When select line is equal to one then MIN operation will be executed. When select line is equal to two then XOR operation will be executed and when select line is equal to three then Full adder operation will be executed. Table 8 provides the description of hybrid quaternary SETMOS operator whereas Table 9 provides comparative information of power dissipation values of hybrid quaternary SETMOS and hybrid binary SETMOS operator [7, 10]. Table 10 provides the comparison of power dissipation of arithmetic logic operator. As for hybridization, single electron transistor and MOSFET are used due to which low power and high speed is achieved as well as interconnects gets reduced with the help of quaternary logic, like for binary 4:1 multiplexer two select lines are required whereas in case of quaternary 4:1 multiplexer only one select line is used which reduces the various interconnections as compared with binary logic (Fig. 14).

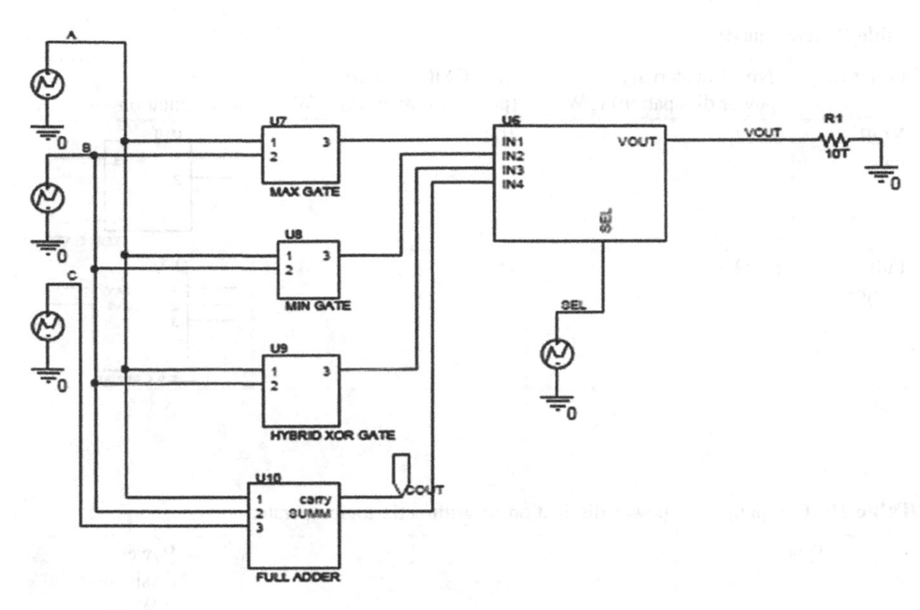

Fig. 13 SETMOS quaternary arithmetic logic operator

Table 8 Description of proposed SETMOS arithmetic logic unit

Sr. no.	Circuit description	Proposed SETMOS quaternary operator
1	Simulator	PSPICE
2	Voltage for logic ‘0’	0 V
3	Voltage for logic ‘1’	1 V
4	Voltage for logic ‘2’	2 V
5	Voltage for logic ‘3’	3 V

Table 9 Summary of quaternary SETMOS gate and full adder

Function	Hybrid quaternary (power dissipation) (μW)	SET-CMOS binary (power dissipation) (μW)	Symbolic representation
MAX gate	0	5.9	U6 1 2 3 MAX GATE
MIN gate	0	5.9	U7 1 2 3 MIN GATE

(continued)

Table 9 (continued)

Function	Hybrid quaternary (power dissipation) (μW)	SET-CMOS binary (power dissipation) (μW)	Symbolic representation
XOR gate	0.214	10.87	U10 1 2 3 XOR GATE
Full Adder	12.92	15.35	U8 1 2 3 caty SUMM FULL ADDER

Table 10 Comparison of power dissipation of arithmetic logic operator

Sr. no	Type	Power dissipation (μW)
1	Single electron transistor ALU (set-arithmetic logic unit)	60.30
2	Hybrid single electron transistor-MOSFET ALU (SETMOS arithmetic logic unit)	70.30
3	Quaternary hybrid single electron transistor-MOSFET ALU (quaternary SETMOS-arithmetic logic unit)	59.40

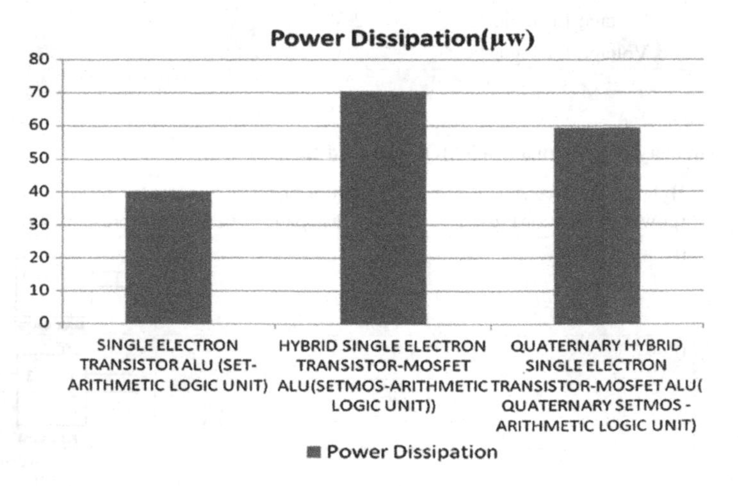

Fig. 14 SETMOS quaternary arithmetic logic operator

4 Conclusion

Hybrid Quaternary SETMOS Operator is proposed with various logical and arithmetic operators and verified according to the truth table. Hybrid SETMOS quaternary gates such as MAX, MIN, XOR gate as well as TSUM. TDIFF. Level 3, Level 0 operator, and full adder is designed and verified. With use of these gates overall power dissipation of quaternary SETMOS operator gets reduced. Design of hybrid quaternary SETMOS operator is simple.

References

1. Lukasiewicz, "On three valued-logic," in *L.* Borkowski, SelectWorks, Amsterdam, North-Holland, 1920, pp. 169–171.
2. E. L. Post, "Introduction to a general theory of elementary propositions," *Amer. J. Math.*, vol. 43, no. 3, pp. 163–185, Jul. 1921.
3. Kenneth C. Smith " Multiple-Valued Logic: A Tutorial and Appreciation" IEEE Trans. April 1998 PP. 17–27.
4. Daniel Etiemble and Michel Israel "Comparison of Binary and Multivalued According to VLSI Criteria" IEEE Trans. Computers April 1998 PP. 28–42.
5. Simulating Hybrid Circuits of Single-Electron Transistors and Field-Effect Transistors Günther LIENTSCHNIG_, Irek WEYMANN and Peter HADLEY, Applied Sciences and DIMES, Delft University of Technology, Lorentzweg 1, 2628 CJ Delft, Netherlands.
6. Günther LIENTSCHNIG_, Irek WEYMANN and Peter HADLEY "Simulating Hybrid Circuits of Single-Electron Transistors and Field-Effect Transistor"Jpn. J. Appl. Phys. Vol. 42 (2003) pp. 6467–6472 Part 1, No. 10 October 2003.
7. Vaishali Raut,Dr.P.K.Dakhole "Design And Implementation Of Single Electron Arithmetic And Logic Unit Using Hybrid Single Electron Transistor And Mosfet At 120 nm Technology" 2015 International Conference on Pervasive Computing (ICPC).
8. Watanabe, T., Matsumoto, M., And Li T., CMOS fourvalued logic circuits using charge-control technique. *18th Int. Symposium on Multiple-Valued Logic*, 1988, pp. 90–97.
9. I. Thoidis D. Soudris, I. Karafyllidis, S. Christoforidis, A. Thanailakis, " Quaternary voltage-mode CMOS circuits for multiple-valued logic" IEEE proceeding 145.1998. pp.
10. Ms. Vaishali Raut, Dr. P. K. Dakhole "Design And Implementation Of Single Electron Transistor N-BIT Multiplier" [ICCPCT]-978-1-4799-2395-3.

Detailed Survey on Attacks in Wireless Sensor Network

A.R. Dhakne and P.N. Chatur

Abstract Now days, Wireless Sensor Networks (WSNs) have widespread applications in the areas of medicine, military etc. So, it is always important to think on the security of this technology. WSN security is having some scientific and technical challenges. Most existing security techniques need a lot of computation, memory and energy which are considered as limitations in WSNs. This paper gives broad overview on all the attacks (vulnerabilities) that can harm to any wireless sensor networking environment such as physical attacks, attacks that can affect the functioning of different layers of networking, attacks related to privacy, Secrecy and Authentication and at the end this paper tried to give overview on some recent attacks that are going to create problems whenever certain nodes wants to calculate trust value of some other node such as Bad mouthing attack, Good mouthing attack and on-off attack.

Keywords Wireless Sensor Network · Attacks · Security · On-Off attack · Wormhole · Sinkhole · Bad mouthing · Good mouthing

1 Introduction

A Wireless Sensor Network (WSN) is generally formed by number of small sensor nodes. These nodes are generally composed of four main components such as processor and memory (Microcontroller), Sender and Receiver (transceiver), power supply and sensor along with digital to analog converter (A/D converter). The simplified architecture of a sensor node is depicted in Fig. 1. As human organs such as eyes and ear have the ability to sense about surrounding environment, sensor nodes are considered similar to them as they gathers the information about

A.R. Dhakne (✉) · P.N. Chatur
Department of CSE, Government College of Engineering, Amravati, India
e-mail: dhakne.amol5@gmail.com

P.N. Chatur
e-mail: chatur.prashant@gmail.com

S.C. Satapathy et al. (eds.), *Proceedings of the International Conference on Data Engineering and Communication Technology*, Advances in Intelligent Systems and Computing 469, DOI 10.1007/978-981-10-1678-3_31

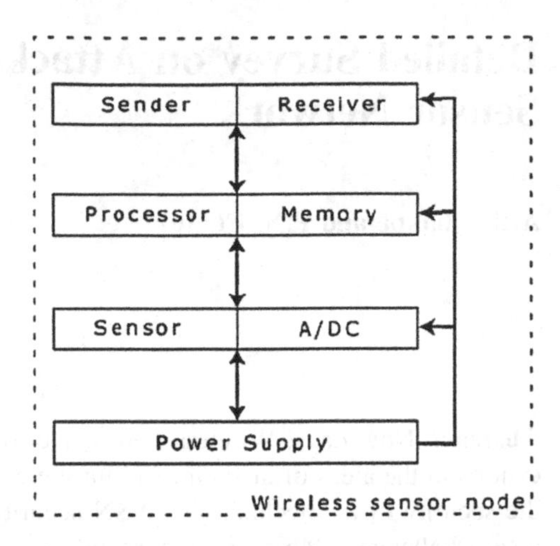

Fig. 1 Simplified architecture of Sensor node

surrounding environment, related to temperature, light, pressure, vibrations, velocity and magnetism. When these sensor nodes are implemented in systematic way, these sensors organize themselves automatically and built a dedicated ad hoc multi hop network and each node can communicate with other node within this network. At the sink, user remotely gives command to nodes via the wireless network and collect data after processing stored into a storage device. These nodes also receive the sensed data from the sink nodes [1].

There are various applications of wireless sensor networking that includes bushfire response, intelligent communications, wildlife monitoring, battlefield surveillance etc. The sensor nodes in network are generally exposed to open environment. Hence Security of such a network is crucial issue [2].

Paper has been organized in different parts. Section 2 will give overview of attacks based on different perspectives. Section 3 classifies attacks according to different layers of network, Sect. 4 focuses on Attacks related to Secrecy and Authentication, Sect. 5 gives overview of attacks that can harm the privacy of network, Sect. 6 gives overview on some additional attacks on wireless sensor network. At the end Sect. 7 conclude the paper and discusses the future issues of research on security in wireless sensor network.

2 Attacks in Wireless Sensor Network

Basically attacks in wireless sensor network can be classified in various ways based on the attacker location, level of damage and attacking devices used [3]. General classification of attacks is given as follows:

2.1 *Outsider Versus Insider Attacks*

In outsider attack, attack is not arranged by network node but external node can be deployed in current network. They are not having access to cryptographic keys or rules as they are not from the network. In Insider attack, insider node is compromised due to some weakness in system. Insider attacks can have some partial keys with them and they are having trust of other sensor nodes. Detection of insider attack is more difficult than outsider attack.

2.2 *Passive Versus Active Attacks*

Passive attacks are somewhat eavesdropping kind and in this, unauthorized user attack tries to track sense and monitor the communication channel; the active attacks are responsible for major modifications of the data or they can create some false stream of data in a WSN.

2.3 *Physical Attack/Node Capture Attack*

In this, attackers get the full control on all the activities going through sensor node. Attackers capture the node itself by having full physical access, so called Physical attack [4, 5]. These attacks harm sensors permanently, so the losses cannot be overcome. Tamper proofing is one of the solutions to avoid physical attack but it is irrelevant in WSN.

3 Attacks According to Layers of Network

These are the attacks that take place at to affect different layers of network such as Physical, Network, Transport and Application layers. This section gives overview of these attacks.

3.1 *Physical Layer Attacks*

In networking, physical layer has different tasks associated with it. Some of the major task of physical layer includes signal detection, selection of frequency for data transfer, data encryption etc. As WSN is deployed in the remote locations, the attackers have a chance to access the physical layer of WSN. Jamming and Tampering are some examples of vulnerability of WSN.

(a) **Jamming**: If the adversaries just have the knowledge about wireless transmission frequency of network then this kind of attack can be done easily by them. The jamming source can either be a powerful or less powerful. Powerful jamming source is able to create traffic in the entire network, whereas less powerful jamming source is able to disrupt only smaller portion of the network. In this, the attacker tries to transmit radio signal arbitrarily with the same frequency as that of other sensor nodes in WSN.

As attacker tries to interfere with signal sent by some another node, receivers that are within radio range of attacker node, will not get any proper message. Thus, due to jamming signal, there no any exchange of message between affected node and other sensor nodes [6].

Frequency hopping can be considered as one of the mechanism for preventing from this kind of attack [7]. Ir is possible to change sequence of frequency in some predetermined way by using frequency hopping technique. But it is not suitable for WSNs because the range of possible frequencies for WSNs is limited and every extra frequency requires extra processing. Another solution to prevent from jamming attack is suggested in [8] called as Ultra Wide Band (UWB) transmission technique. This technique can be called as Anti-jamming solution. It is very difficult to detect jamming attack by this technique but this technique consumes low energy and that's why it is suitable for WSN.

(b) **Tampering**: In this type of attack a node can get altered by a fake node so that attacker can easily get sensitive information and data passing through it. Tamper proofing can be considered as one of the technique to avoid this kind of attack, but it can add additional cost to WSN.
(c) **Path based DoS Attack**: In this type of attack available resources are exhausted to victim node and it prevents original users to access services or resources that are applicable to them to use. DoS attack can undermine the power and authority of network, create disturbance in network, or destroy a network and at same time it diminishes networks capability to provide services.

3.2 Link Layer Attack

Whenever node stops its functioning then this situation is called as Node Outage. Node outage is harmful when there is outage of cluster node, at that time protocols should be designed in such a way that these will provide some alternate route for transmission of messages.

(a) **Collision**: Whenever single transmission channel is overloaded by data from different senders at that time there is possibility of collisions. Whenever packets collide there are chances that it can change some data portion and thus destination will not receive data correctly. There can be collision from attacker in some specific packets such as ACK control message. Such kind of collisions can lead to costly exponential back-off in certain media access control (MAC) protocols [8].

Error correcting codes is one of the defence techniques against collision and these are best suited for the collisions that are happening due to environmental errors. Such kind of codes will need additional processing and communication overhead to overcome the collisions. But, we have to accept the fact that we will not be able to correct more than whatever has been corrupted. Even though it is not impossible to detect these malicious collisions, until now there is no any proper defence technique to completely overcome these attacks.

3.3 Network and Routing Layer Attacks

To improve power efficiency, awareness of location and addressing and to make sensor network more data centric, network and routing layer plays an important role. Major function of network layer is to route messages from one sensor node to another. Attackers can access routing paths to redirect the traffic and provide some wrong information about path to WSN or they can launch Denial-of-Service attacks. Some attacks on this layer are as follows:

(a) **Selective forwarding/Black Hole Attack**: In this type of attack, malicious nodes just drops packets that are to be damaged and selectively forwards other non interesting packets [3]. A black hole attack is one in which node drops all packets that it receives.

In Fig. 2(i) and (ii) selective forwarding attack are described. In Fig. 2(i), S is source node and B is destination, node A forward all packets coming from A but adversary node AD forwards only selected packets D1, D3 and drops D2, D4. In Fig. 2(ii), an adversary AD selectively drop all packets originated from node A and forwards packets that are coming from node B.

One can use multiple paths to end data to defend from selective forwarding attack [3]. Another defense technique is to detect malicious node as early as possible and it should be ignored for transfer of messages by considering some alternative route.

(b) **Sybil Attack**: Rather than having single identity, if node carries different identities to different nodes, then it is a Sybil attack [3]. This attack disrupts functioning of geographic routing protocols by being simultaneously at more than one place.

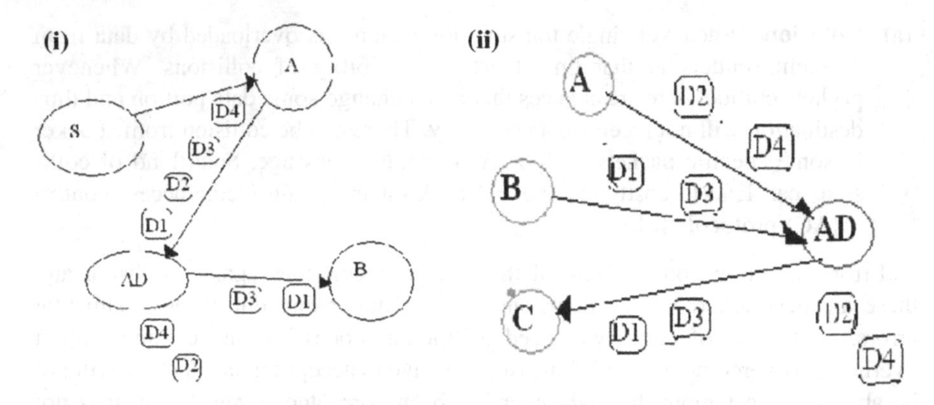

Fig. 2 (**i**) Adversary drops selected packet of a node (**ii**) Adversary drops all packets from selected node

In Fig. 3. Sybil attack is represented by adversary node AD which carries multiple identities. Here node A looks node AD as node F, node D as node A, so when node A wants to communicate with node F, it sends the message to adversary node AD. To defend from Sybil attack one should verify identity of nodes, but as WSN's have some computational limitations where traditional networks Symmetric key and public key algorithm cannot be applied to verify identity in WSN.

(c) **Sinkhole attack**: It is a kind of attack in which compromised node looks more attractive and most of the other surrounding nodes try to forward data to this compromised node as next node [3]. This type of attack is also responsible to boost selective forwarding attack as it is going to attract all traffic from large

Fig. 3 Sybil attack

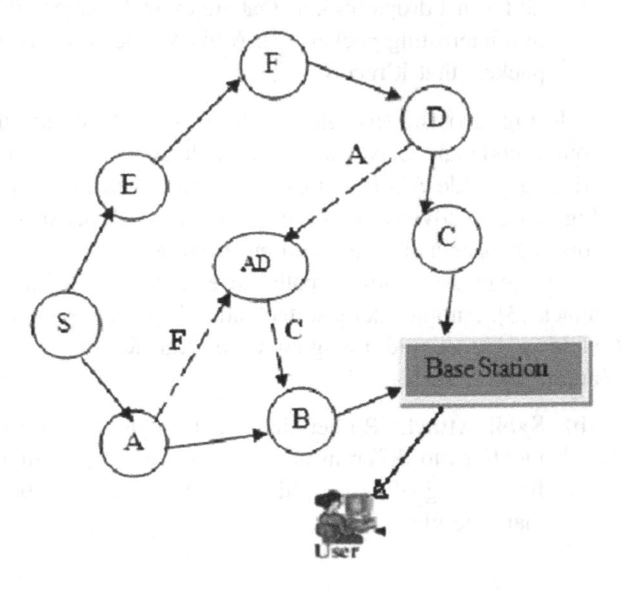

Fig. 4 Sinkhole attack

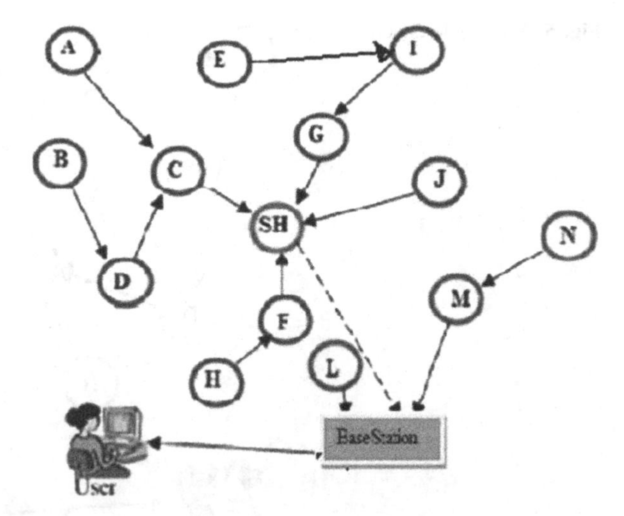

area in network and compromised node can only forward selected packets. In Fig. 4 SH denotes sinkhole attack. This sinkhole tries to attract all traffic from near about all the nodes to route through it except nodes N, L and M.

(d) **Wormhole attack**: In this, node at one end listens packets and transmits them through the tunnel to node at other end, where the packets are replayed to local area [9]. If these two ends forward messages without any change then it is helpful to accomplish faster transmission but most of times packets are dropped at end and selective packets are forwarded. As transfer is fast neighbour nodes can feel that these two nodes are neighbor of each other. If any of the tunnel ends is nearer to base station then it can attract a lot of traffic and it can act as sinkhole attack too.

Figure 5 demonstrates a Wormhole attack where WH is the adversary node that creates a tunnel between nodes E and I. Both Sinkhole and Wormhole attacks are difficult to detect, when routing protocols route packets based on advertisements on remaining energy remained or minimum hop count needed to reach base station. There are some geographic routing protocols that are providing some kind of resilience to these attacks [7].

(e) **Hello Flood attack**: In this attack attacker sends hello message to all surrounding nodes by using powerful transmitter due to which all other nodes that are not in radio range think that sender is in radio range. This makes node to send packets to adversary node rather than to base station.

Figure 6 show an adversary node AD which try to show other nodes that it is neighbor of them. There will be wastage of energy and there will be loss of data in this type because here adversary node will try to pretend other nodes that it is neighbour node, even if that node is far way from those. One partial defence to this attack is to do authentication of such node by some third party.

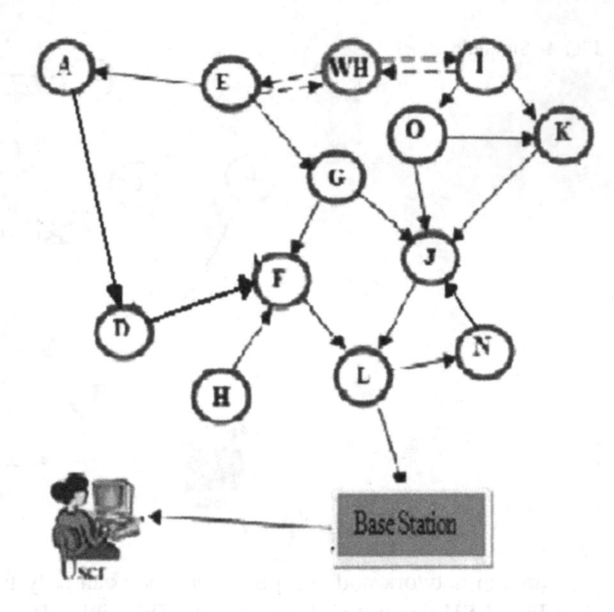

Fig. 5 Wormhole attack

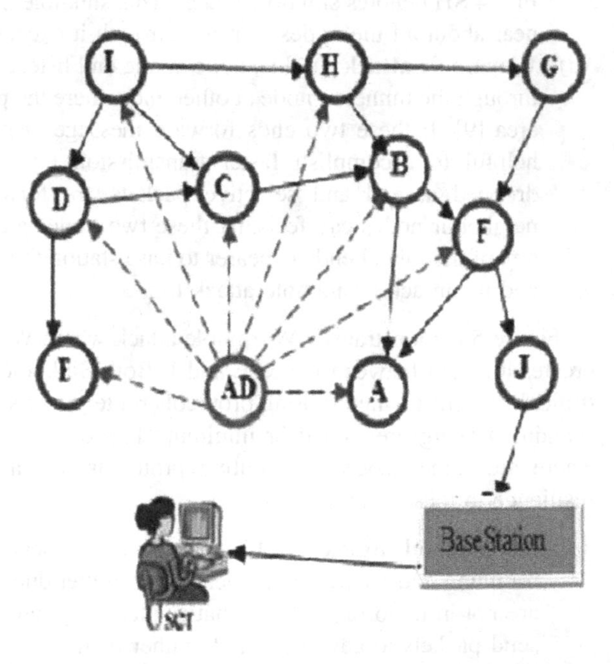

Fig. 6 Hello flood attack

(f) **Spoofed, altered and replayed routing information**: Spoof attack is one in which one person tries to conclude that he is some another one and thus falsifies another to take advantage. Such attacks are responsible to create routing loops to waste time, they can generate false error messages, increase end-to-end latency or they can extend or shorten service routes, [10] etc.

(g) **Node Malfunction**: Node malfunctioning can generate inaccurate data and can expose integrity of data if it is node such as cluster head which is responsible to collect all data [11].
(h) **Acknowledgment Spoofing**: Sometimes acknowledgments are needed by the routing protocols in sensor network [12, 13]. Here attacking node get routing information and provide wrong information to receivers [12].

3.4 Transport Layer Attacks

Functionality of end-to-end communication between source and destination is generally provided by the transport layer. Some possible attacks on this layer are discussed below:

(a) **Flooding**: If source node receives many requests repeatedly, then its memory become exhausted through flooding. An attacker requests for new connection continuously till its maximum limit. In this situation sender decline all request including the genuine request of any node in WSN. Thus, attacker can waste resources of WSN and communication between nodes will stop. In this attack many connection establishment requests can be sent by attacker to victim node to use its resources causing flooding attack. Defence against flooding can include to limit number of connection that can be made by a node as well as one can make compulsory to solve certain puzzle before making connection to another nodes [14].
(b) **Desynchronization**: In this, an existing connection between two nodes get disturbed due to transmission of illegal control flags or fake sequence number messages. By this, other nodes which are transmitting packets will waste their energy due to lack of synchronization. It is better to turn on all control fields in transport header to prevent from this attack. Also, packets exchanged between two nodes can be authenticated to prevent from this attack.

3.5 Application Layer Attacks

This layer basically collets and manages the data. As information is revealed in this layer it can lead to compromise whole network. Exclusion of node is probable solution in this attack if node gets compromised. LEAP (Localized Encryption and Authentication Protocol) [4] can verify if a node has been compromised or not and if it is compromised then it can revoke that node by some efficient keying mechanism.

4 Attacks on Secrecy and Authentication

Following are some attacks which can harm cryptographic techniques to explode secrecy of data and authentication of user:

4.1 Node Replication Attack

In this, attacker tries to have duplicate identity of node by copying identifier of another node. These nodes can disrupt networking by giving wrong information and wrong routes to another node. This can lead to partition the network and it will responsible to communicate false readings of sensor to another nodes. There is always possibility of copying cryptographic keys if attacker gain access to communication channel and these keys can be used by replicated node for communication. Attacker can also place replicated node at some specified location so as to partition network in some groups In Fig. 7, N is the identity of cloned nodes that are mounted in multiple places in the network to confuse the entire network. This kind of attack can be avoided by having single central data gathering mechanism such as base station; This attack can be detected by verifying the identities of these nodes by some trustworthy node [15].

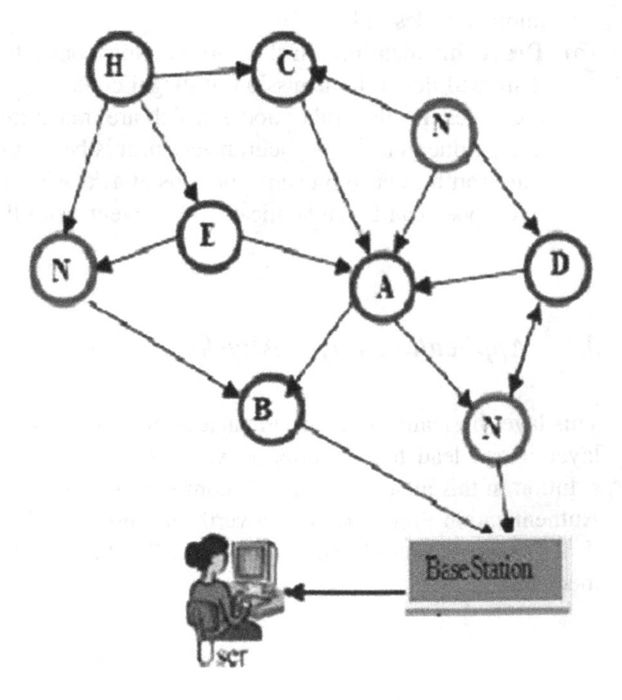

Fig. 7 Node replication attack

5 Attacks on Privacy

Attackers need not be present to gather information, one remote access is sufficient for them to gather information from multiple sites at single instance [16]. Privacy of sensor node can be disturbed by following types of attacks:

5.1 *Eavesdropping and Passive Monitoring*

It is common attack on data privacy. The adversary could easily understand the contents, if cryptographic mechanism is not used for protecting messages. Most of times, more information is revealed by the packets having control information than information available at location server.

5.2 *Traffic Analysis*

Eavesdropping when combined with a traffic analysis, it makes an effective attack on privacy. Due to thorough analysis of traffic for adversary it is possible to understand the roles of nodes. For example, unfortunate increase in message communication can make adversaries to understand that there are some events to happen. Chan.et al., have demonstrated two attacks that can identify the base station just by analysing traffic without having access to contents of message [17].

6 Some Other Security Attacks

There are some less known attacks in WSNs. These mostly concentrate on service availability and sometimes these can occur to misguide the behaviour of neighbour nodes to its head node. We briefly describe them in the following section:

6.1 *Bad Mouthing and Good Mouthing Attack*

In bad-mouthing attack neighbor nodes of sensor node always give some wrong information to degrade trust of sensor node whenever cluster head wants to calculate trust of sensor node from surrounding nodes. For example, they can give less recommendation of other node at the time of calculation of trust value. Thus, these recommendations cannot reflect the real opinions of the recommender. As opposed to this, the sensor nodes which conduct good mouthing attack provide more trust value for malicious nodes intentionally to misguide other nodes.

6.2 *On-Off Attack*

In this attack node can behave good or bad according to situation. When the trust values of malicious nodes are low, nodes can act well for some period to improve trust value of node. Therefore, it is difficult to detect these malicious nodes by conventional trust models.

Basically, an Off-Off attack works with other type of attack. For example, it can work with selective forwarding attack where most of times node can forward all messages but occasionally it can drop all packets too. In On state, attacker attacks node with help of other attack and is called as Attack State. In Off State, malicious nod behaves normally and is called as Normal State [17].

7 Conclusion and Future Scope

As sensor nodes are deployed in open environment they are vulnerable to different attacks. So, security of such network is always important aspect. This paper provides comprehensive survey on attacks in wireless sensor networks with respect to different layers of networking. This paper also gives overview on some new attacks that can harm most of trust models in WSN such as Good Mouthing attack, Bad Mouthing attack and On-Off attack.

This survey brings researchers an view of all attacks in wireless sensor network while designing any security mechanism at different layers of network. Detection of these attacks is important if data to be transmitted to different places in correct way. In future, by taking reference of these vulnerabilities, one can proceed to propose efficient intrusion detection mechanisms to overcome most of these attacks.

References

1. F. Oldewurtel and P. Mahonen. Neural Wireless sensor network ICSNC, 0:28, 2006.
2. J. Yick, B. Mukherjee and D. Ghosal; Wireless Sensor Network Survey; Elsevier's Computer Networks Journal, 52, 2292–2330; 2008.
3. C. Karlof and D. Wagner, Secure routing in wireless sensor networks: Attacks and countermeasures, Elsevier's AdHoc Networks Journal, Special Issue on Sensor Network Applications and protocols, 2003.
4. Z. Tanveer and Z. Albert. Security issues in wireless sensor networks, ICSNC'06: Proceedings of the International Conference on Systems and Networks communication, Washington, DC, USA, IEEE Computer Society, 2006, 40.
5. J. Rehana, Security of wireless sensor networks. In TKKT— 110.5190 Seminar on Internetworking, Helsinki, 2009.
6. J. P. Walters, Z. Liang, W. Shi, and V. Chaudhary. Wireless sensor network security: A survey, Security in Distributed, Grid, and Pervasive Computing, 2006.
7. Y. W. Law and P. Havinga, How to secure a wireless sensor network. December 2005, 89–95.

8. S. Datema. A case study of wireless sensor network attacks. Master's thesis, Delft University of Technology, 2005.
9. I. Khalil, S. Bagchi, and N. B. Shroff, Liteworp: Detection and isolation of the wormhole attack in static multihop wireless networks, International Journal Computer and Telecommunications Networking, Vol. 51, Issue 13, 2007, pp 3750–3722.
10. Jaydip Sen "A survey on wireless sensor networks security" International Journal of Communication Networks and Information Security (IJCNIS) Vol. 1, No. 2, August 2009.
11. Pathan, A.S.K.; Hyung-Woo Lee; Choong Seon Hong, "Security in wireless sensor networks: issues and challenges" Advanced Communication Technology (ICACT), Page(s):6, year 2006.
12. Chris Karlof, David Wagner, "Secure Routing in Wireless Sensor Networks: Attacks and Countermeasures", AdHoc Networks (elsevier), Page: 299–302, year 2003.
13. Yong Wang, Garhan Attebury, and Byrav Ramamurthy "A survey of security issues in wireless sensor networks" 2nd quarter 2006, volume 8, NO. 2 IEEE communication surveys.
14. T. Aura, P. Nikander, and J. Leiwo, Dos-resistant authentication with client puzzles, 2001, 170–177.
15. W. T. Zhu, Node Replication Attacks in Wireless Sensor Networks: Bypassing the Neighbor-Based Detection Scheme, IEEE International Conference on Network Computing and Information Security (NCIS), vol. 2, 2011, 156–160.
16. B. Parno, A. Perrig, and V. Gligor, Distributed detection of node replication attacks in sensor networks, Proceedings of IEEE Symposium on Security and Privacy, 2005, 49–63.
17. J. Deng, R. Han, and S. Mishra, Countermeasures against traffic analysis in wireless sensor networks, Technical Report CU-CS-987-04, University of Colorado at Boulder, 2004.

Comparative Analysis of Frontal Face Recognition Using Radial Curves and Back Propagation Neural Network

Latasha Keshwani and Dnyandeo Pete

Abstract Person identification using face as a cue is one of the most prominent and robust technique. This paper presents 3D face recognition system using Radial curves and Back Propagation Neural Networks (BPNN). The face images used for experimentation are under various challenges like illumination, pose variation, expression and occlusions. The features of images are extracted using Eigen vectors. These features are compared using radial curves on the face starting from center of the face to the end of the face. Each corresponding curve is matched using Euclidean Distance classifier. The BPNN is used to train the features for face matching. The proposed algorithms are tested on ORL and DMCE database. The performance analysis is based on recognition rate accuracy of the system. The proposed radial curve system yields recognition rate accuracy of 100 % for images from the ORL database and 98 % for the images from DMCE database.

Keywords Face recognition • Back Propagation Neural Networks • Radial curves • ORL database

1 Introduction

From the last few decades, instead of PINs and passwords, biometric recognition systems have gained a tremendous significance far and wide. Generally people find it difficult to remember the passwords and PIN numbers. They can even be stolen, forged, or guessed by the intruder. On other hand, biometric recognition systems based on behavioral cues (keystroke, signature, and gait) and physiological cues of

Latasha Keshwani (✉)
Datta Meghe College of Engineering, Airoli, Navi Mumbai, India
e-mail: latashakeshwani@gmail.com

Dnyandeo Pete
Electronics and Telecommunication Department, Datta Meghe College of Engineering, Airoli, Navi Mumbai, India
e-mail: pethedj@rediffmail.com

S.C. Satapathy et al. (eds.), *Proceedings of the International Conference on Data Engineering and Communication Technology*, Advances in Intelligent Systems and Computing 469, DOI 10.1007/978-981-10-1678-3_32

a human (face, finger geometry, fingerprints, palm geometry, palm veins, iris, iris veins, retina, ears, and voice). However an individual's biological features cannot be forgotten, misplaced, duplicated, or stolen.

Biometric-based recognition techniques, like fingerprint detection, palm matching, and iris matching have attained a high level of success in accuracy, but they are not used widely in non cooperative scenario. Besides, face is the non-intrusive method, used in such non-cooperative environment. Face is the natural assurance of someone's identity. Due to this widely accepted representation, face plays an important role in conveying the identity and emotions of a person. Even after many years we can recognize a number of faces that we have seen throughout our lifespan. The human tendency of identifying and recognizing person quickly despite large variations in the looks due to changing condition like aging, beard, glasses, or changes in hairstyle, etc., is the motivation behind development of face recognition systems. With every human, the variation in their fiducial parts such as varying size of nose, eyes, ears, lips, chin, forehead are used as features for perfect matching of the face. A computational framework of face recognition does the same using a proper mathematical calculation to identify a person.

Automatic face recognition technique has numerous applications now-a-days especially the websites hosting images like Picasa, social media like Facebook, Twitter, etc. Systems that can detect and recognize faces can be applied to a wide variety of tasks including criminal/fraud/theft identification, i.e., use of the system by forensic departments and crime departments when the face detected by the surveillance cameras is not so clear, security system by providing access to various control systems such as ATM machine, office entry. It also finds applications in image and film processing. Face recognition system is working at its best level by providing recognition rate of nearly 99 % at a False Acceptance Rate (FAR) of 0.01 %.

A Face Recognition System analyses individual's features and different characteristics to show assertion of one's identity. The generalized block diagram of face recognition system is as shown in Fig. 1.

A face recognition system is basically a two-step procedure in general; they are Training/Enrollment mode and Testing/Identification mode. They can be defined as follows:

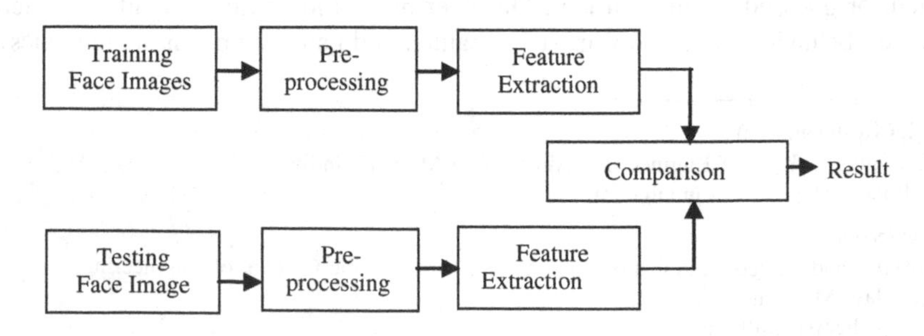

Fig. 1 General schematic of face recognition system

Training/Enrolment mode: In this mode, images of various people are captured and a biometric feature of each person is obtained to be saved in the database. This stage can also be called as database creation stage.
Testing/Identification mode: Where a template is created for an individual and then a match is searched for it in the database of pre-enrolled templates.

The face recognition system starts with acquisition of the face images. If the images are blurred or not so clear, then the preprocessing stage does the necessary action and makes it suitable for further matching. The features of the face are extracted for the database image as well as the testing image. These features of the test image are compared with the features of all the database images. The result is a probable match at the output.

In real-time face recognition system comes across various challenges like pose variation, expression, occlusion, illumination conditions, and clutter. It becomes difficult to identify a person when the face image stored in the database is with a different pose, expression, and illumination conditions; and that for the testing is with a different pose, expression, and illumination condition. Face recognition system also faces difficulty when there is some presence of occlusion like wearing of glasses, growth of beard, change in hair style, etc., it may also happen that the image captured by the camera is not proper. This paper presents an algorithm which has a provision of identifying the person across various challenges viz. pose variation, expression, occlusion, illumination conditions, and clutter.

The content of this paper is organized as follows: Sect. 2 provides the literature survey related to face recognition system. Section 3 describes the in detail system components and algorithms used in proposed face recognition system. The experiment results and discussion is elaborated in Sect. 4. The conclusion and future scope is presented in Sect. 5.

2 Literature Survey

The 3D face recognition has been approached in many ways, leading to different levels of success. Some of the common approaches for face recognition are deformable template-based approach, deals with the matching of entire face, estimates the pose, and occluded areas. Kakadiaris et al. [1] makes use of annotated face model that is deformed elastically to fit each face thus matching different areas such as nose, eyes, mouth.

Local region/features approach handles expression and is based on matching only parts or regions rather than matching full face. Lee et al. [2] uses the ratio of distances/Euclidean distance and angles between fiducial points followed by SVM (Support Vector Machine) classifier. Surface distance-based approach also handles expression variation. It utilizes distance between points on facial surfaces to define features, i.e., the geodesic distance. The open mouth problem is well discussed [3].

Dirira et al. had given emphasis on the elastic models and radial curve that maximally separates the interclass variability from the intraclass variability. Radial curves with the nose-tip detection have tremendous potential in detecting expression, occlusion to at a great extent [3, 4]. Queirolo et al. in [5] performs 3D face recognition using Simulated Annealing (SA) for range image registration and the Surface Interpretation Measure (SIM) as the similarity score between two 3D images. The authentication score is obtained by combining SIM values which corresponds to the matching of four different face regions: circular and elliptical regions around the nose, forehead, and entire face region [5]. Passalis et al. use facial symmetry to handle pose variation [6] in real-world 3D face recognition where automatic landmark detection is employed which estimates poses and detects occluded parts for each face scan. This method makes use of wavelet-based biometric signature as it requires only half of the face to be visible to the sensor [7].

A robust algorithm for 3D face recognition using curvelet transform is proposed in [8]. It detects the fiducial points on specific area of face by examining curvelet coefficient in each subband and builds multiscale local surface descriptors that can capture highly distinctive rotation/displacement invariant local features around the detected keypoints. The Iterative Closest Normal Point (ICNP) [9] method is introduced to find corresponding points between the reference face and every input face. These points are marked across all faces, enabling effective application of discriminant analysis [9]. A survey of various face recognition approaches along with neural networks implementation is presented in [10].

The face recognition is either feature-based, using shape and position of facial features such as eyes, nose, and lips or holistic using overall analysis of facial image. A comparative study of face recognition under varying expression is given in [11]. Erdogmus et al. give performance comparison based on Principal Component Analysis (PCA), Linear Discriminant Analysis (LDA), and Local binary pattern (LBP) [12].

Alyuz et al. [13] introduced a new technique called as masked projection for subspace analysis with incomplete data. Fully Automatic 3D face recognition with occlusion handling is explained. [13]. Ali [14] proposed pose invariant Face Recognition at Distant Framework (FRAD), which introduces an automatic front-end stereo-based system. Once a face is detected by one of the stereo cameras, its 15 facial features are identified using facial feature extraction model. These features are used for steering the second camera to see the same subject. This system performs various steps either in online mode or off-line mode and finally face recognition is performed using nearest neighbor classifier.

The above all the face recognition techniques are implemented on the face database which is available online. This database is constructed using the specialized 3D sensor. While it is challenging in terms of cost to use this 3D sensor for high-resolution images, RGB-D images can be captured by low cost sensors such as kinect [15]. Face recognition with texture and attribute features [16] proposed an algorithm that computes a descriptor based on entropy of RGB-D faces along with the saliency feature obtained from a 2D face. And the attributes are extracted from the depth of the image. A recent advancement in single modal and multimodal face recognition is given in [16] that use various approaches like visual and 3D, visual and IR, visual and IR and 3D.

3 Proposed System

This paper presents a framework for analysis of an individual's face, in the process dealing with various challenges of large expressions, pose variation, and missing parts. The paper shows comparison between two techniques viz. Radial Curves, BPNN. We use the concept of matching radial curves for testing a frontal face. We also make use of variety of tools that help us for the computation of the facial modality. BPNN is used as another technique used for comparison of the system in terms of complexity and efficiency. The image acquisition is the first step, after which the database is created. The system then computes the algorithm in various steps.

3.1 Preprocessing

After the database creation, the frontal face images undergo through the face normalization step. In this step, the preprocessing of an image takes place. First, the RGB image is converted to gray scale image. The preprocessing step also uses Weiner filter for the removal of noise and sharpen the edges of the image which avoids deterioration in the output. The system also uses cropping filter which crops and returns a frontal face portion of face.

3.2 Feature Extraction and Classifiers

Every human face has distinct features. The system extracts the features of every face image. These extracted features play a vital role for matching the perfect face. The classifiers are used for the decision-making process. The classifiers are used to distinguish the image in interclass and intraclass variability. The proposed system uses PCA, Radial Curves, BPNN, and Euclidean Distance for the same. The feature extraction step is carried out during both, Indexing phase and Recognition phase of the face recognition system. The classifiers (based on similarity) are used only in the recognition phase to classify the output whether match found or not. The Indexing phase and Recognition phase is as shown in Figs. 2 and 3 respectively.

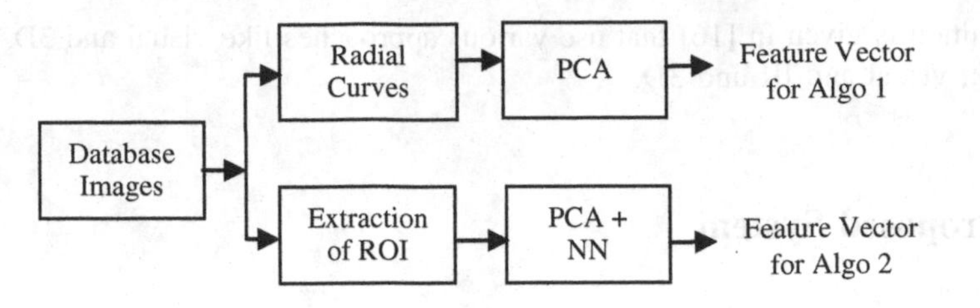

Fig. 2 Indexing scheme for feature vector creating for proposed system

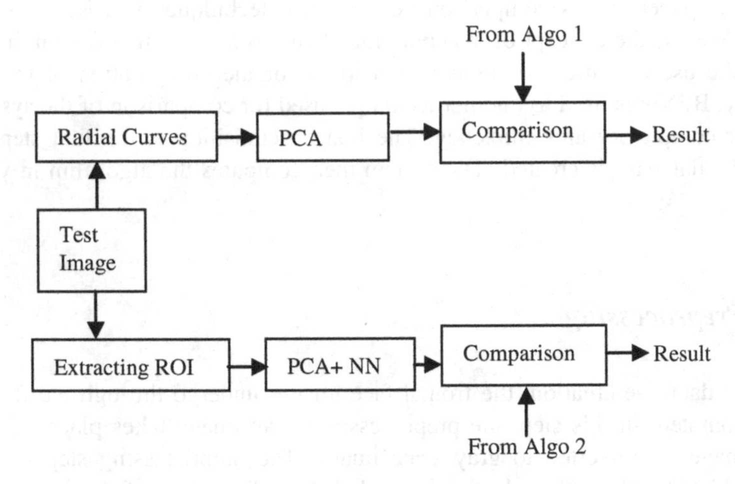

Fig. 3 Recognition scheme for comparing feature vector and obtaining probable matched result for the proposed system

3.2.1 Principal Component Analysis (PCA)

PCA is generally termed as dimensionality reduction technique. PCA also reduces significant redundancies in face images. PCA represents the facial image in the form of vector called feature vector/Eigen vector. This feature representation of image is comparably better than the original distorted images. PCA is applicable to represent the linear variations.

The goal of PCA algorithm is to capture the direction of maximum variance. The algorithm starts by capturing the variance between all images used for training and then creates a set of information in an orthogonal space which constitutes of weights called principal components, i.e., "Eigenfaces". PCA is also called as Eigenface approach. These weights thus obtained become the facial features for classifiers [17, 18].

3.2.2 Radial Curves

Radial Curves are the circular curves drawn on the face of every database image. These curves are drawn by taking into consideration the Iterative Closest Point [19]. The radial curves are drawn during both indexing scheme as well as in the recognition scheme. The proposed system extracts, analyses, and compares the shape of radial curves. The problem of expression, pose variation, and occlusion is solved using radial curves [3]. Figure 4 depicts the face with radial curves.

Algorithm 1:

(a) The images from ORL and DMCE database are taken to form database of the proposed system.
(b) The proposed system draws five radial curves on every face image starting from small circular curve at the center of the face till the end of the face.
(c) The proposed system further uses PCA for feature extraction of each curve.
(d) The PCA used for dimension reduction gives feature vectors for all the database images. One feature vector for each subject
(e) The total number of feature vectors formed by the system is 40.
(f) These vectors constitute as the database for recognition phase.

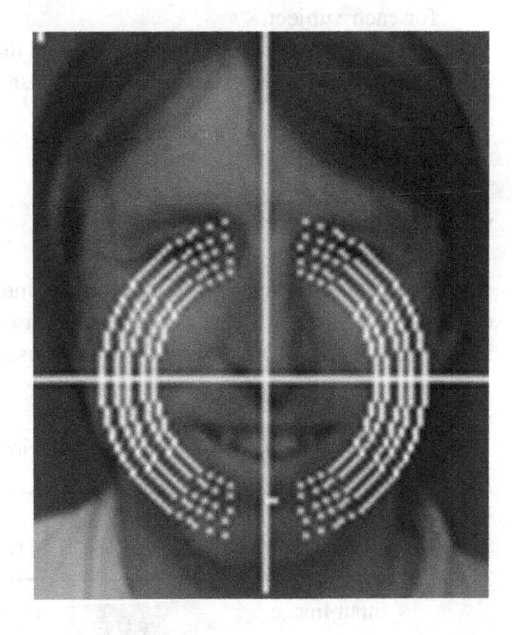

Fig. 4 Facial image with radial curves

3.3 *Back Propagation Neural Network (BPNN)*

Back Propagation Neural Network (BPNN) is used to learn the patterns of PCA features and produce relevant client and imposter scores for verification [18]. BPNN is used to train multilayer feed-forward Neural Network to learn a complex mapping for classification. The BPNN is designed to adapt the weights based on the corresponding error. The goal is to minimize the output error of the network. The network executes the adaptation of weights throughout the epochs until the error reaches an accepted minimum level [18].

BPNN has an input layer, an output layer, and one or more hidden layers in between them. The training of a network by back propagation involves three stages: the feed-forward of the input training pattern, the calculation and back propagation of the associated error, and the adjustment of the weight and the biases [18].

Algorithm 2:

(a) The images from ORL and DMCE database are used for training the Neural Network.
(b) The proposed system extracts Region of Interest (ROI) like nose, eyes and lips from the database image as shown in Fig. 5.
(c) The system further uses PCA for feature extraction process.
(d) The PCA gives feature Vectors for all the database images. One feature vector for each subject.
(e) These images are used for training of the Neural Network.
(f) These feature vectors form the database for recognition phase

3.4 *Euclidean Distance*

The Euclidean Distance [18] is the most commonly used classifier for the matching of faces. The smaller the distance between two vectors, the higher the resemblance in the images. The measure of surface distance is computed by comparing the

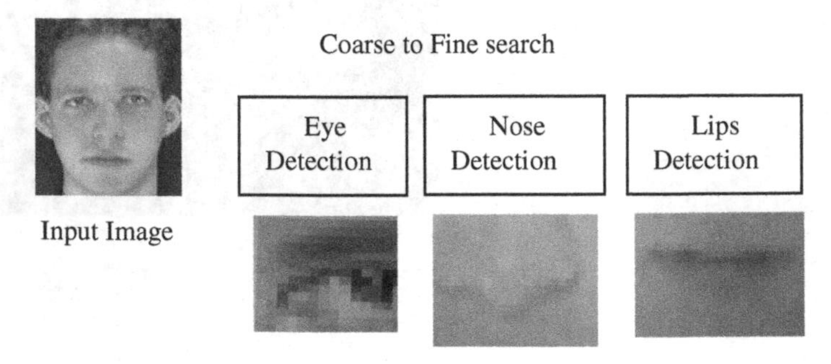

Fig. 5 Block diagram for database creation for Back Propagation Neural Networks

Euclidean distance of radial curves of the all the database images with the test image. Depending on the value of Euclidean distance it is estimated that if there is any resemblance between any one pair of curves then that shows the similarity between the faces. The proposed system shows perfect match is found only when the distance is lower than the threshold.

4 Results and Discussion

The proposed system is evaluated using two databases viz. ORL and DMCE database. The ORL database contains 400 gray scale images of 40 subjects with 10 images per person. The images are taken at different times, varying illumination conditions and expressions (open/closed eyes, smiling/not smiling), varying poses (frontal view, slight left–right rotation) and facial details (glasses/no glasses). The image size is 112 × 92. The DMCE database contains 400 images, of 20 girls and 20 boys, which consist of 10 images for 40 subjects each. These database images are taken under various challenges of expression (smiling, laughter, fear, closed eyes, etc.), pose variation (facing left, right, down), occlusion (wearing spectacles, hat), and different illumination conditions. These database images are resized to 112 × 92 for the DMCE and ORL database.

All the input images are taken from the two databases, i.e., ORL readily available database and DMCE database. These images are selected as database. Then one image is taken as the test image and compared with the remaining all images. First of all, the image is converted from RGB to gray. In the next step, it is passed on to radial curves and extraction of region of interest. Radial curves are drawn on each face emanating from the center of the face to the end of the face in a circular manner. The feature extraction of image is done using PCA. The performance analysis based on different number of curves is carried out. We make the analysis using two, four, five curves. And the corresponding recognition rate is as shown in Table 1. The table shows that as we increase the number of curves, the recognition rate goes on increasing.

These radial curve features of test image are then matched with each database image features (from Algorithm 1) using the Euclidean Distance. Based on the score of Euclidean distance the probably matching faces are recognized as shown in Fig. 6. The same experiments are accomplished with Neural Network-based system.

In case of Neural Networks, the system extracts the ROI same as that done for the database images. The feature extraction done using PCA and the feature vector

Table 1 Comparison of recognition rate for varying number of radial curves

No. of radial curves	Recognition rate (%)
Two	64
Four	96
Five	100

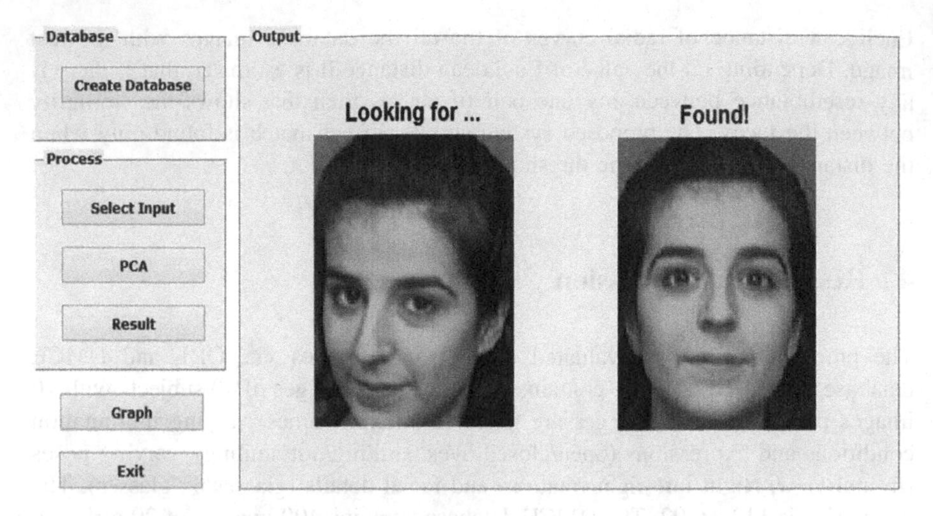

Fig. 6 A snapshot of tested facial image recognition by proposed algorithm

of test image is passed to the comparison block. The other input is given from the Algorithm 2 as shown in Fig. 3. The perfect matching image is thus found.

The resulting recognition rate using five Radial Basis Curves (RBC) and BPNN obtained from the proposed system is as depicted in Table 2. The accuracy is computed by considering the test image once from the database and other time out of the database. When efficiency calculation for image from database is taken into account then 400 images of ORL and DMCE database are used for training, and one out of them is used as query image. Whereas in efficiency calculation for image out of database, 320 images (eight images for each subject) are used for training and the remaining 80 images (two images for each subject) which are not a part of database are used for testing.

The results of the work computed on MATLAB shows that, by implementing Radial Basis Curves with the help of thresholding we can achieve 100 % accuracy for ORL database. The accuracy thus obtained for DMCE database is 98 %. The results show that radial curves prove to be more efficient than the Back Propagation Neural Network.

Table 2 Comparison of recognition rate for proposed system

Method	% Efficiency with image from database		% Efficiency with image out of the database	
	ORL	DMCE	ORL	DMCE
RBC	100	98	99	97
BPNN	99	96	87	86

5 Conclusion

This paper presents a Radial Basis Curves Face Recognition system which can recognize the person even with expression, occlusion, pose variation, illumination, and clutter. The proposed system uses PCA, and Radial curves to achieve face recognition. The complexity and dimensionality is reduced by robust PCA and radial curve results in better performance useful in real-time face recognition systems. The system yields 100 % recognition rate accuracy for images from ORL database and 99 % recognition rate accuracy for images from DMCE database.

Declaration
Author declares that they have the permission for using the human faces for research and publication. Springer will not be held responsible in case of any consequences in future.

References

1. I.A.I. Kakadiaris, G. Passalis, G. Toderici, M.N.M. Murtuza, Y. Lu, N. Karampatziakis, and T. Theoharis, "Three-Dimensional Face Recognition in the Presencc of Facial Expressions: An Annotated Deformable Model Approach," IEEE Trans. Pattern Analysis and Machine Intelligence, vol. 29, no. 4, pp. 640–649, 2007.
2. Y. Lee, H. Song, U. Yang, H. Shin, and K. Sohn, "Local Feature Based 3D Face Recognition," Proc. Audio- and Video-Based Biometric Person Authentication, pp. 909–918, 2005.
3. M H. Drira, B. Ben Amor, A. Srivastava, M. Daoudi, and R. Slama "3D Face Recognition under Expressions, Occlusions, and Pose Variations," IEEE Transactions on Pattern Analysis and Machine Intelligence, vol. 35, no. 9, pp. 2270–2283, 2013.
4. H. Zhou, A. Mian, L. Wei, D. Creighton, M. Hossny, and S. Nahavandi, "Recent Advances on Singlemodal and Multimodal Face Recognition: A Survey," IEEE Transactions On Human-Machine Systems, vol. 44, no. 6, pp. 701–716, 2014.
5. C.C.C. Queirolo, L. Silva, O.R.O. Bellon, and M.P.M. Segundo, "3D Face Recognition Using Simulated Annealing and the Surface Interpenetration Measure," IEEE Trans. Pattern Analysis and Machine Intelligence, vol. 32, no. 2, pp. 206–219, 2010.
6. U. prabhu, J. Heo, and M. Savvides, "Unconstrained Pose-invariant Face Recognition Using 3D Generic Elastic Models," IEEE Transactions on pattern Analysis and Machine Intelligence, vol. 33, no. 10, pp. 1952–1961, 2011.
7. G. Passalis, P. Perakis, T. Theoharis, and I.A.I. Kakadiaris, "Using Facial Symmetry to Handle Pose Variations in Real-World 3D Face Recognition," IEEE Trans. Pattern Analysis and Machine Intelligence, vol. 33, no. 10, pp. 1938–1951, 2011.
8. S. Elaiwat, M. Bennamoun, F. Boussaid, and A. El-Sallam, "3-D Face Recognition Using Curvelet Local Features," IEEE Signal Processing Letters, vol. 21, no. 2, pp. 172–175, 2014.
9. H. Mohammadzade and D. Hatzinakos, "Iterative Closest Normal Point for 3D Face Recognition," IEEE Transactions on Pattern Analysis and Machine Intelligence, vol. 35, no. 2, pp. 381–397, 2013.
10. C.H. Li and I. Gondra, "A Novel Neural Network-Based Approach for Multiple Instance Learning," IEEE International Conference on Computer and Information Technology (CIT 2010), pp. 451–456, 2010.

11. D. Smeets, P. Claes, J. Hermans, D. Vandermeulen, and P. Suetens, "A Comparative Study of 3-D Face Recognition Under Expression Variations," IEEE Transactions on systems, man, and cybernetics—part c: applications and reviews, vol. 42, no. 5, pp. 710–727, 2012.
12. N. Erdogmus and J-L. Dugelay, "3D Assisted Face Recognition: Dealing With Expression Variations," IEEE Transactions on Information Forensics and Security, vol. 9, no. 5, pp. 826–838, 2014.
13. N. Alyuz, B. Gokberk, and L. Akarun, "3-D Face Recognition Under Occlusion Using Masked Projection," IEEE Transactions on Information Forensics and Security, vol. 8, no. 5, pp. 789–802, 2013.
14. A. M. Ali, "A 3D-Based Pose Invariant Face Recognition at a Distance Framework," IEEE Transactions on Information Forensics and Security, vol. 9, no. 12, pp. 2158–2169, 2014.
15. R. Min, N. Kose, and J. Dugelay, "KinectFaceDB: A Kinect Database for Face Recognition," IEEE Transactions on Systems, Man, and Cybernetics: Systems, vol. 44, no. 11, pp. 1534–1548, 2014.
16. G. Goswami, M. Vasta, and R. Singh, "RGB-D Face Recognition with Texture and Attribute Features," IEEE Transactions on Information Forensics and Security, vol. 9., no. 10, pp. 1629–1640, 2014.
17. Y. Woo, C. Yi, and Y. Yi, "FAST PCA-BASED FACE RECOGNITION ON GPUS," IEEE International Conference on Acoustic, Speech and Signal Processing (ICASSP 2013), pp. 2659–2663, 2013.
18. L.-H. Chan, S.-H. Salleh, C.-M. Ting, and A.K. Ariff, "PCA and LDA-Based Face Verification Using Back-Propagation Neural Network," IEEE 10th International Conference on Information Science, Signal Processing and their Applications (ISSPA 2010), pp. 728–732, 2010.
19. J. Yang, H. Li, and Y. Jia, "Go-ICP: Solving 3D Registration Efficiently and Globally Optimally," IEEE International Conference on Computer - Vision, pp. 1457–1464, 2013.

Data Perturbation: An Approach to Protect Confidential Data in Cloud Environment

Dipali Darpe and Jyoti Nighot

Abstract Nowadays there is rapid growth in large-scale databases and cloud computing is becoming tempting key to host data query services due to its advantages in scalability and cost-economy. So organizations are moving towards cloud computing infrastructure. In the commercial areas use of cloud computing has increased due to its features like pay per use, fault tolerance, scalability, elasticity. Despite of these advantages some data owner hesitates to put their data on cloud which is confidential. Unless confidentiality of data and privacy in query processing are not guaranteed, some data owner does not want to move to the cloud. A secured query service should provide efficiency in query processing as well as it needs to reduce the workload of in-house infrastructure to gain the benefits of cloud computing infrastructure. RASP data perturbation is proposed to provide balance between security and efficiency of protected data. RASP method stands for Random Space Perturbation which combines injection of noise, preserving order of encryption. For enhancing performance indexing techniques are also used.

Keywords Confidentiality · Random noise injection · Range query · kNN query

1 Introduction

In cloud computing huge groups of remotely located servers which are networked for centralized data storage are allowed to access computer services or resources online. To accomplish coherence and economies of scale cloud computing rely on resource sharing. It also focuses on maximizing the value of the shared resources. Cloud resources are dynamically reallocated per demand. Cloud computing facil-

Dipali Darpe (✉) · Jyoti Nighot
KJ College of Engineering & Management Research Pune, Pune, India
e-mail: darpedips12345@gmail.com

Jyoti Nighot
e-mail: jyotinighot67@gmail.com

S.C. Satapathy et al. (eds.), *Proceedings of the International Conference on Data Engineering and Communication Technology*, Advances in Intelligent Systems and Computing 469, DOI 10.1007/978-981-10-1678-3_33

itate users to retrieve and update their data without purchasing licenses for different applications by accessing single server. Nowadays organization are moving from traditional approach of buying enthusiastic hardware and decline it over a period of time to the new approach of using shared cloud infrastructure and pay per use it. Cloud data storage is one of the vital service offered by the cloud computing. In this service, subscribers use cloud service providers server to store their own data. Due to the unique advantage in scalability and cost-saving, to host data concentrated query services in cloud becomes popular now. The workload of query services is exceedingly dynamic and serving such workload with in-house infrastructure will be costly and ineffective. On the other hand, service providers do not have control over the data in the cloud query privacy, and data confidentiality becomes the most important issues. Some service providers can see the user's queries or make a copy of the database, which is not so easy to find and prevent in the cloud infrastructure.

Some new approaches are necessary which can ensure privacy of query and protect the confidentiality of data; efficiency in query processing with the advantages of using cloud should also be preserved. As an outcome of security and privacy assurance it is meaningless to provide slow query services. The intension of using cloud resources is to degrade the need of maintaining scalable in-house infrastructure, so it is not useful for the owner of data to utilize a significant amount of in-house resources. There are some approaches which try to solve some aspect of the problem but they do not address all aspect of problem. For example, the crypto-index and Order Preserving Encryption (OPE) are vulnerable to the attacks, the enhanced crypto-index approach puts heavy burden on the in-house infrastructure to improve the security and privacy, the New Casper approach uses cloaking boxes to protect data objects and queries, which affects the efficiency of query processing and the in-house workload [1]. Therefore, there is a complex connection among the data confidentiality, query privacy, the quality of service, and the economics of using the cloud.

A realistic query service in the cloud should satisfy the criterion which includes confidentiality of data, preserving privacy of query, efficiently processing query, and low processing cost of in-house infrastructure. If we consider all these requirements for constructing query services in cloud, it will increase complexity of the system.

2 Related Works

2.1 *Protecting Outsourced Data*

Order Preserving Encryption [2]: These schemes allow comparison operations to be directly applied on encrypted data, without decrypting the operands. After encryption dimensional value order is preserved and in these scheme to break the encryption of the attribute a bucket-based division grouping can be performed by the attacker if he knows the original distribution and manages to spot the mapping between original attribute and its encrypted matching part.

Crypto-Index [3]: It is based on the bucketization approach. In these schemes, an attribute domain is partitioned into a set of buckets. A tag is assigned to each bucket. These bucket tags are referred to as crypto-index for processing queries server uses these crypto-indexes. To protect the access pattern a bucket-diffusion scheme [4] was proposed. But it affects the precision of query results. So client side cost for filtering the query result is increased.

Distance-Recoverable Encryption: To preserve the nearby neighbor DRE is one of the intuitive methods. Many attacks can be applied [5–7] due to the exact preservation of distances. To find kNN instead of preserving distances of nearest neighbor Wong et al. [5] suggested preserving dot products, which is more resilient to distance-targeted attacks. Drawback of this approach is that the search algorithm is limited to linear scan only and we cannot apply any indexing method to enhance the search.

2.2 Preserving Query Privacy

In PIR, it tries to completely preserve the privacy of access pattern, but the data may not be encrypted. PIR schemes are usually extremely expensive. To implement efficient privacy preserving data-block operations Williams et al. [8] use a pyramid hash index which is based on the idea of Oblivious RAM. But it does not give high throughput range query processing. The query privacy problem is addressed by Hu et al. [9] and requires the authorized query users. However, most computing tasks are done in the user's local system, which does not meet the principle of moving computing to the cloud To enhance location privacy Papadopoulos et al. [10] use private information retrieval methods [11]. But in these, confidentiality of data is not considered. To preserve location privacy Space Twist [12] proposes a method to query kNN by providing a bogus user's place. However this method does not consider data confidentiality. Both data confidentiality and query privacy are addressed by Casper approach [13].

3 System Architecture

The reason behind using this architecture is to enlarge the *proprietary database servers* to the public cloud for hosting huge datasets and query services cloud computing infrastructure Amazon EC2 is used. This architecture is used to ensure confidentiality of the data while minimizing the cost and achieving scalable hybrid public-private cloud. Here the database is outsourced to the cloud. Each record in this database has two parts such as some attributes are perturbated using RASP and encrypted original records. To enhance the performance of query processing

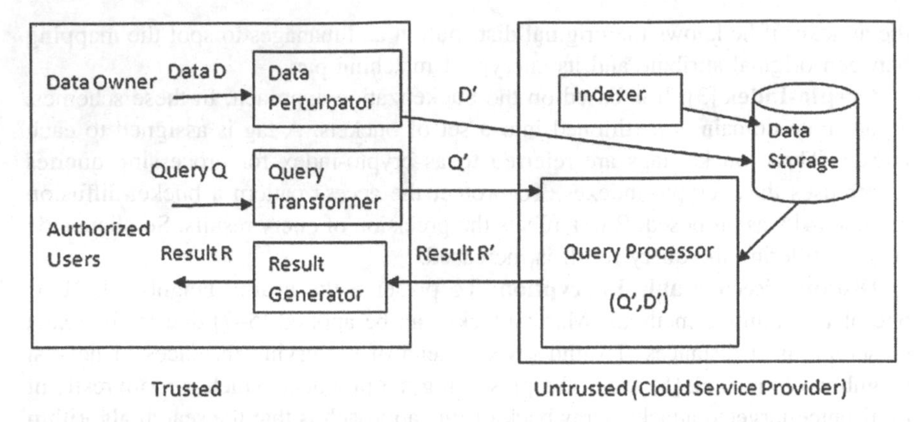

Fig. 1 RASP-based query service

indexing techniques are applied on database. For indexing RASP-perturbed data is used. This system architecture is used to process range queries and kNN queries based on RASP (Fig. 1).

In the system architecture, it has two visibly different parties in which one is trusted and other is not trusted. The trusted parties comprise of the data/service holder, the in-house proxy server, and the authorized users. The data owner store their confidential data on cloud on which RASP perturbation is performed. Users which are authorized only permitted to query the database. To find several records authorized users can submit range queries or kNN queries. The cloud service provider which may hack the confidential data are not trusted hosts the confidential database and query services. To enhance the good query performance indexing is used and to build indices RASP-perturbed data will be used.

3.1 System Module

3.1.1 Perturbation

RASP data perturbation technique is applied to data. After applying RASP original data is transferred to perturbed data. RASP combines different methods together which includes order preserving encryption, random noise injection. In secure transformation the topology of multidimensional range is preserved due to which indexing is applied on perturbated data which helps to efficiently process the query. OPE is encryption technique which allows comparison operation to apply directly on encrypted data where there is no need to decrypt the operands.

OPE: The proposed OPE Scheme is simple to use. It is based on linear expression such as x * y + z where coefficient x and z are kept secret and expression is public. Suppose input value is taken as v, then it is mapped to x * v + z + noise. Noise is some random value. For preserving order of input values noise to be selected properly. To deal with range queries over encrypted databases, OPE is used together with existing encryption algorithms (e.g., AES). For all input values v_1 and v_2 if $v_1 > v_2$ and $x > 0$ then $x * v_1 + z > x * v_2 + z$. The coefficients and input values can be integers or real number. After addition of some noise our expression becomes as

$$x * v_1 + z + noise_1$$
$$x * v_2 + z + noise_2$$

where value selected for noise is randomly sampled from some range. Range of noises is selected such as $v_1 > v_2$ and $x > 0$ then $x * v_1 + z + noise_1 > x * v_2 + z + noise_2$. If noise1 and $noise_2$ are both randomly sample from the range [0, x * 1] then $x * v_1 + z + noise_1 > x * v_2 + z + noise_2$ which holds even when $noise_1$ is minimum of $noise_1$, i.e., 0 and $noise_2$ is its maximum in [0, x * 1].

3.1.2 Query Transformers

In cloud environment, it is difficult to detect and prevent eavesdropped user queries. If the query is in encrypted form it is difficult to eavesdrop. Query transformer do the important task of transforming query, i.e., with the help of key query is encrypted. This transformed query acts on perturbed data.

3.1.3 Query Processor

It processes the encrypted query on protected database.

3.2 Range Query Processing

To minimize client side post-processing cost it is important to process queries efficiently which returns precise results. More common multidimensional tree indices are used for enhancing performance of search. For indexing multidimensional data the construction block used is the axis-aligned minimum bounding region (MBR). Range query on perturbed data is processed in two stages.

Algorithm
It works in two stages.

Proxy server is present at the client side. First it will perform the task of finding MBR of the submitted query. After finding MBR, it will gives this MBR and different query conditions to the server. The records which are enclosed within MBR are returned by the server using tree indexes. In the second stage exact range query result is returned to the server. In the second stage, initial results are filtered by the server using transformed half space conditions. In some cases initial result set is small which is filtered in memory such a case is called as tight ranges. In some cases, linear scan is performed when MBR will include entire dataset. Due to the exact result of the query obtained in second stage the client side post-processing cost is reduced.

3.3 kNN Query Processing

kNN query processing algorithm is based on range query algorithm, so to return faster results indexing is also used to process kNN queries. To find out the kNN candidates which are centered at query points, kNN algorithm is based on square ranges. We cannot directly process kNN query on perturbed database because RASP perturbation does not preserve distances. kNN query processing is combined with range query processing.

Working of kNN-R Algorithm

There are two rounds of interaction between client and server.

(1) The client will send the maximum-bound range and minimum-bound range to the server. Maximum bound range contains more than k points and minimum-bound range contains less than k points. Based on this minimum and maximum range server finds the inner range and it will return it to the client.

(2) Based on inner range client calculates outer range and returns to the server. The records which lie in the outer range are returned to the client by server.

(3) After decrypting the records top k candidates are returned as the final result by the client.

4 Experimental Results

In Fig. 2 it shows the fragment of perturbed database. In the above figure column value of Year_Of_Publication column, we applied OPE scheme which is highlighted. In proposed scheme we combine AES, OPE to secure our database. We considered book dataset from the http://grouplens.org/datasets/ link. On this database we perform equality, range, and kNN queries.

ISBN	Book_Title	Book_Author	Year_Of_Publication	Publisher
//983a4qeeAMAubKbU5voA==	+XANBcIj51JmxD3iVBfarazxVO/KG5LD1lxJtnQpT8Vcl2ToPUXGexFwYpedbdM7	DqJdTkQpxXAYGGVwNIFajA==	400 42	zfy4Wqj9GrVg8SL07
/9LCelvtrE127aIds56Byw==	nzxMD1okz6FjBu2UPo1lT1D5UbB1XHN/7bblayks4ODkPvIGhOxNS+vls0KrMjcD	XxpyRRPah37uZZKbn2IgwA==	396 22	quSkURFMWJY5L7
08z8Pu6d9+x3ZC+IssmdZg==	r7GXZSg9yCpjLBeFb2F8YA==	13bJYxCXYcwFLXl3AGpkZoxD7TtT+Rhom8VhxONFb/w=	399 02	p8Q+wcWAIHDIA8N
10xJjU0EQ8HxW1QqebPhXg==	p7k+zRWUvwvKBgtFEicoPmR6rXgQTAIzBZHODOb+jEQ=	fuRPMcHpHMalOQ/HAvldpA==	400 42	Gp9jUg0o03uguEpZ
1eO/VuLSTMyIQhhryWSjsg==	HvGy/PS9Htb0u8y4QxhsHw==	nQLLUQLnUUIYx0xArRu5kw==	399 82	Cuo1v7PWjiw4hadE
1lPn0l1kUYcv+KRPevzbYA==	4S26XmMBr059SgmNSeuEkw==	3LkEP8U73DWdgFXHgICFsHbeEkkgnQL/8axro9/aOuI=	400 42	u8JiO9Nk9Yzdz3agj
2SZqNrl1CBSmRnrhmjcb3A==	QWFCfX8b3DjPr6Dk1tzbaA==	ERXSvH9JvfLBwMMzn1fJy6Udf3cHymRqcl6WyCAhxEE=	400 02	ebBJSaI95yzF3HPdK
3DAXUI9M2WH5QVqRUahgeg==	PI5kHHUJlQTe2VIHFsaTCFJ2yl0qxEdHLbw8EjQ9vmE=	0C2FaK1bCAPoB01qb2Kn+Q==	398 22	DrqmUe2RL8mkjKB6
4foE4q/kgdhwvgZX+Xlig==	Ve/jVUERIga/MOalJHwyJKUdf3cHymRqcl6WyCAhxEE=	SJvcdAngra5D/1jYlodAnQ==	400 62	zsTnwI9meAZw8nfO
[illegible]	[illegible]	[illegible]	[illegible]	[illegible]

Fig. 2 A fragment of perturbed database

5 Conclusion

RASP is proposed to host query services in cloud. RASP combines noise injection, encryption, and order preserving encryption. RASP satisfies the need of hosting secure query services in cloud. RASP allows indices to be applied to improve performance of query. Range queries and kNN queries are considered. The kNN query is based on range queries. With the help of RASP perturbation, confidential data is protected in cloud environment.

References

1. Huiqi Xu, Shumin Guo, Keke Chen, "Building Confidential and Efficient Query Services in the Cloud with RASP Data Perturbation" IEEE transactions on knowledge and data engineering vol:26 no:2 year 2014.
2. R. Agrawal, J. Kiernan, R. Srikant, and Y. Xu, "Order Preserving Encryption For Numeric Data", in *Proceedings of ACM SIGMOD Conference*, 2004.
3. H. Hacigumus, B. Iyer, C. Li, and S. Mehrotra, "Executing Sql Over Encrypted Data In The Database-Service-Provider Model", in *Proceedings of ACM SIGMOD Conference*, 2002.
4. B. Hore, S. Mehrotra, and G. Tsudik, "A privacy-preserving index for range queries," in *Proceedings of Very Large DatabasesConference (VLDB)*, 2004.
5. W. K. Wong, D. W.-l. Cheung, B. Kao, and N. Mamoulis, "Secure knn computation on encrypted databases," in *Proceedings of ACM SIGMOD Conference*. New York, NY, USA: ACM, 2009, pp. 139–152.
6. K. Liu, C. Giannella, and H. Kargupta, "An attacker's view of distance preserving maps for privacy preserving data mining," in *Proceedings of PKDD*, Berlin, Germany, September 2006.
7. K. Chen, L. Liu, and G. Sun, "Towards attack-resilient geometric data perturbation," in *SIAM Data Mining Conference*, 2007.
8. P. Williams, R. Sion, and B. Carbunar, "Building castles out of mud: Practical access pattern privacy and correctness on untrusted storage," in *ACM Conference on Computer and Communications Security*, 2008.
9. H. Hu, J. Xu, C. Ren, and B. Choi, "Processing private queries over untrusted data cloud through privacy homomorphism," *Proceedings of IEEE International Conference on Data Engineering (ICDE)*, pp. 601–612, 2011.
10. S. Papadopoulos, S. Bakiras, and D. Papadias, "Nearest neighbor search with strong location privacy," in *Proceedings of Very Large Databases Conference (VLDB)*, 2010.

11. B. Chor, E. Kushilevitz, O. Goldreich, and M. Sudan, "Private information retrieval," *ACM Computer Survey*, vol. 45, no. 6, pp. 965–981, 1998.
12. M. L. Yiu, C. S. Jensen, X. Huang, and H. Lu, "Spacetwist: Managing the trade-offs among location privacy, query performance, and query accuracy in mobile services," in *Proceedings of IEEE International Conference on Data Engineering (ICDE)*, Washington, DC, USA, 2008, pp. 366–375.
13. P. Paillier, "Public-key cryptosystems based on composite degree residuosity classes," in *EUROCRYPT*. Springer-Verlag, 1999, pp. 223–238.

Biologically Inspired Techniques for Cognitive Decision-Making

Ashish Chandiok and D.K. Chaturvedi

Abstract In this paper, Artificial Biologically Inspired Techniques based on perceptions related to human cognitive decision-making is presented by proposing a novel BICIT (Biologically Inspired Cognitive Information Theory) architecture. These ideas openly associate neuro-biological brain actions that can explain advanced cognitive meanings comprehended by the mind. This architectural model on the one side elucidates inherent complications in human cognitive sciences, clarifying the behaviour of real-world interaction. On the other hand, points to prototypes of mind in the synergism of neural and fuzzy hybrid cognitive map systems. These structures can contend in forecasting, pattern recognition and classification jobs with neural networks and reasoning tasks with the fuzzy cognitive expert decision making.

Keywords Biological · Cognitive · Neural network · Fuzzy · Cognitive map architecture

1 Introduction

Symbolic and Emergent approach(Biologically Inspired techniques) are two different architectural models for resembling human intelligence and mind conceptual framework [1, 2]. Artificial General Intelligence objectives from starting are for constructing intelligent systems using the handling of significant propositions. Presently, there are critical complications at the groundwork of such an approach for building

Biologically Inspired techniques using BICIT architecture is propose in this paper to have cognitive decision making.

Ashish Chandiok (✉) · D.K. Chaturvedi
Faculty of Engineering, Dayalbagh Educational Institute,
Dayalbagh, Agra, Uttar Pradesh, India
e-mail: a.chandiok@gmail.com

D.K. Chaturvedi
e-mail: dkc.foe@gmail.com
URL: http://www.dei.ac.in

S.C. Satapathy et al. (eds.), *Proceedings of the International Conference on Data Engineering and Communication Technology*, Advances in Intelligent Systems and Computing 469, DOI 10.1007/978-981-10-1678-3_34

model. Starting with the *knowledge acquisition bottleneck problem* (The knowledge components warehoused in the plastic model are in a significant capacity to signify real-world condition). The *symbol grounding problem* (explicit knowledge creates the system due to a symbolic representation of a real-world scenario in a hypothetical way). *Autonomy problem* (The models do not learn by themselves for handling real-world problem). Also the *learning problem* (For every task, insertion of a new knowledge is essential and thus increasing the complexity of data).

In Emergent approach, the system autonomously constructs itself and adapt by interacting with the environment using self-organization process of developing its network. First, in comparison to a Symbolic approach where knowledge acquisition is done by formulating new symbolic representation for every logic. While emergent systems have continuous skill-based learning where knowledge acquires by the impulses taken from the environment [1]. According to which the network, as well as network parameters, are reorganized to match the relationship, maintaining stability between the stimulus perception and the goal achieved. Therefore, the emergent approach can use Neural Network and Fuzzy cognitive mapping combinations for decision-making [3].

In this paper, authors proposes **BICIT** (Biologically Inspired Cognitive Information Technology) architecture for cognitive decision-making. Second, an example on neural-fuzzy system is introduced for knowledge development and Fuzzy Cognitive Map [4] for collaborative expert decision-making [5, 6]. The principal objective is to illustrate how to build a biologically inspired model that can be applied to form a Decision Support System.

Fuzzy Cognitive Map is a meta-rule system [7] such that it not only considers the associations between the causes and effects but also reflects the interactions among the overall cause entities [8–10]. Therefore, it provides a stronger reasoning ability than rule-based reasoning, and it can be used to model complex relationships among different concepts [4, 11].

The traditional decision support expert systems involve the building of a knowledge base; Biological Neural-fuzzy [12–14] cognitive mapping techniques are comparatively faster and easier to obtain knowledge and experience [9, 15]. Mainly using the human approaches that do not typically contemplate in mathematical calculations but in intuitive knowledge represented as natural language. Using the Biological techniques, models can handle distinct and numerous knowledge sources with a range of expertise [13, 16].

2 Biologically Inspired Cognitive Information Technology (BICIT) Architecture

Figure 1 shows the model of the biologically inspired cognitive decision-making system, that author proposes, which reflects that a human being is an intelligent organism cooperating with a society of entities for completing goals. In the equiva-

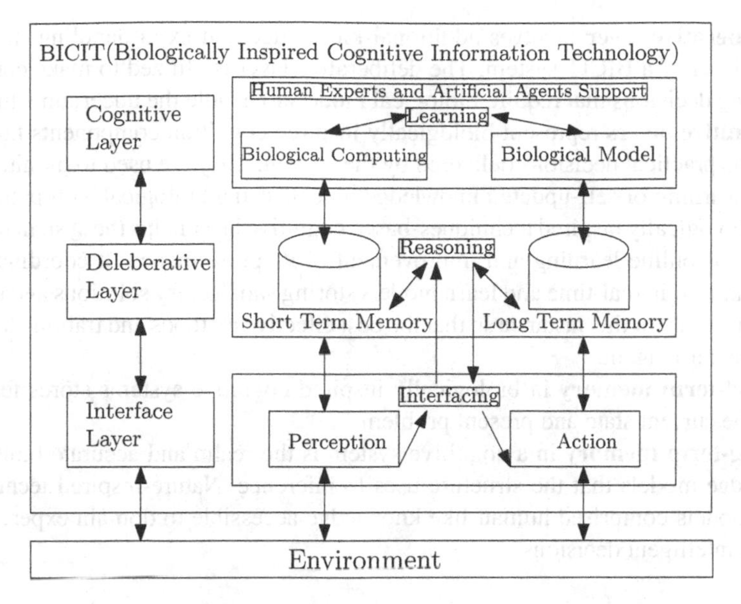

Fig. 1 Biologically Inspired Cognitive Information Technology (BICIT) architecture

lent approach, their artificial counterpart known as Biologically Inspired Cognitive Agents must follow goal completion by the interaction of human experts concious intuitions, machine agents skills, and its internal biological, cognitive powers. Agent trails its goal by interfacing with its external and internal environment. In internal, it interacts with its self-learning ability, emotions, knowledge, reasoning, and internal decision-making states. On the other hand, in the external environment, it cooperates with a human expert or its identical artificial agents. Whatsoever the situation we are bearing in mind, it can be altered, it is active; volatile environment significantly features the rational agent state and, therefore, their actions in reaching goals. Cognitive decision-making methodology shares the features of the symbolic, propositional logic procedures of conventional artificial intelligence, as well as parallel distributed, associative handling of biologically inspired cognitive machine intelligence. Combining the best of both worlds can be done by biological and rule base techniques. There are numerous approaches planned for decision-making like machine learning techniques, fuzzy logic, association rules, clustering, probabilistic networks, neural network, heuristic, and template pattern recognition. Figure 1 illustrates a model of Biologically Inspired Cognitive Information Technology Architecture. It comprises of three layers known as a cognitive, reflective, and interface layer.

Interface layers are unique, rapid decision-making that when at an instance learning performs the action spontaneously. The most common example of such kind of low-level intelligent decision-making is robotic collision avoidance.

Deliberative layer involves additional knowledge and extra handling. It is the central layer of a BICIT system. The deliberative layer is utilized to make complex reasoning decisions that require biological models to handle the uncertain situation.

Cognitive layers represent biologically inspired cognition components that aim about the practical decisions delivered by the system. They are used to monitor biological learning or self-updated knowledge models using biological, computational tools. Biologically inspired techniques-based cognitive layer helps the system to have continuous online learning and improvement of its performance. It coordinates to seize features, in real-time and learn models storing satisfactory solutions and updating the models for this instance so that the best effective methods and training learned from them use again.

Short-term memory in biologically inspired cognitive systems stores features about the current state and present problem.

Long-term memory in a cognitive system is the realm and accurate biological knowledge models that the structure uses to inference. Nature-inspired technique-based models comprised human-like knowledge accessible to domain expert when making intelligent decisions.

2.1 Learning, Reasoning and Interfacing in BICIT Architecture

The biological inspired techniques [1] utilizes Neural Network, Evolutionary algorithm and fuzzy logic. In recent years, the researchers are using hybrid algorithm and techniques to solve problems. Nowadays, fuzzy cognitive map and Neuro-fuzzy [16] are seperately very popular. In this paper, the authors combine Neuro-fuzzy and Cognitive map to solve real-world expert problems based on BICIT architecture.

3 Related Work: Implementation of Biological-Inspired Techniques for Cognitive Decision Making

The selection is a fuzzy cognitive map for expert decision state and neural-fuzzy method for knowledge-based model building. The problem considered is teacher appraisal system.

In recent years, the research area of hybrid cognitive neural and fuzzy has seen a remarkably active development. Furthermore, there has been a massive upsurge in the successful use of hybrid intelligent in many diverse areas. Such as speech/natural language processing, robotics, medical; fault diagnosis industrial equipment, education, e-commerce, recommendation and information outrival. A cognitive neural-fuzzy-based expert system for evaluating teachers on the basis student feedback, examination results, peer feedback, and educational activities is proposed as shown

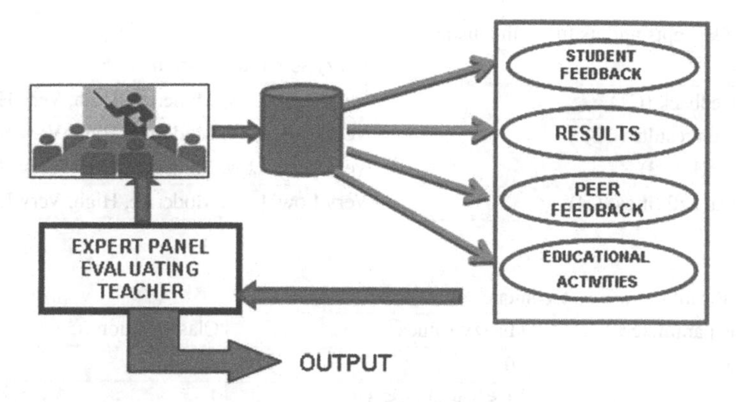

Fig. 2 Generic model of classical teacher evaluation system

in Fig. 2. This generic model of current teacher evaluation based on a static mathematical formula based on experts judgement. The generic model has the following problem that needs elimination. Each expert has intuitions for judgement often not taken into account. Different weights assign to parameters by each expert, but it is not considered. Many experts are getting bashed with one person who might result in wrong outputs. Mathematical computation is not fast and manual entering, and calculations may give an error.

3.1 Procedure for Problem Solving Using Biologically Inspired Techniques

Steps for Biologically Inspired Cognitive Decision

(1) Collect sample input parameter patterns.
(2) Create a cognitive fuzzy map of expert knowledge to get output decision.
(3) Combine input pattern and output decision to get complete training pattern.
(4) Generate the Neural-Fuzzy model and test the result.
(5) Apply to uncertain real-world situation.
(6) Regularly dynamic update fuzzy cognitive knowledge for new pattern obtains.

3.1.1 Sample Collection as Input Pattern

The sample collection is from the annual appraisal lead by the Sharda Group of Institutions, Agra where one of the authors worked as an Assistant Professor. The college holds total staff of 60. The authorized yearly appraisal has numerous sub-inquiries and measures. Therefore, the authors assembled all questions and combined the consequences into four major features as mention below. In facts collection, the

Table 1 Concepts and its fuzzy linguistics

Concepts	Fuzzy semantic relation
Student feedback (C1)	Very Low, Low, Moderate, High, Very High
Examination results (C2)	Very Low, Low, Moderate, High, Very High
Peer feedback (C3)	Very Low, Low, Moderate, High, Very High
Educational activities (C4)	Very Low, Low, Moderate, High, Very High

Table 2 Result of various techniques used

Linguistic parameters	Fuzzy values	Classification
Very Low	0	0
Low	$1 \leq$ variable ≤ 3	1
Moderate	$3 \leq$ variable ≤ 6	2
High	$6 \leq$ variable ≤ 8	3
Very High	$8 \leq$ variable ≤ 10	4

feature data information comprises figurative as well as non-figured features. The handling and model development of such kind of structure is a difficult problem.

3.1.2 Cognitive Fuzzy Map of Expert Knowledge

In this research, the authors uses total four meta-concepts for decision-making of teacher appraisal. These concepts are listed in Table 1 with their linguistics variables. According to domain experts (Directors and Administration) assessments, input concepts representation is with the particular linguistic variables Very Low, Low, Moderate, High, Very High that lies in the fuzzy interval of [0, 10]. On the basis of the doctors intuitive decision, the eight relationships between concepts are shown using semantic variables (Negatively Strong, Negatively Moderate, Negatively Weak, Negatively Very Weak, No Relationship, Positively Weak, Positively Moderate, Positively Strong, and Positively Very Strong) where the values lie in the fuzzy range of [−1, 1]. In accordance to Domain Experts opinion, the Fuzzy values for the concepts variables for our system is shown in Table 2 and relationship between concepts with fuzzy weights in Table 3 below (the fuzzy values are tabulated and classified based on each Director and Management intuitive power). Similarly, Table 3 represent the association fuzzy strength values between concepts. These values are assigned by the experts and use it for representing the weights between two concepts edges.

Table 3 Result of various techniques used

Linguistic parameters	Fuzzy values
Negatively Strong	$-1 \leq$ variable ≤ -0.8
Negatively Moderate	$-0.8 \leq$ variable ≤ -0.6
Negatively Weak	$-0.6 \leq$ variable ≤ -0.3
Negatively Very Weak	$-0.3 \leq$ variable ≤ -0.1
No Relationship	0
Positively Weak	$0.1 \leq$ variable ≤ 0.3
Positively Moderate	$0.3 \leq$ variable ≤ 0.6
Positively Strong	$0.6 \leq$ variable ≤ 0.8
Positively Very Strong	$0.8 \leq$ variable ≤ 1

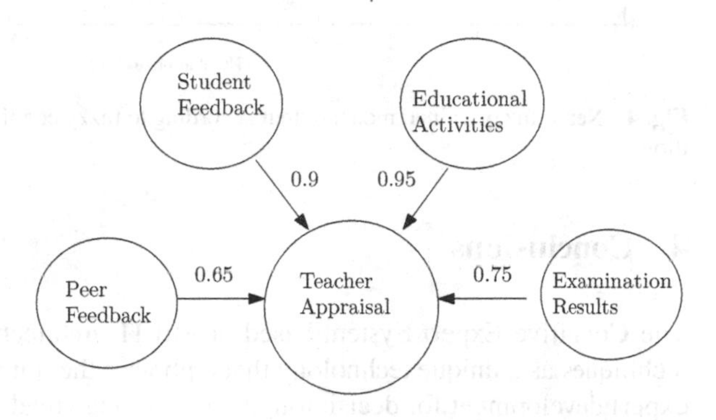

Fig. 3 Fuzzy cognitive map according to expert intuition

The experts Weight Matrix is formed by the intuitions of the each Directors and Management administrators and then authors construct overall weight matrix for the combined cognitive intuitive opinion as represented in the graph shown in Fig. 3.

3.1.3 Neuro-Fuzzy Model Development According to FCM Decision

The model is trained by neuro-fuzzy hybrid technique and can be updated time to time for better performance. The result shown in Fig. 4 concludes that fuzzy cognitive map knowledge combines the artificial cognitive mind of four experts and then training the input parameters and professional expertise output by artificial neural-fuzzy system representing a unique way of cognitive decision-making.

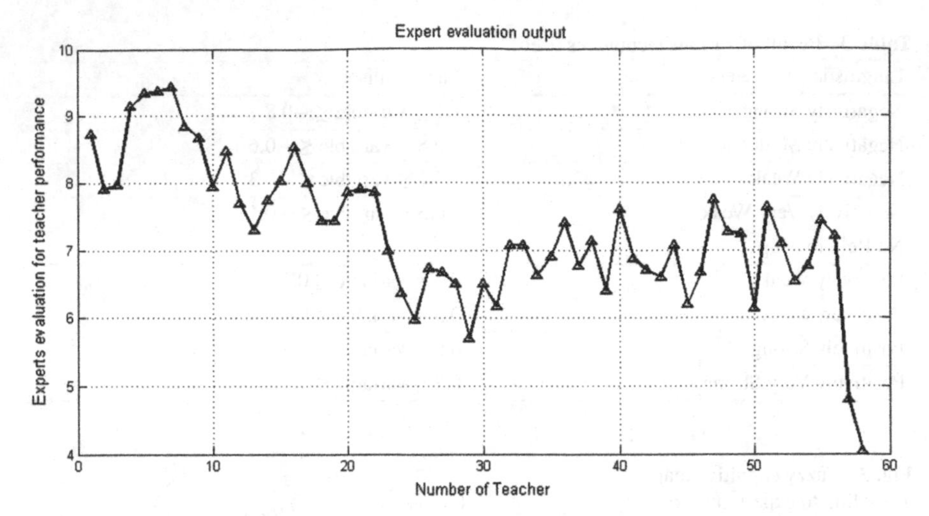

Fig. 4 Neuro fuzzy trained model output according to fuzzy cognitive map following expert intuition

4 Conclusions

The Cognitive Expert System based on BICIT architecture uses neural and fuzzy techniques as a unique technology that embraces the improvement of contemporary expert development for decision-making. The analytical minds of experts are combined qualitatively and quantitatively in a fuzzy cognitive map representing fuzzy and neural aspect respectively. The meta-mind uses to trained hybrid neuro-fuzzy artificial system developing artificial cognitive expert decision system. This era of expert system supports the improved biologically inspired intelligent expert system based on a study of mind.

References

1. Vernon, D., Metta, G., Sandini, G. : A survey of artificial cognitive systems:Implications for the autonomous development of mental capabilities in computational agents. Evolutionary Computation, IEEE Transactions on, 11(2), 151–180, (2007)
2. Chong, H.-Q., Tan, A.-H., Ng, G.-W. (2009). Integrated cognitive architectures: a survey. Artificial Intelligence Review, 28(2), 103–130, (2009)
3. Elomda, B. M., Hefny, H. A., Hassan, H. A.: Fuzzy cognitive map with linguistic values. In Engineering and Technology (ICET), 2014 International Conference, pp.1–6. IEEE, (2014)
4. Kosko, B.: Fuzzy cognitive maps. International Journal of man-machine studies. 24(1), 65–75 (1986)
5. Papageorgiou, E. I.: Fuzzy cognitive map software tool for treatment management of uncomplicated urinary tract infection. Computer methods and programs in biomedicine, 105(3), 233–245, (2012)

6. Wei, X., Luo, X., Li, Q., Zhang, J., Xu, Z.: Online Comment-Based Hotel Quality Automatic Assessment Using Improved Fuzzy Comprehensive Evaluation and Fuzzy Cognitive Map. Fuzzy Systems, IEEE Transactions on, 23(1), 72–84, (2015)
7. Papageorgiou, E. I.: A new methodology for decisions in medical informatics using fuzzy cognitive maps based on fuzzy rule-extraction techniques. Applied Soft Computing, 11(1), 500–513, (2011)
8. Stylios, C. D., Groumpos, P.: Modeling complex systems using fuzzy cognitive maps. Systems, Man and Cybernetics, Part A: Systems and Humans, IEEE Transactions on, 34(1), 155–162, (2004)
9. Papageorgiou, E. I., Stylios, C., Groumpos, P. P.: Unsupervised learning techniques for fine-tuning fuzzy cognitive map causal links. International Journal of Human-Computer Studies, 64(8), 727–743,(2006)
10. Stylios, C. D., Georgopoulos, V. C., Malandraki, G. A., Chouliara, S.: Fuzzy cognitive map architectures for medical decision support systems. Applied Soft Computing, 8(3), 1243–1251, (2008)
11. Miao, Y., Liu, Z. Q., Siew, C. K., Miao, C. Y.: Dynamical cognitive network-an extension of fuzzy cognitive map. Fuzzy Systems, IEEE Transactions on, 9(5), 760–770, (2001)
12. Subramanian, K., Suresh, S., Sundararajan, N.: A metacognitive neuro-fuzzy inference system (McFIS) for sequential classification problems. Fuzzy Systems, IEEE Transactions on, 21(6), 1080–1095, (2013)
13. Kar, S., Das, S., Ghosh, P. K.: Applications of neuro fuzzy systems: A brief review and future outline. Applied Soft Computing, 15, 243–259, (2014)
14. Singh, P., Borah, B.: High-order fuzzy-neuro expert system for time series forecasting. Knowledge-Based Systems, 46, 12–21, (2013)
15. Papageorgiou, E., Stylios, C., Groumpos, P.: Fuzzy cognitive map learning based on nonlinear Hebbian rule. In AI 2003: Advances in Artificial Intelligence (pp. 256–268). Springer Berlin Heidelberg, (2003)
16. Malkawi, M., Murad, O.: Artificial neuro fuzzy logic system for detecting human emotions. Human-centric Computing and Information Sciences, 3(1), 1–13, (2013)

6. W[illegible], [illegible] Zhang, [illegible] Zhang, [illegible] Zhang, [illegible] Three-Component-Based [illegible] Automated Assessment [illegible] Improved Fuzzy Cognitive [illegible] and [illegible] Maps. Fuzzy S[illegible] (2014)
7. Papageorgiou, [illegible] fuzzy cognitive [illegible] Applied Soft Computing [illegible], 300–[illegible] (2013)
8. [illegible] complex [illegible] fuzzy cognitive [illegible] Systems, [illegible] (2010)
9. [illegible]
10. [illegible] 1249–125[illegible] (2005)
11. Miao, Y., [illegible] dynamic cognitive network [illegible] extension [illegible] fuzzy cognitive maps. IEEE Trans. Fuzzy Syst. [illegible] (2001)
12. [illegible] A [illegible] cognitive [illegible] (McFCM) [illegible] 2[illegible] (2015)
13. [illegible] Applied Soft Computing [illegible]
14. [illegible] Knowledge-Based Systems [illegible]
15. [illegible] Human [illegible] Advances in Artificial Intelligence [illegible]
16. [illegible] for [illegible] human [illegible] (201[illegible])

Analysis of Edge Detection Techniques for Side Scan Sonar Image Using Block Processing and Fuzzy Logic Methods

U. Anitha and S. Malarkkan

Abstract Sonar images are generally used for the identification of objects under the sea. The images which are obtained by sonar technology are always gray scale images. For the identification of objects in images, the image processing technique is having some intermediate steps like filtering of noise, edge detection, and segmentation. Edge detection is one of the important steps for image segmentation and object identification. In this paper, edge detection using fuzzy logic technique and the block processing technique is compared. Subjective analysis of the result proves that the fuzzy method is better than the block processing for the identification of objects together with the shadow region in sonar images.

Keywords Side scan sonar image · Edge detection · Segmentation · Object identification · Fuzzy logic method · Block processing

1 Introduction

The technique which is used to navigate under the surface of sea water is called SONAR. Side scan sonar (SSS) is one of the categories of sonar system [1]. The main aim of the side scan sonar image processing is the classification of seafloor and the identification of target object. SSS consist of sensor array on a fish-shaped vessel known as sonar fish which is mounted on a ship hull and is towed deep into the water through some distance above the seabed. It emits fan-shaped sound pulse down toward the sea surface. The reflected pulses from seafloor are again recollected and the sonar imagery is formed finally [2].

U. Anitha (✉)
Faculty of Electronics, Department of Electronics and Control, Sathyabama University, Chennai, India
e-mail: anithaumanath@gmail.com

S. Malarkkan
ManakulaVinayagar Institute of Technology, Puducherry, India
e-mail: malarkkan_s@yahoo.com

S.C. Satapathy et al. (eds.), *Proceedings of the International Conference on Data Engineering and Communication Technology*, Advances in Intelligent Systems and Computing 469, DOI 10.1007/978-981-10-1678-3_35

The quality of raw sonar image is not sufficient for the human intervention. Because it consists of ambient noise, electronic noise, loss due to absorption of sound pulses of the sea surface, reverberation noise, etc. The main components of a side scan sonar imagery system are [3, 4],

Step 1 Positioning of acoustic sensor array system
Step 2 Data acquisition and storage
Step 3 Data processing (Image processing)
Step 4 Interpretation

Image processing is the technique which converts image into a form suitable for specific application based on individual preferences. It consists fundamental steps such as image acquisition, storage, enhancement, restoration, morphological processing, segmentation, object identification, and classification. Many algorithms are available for the processing of each step. Among these edge detection is the primary step in segmentation process for finding the meaningful edges of objects present in an image. There are several operators are available to preserve different types of edges such as horizontal, vertical, and diagonal edges. During the process of edge detection in noisy sonar image, high frequency noise pixel can also interrupt along with edge pixel details. One of the preprocessing techniques called filtration is also included before edge detection in order to reduce the effect of noise pixel [5]. In side scan imagery, an object lying on the seafloor is followed by a shadow zone. But the echoes and shadow zone has different contrast levels [6–8].

In this paper, noisy sonar image filtration is followed by a subjective analysis of image segmentation techniques like fuzzy logic method and block processing are performed.

In Sects. 2 and 3 basics about the fuzzy logic method and block processing operator are discussed, respectively. Sections 4 and 5 converse about the proposed methodology and results obtained correspondingly.

2 Fuzzy Logic Method

In an image, when there is an abrupt change in pixel intensity, it is called as edge pixel [9]. But a minute difference between the pixels is not considered as an edge, because they may be noise effect. In fuzzy logic method, membership function is described in order to define each pixel that how much it belongs to a particular region. The fuzzy inference system designed here is having two inputs and one output that tells whether the pixel is a region pixel or a boundary pixel [10–12].

2.1 Steps

1. Image gradient along x-axis (Ix) and y-axis (Iy) are calculated. The gradient values are in the range of (−1, 1).
2. Create a fuzzy inference system by giving these two (Ix and Iy) as input.
3. Assign membership function for both input and output. In this, Gaussian membership function is used for inputs and the triangular membership function is used for fixing the edge pixel output.
4. Create a set of rule for the system (Fig. 1).

In SSS image, objects are followed by the shadow region. Therefore the rule was formed as given below.

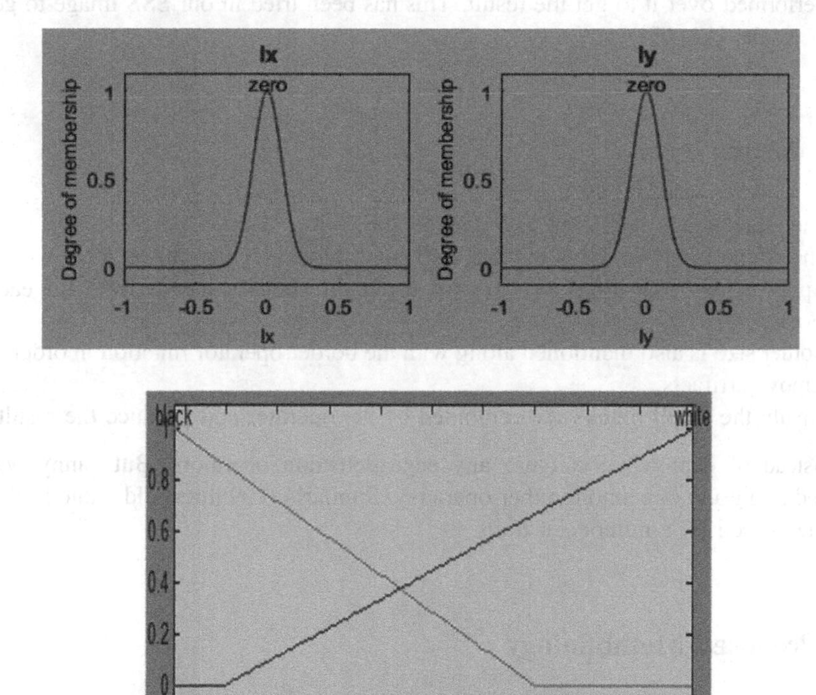

Fig. 1 Membership functions for input and output variables

2.2 Rules

It consists of two main rules.

1. If (Ix is zero) and (Iy is zero) then (Iout is black)
2. If (Ix is not zero) or (Iy is not zero) then (Iout is white)

Further addition rules may improve the performance of edge detection.

3 Block Processing

Usually the larger size images are divided into small blocks and the edge operators are performed over it to get the result. This has been tried in our SSS image to get better results [13–15].

3.1 Steps

1. The input image is resized into 256 × 256.
2. The whole input image is divided into small blocks of size 50 × 50.
3. Apply canny edge detector operator using threshold value of 0.09 for each block.
4. Border size is also mentioned along with the border operator function in order to remove artifacts.
5. Finally the small blocks are combined by the operator and produce the result.

Instead of canny, we can use any edge detection operators. But canny was proved as a good one among other operators. Similarly 0.09 threshold value is also selected based on a number of trails.

4 Proposed Methodology

Algorithm:

1. Get a denoised (preprocessed) sonar image as input
2. Implement a fuzzy logic method of edge detection
3. Apply block processing method for the same input image
4. Subjective analysis of the outputs of both techniques is done.

The above algorithm is given as flow chart below (Fig. 2).

The flow chart of the proposed work is having two tasks. One same image is given as input to two methods which are mentioned above and the results are

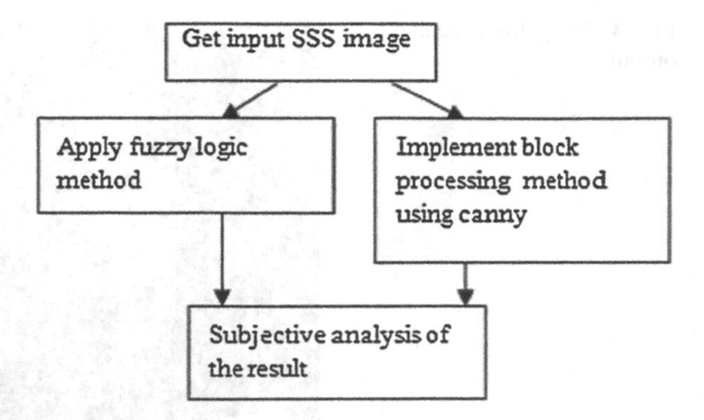

Fig. 2 Flow chart for the proposed work

subjectively analyzed, because there is no qualitative analysis of edge detection techniques.

5 Result and Discussion

In this proposed work, the SSS images along with shadow are taken for the analysis.

Samples of outputs are given below. Figure 3 represents an input image. Fuzzy method works based on FIS rules. The outcome of these methods based on FIS rule is specified in Fig. 4.

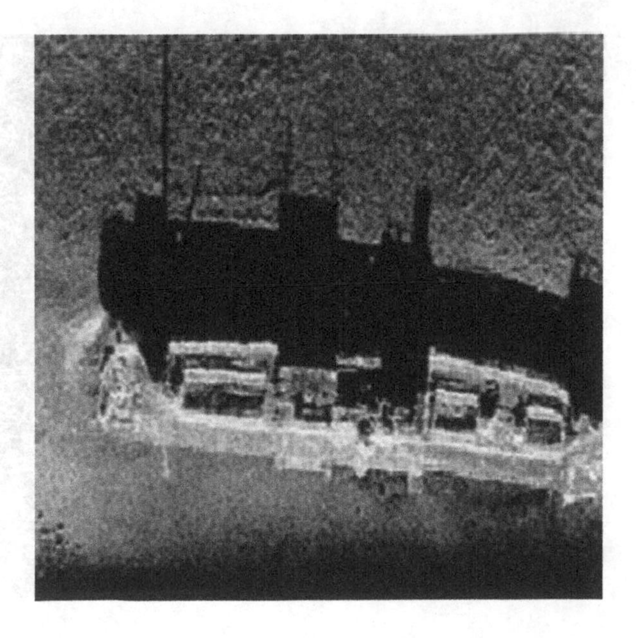

Fig. 3 Input image

Fig. 4 Fuzzy logic method output

The shadow region of the image is represented as dark pixel and the object oriented pixels are looking brighter than the surrounding background pixels.

Figures 5 and 6 are the outputs of the second method called block processing. By comparing these two images, block processing along with canny give many details than conventional canny operator.

Fig. 5 Conventional canny edge detector

Fig. 6 Block processing with canny edge detector

So, using the block processing along with the standard edge detection operators gives better performance than normal edge detection operators. As per the proposed work by comparing fuzzy method with the block processing method, fuzzy gives improved performance than the other and is also more informative. Hence, the subjective analysis of the work tells that, the fuzzy logic method is suitable than block processing method for the SSS image with shadow region.

All this is done with the help of MATLAB software.

6 Conclusion and Future Work

Based on the height of the underwater object and the angle of the sonar devices, signal hit on the object, the width of the shadow region in SSS image varies. So the prediction of one best method for those images is impossible. As per these analyses, fuzzy results show good outcome. It isolates the object, shadow, and background of the image separately using different gray scale intensities. The identification of object area is easier in fuzzy, but in other methods, object area identification is more complicated. In fuzzy method by simply changing the rule or adding more rules can improve the performance. In the future, planned to implement pseudocoloring technique and also wavelet transform for this application.

Acknowledgments I would like to show gratitude to "C-MAX Ltd., 9, Hybris Business Park, Crossways, Dorchester, Dorset DT2 8BF, UK" and also "Geo-Marine Consultants and Technocrats, Chennai" for providing SONAR images for this work and I would like to acknowledge the support of Mr. Livinson, Senior Executive—Business Development, Results Marine Private Ltd., Chennai.

References

1. Punam Thakare (2011) "A Study of Image Segmentation and Edge Detection Techniques", International Journal on Computer Science and Engineering, Vol 3, No. 2, 899–904.
2. Ramadevi, Y & et al (2010) "Segmentation and object recognition using edge detection techniques", International Journal of Computer Science and Information Technology, Vol 2, No. 6, 153–161.
3. R. Gonzalez and R. Woods, Digital Image Processing, Addison Wesley, 1992, pp 414–428.
4. Mee-Li Chiang, Siong-Hoe Lau (2011) "Automatic Multiple Faces Tracking and Detection using Improved Edge Detector Algorithm" IEEE 7th International Conference on IT in Asia (CITA), 978-1-61284-130-4.
5. Lejiang Guo, Yahui Hu Ze Hu, Xuanlai Tang (2010) "The Edge Detection Operators and Their Application in License Plate Recognition, IEEE TRANSACTION 2010, 20978-1-4244-5392-4/10/.
6. Mohammad H. Asghari, et al (2015), "Edge Detection in Digital Images Using Dispersive Phase Stretch Transform", International Journal of Biomedical Imaging, Volume 2015 (2015), Article ID 687819, 6 pages.
7. Kamlesh Kumar, Jian-Ping Li and Saeed Ahmed Khan (2015), "Comparative Study on Various Edge Detection Techniques for 2-D Image", IJCA, vol. 119, no. 22.
8. K.M. Shivani (2015), "Image Edge Detection: A Review", IJEEE, Volume 07, Issue 01, Jan-June 2015.
9. Suryakant, Neetukushwaha, "Edge Detection using Fuzzy Logic in Matlab", IJARCSSE, Vol 2, Issue 4, April 2012.
10. Jaideep Kaur and Poonam Sethi, "An Efficient Method of Edge Detection using Fuzzy Logic", IJCA journal, Vol. 77, No. 15, 2013.
11. Umesh Sehgal (2011) "Edge detection techniques in digital image processing using Fuzzy Logic", International Journal of Research in IT and Management, Vol. 1, Issue 3, 61–66.
12. Kiranpreet Kaur, Vikram Mutenja, Inderjeet Singh Gill" Fuzzy Logic Based Image Edge Detection Algorithm in MATLAB", International Journal of Computer Applications (0975 - 8887) Volume 1 – No. 22, 2010.
13. Geng Xing, Chen ken, Hu Xiaoguang (2012) "An improved Canny edge detection algorithm for colorimage" IEEE TRANSATION, 978-1-4673-0311-8/12/$31.00 ©2012 IEEE.
14. J. F. Canny. "A computational approach to edge detection". IEEE Trans. Pattern Anal. Machine Intell., vol. PAMI-8, no. 6, pp. 679–697, 1986 Journal of Image Processing (IJIP), Volume (3): Issue (1).
15. K. Somasundaram and Veenasuresh, "A Novel Method for Edge Detection using Block Truncation Coding and Convolution Technique for Magnetic Resonance Images (MRI) with Performance Measures", Proceedings of the International Conference in Computational Systems for Health & Sustainability, April 2015.

Leveraging Virtualization for Optimal Resource Management in a Cloud Environment

Dhrub Kumar and Naveen Kumar Gondhi

Abstract Among the various enabling technologies for cloud computing, virtualization stands out high when it comes to optimizing a cloud provider's infrastructure. The success of a cloud environment depends to a large extent on how gracefully its resources are managed. The issue of resource management in cloud computing has been very popular among the researchers. In an IaaS service model, the various parties involved have different goals—the cloud infrastructure provider's objective is to maximize its ROI (return on investment) while the cloud user's intention is to lower its cost while delivering the desired level of performance. This is achieved by managing resources like compute, storage, network and energy in an efficient way. Through this paper, we try to bring out the state-of-the-art in cloud resource management area and also explore how virtualization can be leveraged to perform resource management in data centers during server consolidation, load balancing, hotspot elimination and system maintenance.

Keywords Virtualization · Virtual machine migration · Green computing · Data center

1 Introduction

Cloud computing has been around for a while and nearly every business has embraced it to exploit the benefits offered by it. In cloud computing, virtual machines are utilized to serve the demands of customers in a dynamic fashion by adhering to the service-level agreements (SLA) agreed upon by the service provider and consumers [1]. In the past, various definitions have emerged for cloud

---

Dhrub Kumar (✉) · N.K. Gondhi
School of Computer Science and Engineering, Shri Mata Vaishno Devi University, Katra, Jammu, India
e-mail: dhrubkumar037@gmail.com

N.K. Gondhi
e-mail: naveen.gondhi@smvdu.ac.in

S.C. Satapathy et al. (eds.), *Proceedings of the International Conference on Data Engineering and Communication Technology*, Advances in Intelligent Systems and Computing 469, DOI 10.1007/978-981-10-1678-3_36

Table 1 Cloud service models

Service model	User	Services offered	Vendors/Players
IaaS	System managers/Administrators	VMs, Operating Systems, CPU, Memory, Storage, Network	Amazon EC2, Amazon S3, Joyent Cloud
PaaS	Developers/Deployers	Service/application test, development and deployment	Google App Engine, Windows Azure
SaaS	Business users	Email, CRM, Virtual Desktop, Website testing	Google Apps, salesforce.com

computing but all have certain characteristics common among them viz. virtualization, pay-per-use concept, elasticity and self-service interface. The emergence of cloud computing can be attributed to technologies like virtualization, internet, distributed computing and autonomic computing. Table 1 depicts the various aspects of cloud service models (Infrastructure/Platform/Software as a Service) in terms of its user types, services offered and major service providers.

In the context of cloud computing, resource management process involves managing various resources viz. compute, storage, network and electric energy in such a way so that the objectives of both cloud provider and cloud customer are fulfilled. Failure to do so efficiently can degrade the performance and in turn produces low ROI and make a provider lose its valuable customers. Management of resources in a cloud environment is not a trivial task. The process of resource management often gets tough due to increasing size and complexity of modern-day data centers to cope up with unpredictable workloads and adherence to the signed SLAs.

The rest of the paper is structured as follows. Section 2 presents various concepts related to resource management in cloud environment. Section 3 highlights the major contributions made in the cloud resource management area. Section 4 describes how virtualization can be exploited for resource management in cloud environment.

2 Resource Management in Cloud

To the users it seems as though a data center has unlimited resources to get their jobs done, but in reality it has limited resources. Resource management is an integral key task required of any cloud provider controlling the way common resources are accessed, making changes as and when required to improve performance/efficiency of data center [2]. The different aspects of cloud computing environment that influence the resource management task are discussed below.

2.1 Entities Involved in a Cloud Environment

In a cloud environment, the various entities can be classified according to the roles assumed by them.

Cloud Infrastructure Provider—The responsibility of managing all the data center resources lies with the infrastructure provider who allocates resources to the requesting entity in a way that the SLAs are met.

Cloud Service Provider—This entity requests resources from the cloud infrastructure provider in order to host various application/services that it intends to offer to the end users. In some cases, the same entity assumes the roles of cloud infrastructure as well as cloud service provider.

End User—This entity has no role in resource management. The end user only generates requests either to the cloud infrastructure provider or cloud service provider.

2.2 Goals of Entities

To serve the computational/storage requests of cloud customers, the cloud infrastructure providers maintain large-scale distributed data centers containing thousands of nodes and lease out resources like CPU, memory, network bandwidth, and storage to customers on pay-per-usage basis. Cloud service providers rent resources from cloud infrastructure providers to run various users' applications characterized by unpredictable loads. The sequence of request/response operations between cloud service providers and cloud infrastructure providers is shown in Fig. 1. The client's experience is directly dependent on how efficiently the infrastructure provider manages and allocates its resources.

The ultimate concern for any cloud infrastructure provider is how to make the best use of underlying resources to increase ROI while respecting the signed SLAs. This can be achieved by:

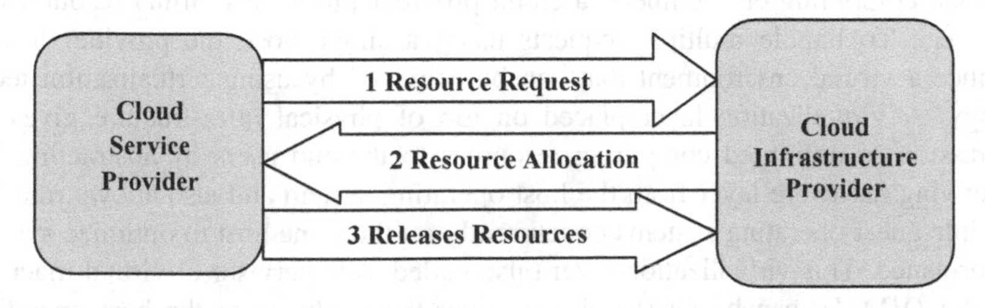

Fig. 1 Request/Response operations between an infrastructure provider and a service provider

(i) Allocating and scheduling the requested resources to customers in an efficient way so that it provides better QoS as per SLAs signed with customers.
(ii) Reducing power consumption by using efficient resource management techniques.

While the target of a cloud service provider is to get various users' jobs done in minimum budget without sacrificing their performance requirements.

2.3 Resources in a Cloud Environment

To serve the unpredictable requests of various user/applications, cloud providers need to ensure that their data centers are equipped with the state of the art devices with a virtualization layer so to avoid any failure or loss of customers. The basic elements that go into the making of a data center primarily include physical servers (compute), storage, network and electric power.

Compute—It is the component that is responsible for the execution of users' task. It consists of physical and logical components. Physical components include hardware devices like Central Processing Unit (CPU), Memory, and Input/output (I/O) devices that communicate with one another using logical components like software/protocols. To further expand the limits of compute resource, features like hyper-threading and multiple cores are added.

Storage—It provides persistent storage of data created by individuals/businesses so that it can be accessed later on. Companies like Google, Amazon depend on data centers for storage, searching and computation purposes [3].

Network—Network is a data path that facilitates communication between clients and compute systems or between compute systems and storage.

2.4 Virtualization Technology

In cloud computing environment, a cloud provider provisions virtual resources to the users. To handle multiple requests using a single host, the provider has to produce a virtual environment that can be achieved by using virtualization technology. A virtualization layer placed on top of physical infrastructure gives an impression of unlimited computing resources to the end users by abstracting the underlying hardware layer from the host operating system and also allows running multiple guest operating systems concurrently on the same host to optimize system performance. This virtualization layer (also called as hypervisor or virtual machine monitor (VMM)) can be placed directly over the hardware or the host operating system. The former is called as Type-1 hypervisor while the latter one is called as Type-2 hypervisor. It is this hypervisor that partitions the physical machine (PM) resources to be allocated to various virtual machines (VMs). Examples of

Type-1 hypervisor VMWare ESX/ESXi, Microsoft Hyper-V and that of Type-2 include VMWare Workstation and Microsoft VirtualPC.

The VMM encapsulates each guest operating system with its corresponding applications to create a logical entity that behaves like a physical machine. This logical entity is called as a virtual machine (VM). Virtualization allows the migration of VMs from one host to another in the same or different data center. A key feature of VMs is that the end-users are unable to distinguish them from actual PMs. Virtualization can be achieved in various ways. This is outlined briefly in Table 2 with their key points and benefits.

Virtual machine technology provides the feature to migrate a running VM across distinct PMs without having to shutting down the VM [4]. This concept of live migration is increasingly utilized in cloud computing environments to deal with these issues—online system maintenance, server consolidation, load balancing, and fault tolerance. In order to optimize resource utilization and save energy consumption, it is necessary to migrate these VMs to some other host in the same data center or an external one.

Table 2 Virtualization types

Type	Key points	Benefits
Server	Masks the physical resources like processor from the users	Server consolidation
	Enables running multiple operating concurrently on a host	Cost minimization
	Enables creation of VMs	Isolated VMs
Storage	Masks the underlying physical storage resources	Minimizes downtime
	Enables creating virtual volumes on the physical storage resources	Supports heterogeneous storage platforms and simplifies management
Network	Masks the underlying physical network hardware and presents multiple logical networks	Enhances performance
	Enables the network to support and integrate better with virtual environments	Enhances security
	Enables creation of overlay networks on top of physical hardware	Improves manageability
Desktop	Detaches OS and applications from physical devices	Enhances accessibility
	Enables desktops to run as VMs which can be accessed over LAN/WAN	Improves data security
	Can be implemented using Remote Desktop Services (RDS) or Virtual Desktop Infrastructure (VDI)	Simplifies backup task
Application	Offers applications to end users without any installation	Simplifies application deployment
	No dependency on the underlying computing platform	Eliminates resource conflicts

3 Related Work

In the past, researchers have optimized data center efficiency by targeting different aspects and goals in mind. Table 3 highlights the main features, strengths and shortcomings of various approaches used by researchers in cloud resource management area.

4 Managing Resources Using Virtualization

A cloud data center must be flexible enough to deal with unpredictable user requests [2]. Virtualization technology provides the solution to most of the issues take place in a data center. The system administrators can utilize it for providing better customer experience and also for efficient management of data center resources. Here, we discuss the various ways in which virtualization can be leveraged for resource management in a cloud environment.

4.1 Server Consolidation

In a data center, server sprawl situation leads to inefficient resource utilization by not exploiting the full capacity of various physical servers. This problem can be overcome by first identifying the under-utilized servers and then migrating VMs from such machines to other machines selected to host the migrated VMs.

In a data center with VMs running on various under-utilized physical machines (PMs), server consolidation will migrate all such VMs on a fewer number of fully utilized machines so that the freed machines can be put to OFF or power-saving state [5]. This will reduce energy consumption and promote green computing. Such a scenario is shown in Fig. 2.

4.2 Load Balancing

In a cloud environment, virtual machines are continuously added and removed to cope up with the varying needs of customers. This dynamism creates imbalance in the resource utilization levels of different PMs across data center with the passage of time resulting in performance degradation. To address this problem, the resource utilization levels of PMs are constantly monitored and if there is a disparity in their utilization levels, load balancing is triggered using live virtual machine migration [6]. This involves migrating VMs from highly loaded PMs to lightly loaded ones. The use of virtualization for load balancing is shown in Fig. 3.

Table 3 Various virtualization-related approaches used for resource management in the past

Author	Features	Strengths	Shortcomings
Zhen Xiao et al. [8]	Addresses the problem of server sprawl by minimizing skewness i.e. uneven utilization of servers	Supports green computing	Doesn't consider all resource types to determine skewness
	Supports dynamic resource allocation	Dynamic in nature	Eliminates hot spots partially
	Adapts to varying application demands		
Ibrahim Takouna et al. [9]	Provides support for detecting over-utilized hosts, selecting and placing VMs by taking into account energy-performance trade-off	Reduces energy consumption significantly	Focuses only on CPU utilization when selecting historical window
	Uses an adaptive window algorithm to minimize undesired VM migrations	Dynamic and robust in nature	More time consuming
Lei Wei et al. [10]	Utilizes Markov Chain model based dynamic resource provisioning approach to form resource groups of different resources types	Provisioning of resources is done according to users' needs	The process of flexible workload provisioning is complex as compared to single resource provisioning scheme
	Develops analytical models using the idea of blocking probability in order to predict suitable number of PMs taking into consideration the QoS requirements	Minimizes energy consumption	
Dong Jiankang et al. [11]	Targets compute and network resources to optimize their utilization to achieve tradeoff between energy efficiency and network performance	Supports green computing	To estimate migration costs, it doesn't take into account memory iterations or migration traffic distances
	To address energy issue, it uses a static VM placement scheme to reduce the number of PMs and network elements	Ensures optimal network resource for saving energy	Considers only maximum link utilization (MLU) network performance
	To address migration costs, it uses a dynamic VM migration scheme in order to optimize the link utilization		

(continued)

Table 3 (continued)

Author	Features	Strengths	Shortcomings
Ning Liu et al. [12]	Provides an optimized task scheduling model based on integer-programming problem to reduce cloud data center energy consumption	Handles heterogeneous tasks	Suffers from energy-response time trade-off
	Uses a greedy task-scheduling scheme to respect the response time constraints of tasks	Consumes 70 times less energy than random-based scheme	
Mahyar Movahed Nejad et al. [13]	Proposes auction-based dynamic virtual machine provisioning scheme based on integer program	Takes into account multiple resource types	Not effective in situations in which VM instances have a high degree of heterogenity
	Uses a truthful greedy technique for multiple resource types, giving incentives to users who disclose true valuations for the requested VMs	Cost beneficial for the cloud providers	
Qingjia Huang et al. [14]	Proposes an elastic consolildation mechanism for scheduling based on VM migration to conserve energy while respecting the SLAs	Increased efficiency by considering the live migration and hotspot elimination overheads	Doesn't initiate migration on machines that are already busy
	Uses resource demand predictor to reserve in advance the minimum amount of resources to be allocated to VMs to satisfy users' requests		

4.3 Hotspot Elimination

Once VMs are provisioned to the requesting customers/application, constant monitoring is required to identify whether the allocated VM is able to satisfy its SLAs as per requirements [7]. When the performance of a VM/PM goes below a certain defined level, a hotspot occurs and must be addressed as it affects the performance of the system. For this, extra resources need to be provisioned to the degrading VMs so as to meet the QoS parameters as per SLAs.

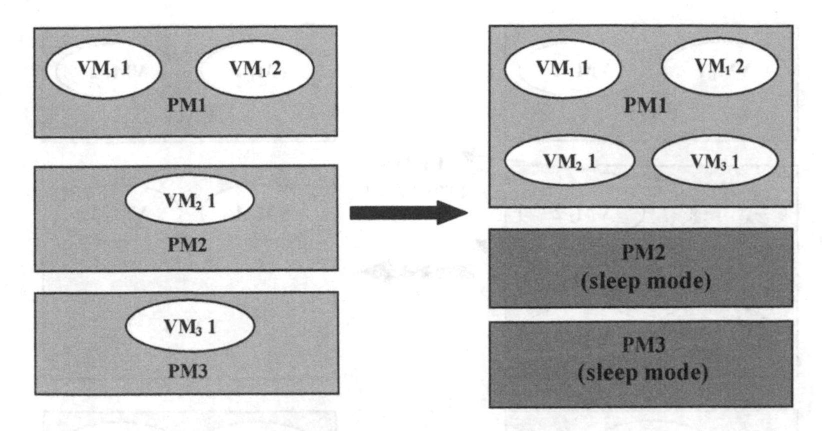

Fig. 2 Pre and post-server consolidation

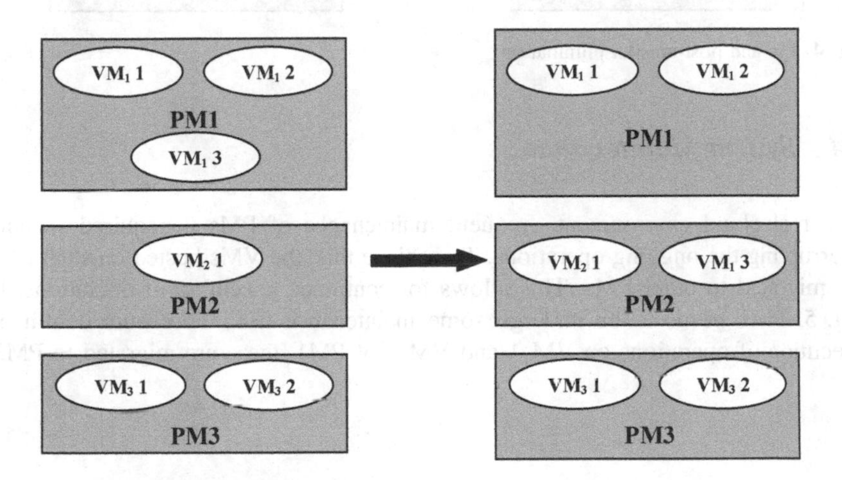

Fig. 3 Pre and post-load balancing

These extra resources can be allocated from within the same PM or from the some other PM. In the latter case, the VM has to be migrated to another PM within the same or different data center to reduce the hotspot. Figure 4 shows a situation where three PMs are hosting VMs on them. Since PM2 is overloaded, it is unable to allocate sufficient resources to its VMs namely $VM_2 1$ and $VM_2 3$. So VM migration takes place to eliminate these hotspots.

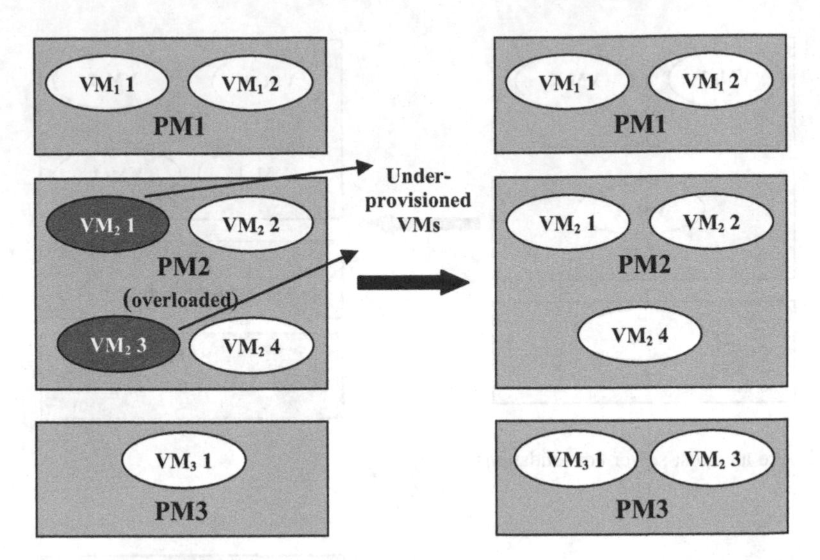

Fig. 4 Pre and post-hotspot elimination

4.4 System Maintenance

In a real cloud environment, frequent maintenance of PMs is required without interrupting the ongoing operations. To achieve this, the VMs hosted on such PMs are migrated to other PMs. This allows for continued execution of operations. In Fig. 5, PM1 needs to be undergo some maintenance task. To ensure continued execution of operations on $VM_1 1$ and $VM_1 2$ of PM1, these are migrated to PM2.

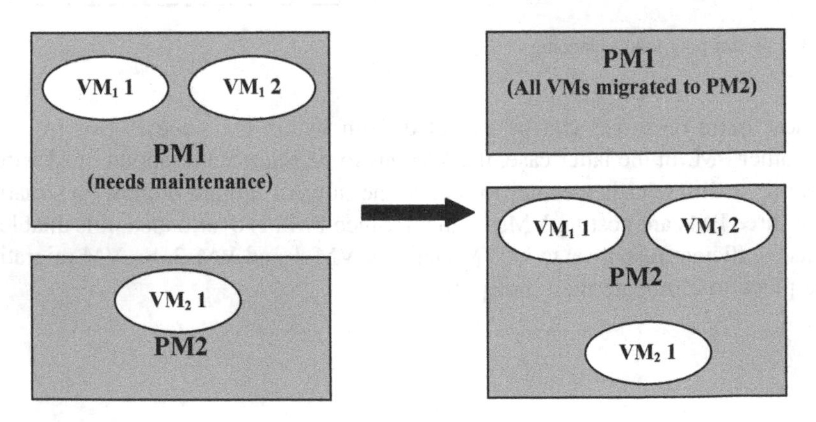

Fig. 5 System maintenance using virtualizaiton

5 Conclusion

Efficient resource management in a cloud to address issues like performance and cost is a key factor that determines its success. Virtualization is the enabling technology for cloud computing. This paper examines the state of the art in resource management area for cloud computing with a focus on improving its performance and addressing the energy concerns. We also discussed how virtualization can be leveraged to perform key tasks like server consolidation, load balancing, hotspot elimination and system maintenance in data centers to exploit the resources to their full potential.

References

1. R. Buyya, C. S. Yeo, S. Venugopal, J. Broberg, and I. Brandic: Cloud computing and emerging IT platforms: Vision, hype, and reality for delivering computing as the 5th utility. In: Future Generation Computer Systems, 25:599_616, 2009.
2. Sunil Kumar S. Manvi and Gopal Krishna Shyam: Resource management for Infrastructure as a Service (IaaS) in cloud computing: A survey. In: Journal of Network and Computer Applications, Elsevier, October 2013.
3. Md. Faizul Bari et al.: Data Center Network Virtualization: A Survey. In: IEEE Communications Surveys & Tutorials, Vol. 15, No. 2, 2013.
4. C. Clark, K. Fraser, S. Hand, J. G. Hansen, E. Jul, C. Limpach, I. Pratt, and A. Warfield: Live migration of virtual machines. In: Proceedings of the Second Symposium on Networked Systems Design and Implementation (NSDI'05), 2005, pp. 273–286.
5. Raj Kumar Buyya et al.: Energy-Efficient Management of Data Center Resources for Cloud Computing: A Vision, Architectural Elements, and Open Challenges. In: Proceedings of the 2010 International Conference on Parallel and Distributed Processing Techniques and Applications, PDPTA 2010, Las Vegas, USA, 2010.
6. C. Clark et al.: Live migration of virtual machines. In: NSDI'05: Proceedings of the 2nd Symposium on Networked Systems Design & Implementation, pp. 273–286, Berkeley, CA, USA, 2005, USENIX Association.
7. A. Meera and S. Swamynathan: Agent based Resource Monitoring System in IaaS Cloud Environment. In: International Conference on Computational Intelligence: Modeling Techniques and Applications (CIMTA), Elsevier, 2013.
8. Zhen Xiao: Dynamic Resource Allocation Using Virtual Machines for Cloud Computing Environment. In: IEEE Transactions on Parallel and Distributed Systems, Vol. 24, No. 6, June 2013.
9. Ibrahim Takouna, Esra Alzaghoul, and Christoph Meinel: Robust Virtual Machine Consolidation for Efficient Energy and Performance in Virtualized Data Centers. In: IEEE International Conference on Internet of Things (iThings), Green Computing and Communications (GreenCom), and Cyber-Physical-Social Computing (CPSCom), 2014.
10. Lei Wei et al.: Towards Multi-Resource Physical Machine Provisioning for IaaS Clouds. In: Selected Areas in Communications Symposium, IEEE, 2014.
11. Dong Jiankang et al.: Energy-Performance Tradeoffs in IaaS Cloud with Virtual Machine Scheduling. In: STRATEGIES AND SCHEMES, China Communications, February 2015.

12. Ning Liu et al.: Task Scheduling and Server Provisioning for Energy Efficient Cloud-Computing Data Centers. In: IEEE 33rd International Conference on Distributed Computing Systems Workshops (ICDCSW), pp. 226–231, July 2013.
13. Mahyar Movahed Nejad et al.: Truthful Greedy Mechanisms for Dynamic Virtual Machine Provisioning and Allocation in Clouds. In: IEEE Transactions On Parallel And Distributed Systems, Vol. 26, No. 2, February 2015.
14. Qingjia Huang et al.: Migration-based Elastic Consolidation Scheduling in Cloud Data Center. In: IEEE 33rd International Conference on Distributed Computing Systems Workshops, 2013.

Reconfigurable Circular Microstrip Patch Antenna with Polarization Diversity

Prachi P. Vast and S.D. Apte

Abstract Single-feed circular microstrip patch antenna with diagonal slot for polarization reconfigurability has been proposed. PIN diode can be used to achieve reconfigurability by placing it within the slot. Ideal diode that is diode size patch is considered in place of PIN diode for simulation and fabrication purpose. When diode is 'ON' and 'OFF', an antenna can switch between linear and circular polarization, respectively.

Keywords Reconfigurable circular microstrip patch · Circular polarization · Linear polarization · PIN diode

1 Introduction

Reconfigurable microstrip antennas due to their properties of adapting to change in environmental and system requirements is motivated a tendency to employ it in recent years [1]. For different wireless communication applications, different frequency bands with different types of radiation and polarization characteristics are required. Conventionally, this can be achieved by designing different antennas. However, practically the physical size of the system does not permit designing separate antennas for each application. Therefore, the design of reconfigurable antennas, which could integrate various operating characteristics in a single antenna, has opened an innovative research area for antenna researchers and engineers. It is well known that microstrip antennas can generate circular polarization (CP) radiation when two fundamental modes with orthogonal property are excited with the same amplitude and a phase difference of 90 [2]. For a single-fed

---

P.P. Vast (✉) · S.D. Apte
Electronics Engineering Department, Rajarshi Shahu College of Engineering Tathwade, Pune, India
e-mail: rupskalsekar@gmail.com

S.D. Apte
e-mail: sdapte@rediffmail.com

S.C. Satapathy et al. (eds.), *Proceedings of the International Conference on Data Engineering and Communication Technology*, Advances in Intelligent Systems and Computing 469, DOI 10.1007/978-981-10-1678-3_37

microstrip antenna, the condition can be achieved by introducing perturbation segments into the antenna structure. With the perturbation segments, the original fundamental mode of the antenna is split into two orthogonal modes, which have different resonant frequencies, and one CP operating frequency can be found between the two resonant frequencies. Several antenna structures offering polarization diversity have been proposed in the literature. Single-feed Square Microstrip patch antenna with single notch for polarization reconfigurability has been proposed [3]. Reconfigurability can be achieved by placing PIN diode within the notch. Four identical notches with parasitic elements are embedded into the circular patch antenna [4]. The antenna can switch between linear polarization (LP), left-hand circular polarization (LHCP) and right-hand circular polarization (RHCP). PIN diodes are used to, respectively, reconfigure the coupling slot and the open stub of the feed line, the polarization of the microstrip antenna can be switched between vertical and horizontal polarizations [5].

In this paper, circular microstrip patch antenna (CMPA) for polarization reconfigurability with rectangular slot on it have been designed and fabricated to resonate at 2.4 GHz. To achieve the reconfigurability, the slot is filled with PIN diode. Switching the diode between ON and OFF positions, changes the polarization from linear to circular. For the fabrication purpose, the PIN diode is considered as ideal diode by placing it with small diode size patch.

2 Proposed Antenna Design

Figure 1 illustrates the geometry of the proposed circular microstrip patch antenna with reconfigurable polarization. The circular patch (radius = 17.1 mm) is implemented on a 1.6-mm thick substrate of relative permittivity 4.4 with loss tangent 0.002. A single slot (3 mm × 8 mm) is made on the circular patch. In Fig. 2, it can be seen that this slot is filled with rectangular parasitic element (1 mm × 4 mm) in place of PIN diode. Simulation is done on HFSS11 (Table 1 and Fig. 3).

Fig. 1 CMPA in 'OFF' state

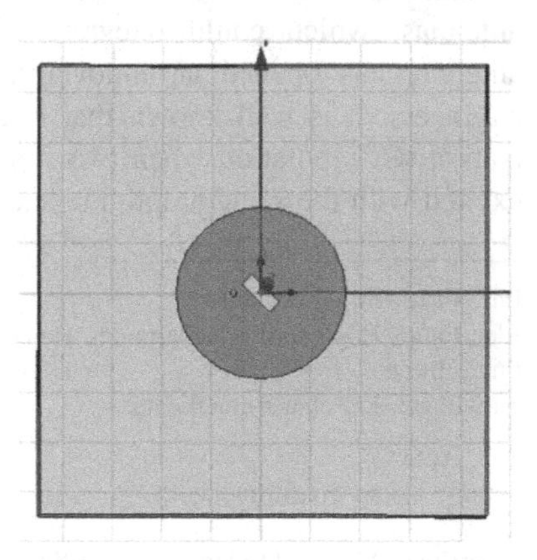

Fig. 2 CMPA in 'ON' state

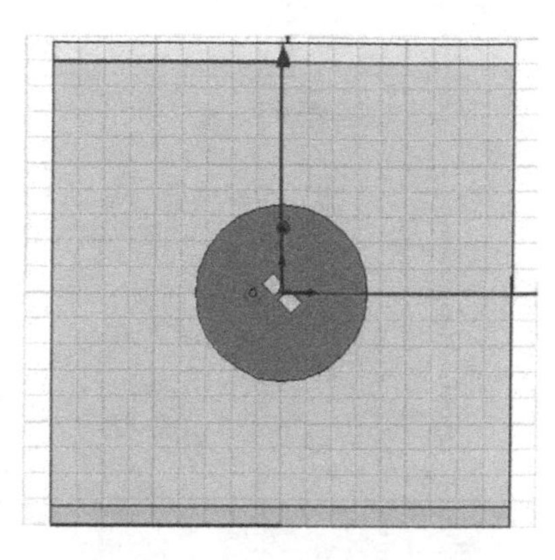

Table 1 Design parameters of SMPA

Sr. no.	Parameter	Value
1	Frequency	2.4 GHz
2	Dielectric constant of substrate	FR4 (4.4)
3	Radius of patch	17.1 mm
5	Slot length	8 mm
6	Slot width	3 mm
7	Parasitic element width	1 mm
8	Parasitic element length	4 mm
9	Height of the substrate	1.6 mm

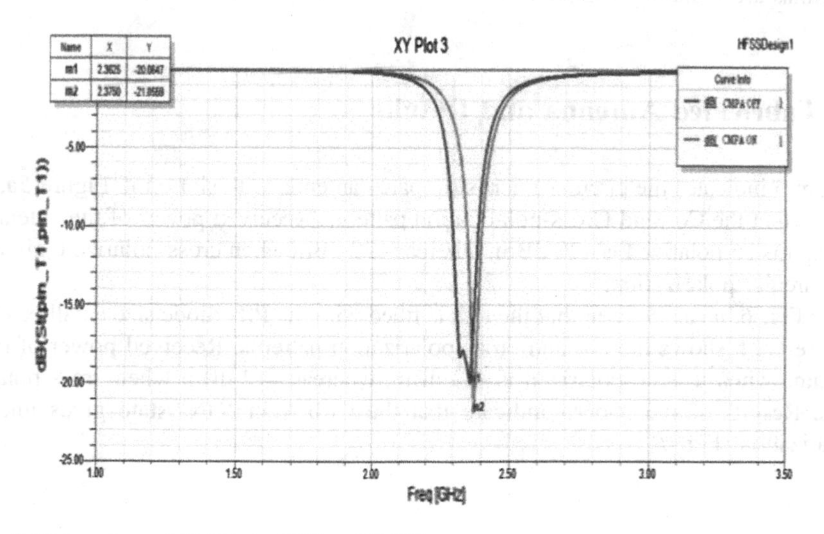

Fig. 3 Return loss of CMPA in 'ON' and 'OFF' state

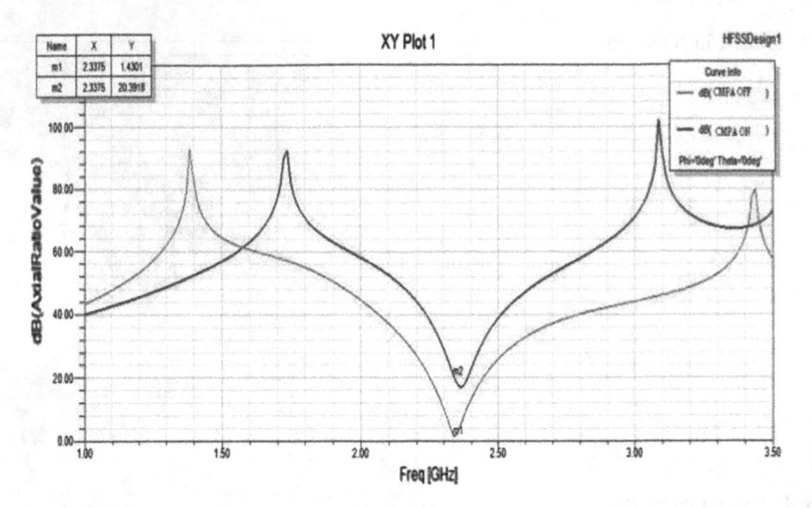

Fig. 4 Axial ratio of CMPA in 'ON' and 'OFF' state

3 Simulation Design and Results

Figures 1 and 2 are the simulated antenna with diagonal slot on CMPA and slot filled with the small patch, respectively. Figure 4 shows that when the diode is 'OFF', the antenna radiates at 2.33 GHz with axial ratio of 1.4301 dB, whereas it is 20.3918 dB when it is ON. That is when SMPA is in 'OFF' state gives Circular polarization, whereas it gives linear polarization when it is in 'ON' state. Both the antennas are radiating at 2.33 GHz.

4 Fabricated Antenna and Results

Figure 5 indicates the circular microstrip patch antenna with slot on it. Figure 5a, b shows that the Co- and Cross-polarization patterns. Received power of the antenna when it is co-polarized is –20 dBm, whereas –22 dBm when cross-polarized, giving the circular polarization.

In Fig. 6, it can be seen that the slot is filled with the PIN diode size small patch. Figure 6a, b shows the co- and cross-polarization patterns. Received power of the antenna when it is co-polarized is -22 dBm, whereas –42 dBm when cross polarized. Results of the pattern indicate that the CMPA in 'ON' state gives linear polarization (Fig. 7).

Fig. 5 CMPA in 'OFF' state
a Co-Polarization of CMPA in 'OFF' state
b Cross-Polarization of CMPA in 'OFF' state

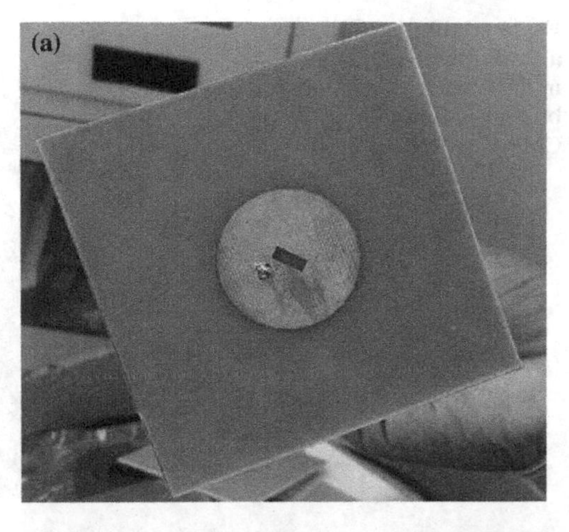

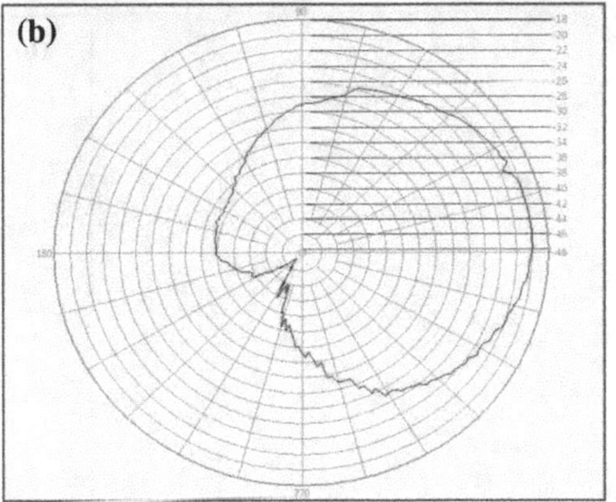

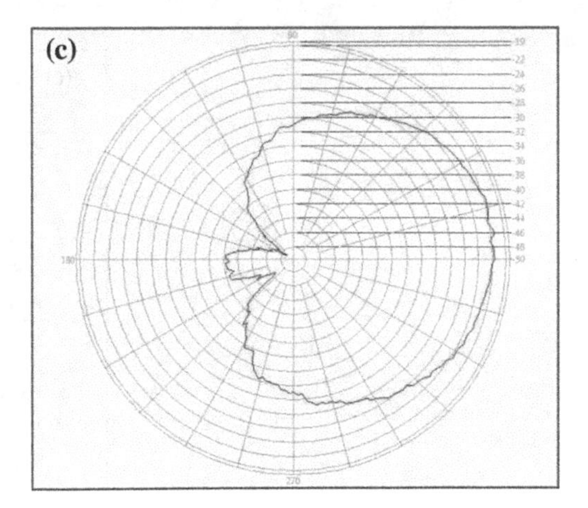

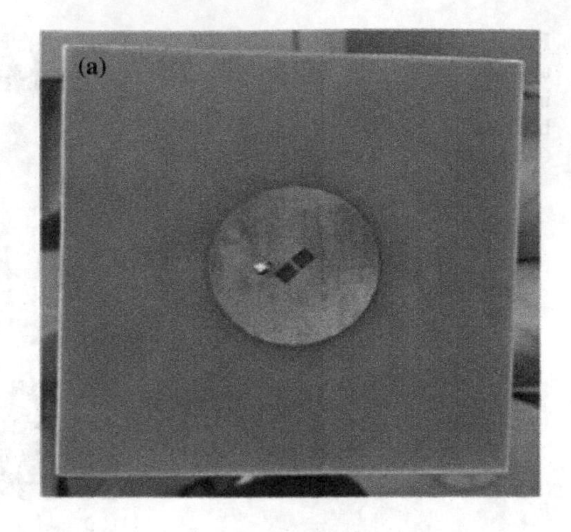

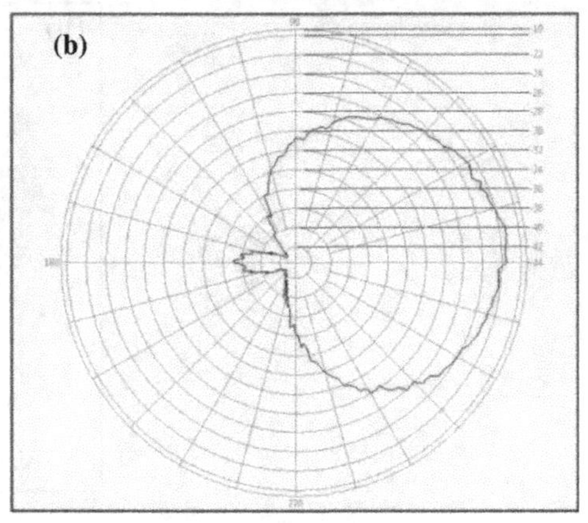

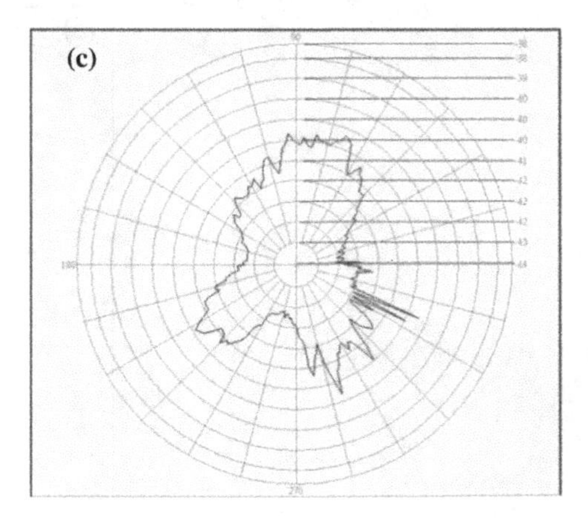

Fig. 6 CMPA in 'ON' state **a** Co-Polarization of CMPA in 'ON' state. **b** Cross-polarization of CMPA in 'ON' state

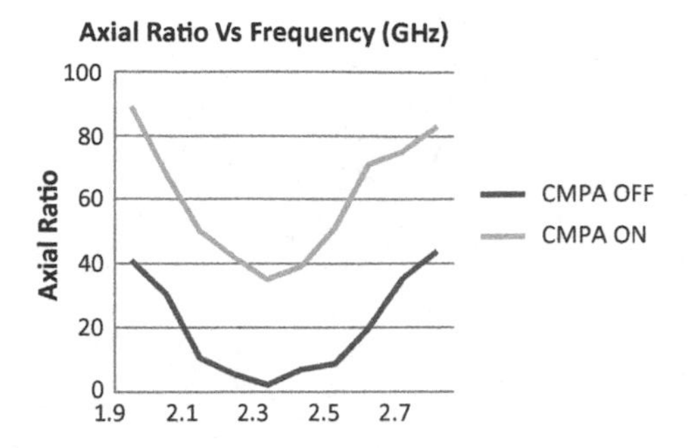

Fig. 7 Axial ratio

5 Conclusion

In this paper, single-feed circular microstrip patch antenna with diagonal slot for polarization reconfigurability has been designed, fabricated, and tested. The antenna can be electronically controlled by controlling the biasing of the PIN diode to switch the polarization mode (linear polarization, circular polarization). Axial ratio found for linear polarization when PIN diode in ON is 38.23 and for circular when PIN diode is OFF is 1.56. From Axial ratio measurement and polarization pattern, it can be seen that proposed antenna shows good polarization reconfigurability.

References

1. Ailar Sedghara and Zahra Atlasbaf.:A New Reconfigurable Single-Feed Microstrip Antenna With Polarization Diversity, 7th European Conference on Antenna And Propagation (EuCAP), 2013.
2. Jia-Fu Tsai and Jeen-Sheen Row.: Reconfigurable Square-Ring Microstrip Antenna, IEEE Transactions On Antennas And Propagation, Vol. 61, No. 5, May 013.
3. Mrs. Prachi P. Vast and Dr. Mrs. S. D. Apte.: Reconfigurable Square Microstrip Patch Antenna With Polarization Diversity, IEEE conference, 2015.
4. N.H. Noordin, Wei Zhou, A.O. El-Rayis, N. Haridas, A.T. Erdogan and Tughrul Arslan,: Single-feed Polarization Reconfigurable Patch Antenna, IEEE Transactions On Antennas And Propagation, 2012.
5. Rui-Hung Chen and Jeen-Sheen Row.: Single-Fed Microstrip Patch Antenna With Switchable Polarization, IEEE Transactions On Antennas And Propagation, Vol. 56, No. 4, April 2008.

Issues with DCR and NLSR in Named-Based Routing Protocol

Rajeev Goyal and Samta Jain Goyal

Abstract Content-based routing, which is under a category of sophisticated routing protocol, is used to propose a routing protocol that can replace Internet protocol. Content-based routing is used to querying, fetching, and finding desired information where in the other routing based system; they forwarded the packets on DNS or data senses. Main issues with the past were constraints of the storage. We have seen in some papers that name-based routing is feasible to achieve content-based routing.

Keywords Content based routing · Information centric network · Named based content routing · Fully qualified domain name · Distance-based content routing · Named-data link state routing protocol · Distributed hash tables

1 Introduction

Content-based routing is a powerful and flexible solution of sending messages, which are addressed to services or clients. It route the message based on the contents of the message [1]. Two types of entities are used to create content-based routing system. One is routers that route messages and other is services that are ultimate consumers of messages (Fig. 1).

This content-based routing has a typical example that represents the information of the content and its name is name-based routing. Based on [2, 3] this routing algorithm, FQDN(**fully qualified domain name**) database is used, which generally takes two entries: One for node ID and other is for routing or forwarding table entries. But sometimes it faces the problem of hardware space and search overhead. Also faces some scalability issues if FQDN is update more frequently.

Rajeev Goyal (✉) · S.J. Goyal
Amity University, Gwalior, Madhya Pradesh, India
e-mail: rgoyal@gwa.amity.edu

S.J. Goyal
e-mail: sjgoyal@gwa.amity.edu

S.C. Satapathy et al. (eds.), *Proceedings of the International Conference on Data Engineering and Communication Technology*, Advances in Intelligent Systems and Computing 469, DOI 10.1007/978-981-10-1678-3_38

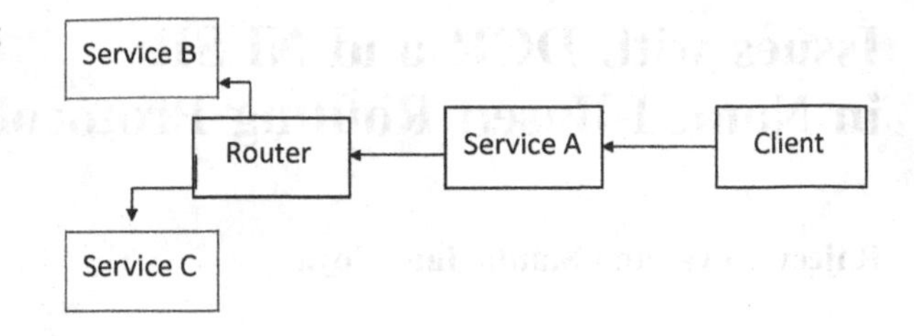

Fig. 1 Content based routing scenario

Basically, FQDN is the complete domain name for a specific machine on the internet, which consists of two parts: (1) host name (2) domain name. Since FQDN [4] is variable length, state-of-the-art memory is used. TCAM is used for routing or forwarding table on routers. It gives high speed during searching the contents. TCAM has excellent performance for the longest prefix match. It has more advantage than a normal RAM [5]. To reduce power consumption in content addressable memory CCAM, more flexible, efficient, and feature rich TCAM has driven by researchers. It is similar to CAM with additional state X, which used to search the given range of values. The TCAM are particularly useful for CIDR (classless inter-domain routing) [6–8]. In a single operation, for the longest prefix matching and for packet classification, CIDR is very particularly useful.

To solve the scalability issues with the FQDN, we implement DHT for storing hierarchically structured FQDN. DHT is used to manage data by using hash function also designed to be update more frequently than DNS-based approach [9]. DHT and cache sharing architectures balances the cost of communication between nodes of the platform and DNS resolution. This provides significant advantage for cache sharing and DHT architectures. DHT nodes routers have more processing ability than regular nodes [10, 11]. Hierarchical DHT specially designed for better fault tolerance, security, effective caching, maximize utilization of bandwidth, and hierarchical access control. In this proposed work, we use hash function to get physical position of node. For this, two measurements landmark clustering and RTT are used to generate proximity information, where landmark clustering is used to get proximity information about the closeness of nodes.

2 Name-Based Content Routing Protocols

For ICN architectures, name-based content routing protocols are essentials. Content-based routing protocols are based on content flooding where requested contents do not know the location [12, 13]. A approach named DIRECT uses for connectivity disruption, whereas direct diffusion was the first approach for the name-based content routing, which used named contents as request on the sensor network. By the means of link state and path vector algorithms, number of approaches is counted to maintain routing table information [14]. NBRP, CBCB, and DONA are some major routing approaches that use content-based routing principles. Those are specifically for multicasting. To accomplish name-based routing, ICN projects use DHT that is used to run over the physical infrastructure [15]. In DHT, destination is assigned to home location for determining namespace of nodes. To improve efficiency, DHT nodes are mapped and built using underlying routing protocols [16–18].

3 Components of DSR and NSLR

In DSR, every node maintains a route cache on the network. While sending data, intermediate node examines header and based on the information on header, forward to the next hop else obtain new route based on the broadcasting a route request within its transmission range [19]. Based on the address of source and destination nodes, route record is created by route request. Based on the received information if destination matches, "sent reply" packet is sent back to originator. Else with the help of intermediate node, "Intermediate node replies" sent to secure to get its destination. After that rebroadcasts the same request [20, 21], "Route error" packet is generated if source route is damaged or corrupted due to any unwanted system. DSR aggressively make use of "route caching". This protocol uses several optimizations like data salvaging, gratuitous replies, route-snooping, etc.

NSLK-This protocol is for named data networking, which is used to identify and retrieve data based on its name prefixes instead of IP prefixes [22]. NSLR protocol has directly benefited of NDN's data authenticity also it has no of forwarding strategies. To exchange sending message, NSLR uses data packet/interest packet of NDN's router to establish connection during failure or recovery of any link/neighbor, NSLR disseminates LSA for the entire network. Advertise name prefix for static or dynamic registrations NLSR provide name prefix to each node on multiple-route as well as it signing and verify of all routes [13]. Each NSLR node builds a network topology based on adjunct LSA's information. To produce multiple next-hopes for destination, use Dijkstra algorithm. To detect failure of processes, NLSR periodically sends INFO to each neighboring node. This "info" if it gets no response, it considered "down" else "active" [23].

4 Security and Performance Issues

DSR faces certain security issues like there is no effective mechanism to remote entries, which causes wastes network bandwidth and loss of data packets also under the guise of data salvaging a malicious node create mistake and gratuitous replies degrades performance of DSR also effects its security system, whereas NLSR is an intra-domain routing protocol. Here network administrator authenticates network with the help of key signing and verification. NSLR [24, 25] strictly follows trust model at the trust anchor. Security with NSLR is more original then the other protocols of content-based routing. NSLR security checks on a key level whether the key indeed belongs to the original routers or not. Once the thing is verified, record this information for future packets.

5 Evaluation

Results of our evaluation and analysis based on its processing time, convergence time, and messaging overhead are shown in Figs. 2, 3, and 4.

In the figures, standard deviation of the distribution of value is given. In each figure, the X-axis resents the number of replicas of name prefix and the quantity presented for each router. The results show that DSR perform operations and events in upper bound indicate our implementation sends all name prefixes instead the changes since the last updates.

Figure 2 shows initialization phase where routers do not have any topology information or prefixes except for those which are locally available. Router sends its

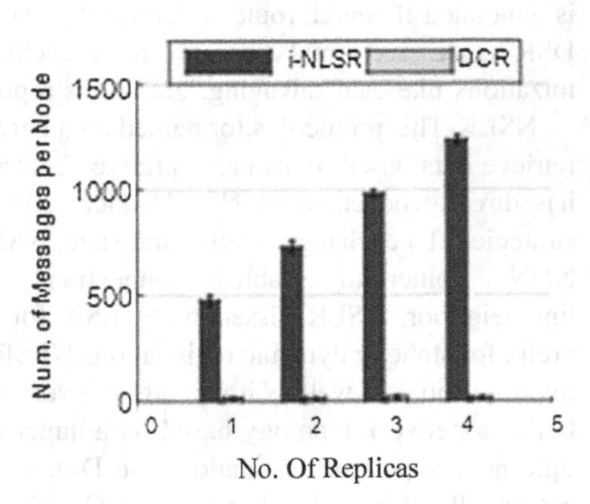

Fig. 2 Evaluation analysis 1

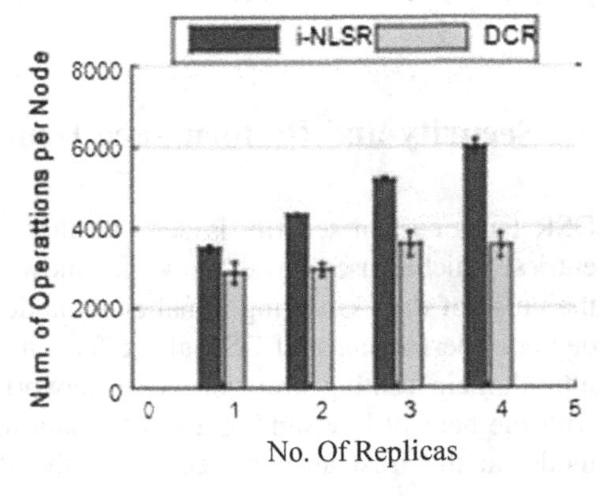

Fig. 3 Evaluation analysis 2

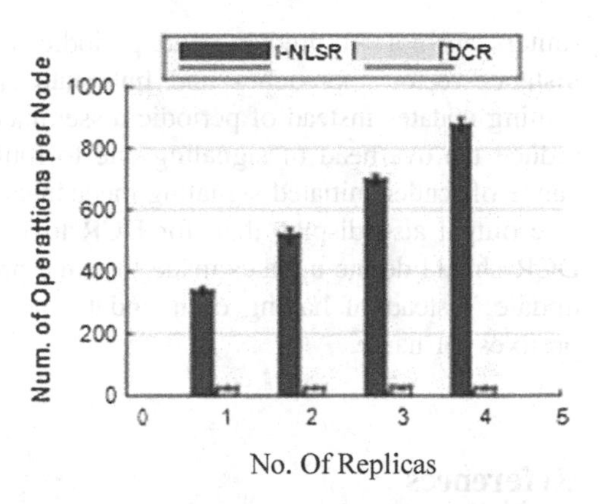

Fig. 4 Evaluation analysis 3

local information for a random number of times at the starting of simulation. In DCR, routers sends the closest anchor to each neighbor to the prefix that is why number of messages does not change for DCR [26] as increases the replicas number. But a number of prefix in NSLR increases the total number of prefixes as average per prefix anchor increase. The operation counts for Dijikstra algorithm are greater than the operation count for DCR.

In Fig. 3, the result shows adding a new prefix into the sensor network. An NLSR uses less number of replicas than DCR uses more messages to exchange. However, in DCR the replicas the number of messages is incomparable and the number of exchanges does not grow with the number of replicas. And Vice versa it happens with NSLR.

In Fig. 4, each router has to process LSA that corresponds to each direction of a link, where the number of messages exchanged per node is smaller than the NLSR. This is independent of the number of replicas of prefixes, maintained in the network.

6 Conclusion

The functioning of an enhanced version of NLSR and DCR was examined. This simulation shows that distances reporting to the closest replicas of prefixes for name is much proficient than preserving whole topology information and the routers information where the numerous replicas of name prefix inhabits. The operating cost in NLSR turns out to be a concern when the average a number of replicas per name prefixes increases further then two.

Our study advises many vital possibilities of fertile research. One, a routing approach based on link-state information in which routers interconnect information about only the prefix that is nearest replicas should be explored and its functioning should be matched against DCR that supports routing to prefix which is nearest replicas with distance information. Two, information routing approaches where routers are not required to send periodic updates should be examined for both distance-vector approaches and link-state approaches. The use of event-driven routing updates instead of periodic dissemination of LSAs or distance updates can reduce the overhead of signaling due to routing considerably. Three, the significance of sender-initiated signaling mechanics in ICN and NDN should be counted. The output also display that, for DCR to work competently, update messages in DCR should define updates made to name prefixes to distances as to send the last update, instead of having each update message comprise information about all prefixes for name.

References

1. J. Choi, J. Han, E. Cho, T. Kwon, and Y. Choi, “A Survey on content-oriented networking for efficient content delivery,” IEEE Communications Magazine, vol. 49, no. 3, pp. 121–127, Mar. 2011.
2. A. Ghodsi, S. Shenker, T. Koponen, A. Singla, B. Raghavan, and J. Wilcox, “Information-centric networking,” in Proceedings of the 10th ACM Workshop on Hot Topics in Networks - HotNets "11. New York, New York, USA: ACM Press, Nov. 2011, pp. 1–6.
3. PARC. CCNx open source platform. http://www.ccnx.org.
4. J. Torres, L. Ferraz, and O. Duarte. Controller-based routing scheme for Named Data Network. Technical report, Electrical Engineering Program, COPPE/UFRJ, December 2012.
5. University Of Arizona. ccnping. https://github.com/NDN-Routing/ccnping.
6. L. Wang, A. M. Hoque, C. Yi, A. Alyyan, and B. Zhang. OSPFN: An OSPF based routing protocol for Named Data Networking. Technical Report NDN-0003, July 2012.
7. C. Yi, A. Afanasyev, L. Wang, B. Zhang, and L. Zhang. Adaptive forwarding in named data networking. SIGCOMM Compute. Common. Rev., 42(3):62–67, June 2012.
8. T. Koponen, M. Chawla, B.-G. Chun, A. Ermolinskiy, K. H. Kim, S. Shenker, and I. Stoica, “A data-oriented (and beyond) network architecture,” ACM SIGCOMM Computer Communication Review, vol. 37, no. 4, p. 181, Oct. 2007.
9. Mobility first project. [Online]. Available: http://mobilityfirst.winlab.rutgers.edu/C. Bian, Z. Zhu, E. Uzun, and L. Zhang. Deploying key management on NDN test bed. Technical Report NDN-0009, Febryary 2013.
10. T. Goff et al., Preemptive Routing in Ad Hoc Networks. In Proceedings of the 7th Annual Int’l Conference on Mobile Computing and Networking (Mobicom), 2001.
11. D. De Couto, D. Aguayo, J. Bicket, and R. Morris. A High-Throughput Path Metric for Multi-Hop Wireless Routing. In Proceedings of MobiCom 2003.
12. Heissenb¨uttel, M., Braun, T., W¨alchli, M., Bernoulli, T.: Evaluating the limitations of and alternatives in beaconing. Ad Hoc Network. 5(5) (2007) 558–578.
13. Mottola, L.Cugola, G.Picc, G.P.: A self-repairing tree topology enabling content-based routing in mobile ad hoc networks. IEEE Trans. on Mobile Computing 7(8) (2008) 946–960.
14. Zorzi, M.Rao, R.R.: Geographic random forwarding for ad hoc and sensor networks: Multihop performance. IEEE Trans. Mob. Compute. 2(4) (2003) 337–348.
15. IEEE Computer Society LAN/MAN Standards Committee, “Part 11: Wireless LAN, medium access control (MAC) and physical layer (PHY) specifications,” IEEE, Inc., Standard ANSI/IEEE 802.11, 1999.
16. M.K. Marina and S. R. Das, Impact of Caching and MAC Overheads on Routing Performance in Ad Hoc Networks, Computer Communications, 2003.

17. Henriksson, D.Abdelzaher, T.Ganti, R.: A caching-based approach to routing in delay-tolerant networks. ICCCN 2007 (Aug. 2007) 69–74.
18. Luo, L. Huang, C. Abdelzaher, T. Stankovic, J.: Envirostore: A cooperative storage system for disconnected operation in sensor networks. (2007).
19. D. Johnson, D. Maltz and Y. Hu. The dynamic source routing protocol for mobile ad hoc networks. IETF MANET Working Group, Internet Draft 2003.
20. T. Jiang, Q. Li, and Y. Ruan. Secure Dynamic Source Routing Protocol. In Proceedings of the Forth International Conference on Computer and Information Technology (CIT), 2004.
21. R. N. Mir and A. M. Wani. Security Analysis of Two on-Demand Routing Protocols in Ad Hoc Networks. In Proceedings of ACM MOBIHOC 2001.
22. Ammari, H.M., Das, and S.K.: Promoting heterogeneity, mobility, and energy-aware voronoi diagram in wireless sensor networks. IEEE TPDS 19(7) (2008) 995–1008.
23. Luo, H., Ye, F. Cheng, J., Lu, S. Zhang, L.: Ttdd: Two-tier data dissemination in large-scale wireless sensor networks. Wireless Networks 11(1–2) (2005) 161–175.
24. Kim, H.S., Abdelzaher, T.F., Kwon, W.H.: Minimum-energy asynchronous dissemination to mobile sinks in wireless sensor networks. In: SenSys. (2003).
25. Somasundara, A., Kansal, A., Jea, D., Estrin, D., Srivastava, M.: Controllably mobile infrastructure for low energy embedded networks. Mob. Comp., IEEE Transactions on 5(8) (Aug. 2006) 958–973.
26. Chatzigiannakis, I., Kinalis, A., Nikoletseas, S.: Efficient data propagation strategies in wireless sensor networks using a single mobile sink. Compute. Commun. 31(5) (2008).

Design and Analysis of Quantum Dot Cellular Automata Technology Based Reversible Multifunction Block

Waje Manisha Govindrao and K. Pravin Dakhole

Abstract In the era of emerging trends quantum dot cellular automata has become very popular because of its extremely small size. In this paper logically reversible 4 × 4 block is proposed. This block is designed and simulated using Quantum Dot Cellular Automata technology. This QCA based 4 × 4 block named Reversible multifunction block is capable to generate different reversible functions like OR, AND, XOR, XNOR, Full Subtractor, Full Adder, Half Subtractor, Half Adder, code Converters like Gray to Binary and Binary to Gray, Pass Gate, Set, Reset and complement, by changing the inputs at different instants. Reversible Multifunction Block constituted of thirteen reversible functions. Performance analysis of this new design shows that using single block or single Gate, multiple functions can be generated whereas if we wish to design all these functions separately too many number of gates are required. This 4 × 4 reversible Multifunction Block is proved to be efficient in the literature. The proposed design is implemented using QCA-Designer 2.3.

Keywords (QCA) Quantum dot cellular automata · Reversible multifunction block · QCA cell, quantum cost, reversible logic

---

W.M. Govindrao (✉)
Department of Electronics Engineering, G.H. Raisoni College of Engineering and Management, Pune, India
e-mail: waje.manisha@gmail.com

K.P. Dakhole
Department of Electronics Engineering, Yeshwantrao Chavan College of Engineering, Nagpur, India
e-mail: pravin_dakhole@yahoo.com

S.C. Satapathy et al. (eds.), *Proceedings of the International Conference on Data Engineering and Communication Technology*, Advances in Intelligent Systems and Computing 469, DOI 10.1007/978-981-10-1678-3_39

1 Introduction

QCA is known as one of the nanotechnology which is efficient in area and power because of the smaller size of Quantum dot [1]. Demonstration of first QCA cell is done in 1997 [2]. Flow of current is not continuous in QCA devices as in case of CMOS devices. As a result power dissipation of QCA devices is very less. Hence QCA technology has the great advantage to scale devices and circuits in the world of quantum.

The QCA technology comprises a clocking means that can be considered same to that of power supply in CMOS technology [3]. Theoretical and experimental demonstration [4] shows that clock in QCA offers power gain and gives a means for memory characteristics in the cells. Different experiments have focused the important features of clocked QCA devices [5, 6]. Employment of reversible logic operations is one of the computing technologies, where information will not be lapsed [7]. Thus, virtually they disperse no heat. Since a QCA circuit maintains information with help of 4 phases of clock [8, 9], the dissipation of heat in a QCA logic circuit can be considerably lesser than kBT ln2. This characteristic favours the introduction of QCA paradigm in reversible logic design.

2 Quantum Dot Cellular Automata Terminologies

2.1 Quantum Cell

A cell comprises of four quantum dots, which are placed at the bend of a square box. It has two numbers of free electrons [1]. These free electrons can tunnel amongst the quantum dots. Two electrons established the polarization values as –1.00 i.e. binary value 0 or +1.00 i.e. binary value 1 as given in Fig. 1.

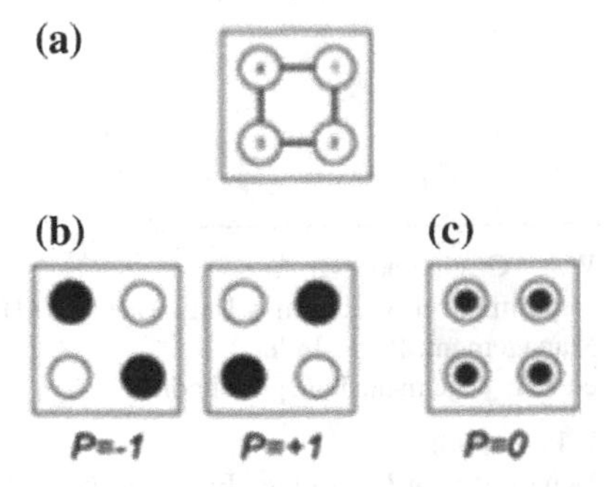

Fig. 1 **a** Structure of a cell **b** Logic '0' and Logic '1' Polarization, **c** Unpolarized cell

2.2 QCA Clock

Four clock phases are there in QCA, namely Relax phase, Switch phase, Release phase, Hold phase [10, 11]. This clocking feature in QCA gives parallel operation also known as pipelining use to provide information in the form of bits for QCA. Clock in QCA technology is used to organize the signal flow. Clock plays a vital role in the QCA circuit for parallel as well as synchronized operation. Loop of QCA cells with different clock zones act as a memory cell (Fig. 2).

2.3 QCA Logic Gates

Basic components of QCA logic circuits are Majority Voter i.e. MV gate and inverter [11]. MV gate can be implemented using only five QCA cells. MV gate has one output and three inputs. With the help of MV gate QCA AND & OR gate can also be implemented just by changing the polarity of third input [12], as shown in Fig. 3. Equations of these gates are,

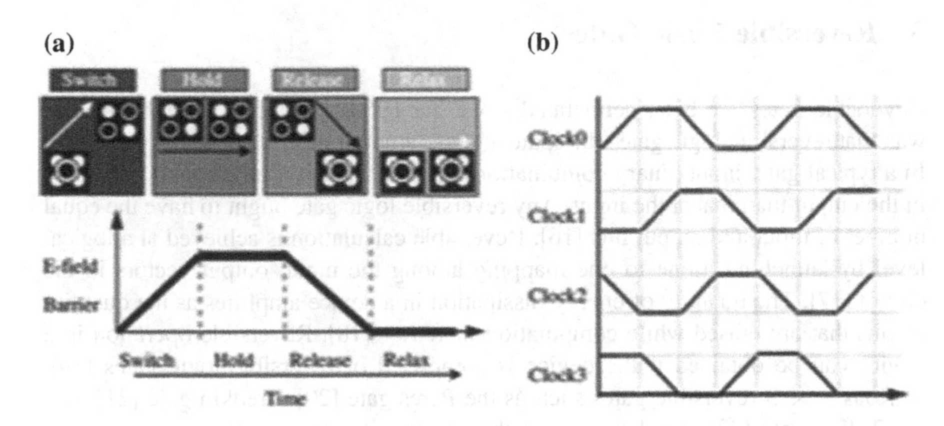

Fig. 2 QCA clock phases 1

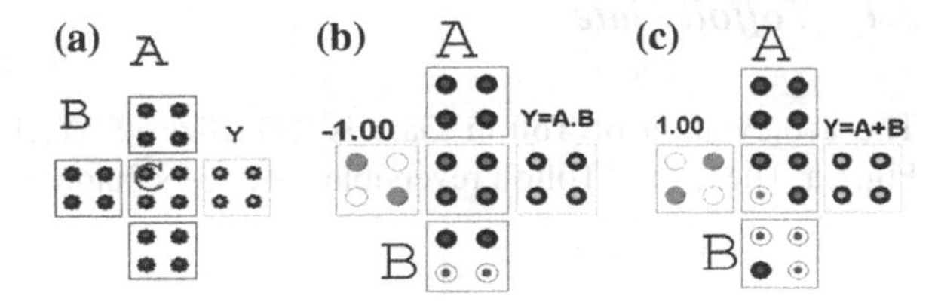

Fig. 3 **a** QCA MV gate; **b** QCA AND gate; **c** QCA OR gate

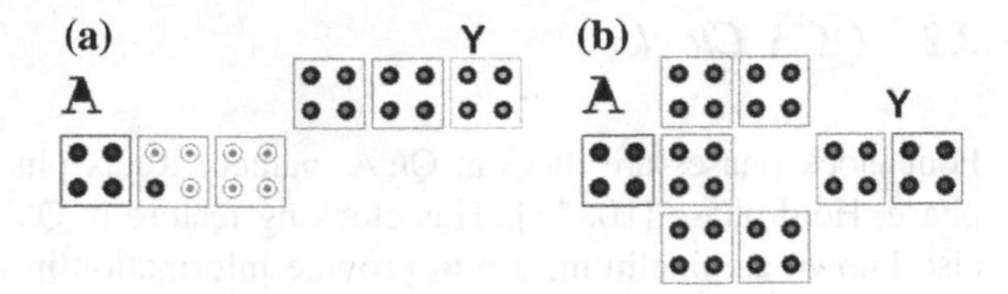

Fig. 4 **a** QCA based NOT gate with 45° rotated cell; **b** QCA based NOT gate

$$\text{MVGate:} \quad Y = \text{MV}(A, B, C) = AB + AC + BC;$$
$$\text{QCA AND Gate:} \quad Y = \text{MV}(A, B, 0) = A \text{ AND } B;$$
$$\text{QCA OR Gate:} \ Y = \text{MV}(A, B, 1) = A \text{ OR } B;$$

QCA based NOT gate can be implemented in two ways as in Fig. 4 [13]. Another important gate in QCA is the NOT gate. NOT gate is obtained with the help of QCA cell which is located 45° to a further QCA cell [14], Consider Fig. 4a, Y will have inverted value of input A. several methods are there to design the QCA inverter, one way of designing is given in Fig. 4b [12].

3 Reversible Logic Gates

Reversible gates are been deliberated since the 1960s [15]. The main motivation was that reversible logic gates dissipate a lesser amount of heat or ideally zero heat. In a typical gate, input binary combinations are lost, since little information is there in the output than that at the input. Any reversible logic gate ought to have the equal number of input and output bits [16]. Reversible calculation is achieved at a logical level by launching a one to one mapping among the input, output vectors in the circuit [17]. The quantity of energy dissipation in a device amplifies as the quantity of bits that are erased while computation increases [18]. Reversible operation in a device can be obtained if the device is comprised of reversible logic gates [19]. Various 3 × 3 reversible gates such as the Peres gate [20], Fredkin gate [21], and the Toffoli gate [22], are discussed in the previous work.

3.1 Toffoli Gate

The abbreviation of Toffoli Gate is TG. Size of TG is 3 × 3 [22] as given in Fig. 5a. The 3 × 3 Toffoli reversible gate gives fanout as 2 and quantum cost value

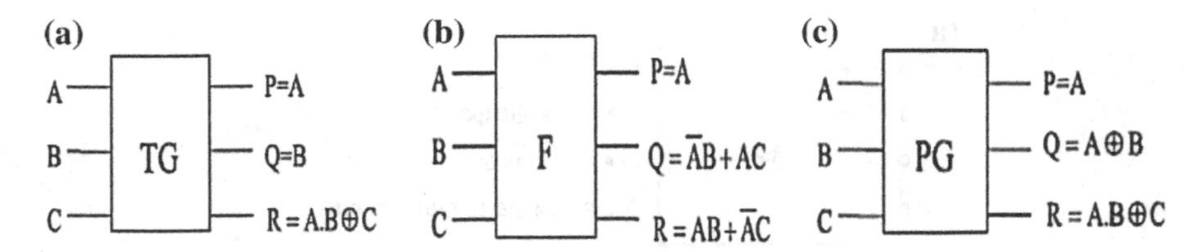

Fig. 5 Reversible Logic Gates **a** Toffoli; **b** Fredkin; **c** Peres

as 5, hence it is widely used gate. The quantum circuit of Toffoli gate is drawn using basic quantum circuits [18].

3.2 Fredkin Gate

Fredkin gate is widely used as it preserves the parity and hence it is also known as conservative as well as parity preserving gate [21]. Parity preserving gate is useful for fault finding thus anyone can design fault tolerant reversible circuit using Fredkin gate. Figure 5b gives the idea about Fredkin gate and it has 5 as a quantum cost.

3.3 Peres Gate

Figure 5c shows the Peres gate [20]. The Peres gate is drawn using two V+ gates, one CNOT gate and one V gate in the design. Peres gate's quantum cost is very less compared to other reversible logic gates.

4 Proposed Work

4×4 Reversible Multifunction Block [RMFB] is proposed in this paper. RMFB has four inputs and four outputs as shown in Fig. 6a; this block is capable of generating 13 functions as mentioned in Table 3.

Calculations for finding the value of quantum cost is done by making use of 1×1 and 2×2 quantum gates in the design [18]. Quantum cost of both the

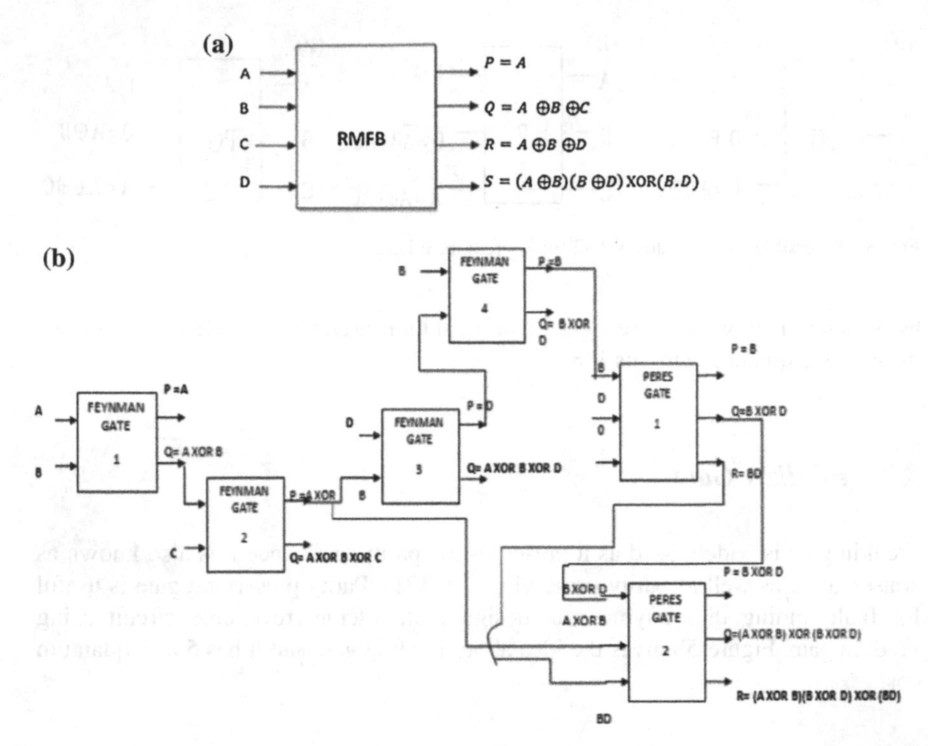

Fig. 6 **a** Block schematic of Reversible multifunction block. **b** Block schematic of reversible multifunction block using Feynman and Peres gates

reversible 1 × 1 and 2 × 2 gates are considered as one [23]. Circuit diagram of RMFB is shown in Fig. 6b

For the implementation of RMFB, 4 Feynman gates and 2 Peres gates are required. Feynman gate has Quantum cost as unity and 4 for Peres gate. Hence the quantum cost of RMFB is 1 + 1 + 1 + 1 + 4 + 4 = 12. By replacing the quantum correspondent of the entire gates and merging the circuit, Quantum cost will optimized to 8.

For the implementation of RMFB, 4 Coupled majority Voter Minority (CMVMIN) gates, 12 Majority Voter (MV) gates and 6 NOT gates are required, that is in total only 22 QCA gates are required for the implementation of all mentioned 13 functions. The quantum cost of RMFB is 8.

While drawing layout of QCA circuits, very important thing is the consideration of number of crossovers required for the design. For the logical schematic of RMFB shown in Fig. 7, the number of crossovers used is zero and the RMFB is known to

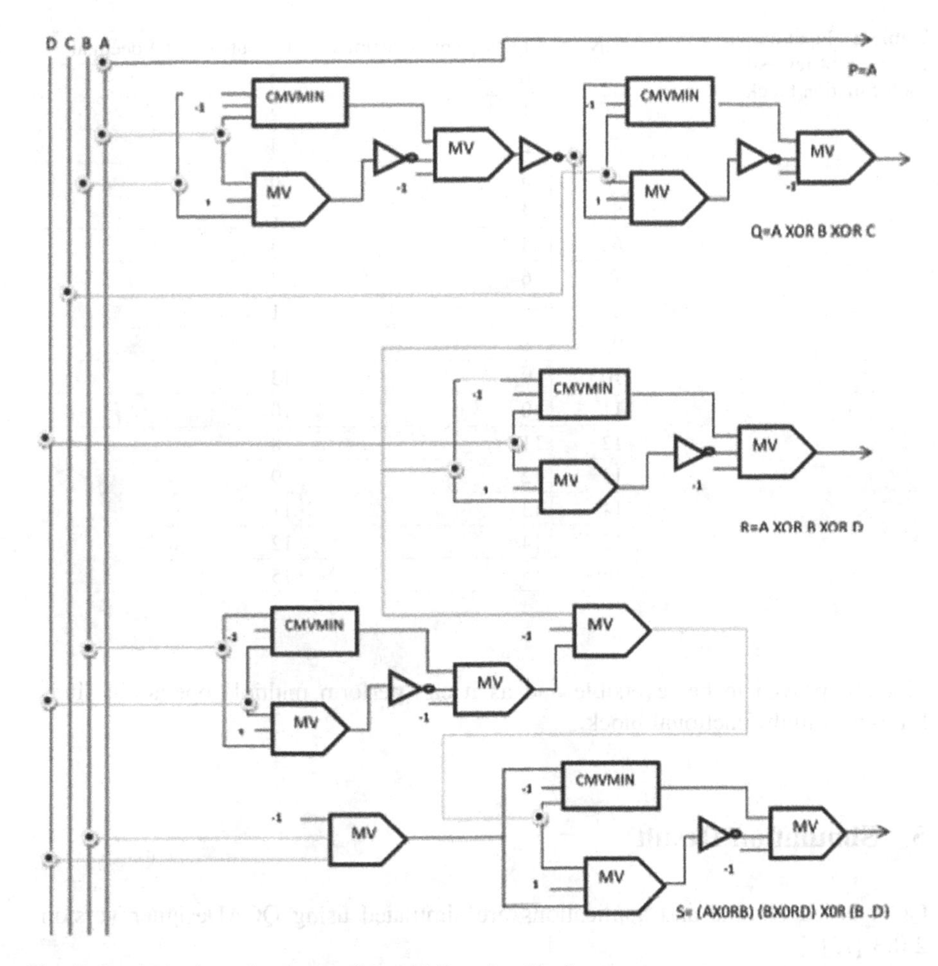

Fig. 7 Logical schematic of reversible multifunction block

be a single layered design. Truth table of Reversible Multifunction Block is as shown in Table 1. Inputs of RMFB are A, B, C, D and P, Q, R, S are the outputs. This block doesn't have information loss as all the input combinations are available at the output side and duplication of sequence is not there. If two RMFBs are connected in cascade, the same input combinations will be available on outputs A, B, C, D. In other words if outputs are considered as inputs, Block is capable to provide outputs where the inputs are located in the circuit. Hence the proposed

Table 1 Input-output mapping of reversible multifunction block

S.N.	Inputs (4 bits) decimal	Outputs (4 bits) decimal
1	0	0
2	1	2
3	2	4
4	3	6
5	4	7
6	5	5
7	6	3
8	7	1
9	8	14
10	9	13
11	10	10
12	11	8
13	12	9
14	13	11
15	14	12
16	15	15

block is proved to be reversible and as it can perform multiple operations, it is known as multifunctional block.

5 Simulation Result

QCA based circuits and applications are simulated using QCADesigner version 2.0.3 [13].

Layout of Reversible multifunction Block is shown in Fig. 8. Instead of using multiple layers in QCADesigner tool, the use of 45° rotated cells are used with normal cells. As the number of layers increased the complexity of designing and manufacturing increases, so these 45° rotated cells are used for drawing the layout of RMFB.

In proposed reversible logic gate, four figure of merits i.e. various parameters of RMFB is studied as shown in Table 2.

Proposed 4 × 4 RMFB gate can perform total 13 functions. In reversible circuits or gate if the output is same as input, it is known as propagate. It can be use as a fanout. Hence propagate is not considered as a garbage output.

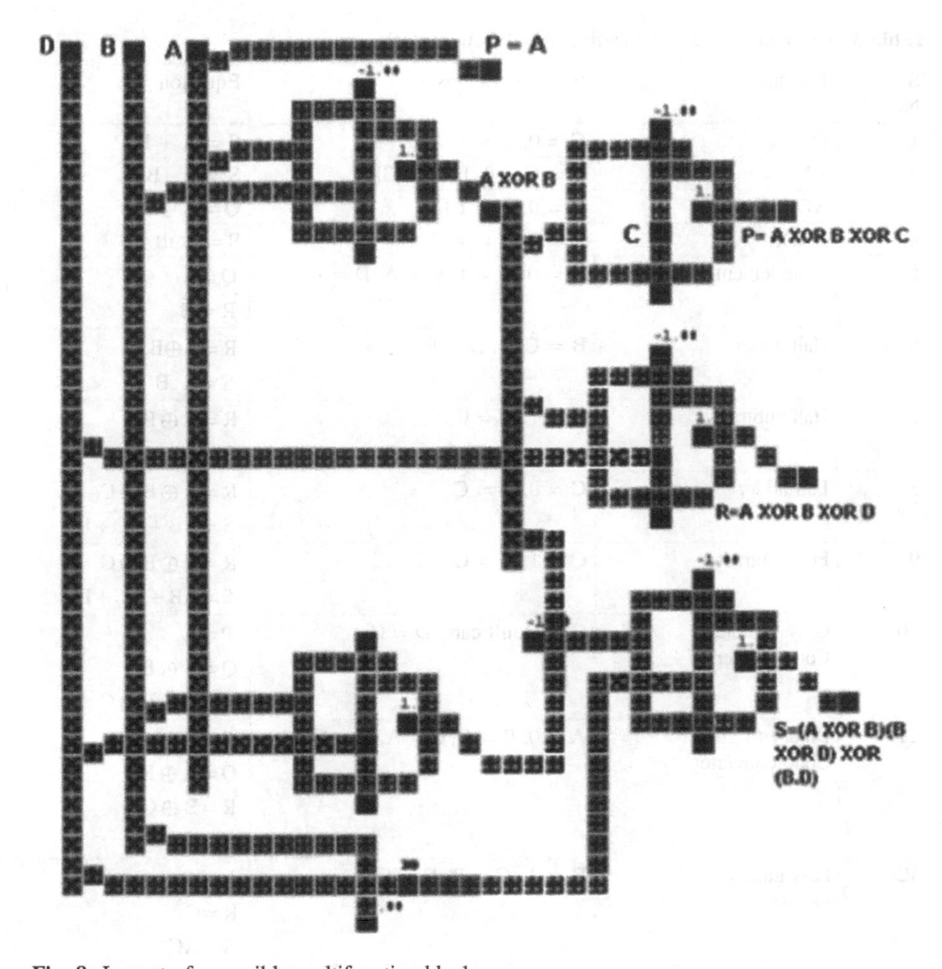

Fig. 8 Layout of reversible multifunction block

Table 2 Simulation parameters and figure of merits of reversible multifunction block

S.N.	Parameters	RMFB
1	Cell count (Complexity)	452
2	Latency (Delay)	1.75
3	Number of basic QCA Gates	22
4	Area	0.93 μm^2
5	Quantum cost	08
6	No. of quantum gates	06
7	Garbage output of RMFB using basic reversible gates	04
8	Constant input	01

Table 3 Functional table of reversible multifunction block

S. No.	Function	Input conditions	Equation
1	OR	C = 0, D = 1	$S = A + B$
2	AND	B = C = 0, D = B [3]	$S = A . B$
3	XOR	C = 0, D = 1 [3]	$Q = A \oplus B$
4	XNOR	C = 0, D = 1 [3]	$R = A \odot B$
5	Complement/Negation	A = 0, B = 1, C = A, D = B	$Q = \bar{A}$ $R = \bar{B}$
6	Half adder	B = C=0, D = B	$R = A \oplus B$, $S = A . B$
7	Half subtractor	C = 1, D = 0	$R = A \oplus B$, $S = \bar{A}B$
8	Full adder	C = 0, D = C	$R = A \oplus B \oplus C$, $S = AB + AC + BC$
9	Full subtractor	C = 1, D = C	$R = A \oplus B \oplus C$, $S = \bar{A}B + \bar{A}C + BC$
10	Gray to binary Code converter	C = Don't care, D = C	$P = A$, $Q = A \oplus B$, $R = A \oplus B \oplus C$
11	Binary to gray Code converter	A = 0, B = B, C = A, D = C [3]	$P = \text{INPUT A}$, $Q = A \oplus B$, $R = B \oplus C$, $S = B$
12	Pass gate/buffer/ Fanout	B = A, C = B, D = C	$P = A$, $Q = B$, $R = C$, $S = AC$
13	Set, reset function	A = 0, B = 1, C = A, D = B	$p = 0$ $S = 1$

6 Conclusion

4 × 4 RMFB is proved to be efficient as discussed in Tables 2 and 3. Various parameters like total count of logic gates required and the number of garbage outputs while designing various digital applications are less. Quantum cost of proposed 4 × 4 reversible logic gate is only 8, which is very less in contrast to the previously presented 4 × 4 Reversible gates. Proposed design shows 66 % improvement in garbage output, 50 % improvement in the constant inputs. QCA layout of 4 × 4 RMFB is capable of performing total 13 functions. As shown in Table 2, for the implementation of 4 × 4 RMFB, area required is only 0.93 μm^2. This block is proved faster as its latency is 1.75. The reversible logic block and

several applications and functions of the block is proposed in this block. Block is designed and implemented in this paper which outlines the basis for a reversible nature of Arithmetic Logic Unit of a quantum Nanoprocessors.

References

1. Orlov, Amlani, H. Lent, G.L, "Realization of a functional cell for quantum-dot cellular automata." Science 277, 928–930 (1997).
2. C.S. Lent, P.D. Tougaw, W. Porod, G.H. Bernstein, Quantum cellular automata, Nanotechnology 4 (1) (1993) 49–57.
3. International Technology Roadmap for Semiconductors, *Process Integration Devices and Structures (PIDS)*, http://www.itrs.net/Links/2011ITRS/Home2011.htm, 2011 Edition.
4. Kai-Wen Cheng and Chien-Cheng Tseng, "Quantum full adder and subtractor", Electronics Letters, 38(22):1343–1344, Oct 2002.
5. K. Kim, K. Wu, R. Karri, 'The Robust QCA Adder Designs Using Composable QCA Building Blocks', IEEE Trans on Computer-Aided.
6. J. Timler and C. S. Lent, "Power gain and dissipation in quantum-dot cellular automata", J. Appl. Phys., vol. 91, no. 2, pp. 823–831, 2002.
7. Vassilios A. Mardiris. "Design and simulation of modular 2n to 1 quantum-dot cellular automata (QCA) multiplexers", International Journal of Circuit Theory and Applications.
8. Manisha G. Waje, and P. K. Dakhole. "Design and implementation of 4-bit arithmetic logic unit using Quantum Dot Cellular Automata." In 3rd IEEE International Advance Computing Conference (IACC), 2013, pp. 1022–1029.
9. J. Timler and C. S. Lent, "Power gain and dissipation in quantum-dot cellular automata", J. Appl. Phys., vol 91, no 2, pp 823–831, 2002.
10. C.S. Lent, P.D. Tougaw, "A device architecture for computing with quantum dots", Proceedings of the IEEE 85 (4) (1997) 541–557.
11. Tougaw PD, Lent CS, "Logical devices implemented using quantum cellular automata", Journal of Applied Physics 1994; 75(3):1818–1825. doi:10.1063/1.356375.
12. Manisha Waje and Dr. Pravin Dakhole. Design and Simulation of Single Layered Logic Generator Block using Quantum Dot Cellular Automata, IEEE, International Conference on Pervasive Computing -2015, Pg. 1–6, doi:10.1109/PERVASIVE.2015.7087101.
13. QCADesigner. http://www.qcadesigner.ca/.
14. Manisha G. Waje, and P. K. Dakhole, "Design and simulation of new XOR gate and code converters using Quantum Dot Cellular Automata with reduced number of wire crossings", 2014 International Conference on Circuit, Power and Computing Technologies (ICCPCT), March 2014, Pages 1245–1250, IEEE.
15. Van R. Landauer, "Irreversibility and heat generation in the computational process", IBM J. Research and Development, 5:183–191, December 1961.
16. C. Bennett., "Logical reversibility of computation", IBM J. Research and Development, 17:525–532, November 1973.
17. D. Maslov and M. Saeedi, "Reversible circuit optimization via leaving the Boolean domain. "Computer-Aided Design of Integrated Circuits and Systems", IEEE Transactions on, 30 (6):806 –816, june 2011.
18. H. Thapliyal and N. Ranganathan. "A new design of the reversible subtractor circuit", 11th International Conference on Nanotechnology, pages 1430–1435, Portland, Oregon, August 2011.
19. M. Mohammadi and M. Eshghi., "On figure of merit in reversible and quantum logic designs", Springer science, Quantum Information Processing, 8(4):297–318, August 2009.
20. Peres, A, "Reversible logic and quantum computers", Phys. Rev. A 32, 3266–3276, 1985.

21. E.Fredkin and Toffoli., "Conservative logic", International Journal of Theor. Physics, 21:219–253, 1982.
22. Tommaso Toffoli, "Reversible computing" Technical Report, MIT Laboratory for Computer Science, 1980.
23. D. Maslov and D. M. Miller, "Comparison of the cost metrics for reversible and quantum logic synthesis", Feb 2008, Pg. 1–10.

Text-Independent Automatic Accent Identification System for Kannada Language

R. Soorajkumar, G.N. Girish, Pravin B. Ramteke, Shreyas S. Joshi and Shashidhar G. Koolagudi

Abstract Accent identification is one of the applications paid more attention in speech processing. A text-independent accent identification system is proposed using Gaussian mixture models (GMMs) for Kannada language. Spectral and prosodic features such as Mel-frequency cepstral coefficients (MFCCs), pitch, and energy are considered for the experimentation. The dataset is collected from three regions of Karnataka namely Mumbai Karnataka, Mysore Karnataka, and Karavali Karnataka having significant variations in accent. Experiments are conducted using 32 speech samples from each region where each clip is of one minute duration spoken by native speakers. The baseline system implemented using MFCC features found to achieve 76.7 % accuracy. From the results it is observed that the hybrid features improve the performance of the system by 3 %.

Keywords Speech processing · Dialect identification · MFCCs · Gaussian mixture models · Regional language processing

1 Introduction

Automated speech recognition (ASR) system recognizes dictation which involves the ability to match a voice pattern against already stored vocabulary. One of the main factors that impacts ASR systems is the accent. When a dialect is identified only in

R. Soorajkumar (✉) · G.N. Girish · P.B. Ramteke · S.G. Koolagudi
National Institute of Technology Karnataka, Surathkal, India
e-mail: soorajkumarbhat@gmail.com

G.N. Girish
e-mail: girishanit@gmail.com

P.B. Ramteke
e-mail: ramteke0001@gmail.com

S.S. Joshi
B.V. Bhoomraddi College of Engineering and Technology, Hubli, Karnataka, India
e-mail: koolagudi@nitk.edu.in

S.C. Satapathy et al. (eds.), *Proceedings of the International Conference on Data Engineering and Communication Technology*, Advances in Intelligent Systems and Computing 469, DOI 10.1007/978-981-10-1678-3_40

terms of pronunciation and prosody, the term accent may be preferable over dialect. Accent plays an important role in the classification of a speaker location, social class, and ethnicity. The dialect identification in Indian regional languages is a challenging task due to multilingual speakers across the country. Kannada is a language spoken in India, predominantly in the state of Karnataka along with the border regions of neighboring states such as Maharashtra, Tamil Nadu, Kerala, Andhra Pradesh, Goa, etc. Hence Kannada language have a dominant influence from these neighboring regions results in accent variation in these regions. Based on this Karnataka have different regions, namely, Mysore Karnataka, Mumbai Karnataka, Karavali Karnataka, and Southern Karnataka, where the language pronunciation properties are influenced by the other states. Even though Kannada is the official language of Karnataka; the accent, lexicon, and prosody of Kannada language varies across these regions.

The variation in accent affects the performance of speech recognition system as there is significant deviation in the pronunciation with respect to actual pronunciation. Hence results in degradation of the performance of the recognition. The regional influence in pronunciation may also affect nonnative language learning such as English, Hindi, etc. Where the pronunciation is affected by the substitution of native phoneme in place of actual phoneme or the actual phonemes are influenced by the native language phoneme. Hence accent identification is more essential in improving the performance of these systems. Based on the accent in pronunciation, it is very easy to identify the identity of a person and his/her native region. Hence this work aims at developing a model for accent identification in Kannada language.

This paper is organized as follows: Sect. 2 briefly outlines the literature in this area. Section 3 explains the proposed approach. The results are discussed in Sect. 4. Section 5 concludes the paper with some research directions.

2 Literature Survey

In recent years, automatic accent recognition and classification systems have become one of the popular research areas. Blackburnn et al. [1] proposed a text-dependent accent classification system for English language using continuous speech. Speakers are chosen from three regions having mother tongue Arabic, Chinese, and Australian, respectively. Hansen et al. [2] explored role of prosodic features in accent classification. Pitch contours and spectral structure for a sequence of accent sensitive words are considered as a feature. An accent sensitive database with four different accents is collected from American English speakers with the foreign accent. Accuracy reported for text-independent and text-dependent system is 81.5 and 88.9 %, respectively. Teixeira et al. [3] used parallel set of ergodic nets with context-independent hidden Markov models (HMMs) for accent classification in English language across six different European countries. The approach claimed 65.48 % accuracy. Hemakumar and Punitha [4] proposed text-dependent approach for speaker accent identification and isolated Kannada word recognition using Baum–Welch algorithm and Normal fit method. The average word recognition rate (WRR) claimed for known

Kannada accent speaker is 95.5 % and unknown Kannada accent speaker is 90.86 %. Another HMM based text-dependent accent classification system classifies the native Australian and English speakers from migrant speakers with foreign accent [5]. The average accent classification accuracy reported are 85.3 and 76.6 % using HMM and HMM+LM, respectively.

Vieru et al. [6] employed machine learning techniques for the analysis of variation in the accent of French native speakers and migrated speakers. The French speech database is collected from the speakers of six different countries speaking French language: Arabic, English, German, Italian, Portuguese, and Spanish are considered for the experimentation. Hamid Behravan et al. proposed a hybrid approach for foreign accent identification in English language using phonetic and spectral features. The extracted speech features modeled with i-vector. The proposed hybrid approach found to outperform the spectral-based systems. Some of the approaches have used Gaussian Mixture Model (GMM) as a classifier for accent identification [7]. Lazaridis et al. [8] proposed an automatic accent identification system using a generative probabilistic framework classification based on GMMs with two different GMM-based algorithms; first is using Universal Background Model (UBM) and Maximum A Posteriori (MAP) and the second is i-vector modeling. Experiments are conducted on Swiss and French language. The accent dataset is collected from four different regions of Switzerland, i.e., Geneva (GE), Martigny (MA), Neuchatel (NE), and Nyon (NY). i-vector-based model is claimed to outperform baseline model. Accent identification systems is also evaluated on English language, other languages like Chinese, Japanese, Korean, and other foreign languages [9]. Accent identification on Indian regional languages such as Hindi, Tamil, Telugu, Punjabi, and Kannada are not explored much. Kumari et al. [10] proposed an accent identification system for Hindi language using MFCC features using hidden Markov model (HMM). An identification system for Telugu language implemented using prosodic and formant features [11]. The classification is performed using nearest neighborhood classifier. Experiments are conducted on speech samples recorded from Coastal Andhra, Rayalaseema, and Telangana regions of Andra Pradesh of India. The system is claimed to achieve 72 % accuracy. In acoustic phonetic-based accent identification system [12], the speech data collected from two regions of Karnataka, namely, South Karnataka often referred as Mysore Karnataka and Coastal Karnataka are used. Where speech recordings consist of 200 Kannada proverbs. HMM is used as a classifier. First and second formant frequencies extracted from vowel mid points are used as features for classification.

From the literature it is observed that most of the approaches are text dependent. These approaches do not consider the variations in the nature of the language due to accent. Also, a large number of word forms are inflected by Kannada language because of variation in accent. An inflected word starts with a root and may have several suffixes added to the right. Hence, it is difficult to consider all these variations for accent identification. In this paper, a text-independent accent identification system is proposed for the Kannada language.

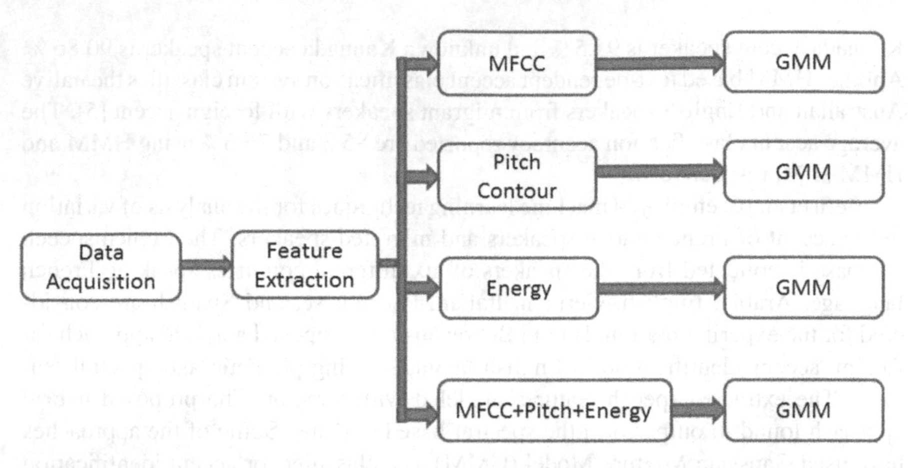

Fig. 1 Block diagram of accent identification system using GMM

3 Methodology

The proposed approach is divided into four steps as database collection, feature extraction, accent classification using GMMs as shown in Fig. 1.

3.1 Database

Accent classification study has been confined to sentences read by male, female, and child speakers from three different regions of Karnataka namely Mysore Karnataka, Mumbai Karnataka, and Karavali Karnataka (refer Table 1). For experimentation, natural database is required, hence the speech is recorded without the knowledge of the speakers. Recording such conversations is a very challenging task due to the background noise and other problem like superposition of voice of two different speakers talking simultaneously. For each region, data samples are collected from 32 speakers where each clip is of 1 min duration.

Table 1 Details of the database considered for accent classification

Sr. no	Region	Number of samples	Age range	Gender
1	Mysore Karnataka	32	15–70	Male, female
2	Mumbai Karnataka	32	10–45	Male, female
3	Karavali Karnataka	32	25–70	Male, female

3.2 Feature Extraction

For the efficient implementation of any system, selection of proper features is an essential task. This section describes the features considered for the experimentation and their significance with respect to accent.

Mel-Frequency Cepstral Coefficient (MFCCs): MFCCs [13] are the most widely used features for the speech recognition. It mimics the human auditory system, i.e., it follows linear scale up to 1 KHz and logarithmic scale beyond it. The region of a person can be easily identified by listening to his/her speech, as there is a variation in the accent of people from different regions. Hence, MFCCs may represent the information in different accents and considered for the classification. 13 MFCC features are considered for the experimentation.

Following are the steps for the computation of MFCC Features:

1. Pre-emphasis: The goal of pre-emphasis is to amplify the high frequency formats and suppress the low frequency formats.
2. Frame blocking: The input speech signal is divided into frames of size 20–30 ms, as speech is assumed to be constant in this range.
3. Hamming windowing: In order to keep the continuity of the first and last frames, each frame is multiplied with a hamming window, i.e., $s(n) * w(n)$, where $s(n)$ is a signal frame and $w(n)$ denotes the hamming window. n varies from $n = 0, 1, 2, \ldots, N-1$,
4. Fast Fourier transform (FFT): The magnitude of frequency and frequency response of each frame is obtained using Fast Fourier Transform.
5. Triangular Bandpass filter: It is used to reduce the size of features involved in signal, and smooth the magnitude spectrum such that the harmonics are flattened to obtain the envelop of the spectrum with harmonics.
6. Discrete Cosine Transform (DCT): By applying DCT on the 20 log energy (E_k) obtained from the triangular bandpass filters to have K Mel-scale Cepstral Coefficients.
7. Log Energy: Log energy is the 13th feature of MFCC.

Energy: Short time energy [13] is one of the important characteristics of speech signal. The language spoken in different accents may reflect variation in the energy, hence it considered as a feature for accent detection. Formula for calculating energy $E(n)$ is given below:

$$E(n) = \sum_{m=-\infty}^{\infty} (s(m) * w(k-m))^2, \quad (1)$$

where $w(n)$ represent the windowing function and k is the shift in number of samples, $s(n)$ denotes the signal frame.

Pitch: Pitch is rate of vocal fold vibration and represents the fundamental frequency 'F0.' Pitch contains the speaker-specific information [13]. It may vary for the pronunciation of the same word in different accent of the language and may capture the accent specific information. Hence pitch is considered as a feature for the classification.

4 Results and Discussion

In this section, the results obtained using the MFCCs, energy, pitch, and combination of these features are presented. The database consists of samples recorded from native Kannada language speakers from three different regions, namely, Mumbai Karnataka, Mysore Karnataka, and Karavali Karnataka) (Refer Sect. 3). Gaussian mixture model(GMM) is used for accent classification. It is an unsupervised classification algorithm which does not require any transcriptions like supervised HMM, hence for text-independent accent identification system it is used as a classifier. Nineteen samples are considered for training the model, and 13 samples are used for testing from each region.

As MFCCs imitates the human auditory system, GMMs are trained to develop a baseline system using MFCCs. Table 1 shows the confusion matrix of classification using MFCC features. Out of 39 samples, 30 samples are correctly identified, hence achieves 76.7 % accuracy. Misclassification in each region is due to the variation in the speech of each speaker. Table 2 shows the confusion matrix of the accent classification using pitch feature. Out of 39 samples, 21 samples are correctly classified. The average accuracy achieved is 53.8 %. Table 3 shows the confusion matrix of classification using energy. Total 17 samples are correctly classified and the accuracy reported is 43.4 %.

From Tables 3 and 4, it is observed that Mysore Karnataka region accent is overlapped with Mumbai Karnataka and Karavali Karnataka region as speakers selected for recording from middle Karnataka usually interact with the people from both the

Table 2 Confusion matrix of the baseline system using MFCC features

	Mysore Karnataka	Mumbai Karnataka	Karavali Karnataka
Mysore Karnataka	9	2	2
Mumbai Karnataka	0	13	0
Karavali Karnataka	2	3	8

Table 3 Confusion matrix of the accent classification using pitch feature

	Mysore Karnataka	Mumbai Karnataka	Karavali Karnataka
Mysore Karnataka	9	3	1
Mumbai Karnataka	2	11	0
Karavali Karnataka	7	5	1

Table 4 Confusion matrix of the accent classification using energy feature

	Mysore Karnataka	Mumbai Karnataka	Karavali Karnataka
Mysore Karnataka	7	4	2
Mumbai Karnataka	3	10	0
Karavali Karnataka	7	6	0

Table 5 Confusion matrix of the accent classification system using MFCCs + energy + pitch feature

	Mysore Karnataka	Mumbai Karnataka	Karavali Karnataka
Mysore Karnataka	9	2	2
Mumbai Karnataka	0	13	0
Karavali Karnataka	2	2	9

regions and have influence of these accent. It is observed that the recognition rate for Mumbai Karnataka accent is higher than other regions, as the energy and pitch of the speakers from these regions is higher than other two regions. The influence of Hindi and Marathi language in these region speakers (e.g., Mumbai Karnataka and middle Karnataka speakers pronounce plate as "taat" which is similar to Marathi language. In case of Mysore Karnataka the speakers from South Karnataka region pronounces plate as "tatte" which is similar to Karvali Karnataka). In Karavali Karnataka region, people speech have low energy and low pitch compared to other regions. It is also influenced by the local languages such as Tulu and Konkani. Hence, results in overlapping of the accent with other regions.

Table 5 shows the confusion matrix of classification using (13) MFCCs + (1) pitch + (1) energy features. The accuracy achieved is 79.5 % accuracy. It is observed that MFCCs achieves better results than pitch and energy features. Using hybrid features (MFCCs, pitch and energy) the accuracy is found to increase by 3 % as it captures the variation in the three regions of Karnataka.

5 Conclusion and Future Work

A text-independent accent classification system for Kannada language is proposed for regions, namely Mysore Karnataka, Mumbai Karnataka and Karavali Karnataka. MFCCs, energy, and pitch are considered as features. GMM is used as a classifier. The hybrid features have improved the accuracy of the classification by 3 %. There is a scope to evaluate the role of various spectral, prosodic, and excitation features in improving the classification accuracy. Also, this work can be extended for the identification of language like Havyaka Kannada in different regions of Karnataka. Such a study will not only enhance our knowledge of dialectal variations of Kannada, but also helps in developing an accent-based speech recognition systems.

References

1. C. Blackburn, J. Vonwiller, and R. King, "Automatic accent classification using artificial neural networks." in *Eurospeech*, 1993.
2. J. Hansen and L. Arslan, "Foreign accent classification using source generator based prosodic features," in *Acoustics, Speech, and Signal Processing, 1995. ICASSP-95., 1995 International Conference on*, vol. 1, May 1995, pp. 836–839 vol.1.
3. C. Teixeira, I. Trancoso, and A. Serralheiro, "Accent identification," in *Spoken Language, 1996. ICSLP 96. Proceedings., Fourth International Conference on*, vol. 3. IEEE, 1996, pp. 1784–1787.
4. G. Hemakumar and P. Punitha, "Speaker accent and isolated kannada word recognition," *American Journal of Computer Science and Information Technology (AJCSIT)*, vol. 2, no. 2, pp. 71–77, 2014.
5. K. Kumpf and R. W. King, "Automatic accent classification of foreign accented australian english speech," in *Spoken Language, 1996. ICSLP 96. Proceedings., Fourth International Conference on*, vol. 3. IEEE, 1996, pp. 1740–1743.
6. B. Vieru, P. B. de Mareil, and M. Adda-Decker, "Characterisation and identification of non-native french accents," *Speech Communication*, vol. 53, no. 3, pp. 292–310, 2011. Available: http://www.sciencedirect.com/science/article/pii/S0167639310001615
7. T. Chen, C. Huang, E. Chang, and J. Wang, "Automatic accent identification using gaussian mixture models," in *Automatic Speech Recognition and Understanding, 2001. ASRU'01. IEEE Workshop on*. IEEE, 2001, pp. 343–346.
8. A. Lazaridis, E. Khoury, J.-P. Goldman, M. Avanzi, S. Marcel, and P. N. Garner, "Swiss french regional accent identification."
9. Q. Yan and S. Vaseghi, "Analysis, modelling and synthesis of formants of british, american and australian accents," in *Acoustics, Speech, and Signal Processing, 2003. Proceedings.(ICASSP'03). 2003 IEEE International Conference on*, vol. 1. IEEE, 2003, pp. I–712.
10. P. Kumari, D. Shakina Deiv, and M. Bhattacharya, "Automatic speech recognition of accented hindi data," in *Computation of Power, Energy, Information and Communication (ICCPEIC), 2014 International Conference on*. IEEE, 2014, pp. 68–76.
11. K. Mannepalli, P. N. Sastry, and V. Rajesh, "Accent detection of telugu speech using prosodic and formant features," in *Signal Processing And Communication Engineering Systems (SPACES), 2015 International Conference on*. IEEE, 2015, pp. 318–322.
12. K. S. Nagesha and G. H. Kumar, "Acoustic-phonetic analysis of kannada accents," *Tata Institute of Fundamental Research, Mumbai*.
13. L. Rabiner and B.-H. Juang, "Fundamentals of speech recognition," 1993.

A Study on the Effect of Adaptive Boosting on Performance of Classifiers for Human Activity Recognition

Kishor H. Walse, Rajiv V. Dharaskar and Vilas M. Thakare

Abstract Nowadays, all smartphones are equipped with powerful multiple built-in sensors. People are carrying these "sensors" nearly all the time from morning to night before sleep as they carry the smartphone all the time. These smartphone allow the data to be collected through built-in sensors, especially the accelerometer and gyroscope give us several obvious advantages in the human activity recognition research as it allow the data to be collected anywhere and anytime. In this paper, we make use of publicly available dataset online and try to improve the classification accuracy by choosing the proper learning algorithm. The benchmark dataset considered for this work is acquired from the UCI Machine Learning Repository which is available in public domain. Our experiment indicates that combining AdaBoost.M1 algorithm with Random Forest, J.48 and Naive Bayes contributes to discriminating several common human activities improving the performance of Classifier. We found that using Adaboost.M1 with Random Forest, J.48 and Naive Bayes improves the overall accuracy. Particularly, Naive Bayes improves overall accuracy of 90.95 % with Adaboost.M1 from 79.89 % with simple Naive Bayes.

Keywords Human activity recognition (HAR) · Smartphone · Sensor · Accelerometer · Gyroscope

K.H. Walse (✉)
S.G.B. Amravati University, Amravati 444601, India
e-mail: walsekh@acm.org

R.V. Dharaskar
DMAT-Disha Technical Campus, Raipur 492001, India
e-mail: rajiv.dharaskar@gmail.com

V.M. Thakare
P.G. Department of CS, S.G.B. Amravati University,
Amravati 444601, India
e-mail: vilthakare@gmail.com

S.C. Satapathy et al. (eds.), *Proceedings of the International Conference on Data Engineering and Communication Technology*, Advances in Intelligent Systems and Computing 469, DOI 10.1007/978-981-10-1678-3_41

1 Introduction

Human activity recognition is active research area since last two decades. Image- and video-based human activity recognition has been studied for a long time [1]. Nowadays, all smartphones are equipped with powerful multiple built-in sensors. People are carrying these "sensors" nearly all the time from morning to night before sleep as they carry the smartphone all the time. These smartphones allow the data to be collected through built-in sensors, especially the accelerometer and gyroscope give us several obvious advantages in the human activity recognition research as it allows the data to be collected anywhere and anytime. Increasingly powerful mobile phones, with built-in sensors such as accelerometer, gyroscope, microphone, GPS, Bluetooth, camera, etc., smartphone becomes sensing platform [2]. The information about mobile device and user activity, environment, other devices, location and time can be utilized in different situations to enhance the interaction between the user and the device [3]. Human activity has enabled applications in various areas in health care, security, entertainment, personal fitness, senior care, and daily activities [4]. Smartphone allows us to collect real data from real-world daily activities. Lab controlled data is not that represent the real world daily activities. The rest of the paper is organized as follows: Sect. 2 talks about the related work done by the earlier researchers. Section 3 emphasizes on the data collection. Section 4 presents the main approach we have utilized in the learning process. This section describes the performance matrix used to present our results. Section 6 presents and discusses our simulation results. Finally, we conclude our work in Sect. 7.

2 Related Work

In this section, we briefly describe the related literature. While studying the related literature within past few decades, many significant research literatures have been contributed, proposing and investigating various methodologies for human activity recognition such as from video sequences, from on body wearable sensors data or from sensors data on mobile device.

Human activity recognition is an active research area since last two decades. Image- and video-based human activity recognition has been studied since a long time [5]. Human activity recognition through environmental sensors is one of the promising approaches. In this approach, it uses motion sensors, door sensors, RFID tags, and video cameras, etc., to recognize human activities. This method provides high recognition rate of human activity and can be more effective in indoor environments (i.e., smart home, hospital and office), but requires costly infrastructure [6, 7]. In another approach, multiple sensors wear on human body which uses human activity recognition. This approach has also proved to be effective sensors for human activity recognition [5]. But this type of sensing has drawbacks that user has to wear lot of sensors on his body.

In the literature, extensive study also found on the approach in which available smart mobile devices used to collect data for activity recognition and adapt the interfaces to provide better usability experience to mobile users. We found few similar studies to the one proposed in the paper [8, 9]. Mobile device becomes an attractive platform for activity recognition because they offer a number of advantages including not requiring any additional equipment for data collection or computing. These devices saturate modern culture, they continue to grow in functionality increasing security and privacy issues [10].

3 Data Collection

In the proposed research work, we have molded a context recognition problem as a six class classification problem. The benchmark dataset considered for this work is acquired from the UCI machine learning repository which is available in public domain [11]. In this dataset, the accelerometer and gyroscope sensor recordings are available for six human activities. The six activities are: walking, sitting, standing, stair-up, stair-down, and lying. After surveying the extensive literature, the methodology is adopted for the system. The system consists of acquisition, preprocessing, feature extraction, feature selection, classification, knowledge base, inference engine, action base, and finally user interface adaptation.

The feature extraction, we have used are statistical and transformed based features extracted. The Classifier is designed with the help of Adaboost.M1 with J.48, Naive Bayes and Random Forest.

3.1 Benchmark Dataset

The benchmark dataset considered for this work is acquired from the UCI machine learning repository which is available in public domain [11]. This dataset is recorded with the help of accelerometer and gyroscope sensors. In this dataset, the experiments were performed with 30 subjects. Each subject wore a smartphone on the waist and performed six activities. Six activities are: walking, sitting, standing, stair-up, stair-down, and lying. With the sampling rate of 50 Hz, the raw data from built-in 3-axial accelerometer and gyroscope were captured. In each record of the dataset following attributes are included:

- Total acceleration from the 3-axis accelerometer and the estimated body acceleration.
- Angular velocity from the 3-axis gyroscope.
- Total 561 features generated from raw data in time and frequency domain.
- Label for corresponding activity

Table 1 Summary of features extraction methods used for accelerometer and gyroscope signals

Group	Methods
Time domain	Mean, SD, MAD, Max, Min, SMA, Energy, Entropy, IQR, Auto regression coefficient, Correlation, Linear acceleration, Angular velocity, Kurtosis, Skewness,
Frequency domain	FFT, mean frequency, index of frequency component with largest magnitude

1. Features are normalized and bounded within [–1,1].
2. Each row is representing feature vector.
3. The unit for acceleration is m/seg2.
4. The gyroscope unit is rad/seg.

3.2 Feature Extraction

The features selected for this dataset come from raw data captured from the accelerometer and gyroscope raw signals. These 3-axial raw signals are time domain signals prefix ‘t’ to denote time. Before selecting features, raw sensor signals were preprocessed by using a median filter and a third-order low pass Butterworth filter to remove noise. Another low pass Butterworth filter with a corner frequency of 0.3 Hz had been used to separate body and gravity acceleration signals from the acceleration signal. Finally, frequency domain signals were obtained by applying a fast Fourier transform (FFT) [11]. To get feature vector for each pattern, these time domain and frequency domain signals were used. The set of variables that were estimated from these signals are shown in Table 1.

4 Classification

The AdaBoost Algo is shown in Fig. 1.

4.1 Adaboost.M1

How to improve the overall accuracy of any learning algorithm will be always an issue of research in soft computing. Boosting is a general technique for improving the accuracy of any learning algorithm [12]. Freund and Schapire [12] introduced adaptive boosting (Adaboost) first time in which boosting is to iteratively learn weak classifiers that focus on different subsets of the training data and combine these into one strong classifier. In Weka Adaboost.M1 is implemented.

Algorithm AdaBoost.M1

Input: Sequence of N examples $\langle (x_1, y_1) .. (x_N, y_N) \rangle$ with labels $y_i \in Y = \{1, ...k\}$

distribution D over the N examples
weak learning algorithm **Weak Learn**

integer T specifying number of iterations

initialize the weight vector: $D_i(i) = 1/m$ for all i.

Do for t= 1,2,..,T

1. Call **Weak Learn,** providing with it the distribution D_t

2. Get back a hypothesis $h_t : X \rightarrow Y$.

3. Calculate the error of $h_t : \varepsilon_t = \sum_{i:h_t(x_i) \neq y_i} D_t(i)$. if $\varepsilon_t \rangle 1/2$, then set T = t-1 and abort loop.

4. Set $\beta_t = \varepsilon_t / (1 - \varepsilon_t)$

5. Update distribution $D_t : D_{t+1}(i) = \frac{D_t(i)}{Z_t} \times \begin{cases} \beta_t & \text{if } h_t(x_i) = y_i \\ 1 & \text{Otherwise} \end{cases}$

where Zt is normalization constant chosen so that D_{t+1} will be distribution

Output the hypothesis:

$$h_{fin}(x) = \arg\max_{y \in Y} \sum_{t:h_t(x)} \log \frac{1}{\beta_t}$$

Fig. 1 A first multiclass extension of AdaBoost.M1 [12]

5 Performance Measures

To assess the neural network performance the following performance measures are used.

5.1 *Confusion Matrices*

The most common way to express classification accuracy is the confusion matrix which is used to display the percentage of accuracy of classification. In the confusion matrix, rows indicate desired or actual classification and columns indicate predicted classifications. Figure 2 shows the confusion matrix for two-class classification problem.

Fig. 2 Sample confusion matrix

True Positive	False Positive
False Negative	True Negative

5.2 *Overall Accuracy*

To access the performance of classifier, the overall accuracy is the important measure. It is the ratio between the total number of correctly classified instances and the overall test set. This is a more informative measure.

$$\text{Accuracy} = \frac{\text{TruePositive} + \text{TrueNegative}}{\text{TruePositive} + \text{TrueNegative} + \text{FalsePositive} + \text{FalseNegative}} \quad (1)$$

5.3 *Precision*

It is the ratio between correctly classified instances to the total number of positive instances.

$$\text{precision} = \frac{\text{TruePositive}}{\text{TruePositive} + \text{FalsePositive}} \quad (2)$$

6 Experiment Results

Finally we used Weka tool to evaluate the performance of the classifiers. Activity recognition on these features was performed using the J.48, Random Forest and Naive Bayes classifiers. Classifier were trained and tested with tenfold cross validation. The confusion matrix was shown in Tables 2 and 3 of NaiveBayes classifier without Adaboost and with Adaboost.M1, respectively.

Tables 2 and 3 show that the misclassification rate was reduced drastically in case Adaboost.M1 with NaiveBayes. So boosting helps to reduce the error and improve the classification performance.

Tables 4 and 5 for J.48 Classifier without and with Adaboost.M1. Similarly, Tables 6 and 7 show the confusion matrix for Random Forest without and with Adaboost.M1 (Table 8).

Table 2 Confusion matrix for NaiveBayes

Output/desired	1	2	3	4	5	6
1	**1250**	283	189	0	0	0
2	37	**1395**	112	0	0	0
3	96	244	**1066**	0	0	0
4	0	25	0	**1575**	143	34
5	2	33	0	1202	**643**	26
6	0	16	1	143	0	**1784**

Table 3 Confusion Matrix for Naive Bayes with Adaboost.M1

Output/desired	1	2	3	4	5	6
1	**1625**	52	45	0	0	0
2	1	**1521**	11	0	0	0
3	12	41	**1353**	0	0	0
4	0	4	0	**1465**	281	27
5	4	11	0	423	**1460**	8
6	0	1	0	0	0	**1943**

Table 4 Confusion Matrix for J.48

Output/desired	1	2	3	4	5	6
1	**1638**	52	31	0	1	0
2	59	**1440**	45	0	0	0
3	35	54	**1317**	0	0	0
4	0	1	0	**1636**	140	0
5	0	1	0	142	**1763**	0
6	0	0	0	1	0	**1943**

Table 5 Confusion Matrix for J.48 with Adaboost.M1

Output/desired	1	2	3	4	5	6
1	**1704**	9	9	0	0	0
2	7	**1527**	9	0	1	0
3	2	17	**1387**	0	0	0
4	0	1	0	**1686**	90	0
5	0	0	0	67	**1839**	0
6	0	0	0	1	0	**1943**

Table 6 Confusion Matrix for Random Forest

Output/desired	1	2	3	4	5	6
1	**1704**	11	7	0	0	0
2	6	**1524**	14	0	0	0
3	7	25	**1374**	0	0	0
4	0	1	0	**1703**	72	1
5	0	0	0	60	**1846**	0
6	0	1	0	0	0	**1943**

Table 7 Confusion Matrix for Random Forest with Adaboost.M1

Output/desired	1	2	3	4	5	6
1	**1706**	7	9	0	0	0
2	3	**1529**	12	0	0	0
3	10	27	**1369**	0	0	0
4	0	1	0	**1703**	71	2
5	0	0	0	57	**1849**	0
6	0	1	0	0	0	**1943**

Table 8 Summary of classifier performance with Adaboost.M1

	RF	RF with Adaboost	J.48	J,48 with Adaboost	NB	NB with Adaboost
% Accuracy	98.01%	98.06%	94.54 %	97.93%	74.89%	90.95%
Kappa statistic	0.9761	0.9766	0.9344	0.9751	0.6987	0.8912
MAE	0.047	0.0466	0.0191	0.0069	0.0836	0.0431
RMSE	0.1056	0.1053	0.1328	0.0778	0.2876	0.1503
RAE (%)	16.95	16.83	6.90	2.48	30.18	15.56
RASE (%)	28.38	28.29	35.69	20.90	77.26	40.39
Coverage (%)	100	100	95.07	98.64	75.72	98.79
MRRS (%)	30.53	30.41	17.01	16.90	16.94	22.28
Total instances	10299	10299	10299	10299	10299	10299
Time taken	13.24	13.25	210	241	1.08	179.8

It was observed from the confusion matrix shown in Tables 4, 5, 6, and 7 that with the Adaboost.M1, misclassifications have reduced. Root mean square error also reduced in case of J.48 and Naive Bayes and percentage of accuracy as shown in Figs. 3 and 4, respectively.

In this paper, we have performed experiment with three classifiers. As shown in Fig. 4 and Table 9, we have received overall accuracy of 74.89 % with Naive-Bayes, 94.54 % overall accuracy with J.48, a decision tee classifier. With random forest, we have received overall accuracy of 98.01 %, random forest. While using adaptive boosting meta-classifier available in Weka tool, we have received overall accuracy of 90.98 % of Adaboost.M1. meta classifier with NaiveBayes. The overall accuracy of 97.93 % from Adaboost.M1 with J.48 decision tree. There is no much improvement of using adaboost.M1 with random forest where we received overall accuracy of 98.06 %. While comparing our results with the existing state of arts, Sharma et al. [13] achieved 84 % overall accuracy using neural network on data from a triaxial accelerometer sensor. Ronao and Cho received overall accuracy of 91.76 % by using two-stage continuous hidden Marko Model (CHMM) approach to recognize human activity using accelerometer and gyroscope sensory data collected

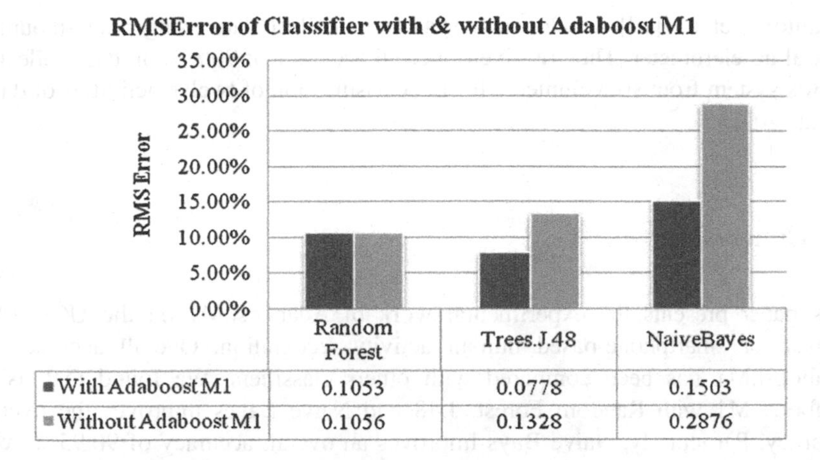

Fig. 3 of Accuracy of classifiers with and without Adaboost.M1

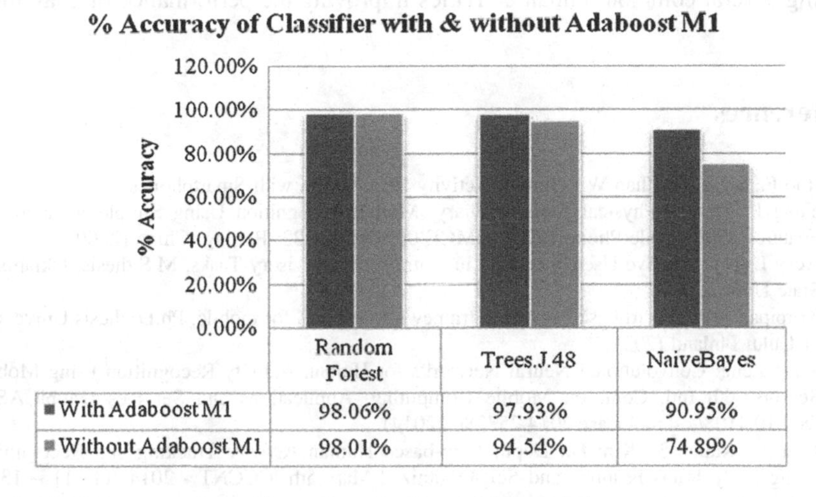

Fig. 4 % of Accuracy of classifiers with and without Adaboost.M1

through smartphone [14]. Anguita et al. [15] achieved an overall accuracy of 96 % for classification of six activities of daily living using the multiclass SVM (MC-SVM). Hanai et al. [16] obtained a recognition performance of 93.9 % for classifying five activities of daily living from data of chest-mounted accelerometer.

Table 9 Summary of classifier performance using experimenter in Weka

Random Forest	Adaboost.M1 with RF	J.48	Adaboost.M1 with J.48	Naive Bayes	Adaboost.M1 with NB
98.05	98.05v	94.40v	97.97v	75.02*	90.95
(v//*)	(0/0/1)	(0/0/1)	(1/0/0)	(1/0/0)	(1/0/0)

Karantonis et al. [17] developed a system to collect data using waist-mounted triaxial accelerometer. They received over-all accuracy of 90.8 % on data collected by this system from six volunteers for the classification of twelve activities of daily living (ADL).

7 Conclusion

This paper presents the experimental work of Adaboost.M1 on the UCI HAR dataset for smartphone-based human activity recognition. Overall accuracy of Adaboost.M1 has been compared with others classifiers. We found that using Adaboost.M1 with Random Forest, J.48 and Naive Bayes improves the overall accuracy. Particularly, Naive Bays improves an overall accuracy of 90.95 % with Adaboost.M1 from 79.89 % with simple Naive Bayes. We found that AdaBoost. M1 algorithm with Random Forest, J.48 and Naive Bayes contributes to discriminating several common human activities improving the performance of Classifier.

References

1. Rao F., Song Y., Zhao W.: Human Activity Recognition with Smartphones
2. Yang J.: Toward Physical Activity Diary: Motion Recognition Using Simple Acceleration Features with Mobile Phones, ACM IMCE'09, October 23; Beijing, China (2009)
3. Acay L. D.: Adaptive User Interfaces in Complex Supervisory Tasks, M.S.thesis, Oklahoma State Univ; (2004)
4. Korpipaa: Blackboard-based software framework and tool for mobile, Ph.D. thesis.University of Oulu, Finland (2005)
5. Ming Zeng: Convolutional Neural Networks for Human Activity Recognition using Mobile Sensors, 6th Intl. Conf. on Mobile Computing, Applications and Services (MobiCASE) DOI:10.4108/icst.mobicase.2014.257786 (2014)
6. Jalal A., Kamal S., Kim D.: Depth Map-based Human Activity Tracking and Recognition Using Body Joints Features and Self-Organized Map, 5th ICCCNT - 2014 July 11 – 13, 2 Hefei, China (2014)
7. Dernbach S., Das B., Krishnan N. C., Thomas B.L., Cook D.J. Simple and Complex Activity Recognition through Smart Phones. *Intelligent Environments (IE) 8th International Conference* 214–21. doi:10.1109/IE.2012.39 (2012)
8. Walse K.H., Dharaskar R.V., Thakare V.M.: Frame work for Adaptive Mobile Interface: An Overview. IJCA Proceedings on National Conference on Innovative Paradigms in Engineering and Technology (NCIPET 2012)14; 2012:27–30 (2012)
9. Walse K.H., Dharaskar R.V., Thakare V.M.: Study of Framework for Mobile Interface. IJCA Proceedings on National Conference on Recent Trends in Computing NCRTC9; 14–16 (2012)
10. Rizwan A., Dharaskar R.V.: Study of mobile botnets: An analysis from the perspective of efficient generalized forensics framework for mobile devices. IJCA Proceedings on National Conference on Innovative Paradigms in Engineering and Technology (NCIPET 2012) ncipet 15; 2012: pp. 5–8 (2012)

11. Anguita D.: A Public Domain Dataset for Human Activity Recognition Using Smartphones. 21th European Symposium on Artificial Neural Networks, Computational Intelligence and Machine Learning, ESANN-2013. Bruges, Belgium (2013)
12. Y. Freund and R. E. Schapire. A decision-theoretic generalization of on-line learning and an application to boosting. In Proceedings of the Second European Conference on Computational Learning Theory, 1995
13. A. Sharma, Y.-D. Lee, and W.-Y. Chung: High accuracy human activity monitoring using neural network," International Conference on Convergence and Hybrid Information Technology, pp 430–435 (2008)
14. Ronao, C.A.; Sung-Bae Cho: Human activity recognition using smartphone sensors with two-stage continuous hidden Markov models.*Natural Computation (ICNC), 2014 10th International Conference on*, vol., no., pp. 681–686, 19-21 Aug. 2014 doi:10.1109/ICNC.2014.6975918
15. D. Anguita, A. Ghio, L. Oneto, X. Parra, and J. L. Reyes-Ortiz: A public domain dataset for human activity recognition using smartphones," European Symposium on Artificial Neural Networks (ESANN), pp. 437–442 (2013)
16. Y. Hanai, J. Nishimura, and T. Kuroda: Haar-like filtering for human activity recognition using 3d accelerometer. In Digital Signal Processing Workshop and 5th IEEE Signal Processing Education Workshop, 2009. DSP/SPE 2009. IEEE 13th,pages 675–678 (2009)
17. D.M. Karantonis, M.R. Narayanan, M. Mathie, N.H. Lovell, and B.G. Celler: Implementation of a real-time human movement classifier using a triaxial accelerometer for ambulatory monitoring. IEEE Transactions on Information Technology in Biomedicine 10(1); 156–167 (2006)

Toward Improved Performance of Emotion Detection: Multimodal Approach

R.V. Darekar and A.P. Dhande

Abstract Emotion detection currently is found to be an important and interesting part of speech analysis. The analysis can be done by selection of an effective parameter or by combination of a number of parameters to gain higher accuracy level. Definitely selection of a number of parameters together will provide a reliable solution for getting higher level of accuracy than that of for the single parameter. Energy, MFCCs, pitch values, timbre, and vocal tract frequencies are found to be effective parameters with which detection accuracy can be improved. It is observed that results with the language are proportional with results with other languages indicating that language will be an independent parameter for emotion detection. Similarly, by addition of an effective classifier like neural network can further yield the recognition accuracy nearly to 100 %. The work attempts to interpret the fact that combining the results of each parameter has improved detection accuracy.

Keywords Feature extraction · Combination · Energy · Pitch · MFCC · Emotion detection

1 Introduction

Speech processing can be divided as speaker identification, speech recognition, and emotion detection [1]. Emotion detection is the part of speech processing in which information about the emotional content of the speaker can be retrieved. The emotional analysis will be helpful for getting the psychological condition of the

R.V. Darekar (✉)
Department of Electronics and Telecommunications Engineering,
Dr. D. Y. Patil Institute of Engineering and Technology, Pune, Maharashtra, India
e-mail: ravirajdarekar@gmail.com

A.P. Dhande
Department of Electronics and Telecommunications Engineering,
Pune Institute of Computer Technology, Pune, Maharashtra, India
e-mail: apdhande@pict.edu.in

S.C. Satapathy et al. (eds.), *Proceedings of the International Conference on Data Engineering and Communication Technology*, Advances in Intelligent Systems and Computing 469, DOI 10.1007/978-981-10-1678-3_42

person who is speaking. Knowing the same, one can communicate more effectively than that of, without knowing the same. It can also lead to an effective communication between man and machine if the machine is capable of understanding the emotions of the person in interaction. Speech recognition can find the exact words spoken by the person [2]. It needs a huge vocabulary which will help for getting the exact words being spoken. Emotion detection will not need such a vocabulary set but will need the threshold levels of range of values of some advanced parameters like MFCCs, energy, duration which can be framed for a certain emotion like happy, angry, sad, neutral, fear, or surprised.

Here some of hidden parameters like MFCCs, timbre, etc., can be found more important than that of the spoken words. The research work for adding emotions to a neutrally spoken speech sentence or removing emotions from sentence with emotion is an example of application of emotion detection. A spoken paragraph in happy emotion can be converted into neutral type or that neutral type can be converted to fear type of emotion content. Emotion detection may also be useful for knowing the psychological condition of the called person with which a company representative would like to communicate for business deal. This is possible only by knowledge about compressing, expanding, or changing attributes or specific parameters for the required change [3].

For the similar type of work with increased accuracy, emotion detection can be considered to be a basic step. The task can be completed with the knowledge of a single or by using a number of parameters together. Definitely, the detection accuracy will be higher if a number of parameters are selected for detection. For instance if energy content is the only parameter under selection, conflicts will be more as higher energy will be indication for happy, surprised, or angry emotions. If two or more parameters are selected for detection, a firm decision can be taken for exact emotion to be detected by virtue of different threshold levels of different parameters under observation. As different parameters will provide information in respective angle, the combination will give rise to higher accuracy. For detection of emotion, there is need of some effective parameters like energy, formant frequencies, vocal tract frequencies, MFCCs, pitch levels, timbre, duration, etc., that can provide efficient information about emotions from speech files. The work presented here is by use of database with one of the Indian language Marathi of 1200 speech files recorded by male and female professional actors for six different emotions happy, angry, neutral, sad, fear, and surprised. In near future, it may also lead to designing a system with capability to work by understanding human emotions.

Emotion detection is found to have increased accuracy level due to a number of factors. The job can be accomplished using image signal, speech signal, or the combination of both for better accuracy level [4]. In image processing, there is minimum scope for work under hidden parameters. The detection will be concentrated on observable parameters. This is not the case with speech signal. Speech signals consist of some additional information apart from the spoken information. Detection of emotion using speech is carried out specifically by retrieval of hidden information using advanced parameters as discussed above. Recently, considerable attention has been drawn by automated emotion detection with human speech and has found huge range of applications. The maximum work of emotion detection is carried over for number of languages like U. S. English, German, and so on, but a very less work is seen on Indian languages. Here, the language under study is Marathi as one of the Indian language. The language under study, i.e., Marathi is found to be the nearest language to south zone consisting of the languages from Dravidian family and can be grouped with Panjabi, Gujarati, and Hindi when grouping Indian languages using machine learning techniques [5]. Out of a number of Indian languages, Marathi is the language which has a special feature observed that, the language differs much by a distance of 20 km. It is also found to have a flexibility on language with a variety of tones and intensities. It can be added here that the language changes the meaning of a single sentence by adding a small pressure on a specific word.

2 Collection of Speech Samples

The database is one of the most important parts of the overall task. For the analysis here, a team of professional actor and actress was defined who have worked on effective emotions in Marathi. A database of 1200 audio files from male and female professional actor and actress is derived and has created 100 speech emotion files of male and female, for six emotions under study, i.e., happy, angry, sad, surprised, fear, and neutral. These speech files are of the duration 3–5 s. The voice segments are recorded in mp3 as well as wave file. The sampling rate is 44100 Hz, with stereo channel and bit rate of 128 kbps.

The reason behind forming such a database is clear that—the strength of database for different emotions will give rise to higher level of accuracy as if speech signal of sad signal may also indicate neutral or fear emotion [6] [7].

Different emotions are created with male and female professional actors and actress and emotions of six types, viz., happy, angry, sad, fear, surprise, and neutral are considered for analysis. The strength and naturalness of the emotions define the accuracy of the system [8].

The summary of implementation done is

1. Preprocessing: the conversion of a single speech input into pair wise speech blocks, noise reduction, and speech splitting.
2. Segmentation—the audio segmentation algorithm employed for splitting input audio into samples ready to be processed.
3. Feature selection: Introduces feature selection algorithm used to reduce number of features.
4. Feature extraction: MFCC, pitch, and energy and vocal tract frequency

The detection of emotion from speech recorded is accomplished by this stage gives features extracted.

Front-end interface: Describes the implementation decisions for the front-end interface for visualizing the output.

Training and classification stage.

2.1 Algorithms in Work Done

The algorithm for the program consists of following stages:

1. Input audio/speech file as input
2. Retrieve effective parameters
3. Calculate MFCC, energy, VT frequency, and pitch value.
4. Store all values in excel sheet format.
5. Define logic to classify emotion
6. Indicate emotion effectively using defined logic.

3 Parametric Analysis

The design can be initiated by the selection of parameters, analyzing their results for different emotions and threshold selection of these parameters effectively so as to get correct level of accuracy in Marathi, which is the language under study. By coding, a set of comparative results can be obtained for different emotions like happy, anger, sad, neutral, etc. [9]. Threshold selection is range definition of minimum and maximum values for any specific parameter under selection. Initially, number of parameters can be considered for analysis of emotion detection like formant frequencies, loudness, timber, etc. Out of these, some of the parameters can be selected for further analysis found to be the effective for emotion detection by results of the analysis. It forms the basis of the good recognition accuracy for emotion detection for any language [10]. Most of the work seen yet today on emotion detection is found to have results derived by a single efficient parameter like MFCC, Pitch, formants, etc. Those systems are found to have an accuracy level up to 70–80 %. The concept of blending or fusion of results of the effective parameters is planned here for getting higher level of accuracy. By scaling of the results of these parameters, the most effective parameter can be scaled higher and the least with least value. Surely it will give rise to an accuracy level above 95 % approaching toward maximum level of accuracy.

The work can be said to help or reference for any other languages also because language is found to have independent over the emotions generated or detected. This is because; any emotion generated in any language may have similar emotion levels for a number of effective parameters. For instance, for angry emotion or sad emotion, though words are different but the energy levels, Pitch values, and similar parameters will be proportional which are the points of our interest here for emotion detection [11]. To get higher level of recognition accuracy, energy, MFCC, and pitch can be fused together. The database formed consists of emotion files of happy, angry, sad, fear surprise, and neutral. Using parameters like Pitch, energy, formant frequencies, speech rate, MFCC and similar, performance of system will increase in terms of accuracy of emotion detection and system will provide higher efficiency. Classifier selection is one of the important tasks. By calculating speech parameters, corresponding results can be provided to the classifier [12]. The database formed extracted feature set and classifier defines the accuracy and effectiveness of the system.

Starting with about 10 parameters, we have concluded to use four effective parameters, viz., energy, Pitch, and MFCC features for detection of emotion. These parameters are found to have good accuracy levels in comparison with other parameters. The strength and the effectiveness of emotions of recorded audio files to be given as input are the main important points for higher level of accuracy as only these files will provide the source input of emotions to be detected. Effectiveness in emotions from audio files will give rise to higher efficiency of system.

4 Comparison of Results

Results of the feature extraction are visualized as

4.1 Vocal Tract Frequency

The results of vocal tract frequencies as shown are as below indicating the emotions with discriminating vocal tract frequencies (Fig 1).

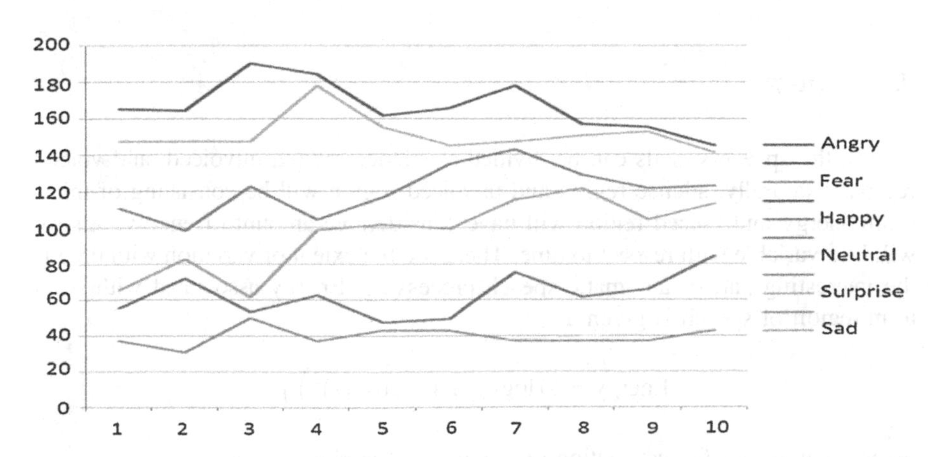

Fig. 1 Graph of vocal tract frequency values for different emotions with emotion files on horizontal axis

The overview of statistical study of analysis indicates that emotions can be found to have distinct values with respect to other emotions in concern with a specific parameter like energy. Mean, variance, and deviation can be found to have a clear distinction for emotions under study.

4.2 Pitch

Basically pitch can be defined as the frequency sensation. Pitch can only be determined in sounds that have a frequency that is stable and clear to separate it from noise. For a sound wave, a lower frequency wave is low pitch wave and higher frequency sound indicates a high pitch wave. Considering loudness, timber, and duration, the major auditory attribute is pitch. Semitones correlate linearly to human's "perceived pitch."

$$\text{Semitone} = 69 + 12\log_2(\text{frequency}/440)$$

For females, the range of pitch values is 100–1000 Hz or 40–80 semitones where as for males, its 60–520 Hz or 30–70 semitones. The human ear has the capability to recognize pitch values and related frequency of the audio signal as longitudinal signals develop variations in pressure at a specific frequency. The frequency of audio signal entering the human ear provides its capability to perceive the pitch value associated with frequency. Figure 2 indicates the ranges of pitch values for different emotions expressing that different emotions will have a spectrum of pitch values. With the frames of speech, recording of the bandwidth and first five formants can be done. Emotional extraction can be done successfully by using formants. To model the change in the vocal tract shape, tracking formants over time can be used.

4.3 Energy

Any of the speech signals can be divided as silence region, unvoiced, and voiced regions. Naturally, silence region and unvoiced region will be consisting of minimum energy and voiced region will have considerable amount of energy. Also it will be a variable with respect to time. Therefore the extent of variation with time is the interesting part of automatic speech processing. Energy associated with short term region of speech is given as

$$\text{Energy} = 20\log_{10}\left(\text{Sum}\left(\text{abs}(y_1)^2\right)\right)$$

where y_1 indicates the correlation of the energy function.

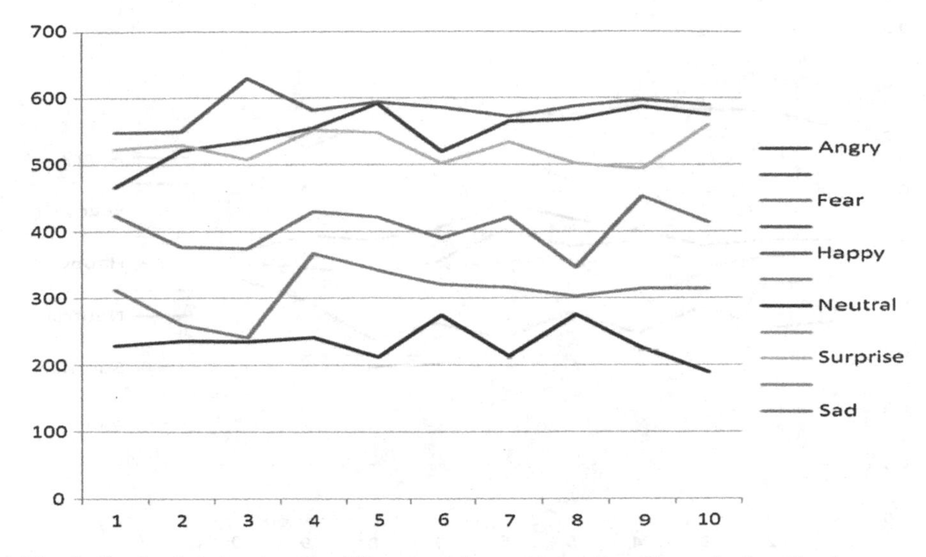

Fig. 2 Graph of pitch values for different emotions with emotion files on horizontal axis

Total energy relation for signal processing as indicated in Fig. 3 provides the relation for finding short time energy. The relation for finding the short term energy can be derived from the total energy relation defined in signal processing as indicated in Fig. 3. For voiced, unvoiced, and silence classification of speech, short term energy can be used.

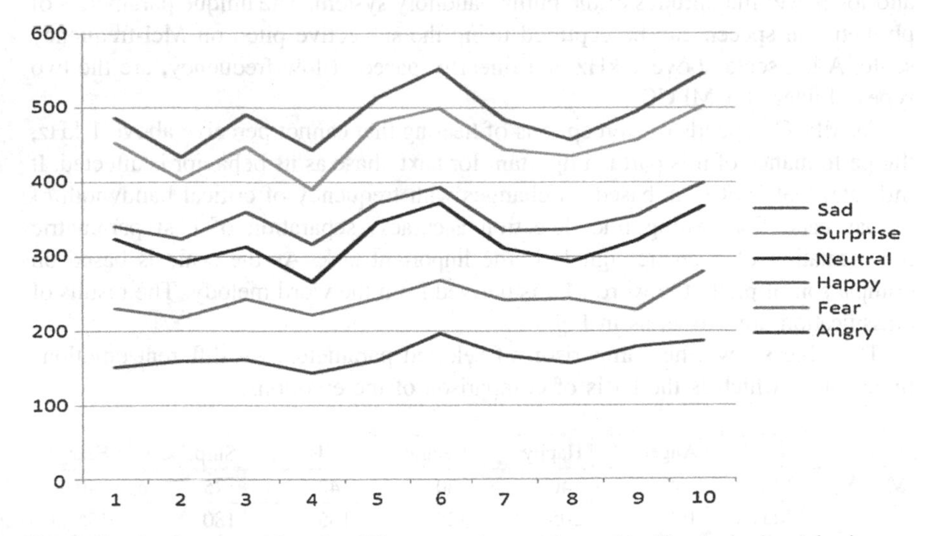

Fig. 3 Graph of energy values for different emotions with emotion files on horizontal axis

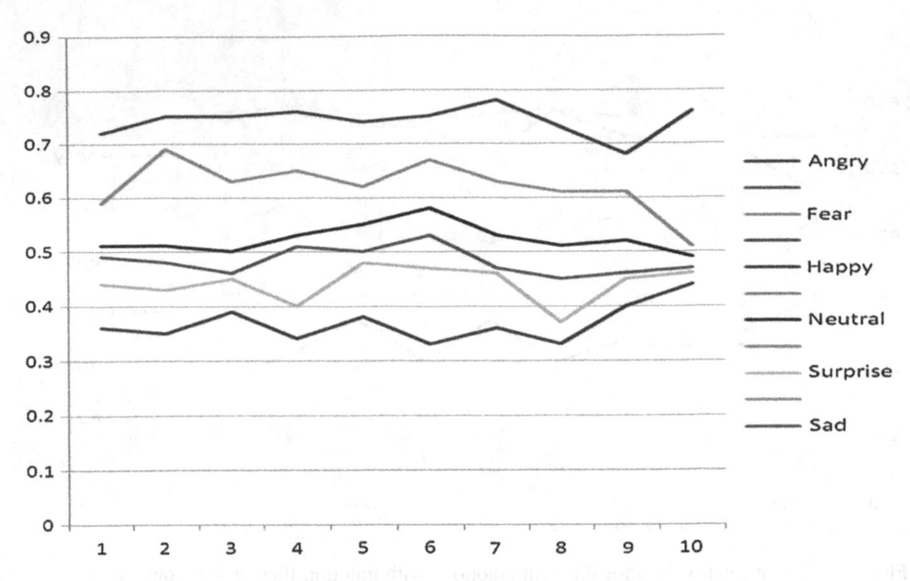

Fig. 4 Graph of MFCC values for different emotions with emotion files on horizontal axis

4.4 MFCC

These model the spectral energy distribution in a perceptually meaningful way, are widely used acoustic feature for speech recognition, speaker recognition, emotion detection, and audio classification. MFCC takes into account a band of frequencies and log power magnitudes of the human auditory system. The unique parameters of phonetics in speech can be captured using the subjective pitch on Mel-frequency scale. A log scale above 1 kHz and linearly spaced at low frequency, are the two types of filters for MFCC.

As MFCC depends on perceptions of hearing that cannot perceive above 1 kHz, the performance of this part is important for next phase as its behavior is affected. It indicates that MFCC is based on changes with frequency of critical bandwidth of human ears. For getting fine detection accuracy, separation of best parametric representation of acoustic signals is the important task. As the scale is based on comparison of pitch, the word Mel is derived from the word melody. The results of classification are shown as in Fig. 4.

The table shows the comparison of selected parameters for different emotions under study which is the basis of comparison of the emotions.

		Anger	Happy	Neutral	Sad	Surprise	Fear
VT. Fr	Min	55	40	30	42	75	40
	Max	190	205	125	155	180	155
Pitch	Min	335	325	225	260	290	145

(continued)

(continued)

		Anger	Happy	Neutral	Sad	Surprise	Fear
	Max	600	590	650	445	560	725
MFCC	Min	0.55	0.25	0.35	0.42	0.4	0.5
	Max	0.8	0.58	0.56	0.58	0.62	0.7
Energy	Min	125	65	-55	50	110	55
	Max	200	130	35	115	145	140

The decibels of 40 db above the perceptual hearing threshold are defined as 1000 Mels, as a reference point the pitch of a 1 kHz tone. We can use the equation below, to compute the Mels for a given frequency in Hz.

$$\mathrm{Mel}(F) = 2595 \times \log 10(1 + f/700)$$

Mel scale is used for measurement of subjective pitch and Hertz is the unit for measurement of actual frequency. A continuous speech signal is transformed in frequency domain after windowing, the logarithm of the Mel spectrum after inverse transform provides MFCC that are utilized here for classification of emotions.

Now, the logarithmic Mel-frequencies can be converted back to time. The result is called MFFC. These coefficients of the audio signal can provide a fine output for spectral properties of the analysis of selected frame as the coefficients are real numbers. By the use of direct cosine transform, these features can be converted to time domain [13]. Matching of patterns and signal modeling are the fundamental operations performed by emotion detection. Speech signal is converted into a set of parameters, is represented by signal modeling whereas the task of selecting the correct parameter from the database that approximately matches with the input pattern is called as pattern matching. The emotional content of the speaker, i.e., aspects of behavior from audio signal can be categorized after these two operations.

Different values of parameters like energy, formant frequencies, etc., provide ranges so that we can correlate them for getting higher accuracies for detection of emotions. For, e.g., if happy emotion indicates a specific range of values, it can be used as the general range for detecting happy emotion from unknown signal. Different threshold values can be defined to assign the specific range of values for a particular parameter.

5 Discussion

The design part is started with the calculation of energy levels, pitch values, MFCCs, and vocal tract frequencies. Parameters have been retrieved with the help of efficient programming. Collection of these values provides the range of values for a particular emotion for specific parameter. These ranges have provided the path forward for classification of different emotions. The basic of the whole research part is to be utilized for further processing. One can easily find here that any specific parameter is having different values of it but possess a specific range which can be utilized for detection of emotion. That is, for energy parameter, different speech signals may have different values in dB but energy levels for happy, sad, angry, fear, surprise, and neutral may have a particular range. This fact can be utilized for emotion distinction [14].

a< energy < b && c < MFCC < d

&& e< pitch < f && g < VT freq. < h

Result ==➔ **Indication of a specific emotion**

Along with the same, it is also observed that happy-angry-surprise and sad-fear-neutral emotions have closer ranges of values but they have distinct pitches or distinct MFCC values or distinct vocal tract frequencies. Using this factor, above sets can be separated using the ANDING of parameters values for different ranges so as to get unique emotion from the analysis.

The actual requirement of such a database is to get sufficient values of coefficients to compare different emotions. Also the huge database will also give rise to get the accurate value mean of the parameter of audio files. To avoid limited database and incorrect mean value, large database will generate effective values for getting desired accuracy for emotion detection. The classification done here is as

From the ANDING conditions, it is clear that a specific emotion will be detected if and only if all the above conditions are satisfied. If any of the above four parameters is not in the given range, it will not indicate the emotion under that specific category. The threshold levels defined are from the minimum and maximum values of the results of speech files for a selected parameter. It can be noted that for the corner values of some speech files, for energy, pitch, MFCC, or vocal tract frequency, about 10 % values are found with incorrect results which will be improved by a suitable classifier. The database newly formed is found to have many variations in different parameters but can be adjusted effectively to get proper results by setting of the threshold levels appropriately [15]. Using the database, minimum and maximum value of the range for any emotion file can be defined. When all the speech file ranges will be defined, we can have the higher accuracy for emotion detection. Also the accuracy can also be improved by addition of neural network.

This will be one of the most important steps of the overall design part as the criteria with which emotions will be detected is building up. Also here, this point can be noted that considering energy as one of the parameter emotions like happy, angry, and surprise will fall under one category and sad, fear, and neutral emotions will be falling under other category. Thus for first level, six emotions can be divided into set of 3–3 emotions. Further similar analysis will help us to distinguish exact emotion out of the set of three with the help of formant frequencies, MFCC, pitch, or similar parameters. All the parameters retrieved from the audio file given as input, collected together for analysis. The suitable coding will provide criteria to retrieve exact emotion.

The threshold values defined are considered from the values of result analysis. They are required to be varied as per the requirement to get higher accuracy. It is also found that independent of any language, emotions generated will be similar though words, sentences are different [16].

From the flowchart and code executed, we have formed tables from all the derived values of speech files. Along with the same, we have noted the minimum and maximum values of the specific range. From the graph shown, though all the emotion values are looking to be mixed, still execution have made it clear that all parameter will fall under a specific emotion result with approx. 90 % accuracy. It is because of the fact that four different parameters selected will not have mixed values. Due to this factor, emotions can be detected efficiently.

From the charts and discussion, it can be concluded that detection efficiency can be increased when different parameters are 'ANDED' together. The threshold levels decided can be retrieved from the tables of values from the analysis. The success of emotion detection is also dependent on the quality or naturalness in audio files recorded, appropriate selection of features and suitable classification range. The combination of results for a single parameter will boost the detection accuracy because for final detected emotion, result will be dependent on the results of 2–4 parameters to yield higher accuracy. This paradigm of combination of speech parameters has improved the detection accuracy by 20 %. By getting higher level of accuracy, research also may forward further for detection of unspoken emotions. It can be the blend of different parameters for emotion detection using speech as well as image to get the desired results [17].

Comparing with other methods, it is seen that most of the work done is by considering a single or two parameters for detection of emotion using speech. Many researchers have worked on the combination of speech and image as the input source. Here, the main emphasis is given by considering about four effective parameters for emotion detection. Also results are fused or combined so as to get higher accuracy. Work accomplished here is concentrated on one of the best Indian language—Marathi in concern with emotion detection.

6 Conclusion

Emotion detection yields good recognition accuracy by selection of suitable parameters like pitch, formant frequencies, energy of signal, MFCCs, and loudness. Accuracy can be increased by combination of results of different parameters together.

By analysis of all the parameter values, using a classifier, if suitable scaling is applied to those parameters which are showing good recognition rates, the detection accuracy can be improved further. The combination of result or suitable classifier can be planned for this purpose which will scale the parameters effectively and improve recognition rate at least by 15 to 20 %. The current recognition rate is approx. 80 % which can be improved by using a scaling method or by selection of suitable classifier, up to 90–95 %. The increase in accuracy is observed because of the fact that multimodal fusion of speech parameters increases the detection accuracy as higher weight can be added to more efficient parameters and least to those having minimum accuracy found. Strong database of speech samples will still improve the recognition accuracy as well as it will make the system for any unknown input signal which can be detected for emotion with improved recognition accuracy.

Energy, pitch, MFCCs, or vocal tract frequencies are independent of spoken words, i.e., spoken words indicate minimum or no impact on results of emotion detection. An effective system will detect emotions of an unknown speaker without consideration of the language he/she is speaking. Language can be considered to be independent for emotion detection as for any language, the parameters of analysis will be unchanged though words are changing.

References

1. L. Rabiner, B. H. Juang, "Fundamentals of Speech Recognition", Pearson Education, 200.
2. Madhavi S. Pednekar, Kavita Tiware and Sachin Bhagwat, "Continuous Speech Recognition for Marathi Language Using Statistical Method", IEEE International Conference on "Computer Vision and Information Technology, Advances and Applications~, ACVIT-09, December 2009, pp. ISSN 2319–7080 International Journal of Computer Science and Communication Engineering Volume 3 issue 1(February 2014 issue) 45 810–816.
3. J Xu, H Zhou, G-B Huang, in *Information Fusion 2012 15th International Conference On* Extreme learning machine based fast object recognition (IEEE, Singapore, 2012), pp. 1490–1496.
4. L.R. Rabiner, "A tutorial on hidden markov models and selected applications in speech recognition", In proc. of the IEEE, Vol. 71, no. 2, pp. 227–286, Feb 1989.
5. A.B. Kandali, A.B. Routray, Basu T.K., "Emotion Recognition From Assamese Speeches Using M FCC And GM M Classifier", IEEE Region Conference TEN CON 2008, India, pp. 1–5.
6. A Khan, A Majid, A Mirza, Combination and optimization of classifiers in gender classification using genetic program., IOS press. Int. J. Knowl. Based Intell. Eng.Syst. 9, 1–11 (2005)
7. Yashpalsing D. Chavhan and M.L. Dhore, "Speech Emotion Recognition using SVM" IEEE International Conference on „Computer Vision and Information Technology, Advances and Applications ~ , ACVIT-09, December 2009, pp. 799–804.
8. Montero, J. M., Gutiérrez-Arriola, J., Palazuelos, S.,Enríquez, E., Aguilera, S., & Pardo, J. M., "Emotional Speech Synthesis: From Speech Database to T-T-S", ICSLP 98, Vol. 3, p. 923–926. Burkhardt, F., "Simulation emotional ersprechweise mit Sprach syntheseverfahren" [Simulation of emotional manner of speech using speech synthesis techniques], PhD Thesis, TU Berlin, 2000.
9. ISCA Workshop on Speech & Emotion, p. 151–156.
10. Vroomen, J., Collier, R., & Mozziconacci, S. J. L., "Duration and Intonation in Emotional Speech", Eurospeech 93, Vol. 1, p. 577–580.

11. Montero, J. M., Gutiérrez-Arriola, J., Colás, J., Enríquez,E., & Pardo, J. M., "Analysis and Modeling of Emotional Speech in Spanish", ICPhS 99, p. 957–960.
12. Murray, I. R., Edgington, M. D., Campion, D., & Lynn., " Rule-based Emotion Synthesis Using Concatenated Speech", ISCA Workshop on Speech & Emotion, 2000, p. 173–177.
13. MA Zissman, Comparison of four approaches to automatic language identification of telephone speech. IEEE Transactions on Speech and Audio Processing. 4(1), 31 (1996).
14. AF Martin, CS Greenberg, in *Odyssey*. The 2009 nist language recog. evaluation (ISCA, Brno, Czech, 2010), p. 30.
15. A.Falaschi, M.Guistianiani, M.Verola, "A hidden markov model approach to speech synthesis", In proc. of Eurospeech, Paris, France, 1989, pp 187–190.
16. S. Martincic- Ipsic and I. Ipsic, "Croatian H M M Based Speech Synthesis," 28th Int. Conf. Information Technology Interfaces ITI 2006, pp. 19–22, 2006, Cavtat, Croatia.
17. Firoz Shah. A, Raji Sukumar. A, and Babu Anto. P, "Discreet Wavelet Transforms and Artificial Neural Networks for Speech Emotion Recognition", International Journal of Computer Theory and Engineering, Vol. 2, No. 3, 1793–8201, June 2010, pp. 319–322.

Priority Dissection Supervision for Intrusion Detection in Wireless Sensor Networks

Ayushi Gupta, Ayushi Gupta, Deepali Virmani and Payal Pahwa

Abstract Wireless sensor networks are prone to failure as a simple attack can lead to damage the entire setup. Ad hoc nature, limited energy, and limited network lifetime are the major reasons that expose WSN's to many security attacks. Intrusion detection is the most important method to intercept all these attacks. In this paper, we propose a new method called Priority Dissection Supervision (PDS) for intrusion detection in WSN. The proposed PDS uses fuzzy logic for prioritizing the impact of distributed intrusion over line of control (LOC). Proposed PDS is a coherent system which models the key elements sensor deployment, intrusion detection, tracking, and priority list. PDS makes use of weight and impact factors which detects the influence field of different types of intruders and accordingly calculates the amount of deterioration. By the use of PDS system the operational and deployment cost can be decreased significantly. Simulation results on MATLAB exhibit the potency and precision of the proposed PDS.

Keywords Priority · Intrusion detection · LOC · Fuzzy · WSN

1 Introduction

Wireless sensor networks (WSN) are used for monitoring physical or environmental conditions like sound, weight, vibrations, velocity, etc. There are defined spatially distributed autonomous which pass their data through the predefined links to the

Ayushi Gupta (✉) · Ayushi Gupta · Deepali Virmani · Payal Pahwa
Bhagwan Parshuram Institute of Technology, Sector-17, Rohini, Delhi 110085, India
e-mail: guptaayushi260894@gmail.com

Ayushi Gupta
e-mail: ayushigupta2995@gmail.com

Deepali Virmani
e-mail: deepalivirmani@gmail.com

Payal Pahwa
e-mail: pahwapayal@gmail.com

S.C. Satapathy et al. (eds.), *Proceedings of the International Conference on Data Engineering and Communication Technology*, Advances in Intelligent Systems and Computing 469, DOI 10.1007/978-981-10-1678-3_43

station and then the data is processed out [1]. A node in a sensor network is capable of information, processing it and communicating with other connected nodes in the network. Nowadays, these are progressively utilized in homeland security, disaster recovery, data logging, area monitoring, and many other disciplines. In accordance with the nature of the military operations, we introduce a set of sensor nodes to collect and operate the data. This acquired data must maintain its accuracy when passed to the evaluation center and must reach in real time. The main contribution of our work is that it demonstrates prioritized list of intrusion activities to be handled and to be responded accordingly.

2 Literature Review

WSN has been utilized for solving challenges like energy efficiency, security, communication, and hardware reliability in military border patrol systems. A variety of different deployment approaches of WSN have been used in border surveillance and intrusions detection. Among the existing prototypes for deployment approach, unmanned aerial vehicles (UAVs) have been used to automatically detect and track illegal border crossing [2]. UAVs provide high mobility and coverage but these have to be paired with the Seismic and Fiber Optic Sensors and are quite costly in terms of deployment. Unlike the wired sensors, there are unattended ground sensors (UGSs) which can detect any kind of vibration/seismic activity or magnetic impact, which indicates crossing of vehicles or infantry at the LOC (line of control) [7–9]. The UGSs provide higher system robustness and are quite effective in detecting intrusion.

The Remotely Monitored Battlefield Sensor System (REMBASS) and Igloo White are the existing radio-based UGS used in intrusion detection [3]. But they have limited networking ability and communicate their sensor readings over relatively long and frequently unidirectional radio links to a central monitoring station.

3 Proposed Approach: Priority Dissection Supervision (PDS)

Comparing with existing intrusion detection techniques, PDS provides the following advantages:

- Ground sensors provide a benefit over multimedia sensors, e.g., [2] intruder hidden behind an obstacle can be detected by ground sensor that is not possible using imaging sensor.
- Underground sensors are less prone to attacks by intruders and provide accurate detection with large coverage area [2].
- Heterogeneous sensors simultaneously give an assurance of intrusion.

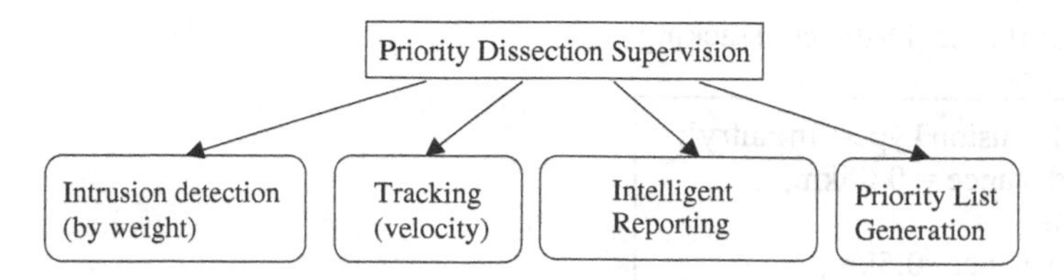

Fig. 1 Phases of proposed approach

Thus, using the proposed PDS system the operational and deployment cost can be decreased significantly [2]. Proposed PDS is divided in four stages as shown in Fig. 1.

3.1 Intrusion Detection

Intrusion detection works in two domains.

(1) **Deployment of Sensors**: The proposed PDS would be established for monitoring a large border area. But due to limited sensing radius of a sensor node, a large number of sensor nodes are required to be deployed to fulfill the coverage requirement. The sensor nodes will be embedded in a strip area at LOC (line of control) [2].

Supposition: Let S be the shortest path length of the region and R be the sensor's sensing range. Then, the required number of sensors to achieve k-barrier coverage in the region is [2].

$$k[D/R] \quad (1)$$

The proof of Supposition has been given in [4]. So, according to Supposition, given a border area with axial distance d, the minimum number of ground/underground sensors required to achieve k-barrier coverage is

$$k[D/R_{UGS}] \quad (2)$$

(2) **Intrusion Detection**: As the intruder enters the sensor coverage area, weight will be recorded [5, 6]. Classification of weight would be done using Algorithm 1.

Algorithm 1: Intrusion Type

```
if weight <= a
 IntrusionType = 'Infantry';
else if weight>a && weight<=b
 IntrusionType='LightWeighted';
else weight>b
 IntrusionType='HeavyWeighted';
```

Algorithm 2: Distance Tracking

```
if IntrusionType='Infantry';
  distance = 0.05km;
else
  distance=0.5km;
Time=distance*speed;
```

3.2 Tracking

Sensor nodes will detect infantry within the range of 50 meters and light weighted/heavy weighted vehicle within the range of 500 m [8]. Accordingly the distance of various intruders and their speed [3] will be recorded.

Using the values of distance and velocity, time taken by intruders to reach the sensor node will be calculated by the Algorithm 2.

3.3 Intelligent Reporting Mechanism

This phase will generate a priority list conveying the severity of intrusion of all sectors on the border. The sector (fixed geographical area) with the most severe intrusion will be at the top of the list followed by other sectors with decreasing severity. Priority list would be depending on two factors:

(1) Time
(2) Impact Factor

where time is calculated as the time taken by the intruder to reach the sensor node from its recorded position. Time will be calculated using the distance and velocity recorded by WSN.

The term impact factor refers the impact or influence of intrusion on the border caused by intruder and depending on its value impact factor is further categorized into following:

(1) Low
(2) Medium
(3) High
(4) Very High

Algorithm 3: Weight Factor

```
if IntrusionType='Infantry'
  WF=x;
else if IntrusionType='LightWeighted'
  WF=y;
else IntrusionType='HeavyWeighted'
    WF=z;
```

Algorithm 4: Impact Factor

```
AvgNo = (∑(Min,Max))/2;
IF=AvgNo*WF;
if IF>=p && IF<=q;
   set IF='Low';
else if IF>q && IF<r
         set IF='Medium';
      else if IF>r && IF<=s
         set IF='High';
else IF>s
         set IF='VeryHigh';
```

The Algorithm 3 assigns the value of weight factor to intrusion and accordingly value of IF will be calculated and categorized using Algorithm 4.

Figure 2 shows the overall flow of data.

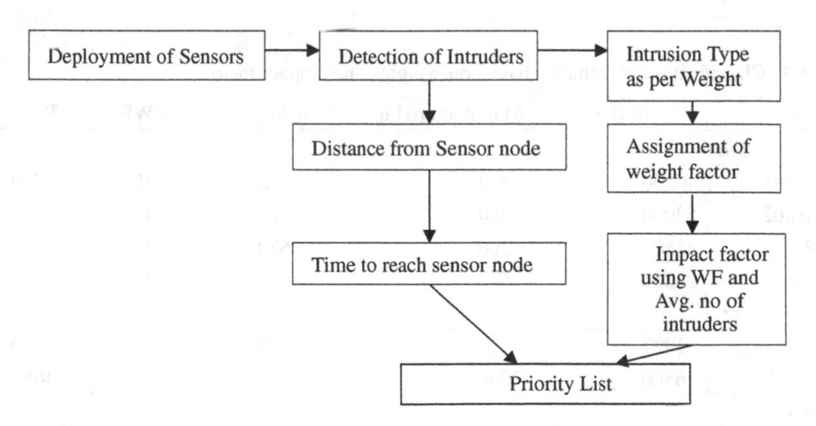

Fig. 2 Overall flow of data

3.4 Priority List

Final output Priority List will be obtained by taking 'AND' operation on two inputs (TIME, IMPACT FACTOR). 'AND' operation gives the minimum value among these two inputs and then fuzzy logic applies the defuzzification to give the final value of the priority list.

$$\sum T_i \wedge IF \tag{3}$$

The Priority List membership function is further being classified into five categories in a range of (0–1):

1. CRITICAL
2. HIGH
3. MEDIUM
4. LOW
5. LOWEST

4 Results

Priority list of the intruders is formulated in accordance to impact of attack. For generation of priority list following parameters are used: time required by the intruder to reach sensor node and impact factor.

Weights of intruders would be tabulated as per the data recorded by BVP [5, 6]. At last the calculated Impact factors are shown in Table 1.

Table 1 Classification of vehicles based on weights and impact factor

Name	Weight (kg)	Manufactured qty.	Avg no.	WF	IF
Heavy weighted					
Leopard1	40200	348	72	10	720
Leopard2	52000	450	225	10	2250
M48	45600	1200	650	10	6500
Jaguar	23200	543	25	10	250
Light weighted					
Wiesel	1927	522	98	5	490
MB1017	6800	7000	800	5	4000
Infantry					
Soldiers	100		16	1	16

Table 2 Weight factors

Type	Heavy weighted	Light weighted	Infantry
Weight factor (WF)	10	5	1

Table 3 Time required to reach Sensor node

	Distance (km)	Speed (km/h)	Time (h)	Time (min)
Vehicle				
	0.500	25.00	0.02	1.20
	0.500	1.00	0.50	30.00
Soldier				
	0.05	20.00	0.0025	0.15
	0.05	1.00	0.05	3.00

Classification of Intruders as per their weight:

- Infantry: < 1700 kg
- Light Weighted: 1700–8000 kg
- Heavy Weighted: > 8000 kg

Weight factors in Table 2 are assigned to the intruders using Algo3 and keeping $x = 1$, $y = 5$ and $z = 10$.

Considering that the intruders are traveling with their maximum/minimum speed and thus time will be calculated using (4) to reach the sensor node and recorded (Table 3).

$$\text{Time} = \text{mean}((\sum \text{Distance}) / (\sum \text{Speed})) \quad (4)$$

5 Simulation Results

Simulation dataset is implemented in MATLAB to generate Priority List (Figs. 3, 4, 5 and 6).

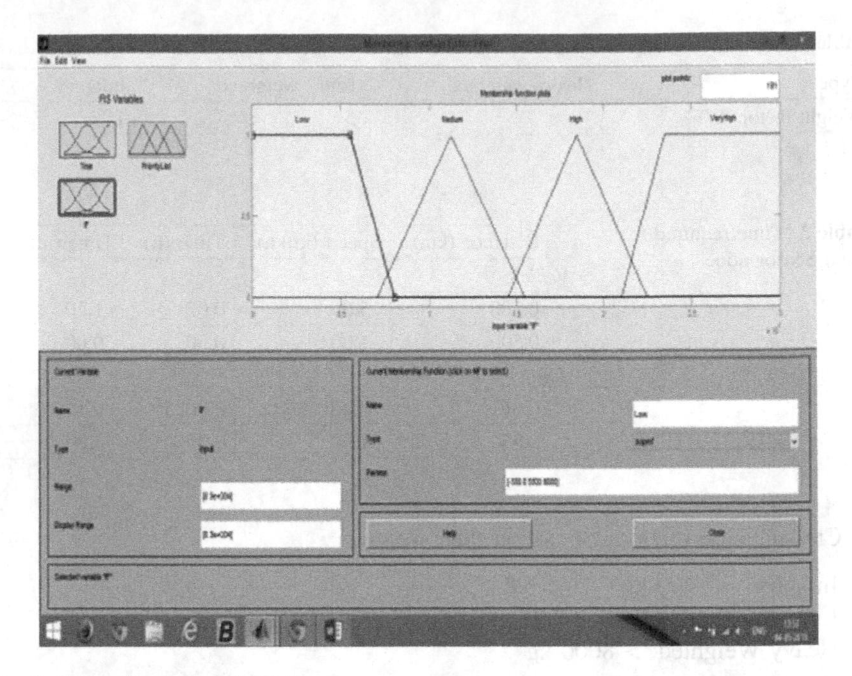

Fig. 3 Impact factor on MATLAB

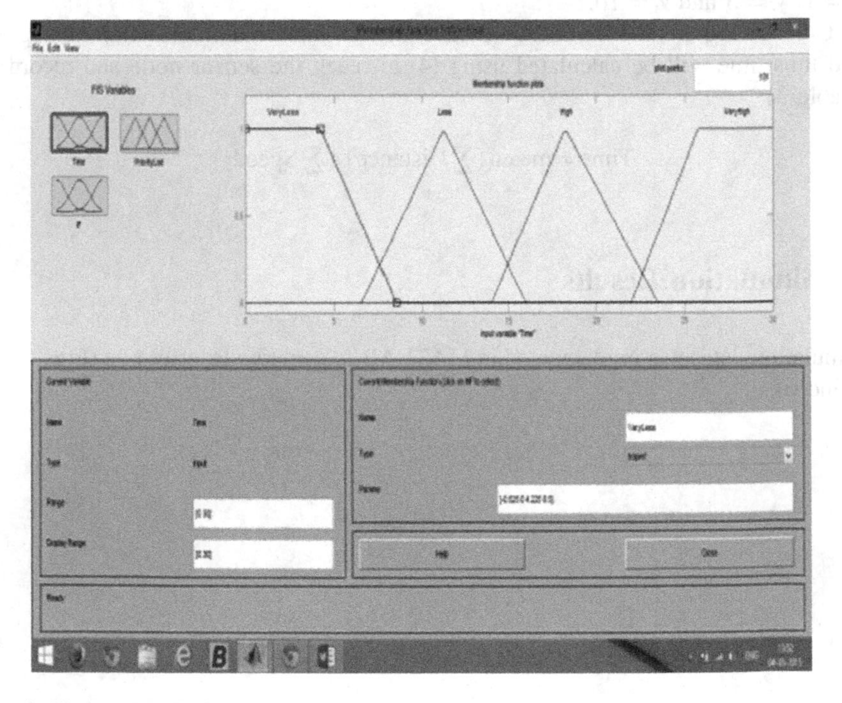

Fig. 4 Time on MATLAB

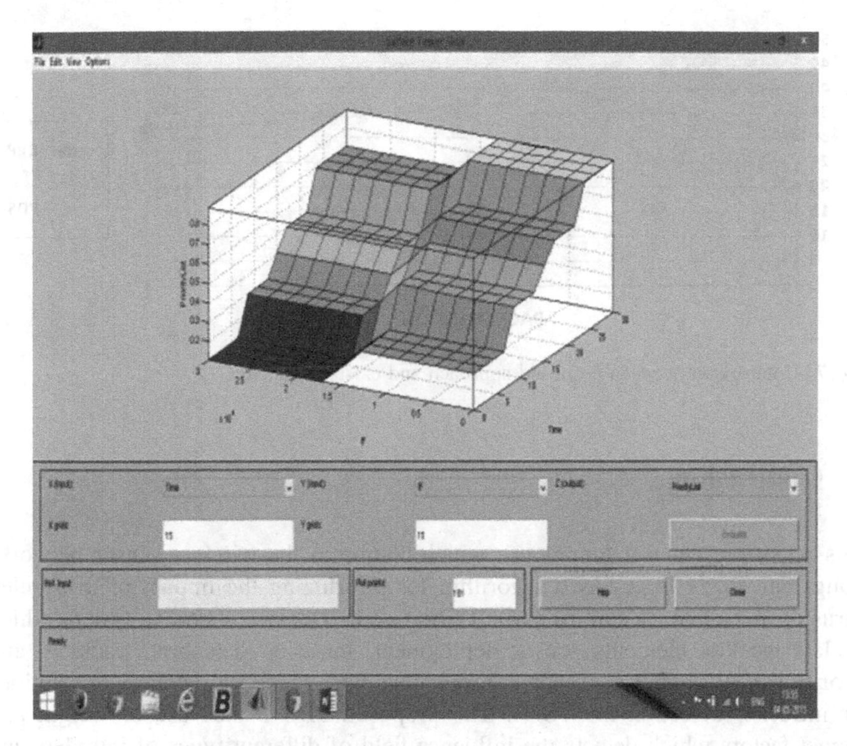

Fig. 5 Surface view of priority list

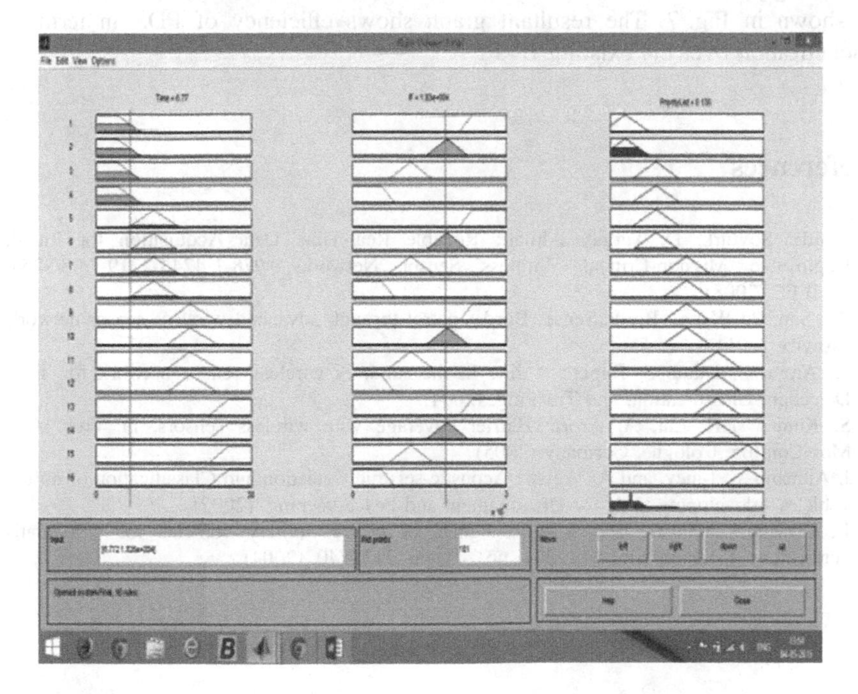

Fig. 6 Rule view of priority list

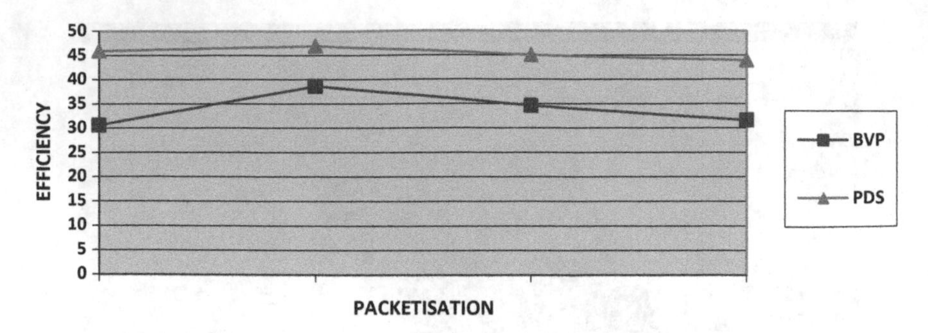

Fig. 7 Comparison between proposed approach and existing system

6 Conclusion

This paper presented a framework which combines the wireless sensor networks along with fuzzy logic-based algorithm for prioritizing the impact of distributed intrusion over line of control (LOC). Proposed PDS is a coherent system which models the key elements sensor deployment, intrusion detection, tracking, and priority list. A number of research efforts have been made for developing solutions for intrusion detection in WSNs. The proposed PDS makes use of weight and impact factors which detects the influence field of different types of intruders and accordingly calculates the amount of deterioration. The efficiency of proposed PDS is shown in Fig. 7. The resultant graph shows efficiency of PDS in terms of packetization over the existing BVP.

References

1. Mujdat Soyturk, D. Turgay Altilar.: Reliable Real-Time Data Acquisition for Rapidly Deployable Mission-Critical Wireless Sensor Networks, 978-1-4244-2219-7/08/$25.00 (c) IEEE (2008)
2. Zhi Sun, Pu Wang.: BorderSense: Border patrol through advanced wireless sensor networks, Elsevier. (2011)
3. A. Arora, P. Dutta, S. Bapat.: A line in the sand: A wireless sensor network for Target Detection, Classification, and Tracking. (2004)
4. S. Kumar, T.H. Lai, A. Arora: Barrier coverage with wireless sensors, in: Proc. ACM MobiCom'05, Cologne, Germany (2005)
5. J. Altmann, S. Linev, and A. Weisz.: Acoustic-seismic Detection and Classification of military vehicles - developing tools for Disarmament and Peace-keeping. (2002)
6. J. Altmann.: Acoustic and seismic signals of heavy military vehicles for co-operative verification. J. Sound Vib., vol. 273, no. 4/5, pp. 713–740. (2004)

7. Crane Wireless, MicroObserver Unattached Ground Sensor System, <http://www.cranewms.com>
8. Elta Systems, Unattended ground sensors network (USGN), <http://defense-update.com/newscast/0608/news/news1506_ugs.htm>
9. Trident Systems Incorporated, Unattended Ground Sensors, <http://www.tridsys.com>

Multi-objective Evolution-Based Scheduling of Computational Intensive Applications in Grid Environment

Mandeep Kaur

Abstract Grid computing has been evolved as a high performance computing to fulfill the demand of computational resources among the geographically dispersed virtual organizations. Grid is used to provide solutions to the complex computational intensive problems. Scheduling of user applications on the distributed resources is an indispensable issue in Grid environment. In this paper, a speed-constrained multi-objective particle swarm optimization (SMPSO) technique-based scheduler is proposed to find efficient schedules that minimizes makespan, flow-time, resource usage cost and maximizes resource utilization in Grid environment. The work is integrated in ALEA 3.0 Grid Scheduling simulator. The results of the proposed approach have been contrasted with Grid's conventional scheduling algorithms like FCFS, EDF, MinMin, and other multi-objective algorithms like NSGA-II and SPEA2. The results discussed in the paper shows that SMPSO outperforms over other scheduling techniques.

Keywords Grid scheduling · Resource utilization · Makespan · Flowtime · Multi-objective · SMPSO

1 Introduction

Grid is a kind of distributed computing where resources are owned by multiple virtual organizations and follows common rules for sharing and selection of resources to execute the user jobs from diverse locations [1, 2]. Grid middleware allows the users to interact with Grid systems. The concept of computational grid has been evolved from power Grid. The computational grids are required to provide reliable and low-cost methods for sharing resources such as computers, storage and working memory, network bandwidth, and software applications across geographically distributed virtual organizations [3]. The resource providers advertise

Mandeep Kaur (✉)
Computer Science Department, Savitribai Phule University, Pune, India
e-mail: mandeep.gondara@gmail.com

S.C. Satapathy et al. (eds.), *Proceedings of the International Conference on Data Engineering and Communication Technology*, Advances in Intelligent Systems and Computing 469, DOI 10.1007/978-981-10-1678-3_44

their resources on Grid using Grid information service and resource consumers execute their jobs on the advertised and available resources with the help of Grid resource broker.

In Grid scheduling, user jobs are to be scheduled on resources that belong to distinct administrative domains. Grid scheduling involves in gathering of information about resources that are capable enough to execute the user jobs, advance reservation of resources, submission of user jobs, and monitoring of user jobs [4]. The Grid scheduler is responsible for sending a user job to the best suitable Grid resource [5]. The grid scheduler matches the optimum resource to the user application. The objective of grid scheduling is to deliver QoS requirements of the grid users like minimum resource usage cost, job execution before predefined deadline and to raise the resource utilization [6].

1.1 Purpose of Research

The motivation behind this research work is to propose a multi-objective approach that is capable to find a diverse set of solutions to address multiple scheduling criteria problem in grid environment. Many researchers have adopted aggregating functions approach where weighted sum of fitness functions is taken to make a single objective function. There is a need of multi-objective optimization technique because grid scheduling criteria considers many conflicting objectives.

1.2 Contribution of the Proposed Work

In this paper, speed-constrained multi-objective particle swarm optimization (SMPSO) [12]-based Grid scheduling approach is proposed, where the user jobs and grid resources are mapped in such a way that maximum user jobs can be successfully scheduled and resource utilization can be increased. The scheduling criteria makespan, flowtime, resource usage cost, resource idle time is minimized in an optimal manner. The SMPSO-based multi-objective approach relies upon three mechanisms; the particle representation, use of leader particles that guide the search and fine tuning of parameters. This approach uses external archive that stores the nondominated solutions separately during the search and these solutions are used as leaders during updation of the positions of particles. The turbulence operator is applied that accelerates the velocity when movement of particles becomes zero to save the swarm from trapping into local minima. The proposed approach produces effective particle positions when the speed of particles becomes too high after applying turbulence to control the velocity of particles. The SMPSO-based scheduler produces evenly distributed solutions in Pareto Optimal. SMPSO produces phenomenal results in accuracy and speed.

In this paper, Sect. 2 highlights State-of-the-art works. Section 3 highlights the proposed approach for job scheduling. Section 4 shows experimental setup and results. The last section provides conclusion and roadmap to future work.

2 The State of the Art

Many researchers have used heuristic-based algorithms to address Grid scheduling problems. Tabu search (TS), simulated annealing (SA), genetic algorithms (GA), ant colony optimization (ACO), bee Colony optimization (BCO) are heuristic-based approaches that have been proposed to tackle Grid scheduling issues. Some of the works have been discussed below that considered multi-objective scheduling criteria. In [7], authors have proposed GA-based heuristic approach that minimizes flowtime and makespan. A single composite objective function is taken for addressing two scheduling criteria. Their work is based on one-to-one job-resource mapping assumption. In [8], authors presented TS algorithm for addressing the batch scheduling problem on computational Grids. The problem is bi-objective, consisting of minimization of makespan and flowtime. This approach is contrasted with GA [7]-based scheduling approach. In [9], DPSO approach is presented which also minimizes makespan and flowtime. However, the approach is new but the algorithm uses aggregation approach where two objective functions are combined into a single objective function and certain weightage is given to both objective functions. In [10], authors have proposed MOEA and minimized makespan and flowtime. The contrast of this approach has been made with single objective algorithms like GA, PSO, and SA. In [11], authors have proposed enhanced ACO-based approach that optimizes time and cost. The paper claims that this approach outperforms over MOACO and basic ACO.

Most of the works have presented grid scheduling problem as a single composite objective function. A few research works have been found in literature with Pareto front and crowding distance-based multi-objective approach that deals with multiple scheduling criteria in Grid environment by giving equal weightage to all criteria. Second, many authors have assumed that a single job is mapped to a single resource, while in real Grid environment; many jobs are mapped to a single resource. The proposed approach in this paper, is addressing four scheduling criteria. The problem formulation is multi-objective and using SMPSO [12] to provide optimum schedule. Many jobs are mapped to single cluster like real-life environment of Grids. The work is integrated in ALEA [13] that is extended from Gridsim, a well-established grid simulator that simulates Grid jobs and Grid resources.

3 Proposed Multi-objective Particle Swarm Optimization-Based Scheduling Approach

In particle swarm optimization, any solution in the search space of the problem is presented by a set of particles (Doctor Kennedy and Eberhart in 1995). The swarm of particles evolves over the generations. The objective function ascertains the optimality of a particle and the fitness of a particle with respect to the near optimal solution. The scheduling problem is formulated as a multi-objective optimization problem in which a group of conflicting objectives are simultaneously optimized. Instead of a single solution, there are diverse set of potential solutions that are optimal in some objectives. An appropriate solution to the problem fulfills the objectives at a moderately good level and the solution is nondominant [14].

3.1 Problem Formulation

Grid scheduling is based on multi-criteria optimization problem. In this work, the focus is on the minimization of makespan, flowtime, resource usage cost and maximization of resource utilization. More precisely, the scheduling problem consists of the following:

- The *N* number of jobs can be represented as independent tasks.
- An *M* number of heterogeneous clusters are available to execute user jobs.
- Many tasks can be scheduled on one resource as one resource is represented as a cluster and one cluster can have many machines.
- The computing capacity of each machine is given in MIPS (million instructions per second).

3.2 Initial Swarm Formation

An initial swarm is used as a starting point to search from. The initial swarm is generated with the following constraints:

- Every job is presented only once in the SMPSO based schedule.
- Each job must be scheduled at one available time slot on a resource.

The first thing is to formulate the particle representation and dimension of a particle [15]. For scheduling, one particle represents the set of jobs that are required to be scheduled on resources; hence, the dimension of a particle is the same as that of total number of jobs. The single particle represents job-resource mapping. The swarm size is taken 50, archive size is 100 and maximum iterations are taken 500 on the basis of empirically driven results of the algorithm.

Job1	Job2	Job3	Job4	Job5	Job6	Job7	Job8
$Resource_1$	$Resource_2$	$Resource_3$	$Resource_5$	$Resource_4$	$Resource_3$	$Resource_2$	$Resource_6$

Fig. 1 Particle representation (job-resource mapping)

Let us say *Ti* be the independent jobs where $i = 1, 2\ldots, n$ and to execute these jobs, there is a need to map them with resources say *Rj* where $j = 1,2\ldots, m$.

If there are eight independent jobs and six potential resources then the initial particle's swarm would be shown below in Fig. 1:

After change in the position of particles, a time slot assignment algorithm is deployed to transfer job assignment string to a feasible schedule.

Algorithm 1: Time Slot Assignment to Jobs
Input: Job Assignment String(Job & Resource mapping)
Output: Job Execution Schedule on particular resource
T←Get the entry job from ready queue of scheduling order
While queue is not empty do
R← obtain the resource allocation from job-assignment string
Compute the execution time of T on R
Check and assign a free time slot on R for executing job T
T← get the next job from the ready queue
end while

3.3 *Exploitation Using Velocity and Position Vector*

PSO is a swarm-based heuristic optimization technique. First of all, it is initialized with a group of random particles. Each particle gets updated in each iteration by its personal best value, i.e., *pbest* and the global best value, i.e., *gbest*. After attaining these two best values, the particle modifies its velocity and position according to Eqs. (1) and (2) to exploit the search space (Doctor Kennedy and Eberhart in 1995). The position vector is represented by *Zi* and updating velocity vector is represented *Vi*. The particles update velocity by moving in the search space.

The velocity vector is represented by

$$v_i^{k+1} \rightarrow \omega v_i^k + C_1 r_1 \left(z_{pbest_i} - z_i^k\right) + C_2 r_2 \left(z_{gbest} - z_i^k\right) \quad (1)$$

where $\omega = 0.95, C_1, C_2 = \text{random}(1.0, 2.5)$, and $r_1, r_2 = \text{random}(0.1, 1.0)$.

The momentum component represents the previous velocity. The cognitive component C_1, which depends heavily on the particle's current distance to the personal best position it has visited so far. The social component C_2, that depends on the particle's distance to the global best position where any of the swarm's particles has ever been.

v_i^k – ith particle's velocity at iteration k
v_i^{k+1} – ith particle's velocity at iteration $k+1$
x_i^k – current position of particle i at iteration k
x_i^{k+1} – position of the particle i at iteration $k+1$
p^{best_i} – best personal position of particle i
g^{best} – position of best fitness particle in a swarm

The position vector is represented by

$$z_i^{k+1} = z_i^k + v_i^{k+1} \tag{2}$$

As shown in Eq. (2), new position of the particle is determined by adding velocity factor to its previous position of the particle.

3.4 Objective Functions

The objective or fitness function is used to evaluate the current swarm to produce quality solutions. Mathematically, MOPs can be defined as

$$\text{Minimize} \quad f(x) = \{f_1(x), f_2(x) \ldots f_n(x)\} \tag{3}$$

Where $x \in S$ and S is a solution space. We can say that the solution is called dominant solution if it is equally good to the other solution and better in at least one fitness function.

Makespan: the time between start and end of all jobs in a schedule.

- Let n be the total number of jobs/tasks $T = \{t_1, t_2, t_3 \ldots t_n\}$
- To execute job t_i on grid, computational node R_j is required.
- Total execution time $TE_{ij} = ET_{ij}$

The makespan is represented as follows:

$$M_{\max} = \max\{TE_{ij}, i = 1, 2, \ldots n, J = 1, 2, \ldots m\} \tag{4}$$

$$\text{The makespan fitness function } f_1(x) = \min_{t \in N} \{M_{\max}\} \tag{5}$$

where each job t belongs to schedule N.

Flowtime: the flow time of a set of jobs is the sum of completion time of all jobs. Let us say S_t is the time in which job t finalizes and N denotes the set of all jobs to be scheduled. The flowtime is represented as follows:

$$\text{The flowtime fitness function} f_2(x) = \min\{\sum_{t \in N} S_t\} \tag{6}$$

Cost: The third fitness function is minimization of cost for job/task t_i for consuming computational node R_j.

- Processing time of t_i on R_j is TE_{ij} and $C(x)$ provides the total cost time for completion.
- The objective is to get the permutation matrix $x = (x_{i,j})$, with $x_{i,j} = 1$ if resource j performs job i and if otherwise, $x_{i,j} = 0$. The cost function $C(x)$ is represented as follows:

$$C(x) = \sum_{i=1}^{m} \sum_{j=1}^{n} TE_{i,j} * x_{i,j} \tag{7}$$

$$\text{The cost fitness function} f_3(x) = \min_{t \in N}(C(x)) \tag{8}$$

Resource Utilization: Maximizing utilization of resources, i.e., minimizing resource ideal time or minimizing $f_4(x)$.

The resource idle time minimizing function

$$f_4(x) = \frac{1}{m} \sum_{j=1}^{m} \{e(l(R_j)) - u(R_j)\} \tag{9}$$

- Function l: provides the last grid job executed on resource R_j.
- Function e: provides the end time of grid job T_i that is mapped on resource R_j.
- Function u: provides the utilized time of resource R_j.

3.5 Turbulence Operator and Velocity Constriction

To avoid the convergence of swarm around local minima; a turbulence operator has been introduced in SMPSO. The turbulence operator is applied in terms of polynomial mutation to the 20 % of swarm particles with 1.0 probability. It allows the particles to explore new solutions in search space. In contrast, SMPSO adopts velocity constriction scheme to control the speed of particles and to let the particles move in bounded search area. The particle's velocity is reversed by multiplying the particle to 0.001 if the resulting position of particle goes out of the bounded search area.

4 Experimental Setup and Results

The proposed approach is implemented in ALEA 3.0 [16] Grid scheduling simulator which is extended from GridSim Toolkit [17]. The jobs are generated in MWF (Metacentrum Workloads Format) and 6 resources (Clusters) are generated with 5 nodes in each of them. The jMetal [20] libraries are used to implement SMPSO [12], SPEA2 [18] and NSGA-II [19]-based metaheuristics. For SPEA2, the population size is 50, the archive size is 100, and maximum iterations are 500. For NSGA-II, the population size is 50 and maximum iterations are 1000. Binary Tournament selection approach is implemented for selecting parents to generate offsprings. Two-point crossover with 0.95 probability, and move mutation with 0.1 probability is implemented for exploring the search space. Grid's conventional algorithm MinMin, EDF, and FCFS have also been integrated to contrast the results.

4.1 Results for Makespan

See Fig. 2.

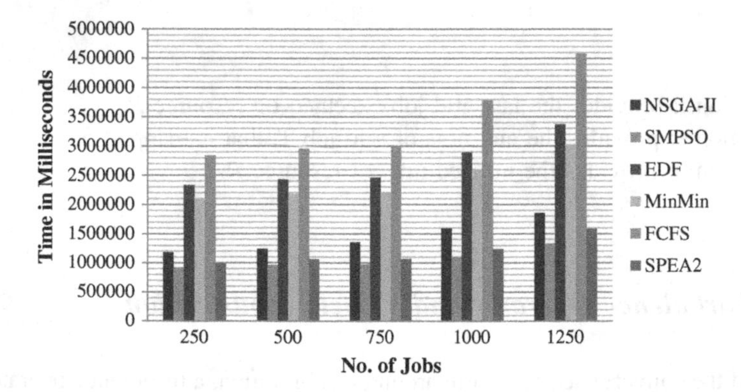

Fig. 2 Shows results for makespan. The jobs are presented along *X*-axis and execution time along *Y*-axis. The six grid scheduling techniques are contrasted with SMPSO-based grid scheduling approach. It is apparent that SMPSO gives minimum makespan or schedule length out of these approaches

4.2 Results for Mean Flowtime

See Fig. 3.

4.3 Results for Resource Usage Cost

See Fig. 4.

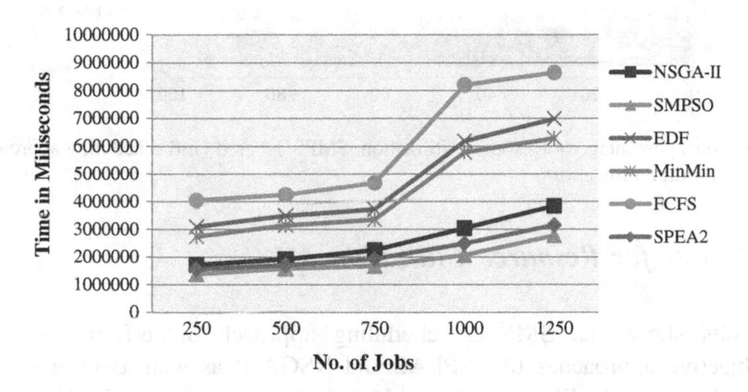

Fig. 3 Depicts that SMPSO gives minimum mean flowtime. The jobs are taken along *X*-axis and time is taken along *Y*-axis in milliseconds

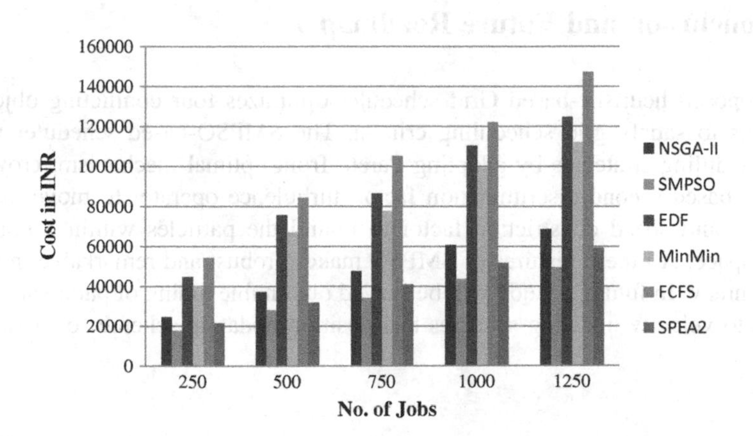

Fig. 4 It shows a comparison of resource usage cost for executing user applications among six Grid scheduling algorithms. The jobs are taken along *X*-axis and resource usage cost is taken along *Y*-axis in Indian rupees. The resource usage cost is also minimized with SMPSO-based proposed scheduling approach as shown in figure

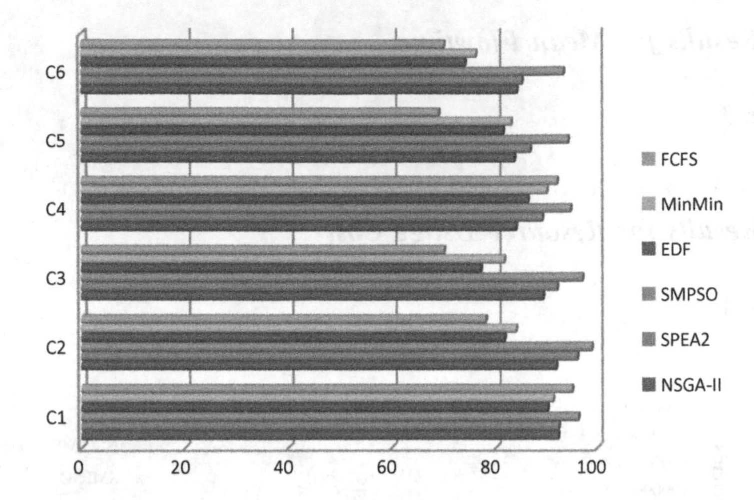

Fig. 5 It shows percentage of Resource Utilization. SMPSO-based Grid scheduling approach also maximizes resource utilization

4.4 Results for Resource Utilization

The results show that SMPSO scheduling approach outperforms over other multi-objective approaches like SPEA2 and NSGA-II as well as Grid's conventional mechanisms like EDF, FCFS, and MinMin by optimizing multiple conflicting scheduling criteria (Fig. 5).

5 Conclusion and Future Roadmap

The proposed heuristic-based Grid scheduler optimizes four conflicting objective functions to satisfy grid scheduling criteria. The SMPSO-based scheduler yields best scheduling strategies by adopting Pareto front optimal mechanism, crowding distance-based second discrimination factor, turbulence operator to move the still particles, and speed constriction factor to bound the particles within a bounded search space. All these features of SMPSO make it robust and remarkable in speed and accuracy. In future, a study will be carried out on fine tuning of parameters with respect to velocity updating schemes and position updating schemes of particles.

References

1. Mandeep Kaur (Computer Science Dept University of Pune), "Semantic Resource Discovery with Resource Usage Policies in Grid Environment," *International Journal of. Compuer Science Issues*, vol. 9, no. 5, pp. 301–307, 2012.
2. R. Buyya and S. Venugopal, "A Gentle Introduction to Grid Computing and Technologies," *Computer Society of India*, no. July, pp. 9–19, 2005.
3. K. Krauter, R. Buyya, and M. Maheswaran, "A taxonomy and survey of grid resource management systems for distributed computing," Software - Practice and Experience, vol. 32, no. September 2001, pp. 135–164, 2002.
4. J. M. Schopf, "Ten actions when Grid scheduling: The User as a Grid Scheduler," *Grid Resource Management: state of the art and Future trends*, pp. 15–23, 2004.
5. A. Pugliese, D. Talia, and R. Yahyapour, "Modeling and Supporting Grid Scheduling," *Journal of Grid Computing*, vol. 6, no. May 2007, pp. 195–213, 2008.
6. P. Varma, U. Grover, and S. Chaudhary, "QoS-Aware Resource Discovery in Grids," *15th International Conference on Advanced Computing and Communications (ADCOM 2007)*, pp. 681–687, 2007.
7. J. Carretero and F. Xhafa, "OPMENT OF ECONOMY TECHNOLOGICAL USE OF GENETIC ALGORITHMS FOR SCHEDULING JOBS IN LARGE SCALE GRID APPLICATIONS," pp. 11–17, 2006.
8. F. Xhafa, J. Carretero, B. Dorronsoro, and E. Alba, "a Tabu Search Algorithm for Scheduling Independent Jobs in Computational Grids," *Computing and Informatics*, vol. 28, pp. 1001–1014, 2009.
9. H. Izakian, B. T. Ladani, K. Zamanifar, and A. Abraham, "A Novel Particle Swarm Optimization Approach for Grid Job Scheduling."
10. H. B. Grosan C, Abraham A, "Multiobjective evolutionary algorithms for scheduling jobs on computational grids.," *International conference on applied, computing*, pp. 459–463, 2007.
11. Q. U. Of, E. Ware, and C. Delimitrou, "Scheduling in Heterogeneous Datacenters With P aragon," vol. 2, no. 5, pp. 17–30, 2014.
12. C. A. C. Coello, F. Luna, and E. Alba, "SMPSO: A New PSO Metaheuristic for Multi-objective Optimization," no. 2.
13. D. Klusacek, L. Matyska, and H. Rudova, "Grid Scheduling Simulation Environment," *Parallel Processing and Applied Mathematics*, vol. 4967, pp. 1029–1038, 2008.
14. C. a. Coello Coello and M. Reyes-Sierra, "Multi-Objective Particle Swarm Optimizers: A Survey of the State-of-the-Art," *International Journal of Computational Intelligence Research*, vol. 2, no. 3, pp. 287–308, 2006.
15. Q. Bai, "Analysis of Particle Swarm Optimization mAlgorithm,"*Computer and Information Science*, vol. 3, no. 1, pp. 180–184, 2010.
16. D. Klusáček and H. Rudová, "Alea 2: job scheduling simulator," *International ICST Conference on Simulation*, pp. 61:1–61:10, 2010.
17. R. Buyya and M. Murshed, "GridSim: a toolkit for the modeling and simulation of distributed resource management and scheduling for Grid computing," *Concurrency and Computation: Practice and Experience*, vol. 14, no. February, pp. 1175–1220, 2002.
18. E. Zitzler, M. Laumanns, and L. Thiele, "SPEA2: Improving the Strength Pareto Evolutionary Algorithm," pp. 1–21, 2001.
19. K. Deb, A. Pratap, S. Agarwal, and T. Meyarivan, "A fast and elitist multiobjective genetic algorithm: NSGA-II," *IEEE Transactions and Evoutionary Computation.*, vol. 6, no. 2, pp. 182–197, 2002.
20. http://jmetal.sourceforge.net/algorithms.html.

Selective Encryption Framework for Secure Multimedia Transmission over Wireless Multimedia Sensor Networks

Vinod B. Durdi, Prahlad T. Kulkarni and K.L. Sudha

Abstract Video compression method such as H.264/AVC often causes huge computing complexity at the encoder that is generally implemented in various multimedia applications. In view of this an article on selective video encryption framework based upon Advanced Encryption Standard (AES) is suggested for prioritized frames of H.264/AVC to improve the multimedia security while maintaining the low complexity at the encoder. The experimental result demonstrates that the recommended technique contributes significant security to video streams while maintaining the original video compression efficiency and compression ratio. Moreover, the proposed scheme provides a good trade-off between encryption robustness, flexibility, and real-time processing.

Keywords Wireless multimedia sensor network · Video compression · Selective encryption · Cross layer · Key frame

1 Introduction

For the military, Internet, real-time video multicast applications the broadcaster needs to get the assurance that unauthorized users will not be able to gain access to its services. Hence multimedia signals need to be broadcast in an encrypted form

V.B. Durdi (✉)
Department of Telecommunication Engineering, Dayananda Sagar College
of Engineering, Bangalore 560078, Karnataka, India
e-mail: vinodduradi@gmail.com

P.T. Kulkarni
Pune Institute of Computer Technology, Pune 411043, Maharastra, India
e-mail: ptkul@ieee.org

K.L. Sudha
Department of Electronics and Communication Engineering, Dayananda Sagar College
of Engineering, Bangalore 560078, Karnataka, India
e-mail: klsudha1@rediffmail.com

S.C. Satapathy et al. (eds.), *Proceedings of the International Conference on Data Engineering and Communication Technology*, Advances in Intelligent Systems and Computing 469, DOI 10.1007/978-981-10-1678-3_45

and can be decrypted by means of authorized user. AES, IDEA, RSA, and DSA are some of the extensively used algorithms for preserving the privacy of the text or any binary data. However lightweight encryption algorithms are suitable for multimedia application as traditional methods such as DES are unable to meet the real-time requirements [1]. Also, it was discovered that the adoption of typical data encryption methods namely Advanced Encryption Standard (AES) and Elliptic Curve Cryptography (ECC) for preserving the privacy of video over wireless channels poses a difficulty, for which the reason is the huge amount of data and the difficulty involved in satisfying the conditions of real-time transmission.

This article introduces a distinct methodology for procuring the safety of multimedia data by regulating the sensitive portion of the multimedia flow, i.e., selective encryption has been proclaimed as a key for safeguarding the content with minimized computational complexity involvement. This paper proposes the framework of protection of key frame using AES from the compressed video. The article is further formulated like this. The multimedia security issues are specified in Sect. 2. In Sect. 3, effective selective encryption approach of multimedia compression is discussed. Various encryption schemes along with simulation scheme are depicted in VI segment. Outcome of the proposed scheme is mentioned in part V. Lastly, VI concludes this paper.

2 Multimedia Security Issues

Once a multimedia stream goes beyond simple public communications, then various issues have to be considered. One important major issue in multimedia wireless network is the security. The acceptable techniques for protected transmission of multimedia message above the networks, a number of cryptographic, steganographic and additional approaches are tried [2–4].

Most of these techniques try to optimize the authentication process in terms of the speed, and the display process. Some of the proposed video encryption schemes are reviewed in the section below. Kankanhalli and Guan [1] mentioned data authentication is very much required in favor of multimedia commerce on the Internet as well as video multicast. The authors proposed joint encryption along with compression. In this framework video data are shuffled into frequency domain.

Either allowing the system performance or defending the computational difficulty selective encryption on the compressed bit stream was proposed by Lookabaugh and Sicker [2] and this technique achieved adequate security.

Yang et al. [3] proclaimed that selective encryption is the key to accomplish multimedia protection and this has been the current topic of research.

The recent works reported in [4] summarizes the latest research related to video encryption and also focused on the widely used H-264 with scalable extension SVC.

The suggested scrambling methods appear to attain a best possible intermediate between many desirable characteristics such as speed, authentication, size of the file, and transcodability. Hence it is very suitable for network video applications.

Podesser and Schmidt [5] propose some research regarding the video/multimedia selective encryption. Grngetto et al. [6] described the multimedia security scheme adopting the arithmetic encoder that is employed for encoding process in global image and video coding formats.

By adopting the method of bits scrambling selectively, shuffling the blocks along with rotating the blocks of the transform coefficients and motion vectors video data are scrambled in frequency domain, which presented encryption in addition to compression scheme jointly [7].

The suggested scrambling methods appear to attain a best possible intermediate between many desirable characteristics such as speed, authentication, size of the file, and transcodability. Hence it is very suitable for network video applications.

A partial encryption method [8] was put forth by Cheng et al. for ciphering a portion of the data which has already been compressed. With the purpose of decreasing the processing time, the complexity of computation and the power dissipation, selective encryption techniques have been put forward [9, 10] which can battle the security problems faced by video applications in real time. This methodology makes use of the characteristics of the MPEG layered structures to scramble various levels of selective portions of the MPEG stream. On the aspects of the frame structure of MPEG (*I*-frame, *P*-frame, also *B*-frames) the conventional selective encryption has been formulated. Only *I*-frame is subjected to scrambling, since *I*-frame comprises of most of the sensitive information. Similarly *P*-frames as well as *B*-frames do not facilitate with the relevant information with the absence of subsequent *I*-frame.

Selective encryption indulges in partial scrambling of compressed video stream to bring down the complexity involved in computation. This approach has already been discussed in several multimedia applications [11, 12]. A novel approach of partial encryption reduces consumption of power through the encryption function in favor of digital content.

On behalf of selective encryption to be influential, the aspects of the compression method are to be essentially taken into account in order to identify the significant information with reference to basic signal, considering a smaller part of the compressed bit stream. Therefore, it is likely to become the main research direction for the technology on video data security based on H.264/AVC.

According to Agi and Gong [13], there exists a fair balance (concession or trade-off) between increasing rate of occurrence of *I*-frames (and thus the level of security) and improving performance of encryption and increased ratio of compression(thereby decreasing the network bandwidth usage/utilization). Experiments usually prove the pattern shown in Fig. 1.

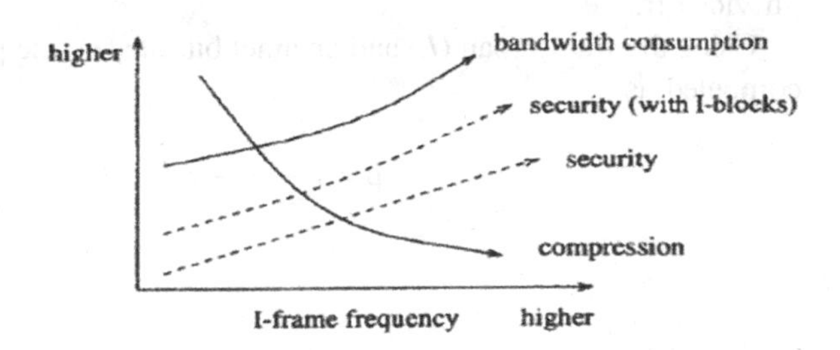

Fig. 1 Trade off between security, performance, and bandwidth consumption

3 Effective Selective Encryption Approach for Multimedia Compression

In this section, a selective encryption approach on the compressed bit stream of the H.264 is discussed. The video sequence is subjected to compression at the encoder for producing *I P B*-frames. The prioritize block is then provided with this compressed data as the input. The prioritizer block has been adopted to judge the priority among the video frames. Therefore, we first carry out detailed examination to learn the dependency on frame level decoding. Analyzing of *I*-frame is not influenced by the decoding of any other video frame within the compressed video sequence.

In a Group of Pictures (GOP), it is *I*-frames that are given utmost prominence, be it in the case of error reduction or decoding dependency dominance. The decoding of *P*-frames relies upon the decoding of preceding *I*-frames as well as/or on *P*-frames, whereas the *B*-frame decoding is reliant on both the previous *I*- and *P*-frames and also the consequent *P*-frames. Let significant weight of a single video frame in a Group of Pictures (GOP) be represented as 'w' and is expressed as follows

$$w = \sum_{\forall i | i \in H} Di \tag{1}$$

where Di is the reduction in distortion, H represents the set of decoding subsequent video frames in GOP [14].

In this method, the video frames are arranged according to their weights which give an insight regarding the number of subsequent frames so as to rely on the interpreting of present video frames and the prominent contribution of those subsequent frames with regard to distortion decrement.

In case of selective encryption, the effective decoding of the encryption blocks has a great impact on the decoding of every frame which is possible on the condition that each bit in the code block succumbs to no fault during propagation. Mathematically the dependency is expressed as

$$E[D] = \sum_{i=0}^{N-1} Di(1-\rho i) \prod_{\forall k | k \in Gi} (1-\rho k) \prod_{\forall j | j \in Si} (1-\rho j) \tag{2}$$

where $E[D]$ is the end-to-end expected distortion reduction after decoding and decryption process [14]. G is the Ancestor frame set for a specific video frame in the decoding dependency graph. S is the association set of encrypted codewords for the *i*th video frame.

Using the frame span (L) and channel bit rate (e), the packet error rate (ρ) can be computed as

$$\rho = 1 - (1 - e)^{L + L_{\text{overhead}}} \tag{3}$$

The term which is the decoding dependency also distortion reduction contribution

$$\sum_{i=0}^{N-1} Di(1-\rho i) \prod_{\forall k \mid k \in Gi} (1-\rho k) \tag{4}$$

are decided with the encoder along with the utilized video compression method like H.264/AVC.

The expected reduction in distortion is possible by maximizing the probability of successful transmission of encryption dependency frames, i.e.,

$$\max_{i} \left\{ \prod_{\forall j \mid j \in Si} (1-\rho i) \right. \tag{5}$$

In cross-layer way, the protection and energy levels are designated at both the application layer and the physical layer on the basis of priority thus ensuring the procurement of maximum number of successful transmissions of vital video frames. The selective encryption and resource allocation are easily compliant to wireless sensor network applications which suffer from restricted energy resources and computation.

4 Encryption Scheme and Simulation Model

Depending on the needs of the applications block ciphers such as AES (Advanced Encryption Standard) or 3DES are used. DES typically uses 56 bit key to encrypt data in 64 bit block size. In each of 16 rounds DES performs bit manipulation and permutation. DES is vulnerable to brute force attack as the 56 key is generating approximately 72 quadrillion possibilities which is not sufficient in today's computing world.

For implementation of content protection when considered together with the combination of authentication and performance AES is chosen as the appropriate algorithm. Similarly AES in terms of efficiency, implementability, and flexibility is made an appropriate selection. With respect to design, AES is also efficient in hardware and faster in software.

In the simulation model for WSN shown in Fig. 2, the H.264 encoder receives the video signal as the input which undergoes compression yielding *I P B*-frames. The prioritize block is then provided with this compressed data as the input. The application layer along with the physical layer with the cross-layer approach, the energy levels are designated at the resource allocator on the basis of priority which assures that the important information is never lost.

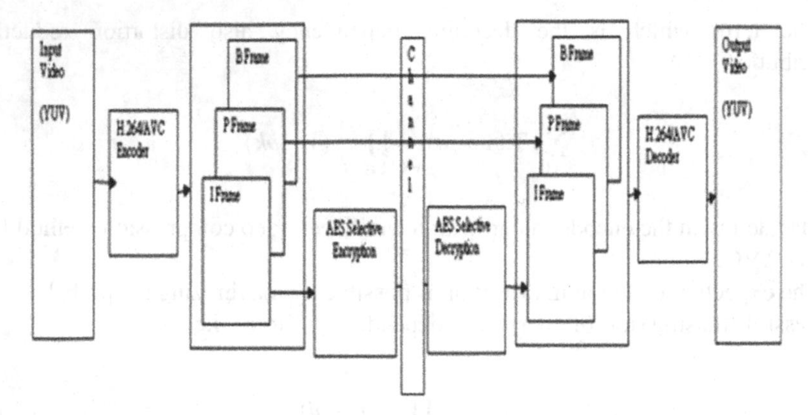

Fig. 2 Simulation model of proposed Selective Encryption method

For the simulation purpose the *I P B*-frames are generated by the H.264 encoder in our work and AES applied to the *I*-frames which are the key frames of the video data.

The kind of input video frame format that has to be given as the input to the encoder is judged with relevance to the specific application. The profoundly used frame format in mobile multimedia applications, QCIF format, has been utilized in our work as this research is concerned to transmission of video over wireless streams. Since H.264 procures greater robustness, efficient encoding, and syntax specifications and supported by most of the networks as well as application is the appropriate choice as the encoder in the implementation.

5 Results and Discussions

The experiments were conducted using JM 18.6 tool to observe the impact of the selective encryption on proposed model in wireless sensor networks using the AES with 128 key size. The simulation specifications are declared as follows. The H.264/AVC reference codec model was adopted at the application layer for compressing and packetizing 300 video frames. The well-known 4:2:0 format of foreman video sequence with preset size of 176X144 QCIF is selected as the sample input signal shown in Fig. 3.

This video signal is encoded with high profile by the frame rate of 30 frames per second (fps), the bit rate 45020 bps so that *I*-frames, *P*-frames, and *B*-frames are generated. Entropy coding method of CABAC with Group of Picture (GOP) of 10 as well as qpIslice and qp slice of 28 is used for the implementation. The encoder compressed the multimedia data with the compression ratio of about 0.0166.

Fig. 3 Original Foreman QCIF video signal: YUV 4:2:0 sub sampled with 176X144

The paper showcases only few images of *I P B*-frames that are obtained as the output of H.264 with regard to the difficulty in displaying many frames in the limited space available.

Using AES selective encryption is performed on the *I*-frames provided by the priority block. Pertaining to earlier discussions, it is the *I*-frames that contribute in recognition of correct information bits in the decoding process. The selectively scrambled *I*-frames when subjected to decoding produces irrelevant codewords leading to erroneous video sequence.

For Encryption key in this framework it is assumed that does not have access to unauthorized users. The encoded bit stream values undergo decoding and the respective decrypted bits are influenced by previous results and hence damage the consequently required decoding steps. In accordance, the reconstructed video poses a pattern resembling the noise pattern.

Figures 4, 5, 6, 7, 8, 9, and 10 show the first, twentieth, thirtieth, fortieth, fiftieth, hundredth, and one hundred nineteenth frames, respectively, show frames decoded

Fig. 4 Frame_1 decoding with encrypted *I*-frame

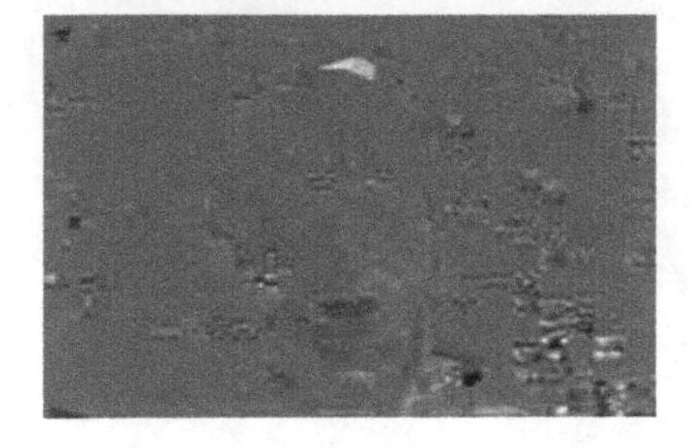

Fig. 5 Frame_20 decoding with encrypted *I*-frame

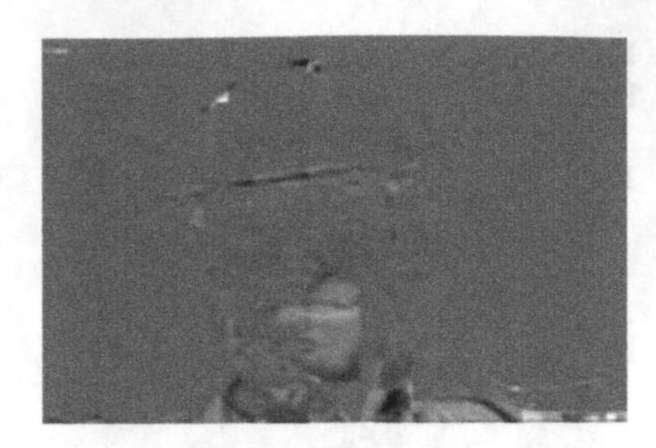

Fig. 6 Frame_30 decoding with encrypted *I*-frame

Fig. 7 Frame_40 decoding with encrypted *I*-frame

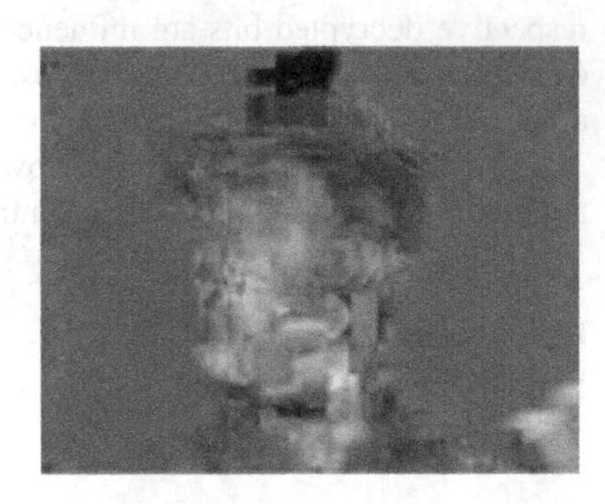

Fig. 8 Frame_50 decoding with encrypted *I*-frame

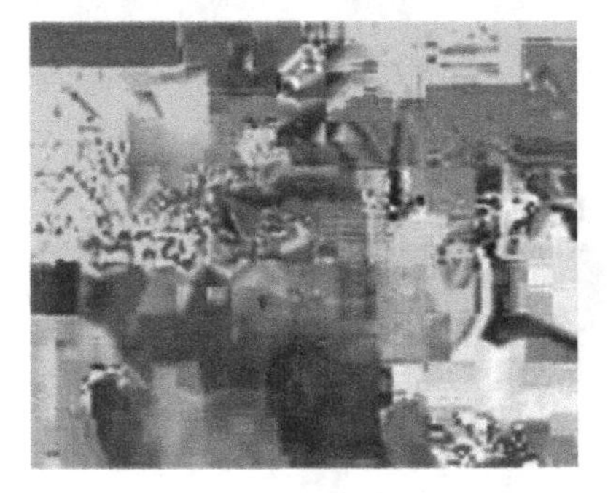

Fig. 9 Frame_100 decoding with encrypted *I*-frame

Fig. 10 Frame_199 decoding with encrypted *I*-frame

with encrypted *I*-frames. From these results, we conclude that the content of every frame within the video clip is protected by ciphering the *I*-frame.

Figure 11 shows the recovered video signal by decoded with decrypted *I*-frames, Figs. 12, 13, 14, 15, 16, 17, 18, and 19 show the first, tenth, twentieth, thirtieth, fortieth, fiftieth, hundredth and one hundred nineteenth frames, respectively, show frames decoded with decrypted *I*-frames.

From these results, we observe that decrypting the *I*-frame with the authentication key generated the original video clip. Also from the proposed method it is observed that only the *I*-Frames bit streams were encrypted instead of the entire video which drastically reduced the processing time as well as it provided adequate security to video stream with no change in compression ratio and do not reduce the original video compression efficiency in applications like wireless sensor networks.

Fig. 11 Decoded received video signal with decrypted *I*-frame

Fig. 12 Frame_1 decoding with decrypted *I*-frame

Fig. 13 Frame_10 decoding with decrypted *I*-frame

Fig. 14 Frame_20 decoding with decrypted *I*-frame

Fig. 15 Frame_30 decoding with decrypted *I*-frame

Fig. 16 Frame_40 decoding with decrypted *I*-frame

Fig. 17 Frame_50 decoding with decrypted *I*-frame

Fig. 18 Frame_100 decoding with decrypted *I*-frame

Fig. 19 Frame_199 decoding with decrypted *I*-frame

6 Conclusion

From the results of selective encryption, it has been found that ciphering *I*-frames of the video provided a meaningless video to the user that is not familiar with the authentication key and also in this event the complex overhead is greatly less compared with total encryption with no change in compression ratio as well as compression efficiency.

Improvement can be accomplished through various approaches. Incrementing the rate of occurrence of *I*-frames attains enhanced protection but at the cost of greater video size, increased consumption of network bandwidth, and intensive complexity involved in computation for encryption and decryption processes.

Overall, it is suitable for secure transmission of videos where the processing time is crucial. Lastly, the proposed model will appear to achieve the solution for the critical issue related to multimedia content protection in wireless sensor networks.

References

1. Kankanhalli, M., Guan, T.: Compressed domain scrambler/descrambler for digital video: IEEE Trans. Consum. Electron. vol. 48, no. 2, pp. 356–365, (2002).
2. Lookabaugh, T., Sicker, D.C.: Selective encryption for consumer applications: IEEE Commun. Mag., vol. 42, no. 5, pp. 124–129, (2004).
3. Yang, M., Bourbakis, N., Li, S.: Data-image-video encryption: IEEE Potentials, vol. 23, no. 2, pp. 28–34, (2004).
4. Stutz, T., Uhi, A.: A Survey of H.264 AVC/SVC encryption: IEEE Trans. Circuits and Systems for Video Technology, Vol. 22, no. 3, pp. 325–339, (2012).
5. Podesser, H. P., Schmidt, Uhl, A.: Selective bitplane encryption for secure transmission of image data in mobile environments: In Proc, 5th Nardic Signal processing Symp. Oct 2002.
6. Grangetto, M., Magli, E., Olmo,: Multimedia selective encryption by means of randomized arithmetic coding: IEEE Trans, Multimedia Vol. 8, no. 5, pp. 905–917, (2006).
7. Zeng, W., Lei, S.: Efficient frequency domain selective scrambling of digital video: IEEE Trans. Multimedia, Vol. 5, no. 1, pp. 118–129, (2003).
8. Cheng, H., Li, X.: Partial encryption of compressed images and videos: IEEE Trans. Signal Process. Vol. 48, no. 8, pp. 2439–2451, (2000).
9. Nithin, M., Damien, L., David, R. B., David, R.: A Novel Secure H.264 Transcoder using Selective Encryption: IEEE International Conference on Image Processing, 2007, (ICIP 2007), 85–88. (2007).
10. Lintian, Q., Nahrstedt, K.: Comparison of MPEG encryption algorithms: Proceedings of the International Journal of Computers and Graphics, special issue Data Security in Image Communication and Network, vol. 28, (1998).
11. Bharagava, B., Shi, C.: A fast MPEG video encryption algorithm: In Proceedings of the 6th International Multimedia Conference, 81–88. Bristol, UKm, (1998).
12. Deniz, T., Cem, T., Nurşen, S.: Selective encryption of compressed video files: International Scientific Conference, 1–4, (2007).
13. Lookabaugh, T., Sicker, D. C., David M., K., Wang, Y. G., Indrani, V.: Security Analysis of Selectively Encrypted MPEG-2 Streams: SPIE proceedings series on Multimedia systems and applications. 10–21, (2003).
14. Wei Wang, Michel Hempel, Dongming Peng, Honggang, Hamid Sharif, Hsiao Hwa Chen,: On energy efficient encryption for video streaming in wireless sensor networks: IEEE Transaction on multimedia, Vol, 12, No 5, pp 417–426, (2010).

Mining Frequent Quality Factors of Software System Using Apriori Algorithm

Jyoti Agarwal, Sanjay Kumar Dubey and Rajdev Tiwari

Abstract Success or failure of any software system mainly depends on its quality. Quality plays a key role for determining the acceptability and adoptability of the software system among the users. Quality is the composite characteristic that depends on its different factors. The researchers have been focusing on different quality factors as per their research work requirement and various quality factors have been proposed by them in different quality models to assess the software quality. These models show that there are differences in the perception of the researchers regarding the important quality factors. Ample research work has been done in the field of software quality but till date no work has been done to prove as which quality factors are frequently used and have high importance. This paper is an attempt to find out the major key factors of the software quality. Various quality models have been considered for this purpose and accordingly data set is prepared. Apriori algorithm is used to generate the frequent quality factors. Experiment is performed in Java programming language. The aim of this research paper is to provide proper focus on the frequently used quality factors, which facilitate to improve the quality of software system. These quality factors will be useful to assess the quality of software system in a prudent manner.

Keywords Quality · Models · Factors · Software system · Apriori algorithm

Jyoti Agarwal (✉) · S.K. Dubey
Amity University, Noida, Uttar Pradesh, India
e-mail: itsjyotiagarwal1@gmail.com

S.K. Dubey
e-mail: skdubey1@amity.edu

Rajdev Tiwari
Noida Institute of Engineering and Technology, Greater Noida, India
e-mail: rajdevtiwari@yahoo.com

S.C. Satapathy et al. (eds.), *Proceedings of the International Conference on Data Engineering and Communication Technology*, Advances in Intelligent Systems and Computing 469, DOI 10.1007/978-981-10-1678-3_46

1 Introduction

In today's world, software has become an important part in all walks of life. Users are capable enough to select the best software as per their requirement. Selection of the best software mainly depends on the quality of the software system. According to IEEE 610.12 standard, quality is defined as "The composite characteristics of software that determine the degree to which the software in use meet the expectations of the customer" [1]. Composite characteristic means that the quality depends on its different factors and these factors can be used to assess the software quality. These quality factors have been defined in different quality models and the researchers have selected different quality factors as per the requirement of their research work. The question arises as to which quality factors should be focused more so that the researchers are able to move in right direction to assess the software quality.

The main objective of this research paper is to enumerate the quality factors that are most frequently used by the researchers in evaluating the software quality and assessing its efficiency. For finding the frequent quality factors data mining is used. Many algorithms are used in data mining for mining the frequent items sets from the transaction database. These algorithms can also be used for determining the frequent quality factors from the quality models. Apriori algorithm is the widely used data mining algorithm [2]. It is used to find the frequent item sets and association among the frequent item sets. In this research paper, the work is limited to generate the frequent quality factors. Association among the frequently used quality factors are not generated in this research paper because the objective is only to identify the frequently used quality factors.

The main contribution of this paper is that the suggested method will help to identify the significant quality factors which play important role to define the quality of the software system and also provide the correct roadmap for the future research on software quality.

This paper is divided into various sections. Research problem identification in Sect. 2. Section 3 is the brief review of the quality models and its different quality factors. Section 4 deals with the experimental work. Section 5 followed by the result of the research questions. The paper is concluded by describing its future scope in Sect. 6.

2 Problem Identification

To find the gap between the present and available research work on software quality, research papers from different sources have been selected and studied. After reading the research papers it is observed that lot of work has been done on software quality and its factors but no research work has been done for finding out the frequent quality factors to provide the right direction for the future research on software quality. It raises a question in the researcher's mind that which quality factor should be taken into account for measuring the quality for software system.

2.1 Inclusion and Exclusion Criteria

Research papers are collected from different digital libraries. Various keywords pertaining to software quality, quality models, quality factors are made used to search relevant papers. This paper includes those research papers and the articles which describe the quality models and the various quality factors. Research papers dealing with the evaluation of the software quality were weeded out, because the aim of this paper is to identify those quality factors which are frequently being used.

2.2 Research Questions

Two research questions are framed on the basis of the study of the research papers.

RQ1: How to determine frequently used quality factors?
RQ2: What all are quality factors being used frequently?

3 Related Work

This section is about the related work on software quality and its different factors. The first quality model was proposed in 1977 by McCall [3]. Research is continued since 1977 for improving the software quality. Number of quality models has been proposed by researchers [4–21].

Table 1 shows the 19 different quality models and the distribution of these quality models as per the nature of software system is shown in Fig. 1.

Table 1 Quality models

Quality model	Symbol	Nature of software system	References
McCall (1977)	Q1	Traditional software system	[3]
Boehm (1978)	Q2	Traditional software system	[4]
ISO-9126 (1986)	Q3	Traditional software system	[5]
FURP (1987)	Q4	Traditional software system	[6]
Ghezzi (1991)	Q5	Traditional software system	[7]
IEEE (1993)	Q6	Traditional software system	[8]
Dromey (1995)	Q7	Traditional software system	[9]
ISO/IEC 9126-1 (2001)	Q8	Traditional software system	[10]
Goulao (2002)	Q9	Component based software system	[11]
Kazman (2003)	Q10	Traditional software system	[12]
Khosravi (2004)	Q11	Traditional software system	[13]
Alvaro (2005)	Q12	Component based software system	[14]
Rawadesh (2006)	Q13	Component based software system	[15]
Sharma (2008)	Q14	Component based software system	[16]
Kumar (2009)	Q15	Aspect oriented software system	[17]
Kumar (2012)	Q16	Aspect oriented software system	[18]
Pasrija (2012)	Q17	Traditional software system	[19]
Roy (2014)	Q18	Traditional software system	[20]
Tiwari (2015)	Q19	Component based software system	[21]

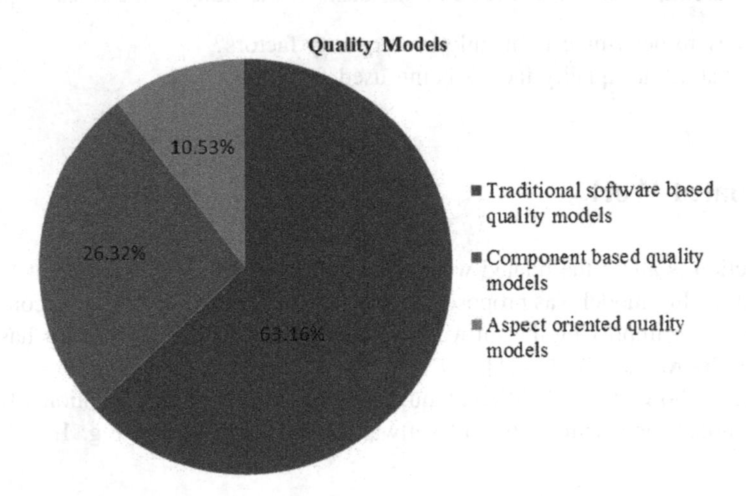

Fig. 1 Distribution of quality models as per the nature of software system

4 Experimental Work

Experiment is performed in Java Programming language. Data set is prepared in two parts according to Table 3. Data set I, contains 19 rows and 25 columns. Row denotes the quality models and column denotes the quality factors of Table 3. Data set I contains two numeric values 1 and 0. Presence of any quality factor in a particular model is represented as 1 and absence of the quality model in the particular quality model is represented as 0. The data set I is shown in Fig. 2. Figure 3 shows the Data set II, in which the first row defines the number of quality factors, i.e. 25, second row defines the number of quality models, i.e. 19 and third row defines the minimum support (min_sup), i.e. 30 % (Table 2).

After preparing the data set, Java code of Apriori algorithm is executed to generate the result. The working of Apriori algorithm is not mentioned in this paper as the objective is to draw the conclusion regarding the frequent quality factors. Agarwal, Srikant [2] have explained the working of Apriori algorithm. The result


```
1 0 1 0 0 0 0 1 0 0 1 0 1 0 1 0 0 0 0 0 0 0 1 1 0
0 0 0 1 0 0 1 0 0 0 1 0 0 0 1 0 1 0 0 0 0 0 0 0 1
0 0 0 0 0 0 1 0 0 0 0 0 0 1 0 0 1 0 0 0 0 1 0 0 1
1 0 0 0 0 1 0 0 1 0 1 0 0 0 1 0 0 1 0 0 0 0 0 0 1
1 0 0 1 0 0 1 0 0 0 1 0 0 0 1 0 0 0 0 0 0 0 0 0 0
1 0 0 1 0 0 1 0 0 0 1 0 0 1 1 0 0 0 0 0 0 0 0 0 0
1 0 0 1 0 0 1 0 0 0 1 0 0 1 1 1 0 1 0 0 0 0 0 0 0
0 0 0 1 0 0 1 0 0 0 1 0 0 0 1 0 1 0 0 0 0 0 0 0 1
1 0 0 0 0 0 0 0 0 0 0 0 1 1 0 0 0 0 0 0 1 0 1 0 1
0 0 0 0 0 1 0 0 0 0 0 0 0 0 0 0 0 1 1 1 0 0 0 0 1
0 0 0 1 0 0 1 0 0 0 1 0 0 0 1 0 1 0 0 0 0 0 0 0 1
0 0 0 1 0 0 1 0 0 0 1 1 0 0 0 0 1 0 0 0 0 0 0 0 1
0 0 0 1 0 0 1 0 0 0 1 0 0 0 1 0 1 0 0 0 0 0 0 0 1
0 0 0 1 0 0 1 0 0 0 1 0 0 0 1 0 1 0 0 0 0 0 0 0 1
0 0 0 1 1 0 1 0 0 0 1 0 0 0 1 0 1 0 0 0 0 0 0 0 1
0 0 0 1 0 1 0 0 0 0 0 0 0 1 0 1 1 0 0 0 0 0 0 0 1
0 1 1 1 0 0 0 0 0 0 0 0 0 0 0 0 0 0 0 0 0 0 0 0 0
0 0 0 1 0 0 1 0 0 0 1 0 0 0 1 0 1 0 0 0 0 0 0 0 1
```

Fig. 2 Data Set I

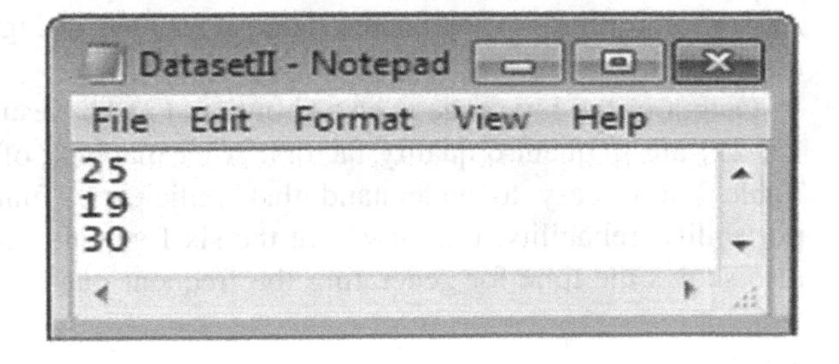

Fig. 3 Data set II

Table 2 Quality Factors

Quality factors	Symbols	Quality factors	Symbols	Quality factors	Symbols
Availability	F1	Interoperability	F10	Reusability	F18
Correctness	F2	Maintainability	F11	Robustness	F19
Effectiveness	F3	Manageability	F12	Scalability	F20
Efficiency	F4	Modifiability	F13	Security	F21
Evolvability	F5	Performance	F14	Suppportability	F22
Flexibility	F6	Portability	F15	Testability	F23
Functionality	F7	Process Maturity	F16	Understandability	F24
Human Engineering	F8	Reliability	F17	Usability	F25
Integrity	F9				

shows the frequent-6 quality factors. These frequent factors are {F4, F7, F11, F15, F17, F25}. From Table 3 it is clear that the frequent quality factors are efficiency, functionality, maintainability, portability, reliability and usability.

5 Results

After applying Apriori algorithm on Data set I (Fig. 2) and Data set II (Fig. 3), answers to the research questionnaire are framed in this section.

RQ1: How to determine frequently used quality factors?

Frequent quality factors can be determined using data mining techniques. Association rule mining is the technique of data mining which describes various algorithms (Apriori, DHP, FP-Growth, etc.) for mining frequent items sets. These algorithms can be applied on the quality models and its factors to find out the frequently used quality factors. Apriori algorithm is not suitable for the large databases so other data mining algorithms can also be used for the same purpose in the future. In this research paper database of quality models is not too large, so Apriori algorithm is applied to find the frequent quality factors.

RQ2: What all are quality factors being used frequently?

Apriori algorithm is used to find out the frequent quality factors. Java code of Apriori algorithm is used on the data set (shown in Figs. 2 and 3) to find out the results.

Output of the Java code is also shown in Fig. 4. Result shows that {4, 7, 11, 15, 17, 25} are frequent-6 quality factors. After mapping of these quality factors from Table 2 it is easy to understand that {efficiency, functionality, maintainability, portability, reliability, usability} are the six frequently used quality factors. Result also shows the time for generating the frequent quality factors.

Table 3 Quality factors in different quality models

Quality models	Q1	Q2	Q3	Q4	Q5	Q6	Q7	Q8	Q9	Q10	Q11	Q12	Q13	Q14	Q15	Q16	Q17	Q18	Q19
Quality factors																			
F1	X	X			X	X	X	X		X									
F2	X																	X	
F3	X	X																X	
F4			X			X	X	X	X			X	X	X	X	X	X	X	X
F5																X			
F6	X				X						X						X		
F7			X	X		X	X	X	X			X	X	X	X	X			X
F8		X																	
F9					X														
F10	X																		
F11	X	X	X		X	X	X	X	X			X	X	X	X	X			X
F12													X						
F13		X								X									
F14				X			X	X		X									
F15	X	X	X		X	X	X	X	X			X		X	X	X	X		X
F16								X											
F17			X	X					X			X	X	X	X	X	X		X
F18	X				X			X			X						X		
F19											X								
F20											X								
F21	X									X									

(continued)

Table 3 (continued)

Quality models / Quality factors	Q1	Q2	Q3	Q4	Q5	Q6	Q7	Q8	Q9	Q10	Q11	Q12	Q13	Q14	Q15	Q16	Q17	Q18	Q19
F22				X															
F23	X	X								X									
F24		X																	
F25	X		X	X	X				X	X	X	X	X	X	X	X	X		X

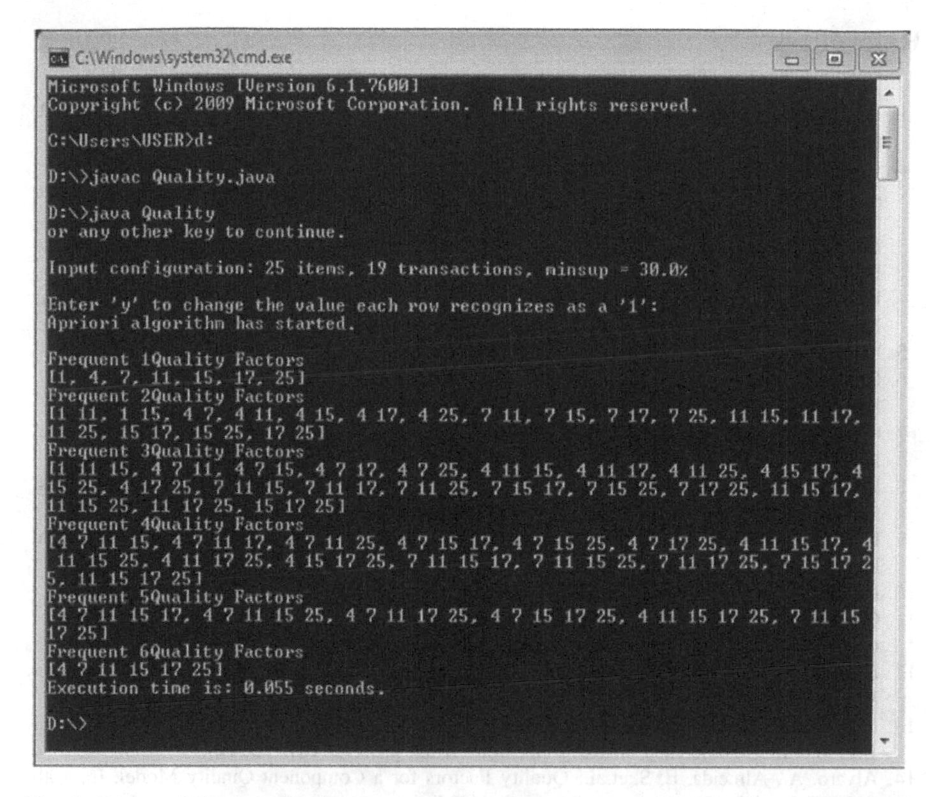

Fig. 4 Output of Frequent Quality Factors

6 Conclusion and Future Scope

Analysis of software quality is a pragmatic approach in the research work. Data mining also has its own importance. This paper presents the combination of data mining along with software quality to work out solution to the research questionnaire framed in this paper. Experiment is performed using Java code of Apriori algorithm and the result of the framed questions show that data mining algorithms can be used to generate the frequent quality factors. Frequent-6 quality factors are identified, viz. efficiency, functionality, maintainability, portability, reliability and usability.

This research work will be helpful to the researchers to measure the software quality in terms of the frequently used quality attributes. In future the objective would be to carry on the research work for the improvement of these frequently used quality factors. These factors will help to measure and improve the software quality, to enhance the acceptability and will also reduce the maintenance cost of the software system. In future work can also be done to identify the subfactors of these frequent quality factors.

References

1. IEEE Standard Glossary of Software Engineering Terminology, IEEE Standard 610.12, pp. 1–84(1990).
2. Agarwal, R., Srikant, R.: Fast algorithms for Mining Association rules. In: Proceedings of the 20th Very Large Data Bases (VLDB) Conference, Vol. 1215, pp. 487–499,Santiago, Chile (1994).
3. McCall, J.A., Richards, P.K., Walters, G.F.: Factors in Software Quality. Griffiths Air Force Base, N.Y. Rome Air Development Center Air Force Systems Command (1977).
4. Boehm, B.W., Brown, J.R., Lipow, M, et. al.: Characteristics of Software Quality. Elsevier, North-Holland (1978).
5. http://www.angelfire.com/nt2/softwarequality/ISO9126.pdf (Last Accessed: 10.09.2015).
6. Grady, R. B. Practical Software Metrics for Project Management and Process Improvement. Prentice Hall, Englewood Cliffs, NJ, USA (1992).
7. Ghezzi, C., Jazayeri, C. M. and Mandrioli, D.: Fundamental of Software Engineering, Prentice-Hall, NJ, USA (1991).
8. IEEE standard: IEEE Recommended Practice for Software Requirements Specifications, Software Engineering Standards Committee of the IEEE Computer Society (1993).
9. Dromey, G. R.: A Model for Software Product Quality: IEEE Transactions on Software Engineering, Vol. 21, No. 2, pp 146–162 (1995).
10. ISO/IEC 9126-1. Software Product Quality – Part 1, Quality model (2001).
11. Goulão, M., Fernando, B.: Towards a Components Quality Models. In: Work in Progress of the 28th IEEE Euromicro Conference, Dortmund, Germany (2002).
12. Kazman, R, Bass, L., Clements, P.: Software Architecture in Practice, 2nd ed., Addison Wesley (2003).
13. Khosravi, K., Guéhéneuc, Y.G.: On issues with Software Quality Models. In: Proceeding of 11th Working Conference on Reverse Engineering, pp. 172–181 (2004).
14. Alvaro, A., Almeida, E. S.,et.al.: Quality Factors for a Component Quality Model. In: 10th WCOP/19th ECCOP, Glasgow, Scotland, (2005).
15. Rawasdesh, A., Matalkah, B.: A new Software quality model for evaluating COTS components. Journal of Computer Science, 2(4), pp. 373–381 (2006).
16. Sharma, A., Kumar, R., Grover, P.S.: Estimation of Quality for Software components: an empirical approach. In: Software Engineering Notes 33, No. 6, pp. 1–10, ACM (2008).
17. Kumar, A., Grover, P.S., Kumar, R.: A quantitative evaluation of aspect oriented software quality model (AOSQUAMO). In: ACM SIGSOFT Software Engineering Notes 34, No.5, pp. 1–9 (2009).
18. Kumar, P.: Aspect oriented Software Quality Model: The AOSQ Model. In: Advanced Computing, An International Journal (ACIJ), Vol.3, No.2, pp. 105–118 (2012).
19. Pasrija, V., Kumar, S., Srivastava, P.R.: Assessment of Software Quality Choquet Integral approach. In: 2nd International Conference on Communication, Computing & Security (ICCCS), pp. 153–162, Elsevier, Procedia Technology (2012).
20. Roy, J., Contini, C., et al.: Method for the Evaluation of Open Source Software Quality from an IT Untrained User Perspective. In: Proceeding of International C* Conference on Computer Science & Software Engineering, pp. 21–26, ACM (2014).
21. Tiwari, A., Chakraborty, S.: Software Component Quality Characteristics Model for Component Based Software Engineering. In: IEEE International Conference on Computational Intelligence & Communication Technology, pp. 47–51(2015).

Algorithm for the Enumeration and Identification of Kinematic Chains

Suwarna Torgal

Abstract The applications of kinematic chains are not limited only to mechanical engineering field rather the kinematic chains are used in almost many areas like mechatronics, robot design, medical applications mainly in orthopedics area, medical instruments, physiotherapy exercise machines, and many more. As the kinematic chain is the base for any mechanism, it becomes necessary to generate and identify the unique, nonisomorphic or distinct feasible kinematic chains and have the digital library of the collection of kinematic chains so that they can be retrieved as and when required during the design phase. The objective being identification and enumeration of each kinematic chain with an identification code (digital), canonical numbering being a unique digital code assigned to identify the kinematic chains. In the proposed paper the identification codes, i.e., maxcode and mincode are explained along with an example of Stephenson's chain.

Keywords Kinematic chain • Adjacency matrix • UTAM (Upper Triangular Adjacency Matrix) • Maxcode • Mincode • Decodability

1 Introduction

Structural enumeration of all the distinct feasible kinematic chains with specified number of links and degrees of freedom enables selection of the best possible chain for specified task at the conceptual stage of design. Kinematicians have synthesized kinematic chains unconsciously since time immemorial, Crossley [1], introduced simple algorithm to solve Gruebler's equation for constrained kinematic chains, the Romanian researchers led by Manolescu [2] have used the method of using the Assur groups, the method is based on visual inspection. Davies and Crossley [3] Franke's condensed notation and synthesized kinematic chains for 10-link, single-freedom

Suwarna Torgal (✉)
Mechanical Engineering Department, Institution of Engineering and Technology, Devi Ahilya Vishwavidyalaya, Indore, India
e-mail: suwarnass@rediffmail.com

S.C. Satapathy et al. (eds.), *Proceedings of the International Conference on Data Engineering and Communication Technology*, Advances in Intelligent Systems and Computing 469, DOI 10.1007/978-981-10-1678-3_47

chains and nine-link, two-freedom chains but the studies showed that Franke's diagrams were constructed manually using visual inspection to discard infeasible molecules and binary-link assignments as well as to discard isomorphic alternatives which is necessary test not the sufficient one. Mruthyunjaya [4] recast, the method of transformation of binary chains, to derive simple-jointed chains in a format suitable for computer implementation. Mruthyunjaya, [5, 6] came up with the first fully computerized approach for structural synthesis. Uicker and Raicu [7] proposed for the first time in kinematics literature that characteristic polynomial adjacency matrix (CPAM) of the graph, but for more number of links to find the solution to polynomials was very difficult. Dube and Rao [8] gave a method for isomorphism check to identify distinct mechanism. Rao and Raju [9], Rao [10] proposed the secondary Hamming Number Technique for the generation of planar kinematic chains which was accepted. A major problem is faced all these years has been the absence of a reliable and computationally efficient technique to pick the nonisomorphic chains. The reason why designers have been plodding through so many new routes instead of sticking to what ought to have been a 'straight-as-an-arrow' path is easy to visualize with an implied requirement of decodability.

Read and Corneil [11] remark that a good solution to the coding problem provides a good solution to the isomorphism problem, though, the converse is not necessarily true. This goes to suggest that a successful solution to the isomorphism problem can be obtained through coding. The concept of canonical numbers provides identification codes which are unique for structurally distinct kinematic chains. One important feature of canonical numbers is that they are decodable, and also promises a potentially powerful method of identifying structurally equivalent links and pairs in a kinematic chain. While describing a method of storing the details of adjacency matrix in a binary sequence, maxcode and mincode are the tools for the identification of kinematic chains. The test of isomorphism then reduces to the problem of comparing max/mincodes of the two chains.

1.1 Advantages of the Proposed Method Over Existing Method

A number of studies are made on the enumeration and identification of kinematic chains. A common limitation of the alternative approaches proposed so far, for kinematic chains is that none uses a graph invariant which is complete. Also, the identification codes, developed on the basis of coefficient of structural similarity are not decodable, i.e., for a given identification code, it is not possible to reconstruct the linkage topology on the basis of identification codes alone.

Thus, while using the existing methods one cannot be sure that all the pairs of cospectral kinematic chains can be differentiated successfully. Also Maxcode, a canonical number is never used till now for the enumeration of kinematic chain with which the digital representation of kinematic chain is possible which is very much convenient to have storage and retrieval.

Hence, in the present paper, concept and algorithm is proposed for the computerized enumeration and identification problem of kinematic chains with the Maxcode, a canonical number. Canonical numbers provide identification codes which are unique for structurally distinct kinematic chains and also they are decodable, and promises a potentially powerful method of identifying structurally equivalent links (Isomorphoic) and pairs in a kinematic chain.

2 Canonical Numbering

According to Ambekar and Agrawal [12, 13] the concepts of maxcode and mincode were introduced as canonical number for the enumeration. For every kinematic chain of n-links, there are $n!$ different ways of labeling the links and hence, $n!$ different binary numbers are possible for the same chain. By arranging these n! binary numbers in an ascending order, two extreme binary numbers can be identified, as significant ones for the same chain. These two binary numbers are: the maximum number and minimum number designated, respectively, as maxcode $M(K)$ and mincode $m(K)$. Since $M(K)$ and $m(K)$ denote two extreme values of binary numbers for a given kinematic chain, and each has a unique position in the hierarchical order and is easily recognized, they are called as canonical numbers. Again each binary number of a given kinematic chain corresponds to a particular adjacency matrix, and hence it also corresponds to a particular labeling scheme.

The unique labeling scheme for the links of a kinematic chain, for which the binary number is in some (either maximum or minimum) canonical form, is called **canonical numbering** (labeling) of the chain. The adjacency matrix, which corresponds to canonical numbering, is said to be in some canonical form.

3 Property of Canonical Numbering

In a binary number an entry of '1' as an $(i + 1)$th digit, counted from the right-hand end, has a contribution to the decimal code equal to 2^i. Also, it follows from the basic property of binary numbers that

$$2^{i} > 2^{0} + 2^{1} + 2^{2} + 2^{3} + \cdots + 2^{(i-1)}. \tag{1}$$

This is obvious because the right-hand side of the above inequality represents summation of terms in geometrical progression and hence, can be shown to be equal to $(2^i - 1)$. This goes to prove that a contribution of any '1', in a binary number, is more significant than even the joint contribution of all the subsequent 'ones' in that binary number. This is fundamental to a basic understanding of any algorithm on maxcode and/or mincode.

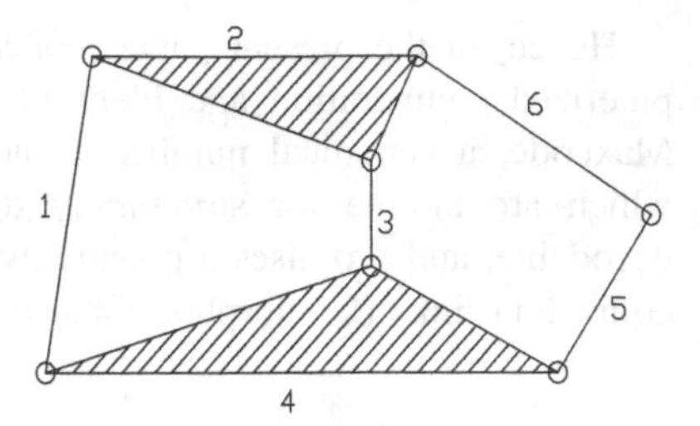

Fig. 1 Stephenson's chain

For the purpose of establishing a binary code one considers upper triangular adjacency matrix for the canonical labeling of the chain. Binary sequence is established by laying strings of zeros and ones in rows, '1' through (i – 1) row, one after the other in a sequence from top to bottom. This binary sequence may be regarded as a binary number illustrated by an example: Consider the **Stephenson's Chain** as shown in Fig. 1 with 6 links: 4 binary and 2 ternary links with single degree of freedom.

According to Bauchabaum and Freudenstein [14] the graphical method to represent kinematic structure consists of polygons and lines representing links of different degrees, connected by small circles representing pairs/joints. It is the powerful tool as it is well suited to computer implementation, using adjacency matrices to represent the graph. The graphical representation of the same chain with arbitrary labeling can be explained in Fig. 2.

For the arbitrarily labeled graph as shown in Fig. 2, the adjacency matrix and the corresponding UTAM (Upper Triangular Adjacency Matrix) is as under

$$A = \begin{vmatrix} 0 & 1 & 0 & 1 & 0 & 0 \\ 1 & 0 & 1 & 0 & 0 & 1 \\ 0 & 1 & 0 & 1 & 0 & 0 \\ 1 & 0 & 1 & 0 & 1 & 0 \\ 0 & 0 & 0 & 1 & 0 & 1 \\ 0 & 1 & 0 & 0 & 1 & 0 \end{vmatrix} \quad \text{Corresponding UTAM} = \begin{vmatrix} 1 & 0 & 1 & 0 & 0 \\ & 1 & 0 & 0 & 1 \\ & & 1 & 0 & 0 \\ & & & 1 & 0 \\ & & & & 1 \end{vmatrix}$$

There are fifteen entries in the UTAM which, if written consecutively by rows as 10100; 1001; 100; 10; 1; results in a binary sequence = 101001001100101.

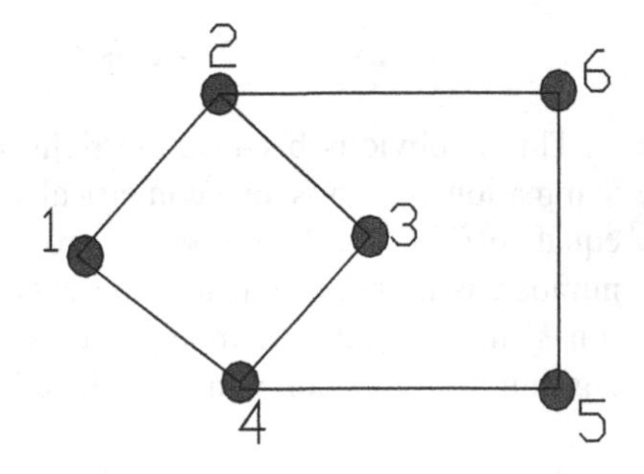

Fig. 2 Graph of chain (arbitrary labeling)

4 Application of Canonical Numbering with Decodability

Intuitively, the labeling of the same graph can be as in Fig. 3, giving the max code $M(G)$ and the corresponding UTAM

Corresponding binary sequence is: $M(G) = 1110000\ 100\ 100\ 11$. One can look at the resulting strings of 'ones' and 'zeros' as representing digits a binary code. And then it is more convenient to express maxcode in corresponding decimal form as

$$M(G) = 1.(2^{14}) + 1.(2^{13}) + 1.(2^{12}) + 0.(2^{11}) + 0.(2^{10}) + 0.(2^{9}) \\ + 0.(2^{8}) + 1.(2^{7}) + 0.(2^{6}) + 0.(2^{5}) + 1.(2^{4}) + 0.(2^{3}) + 0.(2^{2}) + 1.(2^{1}) + 1.(2^{0}) = 28{,}819$$

Intuitively, the labeling of the same graph can be as in Fig. 4 giving the min code $m(G)$ and the corresponding UTAM

In Fig. 4, for min code $m(G)$ corresponding binary sequence is $m(G) = 000110011101100$.

And the corresponding decimal min code is

$$m(G) = 0.(2^{14}) + 0.(2^{13}) + 0.(2^{12}) + 1.(2^{11}) + 1.(2^{10}) + 0.(2^{9}) + 0.(2^{8}) + 1.(2^{7}) + 1.(2^{6}) \\ + 1.(2^{5}) + 0.(2^{4}) + 1.(2^{3}) + 1.(2^{2}) + 0.(2^{1}) + 0.(2^{0}) = 3{,}308$$

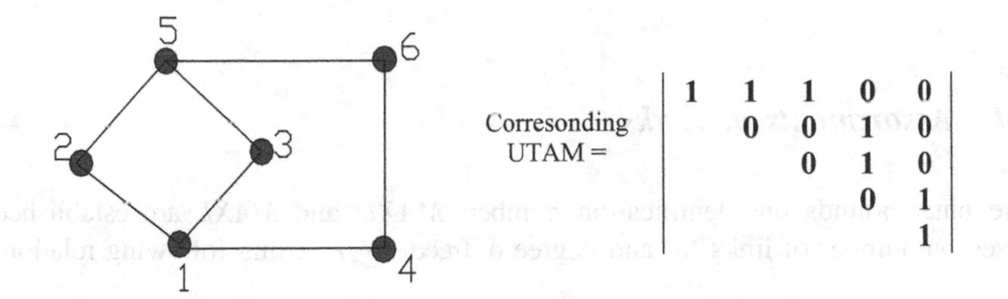

Fig. 3 Graph with maxcode labeling

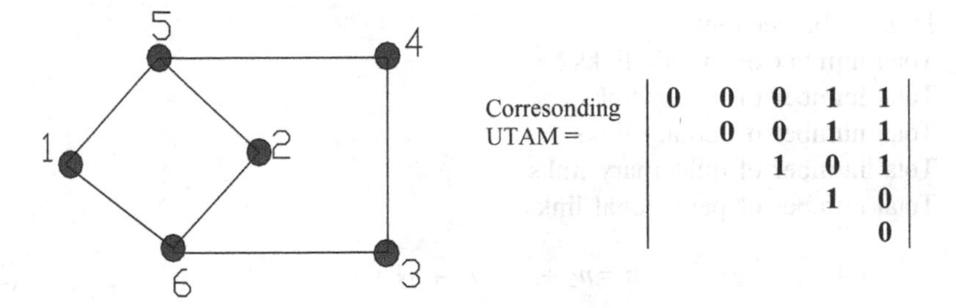

Fig. 4 Graph with mincode labeling

4.1 Decodability

For the given identification code, it is possible to reconstruct the linkage topology on the basis of these identification codes alone. This is made possible by the division of the identification code by 2. The reminders are arranged sequentially to get again the binary number with which the linkage topology can be reconstructed.

5 Algorithm for Structural Enumeration

An algorithm aimed at providing a computerized approach for enumeration of kinematic chains. The proposed method is based on the concept that the same chain can be generated from any of the n! identification codes, possible for an n-link chain. A range of identification codes (*MAXU* ~ *MAXL*) is established on the basis of all possible ways of connecting all k-nary links (k = 2, 3, 4, 5, …), available for a given permutation of links, to a link with highest number of elements. The range so established is then scanned for generating a valid chain, for every integer value in the range. The algorithm uses decoding process for each integer value, in the range of identification codes, to check as to whether or not it corresponds to a feasible chain. Modified maxcode algorithm has been used to verify whether or not the chain, so generated represents a distinct chain.

5.1 Assortments of Links

The outer bounds on identification numbers *MAXU* and *MAXL* are established based on number of links 'n' and degree of freedom 'F' using following relation:

$$F = 3(n - 1) - 2\mathrm{T}, \tag{2}$$

where

T number of simple turning pairs
F Degrees of freedom
n Total number of mobile links
n_2 Total number of binary links
n_3 Total number of ternary links
n_4 Total number of quaternary links
n_5 Total number of pentagonal links

$$n = n_2 + n_3 + n_4 + n_5 + \cdots \tag{3}$$

and

$$2L = 2n_2 + 3n_3 + 4n_4 + 5n_5 + \cdots \quad (4)$$

The link of highest degrees in the chains of this assortment is given by Eq. (4a) and (4b)

$$K = n/2, \quad (\text{when } F \text{ is odd}) \quad (4a)$$

$$K = (n+1)/2. \quad (\text{when } F \text{ is even}) \quad (4b)$$

From Eqs. (3) and (4) the possible combinations of n_2, n_3, n_4, n_5 are established for a given number of links 'n' and given degree of freedom 'F.'

Also it has been shown by Ambekar [15] that minimum number of binary links in a chain of 'n' links and F degree of freedom is given by

$$n_2 = (F+3) + (n_4 + 2n_5 + 3n_6 + \cdots). \quad (5)$$

6 Maxcode Algorithm for Enumeration of Chains

Step 1: Listing all assortments of binary, ternary, etc., links in the chains for given N and F. Assuming any of the link (amongst the available) of highest degree, as link 1 and write down first row of UTAM to extract maximum possible number.

Step 2: The range of Maxcode is established, i.e., upperlimit (MAXU) and lower limit(MAXL) are established.
MAXU—to link 1 connect all links of highest degrees from the remaining so as to produce feasible chains.
MAXL—To link 1 connect all binary links, so as to produce feasible chains.

Step 3: The range is then scanned for every digital number in it for generating feasible chain In the scanning process, the following steps are adopted

Step 3a: MAXL value is selected and its decoding is done to establish a binary code.

Step 3b: From the binary code adjacency matrix is generated which represents the kinematic chain.

Step 3c: To discard the adjacency matrix which does not give feasible chain (by eliminating the adjacency matrix which forms 3-links loop or open chain)

Step 3d: Once the adjacency matrix is selected then maxcode labeling scheme is used to identify the Kinematic Chain.
It is done in following way:

Step $3d_1$: a single Kinematic Chain if number of schemes of labeling are existing they are printed.

Step $3d_2$: Then establishing the decimal code for each scheme of labeling and pick up only those having maximum value of decimal code for the same Kinematic Chain and that scheme of labeling is known as canonical scheme of labeling

Step 3e: A canonical matrix is derived from a canonical scheme of labeling of links of the given Kinematic chain which is unique for a given chain.

Step 3f: The test of isomorphism then reduces to the problem of comparing maxcodes of the two chains.

Step 3g: Then Kinematic Chain is defined completely with the unique code (maximum) which is further checked for its existence in the earlier trails, if not then it is stored.

7 Conclusion

The present work demonstrates the power and potential of canonical number, in identifying kinematic chains and mechanisms. The scheme of numerical representation of kinematic chains gives unique identification codes for all the chains and mechanisms and are decodable. Thus, the canonical numbering (either maxcode or mincode): being unique and decodable holds great promise in cataloguing (storage and retrieval) of kinematic chains and mechanisms.

The paper also provides a computerized methodology whose maxcode (a tool of canonical numbering) algorithm for enumeration of kinematic chains has been elaborated along with the logic involved in each step which has much faster processing time (by reducing the range of identification codes for limiting computation time) in Sect. 5. The advantage with the generated computer program is the enumeration of distinct kinematic chains with automatic check of isomorphism and the catalogue of kinematic chains with their identification codes. Hence, the difficulty in retrieving kinematic chain discarded erroneously as being cospectral but nonisomorphic, during the process of enumeration has been solved.

The paper presents a method aimed at establishing identification codes, in the form of canonical numbers, have the advantage of decodability and also, in that no two nonisomorphic kinematic chains can ever have the same identification code. Thus even due to errors in labeling, no kinematic chain is likely to be discarded unless it is isomorphic to any of the kinematic chains listed already in the process of enumeration. As the "Identification code", is the numeric value used to represent the kinematic chain, the cataloguing of chains, storage and retrieval of kinematic chains is made easier and faster.

References

1. Crossley, F.R.E., A contribution to Gruebler's theory in the number synthesis of planar mechanisms, J. Eng. Indust., ASME Trans., Series B 86, pp. 1–8, (1964).
2. Manolescu, N.I., A unitary method for construction of the kinematic plane chains and plane mechanisms with different degrees of mobility, Revue Roumaine des Sciences Techniques, Serie de Mecanique Appliquee 9, (1964).
3. Davies, T.H., Crossley, F.R.E., Structural analysis of plane linkages by Franke's condensed notation, J. Mechanisms 1, pp. 171–184 (1966).
4. Mruthyunjaya, T.S., Structural synthesis by transformation of binary chains, Mech. Mach. Theory 14, pp. 221–231 (1979).
5. Mruthyunjaya, T.S., A computerized methodology for structural synthesis of kinematic chains: Part 1: Formulation, Mech. Mach. Theory 19, pp. 487–495, (1984).
6. Mruthyunjaya, T.S., A computerized methodology for structural synthesis of kinematic chains: Part 2: Application to several fully or partially known cases, Mech. Mach. Theory 19, pp. 497–505, (1984).
7. Uicker, J. J., and Raicu, A., A Method for Identification and Recognition of Equivalence of Kinematic Chains, Mechanism and Machine Theory, Vol 10, pp. 375–383 (1975).
8. Dube, R.K. and Rao, A.C., Detection of distinct mechanisms of a kinematic Chain, ASME paper 86-DET-172, (1986).
9. Rao, A. C. And Varada Raju, D., Mechanism And Machine Theory, 26(1), Pp. 55–75. (1991).
10. Rao, A. C., Mechanism And Machine Theory, Vol 32, No4 Pp. 489–499 (1994).
11. Read, R. C., And Corneil, D. G., The Graph Isomorphism Disease, J. Graph Theory, Vol. 1, Pp. 339–363, (1977).
12. Ambekar A. G., And Agrawal, V. P., Canonical Numbering Of Kinematic Chains And Isomorphism Problem: Maxcode, Asme, Design Engg. Technical Conference O 5–8, (1986).
13. Ambekar, A. G., and Agrawal, V. P., Canonical Numbering of Kinematic Chains and Isomorphism Problem: Mincode, Mechanism and Machine Theory Vol 22, No. 5, pp. 453–461 (1987).
14. Bauchsbaum, F., And Freudenstein, F., Synthesis Of Kinematic Structure Of Geared Kinematic Chains And Other Mechanisms, J. Mechanisms Vol 5, Pp. 357–392 (1970).
15. Ambekar, A.G., Mechanism and Machine Theory, PHI Learning, 2000, pp 43.

A New Congestion Avoidance and Mitigation Mechanism Based on Traffic Assignment Factor and Transit Routing in MANET

Jay Prakash, Rakesh Kumar and J.P. Saini

Abstract Data communication through wireless channels faces a challenging issue toward congestion and link quality variation. Regardless of mobile or stationary nodes in the network, the link in between the nodes experiences a fluctuation in link quality. Congestion is an intriguing issue, which occurs when there is deterioration in QoS at the link which transports data. This paper describes that an Internet network which is basically a backbone network with wired connection can be utilized for not just Internet services availability but also for mitigating congestion in a network. Data traffic in between nodes in a Mobile Ad hoc Network (MANET) can also be diverted through this backbone network to opt congestion-less environment and higher throughput. This technique is called as transit routing. The proposed protocol is compared with the uniform traffic allocation scheme. The experimental results prove that the proposed method outperforms the existing ones against congestion. It also reduces the packet drop ratio and increases system reliability.

Keywords Congestion control · Multiple paths · Transit routing · MANET

1 Introduction

A MANET is made up of several nodes which are mobile in nature having infrastructure-less decentralize environment [1]. Every mobile node in the MANET having dual nature act as a computing device and a router as well. Routing in

Jay Prakash (✉) · Rakesh Kumar · J.P. Saini
Department of Computer Science and Engineering,
M.M.M. University of Technology, Gorakhpur, India
e-mail: jpr_1998@yahoo.co.in

Rakesh Kumar
e-mail: rkiitr@gmail.com

J.P. Saini
e-mail: jps_uptu@rediffmail.com

S.C. Satapathy et al. (eds.), *Proceedings of the International Conference on Data Engineering and Communication Technology*, Advances in Intelligent Systems and Computing 469, DOI 10.1007/978-981-10-1678-3_48

between a source and a destination node, however, they are not directly in the transmission range of each other, relies on the intermediate routers. MANET nodes lend their hands in data forwarding for special purposes. During route discovery and data transmission phase, a few nodes can find the duplicate data and forward to destination node by using a number of routes. Consequently, more crowd due to the competitions in the nodes and over provisioning in traffic [2–5]. Data can be stored in the node's buffer and hence it results in congestion [6] when insufficient bandwidth is available for data transmission. Thus, it is an area of great concern in MANET to control forwarding of data and congestion control and avoidance techniques. Since we were considering the MANET scenario that already connected with the internet via gateways, the primary goal of an Internet [7] is to offer Internet services to the nodes lies within integrated MANET. Since, the data traffic corresponding to the request and response (request is generated by a host in MANET is responded by a web server placed in the Internet) generated using inherent protocols is routed using the Internet backbone, while the traffic in between two nodes inside the MANET merely routed over wireless connection available in MANET.

This paper illustrates that, how an Internet backbone subnet can also be utilized for routing intra MANET traffic through congestion control. This is called transit routing. This feature has capability to improve the most of the performance metrics and control congestion occurred within MANET, because the Internet subnet is much more trustworthy and carries high bandwidth compared to MANET.

1.1 Theory of Congestion Control System

Congestion control mechanism worries in supervision of control and data traffic [8] within a network. It avoids congestive collapse of a network by preventing unfair allocation of any of either processing or capabilities of a network and also by dropping the rate of sent packets.

(a) **Goals and Metrics of Congestion Control Algorithms**:

Goals for evaluating congestion control mechanisms are:

To achieve fairness and better compatibility with widely used well established protocols.

To achieve maximum bandwidth utilization.

To gain fairness efficiently and effectively.

To sustain high responsiveness.

To reduce the amplitude of oscillations

The rest of this paper is organized as follows. In Sect. 2, related works were discussed. In Sect. 3, we present the congestion avoidance strategy. The simulation results and analysis is given in Sect. 4 and finally in Sect. 5 we conclude the work.

2 Related Works

In wireless sensor network, congestion control mechanism plays an important role not only to ensure the non overflow environment over the reported samples but also to reduce the energy consumption in sensors so that the network lifetime increases.

Four different congestion control mechanism in the wireless sensor network are briefly described in this section.

In [9], a congestion control scheme is illustrated to control the congestion and to achieve efficient bandwidth allocation for data flow. Packet drop rate is used to detect the congestion on sink node. Since Fairness Aware Congestion Control (FACC) mechanism divided the nodes in two parts, near source nodes and near sink nodes based on their proximity to the source and sink nodes in network. Whenever a packet is lost, a near sink node sends a Warning Message (WM) to any node that lies in the near source region and then a Control Message (CM) is generated by that near source node to the source node after CM arrives to it. After getting the CM from the near source node, the source node adjusts its transmission rate based on the present traffic and newly calculated transmission rate.

An adaptive approach which relies on the learning automata for congestion control is discussed in [10]. The prime concern in this paper is to ensure a tradeoff in between data processing rate and transmission rate so that both are equivalent and hence the congestion occurrences decrease gradually. Each node enabled with an automaton that has ability to learn from existing scenario of network and can take decision based on the learnt information and network characteristics.

Another congestion control algorithm based on node priority index is described in [11]. Each node in the sensor network assigned a priority on the basis of their location and function within network. Nodes which lie near the sink assigned higher priority. The congestion detection in this paper relies on the factor calculated as a ratio of packet transmission rate to the packet arrival rate. If transmission rate is lower than the arrival rate then congestion would occur. The congestion-related information and priority index is piggybacked in the packet header of data packet. Adjustment of sending rate on each node is as per of the congestion at the node itself.

3 Congestion Avoidance Mechanism

In this section, we define some parameters that are used to assign traffic to the most prominent path among the number of paths available in between source and destination.

(i) **Buffer Occupancy (BO)**
The buffer occupancy ratio in node *n* is denoted as BUn; it is the ratio of the amount of buffered data to buffer size.

$$\text{BOn} = (\text{Size of buffered data})/(\text{buffer size}) \tag{1}$$

(ii) **Load Factor (LF)**
The load factor is denoted as LFn for node *n*.

$$LF_n = (\text{InFlown}/\text{OutFlown}) \tag{2}$$

Where InFlown = Sum of Inflow velocity at node and OutFn = Sum of Outflow velocity at node *n*.

(iii) **Forwarding Capability Function FC(LF)**
Forwarding capability function is denoted as FC(LFn) for node *n*.

$$FC(LF_n) = \begin{cases} 0, & (1 - LF_n) <= 0 \\ (1 - LF_n), & others \end{cases} \tag{3}$$

where $FC(LF_n) \in [0, 1]$. It infers the node's capability to forward data.

(iv) **Frequency of Node Utilization (FU)**

Frequency of node utilization is represented as FUn, which is the number of paths that uses node *n*. For a number of paths from one source may encompass the node *n* at the same time. Hence, frequency never depends upon number of sources.

Now, suppose a path in MANET encompasses nodes 1, 2, …, *n*, …, *m*, represented as P = {1, 2, …, *n*, …, *m*}. Parameters comprise the occupied ratio of buffers, load factor and frequency of node utilization. The buffer occupancy in path P is denoted as BOP, which means the max of the buffer occupancy Bon

$$\text{BOP} = \max\{\text{BOn} | \text{n} \in \text{P}\} \tag{4}$$

And (1 − BOP) describes the entire path capability to forwarding data.

The value of load factor for path P is represented as LFP, which is the maximum value of the load factor LFn,

$$\text{LFP} = \max\{\text{LFn} | \text{n} \in \text{P}\} \tag{5}$$

Frequency of path utilization is denoted as FUp, Namely

$$\text{FUP} = \max\{\text{FUn} | \text{n} \in \text{P}\} \tag{6}$$

(v) Path Availability Period

Availability period of the path is calculated on the basis of route connectivity extrapolation, where TP denotes the minimum route availability period and Tu is the route availability period in between two intermediate neighborhood nodes in a path from a source (Src) in MANET to the gateway (Gtw).

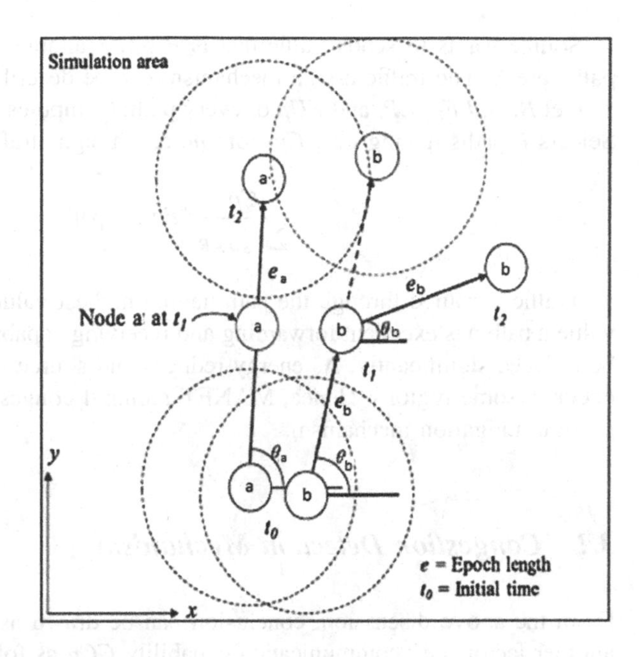

Fig. 1 Mobility pattern in MANET

$$\mathrm{TP} = \min\{\mathrm{Tu}\} \tag{7}$$

where u denotes the link between intermediate nodes in path *P*. In [3], connectivity period of a link Ltu, a link between node m and n, is computed as follows.

In Fig. 1, node *a* and *b* on path *P* are in the transmission range Tr of each other and the current positions of nodes *a* and *b* are (*Xa*, *Ya*) and (*Xb*, *Yb*), respectively.

Suppose θ*a* and θ*b* are moving directions and *va* and *vb* are speeds of node *a* and *b*, respectively. Then, Lu of node *a* and *b* is computed as

$$\mathrm{Lu} = \frac{-(\alpha\beta + \gamma\rho) \pm \sqrt{(\alpha^2 + \gamma^2)tr^2 - (\alpha\rho - \beta\gamma)2}}{\alpha^2 + \gamma^2}$$

where, $\alpha = va\cos\theta a - vb\cos\theta b$, $\beta = xa - xb$, $\gamma = va\sin\theta a - vb\sin\theta b$, and $\rho = ya - yb$.

Finally we compute the *CP* factor, and select the path with highest *CP* value.

$$Cp = \frac{Ep(1 - BOp)FC(LFp)Tp}{FUp} \tag{8}$$

Source wants to sends traffic that is B bit, available paths are M and requisite paths are R. The traffic assign mechanism can be described as algorithm

Get BOp, LFp, EP, and FUp of every path; Computes Cp, and sorts Cp by desc; Selects R paths having max Cp from m, and assigns traffic as

$$\frac{Cp}{\sum_{1 \leq p \leq R}} * B \text{ for } p \text{ path.} \tag{9}$$

Traffic is routed through the path having highest value of Cp. With highest Cp value a path has excellent forwarding and receiving capability hence congestion can be reduced significantly. As energy reduces and source increases, congestion may occur in some regions. Hence, MANET required congestion detection and a congestion mitigation mechanism.

3.1 Congestion Detection Mechanism

From the above discussion, conclusion can be drawn as follows. Here we define another factor, i.e., communication capability CCn as follows:

$$CCn = (1 - BOn) * FC(LFn), CCn \in [0, 1] \tag{10}$$

$BOn = 1$ indicates the buffer is full, $LFn \geq 1$ represents that the outflow velocity is less than the inflow velocity which causes congestion within the network. When $BOn = 1$ or $LFn \geq 1$, $CCn = 0$. This will result congestion in the network which causes packet drops. $BOn < 1$ indicates that the buffer is empty, $LFn < 1$ means the outflow velocity is larger than the inflow velocity, which represents the buffer utilized ratio will shrink more. When $BOn < 1$ or $LFn < 1$, $CCn > 0$ then this will result a non congested scenario.

After forwarding the data, nodes lies in the route calculates CCn and send to upstream with PREP packet. If $CCn = 0$, the node prevents itself to receives data. After getting PREP by upstream nodes (not source), they do:

If CCn value in PREP is 0, it represents that nodes in downstream traversal are congested. The node then blocks to send data to downstream nodes and forwards Path Response (PREP) to upstream.

If $CCn > 0$, it represents that the downstream nodes are placed in order. The node starts forwarding data in downstream fashion, and forwards PREP towards upstream. If the CCn in PREP is greater than the CCn of node, the CCn of PREP is updated by the CCn associated with the node. Finally the source node is aware of the fact of congested node if any.

3.2 Congestion Mitigation Technique

Now we focus on congestion mitigation technique which is based on transit routing. To mitigating the congestion detected in the path, a source node could divert the traffic to the external network via gateway and it is quite efficient and prominent way to mitigating the congestion resides in a particular path.

The basic idea is explored by using a scenario in Fig. 2. As per of scenario in Fig. 2 the actual flow of traffic is passed by internal MANET nodes but as soon as congestion is detected by the source node it make diversion in traffic through the external route, i.e., the internet backbone via gateways. Along with this the source node keep on checking the status of congested node, once the NCi of congested node is greater than zero again the traffic would divert through the inherent way (Internal MANET route). We have shown that the Internet backbone is used for intranet routing efficiently and effectively.

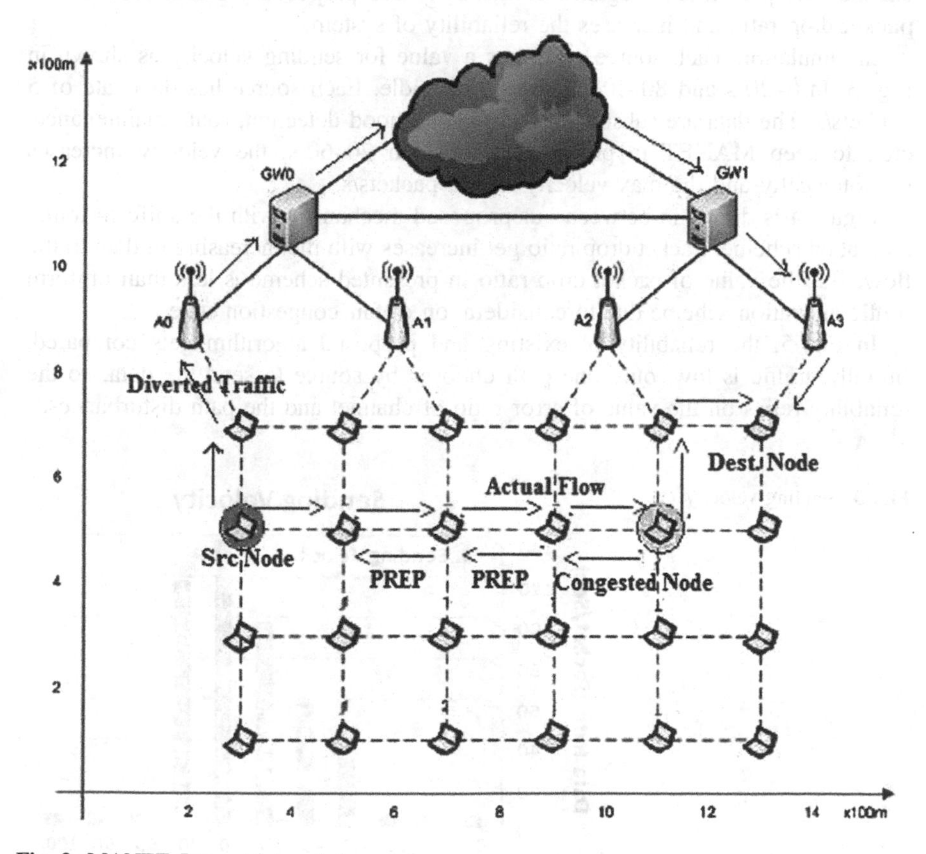

Fig. 2 MANET–Internet integrated scenario

This is the theoretical approach we have followed to mitigate the congestion in MANET. The use of internet backbone not only for accessing internet services but also for routing the data b/w nodes resides in MANET is referred as a transit routing [4].

4 Simulation and Analysis

This work is simulated using NS2. In this, the mobile host's channel bandwidth is set as 2 Mbps. MAC 802.11 protocol layer is used in this simulation with distributed coordination function (DCF) feature, number of mobile nodes are 100 and the canvas size is 1500 * 1000 m^2. Nodes mobility is set for 100 s only. Transmission range of all nodes is 250 m. This protocol is compared with the Uniform traffic allocation [5]. The experimental results prove that the proposed method exceeds the performance against congestion. Our proposed protocol reduces the packet drop ratio and increases the reliability of system.

In simulation, each source node has a value for sending velocity as shown in Fig. 3. In 0–20 s and 80–100 s, sources are idle. Each source has data rate of 5 packets/s. The data are substantial for neighborhood detection, route maintenance, etc., to keep MANET in proper condition. In 20–60 s, the velocity increases monotonically and the max velocity is 150 packets/s.

Figure 4 is drawn in between our proposed mechanism with the uniform traffic allocation scheme. Packet drop ratio get increases with the increasing in data traffic flow. But the value of packet drop ratio in presented scheme is less than uniform traffic allocation scheme due to consideration of fair congestion state.

In Fig. 5, the reliability of existing and proposed algorithm gets compared. Initially, traffic is low; only one path chooses by source to send the data, so the reliability relies on the value of error ratio of channel and the path disturbances.

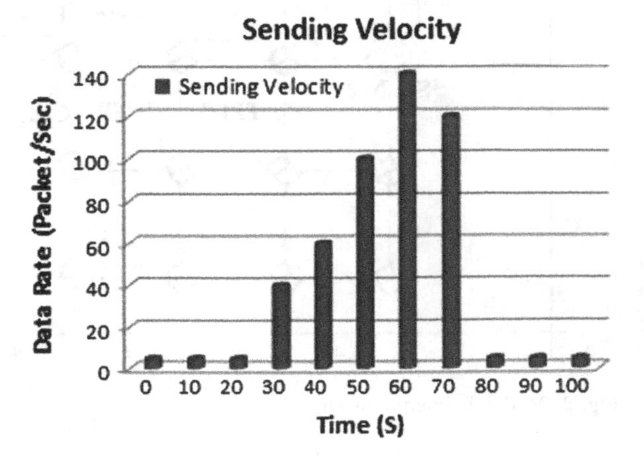

Fig. 3 Sending velocity

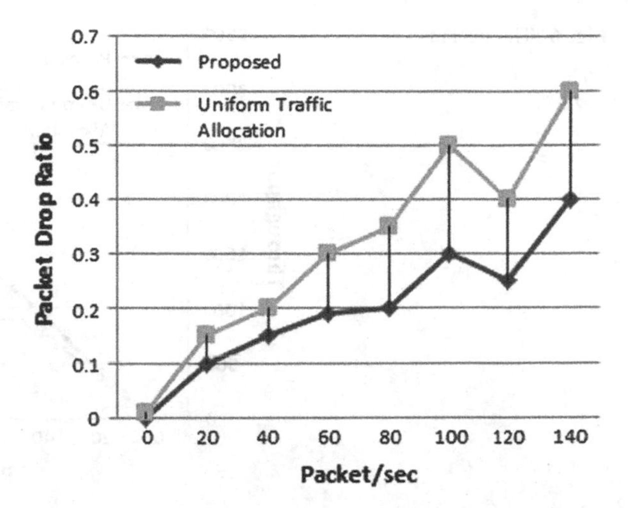

Fig. 4 Packet drop ratio

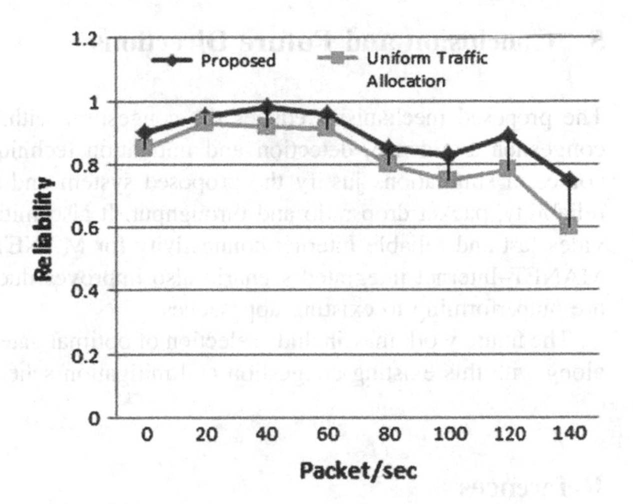

Fig. 5 Reliability

Reliability is increased till data traffic reached 40 packets/s because source adopting different routes in order to send the packets to the destination. Following the traffic is increased and it is reached above 40, the reliability started gradually descending due to the disturbance among paths is increases. Comparatively, the reliability of proposed system is higher.

In Fig. 6, as data rate increases, the throughput increases as well. Moreover, the throughput is higher in proposed system in comparison with existing mechanism.

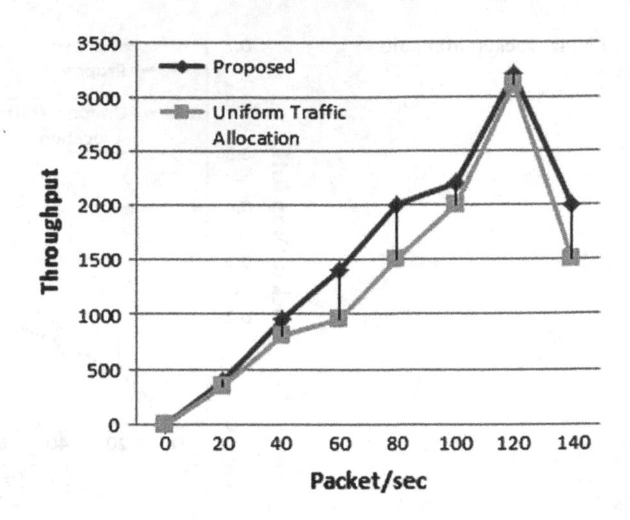

Fig. 6 Throughput

5 Conclusion and Future Directions

The proposed mechanism reduces the congestion within MANET. In this paper, congestion avoidance, detection and mitigation technique is proposed. The outcomes of simulations justify the proposed system and have satisfactory effect in reliability, packet drop ratio and throughput. It also mitigates congestion and provides fast and reliable Internet connectivity for MANET. The performance of the MANET–Internet integrated scenario also improves due to this approach. Results are outperforming to existing approaches.

The future work may include selection of optimal gateway discovery mechanism along with this existing congestion and mitigation scheme.

References

1. Internet Engineering Task Force (IETF) Mobile Ad Hoc Networks (MANETs) Working Group Charter, http://www.ietf.org/html.charters/manet-charter.html.
2. M. Youssef, M. Abrahim, M. Abdelatif and Lin Chen, "Routing Metrics of Cognitive Radio Networks: A Survey", IEEE Communications Surveys and Tutorials, 16(1), pp. 92–109 (2014).
3. Rafi U Zaman, Khaleel Ur Rahman Khan and A. Venugopal Reddy, "Path Load Balanced Adaptive Gateway Discovery in Integrated Internet-MANET", Proceedings of the IEEE International Conference on Communication Systems and Network Technologies, pp. 203–206 (2014).
4. Shahid Asif Iqbal and Humayun Kabir, "Hybrid Scheme for Discovering and Selecting Internet Gateway in Mobile Ad Hoc Network", International Journal of Wireless & Mobile Networks, Vol. 3, No. 4, pp. 83–101 (2011).
5. T. Meng, F. Wu, Z. Yang and G.Chen, "Spatial Reusability-Aware Routing in Multi-Hop Wireless Networks", IEEE Transactions on Computer, Vol. PP 244–255 (2015).

6. Yung Yi, Sanjay Shakkottai "Hop-by-hop congestion control over awireless multi-hop network". IEEE/ACM Transactions on Networking, vol. 15, pp. 133–144 (2007).
7. Z.M Fadlullah, A.V Vasilakos and N. Kato, "On the partially overlapped channel assignment on wireless mesh network backbone: A game theoretic approach", IEEE Journal on Selected Areas in Communications Vol. 30, 1, pp. 119–127 (2012).
8. C. Busch, R. Kannan, and A.V Vasilakos, "Approximating Congestion Dilation in Networks via Quality of Routing Games", IEEE Transaction on Computers vol. 61, 9, pp. 1270–1283 (2012).
9. Xiaoyan, Y., Xingshe, Z., Zhigang, L., Shining, L.: A Novel Congestion Control Scheme in Wireless Sensor Networks. In: 5th International Conference on Mobile Ad-hoc and Sensor Networks, Fujian, pp. 381–387 (2009).
10. S. Misra, V. Tiwari and M. S. Obaidat, LACAS: Learning Automata-Based Congestion Avoidance Scheme for Healthcare Wireless Sensor Networks, IEEE Journal on Selected Areas in Communications, Vol. 27, pp. 466–479 (2009).
11. Wang C, Li B, Sohraby K, Daneshmand M, Hu Y. Upstream congestion control in wireless sensor networks through cross-layeroptimization. IEEE Journal on Selected Areas In Communications, 25(4): 786–795 (2007).

MRWDPP: Multipath Routing Wormhole Detection and Prevention Protocol in Mobile Ad Hoc Networks

Ravinder Ahuja, Vinit Saini and Alisha Banga

Abstract A mobile ad hoc network (MANET) is a continuously self-configuring, infrastructureless network of mobile devices connected without wires. Ad hoc is Latin and means "for this" (i.e., for this purpose). The nodes also act as router. Routing is an important aspect of MANETs which is used to find the route which can be used for communication. Traditionally, routing protocols were focused on performance only. But security is also an important aspect. Due to limited bandwidth, limited storage capacity, dynamic topology, shared medium, open peer-to-peer communication, security in MANETs is difficult to implement compared to wired networks. There are number of attacks on routing protocols like routing table overflow attack, black hole attack, wormhole attack, route cache poisoning, sybill attack, modification, fabrication, location spoofing attacks, etc., which affect the functioning of MANETs and degrade the performance. So, there is need to secure routing protocols so that their functioning is not affected and performance is not degraded due to these attacks. A new way of detection and prevention of wormhole attack is proposed in this paper. Dynamic Source Routing (DSR) is converted into multipath routing protocol by changing the way intermediate node forwards the route request packet. The QualNet 5.0.2 simulator [1] is used to validate the proposed approach. Two new packets Dummy_Request and Dummy_Reply are introduced to check the vulnerability of the path. The format of these packets is same as route reply except the option type. The performance parameters used are packet delivery ratio and throughput. The results show that packet delivery ratio and

Ravinder Ahuja (✉)
Jaypee Institute of Information Technology, Sector-128, Noida, Uttar Pradesh, India
e-mail: ravinder.ahuja@jiit.ac.in

Vinit Saini
Amazon, Banglore, Karnataka, India
e-mail: vinit.saini.ynr@gmail.com

Alisha Banga
Advanced Institute of Technology and Management, Palwal, Haryana, India
e-mail: alishabanga47@gmail.com

S.C. Satapathy et al. (eds.), *Proceedings of the International Conference on Data Engineering and Communication Technology*, Advances in Intelligent Systems and Computing 469, DOI 10.1007/978-981-10-1678-3_49

throughput under wormhole attack is less compared to protocol without wormhole attack and theMultipath Routing Wormhole Detection and Prevention Protocol (MRWDPP) improves performance under wormhole attack.

Keywords MANETs · Wormhole · Split multipath routing · DSR · AODV

1 Introduction

Wireless networks use frequencies to communicate instead of using some physical cables. There exist two types of wireless networks: infrastructure network and infrastructureless network. MANETs is a type of infrastructureless network because there is no base station, no central administration authority, and no fixed routers. Each device acts as a router also. Security is a big issue for wireless networks in today's environment. Wireless devices usually has limited bandwidth, storage space, and processing capabilities due to which it is harder to reinforce security in wireless network. Routing in mobile ad hoc network is a fundamental task which helps nodes to send and receive packets. Traditionally routing protocols focused on the performance but nowadays security is also important. So either new protocols are being designed which includes security also or security feature is being included in the already existing routing protocols. There are number of attacks on the routing protocols like packet dropping, modification, fabrication, impersonation, Sybil attack, wormhole attack, etc. But wormhole attack is more dangerous and very difficult to find out because it may not modify the packets. For wormhole attack two or more nodes are required. Two nodes are connected via secret channel and communicate through this secret channel. In this attack, if two nodes are far apart, then also it seems that they are close to each other because they are connected through secret channel which is not known to other nodes. Whenever we find the route between source and destination these nodes provides shortest route between them. Once they are on the route between source and destination they can easily do anything on the data packets moving between source and destination. Channel between nodes can be specified by two methods.

Packet Encapsulated Channel: It is also called in-band channel. Two malicious nodes $m1$ and $m2$ create path in advance between them and when source node s broadcasts a routing message, it would be received by node $m1$ and then $m1$ encapsulates this message in payload of the data packet and transmit it using prebuilt path between $m1$ and $m2$. $m2$ after receiving encapsulated packet will extract the RREQ message and broadcast it until it reaches the destination node. Path through which this RREQ packet sent is the shortest path and that is why the destination will reply through this path only to source node.

Out-of-Band Channel: A wired network or private channel is used for communication between two malicious nodes.

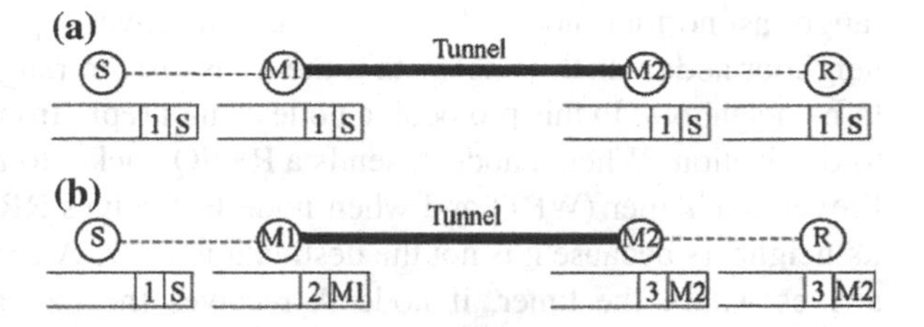

Fig. 1 Types of Wormhole attacks. **a** Hidden attack. **b** Exposed attack

There are two kinds of wormhole attacks [2]. One is hidden attack and the other one exposed attack. In hidden attack malicious node is hidden, legitimate node does not know their presence in the network; and in exposed attack legitimate node knows the existence of wormhole nodes. As shown in Fig. 1.

2 Literature Review

2.1 Packet Leashes [2]

Special information called leash is added in the packet to restrict the packets' maximum allowable distance. Leashes can be geographic leashes and temporal leashes. In geographic leashes each node must know its own location and loosely synchronized clocks. When a node sends packet it includes its location and time, the receiver will check the time and location of packet and calculates the upper bound on the limit between sender and himself. Other techniques of authentication can be used for verification of time and location. In temporal leashes, all nodes are equipped with tightly synchronized clocks. In this sender node, sending the packet includes the time when the packet was sent, *ts*; receiving node compares time t_s (time of sending the packet) with t_r (time at which packet is received). The receiver can calculate how far the packet has traveled. By including the expiration time in the packet temporal leashes can be constructed, after which receiver would not accept the packet. Other techniques of authentication can be used for verification of time.

2.2 WAP: Wormhole Attack Prevention Algorithm in MANETs [3]

This is based on Dynamic source routing (DSR). All nodes monitor their neighbor node behavior while sending RREQ (route request) packet using a special list called neighbor list. Source node receives route reply message from destination and checks whether the routes were under the attack or not. After detecting the wormhole node it is placed in the wormhole node list. In this, radio links are assumed to be bidirectional and wormhole nodes are also having same transmission range as normal node. Through route discovery procedure, node detects its neighbor node which is not within the transmission range of the node but pretends to be neighbors. In this protocol, a node cannot reply from cache if it is having route to destination. When a node A sends a RREQ packet to node B, it starts Wormhole Prevention Timer (WPT) and when node B receives RREQ, it broadcasts to all of its neighbors because it is not the destination. Node A can check whether the RREQ arrives within the timer, if node A receives the message after timer expires, it suspects B or one of its next nodes to be wormhole nodes. Each node is having neighbor node table containing sending time, receiving time, count, RREQ sequence number, and neighbor node ID.

2.3 WARP: Wormhole Avoidance Routing Using Anomaly Detection Mechanism [4]

This algorithm is an extension of AODV for security enhancement. It assumed link disjoint multipath into consideration in route discovery phase. During path discovery in WARP, an intermediate node will try to create a route which is not having hot neighbor on its path, which has a higher rate of route formation than a particular value. It may be that a node is placed at a particular position in the network but due to mobility it will not stay there for long time. In this protocol message format of RREQ has been modified and contains an additional field called first_hop which records the first node receiving the RREQ after leaving the source and a new message called RREP_DEC is also introduced whose format is same as RREP. As WARP is multipath, the node that sends RREQ after receiving back RREP sends a packet (RREP_DEC) on the same route to note down which nodes reside on the path. Every intermediate node creates one forward entry toward destination.

Routing Table Format: Routing table is modified and having additional fields (i) first_hop (ii) RREP count and (iii) RREP_DEC count. Anomaly value of neighbor node can be calculated from last two parameters.

Anomaly value = (number of RREP_DEC)/(number of RREP + 1)

Anomaly value represents that a node is malicious node.

2.4 DelPHI: Wormhole Detection Mechanism for Ad Hoc Wireless Network [5]

Delay information and hop count of paths are collected at the sender and delay/hop information is used as a measure for detection of wormhole attack which provides solution to both types of attacks. Delay under wormhole attack is high compared to normal route delay. Path which has high delay/hop is having high probability of under wormhole attack. It consists of two phases such as data collection phase and data analyses phase. In this, receiver replies to each RREQ received. It consists of two messages DRREQ (DelPHI route request) and DRREP (DelPHI route reply) and includes a timestamp field, hop field, and hop count field previously. Sender broadcasts DRREQ and after receiving the DRREQ, the receiver replies it with the DRREP packet. Sender can receive multiple DRREP packets. Each DRREP contains the hop count information of the path that is associated with it. By calculating the difference between timestamp carried in DRREP and time at which DRREP is received, we can calculate round trip time of the path. Then sender is able to calculate the delay/hop value of corresponding path. DPH value is calculated and arranged in descending order and finds whether there is large gap between the two than a threshold value; then that path is under wormhole attack.

3 Proposed Methodology

This approach is based on DSR protocol.

Algorithm

Step 1: When a source node wants to send data packets to destination, it will check its route cache for the path to destination node; if valid route is available, then it will use valid route to send data packet. Otherwise, source node will broadcast a route request packet to find the routes to the destination; destination receives multiple route requests from the source and replies to each route request with route reply. Sender will receive multiple route reply from the destination and stores multiple paths to the destination in its route cache.

Step 2: After route setup, sender wants to send data packets to the destination before that it sends a Dummy_Request packet to check the vulnerability of the path. Now, sender will choose one of the paths whose hop count is minimum. It will set the number of packets before next Dummy_Request field in Dummy_Request packet, which will represent how many data packets will be sent before sending next Dummy_Request packet. Sender will set a timer and wait for Dummy_Reply packet.

Step 3: Destination node will receive the Dummy_Request packet and form a Dummy_Reply packet and set the number of data packets received till last dummy_request field. Destination node will reverse the path received in Dummy_Request and send Dummy_Reply packet on this path to source node.

Step 4: If source node receives the Dummy_Reply packet before timeout of Dummy_Request packet, then it will compare the number of data packets received till last dummy_request field received in Dummy_Reply; with set number of packets before next Dummy_Request field sent in previous Dummy_Request. If it matches, path is not under wormhole attack and it starts sending data packets. Meanwhile, if sender has to send data packets, then it will start buffering data packets until it receives wormhole free path from source to destination. If sender does not receive Dummy_Reply packet, then it will delete the path from its route cache and selects another path for checking the vulnerability of path.

Step 5: Source node will send fixed number of data packets to destination node which is contained in Dummy_Request packet and then whole process is repeated to check vulnerability of currently using path against wormhole attack.

This approach only checks whether the wormhole nodes are dropping the data packets or not. Detection of wormhole attack is done after few packets have been sent. This approach cannot identify the wormhole link on the route.

4 Simulation and Performance Parameters

The QualNet 5.0.2 simulator [1] is used to simulate and validate the proposed heuristics. Simulation Parameters taken are given below

Scenario-I	
Parameters	Values
Nodes	30
Maximum speed	10 mps
Traffic type	Constant bit rate
Terrain	1000 × 1000 m
Minimum speed	0 mps
Mobility model	Random waypoint
Number of Wormhole links	1
Packet size	512
Packet rate	4 Packets/s
Routing protocol	Multipath routing protocol
Wormhole link	18–19
Simulation time	300 s
Pause time	30, 60, 120, 240, 300
Scenario-II	
Parameters	Values
Nodes	30
Simulation time	100 s
Traffic type	Constant bit rate
Terrain	1000 × 1000 m
Mobility model	Random waypoint
Maximum speed	10 mps

(continued)

(continued)

Scenario-II	
Parameters	Values
Pause time	30 s
Minimum speed	0 mps
Routing protocol	Multipath Routing Protocol, MPR with proposed approach
Packet size	512
No. of Wormhole links	1, 2, 3, 4
Wormhole link	4–19, 16–21, 23–30, 9–20
Packet rate	4 Packets/s

4.1 Performance Metrics

Packet Delivery Ratio (PDR): Packet Delivery is calculated by dividing the total data packets received at destination node to the total data packets sent by source node.

Throughput: Throughput is the successful transmission rate of the network and defined as number of data packets successfully transmitted to destination per time unit. *Throughput* = No. of bytes received * 8 * 100/(Time when last bytes received − Time when first byte received).

5 Results and Discussions

5.1 Results of Multipath Routing Protocol With and Without Wormhole Attack with Varying Pause Time

In this simulation pause time is varied. Pause time is used in random waypoint mobility model in which nodes move in a random direction and stay in a position for certain amount of time. This stay time is called pause time. Previously, performance is measured by varying the node mobility making the network unstable. In this, we have varied the pause time making the network stable. As we increase, the pause time network becomes stable. Threshold value of wormhole is taken 75.

(a) **Packet Delivery Ratio**: In this, only one wormhole link is created between node 18 and node 19. Number of nodes taken is 30. Figure 2a shows that in the presence of wormhole link, packet delivery ratio of multipath routing protocol are decreased. This is because the wormhole nodes drop the data packets passing through them.

(b) **Throughput**: Figure 2b shows that in the presence of wormhole link throughput of multipath routing protocol have been decreased. It is almost half

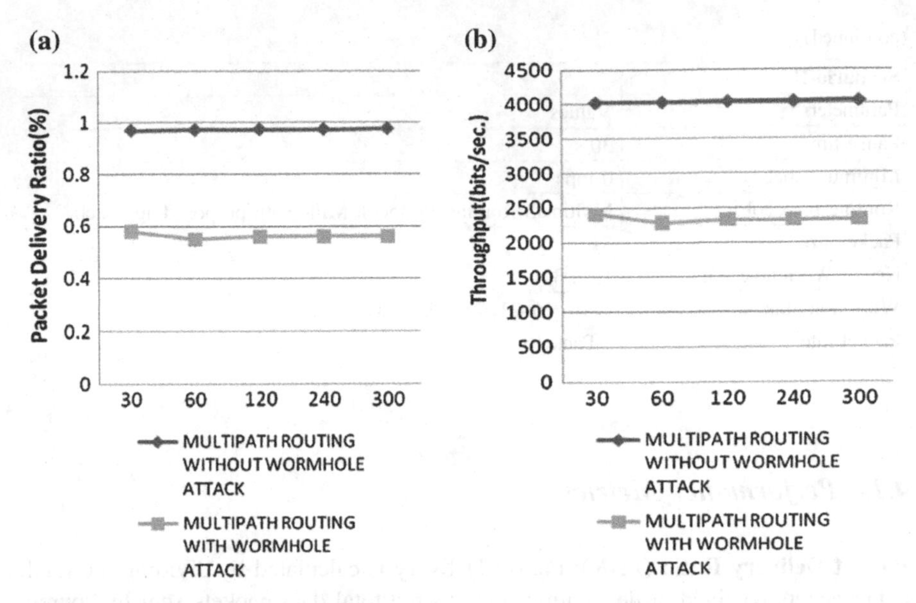

Fig. 2 **a** Packet delivery ratio. **b** Throughput for multipath routing with and without Wormhole attack

the throughput of the network without wormhole attacks. This is because the data packets are dropped by wormhole nodes and bandwidth available is not being used for transmission of data packets. So wormhole attack also decreases the performance of multipath routing protocol.

5.2 *Results of MRWDPP with Multipath Routing Under Wormhole Attack With Variable Wormhole Links*

Scenario-II is taken into consideration. One wormhole link is placed near the source and destination node. Other wormhole links are placed randomly in the network.

(a) **Packet Delivery Ratio**: Figure 3a shows the comparison of MRWDPP with the multipath routing protocol under wormhole attack. Figure 3b shows that multipath routing protocol with MRWDPDP delivers more data packets than multipath routing protocol without new approach; and as there is increase in wormhole links the performance of both is decreased but still MRWDPP delivers more data packets. So effects of wormhole attack on packet delivery ratio of multipath routing protocol are reduced with this approach.

(b) **Throughput**: Figure 3b shows the results of MRWDPP and multipath routing protocol. It shows that in the presence of one wormhole link, throughput of multipath routing is less compared to MRWDPP and as the number of

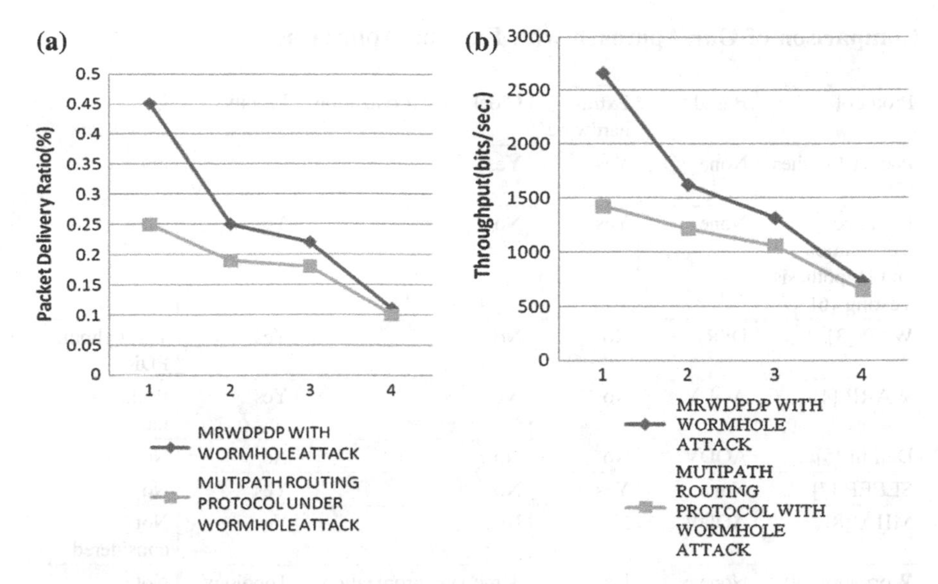

Fig. 3 **a** Packet delivery ratio. **b** Throughput comparison of MRWDPP with multipath routing protocol under wormhole attack with varying number of wormhole link

wormhole links increases throughput of both decreases but MRWDPDP still has higher throughput than the multipath routing protocol. So the solution has eliminated the effects of wormhole on the multipath routing protocol.

6 Conclusions

In this paper, approach for detection and prevention of wormhole attack on multipath routing protocol is proposed called MRWDPP. Two new packets Dummy_Request and Dummy_Reply are introduced. DSR protocol is first converted to multipath routing protocol by changing the way intermediate node forwards the route request. By changing DSR, it becomes multipath routing protocol and during discovery of route from source to destination multiple routes are collected. Performance of multipath routing is measured by changing the pause time (making the network stable) and performance of MRWDPP and multipath routing is evaluated and compared by varying the wormhole links. Results show that (i) Performance of DSR routing protocol is decreased by half under wormhole attack (ii) Performance of DSR routing protocol decreases with the increase in wormhole links (iii) Performance of multipath routing protocol also decreases under wormhole attack (iv) MRWDPP increases the packet delivery ratio and throughput of multipath routing protocol in the presence of wormhole attack compared to multipath routing protocol under wormhole attack.

Comparison of Our Approach with Existing Approaches

Protocol	Based	Extra hardware	Clock synchronization	Mobility	QoS parameters
Packet Leashes [2]	None	Yes	Yes	No	No
Distance Verification and Hypothesis Testing [6]	None	Yes	No	Yes	No
WAP [3]	DSR	No	No	Yes	Throughput, PDR
WARP [4]	AODV	No	No	Yes	Packet loss rate
DelPhi [5]	AODV	No	No	No	No
SEEEP [7]	None	Yes	No	Yes	No
MHA [8]	AODV	No	No	Yes	Not considered
Wormeros [9]	None	No	Time synchronization not considered, RTT between nodes is considered	Topology change is not considered	Not considered
MRWDPP (Our approach)	Multipath routing	No	No	Yes	PDR, throughput

References

1. Qulanet available online http://scaleable-networks.com.
2. Y.-C. Hu, A. Perrig, and D.B. Johnson. "Packet Leashes: A defense against Wormhole Attacks in wireless networks. IEEE INFOCOM, Mar 2003.
3. Sun Choi, Doo-young Kim, Do-Hyeon Lee, Jae-il Jung" WAP: Wormhole Attack Prevention Algorithm in Mobile Ad Hoc Networks" © IEEE 2008.
4. Ming-Yang Su" WARP: A wormhole- avoidance routing protocol by anomaly detection in mobile ad hoc network" Computers and Security 29(2010) 208–224 © 2009 Elsevier Ltd.
5. Hon Sun Chiu and King-Shan Lui "DelPHI: Wormhole Detection Mechanism for Ad Hoc Wireless Networks" © 2006 IEEE.
6. M Yifeng Zhou Louise Lamont Li "Wormhole Attack Detection Based on Distance Verification and the Use of Hypothesis Testing for Wireless Ad Hoc Networks" © 2009 Crown.
7. Neelima Gupta and Sandhya Khurana "SEEEP: Simple and Efficient End-to-End protocol to Secure Ad Hoc Networks against Wormhole Attacks" © 2008 IEEE.
8. Shang-Ming Jen 1, Chi-Sung Laih 1 and Wen-Chung Kuo "MHA: A Hop count analysis Scheme for Avoiding Wormhole Attacks in MANETs" *Sensors* 2009.
9. Hai Vu, Ajay Kulkarni, Kamil Sarac, and Neeraj Mittal "WORMEROS: A New Framework for Defending against Wormhole Attacks on Wireless Ad Hoc Networks" WASA 2008, LNCS 5258, pp. 491–502, 2008

Self-coordinating Bus Route System to Avoid Bus Bunching

Vishal B. Pattanashetty, Nalini C. Iyer, Abhinanadan Dinkar and Supreeta Gudi

Abstract This paper elucidates the problem on bus bunching and the proposed solution to the problem. The motive is to consider the density of passengers and schedule the bus based on the density. The system is capable of considering the density by user-friendly mobile application. The mobile application is accessible within the personal area communication range of the bus terminus. The whole system is controlled by a low power, highly sustainable, and secured wireless technology. The main advantage of this proposal is to reduce unnecessary running of the buses, by which the fuel consumption is reduced and passengers need not wait for a longer time for the buses. The outcome of this result is to establish a well-maintained public transportation by reducing the problem of bus bunching.

Keywords Ad Hoc networks · WSN · ZigBee · IOT · Public transport · Bus bunching

1 Introduction

Traveling through buses is one of the main means of transport in our country. Population of over one million people chooses bus as the means to travel to different places in urban as well as rural areas. Buses take around 90 % of public transport in Indian cities, and serve a cheap and convenient mode of transport for all

V.B. Pattanashetty (✉) · N.C. Iyer · Abhinanadan Dinkar · Supreeta Gudi
Department of Instrumentation Technology, B.V. Bhoomaraddi College of Engineering & Technology, Vidyanagar, Hubli, Karnataka, India
e-mail: vishalbps@bvb.edu

N.C. Iyer
e-mail: nalini_c@bvb.edu

Abhinanadan Dinkar
e-mail: abhidnkr07@gmail.com

Supreeta Gudi
e-mail: gsupreeta@gmail.com

S.C. Satapathy et al. (eds.), *Proceedings of the International Conference on Data Engineering and Communication Technology*, Advances in Intelligent Systems and Computing 469, DOI 10.1007/978-981-10-1678-3_50

classes of society. To make the public bus service helpful for passengers, the proper bus schedule is maintained. However, it is impossible to maintain equal headways between buses because of variability in traffic and in the boarding and de-boarding of passengers, which leads to bunching of buses. This paper provides a solution to the problem of bus bunching, which is bothering both passengers and operators in public transportation system. The bus bunching problem typically means buses of the same service route getting too close together so that they "bunch" each other. It is undesired because it usually increases the waiting time of passengers before and after the arrival of the bunched buses. Several approaches have been proposed to reduce bus bunching, in particular bus routing, bus stop planning and bus scheduling. One of them is using GPS system [1] and another one is using IR sensors and Microcontroller [2]. In this paper cause of the bus bunching problem is analyzed, and proposed a new solution taking the density of people into consideration in scheduling the buses to different areas. This paper facilitates the use of reliable low power, self-healing, and highly secured personal area network wireless communication protocol [3].

2 Project Organization and Framework

The whole working system is organized in three stages: refer Fig. 1.

Stage A: Monitoring Department
This is the initial stage and here the role of monitoring department is to keep track of the information acquired from the other stages, until the assignment of locations to the bus, arrived at the terminal. The bus is detected, as it arrives at the bus terminus and the previous location tagline on the bus is erased and in concert the passenger density information from the successive stage is acquired.

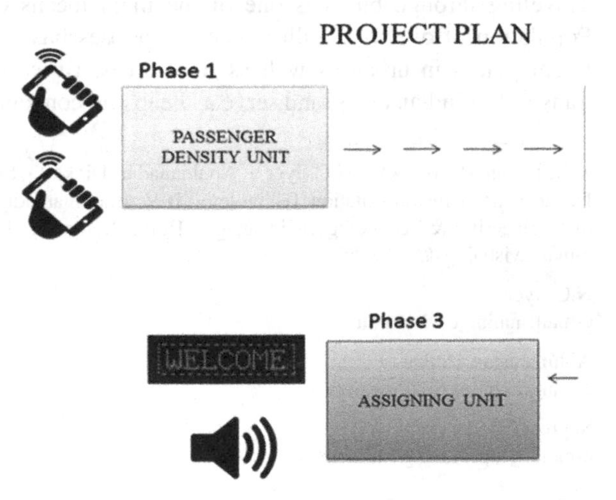

Fig. 1 Project framework

Stage B: Bus terminus Wi-Fi range
This is the eminent stage, as the density of the passenger is conveyed to the stage A. The passenger density is monitored using an app, which is easily accessible by all the passengers. The mobile application can be easily accessible within the Wi-Fi range of bus terminus. The application has the list of locations and the information of lanes. The passengers have to select the locations. Based on the density of the passengers the buses are rescheduled.

Stage C: Assignment of locations
In this stage the information regarding the density of passengers is considered from the preceding stages and based on that the assignment of locations to the bus is done. The information of the passenger density from the different lanes is compared and the lane with the higher density is given the priority and the bus is allocated to that particular lane.

3 Proposed Methodology

3.1 ZigBee

ZigBee is a specification for set of high-level communication protocols using tiny, low-power digital radios based on IEEE 802 standard for personal area networks. For controlling of traffic management ZigBee has been implemented. ZigBee devices are suitable for providing wireless capability to any product with serial data

Table 1 ZigBee specification

Specifications	Range
Maximum transmit power	1 mW
RF data rate	250 kbps
Receiver sensitivity	−92 dBm
Serial interface data rate	Maximum 115200 baud
Antenna options	Chip & wire antenna
Operating temperature	−40 to 85 °C

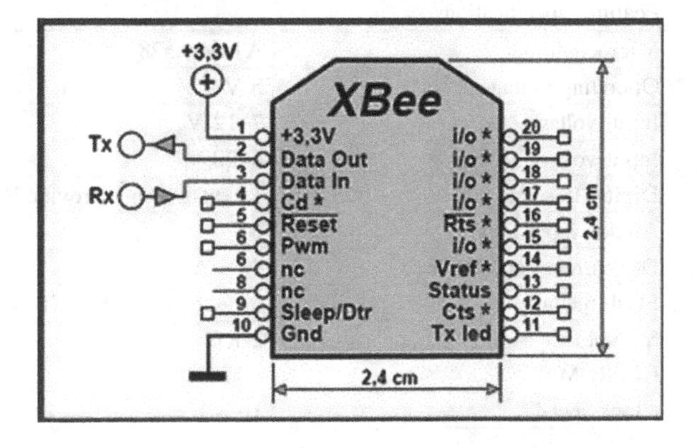

Fig. 2 Pin configuration of ZigBee

interface. ZigBee is a low-cost, low-power wireless mesh network standard. The low cost allows the technology to be widely used deployed in wireless control and monitoring system. Low-power usage allows longer life with smaller batteries. The specifications of ZigBee are as follows (Table 1 and Fig. 2).

3.2 Arduino

The microcontroller is heart of the system which acts as a central controller for the different phases. In search of microcontroller which has low cost, high speed, reliable and is efficient in performing the defined task, we came across a flexible software programming, easy-to-use open-source prototyping Arduino board. Ardunio board is one type of a microcontroller board which comes with an easy provision of connecting with the CPU of the computer using serial communication over USB as it contains built-in power and reset circuitry. The reason for the selection of Arduino is in view of its attractive features and simpler process of work with microcontrollers. Arduino board basically includes the hardware architecture where the program code and program data have separate memory. Programs are stored in flash memory and data is stored in SRAM.

The hardware architecture has the best applications in embedded system, when things get complicated the process start getting quicker in such systems. The Arduino has a series of development boards of which the Arduino UNO is best suited for our purpose. The technical specifications of Arduino UNO board are as shown in Table 2.

3.3 Android Mobile Application

Nowadays number of passengers possessing and using smartphones is increased rapidly. While their interest for mobile applications that enable the storage, analysis, and visualization of the collected space time information is apparent.

Table 2 Arduino specification

Features specifications	
Microcontroller	ATmega328
Operating voltage	5 V
Input voltage	7–12 V
Input voltage (limits)	6–20 V
Digital I/O Pins	14 (of which 6 provide PWM output)
Analog input pins	6
DC current per I/O Pin	40 mA
Flash memory	32 KB of which 2 KB used by boot loader
SRAM	2 KB
EEPROM	1 KB
Clock speed	16 MHz

There are many apps used for wireless technology. Some of the general purpose apps are "My Tracks" [4], "And Ando" [5], "GPS Tracker" [6], and "Every Trail" [7]. In our proposal we are going to develop an app with eclipse software as open source software development project, Eclipse focuses on providing a full functional platform for the integrated tool development. In this paper we develop an app based on the eclipse platform because it can work well in different systems, and it is an outstanding simulator management console. Eclipse uses plug-ins to provide all the functionality within and on top of the runtime system. Its runtime system is based on Equinox, an implementation of the core framework specification. In addition to allowing the Eclipse Platform to be extended using other programming languages, such as C and python, the plug-in framework allows the Eclipse Platform to work with typesetting languages like LaTeX and networking applications such as telnet and database management systems (Fig. 3).

Fig. 3 Home screen of an application

4 System Implementation

The project proposal is implemented using a low-power wireless technology ZigBee and Arduino UNO R3, and a mobile application which is accessible within the Wi-Fi range of the bus terminus. The Mobile App has the display of different locations; the passenger has to click on the destination location. One click depicts one passenger at the lane. One click through the app, the proxy of Ethernet gets connected in turn to the Arduino and the count gets incremented. The passenger gets the confirmation voice notification once his density count is considered. As explained earlier, the whole procedure is divided into three different stages, the flow of the system procedure is shown [8]. The passenger density information acquired from the mobile app is received by the Ethernet present at the monitoring department, the Arduino compares the densities of the different lanes, and the pin Tx [1] of Arduino triggers and the output signal from Rx is received by pin 3 of ZigBee that is data-in. From ZigBee data-out pin2 (Fig. 1), the information is sent to pin3 of the other ZigBee transmitter fit in the bus. The ZigBee is in turn interfaced with Arduino board which is connected to LCD. The location of the place is displayed on the LCD screen (Figs. 4, 5 and 6).

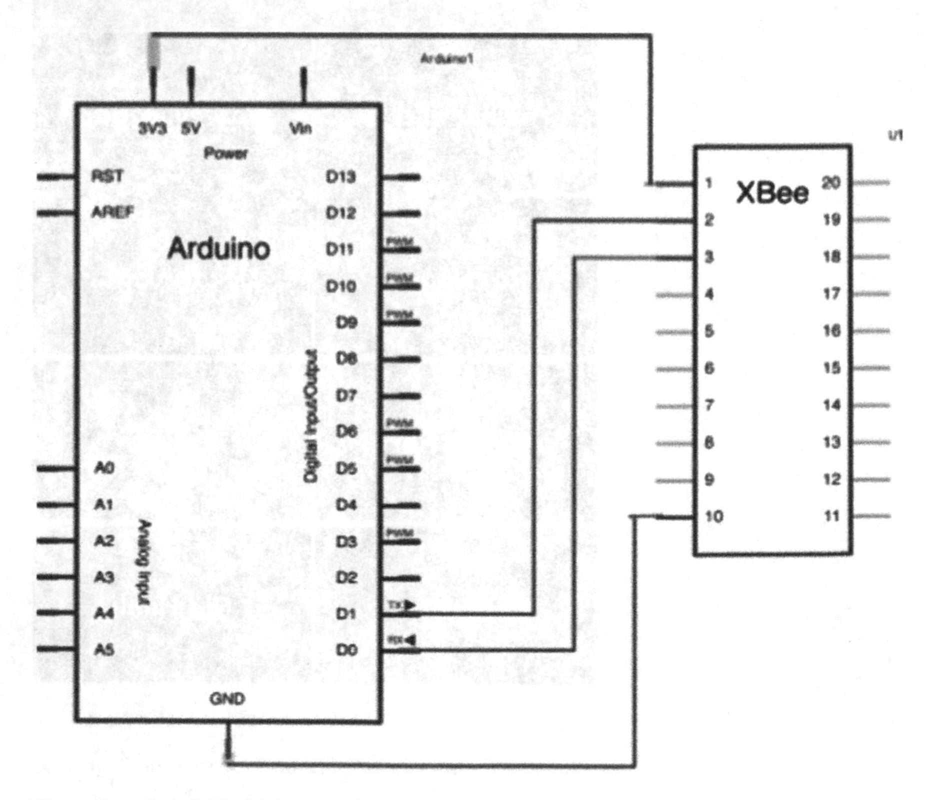

Fig. 4 Interfacing of Arduino and Xbee

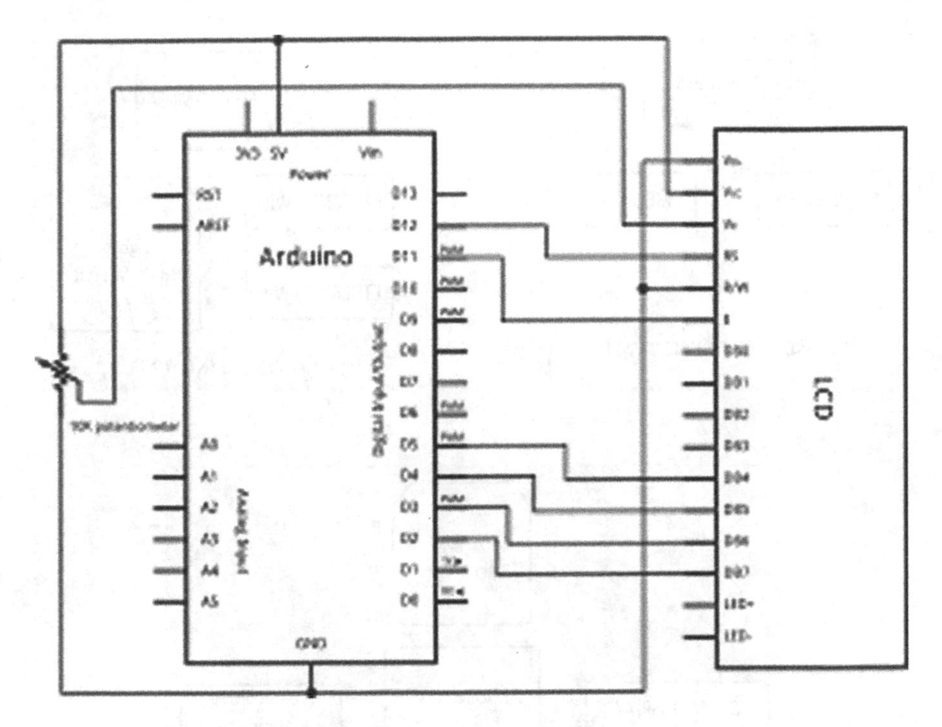

Fig. 5 Interfacing of Arduino and LCD

5 Results and Conclusions

- Delay time in the running of buses is reduced.
- Passengers need not wait for a long time at the terminus for buses, as buses are allocated based on the passenger density.
- Unnecessary running of the buses is reduced.
- Well-managed public transport system in the society. The traffic congestions, road blockages due to traffic and road accidents can be reduced.
- Circulation of buses in cities only when required reduces the amount of the fuel energy consumed.

This project proposal has a very promising future, the system can be extended to an extent where the passengers can get the tickets automatically at the lanes by providing the fixed fare, and even their bank accounts can be linked to the ticket generating system, so that the transactions take place automatically. It is efficient in terms of security also, as the monitoring or the controlling systems are

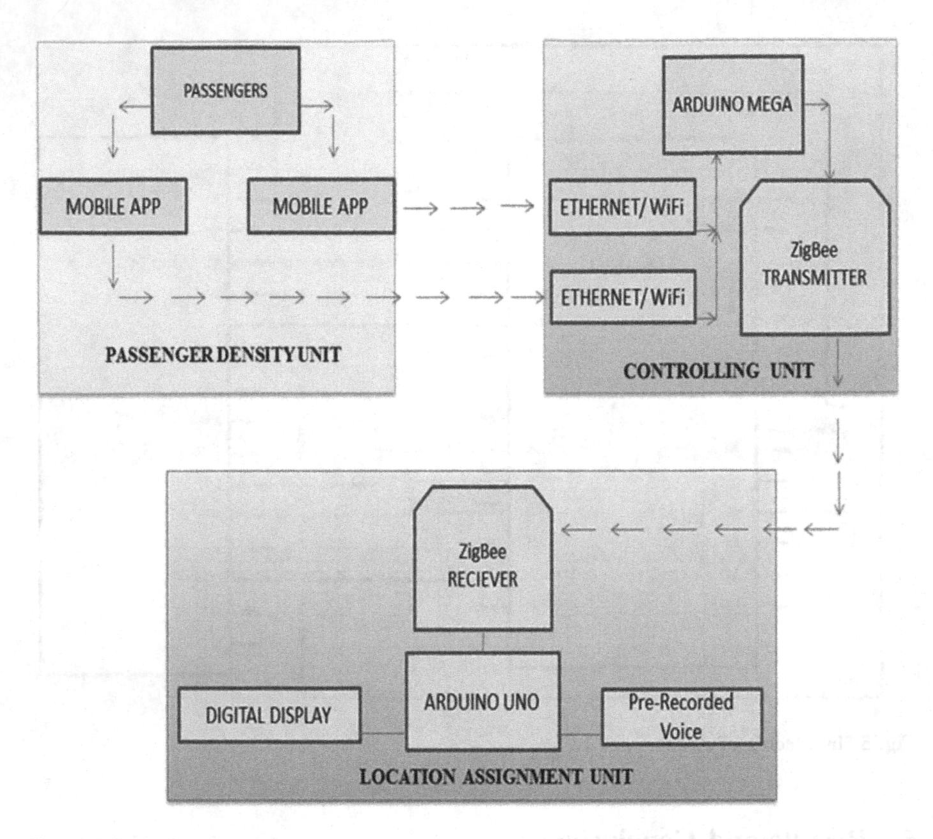

Fig. 6 Block diagram of the system

interconnected and in case of any consequences the passenger can contact the controlling department.

This paper provides the very efficient way of transport to public. It will play a very vital role in the field of Traffic and Transportation. In India bus transportation segment accounts for about 70 % of fuel consumption. In cities the Buses circulate according to the scheduled time, irrespective of the passenger density. Our project aims on scheduling the buses only when the density of passengers increases above the specified passenger density limit. Hence, the unnecessary running of scheduled buses and fuel consumption by them is reduced. The passengers can access the mobile application which is user friendly to know the status of buses. The density of passengers is obtained from the mobile application. The app is developed such that it displays different locations (Figs. 7 and 8).

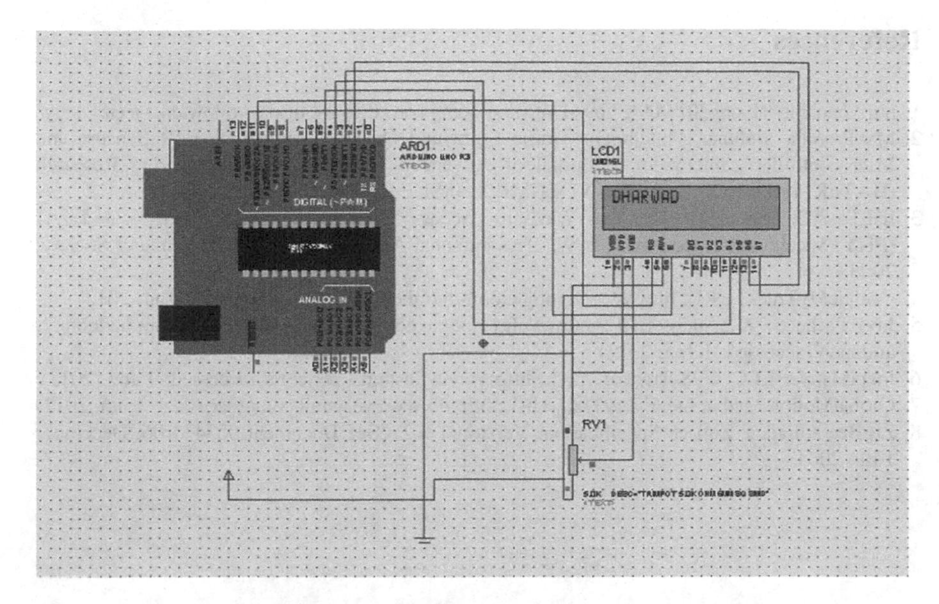

Fig. 7 Simulation results of Interfacing Arduino with LCD

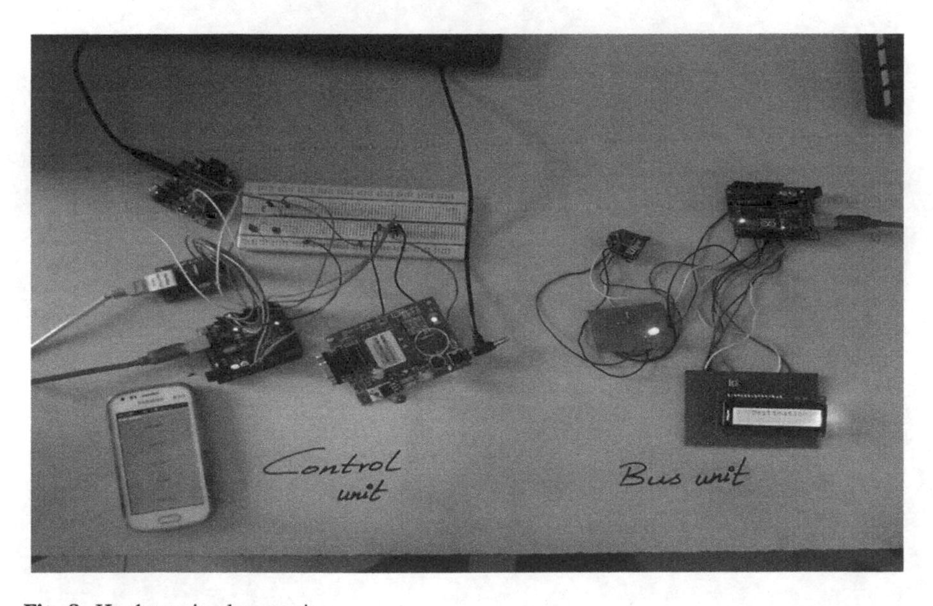

Fig. 8 Hardware implementation

References

1. http://mobilitylab.org/2015/09/17/bus-bunching-being-solved-in-d-c-thanks-to-low-cost-technology/
2. Ms Promila Sinhmar, "Intelligent Traffic Light and Density Control using IR Sensors and Microcontroller", International Journal of Advanced Technology & Engineering Research (IJATER) ISSN NO: 2250-3536 VOLUME 2, ISSUE 2, MARCH 2012
3. IEEE 802 Part 15.4: Wireless Medium Access Control (MAC) and Physical Layer (PHY) Specifications for LowRate Wireless Personal Area Networks, IEEE Computer Society, 2003
4. GoogleMyTrack. URL: http://www.google.com/mobile/mytracks/ (accessed: 1 Jun. 2012)
5. Javi Pacheco, AndAndo. URL: https://play.google.com/store/apps/details?id=com.javielinux.andando (accessed: 1 Jun. 2012)
6. InstaMapper LLC, GPS Tracker. URL: http://www.instamapper.com/ (accessed: 1 Jun. 2012)
7. GlobalMotion Media, Inc, EveryTrail. URL: http://www.everytrail.com/ (accessed: 1 Jun. 2012)
8. ZigBee Alliance, ZigBee Specification. Version 1.0 ZigBee Document 053474r06, December 14th, 2004

Review on Data Hiding in Motion Vectors and in Intra-Prediction Modes for Video Compression

K. Sridhar, Syed Abdul Sattar and M. Chandra Mohan

Abstract In this paper, the main objective is an inter-frame data hiding in motion vectors and data hiding through intra-frame prediction method. Data hiding in the compression video is through motion vectors, i.e., embedding data in phase angle between two successive CMVs, with the best matching macroblock-based search. Another method is through 4 × 4 intra-prediction macroblocks. In this paper, we proposed two new advanced embedding scheme and steganography. By the inter-frame motion vectors and through intra-prediction modes, we can embed with more efficiency. The main advantage of the proposed method is that data hiding video is stego video and cannot understood by any hackers and maintains good perceptual quality of video.

Keywords Data hiding · Motion vectors · CMV · Intra-prediction · Steganography

1 Introduction

For secrete communication copy right protection and authentication for new applications, data hiding is required for image and video processing, to avoid illegal activity or reproduce or manipulate them in order to change their owners identity. Digital water marking and data hiding are the techniques providing best embedding and copy right information in images and videos [1]. The aim of water marking is

K. Sridhar (✉)
Department of ECE, VREC, Naziabad, India
e-mail: rahulmani_147@yahoo.com

S.A. Sattar
Department of ECE, RITS, Chevella, India
e-mail: syedabdulsattar1965@gmail.com

M.C. Mohan
Department of CSE, SDC, JNTUH, Hyderabad, India
e-mail: c_miryala@yahoo.com

S.C. Satapathy et al. (eds.), *Proceedings of the International Conference on Data Engineering and Communication Technology*, Advances in Intelligent Systems and Computing 469, DOI 10.1007/978-981-10-1678-3_51

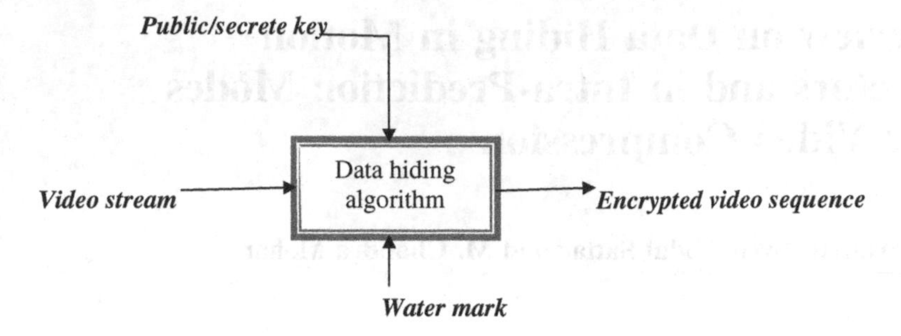

Fig. 1 Model of embedding technique

not to restrict to the original image to access and ensure that embedding data again recoverable [1]. Developers seeking for providing protection on top of data encryption and scrambling for content protection. Developers seeking to protect owner information (water mark) embedded in original image or video. Cryptography converts into cipher text and it cannot be understood as shown in Fig. 1. A basic model of embedding and steganography is shown below.

Data hiding techniques have become more important since two decades in various applications. For data hiding, so many techniques and algorithms introduced for digital images and videos. Nowadays, audio and video and picture frames with hidden copy right symbol help to avoid un authorized copying directly [2]. Hackers may modify the script of particular organization or revel the information to others. A lot of researches have been done in this field and many techniques are introduced, but due to drawbacks in techniques quality of video like sudden change in frames or noise is disturbed. All the above problems solved by high hiding techniques and steganography. Even though the host data is damaged, this steganography make more complexity to obtain the data.

2 Related Work

VIDEO COMPRESSION: Video compression techniques are used to minimize redundancy in video data with unchanged video quality, it mostly used in video conferences and military communication applications and real-time applications [3]. For motion-based video compressed process, motion estimation and motion compensation techniques are used for temporal redundancy in frames. We target the motion vectors to encode and reconstruct the video, both the predictive (p)-frame and bidirectional (b)-frames used for getting motion vectors to embed the secrete data by LSB method.

For motion vectors, we have two successive methods: one is associated macroblock prediction error and other is using magnitude and phase angle in motion vectors and before this the most common method is temporal differencing is used [4].

It compares the current frame with previous frame then the current frame is threshold to segment out four ground object, but this technique also has disadvantages. Due to these reasons, authors suggested video coding by motion vectors. Basic idea behind these techniques is all consecutive frames are having same similarities both before and after frames.

The aim is to reduce this redundancy by block-based motions, for this approach motion estimation is required. The consecutive frame is similar but except some objects moving the within the frames. The most accurate motion is estimated and matched for residual error nearer to zero and coding efficiency will be high. In the motion vectors the predicted frame is subtracted from current frame, the data bits are hidden in some of motion vectors that too in CMVs. These CMVs magnitude is having an above a predefined threshold value.

Here we have two approaches,

1. ***Embedding in motion vectors (i.e., candidate motion vectors)***.
2. ***Intra-prediction modes***.

3 Embedding in Motion Vectors

We hide the data in video using phase angle between two consecutive CMVs, the data bit code is embedded depend up on phase angle criteria, depends on phase angle sectors [5]. This method is used for entire CMVs in all frames and also at data retrieving place.

The below figure explains each macr block in a predicted frame can be encoded as motion vector (Fig. 2).

Up to now, we know data hiding and water marking in digital [6] Images and raw videos data hiding in motion vector for compression. The message should be surviving video lossy compression and without loss extracted. A novel video water mark technique in motion vectors explained with simple diagram is mentioned in (Fig. 3).

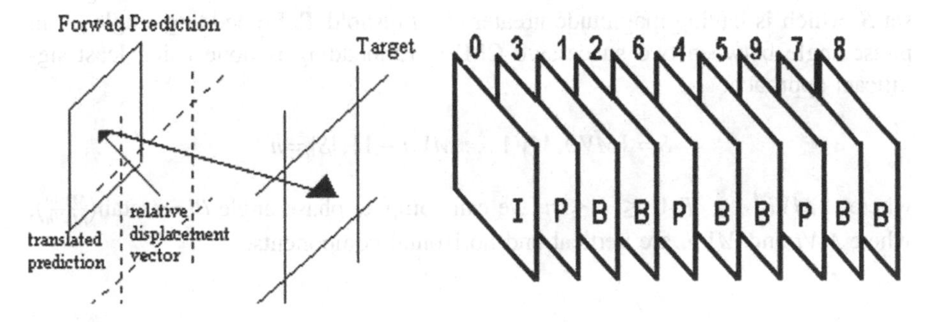

Fig. 2 Forward prediction and GOP in compressed video

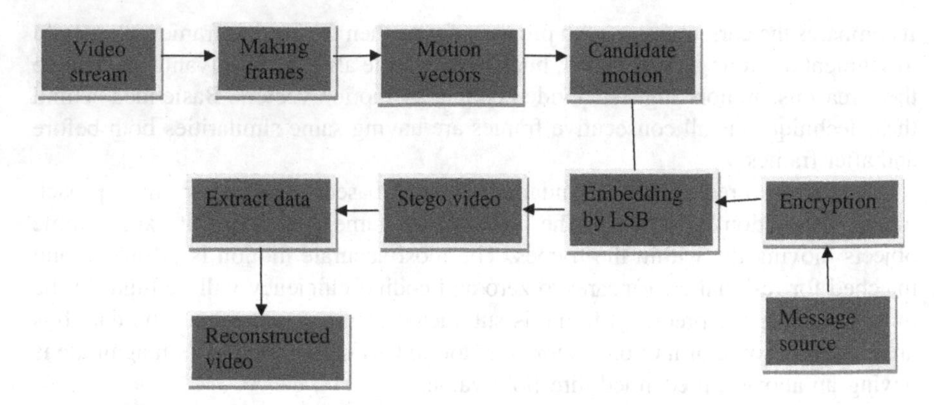

Fig. 3 Advanced data hiding scheme

Figure 3 explains that for video compression basic step is continue video and is converted into number of frames, which depends up on mpeg format; we have predictive (p)-frame and bidirectional (b)-frame. After conversion of frames with motion estimation process, we can reduce temporal redundancy. Motion estimation compares adjacent frames, the displacement of macroblock from reference frame to current frame called motion vector [7] by motion compensation an algorithm employed for video compression.

In general, the video format have 'IPBBPBBPI' of frames, we select p-frame and b-frame for embedding. In intra-frame, we need no remember search area that is limited for best match. In general MBs are usually multiple of 4(16 × 16, 16 × 8, 8 × 16, 8 × 8), and further dividing into transform blocks, and again subdivide into predictive blocks, after making fixed transform blocks, block search starts in reference frame with current frame, search is continues for best matching for least prediction error. If matched block is found, it is it is encoded by vector called motion vector.

1. ***Candidate motion vectors*** (***CMVs***): candidate motion vector means encoder find best matched block for encode by vector for motion vector [8], in those motion vectors some of the motion vectors having magnitude greater than threshold value called candidate motion vectors. We find that best motion vectors are arranged as set ***S***, which is having magnitude greater the threshold ***T***. Embedding single bit in phase angle between two successive CMVs. Embedding is done using least significant approach.

$$S = \{MV0, MV1 \ldots .MVn-1\}, |S| = n$$

where $\mid MVi \mid \; \geq T, 0 \leq i \leq n$, we can compute phase angle $\theta i = \arctan(\frac{MViv}{MVih})$, where MVi and $MVih$ are vertical and horizontal components.

ALGORITHM:

Step 1: *Video is converted into frames (frame separation).*
Step 2: *Select P-frame and B-frames for embedding.*
Step 3: *Performing motion estimation and motion selecting best match block motion.*
Step 4: *Embedding bits in phase angle between two successive CMSs.*
Step 5: *Apply secrete key for encryption.*
Step 6: *Generate stego video.*

For encryption, steganography is the process to convert message into cipher text, [9]. The advantage of steganography is that data keeps in secrecy key is need in encryption process. RSA algorithm is the best suitable for data secrecy we get good PSNR value after extraction.

2. ***Another Advanced Method***: Because of data size increased at extraction time for video [2], the new method explains as up to now we are considering p-frame and b-frame for embedding now I-frame is also encoded for using regular method like jpeg, At decoder will extracted independently I-frame, p-frame, and b-frame. We know that video is making like no of group of pictures in each GOP I-frame-frame and P-frame will be there as for mpeg format. The relative information or redundancy is employed by temporal redundancy by using block-based motion estimation, here also single bit hidden in CMVs. In this scheme, data bits will increase by block size decreasing, here code efficiency is increasing because microblocks increased in intra-frame, it is increasing the more searching option for best match, it increases best video quality.

4 Intra-PredictionModes

Another important data hiding technique is data hiding using intra-prediction modes. For video coding, several methods have been proposed in H.264/AVC compressed video standard. Coding efficiency is compare more with previous standards. For this standard, the 4×4 spatial intra-predictions provide good quality at decoder.

Some works have been done [10]' on data hiding with the Intra-prediction modes. Authors have penned the data hiding methods by modifying the intra 4×4 prediction modes with a technique that is mapping between the modes and hidden private data. For this in the H.264/AVC standard, the intra-prediction uses in the spatial correlation method. A desired macroblock can be predicted from macroblocks, which have been encoded and decoded [11]. The subtraction from the current macroblock is coded and represented with a few number of bits analyzed with the macroblock used for the processing of the macroblock itself. In all modes, we need to select one 4×4 mode for embedding and coding.

This intra-prediction mode shows convenient for area of significant details in the picture. One block divide into 16 4 × 4 macroblocks in that A TO M are the already encoded samples and other are to be encoded. The modes are decided by formula called rate distortion optimization (RDO).

$$J = d + \lambda\,\mathbf{mode}^{*}R$$

where R and D represent bit rate and distortion, λ mode represents lagrangian multiplier which represents quantization parameter (Fig. 4).

The modes classified an groups listed below

In Group one: Mode 1 and mode 8.

In Group two: Mode 3 and mode 7.

In Group three: Mode 2, mode 4, mode 5 and mode 6.

In Group four: Mode 0 and mode 2.

The embedding process as follows for group 1 the horizontal mode unchanged by embedding 1. Same way if we embed 0 to mode 8, its remains same. Similar way by modification '00' to mode 4, it becomes 2, '01' to mode 4 it is equal to 6. In this way embedding and extraction takes place.

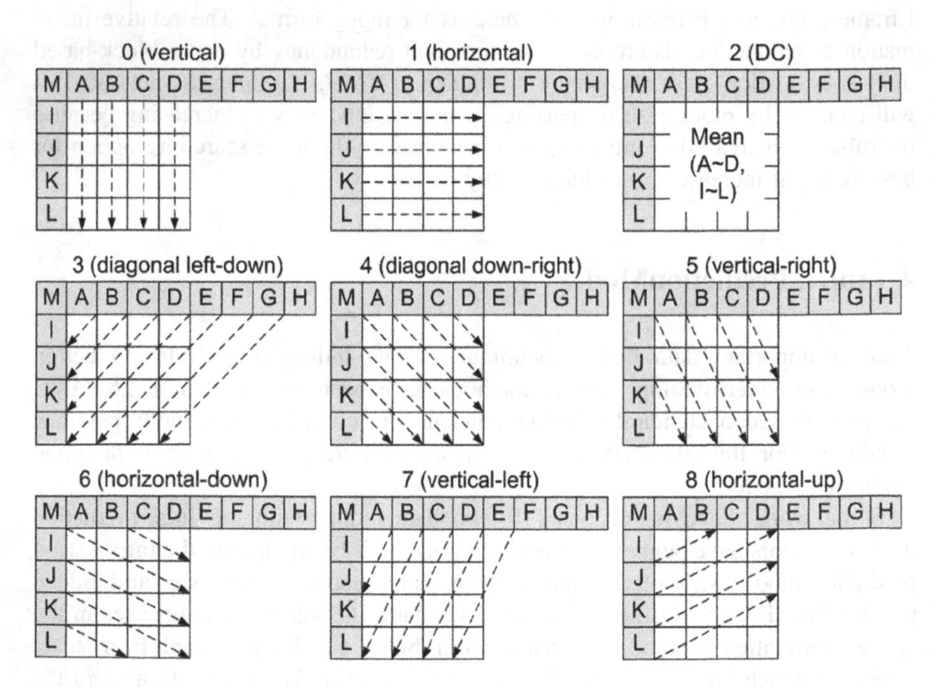

Fig. 4 Types of modes in intra-prediction

5 Conclusion

We discussed in this chapter the improved methods for data hiding. Depending up on macro block size and best matched motion vectors, we can embed single bit in phase angle between two successive CMVs and also suggested popular steganogrphy method for data secrecy. We can decrease bandwidth of video by small microblock intra-frame for the best match block. This algorithm can extend in future for different transform domain with embedding techniques for good perceptual quality.

References

1. S. Jayarajan, S. esakkirajan, T. veerakumar, 2012, "digital image processing" TAT Mc GraHill.
2. Aditya pothan raj, 2013, "compacted video based data hiding for motion vectors" elixir.
3. LI Man H, 2006 "Variable block based size motion estimation hardware for video encoders".
4. R. Jain and H.H. Nagel, april 1979, "On the analysis of accumulative difference pictures from Image sequences of real world sence" IEEE.
5. U. Manjula, R. Rajani, K. Radhika, 2012, "A Secured data hiding techniques in compressed video using a secrete key" IJCSIT.
6. F. Jordan, M. Kutter, and T. Ebrahimi, "Proposal of a watermarking technique for hiding data in compressed and decompressed video," ISO/IEC Document, JTC1/SC29/WG11, Stockholm, Sweden, Tech. Rep. M2281, Jul. 1997.
7. M. Usha, 2014. "Motion detection in compressed video using macro block classification", ACIJ.
8. Ding-yu fang, Long-wen chang "Data hiding for digital video with phase of motion vector" IEEE, 2006.
9. X. He and Z. Luo, "A novel steganographic algorithm based on the motion vector phase," in Proc. Int. Conf. Comp. Sc. and Software Eng., 2008, pp. 822–825.
10. Y. Hu, C. Zhang, and Y. Su, "Information hiding based on intraprediction modes for H.264/AVC," in *2007 IEEE Int. Conf. onand Expo (ICME)*, pp. 1231–1234.
11. Li L, Wang P, Zhang ZD, "A video watermarking scheme based on motion vectors and mode selection". In: Proceedings of IEEE International Conference Computer Science and Software Engineering, vol. 5. 2008.

Generation of Product Cipher for Secure Encryption and Decryption Based on Vedic Encryption Ideology and Using Variable and Multiple Keys

Vaishnavi Kamat

Abstract Information Security in the recent years has become an important area. There is always a necessity for exchanging information secretly between the sender and the receiver. Cryptography reveals its deep roots of history in providing a way to transmit the sensitive information across networks, so that only the intended recipient can read it. There are several methods of encryption and decryption developed to improve the security of the information being transmitted over last few years. Evidence of encryption is evidently present in the Ancient-Vedic Indian period for example in Sri Ram Shalakha, the placement of couplets. In this paper, a contemporary approach is developed, i.e., a "product cipher" to harness the dominant features of Vedic encryption style as well as incorporate the classic behavior and feature of modern cryptographic algorithms.

Keywords Encryption • Decryption • Sri Ram Shalakha • Random number • Substitution • Transposition • Product • Symmetric key

1 Introduction

Cryptography is the practice of mathematical scrambling of word. Encryption being one of its special case, which transforms data into cipher text, so that it is difficult or impossible for the unauthorized users to decipher it. Ancient India, especially during its Vedic period, has witnessed use of encryption for its scriptures. In this paper, an attempt has been made to utilize the Vedic encryption ideas and modern cryptographic algorithms, to produce a secured cipher. The book titled "Ram Charit Manas" authored by Saint Tulsidas, depicts the life of Lord Shri Ram. In this book, we will find a "Sri Ram Shalakha Prashnavali". The "Prashanavali" is an extract from Shri Ram Charit Manas. "Ram-shalaka Prashnavali" used to get an answer or

Vaishnavi Kamat (✉)
Department of Computer Engineering, Agnel Institute of Technology and Design, Assgao, Bardez, Goa, India
e-mail: vaishnavi.kunkoliker@gmail.com

S.C. Satapathy et al. (eds.), *Proceedings of the International Conference on Data Engineering and Communication Technology*, Advances in Intelligent Systems and Computing 469, DOI 10.1007/978-981-10-1678-3_52

indication to the question proposed. The answers/indications are based on Chopais (couplets) from the book "Shri Ram Charit Manas". The technique encrypts as well as encodes the information from the book into the "Prashnavali" [1]. An overview of this technique is given in the sections below. This technique in this paper is modified and tried for different numbers. Apart from the encryption and decryption algorithm, this technique uses multiple keys for the ciphering and deciphering, which makes the cipher text passed through the channel more secured. As we mentioned in the abstract, both privacy of communication and authenticity of entities involved can be achieved with symmetric cryptography, so this technique falls under symmetric cryptography category and it processes plain text as a stream cipher. Here, a combination of substitution and transposition techniques is used to generate the cipher. In the concluding part we can see that, it is difficult to decrypt the cipher by unauthorized users, without the appropriate keys.

2 Cryptosystem

Cryptosystem is a complete package of all the algorithms required to implement the security for the message transferred. As shown in Fig. 1, a message source stores the message which is to be encrypted, i.e., X. This message X, is sent as input to encryption algorithm block, where different techniques like substitution are used to encrypt the data; that message is termed as Y. Y travels through the channel and reaches the destination side, where it has to be decrypted; decryption algorithm blocks and is made visible as X to the receiver. Meanwhile, the keys are transmitted over secured channel as per the agreement between sender and receiver.

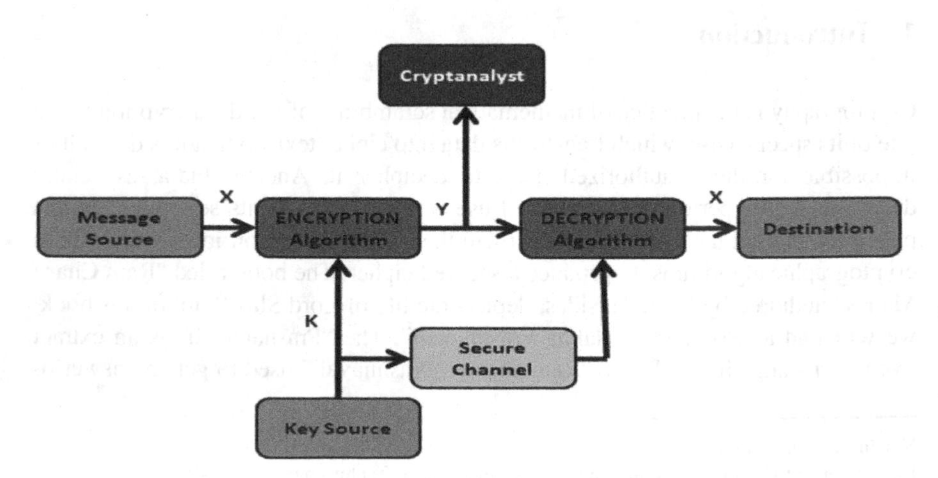

Fig. 1 Model of conventional cryptosystem

2.1 Symmetric Cryptosystem

In this paper, a symmetric cryptosystem model is implemented. A symmetric encryption model has five components [2]

(1) **Plain Text**: the original message that is given as input.
(2) **Encryption Algorithm**: it performs various substitutions and transformations on the plain text.
(3) **Secret Key**: is a value independent of the plain text and of the algorithm. The exact substitutions and transformations performed by the algorithm depend on the key.
(4) **Cipher text**: it is the scrambled message produced as output. It depends on the plain text and the secret key.
(5) **Decryption algorithm**: it takes the cipher text and the secret key and produces the original plain text Fig. 2.

In symmetric cryptography, first each pair of communicating entities needs to have a shred key. Second these keys must be transmitted securely

2.2 Characteristics of Cryptographic Model with Respect to the Proposed Algorithm

A cryptographic model can be characterized by describing **three** parameters [3]

(1) **The type of operations used for transforming plain text to cipher text.**

- *In the encryption algorithm used in this paper, first applies transposition technique to the plain text and then performs substitution on it.*

(2) **The number of keys used.**

- *This algorithm uses two keys of variable length and a single value key.*

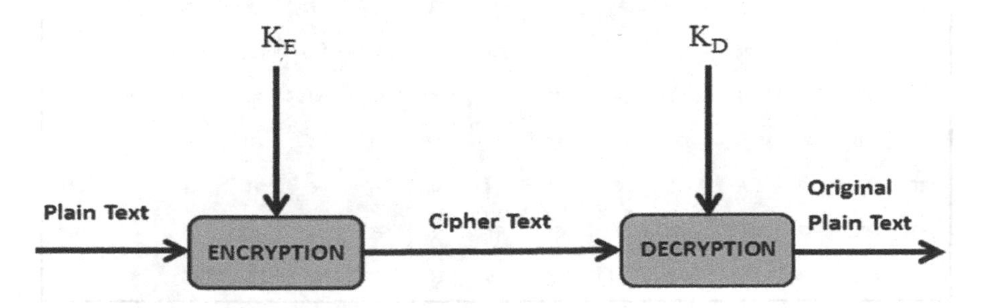

Fig. 2 Symmetric cryptosystem: KE = KD

(3) **The way in which the plain text is processed.**

- *The plain text is processed as a stream cipher.*

3 Sri Ram Shalaka Prashnavali

Sri Ram Shalaka Prashnavali is a matrix of 15 × 15 squares. Each square in the grid has one Akshar (alphabet) from nine Chopais (couplets) of Shri Ram Charit Manas.

To use the Prashnavali, think of a question in mind and select any one of the squares randomly Fig. 3.

Now proceeding further, count the ninth square from the one initially picked and then another ninth, i.e., counting every ninth square till all the letters collected from the squares complete a chaupai. The couplet thus formed will provide the answer to the question in mind [4]. Couplet formed is given in the Figs. 4 and 5.

One should remember that in certain squares only the sign standing for a vowel sound has been given. Such squares must not be left out nor should those squares be counted twice which contain two letters.

सु	प्र	उ	बि	हो	मु	ग	ब	सु	नु	बि	घ	धि	इ	द
र	रु	फ	सि	सि	रहिं	बस	हि	मं	ल	न	ल	य	न	अं
सुज	सो	ग	सु	कु	म	स	ग	त	न	इ	ल	धा	बे	नो
त्य	र	न	कु	जो	म	रि	र	र	अ	की	हो	सं	रा	य
पु	सु	थ	सी	जे	इ	ग	म	सं	क	रे	हो	स	स	नि
त	र	त	र	स	हुँ	ह	ब	ब	प	चि	स	हिं	स	तु
म	का	ा	र	र	म	मि	मी	म्हा	ा	जा	हू	हीं	ा	ा
ता	रा	रे	री	ह्र	का	फ	खा	जू	ई	र	रा	पू	द	ल
नि	को	जो	गो	न	मु	ज	य	ने	मनि	क	ज	प	स	ल
हि	रा	मि	स	रि	ग	द	न्मु	ख	म	खि	जि	म	त	जं
सिं	ख	नु	न	को	मि	निज	र्क	ग	धु	ध	सु	का	स	र
गु	ब	म	अ	रि	नि	म	ल	ा	न	ढ़	ती	न	क	भ
ना	पु	व	अ	ा	र	ल	ा	ए	तु	र	न	नु	वै	थ
सि	हु	सु	म्हा	रा	र	स	स	र	त	न	ख	ा	ज	ा
र	ा	ा	ला	धी	ा	री	ा	हू	हीं	खा	जू	ई	रा	रे

Fig. 3 Sri Ram Shalaka Prashnavali

<table>
<tr><td>सु</td><td>प्र</td><td>उ</td><td>बि</td><td>हो</td><td>मु</td><td>ग</td><td>ब</td><td>सु</td><td>नु</td><td>बि</td><td>घ</td><td>धि</td><td>इ</td><td>द</td></tr>
<tr><td>र</td><td>रु</td><td>फ</td><td>सि</td><td>सि</td><td>रहिं</td><td>बस</td><td>हि</td><td>मं</td><td>ल</td><td>न</td><td>ल</td><td>य</td><td>न</td><td>अं</td></tr>
<tr><td>सुज</td><td>सो</td><td>ग</td><td>सु</td><td>कु</td><td>म</td><td>स</td><td>ग</td><td>त</td><td>न</td><td>इ</td><td>ल</td><td>धा</td><td>बे</td><td>नो</td></tr>
<tr><td>त्य</td><td>र</td><td>न</td><td>कु</td><td>जो</td><td>म</td><td>रि</td><td>र</td><td>र</td><td>अ</td><td>की</td><td>हो</td><td>सं</td><td>रा</td><td>य</td></tr>
<tr><td>पु</td><td>सु</td><td>थ</td><td>सी</td><td>जे</td><td>इ</td><td>ग</td><td>म</td><td>सं</td><td>क</td><td>रे</td><td>हो</td><td>स</td><td>स</td><td>नि</td></tr>
<tr><td>त</td><td>र</td><td>त</td><td>र</td><td>स</td><td>हुँ</td><td>ह</td><td>ब</td><td>ब</td><td>प</td><td>चि</td><td>स</td><td>हिं</td><td>स</td><td>तु</td></tr>
<tr><td>म</td><td>का</td><td>ा</td><td>र</td><td>र</td><td>म</td><td>मि</td><td>मी</td><td>म्हा</td><td>ा</td><td>जा</td><td>हू</td><td>हीं</td><td>ा</td><td>ा</td></tr>
<tr><td>ता</td><td>रा</td><td>रे</td><td>री</td><td>इ</td><td>का</td><td>फ</td><td>खा</td><td>जू</td><td>ई</td><td>र</td><td>रा</td><td>पू</td><td>द</td><td>ल</td></tr>
<tr><td>नि</td><td>को</td><td>जो</td><td>गो</td><td>न</td><td>मु</td><td>ज</td><td>य</td><td>ने</td><td>मनि</td><td>क</td><td>ज</td><td>प</td><td>स</td><td>ल</td></tr>
<tr><td>हि</td><td>रा</td><td>मि</td><td>स</td><td>रि</td><td>ग</td><td>द</td><td>न्मु</td><td>ख</td><td>म</td><td>खि</td><td>जि</td><td>म</td><td>त</td><td>जं</td></tr>
<tr><td>सिं</td><td>ख</td><td>नु</td><td>न</td><td>को</td><td>मि</td><td>निज</td><td>र्क</td><td>ग</td><td>धु</td><td>ध</td><td>सु</td><td>का</td><td>स</td><td>र</td></tr>
<tr><td>गु</td><td>ब</td><td>म</td><td>अ</td><td>रि</td><td>नि</td><td>म</td><td>ल</td><td>ा</td><td>न</td><td>ढ</td><td>ती</td><td>न</td><td>क</td><td>भ</td></tr>
<tr><td>ना</td><td>पु</td><td>व</td><td>अ</td><td>ा</td><td>र</td><td>ल</td><td>ा</td><td>ए</td><td>तु</td><td>र</td><td>न</td><td>नु</td><td>वै</td><td>थ</td></tr>
<tr><td>सि</td><td>हु</td><td>सु</td><td>म्हा</td><td>रा</td><td>र</td><td>स</td><td>स</td><td>र</td><td>त</td><td>न</td><td>ख</td><td>ा</td><td>ज</td><td>ा</td></tr>
<tr><td>र</td><td>ा</td><td>ा</td><td>ला</td><td>धी</td><td>ा</td><td>री</td><td>ा</td><td>हू</td><td>हीं</td><td>खा</td><td>जू</td><td>ई</td><td>रा</td><td>रे</td></tr>
</table>

Fig. 4 Result from Sri Ram Shalaka for a couplet after clicking on the character with red background

चौपाई: सुनु सिय सत्य असीस हमारी। पूजहि मन कामना तुम्हारी॥

Fig. 5 Couplet formed after selecting a random character from the grid

The vowel symbol when reached in the relevant square should be added to the preceeding letter, and where the square with two letters is reached both letters should be written together as part of the couplet to be formed.

4 Substitution and Transposition Techniques

A substitution cipher changes characters in the plain text to produce to cipher text, i.e. a substitution technique is one in which the letters of plain text are replaced by other letters or by numbers or symbols, for example, Caesar cipher. A transposition cipher, rearranges in the plain text to form the cipher text. The letters are not changed. Example: Rail fence cipher. Substitution and Transposition technique together are the building blocks of several encryption and decryption algorithms.

When Substitution and Transposition Techniques are used together, the resultant in called as **Product cipher**.

5 Character Set Used in the Proposed Algorithm

Character set used in the algorithm is illustrated in the Fig. 6

6 Proposed Encryption Algorithm

Step 1 **Take input string to be encrypted and find its length in terms of characters, excluding blank spaces** Figs. 7 and 8.

Step 2 **Select nearest possible symmetric matrix.**
In the given example, total number of characters is 59, and the nearest possible symmetric matrix will be 8 × 8, that gives a total of 64 squares. Row wise we arrange the input text in this symmetric matrix. 8 × 8 is the initial size of the matrix. As we proceed to step 3, we will notice that, the size of the matrix increases.

Step 3 **Randomly select any number between 2 and 10.**
Since the size of the text transmitted is small, we select any number between the above given range. If the limit of text size is increased, this

a	0	k	10	u	20	#	30	[	40	+	50	2	60
b	1	l	11	v	21	$	31	]	41	-	51	3	61
c	2	m	12	w	22	%	32	|	42	*	52	4	62
d	3	n	13	x	23	^	33	\	43	/	53	5	63
e	4	o	14	y	24	&	34	:	44	<	54	6	64
f	5	p	15	z	25	(	35	;	45	>	55	7	65
g	6	q	16	`	26	)	36	"	46	=	56	8	66
h	7	r	17	~	27	_	37	’	47	'	57	9	67
i	8	s	18	!	28	{	38	?	48	0	58		
j	9	t	19	@	29	}	39	,	49	1	59		

Fig. 6 Character set used for substitution

The world is the great gymnasium where we come to make ourselves strong

Fig. 7 Input text

Character Count per word :

The = 3 world = 5 is = 2 the = 3 great = 5
gymnasium = 9 where = 5 we = 2 come = 4 to = 2
make = 4 ourselves = 9 strong = 6

Total no. of characters = 59

Fig. 8 Character count

range also can be increased. This range is fixed currently using trial and error method. For transposing text of small size, any number between 2 and 10 is sufficient.

Step 4 **Transpose the characters from the input string by 'n' places as selected in Step 3.**
Consider Fig. 9, first word to be placed is "the" from plain text. We place character 't' in the first square of the matrix. Leaving the next six squares from 't', we place character 'h' and again leaving six square from 'h', we place character 'e'. The placement of the characters 't', 'h', 'e' are highlighted with yellow background color. Now while placing the next word, i.e., "world", look out for the first empty space from start of the matrix. We see that second square is free, so we insert character 'w' in that square and accordingly space the remaining characters of the word in the matrix with a space of six squares in between.
Next we place word "is" → "the" → "great" → "gymnasium" → "where" → "we" → "come" → "to" → "make". After placing all these words we see that, when we insert the characters from the next word, the size of the matrix starts increasing row wise, as all the spaces in 8 × 8 matrix have been full or it cannot accommodate the character at the location where it is free. We observe that, when we place the word "ourselves", it starts from one of the free locations inside 8 × 8 matrix, but ends up increasing the number of rows, so that all its characters can be placed by a gap of six squares in between. This is one of the important

Say Randomly choose number : 6

t	w	i	t	g	g	h	o
s	h	r	y	e	r	w	e
e	m	c	l	h	w	a	n
o	d	e	e	t	a	m	t
r	m	o	s	e	o	e	a
u	i	s			k	r	u
t			e	s	m	r	
		e		o			
l		n				v	
g				e			
		s					

Fig. 9 Transposed text with number 6

points, which highlight the dynamic nature of the algorithm. Because, for every text wordings will be different, so matrix size also will be different. This hides the actual length of the text. And based on the size of the final matrix, key size is determined. Every time we encrypt, we will get a variable length key. This is Key number 1. Similarly, we place all the characters in the matrix and final matrix size is 8 × 11.

Step 5 **The Substitution step**.
Substitute each character, with its equivalent substitution code from the given set of data characters in Fig. 6. This substitution process will be carried out on the matrix received as output form Step 4. Figure 10, displays the substituted output for the transposed text. Squares which are empty will be substituted with 0(zero).

Step 6 **Generate the Key number 1**, of variable size, i.e., in this example, size of the key will be "8 × 11". For generation of key, numbers are randomly selected (Fig. 11).

Step 7 **Add the key generated in Step 6 to the product matrix**, which is resultant of transposition and substitution methods. New matrix after result of addition is given below in Fig. 12.

Step 8 **Since, the cipher text generated has to fit within the character set given, we perform "mod" operation on the matrix received from** Fig. 12. Result of "mod" operation is shown in Fig. 13 below.

Step 9 **Substitute in the given matrix, whatever the numbers are pointing**. i.e., characters, alphabets, symbols, etc., to generate the cipher text. Matrix with substituted characters, symbols, etc. is illustrated in Fig. 14 below.

19	22	8	19	6	6	7	14
18	7	17	24	4	17	22	4
4	12	2	11	7	22	0	13
14	3	4	4	19	0	12	19
17	12	14	18	4	14	4	0
20	8	18			10	17	20
19			4	18	12	17	
		4		14			
11		13				21	
6				4			
		18					

Fig. 10 Result of substitution

93	70	115	153	73	86	197	145
214	156	216	126	310	471	287	469
387	411	374	455	184	89	163	99
236	241	322	379	419	487	394	463
488	361	383	242	251	144	490	364
94	75	111	158	142	357	455	476
257	380	280	484	366	183	169	279
374	316	418	435	356	274	299	444
72	93	447	164	196	348	406	204
402	500	470	243	248	371	382	455
431	331	231	320	347	281	161	113

Fig. 11 Key number 1

112	92	123	172	79	92	204	159
232	163	233	150	314	488	309	473
391	423	376	466	191	111	163	112
250	244	326	383	438	487	406	482
505	373	397	260	255	158	494	364
114	83	129	158	142	367	472	496
276	380	280	488	384	195	186	279
374	316	422	435	370	274	299	444
83	93	460	164	196	348	427	204
408	500	470	243	252	371	382	455
431	331	249	320	347	281	161	113

Fig. 12 Resultant matrix after addition of key 1 and product matrix

Step 10 **Now read the matrix row wise, to generate the cipher text**. Cipher Text to be sent over the insecure channel is displayed below.

```
:y >)lyax! ~ @o|m_7-p)0 >\~:”[<\#l8 g@^’ = -wsy”p3wg ~ 6ue[im:l + h&:
o ~#c ~)pz*!2itaay4}?$|’xl;?hjz;
```

Step 11 Now cipher text is ready to be sent over the channel, but for the receiver to decrypt, the cipher text, an additional key is required. That key will be based on the length of each word present in the input string. We know that, we have 13 words in the plain text and the nearest possible symmetric matrix used to accommodate length of each word is 4 X 4 matrix, which contains maximum of 16 squares (Fig. 15).

44	24	55	36	11	24	0	23
28	27	29	14	42	12	37	65
51	15	36	58	55	43	27	44
46	40	54	43	30	11	66	6
29	33	57	56	51	22	18	24
46	15	61	22	6	27	64	20
4	40	8	12	44	59	50	7
34	44	14	27	30	2	27	36
15	25	52	28	60	8	19	0
0	24	62	39	48	31	42	47
23	59	45	48	7	9	25	45

Fig. 13 Resultant matrix after mod operation

:	y	>	)	1	y	a	x
!	~	@	o	|	m	_	7
-	p	)	0	>	\	~	:
"	[	<	\	#	1	8	g
@	^	'	=	-	w	s	y
"	p	3	w	g	~	6	u
e	[	i	m	:	1	+	h
&	:	o	~	#	c	~	)
p	z	*	!	2	i	t	a
a	y	4	}	?	S	|	'
x	1	;	?	h	j	z	;

Fig. 14 Substituted back to character set matrix: Cipher Text Matrix

Fig. 15 Length of each word from plain text

3	5	2	3
5	9	5	2
4	2	4	9
6	0	0	0

We use simple Caesar Cipher Substitution technique to encrypt the key. This is Key number 2. This key will also be of variable length, as number of words in each plain text varies. Using Caesar cipher to encrypt the key values provides it with additional security. Here, we select Caesar cipher Key to be 14. New encrypted Key 2 is shown in Fig. 16.

17	19	16	17
19	33	19	16
18	16	18	33
20	14	14	14

Fig. 16 Encrypted Key number 2

Step 12 Now both the Keys are agreed upon and transferred to the receiver through a secured channel. Then cipher text is sent over the insecure channel to the receiver.

7 Proposed Decryption Algorithm

Step 1 **Receive the Cipher Text**. Based on the first key size receiver we will know that size of matrix to be created.

In our example, size of the matrix is 8 × 11. Once matrix is created arrange the characters row wise in the matrix known as the Cipher Text Matrix as displayed in Fig. 14

Step 2 **Replace the characters by numbers from the data set**.

We will see matrix formed as in Fig. 13.

Step 3 **Now to the matrix obtained from Step 2, add the total count of the dataset, to each and every element in the square**.

In this example, since our dataset size is 68, it will be added to each and every element from the square (Fig. 17).

112	92	123	104	79	92	68	91
96	95	97	82	110	80	105	133
119	83	104	126	123	111	95	112
114	108	122	111	98	79	134	74
97	101	125	124	119	90	86	92
114	83	129	90	74	95	132	88
72	108	76	80	112	127	118	75
102	112	82	95	98	70	95	104
83	93	120	96	128	76	87	68
68	92	130	107	116	99	110	115
91	127	113	116	75	77	93	113

Fig. 17 Matrix after adding 68, to matrix from Step 2

Step 4 **Next we subtract matrix obtained from Step 3 by the matrix containing Key 1**. Key 1 matrix is given in Fig. 11. By performing this subtraction, we are trying to get back the original substituted matrix for transposed characters.

If we look at the contents of Middle Matrix in Fig. 18, we see that some of its squares are matching matrix from Fig. 10, as a result of substitution. Comparing the squares, squares which are empty in Fig. 10, are filled with 0

We see that, the first, second and third squares are matching, while the fourth square is not. Similarly, we can see that, there are many squares where the value is negative. All the squares with the negative values have to be worked upon to get the right substituted text. So, we go on adding 68, as many times it is required to get the number in the square to a positive value. We need to add 68 only once for the fourth square to get a positive number in that square. When we add 68, as many time required, we finally get the original substituted text back. In this case, fourth square value will be 19. Similarly, if we consider the seventh square, we need to add 68 twice, to get the seventh square on the positive side. The procedure looks simple, but it is difficult to do so, as Key 1 value is required and length of the dataset is required. Matrix displayed in Fig. 19 shows all the corrected values.

Step 5 **Subtract the corrected value matrix by Key 1 again. Result is shown in** Fig. 20.

Now perform a mod operation on the matrix from Fig. 20, to get the original substituted matrix as shown in Fig. 10 earlier and from that we replace them by the corresponding characters from the dataset as illustrated in Fig. 9. As we decrypt, we can see that there are lot of character 'a' present in the matrix. Some are genuine and some are placed to fill up the empty square. Next task is to identify the genuine text square and at the same time get back the original plain text. For this we require, the value of second key that will tell us about the length of each word in the text. Remember, it is also in the encrypted form, so first decrypt the key and then apply it. (Refer to Figs. 15 and 16 for Key 2). Next the receiver has to compute what will

19	22	8	-49	6	6	-129	-54
-118	-61	-119	-44	-200	-391	-182	-336
-268	-328	-270	-329	-61	22	-68	0
-122	-133	-200	-268	-321	-408	-260	-389
-391	-260	-258	-118	-132	-54	-404	-272
20	8	18	-68	-68	-262	-323	-388
-185	-272	-204	-404	-254	-56	-51	-204
-272	-204	-336	-340	-258	-204	-204	-340
11	0	-327	-68	-68	-272	-319	-136
-334	-408	-340	-136	-132	-272	-272	-340
-340	-204	-118	-204	-272	-204	-68	0

Fig. 18 Middle matrix

112	92	123	172	79	92	204	227
232	163	233	150	314	488	309	473
391	423	376	466	191	111	163	112
250	244	326	383	438	487	406	482
505	373	397	260	255	158	494	364
114	83	129	158	142	367	472	496
276	380	280	488	384	195	186	279
374	316	422	435	370	274	299	444
83	93	460	164	196	348	427	204
408	500	470	243	252	371	382	455
431	331	249	320	347	281	161	113

Fig. 19 Corrected values

19	22	8	19	6	6	7	82
18	7	17	24	4	17	22	4
4	12	2	11	7	22	0	13
14	3	4	4	19	0	12	19
17	12	14	18	4	14	4	0
20	8	18	0	0	10	17	20
19	0	0	4	18	12	17	0
0	0	4	0	14	0	0	0
11	0	13	0	0	0	21	0
6	0	0	0	4	0	0	0
0	0	18	0	0	0	0	0

Fig. 20 Result of subtraction of corrected value by Key 1

be value of ‘n’ for the given matrix, using the formula mentioned below (x in the formula represents n).

$$(((1 + x)\text{No.of Rows}) + \text{No.of Columns}) - \text{Total count of characters in the input text} = \text{total no.of squares in the actual cipher text matrix.}$$

$$(((1 + x)11 + 08) - 59 = 88 \quad \text{therefore, } x = 06$$

Once we have the decrypted key and value of n, we can reconstruct the original plain text. ***“The world is the great gymnasium where we come to make ourselves strong”*****. Note: For decryption, we require Key 1 twice, Key 2 once, Key to decrypt Key 2 and value of n.**

8 Results and Discussion

The method proposed in the paper uses the combination of two classic techniques, i.e., substitution and transposition, which generates a product cipher, which is more secure and strong cipher, which makes it difficult to break. From the initial matrix, we generate a transposed matrix and from the transposed matrix we generate the substituted matrix.

To convert initial matrix to transposed matrix, we require the knowledge of "spacing of characters", i.e., given by variable 'n'. Once, the transposed matrix is completely ready, it is substituted with corresponding characters form the dataset. Key 1 is used to encrypt the text once, and that matrix which stores the cipher text is the "Product Cipher Matrix". Key 2 is also generated depending upon the input text.

It is difficult to break the cipher generated by this technique. With respect to the example illustrated in the paper, if Brute force attack is tried, and since the unauthorized user does not have direct knowledge of mapping of character substitutions, he will have to try first **68!** combinations. Suppose, he is able to get the corresponding mappings right (practically very difficult to achieve), then at the next level, the unauthorized user will have to guess all numbers between 2 and 10, to find out 'n'. It is easy to calculate 'n' if we know what is the actual length of the input string, from the cipher text actual length of the input string cannot be known. So, the unauthorized user has to make **08 attempts**. If say value of 'n' is known to him, he will have the knowledge how it is spaced. But **without the key 1, it is difficult to get the substituted text for transposition because (1) key 1 size is variable (2) key 1 is randomly generated, so it is difficult to guess the values of the key and without this key we will not know whether the substitutions are correct or incorrect as a two-level check is done with this Key 1 in the proposed algorithm.** If key 1 values are known, then without key 2 values it is difficult to get what it is made up of, as key 2, first has to be decrypted using another key which can take another **n!** attempts to break it. Even after getting the substituted text using key 1, and if Key 2 is not known, say for example input string in this example has 13 words made up of 59 characters, then the unauthorized user will have to try and find out combinations of words that make up 59 characters with the help of a dictionary and if we consider even the latest edition of the Concise Oxford Dictionary, it contains more than 2,40,000 entries.

9 Conclusion

From the Sect. 8, we can observe that we make use of multiple keys both having variable length. Key 1, since it contains randomly chosen numbers is difficult to guess and without use of that key, we cannot get the substituted matrix during decryption.

Future improvements can be made to improve the encryption provided to the Key 2 as well as keys can be made more stronger by performing permutations on them. This method will perform well on small text size, but can be improved to work on block ciphers. "Prashnavali" displayed one of the techniques how a large text was encrypted into a small grid of 15×15 matrix. Similarly, other techniques in Vedic Scriptures also can be explored and mixed with modern cryptographic algorithms to generate a strong cipher and provide a secure communication.

References

1. Rajkishore Prasad: *Sri Ramshalaka*: A Vedic Method Of Text Encryption And Decryption. IJCSE, Vol. 4 No.3 Jun-Jul, pp. 225–234 (2013).
2. I.A. Dhotre, V.S. Bagad: Cryptography And Network Security, Technical Publications, pp. 1–18, 34–35(2008).
3. V. K. Pachghare: Cryptography And Information Security, PHI Learning Pvt. Ltd, pp. 17 (2015).
4. Tulsidasa, Edited by Ramchandra Prasada,: Gosvami Tulsidasakrta Sriramcharitramanasa. Motilal Banarsidass Publisher, pp. 853. (1989).

A Novel Integer Representation-Based Approach for Classification of Text Documents

S.N. Bharath Bhushan, Ajit Danti and Steven Lawrence Fernandes

Abstract Text Classification approaches are gaining more and more attention due to the exponential growth of the electronic media. Text representation and classification issues are usually treated as independent problems, but this paper illustrates combined approaches for text classification system. Integer representation is achieved using ASCII values of the each integer and later linear regression is applied for efficient classification of text documents. An extensive experimentation using nearest neighbor supervised learning algorithms on four publically available corpuses are carried out to reveal the efficiency of the proposed technique.

Keywords Document classification · Integer representation

1 Introduction

Due to the exponential increase in the popularity of the Internet and the World Wide Web text data has became the most common types of information store house. Most common sources are web pages, emails, newsgroup messages, internet news feeds, etc., [1]. Many real-time text mining applications have gained a lot of attention due to large production of textual data. Many applications of text classification are spam filtering, document retrieval, routing, filtering, directory maintenance, ontology mapping, etc.

S.N.B. Bhushan (✉) · Ajit Danti · S.L. Fernandes
Karnataka Government Research Center, Sahyadri College of Engineering and Management, Mangalore, Karnataka, India
e-mail: sn.bharath@gmail.com

Ajit Danti
e-mail: ajitdanti@yahoo.com

S.L. Fernandes
e-mail: steven.ec@shayadri.edu.in

Ajit Danti · S.L. Fernandes
Department of Computer Science, King Khalid University, Abha, Saudi Arabia

S.C. Satapathy et al. (eds.), *Proceedings of the International Conference on Data Engineering and Communication Technology*, Advances in Intelligent Systems and Computing 469, DOI 10.1007/978-981-10-1678-3_53

The main objective of the text classification algorithm is to identify text documents with the ontology of domains defined by the subject experts. A classifier can be designed systematically by training it using a set of training documents. Generally, literary information being unstructured in nature, contains more number of issues like selection of desired representation model, high dimensionality, semanticity, volume, and sparsity. Solutions for these issues are found in [2].

In this article, an integer representation model for text document which requires minimum amount of memory to store a word which in turn reduces the processing cost is presented. Text representation algorithm works on the principle that, an integer number requires minimum memory when compared to store a word. Integer representation-based classification of text documents is an unconventional approach for classification of text documents.

The remaining part of the paper is planned as follows. In Sect. 2 a brief literature survey on the text classification is provided. In Sect. 3, a proposed model for the compression-based classification of text document. Section 4 discusses about experimentation and comparative analysis performed on the proposed models. Paper will be concluded in Sect. 5.

2 Literature Survey

In literature, few works of compression-based text classification can be seen. Generally, text compression involves context modeling which assigns a probability value to new data based on the frequencies of the data that appeared before. But, context modeling algorithms requires more running time and large amount of main memory for processing. Mortan [3] proposed a modeling-based method for classification of text documents. A cross entropy-based approach for text categorization is presented in [4]. It is based on the fact that the entropy is the measure of information content. Compression models are derived from information theory, based on this theoretical fact, language models are constructed for text categorization problem. Authors have also illustrated that, character-based prediction by partial matching (PPM) compression schemes have considerable advantages over word-based approaches. Frank [5] considered classification task as a two class problem. Different language models such as Ma and Mb are constructed for each class using PPM methods. Test document will be compressed according to different models and gain per model is calculated. Finally, class label will be assigned based on the positive and negative gain. Modeling-based compression for low complexity devices is presented in [6]. This method is based on the fact that, PPM-based approaches require high computational effort which is not practically advisable for low complexity devices such as mobile phones. Algorithm makes use of static context models which efficiently reduce storage space. Similar type of work for low complexity devices are found in [7]. This approach split the data into 16 bit followed by the application of Quine-McCluskey Boolean minimization function to find the minimized expression. Further static Huffman encoding method is used for text compression. Dvorski [8] proposed an indexed-based compression scheme for text retrieval systems. A document is considered as a combination of words and non words. Similarly, Khurana and Koul [9] considered English text as a dictionary where each word is identified by a unique number in which a novel algorithm is proposed which consists of four phases, where each phase is for different type of input conditions along with a technique to search for a word in the compressed dictionary based on the index value of the word. Word-based semi-static Huffman compression technique is presented [10] in which algorithm captures the words features to construct byte oriented Huffman tree. It is based on the fact that byte processing is faster than bit processing. Automaton is constructed based on the length of search space. End tagged dense code compression is proposed in [11]. Though the proposed method looks similar to tagged Huffman technique, the algorithm has the capacity in producing better compression ratio, constructing a simple vocabulary representation in less computation time. Compressed strings are 8 % shorter than tagged Huffman and 3 % over conventional Huffman. Another word-based compression approach is found in [12]. This method compresses the text data in two levels. First level is the reduction which is through word lookup table. Since word lookup table is operated by operating system, the reduction is done by operation system only. In this technique, each term will be represented by its address index. Next stage is the compression stage. Four different Huffman, W-LZW, word-based first-order and first-order context modeling methods for text compression are presented in [13]. All the word-based compression techniques discussed above, maintain two different frequency tables for words and nonwords.

In the literature, many approaches are found for classification of text documents. They are naïve bayes [14, 15], nearest neighbor [16, 17], decision trees [18], support vector machines [19], and neural network [20] approaches.

3 Proposed Model

The proposed model can be categorized into two different stages, such as text representation stage and regression-based searching stage.

3.1 Text Representation

It is theoretically verified that a sequence of characters requires more memory than an integer number. Based on this, a novel text representation algorithm is proposed, which has the facility of representing character string (a word) by a unique integer number. It is verified that, terms are the set of alphabets, which denote a specific meaning. Similarly, a document is collection of such strings which represent a specific domain. Terms from each document are extracted and then subjected for compression algorithm and then it is represented by an integer number. The whole procedure is algorithmically represented in the algorithm 1 and pictorially represented in Fig. 1 and the proposed method are explained in illustration 2.

Cumulative sum of ASCII value is determined for given textual word using the equation (1)

$$C_w = \sum_{k=1}^{n} a_k\, b^k \qquad \ldots(1)$$

Where,

C_w = Cumulative sum of ASCII values

w = length of the document. (Number of words in the documents).

k = number of alphabets in the word.

a = ASCII value of alphabet.

b = base.

Illustration : 1

Input word: **heart**.

ASCII values: 104,101,97,114,116.

= $104 \times 2^4 + 101 \times 2^3 + 97 \times 2^2 + 114 \times 2^1 + 116 \times 2^0 =$ **3204.**

Data	Before Compression	After Compression
heart	5 bytes	2 bytes

heart will be represented by an integer number **3204**.

Fig. 1 Pictorial representation of compression algorithm

3.2 Regression-Based Searching Stage

Now, the text document can be viewed as a collection of integer vales. As a result text classification problem got reduced into integer searching problem. Once the data is represented by an integer value, it is sorted and linear regression is applied using the Eq. 2.

$$a_0 = \frac{\sum x_i y_i \sum x_i - \sum x_i^2 \sum y_i}{\left(\sum x_i^2 - n \sum x_i^2\right)}$$
$$a_1 = \frac{n \sum x_i y_i - \sum x_i \sum y_i}{n \sum x_i^2 - \left(n \sum x_i^2\right)} \qquad (2)$$

where, x = Positional Value, and y = word.

Once the regression algorithm is applied, the data will represented by a straight line, as shown in Fig. 2 using the Eq. (3)

Fig. 2 Regression-based searching

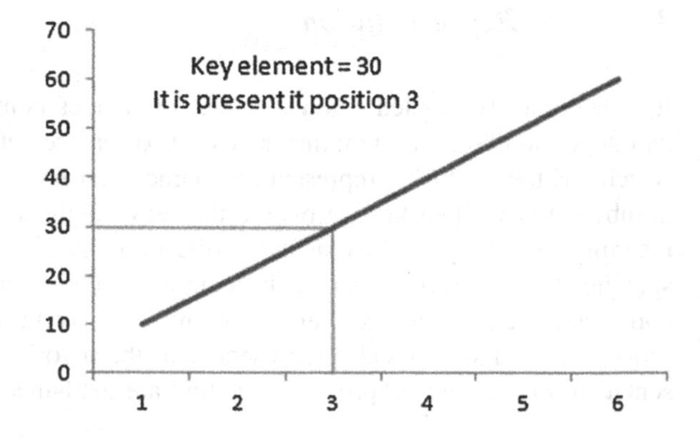

$$y = a_1 x + a_0. \tag{3}$$

where x gives the appropriate position of the search key element as shown in Fig. 2.

Classification of text documents will be accomplished by subjection training documents to representation algorithm. Once the data representation stage is accomplished linear regression is calculated for each class. Effect of this process, training phase will be seen as collection linear regression values. Then, query documents will be subjected to the compression algorithm. As a result, test documents will be collection of integer number. Each integer number from testing document is considered and is fed to regression-based searching algorithm. The main advantage of the regression is that, it takes minimum computational unit to search an integer number from the database. Similarly, the procedure is carried out and class label will be given to all the integer values. Test document will be assigned a class label based on the maximum class label assigned each integer. Figure 3 present the block diagram of the proposed method.

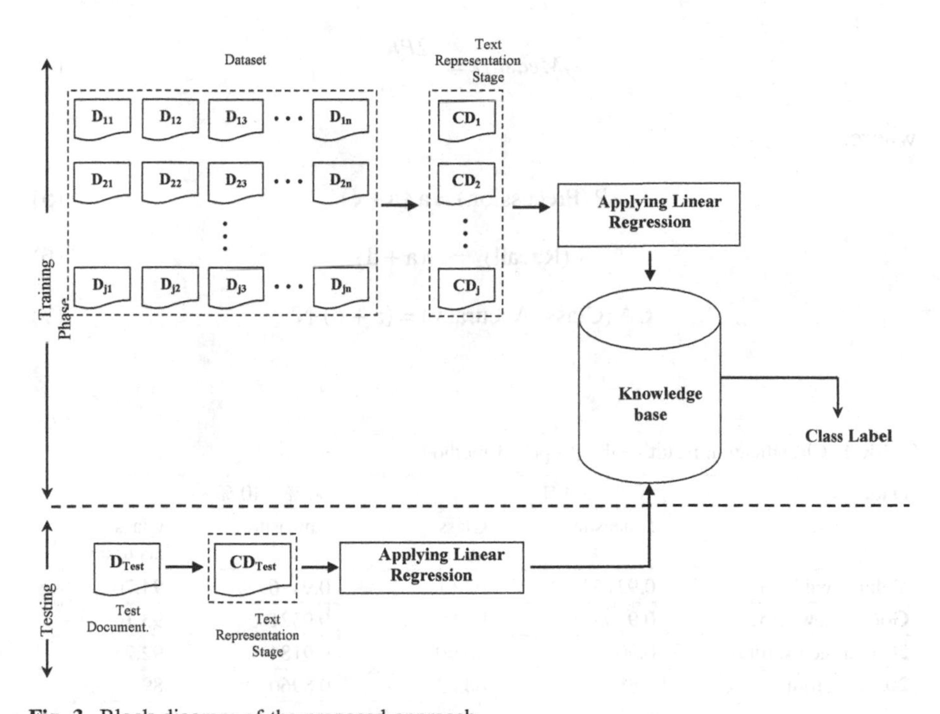

Fig. 3 Block diagram of the proposed approach

4 Experiments

To assess the efficiency of the proposed technique a well-known classification algorithm namely nearest neighbor classification technique is considered. Four publicly available corpuses are considered as databases where the first dataset is from Wikipedia pages which contain the characteristics of the automobiles. The second dataset includes 10 different classes for 1000 documents are from Google newsgroup data. The third and fourth is from 20 Mini newsgroup and 20 newsgroup datasets. Two different set of experiments are conducted are demonstrate the performance of the proposed technique. First set of experimentation consists of 40 % training and 60 % testing on all the four datasets. Second set of experimentation consists of 60 % training and 40 % testing on all the four datasets. The details of the first and second set of experiments are shown in Table 1.

F measure is considered for the evaluation of the proposed methods, precision, recall and class accuracy (CA) for each set of experiments using the Eq. (4)–(7). Let a, b, c, and d, respectively, denote the number of correct positives, false negatives, false positives, and correct negatives.

$$fMeasure = \frac{2PR}{P+R} \tag{4}$$

where,

$$\text{P (Precession)} = \text{a}/(\text{a}+\text{c}) \tag{5}$$

$$\text{R (Recall)} = \text{a}/(\text{a}+\text{d}) \tag{6}$$

$$\text{CA (Class Accuracy)} = (\text{a}+\text{d})/\text{N} \tag{7}$$

Table 1 Classification result of the proposed method

Datasets	40 % : 60 %		60 % : 40 %	
	f measure	Class accuracy	f measure	Class accuracy
Vehicle wikipedia	0.9273	92.61	0.9476	94.70
Google newsgroup	0.9205	92.00	0.9321	93.16
20 mini newsgroup	0.9003	90.00	0.9184	92.25
20 newsgroup	0.8873	88.62	0.8960	89.47

5 Conclusion

This paper illustrates a method of providing an integer representation of text documents and regression for classification of documents. Proposed model has the facility of representing character string (a word) by a unique integer number representation. It is verified that, terms are the set of alphabets, which denote a specific meaning. Similarly, a document is collection of such strings which represent a specific domain. Later on the concept regression is explored for classification of text documents. An extensive experimentation is carried on four publically available datasets to demonstrate the efficiency of the proposed models. The performance evaluation of the proposed method is carried out by performance measures such as f-measure and class accuracy (CA). The proposed model is very simple and computationally less expensive. One can think of exploring the proposed model further for other applications of text mining. This can be one of the potential directions which might unfold new problems.

References

1. Rigutini L., Automatic Text Processing: Machine Learning Techniques. Ph.D. Thesis, University of Siena 2004.
2. Sebastiani F., Machine learning in automated text categorization. ACM Computing Surveys. 2002, vol. 34, pp. 1–47.
3. Yuval M, Nign Wu., and Lisa Hellerstein., On compression based text classification. Advances in information retrieval in Advances in information retrieval, 2005, pages 300–314.
4. Teahan W., and Harper D., Using compression based language models for text categorization. Proceedings of 2001 workshop on language modeling and information retrieval.
5. Frank E., Cai C., and Witten H., Text Categorization using compression models. In proceedings of DCC-00, IEEE Data compression conference 2000.
6. Clemens S and Frank P. Low complexity compression of short messages. In proceedings of IEEE Data Compression Conference, 2006, pp 123–132.
7. Snel V., Plato J., and Qawasmeh E. Compression of small text files. Journal of Advanced Engineering Informatics Information Achieve, 2008, vol. 20, pages 410–417.
8. Dvorski J., Pokorn J and Snsel V. Word-based compression methods and indexing for text retrieval systems. Proceeding third east European conference on advances in databases and information systems, 1999, pages 75–84.
9. Khurana U and Koul A, 2005. Text compression and superfast searching. Proceedings of the CoRR., 2005.
10. Moura E., Ziviani N and Navarro and Yates RB. Fast searching on compressed text allowing errors. Proceedings of the 21st annual international ACM Sigir conference on Research and Development in Information retrieval, 1998, pp 298–306.
11. Nieves G., Brisaboa, Eva L, and Param J, An efficient compression code for text databases. Proceedings of the 25th European conference on IR research, 2003. pp 468–481.
12. Azad AK., Ahmad S., Sharmeen R., and Kamruzzana SM, An efficient technique for text compression. In proceedings of International Conference on Information Management and Business, 2005, pp. 467–473.
13. Horspool R.N., and Cormack G.V. Constructing word based text compression of short messages. In Proceedings of the IEEE Data compression conference, 1992, pages 62–71.

14. Hava O., Skrbek M., and Kordík P., Supervised two-step feature extraction for structured representation of text data. Journal of Simulation Modelling Practice and Theory, 2013, vol. 33, pp. 132–143.
15. Rocha L., Mourao F., Mota H., T.Salles, M. A. Gonc-alves, and W. Meira., Temporal contexts: Effective text classification in evolving document collections. Journal of Information Systems, 2013, vol. 38, pp. 388–409.
16. Ur-Rahman N. and Harding J.A. Textual data mining for industrial knowledge management and text classification a business oriented approach. Journal of Expert Systems with Applications, 2012, vol. 39, pp. 4729–4739.
17. Ajit Danti and S. N. Bharath Bhushan. Document Vector Space Representation Model for Automatic Text Classification. In Proceedings of International Conference on Multimedia Processing, Communication and Information Technology, Shimoga.2013, pp. 338–344.
18. Lewis D. D and M. Ringuette, A comparison of two learning algorithms for text classification. Proceedings of the 3rd Annual symposium on Document Analysis and Information Retrieval, 1998, pp. 81–93.
19. Joachims T., Text categorization with support vector machines: Learning with many relevant features. Proceedings of European Conference on Machine Learning (ECML), 2000, no. 1398, pp. 137–142.
20. Patra A and Singh D., 2013. Neural Network Approach for Text Classification using Relevance Factor as Term Weighing Method. International Journal of Computer Applications, 2013, vol 68, no. 17.

Communication Device for Differently Abled People: A Prototype Model

Rajat Sharma, Vikrant Bhateja, S.C. Satapathy and Swarnima Gupta

Abstract The process of communication between marginalized communities like deaf-blind-dumb people has always been a matter of great concern and these differently abled people are not able to easily communicate their thoughts and talks with other people as normal people does by using mobile phones, etc. So, it is the greatest need of this hour to think and act upon the development of such people as they are also the equal part of our society. The proposed model in this paper, proposes a finely tuned solution to mitigate this problem of ever increasing communication gap between differently abled people and normal people. The architecture of this portable device is presented and its operations are discussed via three embedded algorithms for faster, easier, and accurate message communication.

Keywords Differently abled people • Hardware prototype • Communication • Machine model

Rajat Sharma (✉)
Sopra Steria Pvt. Ltd., Noida, U.P., India
e-mail: march10rajat@gmail.com

Vikrant Bhateja
Department of Electronics and Communication Engineering,
Shri Ramswaroop Memorial Group of Professional Colleges (SRMGPC),
Lucknow 227105, U.P., India
e-mail: bhateja.vikrant@gmail.com

S.C. Satapathy
Department of Computer Science and Engineering, ANITS,
Visakhapatnam, A.P., India
e-mail: sureshsatapathy@gmail.com

Swarnima Gupta
ABES Engineering College, Ghaziabad, U.P., India
e-mail: swarni.smile@gmail.com

S.C. Satapathy et al. (eds.), *Proceedings of the International Conference on Data Engineering and Communication Technology*, Advances in Intelligent Systems and Computing 469, DOI 10.1007/978-981-10-1678-3_54

1 Introduction

The existing ways followed by differently abled people to communicate with each other till date has been by mere gestures, physical touch, finger sensations and stimulations on the skin of the sufferer and a multitude of techniques that did not find its existence on the grounds of technicality [1]. When a deaf-dumb person speaks to a normal person, the normal person seldom understands and asks the deaf-dumb person to show gestures for his/her needs [2]. While many deaf persons communicate effectively using a form of sign language or the fingerspelling alphabet, problems arise when a hearing person, who does not know sign language, attempts to interact with a deaf individual. For deaf-blind persons, they must be able to touch the hand of the person with whom they wish to interact. Thus, there is a great need for a portable communication aid which would permit deaf, deaf blind, and nonvocal individuals to engage in conversation among themselves and with hearing, vocal persons [3, 4]. The existing works done in this domain for differently abled people include: an instrumented glove by Rehman et al. [4], Choudhary et al. [5], developed a smart glove translating braille alphabet into text and vice versa. U. Gollner et al. [6] developed a Mobile Lorm glove which will act as a common form of communication to be used by people with both hearing and sight impairment. Ohtsuka et al. [7] developed a jacket consisting of vibrators which will be worn by deaf-blind person and non-disabled person will be having electronic device with LCD display. Therefore, there is a requirement of such a model which would be faster, easier, and economical to develop for facilitating a faster communication between disabled people. The prototype model consists of RF transmitters and receivers working on WI-FI protocol for establishing on-spot network for faster transference of messages. We have used parallel working methodology technique of three algorithms to establish a faster and more accurate connection link for easy transference.

2 Methodology and Implementation

The hardware architectural prototype is shown below in Fig. 1 along with component description as:

2.1 Component Description of Hardware Model

The machine architecture is described with the following components and functionalities as is shown in Table 1.

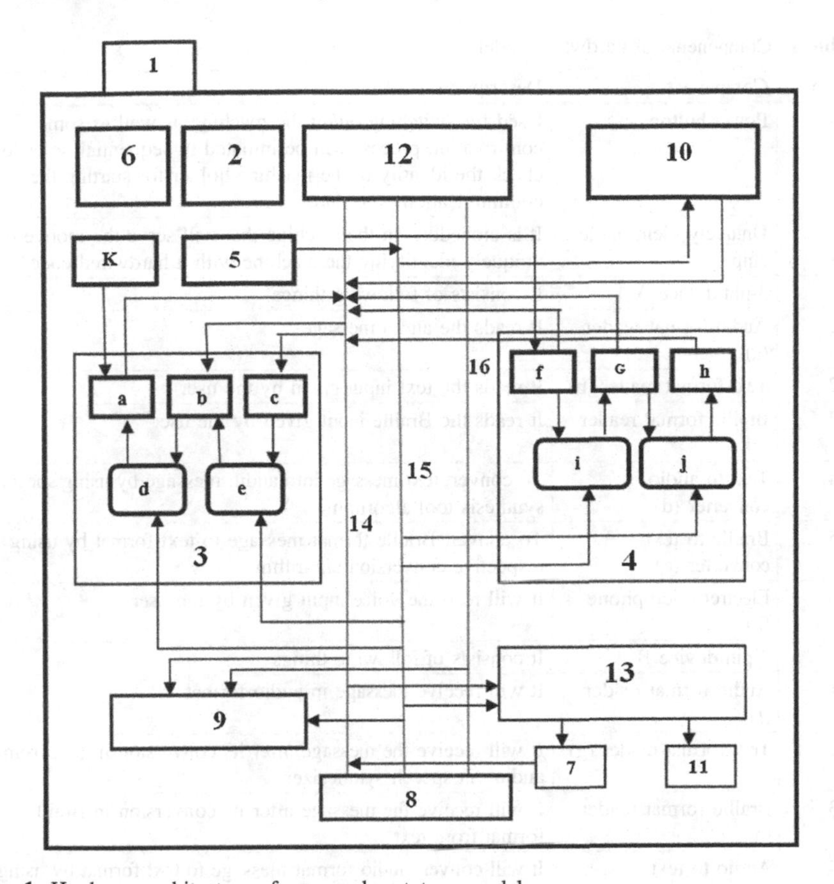

Fig. 1 Hardware architecture of proposed prototype model

2.2 Proposed Methodology

When machine model will be purchased from manufacturer then for the very first time it will be started by switching on the power button, provided in machine architecture. As soon as the machine will start-up, the uniquely identifiable chip embedded in the machine will get activated and will set a unique code for all time recognition purposes of the machine. Once machine is set in active mode for the first time, Algorithm 1 will be executed so as to set the formats to input the message on machine by the holder and to set receiving message format coming on the machine in an understandable format depending on the disability, respectively. Now, using this unique identifiable code, the process of connection establishment between sufferers will be initiated as—The communication link between the two machines will be set up by using RF transmitter and receiver working on WI-FI protocol embedded with IEEE 802.1X for establishing a secured network for data communication. The uniquely identifiable code will be used to form a network

Table 1 Components of hardware model

Sr. no.	Component	Description
1	Power button	Used for switching on/off the machine as well as some confirmation process will be initiated in sequential order to check the identity of the machine holder for starting the communication procedure
2	Uniquely identifiable chip	It is embedded in the machine that will serve the motive of uniquely identifying the machine with a hardwired code
3	3-pin device A	It consists of following things
3.1	Audio format reader (a)	It reads the audio message
3.2	Text format reader (b)	It reads the text input given by the user
3.3	Braille format reader (c)	It reads the Braille input given by the user
3.4	Text to audio converter (d)	To convert text message into audio message by using speech synthesis tool algorithm
3.5	Braille to text converter (e)	To convert Braille format message to text format by using respective conversion algorithm
3.6	Electret microphone (k)	It will read the voice input given by the user
4	3-pin device B	It consists of following things
4.1	Audio format reader (f)	It will receive message in audio format
4.2	Text format reader (g)	It will receive the message after its conversion in text from audio via speech synthesizer
4.3	Braille format reader (h)	It will receive the message after its conversion in Braille format from text
4.4	Audio to text converter (i)	It will convert audio format message to text format by using respective conversion algorithm
4.5	Text to Braille converter (j)	It will convert text format message to Braille format by using respective conversion algorithm
5	Modem	It is assembled so as to modulate the signal during the time of transmission process and to demodulate the signal while receiving process
6	GPS tracker	It is embedded in the machine that will be used to track the machine by the use of uniquely identifiable chip already inserted into the machine
7	Display screen	It will be partitioned into two halves—one half showing virtual keypad and the other half showing Braille keypad
8	Battery slot	The batteries will be fitted to operate the whole functional mechanism of the machine
9	Transmitter/modulator	It is used to carry forward the message in audio format to the receiver/demodulator on the receiver's machine
10	Receiver/demodulator	It is used to receive the message in audio format from the transmitter/modulator of the sender's machine

(continued)

Table 1 (continued)

Sr. no.	Component	Description
11	Speaker	It will convert the audio signal received by the sender's machine into sound that can be heard at specified volume rate
12	Microcontroller	It is CONTROL UNIT of the machine included to perform all the operations on the three algorithms hardcoded
13	Interface unit	It will take the data from 3-pin device B in the form of either audio, text or Braille. It will be connected to two output devices of the machine, speaker and the screen
14	Data bus	All the inputs (in any form i.e., either in Braille/text/audio) coming from the user are fed into the data bus. The message that will be received by the receiver will also be fed into the same data bus
15	Control bus	All the algorithm that will be responsible for data conversion purpose(s) into required message formats either into audio or into text or into Braille are embedded into the control bus
16	Address bus	It will be used to transfer data to desired address (uniquely identifiable code received from other machine)

created on the spot by using this protocol and the connection will be established as the address bus of the machine will receive that particular address for connecting link. Once the connecting link is established the signals can be easily transferred. Once the input formats (speak, type in text, type in braille) are set and connection is established then Algorithm 2 will be executed for internal format conversions to make the message ready for transmission. Now, discussing about the sending phase of message by machine as—for the all purposes required to do the work of generating a message, are done by 3-pin device A and by using its five components it can be used as—there are three cases to be considered for better understanding—Case 1: The user will give the audio input to the electret microphone present on top surface of the machine model which would be then sent to the audio format reader to recognize the audio input. Now the received audio message will be directed to data bus of the controller so as to redirect the signal to modem for message modulation purposes. Case 2: The user will give the input in text format by directly typing on the virtual keypad and this input will be received by the text format reader present in the device from data bus and it will be sent into text to audio converter (speech synthesizing tool) by control bus as final message sending mode is audio format as output. Now, this audio message will be directed to data bus of the controller. Case 3: The user will give the input in braille format by typing on the braille keypad which will be present on the architectural model screen and this input will be received by the braille format reader from data bus and it will be sent into braille to text converter (first) by control bus and then again with help of data and control bus, message is converted from text to audio format (second) as discussed in previous case. Now, the received audio message will be directed to data bus of the controller. To bring these tasks of message format conversion easily by microcontroller [8], Algorithm 4 is used for above three cases and by undergoing the process executed by Algorithm 4, message is finally transmitted by transmitter on the destination machine as It receives the message to be sent from the data bus of the controller and the address (uniquely identifiable code) of the receiver's machine from the address bus. Now, once the message is sent by sender, then it is received by receiver machine as—the demodulation activity of modem is performed on received message by using Algorithm 5 so as to remove all the extra noise and attenuation from speech signal and to maintain its quality without any degradation, after which it is redirected to data bus. Now, Algorithm 3 is executed so as to fulfill the purpose of showing a message into the understandable format set as default on start-up(may be as audio on speaker or text on screen or braille on screen). As soon as he message is received by receiver, it is redirected to data bus for taking the message to audio reader of 3-pin device B comprising of five components as were present in above discussion of 3-pin device A. By checking the conditions of understandable format of suffer in Algorithm 3, data bus with help of control bus will send the data (may be to amplifier speaker in case of audio) or audio will go in one conversion step of text by audio to text converter or audio will first be converted to text and then to braille by using text to braille converter depending on the condition checks. The final format of the message will travel via data bus to interface unit and by help of control bus it will be either listened on speaker (if audio) otherwise it will be shown on screen (if text or braille). All the components of machine will work finely as battery will complete the purpose of providing current.

2.3 *Description of Algorithms*

The below section will give the information of all the algorithms hardcoded in the microcontroller. Algorithm 1 will be executed at the first time start-up of the machine only. Here, we check the condition of the sufferer and accordingly set the format for inputting the message as well as the format in which the sufferer wants to receive the message.

2.3.1 Algorithm 1

BEGIN:

Step 1:- Operate the function of "ON" to the power button of the machine for the first time (only applicable for the machine which is newly purchased from the market).

Step 2:- There are some required settings that are to be done by the helper for the first time to give the required instructions at the time of initialization process.

Step 3:- Initialize the value of flag variable with zero i.e., FLAG<- 0.

Step 4:- Set the date, time and format of the input to be given and output to be seen on the machine screen after receiving the message from sender machine as –

IP is a variable whose value is set according to the disability of the sufferer for inputting the message.

If (sufferer==blind)
    Set : IP<-1
Else if (sufferer==dumb ||sufferer==deaf && dumb)
    Set : IP<-2
Else if (sufferer==blind && dumb ||sufferer==blind && deaf && dumb)
    Set : IP<-3

Step 5:- OP is a variable whose value is set according to the disability of the sufferer for receiving the message.

If (sufferer==blind|| sufferer==dumb ||sufferer==blind && dumb)
    Set : OP<-1
Else if (sufferer==deaf && dumb)
    Set : OP<-2
Else if (sufferer==blind && deaf && dumb)
    Set : OP<-3

Step 6 :- Check that

If (OP==1)
    Set : audio format as default output message mode.
Else if (OP==2)
    Set : text format as default output message mode.
Else
    Set : braille format as default output message mode.

Step 7:- After setting the required input value in input variable and output value in output variable set the value of FLAG variable as one

i.e. FLAG<-1.

Step 8:- Check that

If (IP==1)
    Set : audio format as default input format.
Else if (IP==2)
    Set : text format as default input format.
Else
    Set : braille format as default input format.

END

Algorithm 2 will be used during the sending phase of the machine i.e. when the user's machine is sending the message to another machine.

2.3.2 Algorithm 2

```
BEGIN:
Step 1:- Switch 'ON' machine.
Step 2:- if (flag==0)    [Input, output formats are not set]
         {
          Execute Algorithm 1;
         }
Step 3:- if (IP==1)
         {
              Disable  :   Input pins of text and braille format reader.
              Direct   :   Message from microphone to audio format reader.
              Direct   :   Audio message from audio format reader to data bus.
              Send     :   Audio signal from data bus to transmitter.
         }
         Else if (IP==2)
         {
              Disable  :   Input pin of braille format reader.
              Disable  :   Microphone input pin.
              Direct   :   Message from screen (interface unit) to text format reader.
              Take     :   Input message from text format reader.
              Activate :   Text to audio converter.
              Pass     :   Message through text to audio converter.
              Direct   :   Output message to audio format reader.
              Direct   :   Audio message from audio format reader to data bus.
              Send     :   Audio signal from data bus to transmitter.
         }
         Else
         {
              Disable  :   Input pin of text format reader.
              Disable  :   Microphone input pin.
              Direct   :   Message from screen (interface unit) to braille format reader.
              Take     :   Input message from braille format reader.
              Activate :   Braille to text converter.
              Pass     :   Message through text to audio converter.
              Direct   :   Output message from braille format reader to text format reader.
              Take     :   Input message from text format reader.
              Activate :   Text to audio converter.
              Pass     :   Message through text to audio converter.
              Direct   :   Output message to audio format reader.
              Direct   :   Audio message from audio format reader to data bus.
              Send     :   Audio signal from data bus to transmitter.
         }
Step 4:- Direct: address (uniquely identifiable code) of receiver's machine from address bus to modem
         (transmitter).
Step 5:-Transmit by using modulator function of modem.
END.
```

Algorithm 3 will be used for directing message from 3-pin device B to the interface unit and from interface unit to the output devices, i.e., speaker and screen.

2.3.3 Algorithm 3

```
BEGIN:
Step 1:- Switch 'ON' machine.
Step 2:- if (flag==0)  [Input, output formats are not set]
        {
        Execute Algorithm 1;
        }
Step 3:- Demodulate: received signal from receiver.
         Send: demodulated signal to data bus.
Step 4:- Send: signal from the data bus to the audio format reader of the 3-pin device B.
Step 5:-if (OP==1)
        {
            Disable :   Output pin of text format reader.
            Disable :   Output pin of braille format reader.
            Send    :   Signal from audio format reader to data bus.
            Direct  :   Signal from data bus to interface unit.
            Direct  :   Signal from interface unit to speaker.
        }
        Else if (OP==2)
        {
            Take     :  Input from audio format reader.
            Disable  :  Output pin of audio format reader to data bus.
            Activate :  Audio to text converter.
            Pass     :  Message through audio to text converter.
            Direct   :  Output from converter to text format reader.
            Send     :  Signal from text format reader to data bus.
            Direct   :  Signal from data bus to interface unit.
            Direct   :  Signal from interface unit to screen.
            Show     :  Text message on selected area of screen.
        }
        Else
        {
            Take     :  Input from audio format reader.
            Disable  :  Output pin of audio format reader to data bus.
            Activate :  Audio to text converter.
            Pass     :  Message through audio to text converter.
            Direct   :  Output to text format reader.
            Disable  :  Output pin of text format reader to data bus.
            Activate :  Text to braille converter.
            Pass     :  Message through text to braille converter.
            Take     :  Input from text format reader.
            Direct   :  Output to braille format reader.
            Send     :  Signal from braille format reader to data bus.
            Direct   :  Signal from data bus to interface unit.
            Direct   :  Signal from interface unit to screen.
            Show     :  Braille message on selected area of the screen.
        }
END.
```

So, in this manner all three algorithms will be loaded so as to perform the function of communication between two devices.

3 Conclusion and Future Scope

The proposed model can be successfully used for communication purpose by differently abled persons with mono dual or multiple defects. The fundamental model of the proposed machine along with running algorithms has been designed. A differently abled person carrying the machine could send message in pre-set input format to another person (carrying the machine and linked with the sender machine) with similar or dissimilar disability and the other person would receive the message in pre-set output format. The proposed machine serves as easy-to-use and with negligible time delay communication medium between disabled people. The work can be further extended to facilitate some of the applications like—a section for storing the chats could be added up and an android application can be designed to be used by the guardian of the disable person to track his path so that security of the person can be ensured at every point.

References

1. Sharma R., Gupta S.: A One-to-One Communication Model to Facilitate Conversation Between Differently-Abled People by a Portable and Handy Machine. In: Proceedings of the Second International Conference on Computer and Communication Technologies, pp. 651–659, Springer India (2016).
2. Radha, H. G., Shruti S. D., and Kanya B. S.: Design And Development of an Assistive Device For Speech and Hearing Impaired. In: IJITR 2, no. 2, pp. 859–862 (2014).
3. Kramer, J. P., Lindener P., and George W. R.: Communication system for deaf, deaf-blind, or non-vocal individuals using instrumented glove. In: U.S. Patent 5,047,952, (1991).
4. Rehman A. U., Rehman A. U., Afghani S., Akmal M., and Yousaf R.: Microcontroller and sensors based gesture vocalizer. In: Proceedings of the 7th WSEAS International Conference on Signal Processing, Robotics and Automation, pp. 82–87. World Scientific and Engineering Academy and Society (WSEAS) (2008).
5. Choudhary, T., Kulkarni S., and Reddy P.: A Braille-based mobile communication and translation glove for deaf-blind people. In: International Conference on Pervasive Computing (ICPC), pp. 1–4. IEEE (2015).
6. Gollner, U., Bieling T., and Joost G.: Mobile Lorm Glove: introducing a communication device for deaf-blind people. In: Proceedings of the Sixth International Conference on Tangible, Embedded and Embodied Interaction, pp. 127–130. ACM (2012).
7. Ohtsuka, S., Hasegawa S., Sasaki N., and Harakawa T.: Communication system between deaf-blind people and non-disabled people using body-braille and infrared communication. In: 7th IEEE Consumer Communications and Networking Conference (CCNC), pp. 1–2. IEEE (2010).

8. Sharma R., Bhateja, V., Satapathy, S. C.: GSM based Automated Detection Model for Improvised Explosive Devices. In: Proceedings of the Third International Conference on INformation systems Design and Intelligent Applications, pp. xx, Springer India (2016).

Combination of PCA and Contourlets for Multispectral Image Fusion

Anuja Srivastava, Vikrant Bhateja and Aisha Moin

Abstract Multispectral image fusion proceeds to combine images from diverse modalities to congregate the expedient information, while abandoning the redundant information from the input images. The aim is to realize a single image holding anatomical information without compromising on the functional information. This paper presents a principal component analysis (PCA)-based multispectral fusion framework for medical images acquired from two different sensor modalities using the contourlet transform. During the process, the input multispectral image is transformed from the *RGB* to *YIQ* color space in order to provide better retention of functional information. Subband decomposition of source images is employed with contourlet transform; to impart anisotropy and seizing of better visual geometrical structures. Further, PCA provides the dimensionality reduction followed by application of min-max fusion of subbands. The superiority of the fusion response is depicted by the comparisons made with the other state-of-the-art fusion approaches.

Keywords PCA · Multispectral · Contourlet · *YIQ* color space

1 Introduction

With the advent of diverse sophisticated imaging modalities, image fusion is gaining considerable importance in both medical and non-medical domains. Image fusion in medical domain has enticed researchers from all over the world as it

Anuja Srivastava (✉) · Vikrant Bhateja · Aisha Moin
Department of Electronics and Communication Engineering, Shri Ramswaroop Memorial Group of Professional Colleges (SRMGPC), Lucknow 227105, Uttar Pradesh, India
e-mail: anuja.srivastava009@gmail.com

Vikrant Bhateja
e-mail: bhateja.vikrant@gmail.com

Aisha Moin
e-mail: aishamn04@gmail.com

S.C. Satapathy et al. (eds.), *Proceedings of the International Conference on Data Engineering and Communication Technology*, Advances in Intelligent Systems and Computing 469, DOI 10.1007/978-981-10-1678-3_55

facilitates in accurate and easier clinical diagnosis of numerous diseases. Few of the common imaging modalities used for medical imaging are: computer tomography (CT) scan, MRI (magnetic resonance imaging) scan, SPECT (single positron emission tomography) scan, and PET (positron emission tomography) scan. But these modalities often provide complementary and redundant information about the same Region of interest (ROI). Hence, there comes a need for an efficient fusion process that not only integrates the complementary and disparate information but is also effective in discarding the superfluous information in order to confer a final fused image which is more reliable and informative than the individual source images. Out of the aforementioned modalities, CT and MRI scans provide panchromatic images detailing only into the anatomical information (spatial features), whereas PET and SPECT scans provide multispectral images emphasizing only on functional information (spectral features). Hence, for a compendious view in neuroimaging, multispectral fusion of SPECT/PET with CT/MRI is considered to give a final fused image which details into both anatomical as well as functional information [1]. Moreover, it is visually convenient to interpret features with high spatial resolution and multispectral content than a single high resolution panchromatic image. Further, PET/SPECT scans are considered as advancement to other scans, thus, fusion of CT/MR with PET/SPECT provides better insight into various diseases [2]. But, medical images are often superimposed by noises during acquisition or transmission. Thus, image prefiltering is necessary to suppress the erroneous intensity fluctuations caused due to imperfection of imaging devices or transmission channels [3–5]. Fusion algorithms can be cataloged into three levels: pixel, feature, and decision level. In this paper, pixel level fusion is used [6, 7]. Among different types of transform bases, the most prominent ones are discrete wavelet transform (DWT) [8–11], ridgelet transform [12], curvelet transform [13], and contourlet transform [14–17]. Xu [6] performed pixel level fusion based on local extrema of CT scan and MR source images. The authors' in this work decomposed source images into coarse and detailed layers followed by application of local energy and contrast based fusion rules. Liu et al. [8] proposed a compressive sensing approach to multimodal fusion. During the fusion process, the sparse coefficients were obtained by DWT and then two different fusion rules were applied to fuse the low and high frequency components. But, this leads to non-satisfactory robustness of the fusion method due to the constraint of uncertainty as it utilizes the random measurement matrices. Contourlet transform proves to stand out among these transforms in representing signals by capturing the 2D geometrical structures in visual information. It captures sharp transitions like edges, geometrical structures more efficiently than any other transform. Moreover, to enhance the quality of image fusion, PCA algorithm is applied followed by min-max fusion rule. The remaining part of the paper is organized as follows: Sect. 2 details the proposed fusion methodology. Section 3 discusses the obtained results and conclusions are drawn in Sect. 4.

2 Proposed Fusion Methodology

This section details about the steps required for multispectral fusion of CT/MR images with PET/SPECT images in Contourlet domain. The processing has been carried out in specifically in Contourlet domain due to the aforesaid advantages. But as Contourlet transform captures limited directional information, it has been hybridized with principle component analysis [18]. PCA not only counters the limited directionality limitation of Contourlet transform, but also enhances the fusion of medical images. It is important to note that as a multispectral source image is used for fusion process, color transformation is an integral part of the whole fusion methodology. The fusion methodology can be better explained with the help of proposed block diagram (refer Fig. 1). Here, it is assumed that the input images are noise free and preprocessed.

2.1 Color Transformation

Multispectral Image fusion takes into account the fusion of a panchromatic and a multispectral image. Since, one of the input image is multispectral, color transformation becomes an integral part of the fusion process. SPECT and PET images are inherently *RGB* but medical image fusion in *RGB* color space is not desirable because in *RGB* model it is difficult to determine specific color and it is not useful for object specification and recognition of colors. Thus, there is a need to transform *RGB* to a different color model. One of the widely used transform in image fusion is IHS transform, but there are certain problems while using it in the medical domain. Hence, to preserve both spatial as well as spectral features *YIQ* color model is considered over IHS color model. In *YIQ* color model *Y* corresponds to

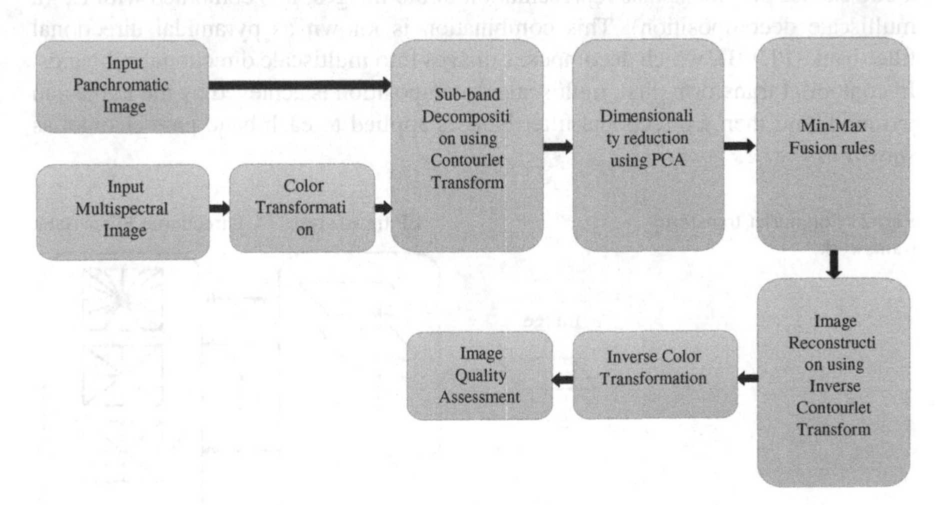

Fig. 1 Block diagram of proposed fusion methodology

luminescence (lightness), I and Q contains the chrominance information. I corresponds to orange-cyan axis, whereas Q corresponds to magenta-green axis and represent the hue and saturation, respectively. The relation between *YIQ* and *RGB* is shown by Eqs. (1) and (2) [19]. Once the input images are preprocessed, the multispectral image is converted to *YIQ* color space and its Y channel is fused with the input panchromatic image.

***RGB* to *YIQ* conversion**:

$$\begin{bmatrix} Y \\ I \\ Q \end{bmatrix} = \begin{bmatrix} 0.299 & 0.587 & 0.114 \\ 0.586 & -0.275 & -0.321 \\ 0.212 & -0.528 & 0.311 \end{bmatrix} \begin{bmatrix} R \\ G \\ B \end{bmatrix} \tag{1}$$

***YIQ* to *RGB* conversion**:

$$\begin{bmatrix} R \\ G \\ B \end{bmatrix} = \begin{bmatrix} 0.300 & 0.600 & 0.210 \\ 0.590 & -0.280 & -0.520 \\ 0.110 & -0.320 & 0.310 \end{bmatrix} \begin{bmatrix} Y \\ I \\ Q \end{bmatrix} \tag{2}$$

2.2 Decomposition into Subbands Coefficients

The contourlet transform consists of two steps namely subband decomposition and the directional transform. First, Laplacian pyramid (LP) performs subband decomposition and generates a low-pass version of the original image. Implicit oversampling occurs in LP decomposition thus directional filter bank (DFB) is applied to capture the high frequency components of the source image. But as DFB alone cannot provide sparse representation of the images, it is combined with LP (a multiscale decomposition). This combination is known as pyramidal directional filter bank (PDFB), which decomposes images into multiscale directional subbands. In contourlet transform, first, multiscale decomposition is achieved by the Laplacian pyramid, and then a directional filter bank is applied to each band pass channel as shown in Fig. 2.

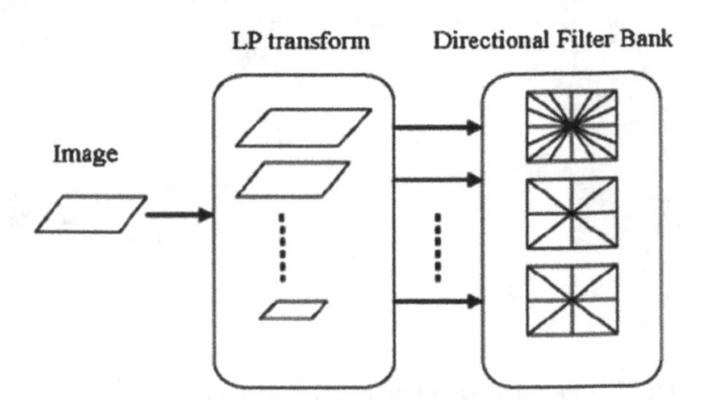

Fig. 2 Contourlet transform framework

2.3 *Dimensionality Reduction Using PCA*

Real-world data such as speech signals, medical images, etc., generally have very high dimensionality or redundancy. Various methods therefore are used to reduce the dimensions of these high-dimensional data. Most popular among them is PCA. PCA being an orthogonal transform helps to reduce the redundancy present in both the source image. PCA is a classical method that provides a sequence of best linear approximations to a given high-dimensional observation. It is one of the most popular techniques for dimensionality reduction. The subspace modeled by PCA captures the maximum variability in the data, and can be viewed as modeling the covariance structure of the data [20–23].

3 Simulation Results and Discussions

To evaluate the effectiveness of the proposed fusion methodology, this section analysis the fused image both quantitatively and qualitatively.

3.1 *Fusion Metrics for IQA*

The main aim of image fusion is to preserve the complementary and discard the superfluous information and distortion present in the individual source images. Performance measures are thus essential to measure the fruitfulness of the fusion methodology and compare the results from different algorithms. Relevant studies show that a single quality measure cannot be fully sufficient to quantify the performance of the fusion approaches as a particular metric can assess a particular aspect of the fused image [24, 25]. Therefore, IQA metrics like entropy (E), edge strength ($Q^{AB/F}$), and SSIM are employed to assess the effectiveness and efficiency of the fusion methodology. Higher values of each of these indices demonstrate the effectives of the fusion. These metrics are explained in Table 1.

Table 1 IQA of proposed fusion method using different metrics

Metrics	Formulae
Entropy	$E = \sum_{l=0}^{L-1} P_1 \log_2 P_1$ Higher value indicates more information in the fused image
Edge strength	$Q^{AB/F} = \frac{\sum_{m,n} Q^{AF}(m,n)_{W_A}(m,n) + Q^{BF}(m,n)_{W_B}(m,n)}{\sum_{m,n} W_A(m,n) + W_B(m,n)}$ Higher value indicates higher degree of edge preservation
Structural similarity index	$\text{SSIM} = \left(\frac{\sigma_{xy}}{\sigma_x \sigma_y}\right)\left(\frac{2\overline{xy}}{(\bar{x})^2 + (\bar{y})^2 + K_1}\right)\left(\frac{2\sigma_x \sigma_y}{(\sigma_x)^2 + (\sigma_y)^2 + K_2}\right)$ Higher value indicates preservation of luminance, contrast and structural content

3.2 Simulation Results

The test images in the present work include three sets of CT/MR and SPECT/PET images (namely Test Image 1, Test Image 2, and Test Image 3). Figure 3 shows the fusion response on these three test image sets. It can be inferred from obtained results that both the anatomical (soft and hard tissues) as well as functional (blood flow activity depicted by different colors) are visible in the composite fused images. It can be therefore concluded that the fusion response in Fig. 3 effectively demonstrates the features of both the modalities. High values of the IQA metrics as shown in Table 2 denote the high quality of the obtained fusion responses. The high entropy values indicate that high amount of information content is present in the fused images. On the other hand, higher values of *SSIM* indicate the preservation of luminance, contrast and structural content. Above all, the amount of edge

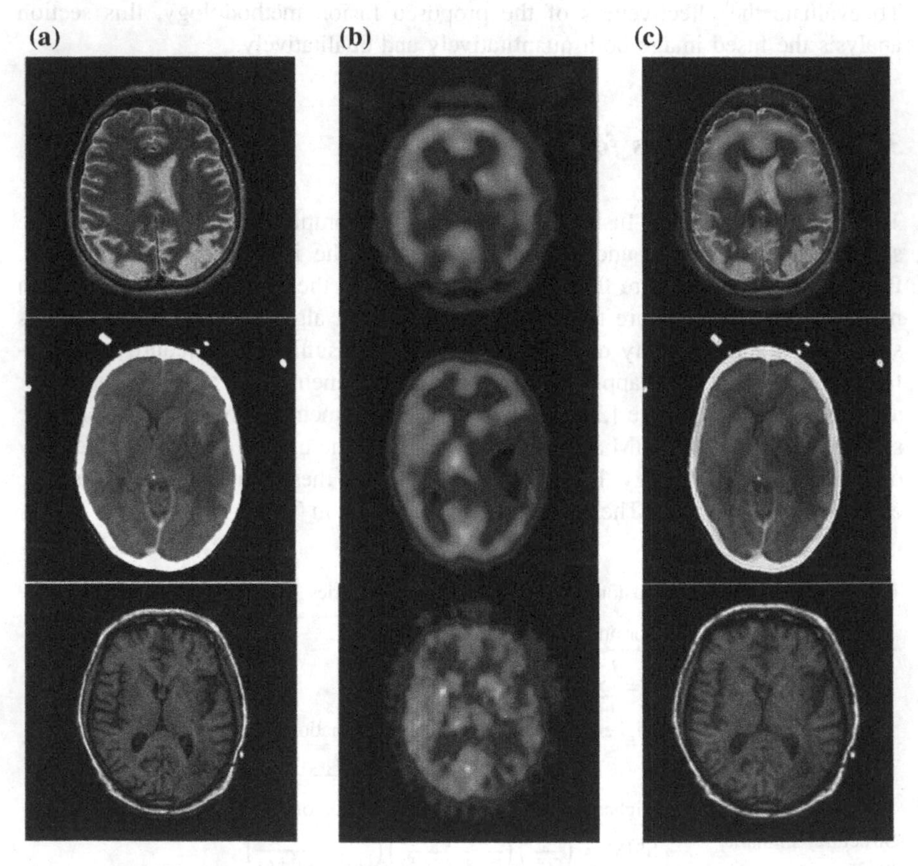

Fig. 3 An Illustration showing fusion of different modalities. **a** Panchromatic images. **b** Multispectral images. **c** Fused images

Table 2 Image quality evaluation of proposed method using different metrics

Data set	E	SSIM	$Q^{AB/F}$
Image set 1	5.7389	0.8388	0.6276
Image set 2	5.8711	0.8259	0.6246
Image set 3	5.6885	0.8458	0.6337

preservation in the fused images determines the major quality aspect of image fusion. Thus, higher value of $Q^{AB/F}$ indicates higher degree of edge preservation.

The proposed fusion response is further compared with other state-of-art methods (refer Fig. 4). It is clearly evident that the proposed fusion response is superior to other methods as it is capable of preserving both spatial as well as spectral features. The high values of entropy (refer Table 3) indicate that the final fused image contain high information and is more beneficial than the individual source images.

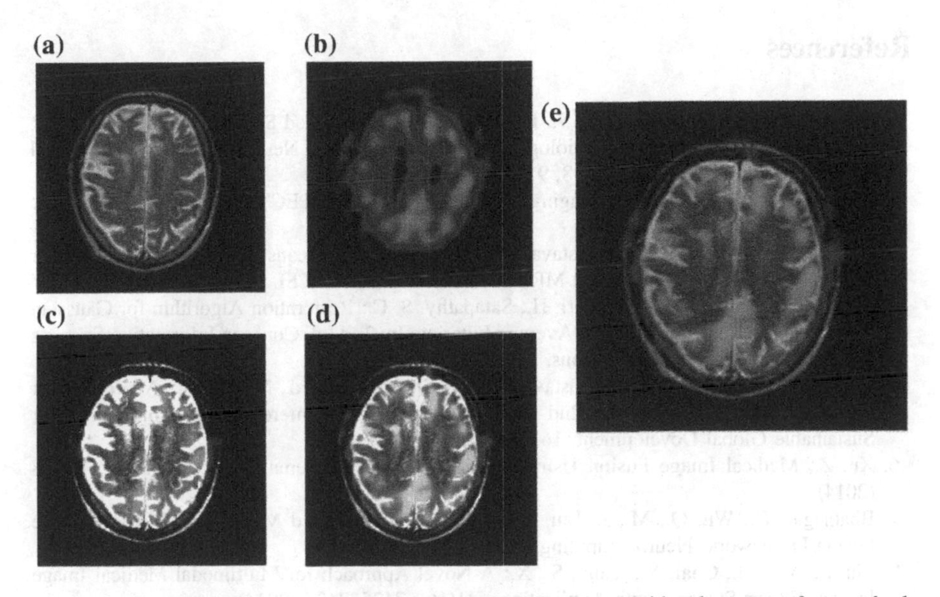

Fig. 4 A comparative analysis of proposed fusion methodology with other state-of-art method. **a** Input MR image. **b** SPECT-image. **c** Curvelet [13]. **d** Contourlet [14]. **e** Proposed method

Table 3 IQA of proposed fusion method compared with other state-of-art approaches

Approaches	E
Curvelet [13]	5.326
Contourlet [14]	5.481
Proposed method	5.723

4 Conclusion

A contourlet-based approach in combination with PCA for the fusion of multi-spectral medical images is presented in this paper. Contourlet is selected as the decomposition transform owing to its properties like anisotropy and better visualization of the geometrical structures present in the input images. The redundancy of the input medical images is reduced using PCA as the technique. Min-max fusion rules are employed to fuse the decomposed coefficients. The final fused image obtained from the proposed fusion approach shows both anatomical and functional information and provide a better visual perception. The values of the quality parameters are indicative of an effective fusion with all the desirable information present in the fused image. The results achieved are better than the previously proposed fusion approaches both visually and quantitatively.

References

1. Singh, S., Gupta, D., Anand, R., S., Kumar, V.: Nonsubsampled Shearlet based CT and MR Medical Image Fusion using Biologically Inspired Spiking Neural Network. Biomedical Signal Processing and Control. 18, 91–101 (2015)
2. Cherry, S., R.: Multimodality Imaging: Beyond PET/CT and SPECT/CT. Seminars in Nuclear Medicine. 39(5), 34–353 (2009)
3. Bhateja, V., Tiwari, H., Srivastava, A.: A Non-Local Means Filtering Algorithm for Restoration of Rician Distributed MRI. Annual Conv. of the CSI. 2, 1–8 (2014)
4. Srivastava, A., Bhateja, V., Tiwari, H., Satapathy, S. C.: Restoration Algorithm for Gaussian Corrupted MRI using Non-Local Average Filtering. In: 2nd Int. Conf. on Information Systems Design and Intelligent Applications. 2, 831–840 (2015)
5. Tiwari, H., Bhateja, V., Srivastava, A.: Estimation Based Non-Local Approach for Pre-Processing of MRI. In: 2nd IEEE International Conference on Computing for Sustainable Global Development. 1622–1626 (2015)
6. Xu, Z.: Medical Image Fusion Using Multi-level Local Extrema. Info. Fusion. 19, 38–48 (2014)
7. Bhatnagar, G., Wu, Q., M., J., Liu, Z.: A New Contrast based Multimodal Medical Image Fusion Framework. Neurocomputing. 157, 143–152 (2015)
8. Liu, Z., Yin, H., Chai, Y., Yang, S., X.: A Novel Approach for Multimodal Medical Image Fusion. Expert Systems with Applications. 41(16), 7425–7435 (2014)
9. Bhateja, V., Patel, H., Krishn, A., Sahu, A., Lay-Ekuakille, A.: Multimodal Medical Image Sensor Fusion Framework using Cascade of Wavelet and Contourlet Transform Domains. IEEE Sensors Journal. 16(8), 1–8 (2015)
10. Bhateja, V., Verma, R., Mehrotra, R., Urooj, S., Lay-Ekuakille, A., Verma, V., D.: A Composite Wavelets and Morphology Approach for ECG Noise Filtering. In: 5th Int. Conf. on Pattern Recognition and Machine Intelligence. 8251, 361–366. Springer (2013)
11. Shrivastava, A., Alankrita, Raj, A., Bhateja, V.: Combination of Wavelet Transform and Morphological Filtering for Enhancement of Magnetic Resonance Images. In: International Conference on Digital Information Processing and Communications, Part-I. 460–474 (2011)
12. Do, M., N., Vetteli, M.: The Finite Ridgelet Transform for Image Representation. IEEE Transactions on Image Processing. 12, 16–28 (2003)

13. Himanshi., Bhateja, V., Krishn, A., Sahu, A.: Medical Image fusion in Curvelet Domain Employing PCA and Maximum Selection Rule. In: Proceedings of Second International Conference on Computers and Communication Technologies. Springer. (2015)
14. Yang, L., Guo, B., Ni, W.: Multimodality Medical Image Fusion based on Multiscale Geometric Analysis of Contourlet Transform. Neurocomputing. 72(1), 203–211 (2008)
15. Ganasala, P., Kumar, V.: CT and MR Image Fusion Scheme in Non-Subsampled Contourlet Transform Domain. J. Digit. Imag. 27(3), 407–418 (2014)
16. Da Cunha, A., L., Zhou, J., Do, M., N.: The Nonsubsampled Contourlet Transform: Theory, Design, and Applications. IEEE Tran. Image Proc.. 15(10), 3089–3101 (2006)
17. Gupta, A., Tripathi, A. and Bhateja, V.: Despeckling of SAR Images in Contourlet Domain using a New Adaptive Thresholding. In: Proceedings of 3rd *International Advance Computing Conference. IEEE. Ghaziabad (U.P.),* India. 1257–1261 (2013)
18. Krishn, A., Bhateja, V., Himanshi., Sahu, A.: Medical Image Fusion using Combination of PCA and Wavelet Analysis. In: Proceedings of Third International Conference on Advances in Computing. Communications and Informatics. 986–991 (2014)
19. Bhatnagar, G., Wu, Q., M., J., Liu, Z.: Directive Contrast Based Multimodal Medical Image Fusion in NSCT Domain. IEEE Tran. on Multimedia. 15(5), 1014–1024 (2013)
20. Sahu, A., Bhateja, V., Krishn, A., Himanshi.: Medical Image Fusion with Laplacian Pyramids. An Improved Medical Image Fusion Approach Using PCA and Complex Wavelets. In: IEEE International Conference on Medical Imaging, m-Health & Emerging Communication Systems. 448–453 (2014)
21. Krishn, A., Bhateja, V., Himanshi, Sahu, A.: PCA based Medical Image Fusion in Ridgelet Domain. In: 3rd International Conference on Frontiers in Intelligent Computing Theory and Applications. Springer 328, 475–482 (2014)
22. Himanshi, Bhateja, V., Krishn, A., Sahu, A.: An improved Medical Image Fusion Approach Using PCA and Complex Wavelets. In: IEEE International Conference on Medical Imaging, m-Health & Emerging Communication Systems. 442–447 (2014)
23. Moin, A., Bhateja, V. and Srivastava, A.: Weighted-PCA based Multimodal Medical Image Fusion in Contoulet Domain. In: Proceedings of 2nd International Congress on Information and Communication Technologies. x, xx–xx (2015) (in-press)
24. Gupta, P., Srivastava, P., Bharadwaj, S. and Bhateja, V.: A Novel Full-Reference Image Quality Index for Color Images. In: Proceedings of the International Conference on Information Systems Design and Intelligent Applications Springer 253–245 (2012)
25. Bhateja, V., Krishn, A., Himanshi and Sahu, A.: Medical Image Fusion in Wavelet and Ridgelet Domains: A Comparative Evaluation, IJSDA 2(2), 78–91 (2015)

A Behavioral Study of Some Widely Employed Partitional and Model-Based Clustering Algorithms and Their Hybridizations

D. Raja Kishor and N.B. Venkateswarlu

Abstract The present work experiments with the K-means (partitional), expectation-maximization, and fuzzy C-means (model-based) techniques, and some of their hybridizations to study their behavior. Experiments are carried out on three different datasets of which one is synthetic dataset. On each dataset, experiments are carried out with varying number of clusters. To measure the clustering performance, the experiments compute *clustering fitness* and *sum of squared errors* (SSE). Execution time is also taken into consideration for all the algorithms with all the datasets. Though the algorithm for K-means (StKM) is taking less execution time as it involves in computing only one parameter, i.e., cluster means, the algorithm for K-means followed by standard fuzzy c-means (KMFCM) may be preferred when we consider high intracluster similarity with a good separation of clusters also.

Keywords Clustering · K-means · EM algorithm · Fuzzy C-means · Hybridization · Clustering fitness · Sum of squared error

1 Introduction

Cluster analysis is the identification of groups of observations that are cohesive within the same group and are separated from other groups [1]. Clustering methods can be partitional [2] (like K-means, PAM, CLARA, CLARANs, etc.), or model-based [3] (like expectation-maximization, fuzzy c-means, SOM, and mixture model clustering) classes. Partitional clustering methods attempt to break a population of data into k clusters such that the partition optimizes a given criterion [2].

D.R. Kishor (✉)
Department of CSE, JNTU, Hyderabad, Andhra Pradesh, India
e-mail: rajakishor@gmail.com

N.B. Venkateswarlu
Department of CSE, AITAM, Tekkali, Andhra Pradesh, India
e-mail: venkat_ritch@yahoo.com

S.C. Satapathy et al. (eds.), *Proceedings of the International Conference on Data Engineering and Communication Technology*, Advances in Intelligent Systems and Computing 469, DOI 10.1007/978-981-10-1678-3_56

Model-based or prototype-based clustering methods attempt to optimize the fit between the given data and some mathematical model [1].

This work experiments with the algorithms for K-means, expectation-maximization (EM), fuzzy C-means (FCM), and some of their hybridizations. While clustering the data, the K-means algorithm aims at the local minimum of the distortion [4]. EM is a model-based approach, which aims at finding clusters such that maximum likelihood of each cluster's parameters is obtained. Introducing the fuzzy logic in K-means clustering algorithm is the Fuzzy C-means algorithm [5]. In Fuzzy C-means, every object belongs to every cluster with a membership weight that ranges from 0 to 1. Clusters are treated as fuzzy sets. The EM and Fuzzy C-means techniques belong to the same category of clustering techniques. However, they differ in the way they assign data points to the clusters. The EM technique assigns the data points to the clusters based on their probabilities belonging to the clusters, whereas the Fuzzy C-means technique assigns them based on their cluster memberships [6, 7].

Along with the standard K-means, EM, and fuzzy C-means algorithms, experiments are carried out with the proposed hybridizations and finally performance comparison is made among all the algorithms. The same termination condition (8) is used for all the algorithms. Though the algorithm for K-means (StKM) is taking less execution time as it involves in computing only one parameter, i.e., cluster means, the algorithm for K-means followed by FCM (KMFCM) is showing better performance with acceptable clustering fitness value and the algorithm for standard FCM (StFCM) is showing better performance with acceptable SSE.

2 Standard K-means (StKM)

The k-means algorithm is a simple iterative method to partition a given dataset into a prespecified number of clusters, k [8]. The K-means algorithm can be used to simplify the computation and accelerate convergence as it requires only one parameter to compute, i.e., cluster means. The algorithm for conventional K-means is given below.

Select k vectors randomly from the dataset as the initial cluster means, μ. Set the current iteration $t = 0$.

Repeat

Assign each vector X_i from the dataset to its closest cluster mean using Euclidean distance.

$$\mathrm{dist}(X_i, \mu_j) = \sqrt{\sum_{l=1}^{d} (x_{il} - \mu_{lj})^2} \quad (1)$$

where X_i is the ith vector in the dataset, μ_j is the mean of the cluster j, and d is the number of dimensions of a data point.

Recompute the cluster means and set $t = t + 1$.
Compute *percentage change* using (8).
Until *percentage change* is <3.
End of K-means

3 Standard EM (StEM)

The expectation-maximization (EM) is a probabilistic clustering method, where each observed vector is probabilistically assigned to the k clusters by estimating the respective parameters [9]. The EM algorithm iteratively refines initial mixture model parameter estimates to better fit the data and terminates at a locally optimal solution. For each of the input vectors X_i, $i = 1, \ldots, N$, the algorithm calculates the probability $P(C_j|X_i)$. The highest probability will point to the vector's class. The EM algorithm works iteratively by applying two steps: the Expectation step (E-step) and the Maximization step (M-step).

Formally, $\hat{\theta}(t) = \{\mu_j(t), \Sigma_j(t), W_j(t)\}; j = 1, \ldots k$ stands for successive parameter estimates. Given a dataset of N, d-dimensional vectors, the EM algorithm has to cluster them into k groups.

The multidimensional Gaussian distribution for the cluster C_j is parameterized by the d-dimensional mean column vector μ_j and d x d covariance matrix Σ_j is given as follows [9]:

$$P(X_i \mid C_j) = \frac{1}{\sqrt{(2\Pi)^d \left|\Sigma_j\right|}} e^{-\frac{1}{2}(X_i - \mu_j)^T (\Sigma_j)^{-1} (X_i - \mu_j)} \quad (2)$$

where X_i is a sample column vector, the superscript T indicates transpose of a column vector, $|\Sigma_j|$ is the determinant of Σ_j and $(\Sigma_j)^{-1}$ is its matrix inverse of covariance matrix Σ_j.

The mixture model probability density function is

$$p(X_i) = \sum_{l=1}^{k} W_l P(X_i \mid C_l) \quad (3)$$

where W_l is the weight of cluster C_l.

In E-step, the algorithm estimates the probability of each class C_j ($j = 1, 2, \ldots, k$), given a certain vector X_i ($i = 1, 2, \ldots, N$) for current iteration t using the following formula and assign X_i to the cluster with the maximum probability.

$$P(C_j \mid X_i) = \frac{W_j P(X_i \mid C_j)}{P(X_i)} = \frac{\left|\sum_j (t)\right|^{-1/2} \exp^{\eta_j}.W_j(t)}{\sum_{l=1}^{k} \left|\sum_l (t)\right|^{-1/2} \exp^{\sigma_l}.W_l(t)} \quad (4)$$

where

$$\eta_j = -\frac{1}{2}(X_i - \mu_j(t))^T \sum_j^{-1} (t)(X_i - \mu_j(t))$$
$$\sigma_l = -\frac{1}{2}(X_i - \mu_l(t))^T \sum_l^{-1} (t)(X_i - \mu_l(t))$$

Each of the k clusters has its mean (μ_j) and covariance (Σ_j); $j = 1, 2, \ldots, k$. W_j is the weight of jth cluster.

In M-step, for jth cluster, the algorithm updates the parameter estimation for the iteration $t + 1$ as follows:

$$\mu_j(t+1) = \frac{\sum_{i=1}^{N} P(C_j \mid X_i) X_i}{\sum_{i=1}^{N} P(C_j \mid X_i)} \quad (5)$$

$$\sum_j (t+1) = \frac{\sum_{i=1}^{N} P(C_j \mid X_i)(X_i - \mu_j(t))(X_i - \mu_j(t))^T}{\sum_{i=1}^{N} P(C_j \mid X_i)} \quad (6)$$

$$W_j(t+1) = \frac{1}{N} \sum_{i=1}^{N} P(C_j \mid X_i) \quad (7)$$

The algorithm repeats through the E-step and M-step till the termination condition.

A. *Termination Condition*

As the termination condition, percentage change is computed using the following formula:

$$\text{Percentage change} = \frac{|\Psi_t - \Psi_{t+1}|}{\Psi_t} \times 100 \quad (8)$$

where Ψ_t is the number of vectors assigned to new clusters in tth iteration and Ψ_{t+1} is the number of vectors assigned to new clusters in $(t + 1)$th iteration. The algorithm terminates when the percentage change <3.

4 Fuzzy C-means (StFCM)

Given a dataset, the aim of Fuzzy C-means algorithm is to produce fuzzy c-partitions [5]. A fuzzy c-partition of the dataset is the one which characterizes the membership of each data point in all the clusters by a membership function which ranges between 0 and 1. The sum of the memberships for each data point must be unity.

The FCM algorithm for Gaussian Mixture Models [6] proceeds as follows:

1. Initialize parameters: set current iteration $t = 0$; set membership weight $m = 1.25$; select k vectors randomly as cluster means; compute global covariance matrix and set it to be the covariance matrix for all clusters; set initial membership matrix $U^{(0)}_{k\times N}$ with all $u_{ji} = 1/k$ for $j = 1, \ldots, k$, $i = 1, \ldots, N$.
2. Initially, assign each data vector X_i to clusters using the initial membership matrix.
3. Compute mean of jth cluster as follows:

$$\mu_j^{(t+1)} = \frac{\sum_{i=1}^{N} (u_{ji})^m X_i}{\sum_{i=1}^{N} (u_{ji})^m} \tag{9}$$

4. Compute new membership matrix using

$$u_{ji}(t+1) = \left[\sum_{l=1}^{k} \left(\frac{||X_i - \mu_j(t)||^2}{||X_i - \mu_l(t)||^2} \right)^{2/m-1} \right]^{-1} \tag{10}$$

5. Assign points to clusters based on their memberships.
6. Compute new covariance matrix for each cluster.
7. Compute percentage change using (8).
8. Stop the process if the percentage change is <3. Otherwise, set t = t + 1 and repeat the steps 3–7 with the updated parameters.

5 Hybridization of K-means, EM and FCM Algorithms

Along with the standard K-means, EM, and Fuzzy C-means algorithms, this work experiments with several hybridizations of those algorithms. The algorithms for hybridizations are briefly discussed below. In all the experiments, same termination condition, Eq. 8, is used.

A. *KMFCM*
This method performs clustering, which first runs the standard K-means algorithm completely on the given dataset and then runs the standard Fuzzy C-means algorithm till termination on the results of K-means.

B. *KMandFCM*
This method performs clustering running K-means and fuzzy C-means techniques in the alternative iterations till termination. For Fuzzy C-means step, the membership matrix and cluster means are calculated using the results of K-means step.

C. *EMFCM*
This method performs clustering, which first runs the standard EM algorithm completely on the given dataset and then runs the standard fuzzy C-means algorithm till termination on the results of EM.

D. *FCMEM*
This method performs clustering, which first runs the standard fuzzy C-means algorithm completely on the given dataset and then runs the standard EM algorithm till termination on the results of Fuzzy C-means.

E. *EMandFCM*
This method performs clustering running expectation-maximization and fuzzy C-means techniques in the alternative iterations till termination. For maximization step for EM, cluster weights, means, and covariance matrices are calculated using the results of Fuzzy C-means step, and for Fuzzy C-means step, membership matrix is calculated using the results of EM step.

6 Clustering Performance Measures

As a measure of clustering performance, the *Clustering Fitness* [10] and the sum of squared errors are computed for all the algorithms. The calculation of clustering fitness involves in computing intracluster similarity and intercluster similarity.

A. *Intracluster Similarity for the Cluster C_j*
It can be quantified via some function of the reciprocals of intracluster radii within each of the resulting clusters. The intracluster similarity of a cluster C_j $(1 = j = k)$, denoted as $S_{tra}(C_j)$, is defined by

$$S_{tra}(C_j) = \frac{1+n}{1+\sum_1^n \text{dist}(I_l, \text{Centroid})} \tag{11}$$

Here n is the number of items in cluster C_j, $1 = l = n$, I_l is the lth item in cluster C_l, and dist(I_l, Centroid) calculates the distance between I_l and the centroid of C_j, which is the intracluster radius of C_j. To smooth the value of $S_{tra}(C_j)$ and allow for possible singleton clusters 1 is added to the denominator and numerator.

B. *Intracluster Similarity for One Clustering Result C*
Denoted as $S_{tra}(C)$, it is defined by

$$S_{tra}(C) = \frac{\sum_1^k S_{tra}(C_j)}{k} \tag{12}$$

Here k is the number of resulting clusters in C.

C. *Intercluster Similarity*
It can be quantified via some function of the reciprocals of intercluster radii of the clustering centroids. The intercluster similarity for one of the possible clustering results C, denoted as $S_{ter}(C)$, is defined by

$$S_{ter}(C) = \frac{1+k}{1+\sum_1^k \text{dist}(\text{Centroid}_j,\ \text{Centroid}^2)} \tag{13}$$

Here k is the number of resulting clusters in C, $1 = j = k$, Centroid_j is the centroid of the jth cluster in C, Centroid^2 is the centroid of all centroids of clusters in C. We compute intercluster radius of Centroid_j by calculating dist (Centroid_j, Centroid^2), which is distance between Centroid_j, and Centroid^2. To smooth the value of $S_{ter}(C)$ and allow for possible all-inclusive clustering result, 1 is added to the denominator and the numerator.

D. *Clustering Fitness*
The clustering fitness for one of the possible clustering results C, denoted as CF, is defined by

$$\text{CF} = \lambda \times S_{tra} + \frac{1-\lambda}{S_{ter}} \tag{14}$$

Here $0 < \lambda < 1$ is an experiential weight. To make the computation of Clustering Fitness unbiased, the value of λ is taken as 0.5.

E. *Sum of Squared Errors*
The experiments also compute the Sum of Squared Error (SSE) for the results of all the algorithms as a measure of performance [3]. The SSE is computed using the following formula.

$$SSE = \sum_{j=1}^{k} \sum_{X_i \in C_j} (X_i - \mu_j)^2 \tag{15}$$

Here X_i is a vector in the dataset, μ_j is the means of the cluster C_j, k is the number of clusters, and N is the number of vectors in the dataset. The objective of clustering is to minimize the within-cluster sum of squared errors. The lesser the SSE, the better the goodness of fit is.

7 Experiments and Results

Experimental work has been carried out on the system with Intel(R) Core i7-3770 K with 3.50 GHz processor speed, 8 GB RAM with 1666FSB, Windows 7 OS and using JDK1.7.0_45. Letter recognition and magic gamma datasets are used for the present work from UCI ML dataset repository [11]. The present work also uses one synthetic dataset that is generated by an algorithm [12] that generates multivariate normal random variables.

Table: Dataset Description

Data set	No. of points	No. of dimensions
Magic Gamma data	19020	10
Poker Hand data	1025010	10
Synthetic data-1	50000	10

All the algorithms are studied by executing on each dataset by varying number of clusters (k = 10, 11, 12, 13, 14, 15). The details of execution time, clustering fitness, and SSE of each algorithm are separately given in the tables below for each dataset (Tables 1, 2, 3, 4, 5, 6, 7, 8 and 9).

A. *Observations on Magic Gamma dataset*
See Figs. 1, 2 and 3.

B. *Observations on Poker Hand dataset*
See Figs. 4, 5 and 6.

Table 1 Execution time of each clustering method in seconds (Magic Gamma dataset)

K	StKM	StEM	StFCM	KMFCM	KMandFCM	EMFCM	FCMEM	EMandFCM
10	0.0420	2.8100	0.2440	0.4580	0.3960	3.2120	4.9760	0.4160
11	0.0820	3.7480	0.1070	0.5400	0.0990	4.3610	1.9920	0.4210
12	0.0630	3.5110	0.1150	0.5710	0.1010	4.1750	1.2920	0.3960
13	0.0370	4.7470	0.1240	0.5890	0.1120	5.4530	3.6350	0.4220
14	0.0470	5.1190	0.1350	0.6400	0.1160	5.8870	5.2640	0.4550
15	0.0800	7.3250	0.1400	0.7210	0.1230	8.1230	4.8980	0.4960

Table 2 Clustering fitness of each clustering method (Magic Gamma dataset)

K	StKM	StEM	StFCM	KMFCM	KMandFCM	EMFCM	FCMEM	EMandFCM
10	47.8101	37.8917	47.7119	53.4277	39.0281	61.3841	37.5678	35.1167
11	59.3189	38.1009	59.3537	69.4608	47.7191	64.3828	42.3031	41.8048
12	63.3480	39.7301	63.4965	72.5162	49.0666	62.1473	53.3417	35.6841
13	58.5164	39.9926	58.5447	69.1421	48.5111	59.1161	40.8975	37.6999
14	58.2683	36.5958	58.9229	70.2297	49.6919	61.6858	43.5824	41.2995
15	59.2917	38.9988	59.4293	67.5864	41.2602	66.7143	43.1607	34.3483

Table 3 SSE of each clustering method (Magic Gamma dataset)

K	StKM	StEM	StFCM	KMFCM	KMandFCM	EMFCM	FCMEM	EMandFCM
10	0.2709	0.4689	0.2704	0.4863	0.3426	1.2949	0.4355	0.5449
11	0.2375	0.4703	0.2374	0.7438	0.3182	1.5331	0.3887	0.4938
12	0.2210	0.4930	0.2209	0.6285	0.3418	1.5624	0.2931	0.5131
13	0.2241	0.4698	0.2240	0.6725	0.2983	1.1293	0.4085	0.5172
14	0.2204	0.4945	0.2196	0.6985	0.2724	1.4690	0.3893	0.4386
15	0.2035	0.4856	0.2034	0.7124	0.3195	1.7427	0.3483	0.5400

Table 4 Execution time of each clustering method in seconds (Poker Hand dataset)

K	StKM	StEM	StFCM	KMFCM	KMandFCM	EMFCM	FCMEM	EMandFCM
10	6.9080	136.1450	7.0590	34.4880	6.2200	170.3420	127.8200	20.3520
11	2.2410	131.8070	7.3520	32.2050	6.3400	168.2180	316.2970	21.9780
12	5.6940	90.8490	7.7300	38.6630	7.1280	133.3780	154.8670	24.5970
13	2.4580	176.5590	8.1080	38.0740	7.8100	221.7310	185.9840	26.3670
14	8.7340	126.0720	8.5870	46.3580	8.0800	174.7470	52.0460	28.2390
15	8.6800	133.2020	8.8500	48.4580	8.1750	183.7710	187.4290	29.6410

Table 5 Clustering fitness of each clustering method (Poker Hand dataset)

K	StKM	StEM	StFCM	KMFCM	KMandFCM	EMFCM	FCMEM	EMandFCM
10	2.9468	1.6980	2.9468	3.0560	2.7986	1.8286	2.6911	2.1501
11	2.9783	1.4581	2.9805	3.1157	2.9024	1.5524	2.7246	2.1078
12	3.0783	1.4558	3.0784	3.2057	3.0428	1.6016	2.8647	2.2930
13	3.0997	1.7492	3.1010	3.2455	3.0469	1.8026	2.8564	2.3098
14	3.1992	1.4818	3.1993	3.3436	3.0328	1.6388	3.0577	2.1669
15	3.2574	1.6981	3.2574	3.4091	3.1035	1.8455	3.0276	2.3583

Table 6 SSE of each clustering method (Poker Hand dataset)

K	StKM	StEM	StFCM	KMFCM	KMandFCM	EMFCM	FCMEM	EMandFCM
10	38.4061	60.3787	38.4061	38.7991	45.5360	62.3060	41.3860	58.6724
11	38.1672	65.1422	38.1304	38.8203	41.9040	66.2561	41.3724	59.4302
12	35.9512	63.0996	35.9499	36.3821	41.8733	66.5395	38.4630	57.9042
13	35.7055	61.4338	35.6877	36.3774	40.9324	63.9005	38.6120	57.0821
14	33.4674	61.8512	33.4660	33.9366	41.1059	63.6446	34.2793	58.9143
15	32.1261	59.0757	32.1256	32.6212	38.7390	61.2613	34.9232	55.3141

Table 7 Execution time of each clustering method in seconds (Synthetic dataset)

K	StKM	StEM	StFCM	KMFCM	KMandFCM	EMFCM	FCMEM	EMandFCM
10	0.1370	6.4000	0.2680	1.2980	0.2340	7.8620	2.8460	0.8690
11	0.1000	6.3360	0.2950	1.3610	0.2540	7.9500	3.1280	0.9550
12	0.1360	5.3840	0.3080	1.5080	0.2790	7.1280	2.6300	1.0320
13	0.1490	4.9920	0.3310	1.6400	0.2890	6.8870	3.6700	1.1100
14	0.1470	8.0410	0.3600	1.6990	0.3170	10.0690	4.8530	1.1990
15	0.1850	8.6250	0.3730	1.8840	0.3260	10.8060	4.2190	1.2840

Table 8 Clustering fitness of each clustering method (Synthetic dataset)

K	StKM	StEM	StFCM	KMFCM	KMandFCM	EMFCM	FCMEM	EMandFCM
10	1076.0893	775.4051	1076.6322	1124.1167	1126.0081	852.0542	977.2159	1049.3886
11	1088.8953	661.5773	1090.2945	1134.7413	1077.0396	756.1819	992.7452	981.0285
12	1122.4485	819.4942	1123.2434	1186.0276	1137.7077	898.2294	1035.3235	1080.4385
13	1151.4744	695.0268	1151.8879	1195.7001	1156.3846	835.6280	1058.7946	1068.5140
14	1167.9671	841.5911	1168.9156	1213.7886	1241.8767	925.9731	1067.6871	1145.8648
15	1195.8812	780.9271	1196.1747	1235.8618	1197.1809	844.9339	1094.3546	1101.4708

Table 9 SSE of each clustering method (Synthetic dataset)

K	StKM	StEM	StFCM	KMFCM	KMandFCM	EMFCM	FCMEM	EMandFCM
10	229.5530	272.0786	229.5315	234.8056	247.3580	277.4283	233.9949	264.3630
11	226.9850	289.0519	226.9133	231.9029	252.8353	291.9495	231.3350	274.9175
12	223.1056	271.9045	223.0528	227.3112	245.0971	277.9409	226.6494	262.4669
13	219.0802	272.4666	219.0384	225.0529	236.5526	273.6704	223.8112	256.1266
14	216.5148	273.5514	216.4372	220.0887	239.1548	275.1162	223.2097	263.3026
15	213.9715	269.8829	213.9480	218.5386	234.9775	273.5891	219.3681	257.1976

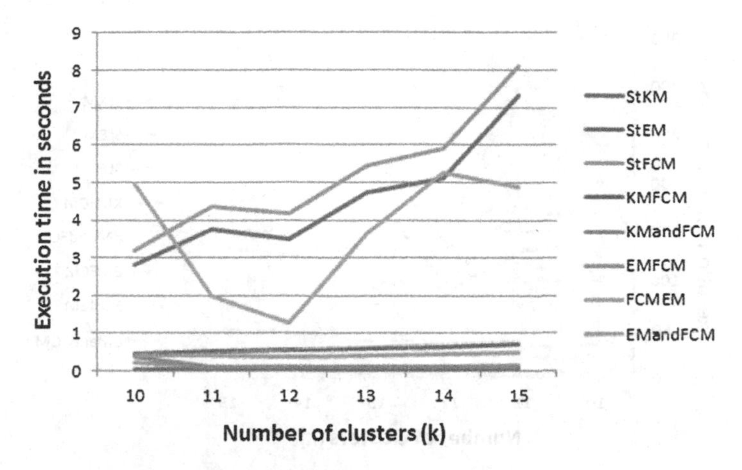

Fig. 1 Execution time (Magic Gamma dataset)

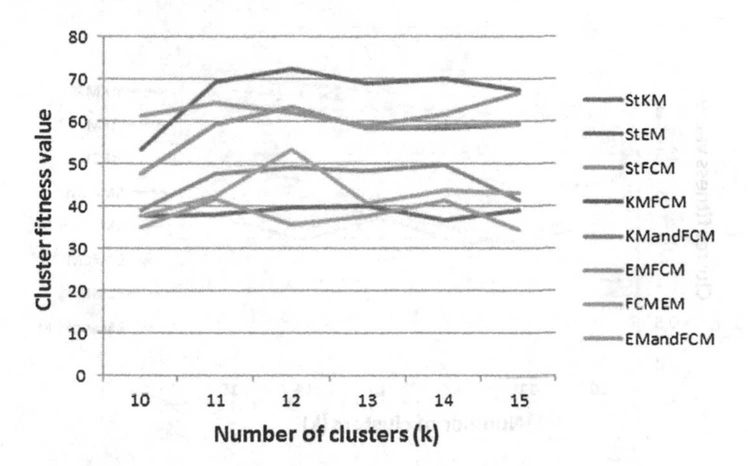

Fig. 2 Clustering fitness (Magic Gamma dataset)

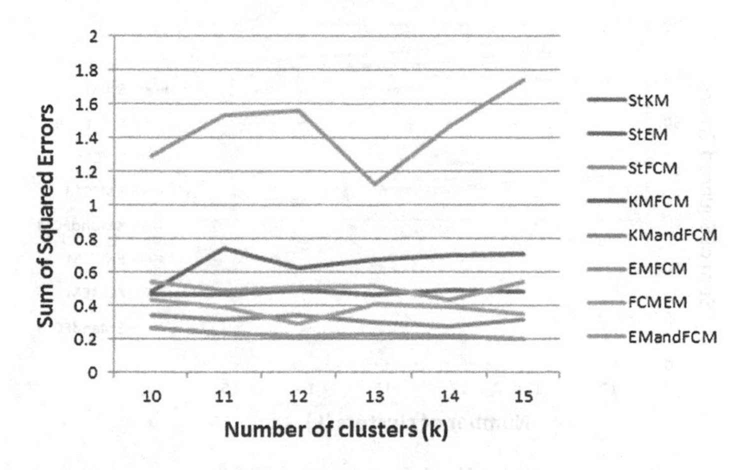

Fig. 3 Sum of squares errors (Magic Gamma dataset)

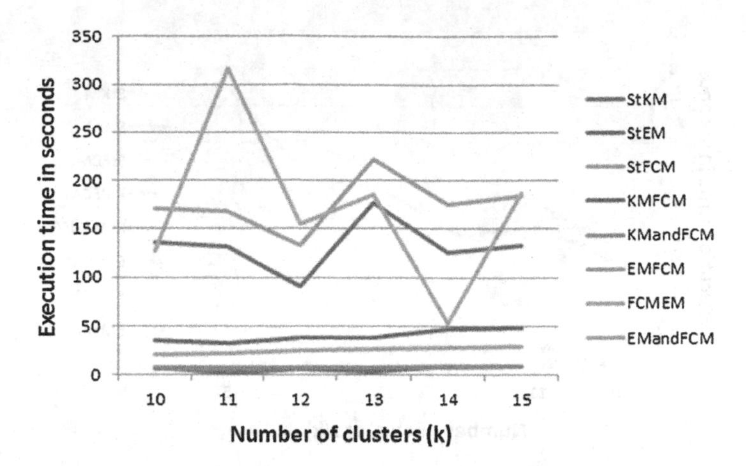

Fig. 4 Execution time (Poker Hand dataset)

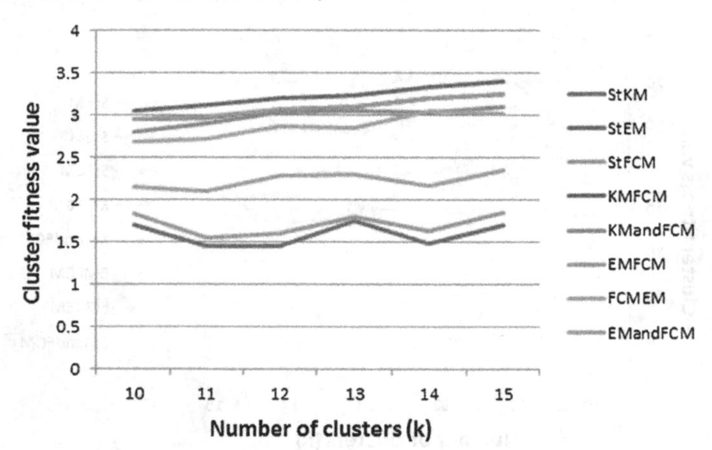

Fig. 5 Clustering fitness (Poker Hand dataset)

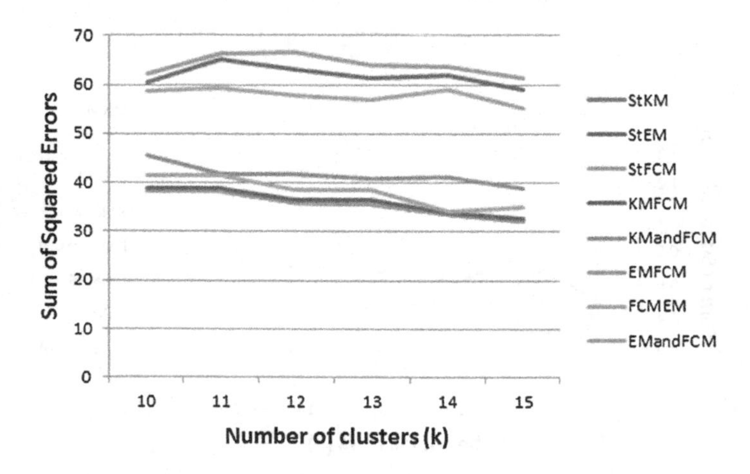

Fig. 6 Sum of squares errors (Poker Hand dataset)

C. *Observations on Synthetic dataset*
See Figs. 7, 8 and 9.

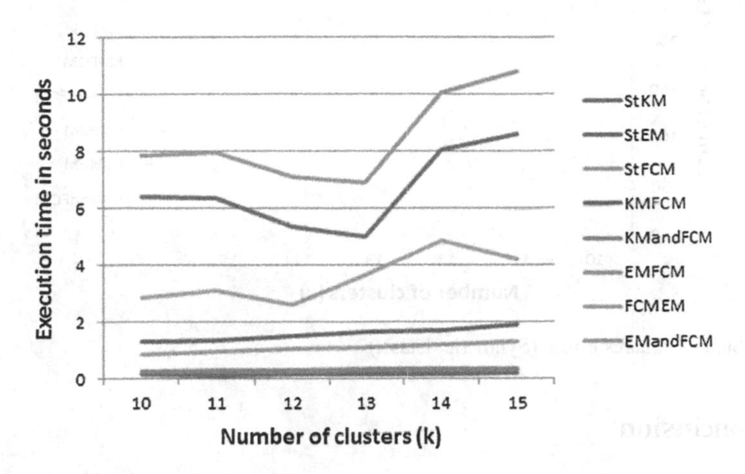

Fig. 7 Execution time (Synthetic dataset)

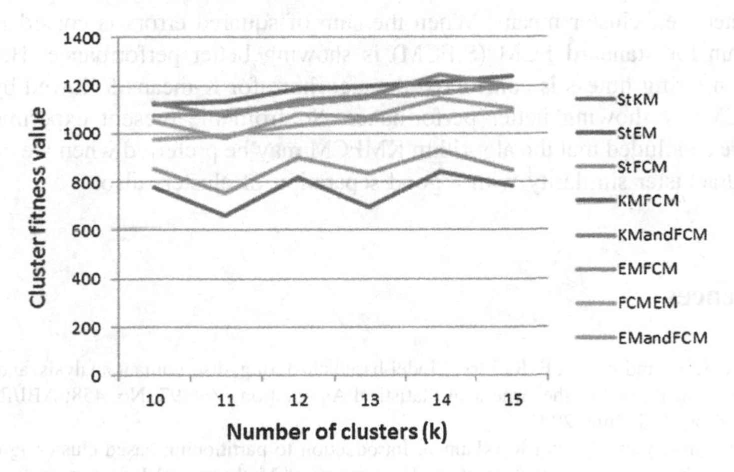

Fig. 8 Clustering fitness (Synthetic dataset)

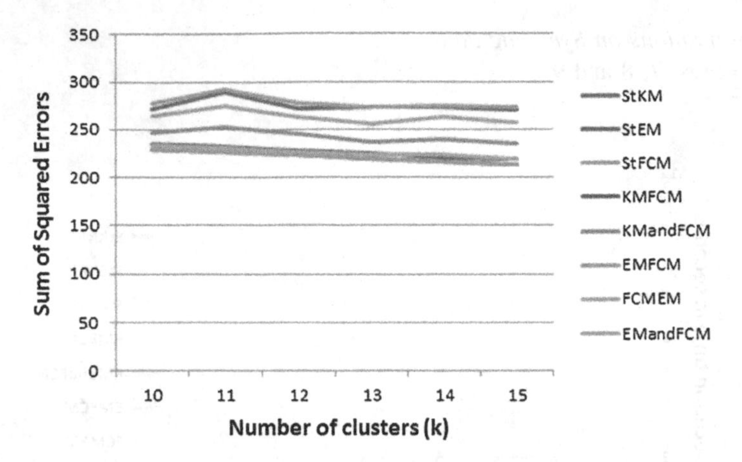

Fig. 9 Sum of squares errors (Synthetic dataset)

8 Conclusion

In all the experiments, it is observed that the algorithm for standard K-means (StKM) is taking less execution time as it involves in computing only one parameter, i.e., cluster means. When the sum of squared errors is considered, the algorithm for standard FCM (StFCM) is showing better performance. However, when clustering fitness is considered the algorithm for K-means followed by FCM (KMFCM) is showing better performance. So from the present experiments, it could be concluded that the algorithm KMFCM may be preferred when we consider high intracluster similarity with a good separation of clusters also.

References

1. Chris Fraley and Adrian E. Raftery, Model-based clustering, discriminant analysis, and density estimation, Journal of the American Statistical Association; Vol. 97, No. 458; ABI/INFORM Global pg. 611, June 2002.
2. Sami Ayramo and Tommi Karkkainen, Introduction to partitioning-based clustering methods with a robust example, Reports of the Department of Mathematical Information Technology Series C. Software and Computational Engineering, No. C. 1/2006, 2006.
3. Jiawei Han and Micheline Kamber, Data Mining Concepts and Techniques, 2/e, Elsevier Inc, 2007.
4. Adigun Abimbola Adebisi, Omidiora Elijah Olusayo and Olabiyisi Stephen Olatunde, An exploratory study of K-means and expectation maximization algorithms, British Journal of Mathematics & Computer Science, Vol. 2, No. 2, pp. 62–71, 2012.
5. Soumi Ghosh and Sanjay Kumar Dubey, Comparative analysis of K-means and Fuzzy c-means algorithms, international Journal of Advanced Computer Science and Applications, Vol. 4, No. 4 pp. 35–39, 2013.

6. George J. Klir and Bo Yuan, Fuzzy Sets and Fuzzy Logic: Theory and Applications, Prentice Hall of India Private Limited, 2005.
7. Thales Sehn Körting, Luciano Vieira Dutra, Leila Maria Garcia Fonseca, Guaraci Erthal, Felipe Castro da Silva (2007), Improvements to Expectation-Maximization approach for unsupervised classification of remote sensing data, GeoINFO. Campos do Jordão, SP, Brazil, 2007.
8. Neha Aggarwal and Kirti Aggarwal (June 2012), A mid-point based K-mean clustering algorithm for data mining, International Journal on Computer Science and Engineering, Vol. 4, No. 06, pp. 1174–1180, June 2012.
9. Nagendra Kumar D.J., Murthy J.V.R., Venkateswarlu N.B., Fast expectation maximization clustering algorithm, International Journal of Computational Intelligence Research, ISSN 0973-1873 Vol. 8, No. 2, pp. 71–94, 2012.
10. Xiwu Han and Tiejun Zhao, Auto-k dynamic clustering algorithm, Journal of Animal and Veterinary Advances 4 (5), pp. 535–539, 2005.
11. UCL Machine Learning Repository http://archive.ics.uci.edu/ml/datasets.html.
12. Amitava Ghosh and Pinnaduwa H.S.W. Kulatilake, A FORTRAN program for generation of multivariate normally distributed random variables, Computers & Geosciences, Vol. 13 No. 3, pp. 221–233, 1987.

An Adaptive MapReduce Scheduler for Scalable Heterogeneous Systems

Mohammad Ghoneem and Lalit Kulkarni

Abstract Hadoop MapReduce has been proved to be an efficient model for distributed data processing. This model is widely used by different service providers, which create a challenge of maintaining same efficiency and performance level in different systems. One of the most critical problems for this model is how to overcome heterogeneity and scalability in different systems. The decreases of performance in heterogeneous environment occur due to inefficient scheduling of Map and Reduce tasks. Another important problem is how to minimize master node overhead and network traffic created by scheduling algorithm. In this paper, we introduce a lightweight adaptive scheduler in which we provide the classifier with information about jobs requirement and node capabilities. The scheduler classifies jobs into executable and nonexecutable according to the nodes capabilities. Then the scheduler assigns the tasks to appropriate nodes in the cluster to get highest performance.

Keywords Hadoop · MapReduce · JobTracker · TaskTracker · Heartbeat message · Heterogeneous environment · Task assignment · Classifier

1 Introduction

Hadoop is a MapReduce-based paradigm that support distributed data processing. This model is widely used in cloud computing to process big data applications in the cloud. The MapReduce receive the jobs from the users and divide each job into a number of Map and Reduce tasks as shown in Fig. 1. The Map tasks process the data blocks at each node in the cluster and produce a sub result. These sub results are combined using Reduce tasks to get final result [1].

Mohammad Ghoneem (✉) · Lalit Kulkarni
Department of Information Technology, Pune University, Pune, India
e-mail: Mhd89aiu@gmail.com

Lalit Kulkarni
e-mail: lalit.kulkarni@mitcoe.edu.in

S.C. Satapathy et al. (eds.), *Proceedings of the International Conference on Data Engineering and Communication Technology*, Advances in Intelligent Systems and Computing 469, DOI 10.1007/978-981-10-1678-3_57

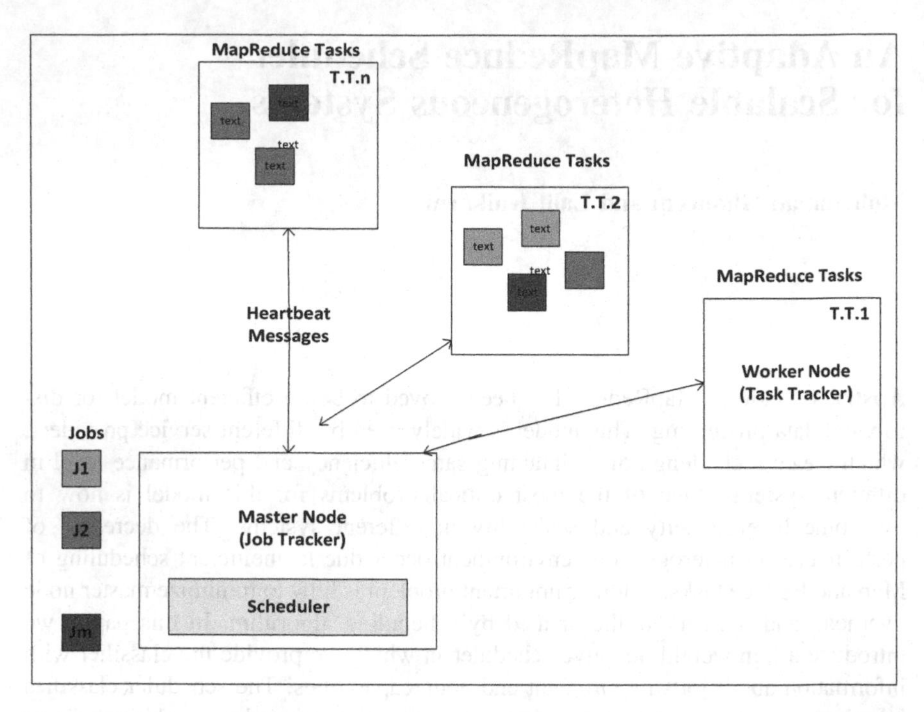

Fig. 1 MapReduce scheduler

MapReduce scheduler is considered as one of the critical elements in Hadoop environment, which can significantly increase or decrease the performance. The scheduler performance might get affected by different factors such as heterogeneity of the cluster, scalability, and variety of Hadoop applications. The current Hadoop schedulers perform well in homogeneous Hadoop environment [1]. However, these schedulers do not take into consideration the heterogeneity and scalability of the cluster [2]. As a result, these issues might significantly degrade the performance of MapReduce model when it is used in the cloud [3]. The performance decreases in such a case due to mismatch between job requirement and node capabilities, which led to a task failure at that node and relaunch of the task at some other node [4]. As a result, the execution time of a job will increase, and node utilization will decrease [5].

MapReduce scheduler is considered as one of the critical elements in Hadoop environment, which can significantly increase or decrease the performance. The scheduler performance might get affected by different factors such as heterogeneity of the cluster, scalability, and variety of Hadoop applications. The current Hadoop schedulers perform well in homogeneous Hadoop environment [1]. However, these schedulers don't take into consideration the heterogeneity and scalability of the cluster [2]. As a result, these issues might significantly degrade the performance of MapReduce model when it is used in the cloud [3]. The performance decreases in such a case due to mismatch between job requirement and node capabilities which led to a task failure at that node and relaunch of the task at some other node [4]. As a result the execution time of a job will increase, and node utilization will decrease [5].

One of the emerging problems in Hadoop environment is to make scheduler a ware about node capabilities and job requirement in order to increase utilization of the resources in the cluster [6]. This can be achieved by providing the scheduler with a classifier. The input for this classifier will be the node capabilities and job features. The output of the classifier will be whether this job is executable or nonexecutable at a specific node.

2 Background

2.1 *Heterogeneity and Scalability in MapReduce Environment*

Heterogeneity and scalability in MapReduce paradigm can be summarized in three main categories: Cluster, jobs workload, and users.

- Cluster resources might have different capabilities in heterogeneous environment such as node processing unit, RAM size, and storage unit [7].
- Jobs workload including the arrival rate of jobs, number of task in each, computation requirement, and execution time [7].
- Users' priorities and number of slots assign for each user which is known as minimum share.

2.2 *MapReduce Schedulers*

Different schedulers have been introduced by different companies to meet specific requirement. Here we briefly described these schedulers and find out the common problems between these schedulers.

FIFO scheduler is the default Hadoop scheduler, which is provided by apache Hadoop. The scheduler assigns jobs to the nodes according to its arrival time. This scheduler performs well in low load cluster [5].

Fair scheduler introduced by Facebook. Fair scheduling tries to assign a pool for each job. This pool consists of a number of Map and Reduce slots [7]. The job can use its pool slots and whenever the pool is free these slots can be used by other jobs.

Longest approximate time to end (LATE) scheduler uses the speculative execution concept [8]. Speculative execution tries to detect the slow running tasks and create a backup copy of this task to run at different node. If the backup copy completes before the original copy then the execution time is improved and the result of original copy ignored.

2.3 Performance Metrics

Performance of MapReduce model can be measured using different metrics. Here we cover some of the most popular metrics. These metrics are:

- Completion Time: is the time to complete all tasks belonging to same job [9]. Average completion time is the average time to complete all jobs belongs to same category.
- Scheduling time: is the time taken by master node to schedule job submitted by users [10]. It also measures the overhead on master node to schedule all jobs in term of RAM usage, and processor usage.
- Locality: means the number of tasks that are scheduled at the nodes which contain the block of data to be processed.

2.4 Performance Issues

All of the current schedulers have been designed by different companies to meet different goals. But each one of these schedulers has its own drawbacks as follow:

- Small job delay: in FIFO scheduler, the tasks to be executed are scheduled in the cluster according to their arrival time. This means that small jobs might get stack until large jobs are executed successfully which degrade the performance [9].
- Sticky node: Fair scheduler assigns a pool of slots for each job. These jobs will be dedicated to this job only and cannot be used by other jobs [11]. The problem happens when node capabilities are less than job requirement which increase average completion time [5].
- LATE scheduler issues: the default speculative execution in LATE scheduler makes assumption that the jobs and clusters are homogeneous which is not true in case of cloud computing. As a result these wrong assumptions will degrade the performance when the cluster is heterogeneous [5].
- One of the critical problems common for all current schedulers is the mis-matching between job and resource in heterogeneous environment [12]. To overcome these problems we propose an adaptive scheduler that can schedule jobs at proper nodes according to job requirements.

3 Proposed Implementation

Our implementation will include three nodes that will form our cluster. One of the nodes will play the role of master node (JobTracker) and the other two nodes will be the slave nodes (TaskTracker). Our adaptive scheduler will be implemented throughout three steps:

- The scheduler will use a classification algorithm to classify jobs into executable and nonexecutable according to job processing requirement and node available resources.
- To classify jobs we need to provide the classifier with job processing requirement which will be stored as a log file at JobTracker as shown in Fig. 2.
- Also we have to provide the classifier with the nodes current available resources to choose the TaskTracker which will minimize execution time.

To classify tasks we will use support vector machine classifier (SVM) which will differentiate tasks into two classes executable and nonexecutable as shown in Fig. 2. SVM is a single layer neural network algorithm for supervised learning of task classification. The advantage of this classifier is that the prediction accuracy is generally high with less computation. This implies less overhead on master node. Also SVM is considered as online algorithm which can accept input at real time and adapt the classification of the tasks based on results of execution.

The jobs processing requirement will be first input to the SVM classifier. This information will be stored as a log file at JobTracker. The JobTracker maintain this information by receiving thc heartbeat message from the TaskTrackers. This message contains the task processing requirement, execution time, current status of the node, and some other information.

The second input to the classifier is the available resources of nodes. These resources have two types: static resources and dynamic resources. Static resources can be the number of processors in each node, total memory, total storage, and processing speed. The dynamic resources include CPU usage, free memory, and

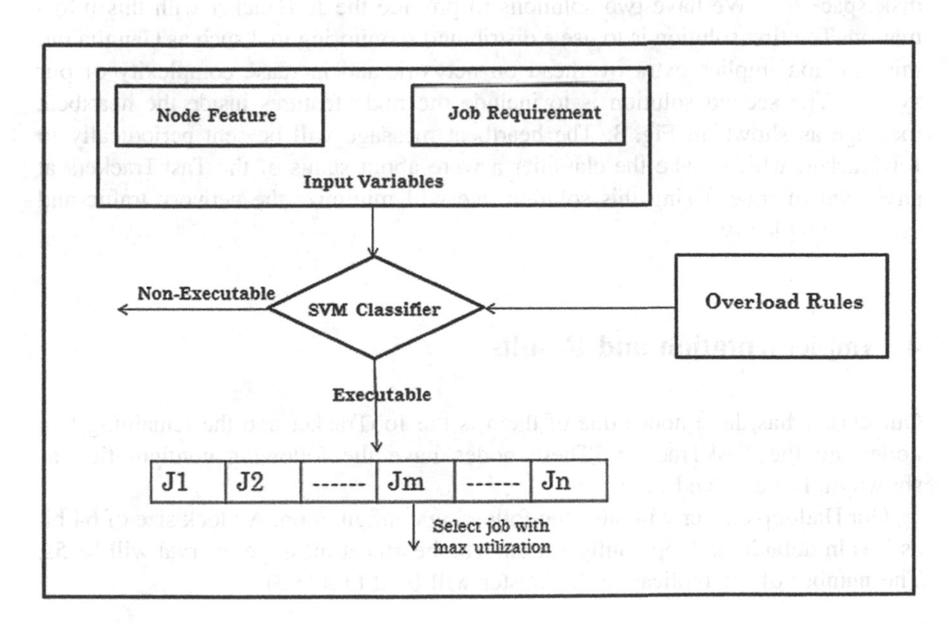

Fig. 2 Task assignment in Hadoop scheduler

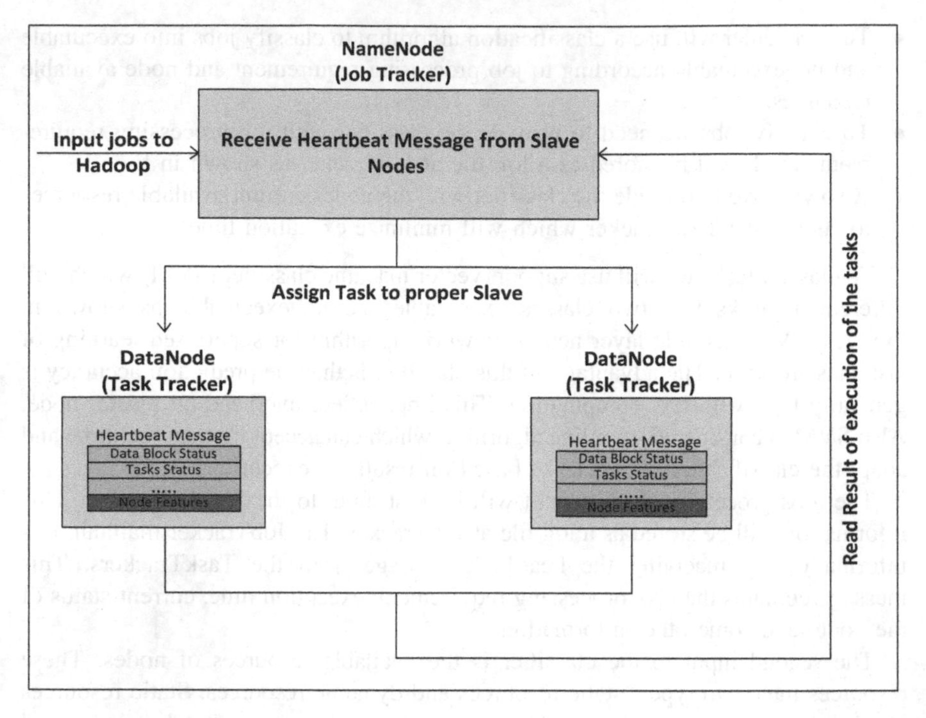

Fig. 3 Heartbeat message details

disk space left. We have two solutions to provide the JobTracker with this information. The first solution is to use a distributed monitoring tool such as Ganglia but this solution implies extra overhead on network and increase complexity of our system. The second solution is to include the node features inside the heartbeat message as shown in Fig. 3. The heartbeat message will be sent periodically to JobTracker, which make the classifier a ware about status of the TaskTrackers at any point of time. Using this solution, we will minimize the network traffic and system complexity.

4 Implementation and Results

Our cluster has three nodes one of them is the JobTracker and the remaining two nodes are the TaskTrackers. These nodes have the following configuration as shown in Tables 1 and 2.

Our Hadoop cluster will have the following configuration. A block size of 64 bit as it is in default Hadoop configuration. The heartbeat message interval will be 5s. The number of the replicas in the cluster will be 2 (Table 3).

Table 1 Master node configuration

Master node	
Hard disk	50 GB
RAM	4 GB

Table 2 Slave nodes configuration

Slave node 1		Slave node 2	
Hard disk	50 GB	Hard disk	50 GB
RAM	4 GB	RAM	2 MB

Table 3 Hadoop configuration

Hadoop configuration	
Replication	2
HDFS block size	64 MB
Heartbeat interval	5 s

In our cluster implementation, we will run WordCount example and monitor the results of execution with the new scheduler. Four jobs will be under execution while monitoring the system. These jobs will be processed in throughout two phases. In the first phase, the four jobs will be further subdivided into N number of Map tasks. These Map tasks will be processed by TaskTrackers to get the sub results. In the second phase, the Reduce tasks will merge the sub results of all Map tasks in the Cluster to get the final result.

The overload rule for our scheduler is that the CPU usage should not cross 80 %. After the execution, we can see that we have achieved a very high CPU and memory utilization as shown in Fig. 4.

Some exceptions are there during the execution when the CPU usage slightly crosses 80 % as shown in Fig. 4. This is normal because our SVM classifier still in the learning phase. When the usage of CPU cross the threshold the monitoring system should update thc logs at master node so that the classifier can adapt itself and change the weights of the tasks to take appropriate assignment decision in the next phase.

During the execution of jobs in the default Hadoop system, some of the tasks might failed or straggle due to wrong assignment and it should be relaunched again at some other nodes. This method known as speculation technique used to minimize average completion time of the jobs. This failure occur because the scheduler is not aware a bout node capabilities and its current available resources.

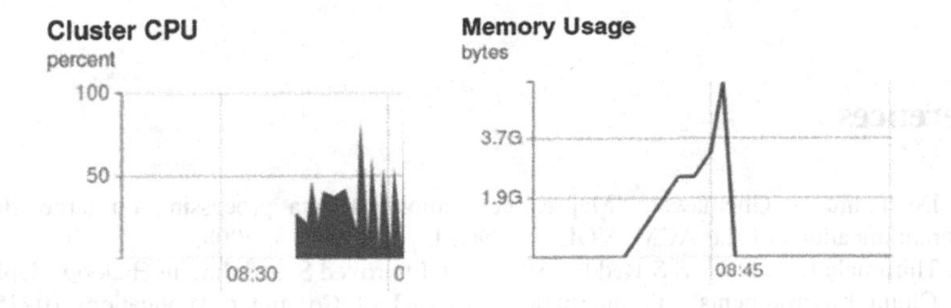

Fig. 4 Cluster CPU utilization and JobTracker memory utilization

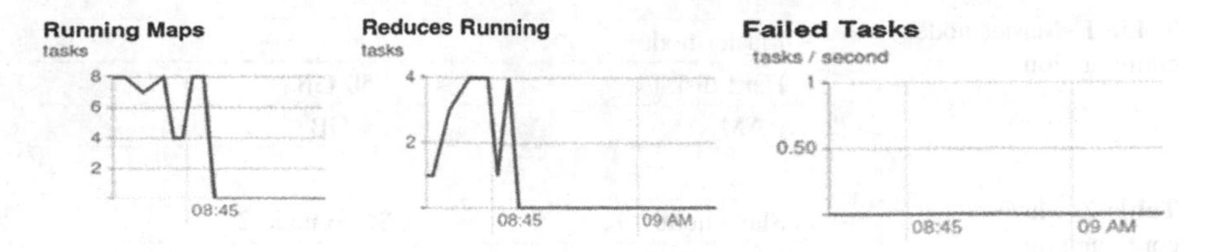

Fig. 5 Tasks execution and failure in the cluster

In our system, the scheduler assigns tasks according to node current available resources and task processing requirement. As a result all tasks will be executed successfully without any failure as shown in the Fig. 5. By following this scheduling technique, we will achieve a high level of resources utilization and improve the average completion time of jobs as shown in Fig. 5.

5 Conclusion and Future Work

In this paper, we introduce an adaptive scheduling technique for MapReduce scheduler to increase efficiency and performance when it is used in heterogeneous environment. In this model, we make the scheduler aware about cluster resources and job requirement by providing the scheduler with a classification algorithm. This algorithm classifies jobs into two categories executable and nonexecutable. Then the executable jobs are assigned to the proper nodes to be executed successfully without failures which increase execution time of the job. This scheduler overcomes the problems of previous schedulers such as small job starvation, sticky node in fair scheduler, and mismatch between resource and job. The adaptive scheduler increase performance of MapReduce model in heterogeneous environment while minimizing master node overhead and network traffic.

In the feature, we plan to improve performance by including multiple performance metrics such as average completion time, scheduling time, locality, and fairness. Also we plan to run different benchmarks such as sorting and searching on a bigger cluster and make a comparison with the default Hadoop system. The analysis of our system with default system will give us the pros and cons of our model. This will allow to do further improvement on our system.

References

1. J. Dean and S. Ghemawat, “Mapreduce: Simplified data processing on large clusters”, Communications of the ACM, VOL. 51, NO. I, pp. 107–113, 2008.
2. B. Thirumala Rao, Dr. L S S Reddy “Survey on Improved Scheduling in Hadoop MapReduce in Cloud Environments”, in International Journal of Computer Applications (0975-8887) Volume 34. No. 9, November 2011.

3. B. Thirumala Rao, V. Krishna Reddy. "Performance Issues of Heterogeneous Hadoop Clusters in Cloud Computing", Global Journal of Computer Science and Technology, Volume XI, Issue VIII, May 2011.
4. Dr. J. Aghav and Shyam Deshmukh (2013),"Job Classification for MapReduce Scheduler in Heterogeneous Environment", IEEE Cloud & Ubiquitous Computing & Emerging Technologies (CUBE), 15–16 Nov. 2013, **Page:** 26.
5. Dhok J, Varma V (2010), "Using pattern classification for task assignment in MapReduce", Proceedings of the 8th IEEE International Conference on Grid and Cooperative Computing, Volume 34. No. 9, November 2011.
6. M. Zaharia, A. Konwinski, A.D. Joseph, R. Katz, and I. Stoica. "Improving mapreduce performance in heterogeneous environments",. In Proc. Of USENIX OSDI, 2008.
7. Y. Yao, J. Tai, B. Sheng, and N. Mi, "Scheduling heterogeneous mapreduce jobs for efficiency improvement in enterprise clusters", Integrated Network Management (1 M 2(13), 2013 IFIPIIEEE International Symposium on, pp. 872–875, 2013.
8. K. Kc and K. Anyanwu, "Scheduling Hadoop Jobs to Meet Deadlines", in Proc. CloudCom, 2010, pp. 388–392.
9. Rasooli and D. G. Down, "A hybrid scheduling approach for scalable heterogeneous hadoop systems", IEEE Computer Society, 2012, pp. 1284–1291.
10. J. S. Manjaly and V. S. Chooralil, ''Tasktracker aware scheduling for hadoop mapreduce", 2013 Third International Conference on Advances in Computing and Communications, pp. 278–281, Aug. 2013.
11. M. Hammoud and M. F. Sakr, "Locality-aware reduce task scheduling for Mapreduce", in Proceedings of the 2011 IEEE Third International Conference on Cloud Computing Technology and Science, ser. CLOUDCOM '11. Washington, DC, USA: IEEE Computer Society, 2011, pp. 570–576.
12. S. Humbetov, "Data-intensive computing with map-reduce and hadoop", IEEE International Conference on Application of Information and Communication Technologies, 17–19 Oct. 2012, pp. 1–5.

Enhancement in Connectivity by Distributed Beamforming in WSN

Vandana Raj and Kulvinder Singh

Abstract A pair of node is said to be connected if they lie in communication range of each other or more precisely in terms of graph theory, we can say that a pair of node is said to be connected if at least, there is a single path exist between them. This connectivity is affected by displacement, dying node, and communication blockage. Due to this reason, the network topology also changes dynamically. Disconnection results in nonfunctional WSN, although other nodes remain operational. Even owing to one node failure can results in end of whole network and we have to deploy the whole WSN again. We will show an improvement in connectivity and increase in lifetime of WSN and better quality signal by using master–slave architecture of distributive beamforming, if a node failure occurs. This approach requires no receiver feedback, other approaches such as one-bit feedback requires feedback from receiver for correctly superimposition of signals from two or more sources on receiver and therefore not an energy-efficient approach. The simulations of the proposed approach are performed and the acquired results highlight the benefits of this proposal.

Keywords Wireless sensor network · Distributed beamforming · Master–slave architecture · Directional antenna

1 Introduction

Connectivity is a major concern in many applications of WSNs such as monitoring, surveillance, battlefield, potentially terrorist attack detection, disaster management, target monitoring, border protection, and more. Connectivity is one of the crucial

---

Vandana Raj (✉) · Kulvinder Singh
University Institute of Engineering and Technology Kurukshetra University, Kurukshetra, India
e-mail: raj.vandana.085@gmail.com

Kulvinder Singh
e-mail: kshanda@rediffmail.com

S.C. Satapathy et al. (eds.), *Proceedings of the International Conference on Data Engineering and Communication Technology*, Advances in Intelligent Systems and Computing 469, DOI 10.1007/978-981-10-1678-3_58

requirements of WSN, as information collected needs to be transmitted to the sink node or processing centers, if it fails to do so, then network life time ends no matter if other nodes are still operational. It is an open research problem. Connectivity depends on link existence. Connectivity can be modeled as graph, G(V, E), where vertices V are sensor nodes, and link between them is an edge. Graph is said to be connected, if there is at least one path between each pair of nodes. With the help of equation, connectivity can be defined as follows:

$$\mu(r) = N\pi r^2/A \tag{1.1}$$

where N is the number of sensors in area A, and r is the radius of transmission [1]. Due to a wide-range of potential applications, the concept of WSN has attracted a great deal of research attention. A wireless sensor network is composed of small sensing devices having processing and wireless communication capabilities, which are deployed in a region of interest. They gather information about the environment, generate and deliver messages to the remote base station. The issues related to WSN connectivity are: critical battery, environmental changes dying node leading to sparse amount of network or disconnected network, and security attacks on nodes such as denial of service [2]. Due to their limited range of communication, holes occur in the network. After operating for a long period, sensor nodes battery starts exhausting and thereby results in disconnection of other nodes from sink node. Figure 1a shows an example of a connected WSN. Communication here is multihop, therefore, if sensor 1 fails then it will result in no data transfer at sink node at all. We know that the purpose of WSN is to deliver data to sink node. But only due to sensor 1 failure can bring down whole WSN. Figure 1b shows an example of disconnected WSN containing isolated nodes. These sensor nodes are not able to send their data to the base station as sensor 1 has stopped functioning. Here one node failure resulted in nonfunctional WSN (Fig. 2).

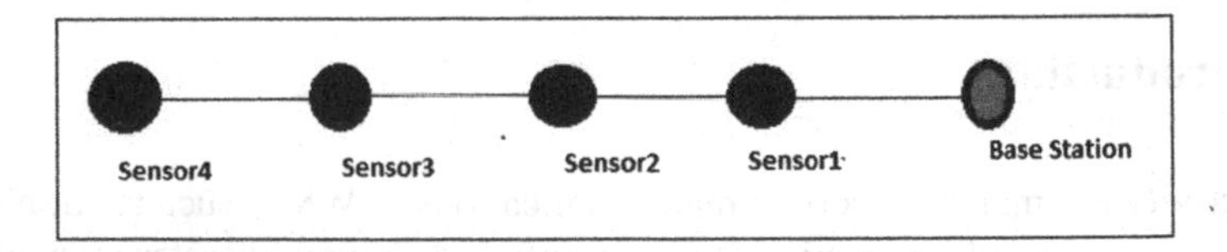

Fig. 1 This figure, shows linear topology WSN with four sensor nodes, and one base station or sink node, communication is multihop

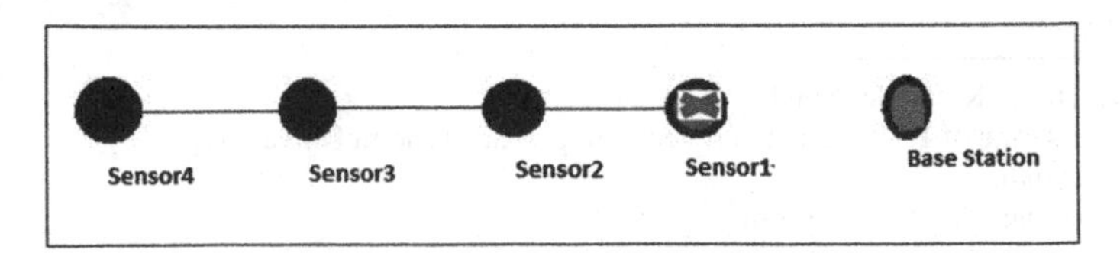

Fig. 2 This figure shows disconnected WSN, sensor 1 has stop functioning, there is no communication of data to base station which results in nonfunctional WSN

To tackle the problem of disconnected WSN, we are using master–slave architecture of DBF, where master node calibrates and synchronizes the carrier signal of slave sensor to achieve beamforming. In paper [3], the author showed that this approach is feasible and is energy-efficient too. The remaining text is organized as follows: Sect. 2 presents related work in this field. Section 3 presents the proposed work. Simulation results are presented and discussed in Sect. 4. Section 5 concludes the paper.

2 Related Work

Advancement in directional antenna technologies had made it possible for sensor nodes to be equipped with directional antennas. Kranakis et al. [4] showed that using directional antenna, energy saving could be achieved as compared to omnidirectional antenna. The author here described that directional antenna provides same amount of connectivity as omnidirectional antenna by giving a condition on beam-width of unidirectional antenna. Saha and Johnson [5] bridged partition of network using unidirectional antenna, for maintaining connectivity. The conversion of sensor network of omnidirectional antenna to strongly connected sensor network of unidirectional antenna was first addressed by Caragiannis et al. [6]. The author here presented a polynomial time algorithms for linear case and two-dimensional case when the sector angle of the antennas was at least $8\pi/5$. When the sector angle is smaller than $2\pi/3$, it was shown that the problem of determining the minimum radius to achieve connectivity was NP-hard. To maintain connectivity, Pignaton de Freitas et al. [7] proposed mobile sinks (UAV). While moving, these mobile sinks cover up the entire region deployed of sensor nodes and thus connectivity and availability of information was achieved. Kranakis et al. [8] studied how to maintain network connectivity when antennae angles were being reduced, while at the same time the transmission range of the sensors was being kept as low as possible. It was showed in [9] that directional antenna could reduce security risk. It was shown that with narrower beam-width of directional antenna, the signal was less likely to be intercepted by enemy. The connectivity recovery technique was proposed by Abbasi et al. [10]. The author proposed "Distributed Actor Recovery Algorithm" for maintenance of connectivity. Cheng et al. [11] proposed connectivity restoration by node rearrangement. Younis et al. [12] proposed an approach "Recovery by Inward Moving" to recover connectivity owing to one node failure. Ahmed et al. [13] proposed another approach of "Least Destructive Topology Repair" for connectivity recovery. To enhance connectivity, by increasing lifetime of the sensor nodes an algorithm was proposed by Guha and Khuller [14]. The Marinho and Marco [15] considered cooperative MIMO technique and UAV relay network together to support connectivity. It was shown here that by utilizing CMIMO and UAVs together increases the connectivity than using them alone. Heimfarth et al. [16] proposed another method utilizing UAV, for connecting disjoint segment in the WSN. In this method, UAV and message ferrying concept was utilized to join isolated network segment.

3 Proposed Work

To overcome this problem of disconnected WSN owing to node failure, and other issues like limited communication range of sensors, dying node problem and critical battery of sensor nodes, the master–slave architecture approach of distributed beam forming is proposed here. We will show that using this approach good connectivity, longer lifetime of WSN and quality of links could be achieved with inexpensive local coordination with a master transmitter. We will give a brief introduction of master slave architecture, after that simulation results achieved will be shown.

3.1 Master–Slave Architecture

Master sensor has a local oscillator that generates a sinusoid that serves as the reference signal for the network. The master sensor broadcasts reference signal to all the slaves. The slave sensor uses this signal as input to a second-order phase-locked loop, driven by a VCO with a quiescent frequency close to carrier frequency. From PLL theory, we know that the steady-state phase error between VCO output and signal received at slave sensor is zero, and therefore, the steady-state VCO output can be used as a carrier signal consistent across all sensors provided that the offset phase offset can be corrected for.

3.2 Implementation

The sensor data that is transmitted is a binary pulse train of random bits. Modulation of carrier wave is done by multiplying the carrier wave with the sensor data which is equivalent to BPSK modulation. Assumption here is that nodes has knowledge of their location and another standard assumption is that receiver has perfect knowledge of channel. This ensures phase synchronization, and therefore coherent demodulation of signal is achieved at the receiver.The base station or sink node first converts the signal to baseband using a mixer to multiply the incoming carrier signal with a local oscillator. The local oscillator is assumed to be frequency synchronized with the transmitting sensors. The implementation above satisfies the results, i.e., for integral wavelength $\lambda, 2\lambda, 3\lambda \ldots$ there is constructive interference and for fractional λ values we got destructive interference. Therefore, implementation satisfied the condition of constructive and destructive interference.

4 Simulation Results

We performed simulation on SIMULINK. Simulation time was 0.3 s. Communication between nodes is multihop. We experimented with 4 sensor nodes and 1 sink node as shown in Fig. 3. By varying distances, between sensor nodes, we have shown that connectivity could be achieved by using master–slave architecture. We have defined distance using the time delay, i.e., we have frequency = 10 kHz, c = 3×10^8 m, by this we calculated the wavelength, λ which equals to 3×10^4 m. We calculated time period that amounts to 0.00001 s. Therefore, 1 m distance time delay amounts 0.3×10^{-8} s. In every sensor node, we created the delay of 0.3×10^{-8} s after that we started the simulation. The result of simulation after Distributed Beamforming on sink is shown in Fig. 4. It shows observed result with N = 4, where N is the number of nodes used in distributed Beamforming. These nodes cooperatively direct their data to sink node. Sink node has the same frequency of 10 KHz, i.e., it is frequency synchronized with transmitting nodes. The distance taken between each node is 1 m, i.e., a delay of 0.3×10^{-8} s and after performing simulation we gained signal-to-noise ratio = 3.85 dB. Here x-axis represents time and y-axis represents achieved signal-to-noise ratio. After that we experimented with N = 3, the SNR achieved here is 2.8 dB with 1 m distance between each node. Achieved result is shown in Fig. 5. We took N = 2 and performed simulation we got SNR = 1.9 dB. We have created

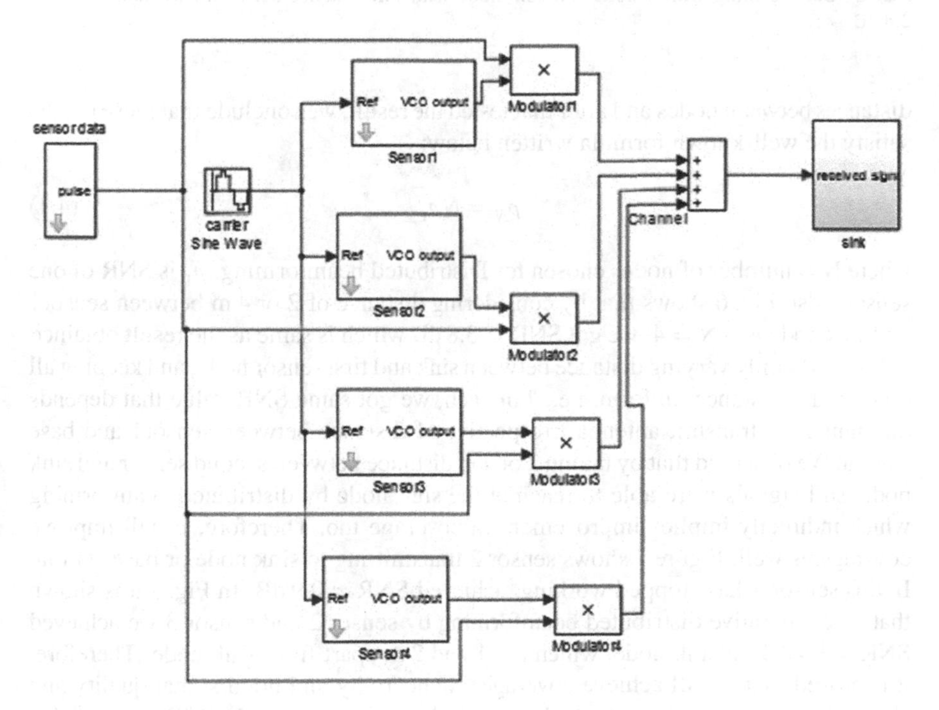

Fig. 3 Designed model of master slave architecture in SIMULINK

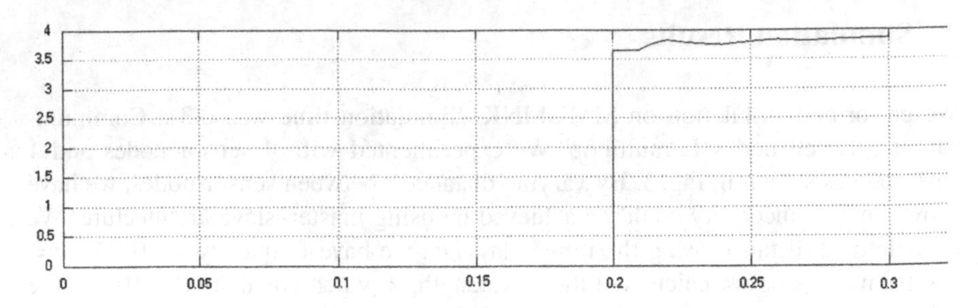

Fig. 4 Beamforming with 4 nodes on sink node with 1 m distance between each node. SNR = 3.85 dB

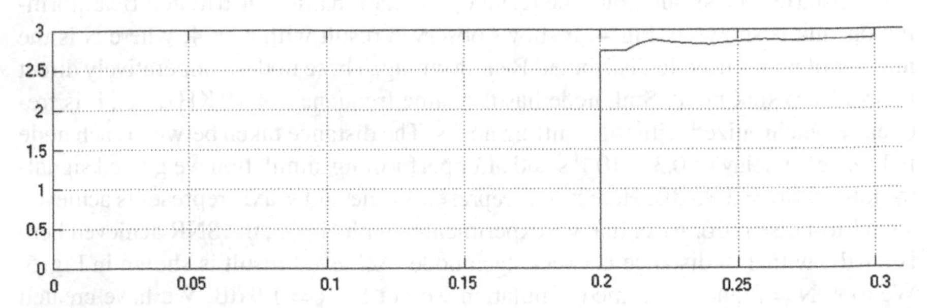

Fig. 5 Beamforming with 3 nodes on sink node with 1 m distance between each node. SNR = 2.8 dB

distances between nodes and after that tested the result, we conclude that these results satisfy the well-known formula written below:

$$\rho_N = N\rho_1 \tag{4.1}$$

where N = number of nodes chosen for Distributed beamforming, ρ_1 is SNR of one sensor. Also Fig. 6 shows that by considering distance of 2 or 4 m between sensor1 and sink node and N = 4, we got SNR = 3.8 dB which is same as the result obtained in Fig. 4. By only varying distance between sink and first sensor node, and keeping all other nodes distance uniform, i.e., 1 m each, we got same SNR value that depends on number of transmit antenna irrespective of distance between sensor1 and base station. We observed that by taking 2 or 4 m distance between second sensor and sink node, still signals were able to reach at the sink node by distributed beamforming which indirectly implies improvement in coverage too. Therefore, it will improve coverage as well. Figure 7 shows sensor 2 transmitting to sink node or base station. In this sensor 1 has stopped working, achieved SNR = 0.9 dB. In Fig. 8 it is shown that by cooperative distributed beamforming by sensor 2 and sensor 3 we achieved SNR = 1.9 dB on sink node, which are 1 and 2 m apart from sink node. Therefore, it is proved that it will achieve coverage, connectivity and good signal quality and also at the same time elongate the life of wireless sensor network (WSN). Using this

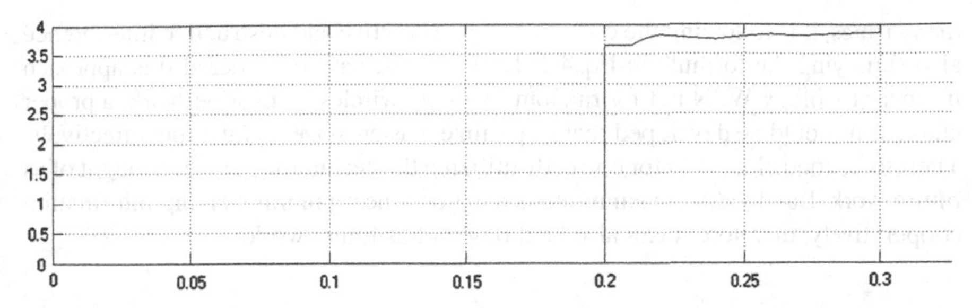

Fig. 6 Beamforming with 4 nodes on SINK by varying distance between sink and sensor. We have taken 2 and 4 m distance between SINK and sensor1 and in all 1 m distance SNR = 3.8 dB

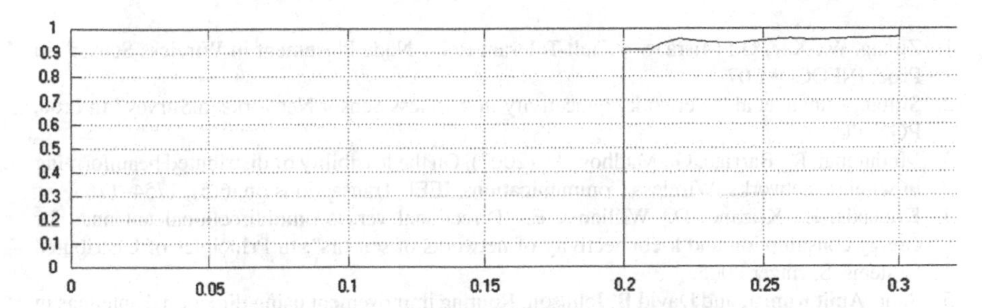

Fig. 7 Without beamforming, when sensor1 stop working, we got SNR = 0.9 dB by sensor 2

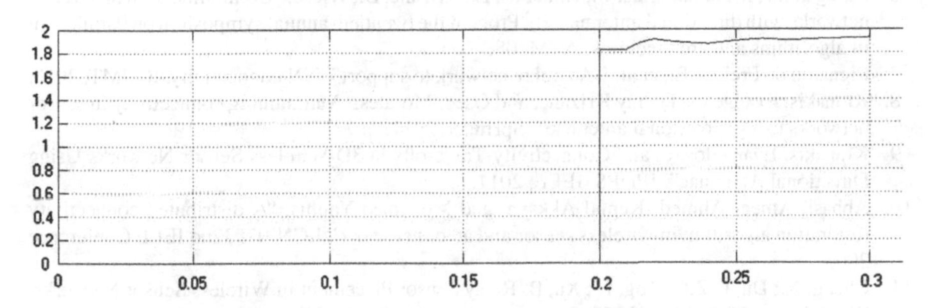

Fig. 8 Beamforming with sensor 2 and sensor 3 when sensor1 stop working. SNR = 1.9 dB

approach of connectivity, we will decrease the necessity of redeploying the nodes in the region of interest.

5 Conclusion and Future Directions

With the help of master–slave architecture of distributive Beamforming, we showed that by increasing distance between nodes and when a node fails still we are achieving good connectivity, and quality of signals. We have tested master–slave architecture many times, it is following the condition of constructive and destructive interference also satisfying the formula in Eq. 4.1. However, we have considered this approach in linear topology WSN but for random topology wireless sensor network, a proper algorithm should be developed that can utilize master–slave architecture effectively. Therefore, good algorithm for correctly utilizing this technology can be thought of as future work. Bandwidth consumption is a major concern in transferring information cooperatively, therefore it can also be thought of as future work.

References

1. Zhang, W.; Xue, G.; Misra, S. “Fault-Tolerant Relay Node Placement in Wireless Sensor” In Proc. INFOCOM’07.
2. Sultan, Adnan, et al. “Network connectivity in Wireless Sensor Networks: a Survey” In Proc. PGNet’09.
3. Mudumbai, R., Barriac, G., Madhow, U. (2007). On the feasibility of distributed beamforming in wireless networks. Wireless Communications, IEEE Transactions on, 6(5), 1754–1763.
4. Kranakis, E., Krizanc, D., Williams, E. “Directional versus omnidirectional antennas for energy consumption and k-connectivity of networks of sensors”. In Principles of Distributed Systems. Springer 2005.
5. Saha, Amit Kumar, and David B. Johnson. Routing improvement using directional antennas in mobile ad hoc networks. In Proc. of GLOBECOM 04.
6. Caragiannis, I., Kaklamanis, C., Kranakis, E., Krizanc, D., Wiese. “Communication in wireless networks with directional antennas”. In Proc. of the twentieth annual symposium on Parallelism in algorithms and architectures. ACM,’08.
7. Pignaton de Freitas, E., et al. UAV relay network to support WSN connectivity. ICUMT, 2010.
8. Kranakis, Evangelos, Danny Krizanc, and Oscar Morales. “Maintaining connectivity in sensor networks using directional antennae”. Springer, 2011.
9. Kranakis, Evangelos, et al. “Connectivity Trade-offs in 3D Wireless Sensor Networks Using Directional Antennae”. IPDPS. IEEE, 2011.
10. Abbasi, Ameer Ahmed, Kemal Akkaya, and Mohamed Younis. “A distributed connectivity restoration algorithm in wireless sensor and actor networks”. LCN’07. 32nd IEEE Conference 2007.
11. Cheng, X.; Du, D.Z.; Wang, L.; Xu, B. Relay Sensor Placement in Wireless Sensor Networks. Wirel. Netw. 2008, 14, 347355.
12. Younis, Mohamed, Sookyoung Lee, and Ameer A. Abbasi. A localized algorithm for restoring internode connectivity in networks of moveable sensors. Transaction of Computers 2013.
13. Ameer Ahmed, Mohamed F. Younis, and Uthman A. Baroudi. Recovering from a node failure in wireless sensor-actor networks with minimal topology changes. Transaction of Vehicular Technology, 2013.
14. Sharma, Lokesh, Jaspreet Singh, and Swati Agnihotri. “Connectivity and Coverage Preserving Schemes for Surveillance Applications in WSN”. International Journal of Computer Applications, 2012.
15. Marinho, Marco AM, et al. Using cooperative MIMO techniques and UAV relay networks to support connectivity in sparse Wireless Sensor Networks. In Proc. of Computing, Management and Telecommunications, 2013.
16. Heimfarth, Tales, and Joo Paulo de Araujo. Using unmanned aerial vehicle to connect disjoint segments of wireless sensor network. In Proc. of Advanced Information Networking and Applications, 2014.

Sparse Representation Based Query Classification Using LDA Topic Modeling

Indrani Bhattacharya and Jaya Sil

Abstract In recent years, tremendous growth of documents provides scope and challenges to the interdisciplinary research community in text processing for retrieving information. Text analytics reveals high-quality information by identifying patterns and its trends using statistical methods. In this paper, we propose a novel approach to classify user query in a reduced search space by considering the query as a collection of words distributed over different topics. Latent Dirichlet allocation (LDA) has been used for topic modeling and a collection of topics containing words are obtained following Dirichlet distribution. We construct a sparse matrix called topic-vocabulary matrix (TVM) using probability distribution of words appearing in the topics. Finally, sparse representation based classifier (SRC) has been applied for classifying query using TVM consisting of training patterns. Here, we have analyzed the effect of number of patterns in classifying the queries and achieved 90.4 % accuracy.

Keywords Topic modeling • LDA • Sparse classifier • Statistical methods

1 Introduction

Huge collection of electronic documents posed new challenges to the researchers for developing automatic techniques to visualize, analyze and summarizing the documents in order to retrieve information accurately and computationally efficient manner. Topic models identify patterns which reflect underlying semantic embedded in the documents [1], needed to classify the documents based on the requirement of the users. Topic modeling is applied to index the documents using relevant terms whereas a topic is a probability distribution over words [2].

Indrani Bhattacharya (✉) · Jaya Sil
Indian Institute of Engineering Science and Technology, Shibpur, Howrah, India
e-mail: Indrani.84@hotmail.com

Jaya Sil
e-mail: jayaiiests@gmail.com

S.C. Satapathy et al. (eds.), *Proceedings of the International Conference on Data Engineering and Communication Technology*, Advances in Intelligent Systems and Computing 469, DOI 10.1007/978-981-10-1678-3_59

In classical document clustering approaches [3] documents are represented by bag-of-word (BOW) model based on raw term frequency and therefore, poor in capturing the semantics. Topic models are able to combine similar semantics into the same group, called topic.

Term frequency inverse document frequency (tf-idf) method [4] can be able to identify discriminative words for a document, but pays little attention to inter- or intra-document statistical structure [5]. To address the problem, first latent semantic indexing (LSI) [6] and latter a generative probabilistic model [7] of text corpora were proposed. A significant step toward probabilistic modeling of text is probabilistic latent semantic analysis (PLSA) reported in [8]. LDA model has been developed to improve the process of forming the mixture models by capturing the exchangeability of both words and documents previously not explored in PLSA and LSA [9]. There are many LDA-based models including temporal text mining, author-topic analysis, supervised topic models, latent Dirichlet co-clustering, and LDA-based bioinformatics [10]. Several improvements have been proposed on LDA, such as the hierarchical topic models [11] and the correlated topic models [12].

A query consisting of several keywords can be viewed as distribution of words with probability over topics. Challenge is to develop more efficient retrieval mechanism for searching related topics from the corpus similar to the query submitted by the user. In this paper, we use LDA method to extract the topics from a large corpus of documents [2]. Then we propose a sparse representation-based classifier [13] for classifying the query, which is distribution of words with probability among the topics. A term vocabulary (TRV) has been constructed using unique terms in the topic corpus, representing the document repository. Since the number of terms present in the query is very specific, the query vector is highly sparse with respect to the TRV. In Sparse representation based classifier (SRC), query is represented in an overcomplete dictionary whose base elements are the training samples. In this paper, the topics are encoded using TRV and considered as training samples in topic-vocabulary matrix (TVM). Finally, we apply SRC to classify the query vector in a reduced search space.

This paper is divided into four sections. Section 2 describes detailed methodology using statistical topic modeling and sparse representation based classifier while results are summarized in Sect. 3 and conclusions are arrived in Sect. 4.

2 Methodology

In the proposed method, first we obtain the topics and the word distribution among the topics using traditional statistical topic models (LDA model). In the second part, we apply SRC to classify the users' query.

2.1 Statistical Topic Modeling

LDA is a basic probabilistic model that describes a generative topic model for a large corpus of documents. In this model, each document is defined as a mixture of words with a given probability. The most dominant topics in the document are with highest probabilities.

Basic LDA model is built on certain assumptions. It assumes that (i) k number of topics is in the corpus, (ii) each document has topic proportions of θ, (iii) a word w is generated from a topic z, and (iv) each topic is defined as the word proportions β over the number of existing words.

The generative process is described below:

1. For each document d in the corpus
   (a) Generate $\theta_d \sim Dir(\alpha)$
   (b) For each word
      i. Draw a topic $z_{d,n} \sim Mult(\theta_d)$
      ii. Draw a word $w_{d,n} \sim Mult(\beta_{z_{d,n}})$

where n is the number of words and α is Dirichlet prior vector for θ and β is the topic probability. The joint distribution of topic mixture θ for given parameter α and β, a set of N topics zd and a set of n words wd is given by Eq. (1):

$$p(\theta, z_d, w_d \mid \alpha, \beta) = p(\theta \mid \alpha) \prod_{n=1}^{N} p(z_{d,n} \mid \theta).p(w_{d,n} \mid z_{d,n}, \beta) \tag{1}$$

where $p(z_{d,n} \mid \theta)$ is $\theta_{d,i}$ for the unique i such that $z_n^i = 1$.

Equation (2) shows the marginal distribution of a document:

$$p(w_{d,n} \mid \alpha, \beta) = \int p(\theta \mid \alpha).(\prod_{n=1}^{N} \sum p(z_{d,n} \mid \theta).p(w_{d,n} \mid z_{d,n}, \beta))\, d\theta \tag{2}$$

Finally, we obtain the probability of a corpus as given in Eq. (3):

$$p(D \mid \alpha, \beta) = \prod_{d=1}^{M} \int p(\theta_d \mid \alpha)\, (\prod_{n=1}^{N_d} \sum p(z_{dn} \mid \theta_d) p(w_{dn}, \beta))\, d\theta_d \tag{3}$$

In the learning method, latent variables z and θ are searched using LDA with an objective to maximize log-likelihood of the data and this problem is NP hard. Several approximate inference algorithms include Gibbs sampling [14] and variation inference [15] have been used for learning purpose. Figure 1 shows graphical representation of LDA model where each node is represented as a random variable and labeled according to the generative process.

In our experiment, we apply LDA considering variable number of topics. We set initial value of parameter α in the range [0, 1] and obtain its effect on distribution of number of topics. We choose 10^{-2} as threshold of difference in α varied over number of topics for the proposed retrieval method. We choose 25 topics as threshold because no significant change in α has been observed further.

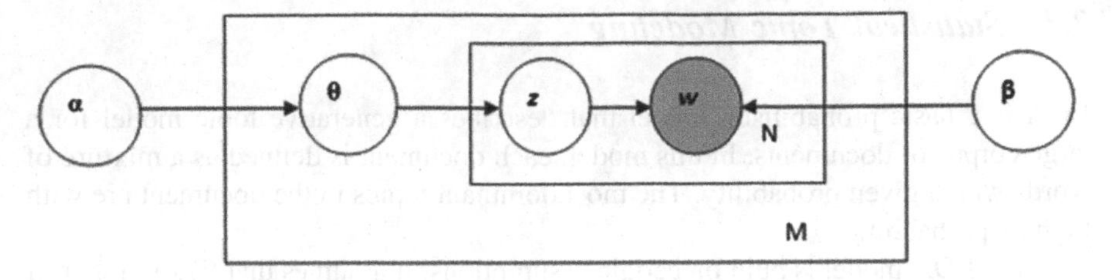

Fig. 1 Graphical representation of LDA model

2.2 *Sparse Classifier*

We obtain n topics containing words using LDA from the corpus of documents and a vocabulary TRV is prepared using unique terms present in different topics. Let us assume the number of unique words in n no. of topics is k, so the dimension of TRV is $1 \times k$. On the basis of TRV a feature vector (FV) of dimension $1 \times k$ for each topic is defined in Eq. (4):

$$\begin{aligned} \mathbf{FV}[i] &= P(w_i \mid T) \cdot P(w_i), \quad \text{if } w_i \text{ presents in the topic} \\ &= P(w_i), \quad \text{otherwise} \end{aligned} \tag{4}$$

where $P(w_i|T)$ denotes the probability of word w_i in topic T and $P(w_i)$ gives the probability distribution of the word in the corpus.

Let us assume that there are c known pattern classes. Let Ai be the matrix obtained using the training samples of class i, i.e., $\boldsymbol{A}_i = [\boldsymbol{y}_{i1}, \boldsymbol{y}_{i2}, \ldots, \boldsymbol{y}_{iRi}] \in \mathbb{R}^{d \times Si}$ where M_i is the number of training samples of class i.

Let us define a matrix $\boldsymbol{A} = [\boldsymbol{A}_1, \boldsymbol{A}_2, \ldots \boldsymbol{A}_c] \in \mathbb{R}^{d \times S}$, where $S = \sum_{i=1}^{c} S_i$. The matrix $\boldsymbol{A}$ is built for the entire training samples. Given a query test sample $\boldsymbol{y}$, we represent $\boldsymbol{y}$ in an overcomplete dictionary whose basis are training samples, so $\boldsymbol{y} = \boldsymbol{A}\boldsymbol{w}$ If the system of linear equation is underdetermined (P < S), this representation is naturally sparse.

The sparsest solution can be obtained by solving the following L_1 optimization problem given in Eq. (5),

$$(L_1)\ \widehat{\boldsymbol{w}_1} = \arg\min \|\boldsymbol{w}\|_1, \quad \text{subject to } \boldsymbol{A}\boldsymbol{w} = \boldsymbol{y} \tag{5}$$

This problem can be solved in polynomial time by standard linear programming algorithm [16]. After the sparsest solution $\widehat{\boldsymbol{w}_1}$ is obtained, the SRC can be done in the following way [17].

For each class i, let $\partial_i: \mathbb{R}^S \rightarrow \mathbb{R}^S$ be the characteristic function that selects the coefficient associated with the ith class.

For $\boldsymbol{w} \in \mathbb{R}^S$, $\partial_i(\boldsymbol{w})$ is a vector whose nonzero entries are in w associated with class i. Using only the coefficient associated with the ith class, reconstruction can be done on a given test sample y as $v^i = A\partial_i\widehat{w_1}$; v_i is called the prototype of class i with respect to the sample y. Equation (6) shows the residual distance between y and its prototype v_i of class i,

$$r_i(y) = \|y - v^i\|_2 = \left\|y - A\partial_i(\widehat{w_1}).v^i\right\|_2 \tag{6}$$

The SRC decision rule is
if $r_l(y) = \min_i r_i(y)$, y is assigned to class l.

In the experiment, we consider the **TVM** as the training set. **TVM** has been built by considering the feature vectors $\mathbf{FV}_i$ for each topic i as described below.

$$\mathbf{TVM} = [\mathbf{FV}_1, \mathbf{FV}_2 \ldots \mathbf{FV}_n]^T$$

The dimension of **TVM** is $n \times k$. Now, we consider a user query q as a test sample and convert it into feature vector $\mathbf{FV_q}$ of dimension $1 \times k$ as described in Sect. 2.2.

We apply SRC on **TVM** for reconstruction of $\mathbf{FV_q}$ and assigned nearest topic to $\mathbf{FV_q}$. The procedure is described in Algorithm 1.

Algorithm 1 Query classification using SRC

```
Input: Set of topics T , query Q, Set of unique keywords TRV, Number of
topics n
Output: Topic related to the query,
  Topic-SRC (T, Q, TRV, n)
  1. TVM ← Φ; FV ← Φ   // Topic-vocabulary matrix & Feature vector
  3 For each t ϵ T
  4.      FV = Feature-Vector (T, TRV)
  5. i ← 0
  6. For i < n
  7.      TVM[i] = FV [i]^T
  8. FV_q = Feature-Vector (Q, TRV)
  9. t-Class = SRC (FV_q, TVM, n)   // Topic of the query
  10. Return t-Class
  Procedure1: Feature-Vector (T, TRV)
  1. For each w ϵ T
  2.      Calculate P(w | T)
  3. FV ← Φ; i ← 0
   6. For i < length (TRV)
  7.      If T[i] = = TRV[i]
  8.         FV[i] ← P(w | T) .P(w)
  9.      Else
  10.        FV[i]  ← P(w)
  11. Return FV
  Procedure 2: SRC (FV_q, TVM, n)
  1.  W ← Φ; D ← Φ          //Sparse co-efficient vector & Set of distance
  3.  W = pinv(TVM) * FV_q^T // Construction of W
  4. Find sparsest solution W_s for W by equation (5)
  5.  i ← 0
  6. For each i < n
  7.      (FV_q)^i_new = TVM * W_s  // Reconstruction of sample vector
  8.      D[i]   ← Norm (FV_q , (FV_q)^i_new )
  9. t-Class = min(D)          //Finding minimum residue distance
  10. Return t-Class //Returning nearest topic class of  Query
```

3 Results and Discussion

In the experiment, we used a subset of TREC AP (Academy Press) corpus containing 2246 news articles with 10473 unique terms. After preprocessing the document, we apply LDA and obtain 25 topics that are optimum for this experiment. Each topic is visualized as a set of words where each element of the set is assigned with the posterior topic measures. Table 1 shows five different topics with most frequent 10 keywords. Test samples as users' query are classified by executing algorithm1.

The performance of the classifier is evaluated using different statistical measures [18] considering 10 queries for each 25 topics, i.e., $10 \times 25 = 250$ queries. We considered varied length of query up to five keywords and no significant change is

Table 1 Topics with top 10 keywords

Economy	Administration	Judiciary	Healthcare	Aviation
Oil	Police	Court	Aids	Air
Cents	People	Trail	Health	Space
Price	Killed	Case	Hospital	Flight
Futures	Authorities	Charges	Medical	Plane
Cent	Army	Attorney	Disease	Two
Lower	City	Prison	Drug	Aircraft
Market	Man	Judge	Patients	Planes
Higher	Government	Two	Care	Accident
Million	Officials	Guilty	Federal	Navy
Farmers	Reported	Years	Doctors	Ship

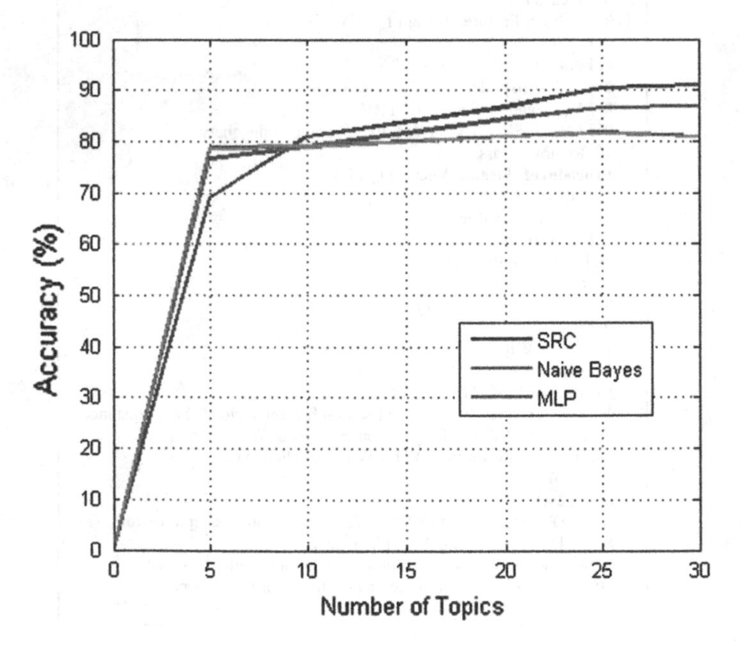

Fig. 2 Accuracy versus number of topics

Table 2 Comparison with different classifiers

Classifiers	Statistical measures							
	Accuracy (%)	Misclassification (%)	TP rate	FP rate	Precision	Recall	F-measure	Specificity
Naive-Bayes	86.6	13.3	0.86	0.015	0.888	0.867	0.868	0.985
Multilayer perceptron	81.66	18.33	0.82	0.02	0.823	0.817	0.819	0.98
SRC	90.4	9.6	0.9	0.09	0.978	0.9	0.94	0.91

reported in the retrieved information. It has been observed statistically that maximum five keywords are provided by most of the users for their query [19]. For instance, the query vectors [judge trial charges guilty]^T, [oil higher market price]^T, [government authorities reported official]^T and [aids patients doctors care]^T of length 4 are classified as 'Judiciary', 'Economy', 'Administration', and 'Healthcare'.

Accuracy has been improved with the number of topics indicating appropriateness of applying sparse classifier in the proposed method. Figure 2 shows improvement in accuracy with increasing number of topics, not remarkable for other classifiers unlike SRC. High accuracy and high TP rate ensures good precision and recall for retrieval method that guarantees lower misclassification too. Detailed comparison of different classifiers using statistical measures summarized in Table 2 and ROC curve for SRC is shown in Fig. 3.

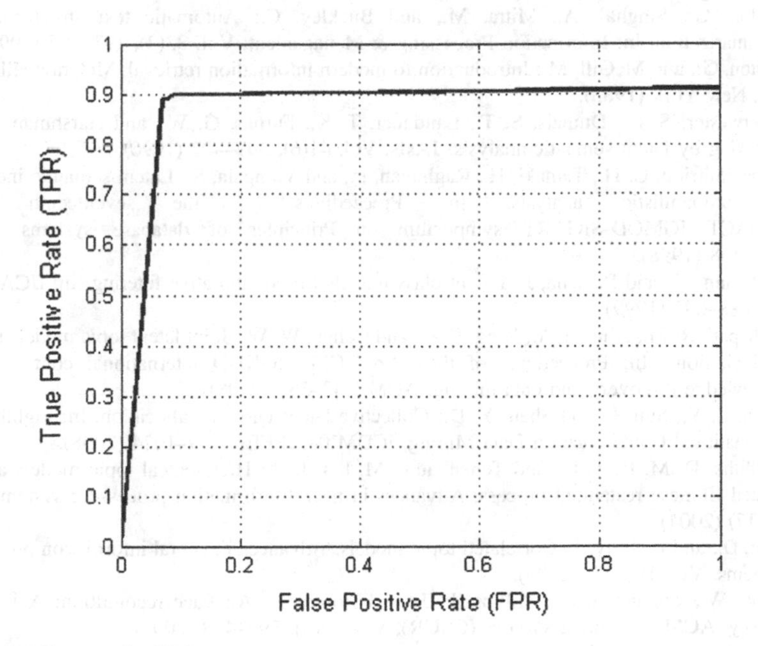

Fig. 3 ROC curve for SRC

4 Conclusion

LDA is a basic and generative model for topic modeling used in the paper for initial latent topic identification. We consider any query as a distribution of words obtained from topics which is sparse with respect to high dimensional input space represented using vocabulary. Proposed approach gives satisfactory results in terms of statistical measures and improves with the size of the training set. For Naïve Bayes classifier performance is better for less number of topic, however, proposed approach using SRC outperforms when size of the training set increases, shown in Fig. 2.

Acknowledgments This research was partially supported by grants from Information Technology Research Academy (ITRA), under the Department of Electronics and Information Technology (DeitY), Government of India.

References

1. Baeza-Yates, R., and Ribeiro-Neto, B.: Modern information retrieval (Vol. 463). ACM press, New York (1999).
2. Blei, D. M., Ng, A. Y., and Jordan, M. I.: Latent dirichlet allocation. The Journal of machine Learning research, Vol. 3, 993–1022 (2003).
3. Manning, C. D., Raghavan, P., and Schütze, H.: Introduction to information retrieval (Vol. 1). Cambridge university press, Cambridge (2008).
4. Salton, G., Singhal, A., Mitra, M., and Buckley, C.: Automatic text structuring and summarization. In: Information Processing & Management, Vol. 33(2), 193–207 (1997).
5. Salton, G., and McGill, M.: Introduction to modern information retrieval. McGraw Hill Book Co, New York (1983).
6. Deerwester, S. C., Dumais, S. T., Landauer, T. K., Furnas, G. W., and Harshman, R. A.: Indexing by latent semantic analysis. JAsIs, Vol. 41(6), 391–407 (1990).
7. Papadimitriou, C. H., Tamaki, H., Raghavan, P., and Vempala, S.: Latent semantic indexing: A probabilistic analysis. In: Proceedings of the seventeenth ACM SIGACT-SIGMOD-SIGART symposium on Principles of database systems. ACM, 159–168 (1998).
8. Hofmann, T., and Puzicha, J.: Latent class models for collaborative filtering. In: IJCAI, Vol. 99, 688–693 (1999).
9. Nallapati, R. M., Ahmed, A., Xing, E. P., and Cohen, W. W.: Joint latent topic models for text and citations. In: Proceedings of the 14th ACM SIGKDD international conference on Knowledge discovery and data mining. ACM, 542–550 (2008).
10. Shen, Z. Y., Sun, J., and Shen, Y. D.: Collective latent Dirichlet allocation. In: Eighth IEEE International Conference on Data Mining, ICDM'08. IEEE, 1019–1024 (2008).
11. Griffiths, D. M. B. T. L., and Tenenbaum, M. I. J. J. B.: Hierarchical topic models and the nested Chinese restaurant process. Advances in neural information processing systems, Vol. 16(17) (2004).
12. Blei, D., and Lafferty, J.: Correlated topic models. Advances in neural information processing systems, Vol. 18(147) (2006).
13. Zhao, W., Chellappa, R., Phillips, P. J., and Rosenfeld, A.: Face recognition: A literature survey. ACM computing surveys (CSUR), Vol. 35(4), 399–458 (2003).

14. Porteous, I., Newman, D., Ihler, A., Asuncion, A., Smyth, P., and Welling, M.: Fast collapsed gibbs sampling for latent dirichlet allocation. In: Proceedings of the 14th ACM SIGKDD international conference on Knowledge discovery and data mining. ACM, 569–577 (2008).
15. Jordan, M. I., Ghahramani, Z., Jaakkola, T. S., and Saul, L. K.: An introduction to variational methods for graphical models. Machine learning, Vol. 37(2), 183–233 (1999).
16. Chen, S. S., Donoho, D. L., and Saunders, M. A.: Atomic decomposition by basis pursuit. SIAM journal on scientific computing, Vol. 20(1), 33–61 (1998).
17. Yang, J., Chu, D., Zhang, L., Xu, Y., and Yang, J.: Sparse representation classifier steered discriminative projection with applications to face recognition. Neural Networks and Learning Systems, IEEE Transactions on, Vol. 24(7), 1023–1035 (2013).
18. Davis, J., and Goadrich, M.: The relationship between Precision-Recall and ROC curves. In: Proceedings of the 23rd international conference on Machine learning. ACM, 233–240 (2006.).
19. http://www.keyworddiscovery.com/keyword-stats.html.

Multiple Home Automation on Raspberry Pi

Krishna Chaitanya, G. Karudaiyar, C. Deepak and Sainath Bhumi Reddy

Abstract A wireless home automation is an emerging trend in this era. Smart home automation finds itself a place that cannot be neglected, as more physical components can be connected through the internet, i.e., wirelessly. The basic elements in a wireless home automation network consist of an embedded sensors accompanied with actuators. This automation network can be used to control home appliances. WebIOPi is a web-based interface and browser-based software that is similar to an application in android phone. To automate multiple homes, the users need a common platform like Raspberry Pi, so that all homes can be managed through WebIOPi with minimal cost using Raspberry Pi. This paper mainly focused the emerging solutions that are suitable for Raspberry Pi and ZigBee.

Keywords ZigBee · WebIOPi · Raspberry Pi

1 Introduction

In traditional communication network all devices are connected through wires. This resulted in a major impediment in network advancements as more money has to be spent in installing wires and establishing connections between networks. Switch played a vital role in these wired connectives. As air space was used as a medium for communication, research and further development in technologies aiding this air

Krishna Chaitanya (✉) · G. Karudaiyar · C. Deepak · S.B. Reddy
Master of Engineering, Embedded System, Sathyabama University, Chennai, India
e-mail: Chaithu11@gmail.com

G. Karudaiyar
e-mail: kallanaikaru@gmail.com

C. Deepak
e-mail: deepakavn@gmail.com

S.B. Reddy
e-mail: saibhumireddy@yahoo.com

S.C. Satapathy et al. (eds.), *Proceedings of the International Conference on Data Engineering and Communication Technology*, Advances in Intelligent Systems and Computing 469, DOI 10.1007/978-981-10-1678-3_60

medium began to slowly emerge. This led to the advent of wireless technology. Internet was used as a base platform for all wireless communications. All physical devices that are connected to the internet are assigned a separate internet protocol (IP) address. This IP address is the physical representation for the device. In recent years wireless sensor network (WSNs) receives significant attention from diversified spectrum of industries, from health care, video surveillance, military, etc, with the emphasis being placed in the field of home appliances, where there is a huge market for consumer electronic goods and a considerable population using these home appliances [1].

Wireless home automation is implemented through a network of interconnected devices which can monitor and control applications for home. A wireless home automation typically includes several embedded devices, which are battery powered and furnished with low-power radio frequency (RF) transceivers. One of the greatest advantages of using a radio frequency communication is its flexibility in allowing new devices to be added and removed as per the user's convenience. Moreover, it minimizes the installation costs since wired networks require ducts or cable trays [2]. Despite these advantages, the wireless home automation also poses some challenges like dynamics of radio propagation, resource limitations, and mobility of devices. A considerable advancement in architectural developments has been made by several organizations and companies across the length and breadth of the world. These architectural designs have been established as per the requirements of the companies personally. The main objective of home automation is to monitor and control devices in home with precision and with minimal direct human interference [3].

2 Main Features of Home Automation

2.1 Light Control

Light can be controlled from outside the place by passing commands. It will reduce the usage of wired connection. It does not automatically on and off. The human intervention is must. The sensor will decide whether the light will be on and off [4].

2.2 Remote Control

Infrared technologies are used for controlling television, heating, and ventilating equipment, but the infrared wavelength is only suitable for short distances. So the radio frequency waves are used to overcome the drawback of infrared solution [5].

2.3 Smart Energy

The sensor only gives the information about temperature, humidity, and light, so smart technology is used to avoid the unnecessary waste of energy. The smart utility meters can be used to detect usage peaks and alert the household devices that may be causing them. Energy supply companies may also use wireless home automation mainly used to perform energy load management [2].

2.4 Remote and Care

Internet of things pills can be used to detect patient health condition and inform to the doctor. Wearable wireless sensors can continuously indicate the condition of several body parameters (e.g., temperature, blood pressure, and insulin) for a correct diagnosis. If acceleration sensors indicate that a person has health problem, alarms can be activated promptly [6].

3 Single Home Automation

Automation is reducing human intervention from the physical devices. Single home automation can be defined as an ideal environment where all appliances and devices inside the home can be controlled with a tap of a finger. Single home automation using the embedded browser of two-way communication is made possible. For all home automations, remote control is the main aim [7] (Figs. 1, 2, 3, 4, and 5).

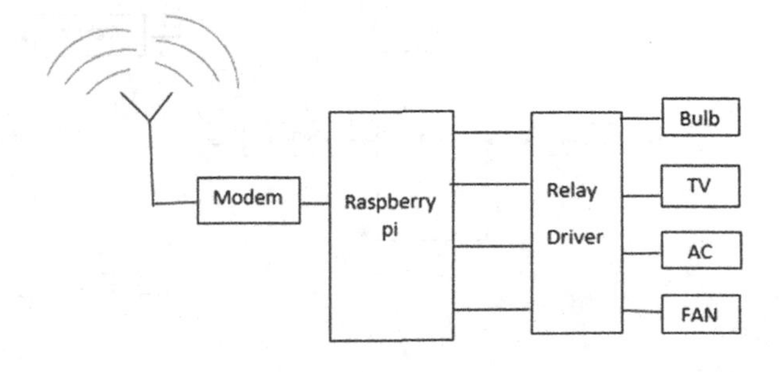

Fig. 1 Single home automation

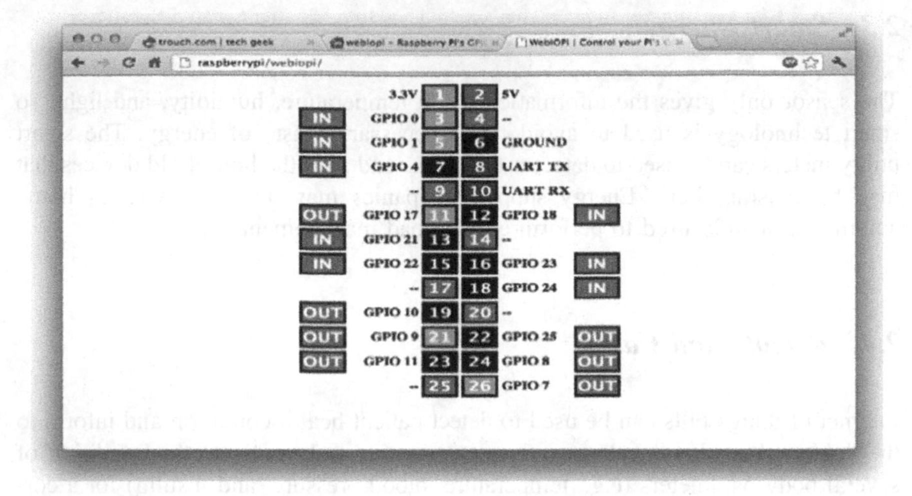

Fig. 2 WebIOPi

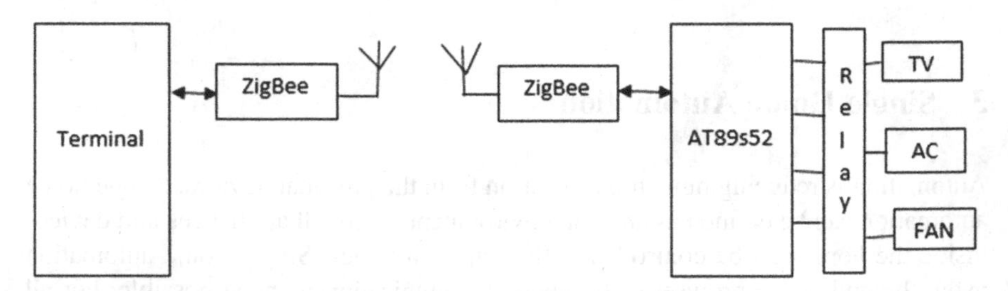

Fig. 3 Single home automation in ZigBee

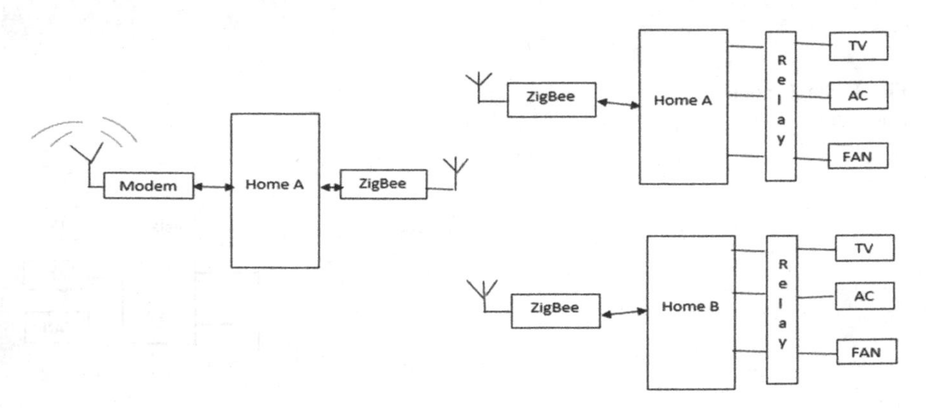

Fig. 4 Multiple home automation using Raspberry Pi and ZigBee

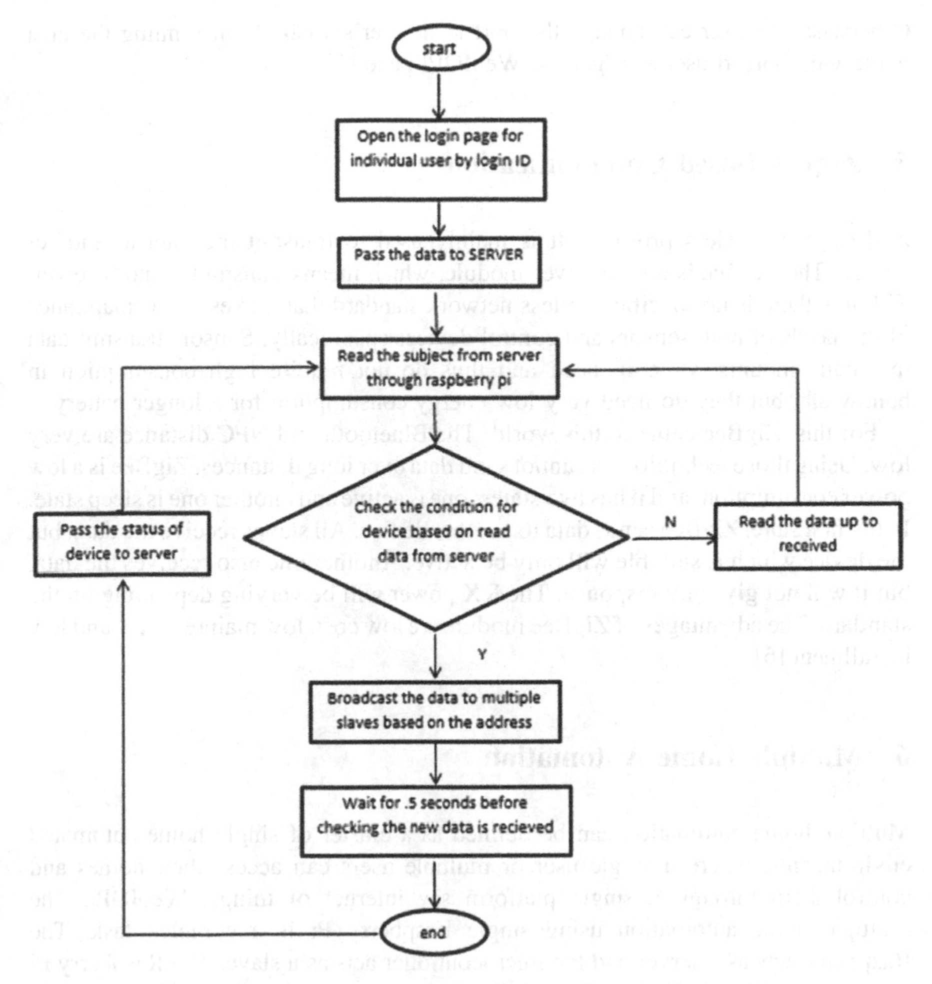

Fig. 5 Flow chart for the proposed method

4 Proposed Work

4.1 WebIOPi

WebIOPi can be defined as a REST framework that allows the user to control GPIO pins in Raspberry Pi from a browser. It is a client–server model of communication. The client is written in JavaScript and the server is coded in Python. The benefit of the WebIOPi is that it can perform the two-way communication [3]. In WebIOPi control, the air conditioner depends on the temperature value, because WebIOPi will give the options for digital monitor. For changing the Raspberry Pi as a server, Raspberry Pi will provide one hostname with port id, and will assign a new port.

Otherwise, the user can change the port id as user's wish. Then running the host name with port id user will get the WebIOPi page [1].

5 ZigBee-Based Communication

ZigBee is a wireless protocol. It is mainly used to transmit the data to another device. The ZigBee is a transceiver module, which means transmitter and receiver. Till now there is no specific wireless network standard that makes the acquaintance of the needs of both sensors and control devices specifically. Sensors transmit data in small amounts while in need and thus do not require high consumption in bandwidth, but they do need very low energy consumption for a longer battery.

For this, ZigBee came to this world. The Bluetooth and NFC distance are very low. Using those technologies cannot send data over long distances. ZigBee is a low power consumption, and it has two states: one is active and another one is sleep state. In the first state, ZigBee sends data to another device. All slaves receive the data, but the device which is suitable will only be active. Another one also receives the data, but it will not give any response. The RX power will be varying depending on the standard. The advantages of ZigBee module are low cost, low maintenance, and low installment [6].

6 Multiple Home Automation

Multiple home automation can be defined as a cluster of single home automated environments, where a single user or multiple users can access their homes and control them through a single platform say internet of things, WebIOPi. The multiple home automation using single Raspberry Pi is a complex task. The Raspberry acts as a server and the microcontroller acts as a slave. The Raspberry Pi sends a data through ZigBee. One ZigBee is connected to Raspberry Pi and another ZigBee is connected to microcontroller. While pressing the button in Raspberry Pi, the string data will come out. The ZigBee receives the data and sends to microcontroller. The slave ZigBee receives the data and gives to that slave host. The slave host is a microcontroller. All microcontrollers will receive the data, but the microcontroller where the data is suitable will move to active state, while other microcontrollers will be in sleep state only.

7 Algorithm for Proposed Method

The above flowchart says the individual user gets a separate login id. When the user enters the login and presses the button, data is sent to the server. Then the server reads the data through Raspberry Pi, after that check the condition for device based on read data from server. If the condition false, then it will move to read data up to received. If the condition is true, it will broadcast the data to multiple slaves based on the address. The server will give the service which depends on the priority, wait for 5 s before checking the new data which is received. The 8051 microcontroller acts as a slave in this work, and has 40 pins and 4 ports. The user can access only two ports at a time. The microcontroller will be host for the ZigBee. User already dumbed the program in microcontroller. The serial communication program will work perfectly. There is no communication between the microcontrollers; the baud rate of this microcontroller is 9600. The parity bits also have to give [8].

8 Results and Discussions

Figure 6 shows the output of serial communication in AT89s52. Here, the Raspberry Pi is connected to microcontroller [3]. First, the Raspberry Pi is changed into local server. The frame name is WebIOPi. The Raspbian OS is running on the Raspberry Pi. The Raspberry Pi did not have an internal memory. The SD card is used for the memory of Raspberry Pi. For running the WebIOPi framework user has to run the following code:

Sudo webiopi -d -c/etc./webiopi/config

After running the above code, the WebIOPi framework page will be opened. Then the users have to type the username and password. The default username is WebIOPi and password is Raspberry, but the user can change the user name and password they like. After running the code the below page will be opened. The Linux kernel has to upload all the details into the configuration file. The user copies

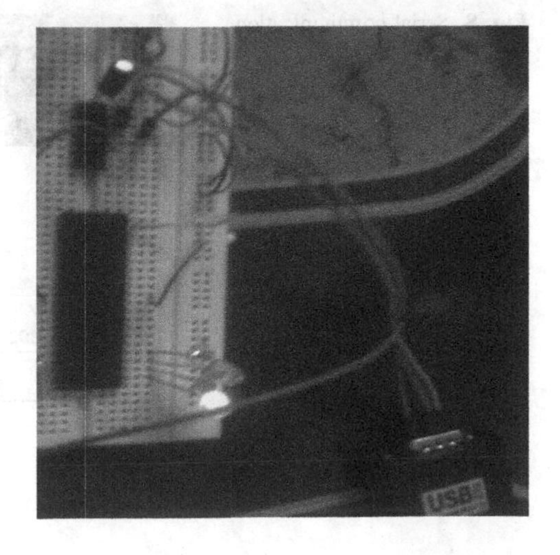

Fig. 6 Serial communication in AT89s52

the internet protocol address with port id which is generated from the WebIOPi. The format is

http://ipaddress of Raspberry pi: Port id

By default, the Raspberry Pi will communicate on the port 8000, but the user wants to change the port id that is possible (Figs. 7, 8, 9 and 10).

When the user presses the above button, the character 'A01' will be sent to microcontroller through ZigBee. The ZigBee is a wireless protocol. The second part microcontroller will receive the string 'A01'. The character A means that it will select the first home. The character 'B' means it will select the home B.01 means the first light will glow and 02 means the second light will glow. The microcontroller will receive the code and will react depending on the code. The figure shows the output of serial communication in Raspberry Pi [5].

The above diagram shows how the relay circuit is connected with Raspberry Pi, because relay circuit is very important one. The need of relay converts the controller output into commercial output because all the home appliances needed 230 V power supply, but the Raspberry Pi cannot able to give that much of power. Our Raspberry Pi input voltage is 3–5 V. The relay circuit is used to connect with another end switch: the TTL output to commercial output (230 V, 50 Hz) [7].

Fig. 7 Output of Raspberry Pi server

Fig. 8 Serial communication in Raspberry Pi server

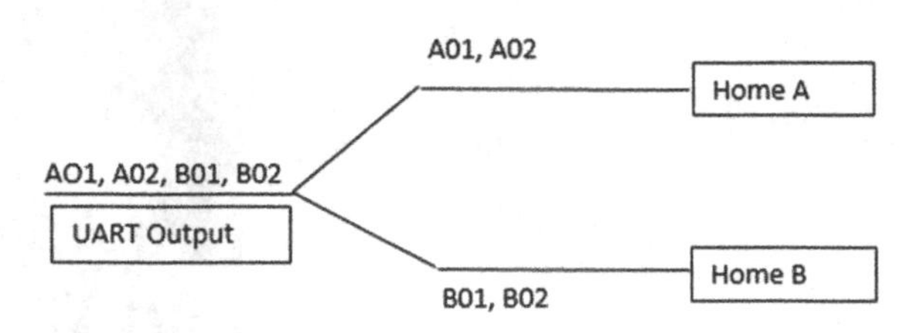

Fig. 9 Communication structure

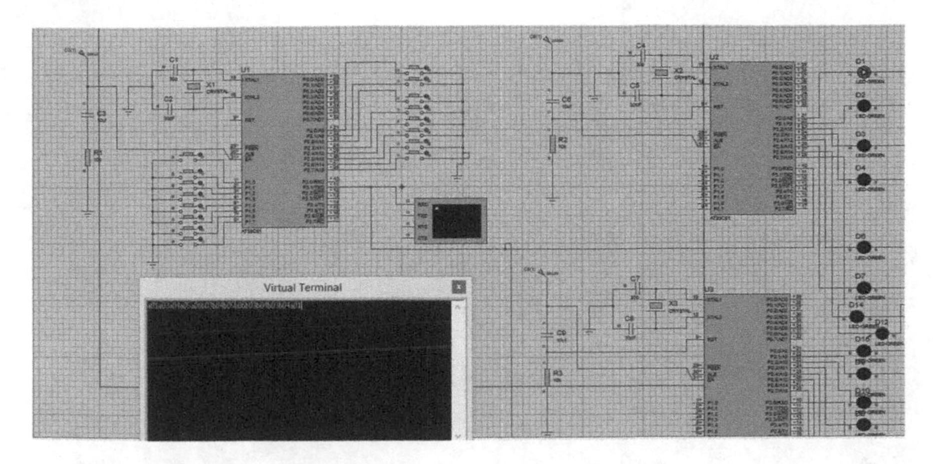

Fig. 10 Microcontroller simulation result

9 Conclusions

An approach for implementing multiple home automation is done using single Raspberry Pi. But the performance can be further enhanced using ZigBee module, as ZigBee reduces power consumption and its ability to establish communication even in the presence of barriers. Data is sent to the server; this minimizes the problem of transmission of data. Hence, using both internet of things (IOT) and ZigBee protocol, this approach could satisfy all hindrances in home automation like power consumption and cost issues. This approach ensures home automation with greater efficiency at lower cost.

References

1. A.R. Al-Ali and M. AL-Rousan, "Java Based Home Automation System" IEEE Trans. Consumer Electron. Vol. 50, No. 2, pp. 498–504. May 2004.
2. Hiroki sawda, hase tomohiro (2013) "A Remote Contoller With Embedded Browser as a Platform" ieee second international conference on consumer electronics.
3. Home gateway based on open service gateway initiative platform, The 8th International Conference on Advanced Communication Technology, pp. 1517–1520, 2006.
4. S.R. Bharanialankar, C.S. Manikandababu (2014) "Intelligent Home Appliance Status Intimation Control And System Using Gsm" International Journal of Adavnce Research in Computer Science And Software Engineering volume4, issue4.
5. Jong-hyuk Roh and Seunghun Jin (2014) "Device Control Protocol Using Mobile Phone" published a paper on ICACT 2014.
6. Van der weff M. Gui X. Xu, W.L, "a mobile based home automation system" second international Nov 2005 pp. 40.
7. Shaiju Paul, Ashlin Antony (2014), "Android Based Home Automation Using Android", International journal Of Computing Technology, volume 1, issue1.
8. J. Bray, C. F. Sturman, "Bluetooth 1.1: Connect without Cable", Pearson Education, edition 2, 2001.

Sentiment Analysis Based on A.I. Over Big Data

Saroj Kumar, Ankit Kumar Singh, Priya Singh, Abdul Mutalib Khan, Vibhor Agrawal and Mohd Saif Wajid

Abstract Area of interest over big data is a basic problem of data management system. In this paper we elaborate a methodology for sentiment analysis based on artificial intelligence. In any AI (Nicole, IEEE Trans Inf Theory, IT-9:248–253 (1963) [1]) systems think like human, think like rationally, act like human, and act like rationally. This type of AI system is imposed on big data to find the common sentiment as per user recommendation. In this paper we present a recommended technique for sentiment analysis. Recommended technique is based on AI approaches. In this paper we elaborate a matrix for user recommended data group for big data which is reduced by dimension reduction technique.

Keywords Cloud · Artificial intelligence · Recommended system · Collaborative filtering · Single value decomposition (SVD) · Big data

Saroj Kumar (✉)
CSE Department, BBDNITM, Lucknow, India
e-mail: saroj.kumar999@gmail.com

A.K. Singh
IT Department, BBDEC, Lucknow, India
e-mail: ankit.singh.mails@gmail.com

Priya Singh
CSE Department, ASET, Lucknow, India

A.M. Khan · Vibhor Agrawal
CSE Department, BBDEC, Lucknow, India
e-mail: abdul.khan619@gmail.com

Vibhor Agrawal
e-mail: vibhor_vaibhav@hotmail.com

M.S. Wajid
CSE Department, BBDU, Lucknow, India
e-mail: mohdsaif06@gmail.com

S.C. Satapathy et al. (eds.), *Proceedings of the International Conference on Data Engineering and Communication Technology*, Advances in Intelligent Systems and Computing 469, DOI 10.1007/978-981-10-1678-3_61

1 Introduction

The user opinion about the EaaS (Everything as a Service) can be based upon the interest of the downloading or viewing the data over the Cloud. Cloud [2] is also known as a one-time big data. The relevant data, large data storage, high virtualization different networking equipments, and large platform are the area of interest for big enterprises as well as single user. Public cloud permits a large amount of availability of public data (texts, audio, video, images) which can be shared between large number of end consumers. These shared data help to find out the interest of the users. By the interest of cloud service consumers, cloud service providers can easily maintain a group of user recommended data group (Big data). This group contains a common data which is accessed by a number of service consumers.

Advantages of user recommended data group (Big data) [3] are as follows:

Each time when the service consumer logins in, he logins as a member of this group and finds out the people who have similar interest as him.
Cloud service provider can save himself from the overhead of data availability and data security.
Public Cloud becomes a platform where the service consumers with same interest can interact.

2 Problem Statement

Challenges for making a user recommended data group (Big Data) are as follows:

A large public Cloud can have huge amount of data, tens of millions of service consumers, and millions of different catalog groups.
New service consumer can have extremely limited information, based on very limited data which is accessed.
Old service consumers can have a glut of information, based on thousand of data which are accessed.
End-user consumers are volatile.

3 Proposed System

In this paper we are going to introduce the methodology for creating user recommended data group (Big data). In the present scenario for making data group, public Cloud is either done by the Cloud service provider or Cloud server. Manually, making these groups takes large amount of processing time, so it is important to make it using Cloud servers. Hence, using Cloud servers introduces artificial

Table 1 Categories of AI [16]

Systems that assume like humans	Systems that assume rationally
"The exciting new effort to make computers think …. Machines with minds, within the full and literal sense"	"The study of mental faculties through the use of computational models"
Systems that act like humans	Systems that act rationally
"The art of creating machines that perform functions that require intelligence when performed by people"	"Computational Intelligence is the study of the design of intelligent agents"

intelligence in the cloud. For making these types of group we involve the search methods of artificial intelligence known as recommender system. In recommender system we use collaborative filtering to make user recommended data group (Big data) (Table 1).

Artificial Intelligence: Following table defines the artificial intelligence, organized in four categories:

Search Methods: The process of looking for sequence from problem formulation to solution is known as search methods. In artificial intelligence there are two methods:

Uninformed search: Sometimes we may not get much relevant information to solve a problem. This type of search is called uninformed search.

Informed search [4]: Informed search is also decision heuristic search. Rather than looking one path or several ways, similar to that informed, search uses the given heuristic info to make a decision whether or not to explore the present state further.

Recommender systems: Recommender systems were created to assist in sorting through the vast amount of information that the internet can provide. These systems function by taking in some types of user information, such as preferred music artists, etc., and provide recommendations for new data based on the user's previous choices.

4 Research Methodology

Detailed description of research methodology is as follows:

1. Cloud consumer or user logins access the data from the Cloud. Each time a log file is created corresponding to access data and forwarded to Cloud server.
2. Cloud server takes current log files and all the previous log files and applies traditional collaborative filtering [5] over them.
3. In this way a large cluster [6] is created over big data based on user interest or user recommendation.
4. This large cluster contains thousands of data interest over user interest so it is impossible or too difficult for a cloud server to match these files with other user interest files. So cloud server applies size reduction policy.

5. In size reduction policy cloud server uses single value decomposition mechanism.
6. In this way size reduction policy is processed and user interest small cluster is created.
7. Cloud server matches small cluster of user interest of each user with other users and creates sentimental group.
8. In this way Cloud-based sentimental group is created.

5 Solution Approach

5.1 Traditional Collaborative Filtering [7, 8]

Collaborative filtering has two senses, a slim one and a lot of general one. In general, collaborative filtering is that the method of filtering for data or patterns and victimization techniques involving collaboration among multiple agents, viewpoints, information sources, etc. Applications of collaborative filtering usually involve terribly massive information sets. Collaborative filtering ways are applied to several totally different styles of information (Fig. 1).

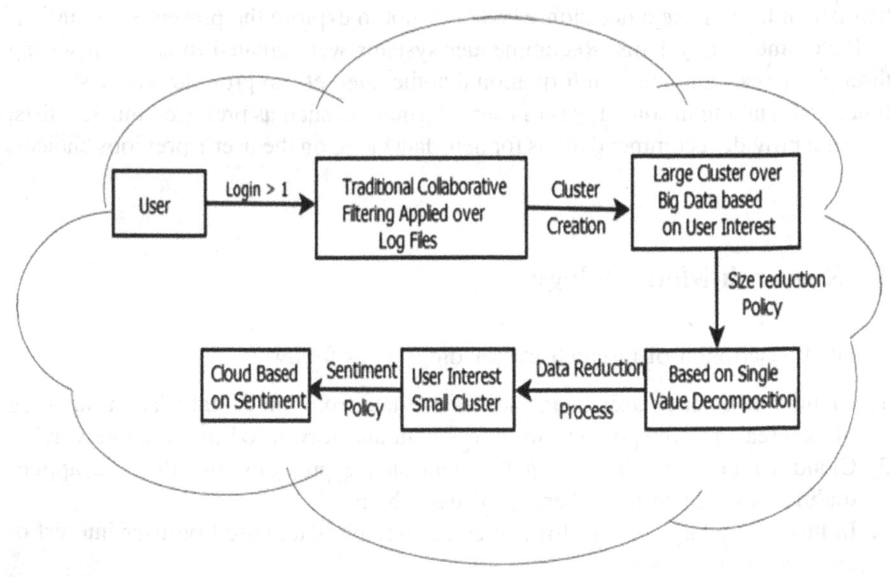

Fig. 1 Research methodology

1. Prediction: Prediction is a mathematical value, and expresses the predict likeliness of the item for the active user.
2. Recommendation: It is a list of N items that the active user will like the most. The recommended list contains the items not already purchased by user.

Within the newer, narrower sense, collaborative filtering [9] may be a technique of constructing automatic predictions (filtering) regarding the interests of a user by aggregation preferences or style data from several users (collaborating). A collaborative filtering [10] formula represents a client as an N-dimensional vector of data, wherever N is that the variety of distinct catalog data. The elements of the vector are positive for accessed or absolutely rated information and negative for solely viewed information and negatively rated data. For almost all service customers, this vector is extraordinarily thin. The formula generates recommendations based on many service customers who are most similar to the finish user. It will live the similarity of two service customers, A and B, in numerous ways; a common method is to live the cos of the angle between the two vectors (Figs. 2, 3, 4, and 5).

Using collaborative filtering to get recommendations is computationally dearly won. It is O(MN) within the worst case, wherever M is the range of service customers and N is the number of product catalog data, since it examines M service customers and up to N knowledge for every client. However, because the average client vector is extraordinarily thin, the algorithm's performance tends to be nearer to O(M + N). Scanning each client is approximately O(M), not O(MN), as a result of the majority customer vectors contain a tiny low range of information, regardless of the dimensions of the catalog. However, there is a space for few service customers who have accessed a significant share of the catalog, requiring O(N) time interval. Thus, the ultimate performance of the algorithmic rule is roughly O

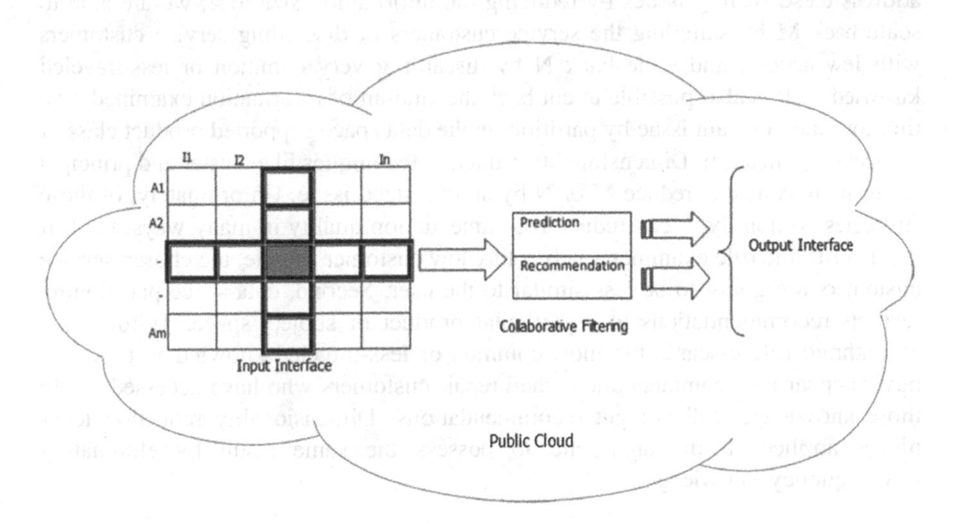

Fig. 2 The collaborative filtering process on public cloud

$$\textbf{similarity} = \cos(\theta) = \frac{A \cdot B}{\|A\|\|B\|} = \frac{\sum_{i=1}^{n} A_i \times B_i}{\sqrt{\sum_{i=1}^{n} (A_i)^2} \times \sqrt{\sum_{i=1}^{n} (B_i)^2}}$$

Fig. 3 Cosine similarity rule [15]

$$\underset{m \times n}{\overset{X}{\begin{pmatrix} x_{11} & x_{12} & \cdots & x_{1n} \\ x_{21} & x_{22} & \cdots & \\ \vdots & \vdots & \ddots & \\ x_{m1} & & & x_{mn} \end{pmatrix}}} = \underset{m \times r}{\overset{U}{\begin{pmatrix} u_{11} & \cdots & u_{1r} \\ \vdots & \ddots & \\ u_{m1} & & u_{mr} \end{pmatrix}}} \underset{r \times r}{\overset{S}{\begin{pmatrix} s_{11} & 0 & \cdots \\ 0 & \ddots & \\ \vdots & & s_{rr} \end{pmatrix}}} \underset{r \times n}{\overset{V^{\mathsf{T}}}{\begin{pmatrix} v_{11} & \cdots & v_{1n} \\ \vdots & \ddots & \\ v_{r1} & & v_{rn} \end{pmatrix}}}$$

Fig. 4 Single value decomposition matrix

$$\underset{m \times n}{\overset{\hat{X}}{\begin{pmatrix} x_{11} & x_{12} & \cdots & x_{1n} \\ x_{21} & x_{22} & \cdots & \\ \vdots & \vdots & \ddots & \\ x_{m1} & & & x_{mn} \end{pmatrix}}} \approx \underset{m \times r}{\overset{U}{\begin{pmatrix} u_{11} & \cdots & u_{1r} \\ \vdots & \ddots & \\ u_{m1} & & u_{mr} \end{pmatrix}}} \underset{r \times r}{\overset{S}{\begin{pmatrix} s_{11} & 0 & \cdots \\ 0 & \ddots & \\ \vdots & & s_{rr} \end{pmatrix}}} \underset{r \times n}{\overset{V^{\mathsf{T}}}{\begin{pmatrix} v_{11} & \cdots & v_{1n} \\ \vdots & \ddots & \\ v_{r1} & & v_{rn} \end{pmatrix}}}$$

Fig. 5 Reduced matrix for user recommended data group

(M + N). Even so, for terribly massive knowledge sets—like ten million or more service customers and one million or additional catalog data—the algorithmic rule encounters severe performance and scaling problems. It is attainable to partly address these scaling issues by reducing the information size to 4; we are able to scale back M by sampling the service customers or discarding service customers with few access, and scale back N by discarding very common or less-traveled knowledge. It is also possible to cut back the amount of information examined by a tiny low and constant issue by partitioning the data space supported product class or subject classification. Dimensionality reduction techniques like cluster and principal element analysis can reduce M or N by an oversized issue. Unfortunately, of these strategies conjointly, it can reduce recommendation quality in many ways. First, if the algorithmic rule examines solely a tiny low customer sample, the chosen service customers are going to be less similar to the user. Second, data-space partitioning restricts recommendations to a particular product or subject space. Third, if the algorithmic rule discards the most common or less-traveled knowledge, they will never appear as recommendations, and repair customers who have accessed solely those knowledge will not get recommendations. Dimensionality reduction techniques applied to the area tend to possess the same result by eliminating low-frequency knowledge.

5.2 Dimension Reduction [11] and Single Value Decomposition [12]

One common thanks to represent datasets is as vectors during a feature area. As an example, if we have a tendency to let every dimension be a film, then we are able to represent users as points. One natural question to request this setting is whether or not it is attainable to cut back the amount of dimensions we want to represent the information. As an example, if each user who likes the matrix conjointly likes Star Wars, then we are able to cluster them along to make an agglomerated film or feature. We are able to then compare two users by watching their ratings for various options instead for individual movies. There are unit many reasons we would need to try to this. The primary is quantifiability. If we have got a dataset with 20000 movies, then every user may be a vector of 20000 coordinates, and this makes storing and examination users comparatively slow and memory-intensive. It seems, however, that employing a smaller range of dimensions will really improve prediction accuracy. As an example, suppose we have got two users who like phantasy movies. If one user has rated Star Wars extremely and also the different has rated Empire Strikes Back extremely, then it is smart to mention the users' area unit similar. If we have a tendency to compare the users' supported individual movies, however, solely those movies that each user has rated can have an effect on their similarity. This can be an extreme example; however, one will actually imagine that there area unit numerous categories of films that ought to be compared. They were making an attempt to check document exploitation of the words they contained, and that they planned the thought of making options representing multiple words and so examination those. To accomplish this, they created use of a mathematical technique referred to as singular value decomposition. In addition, recently, Sarwar et al. created use of this method for recommender systems. The singular value decomposition (SVD) [13] may be a documented matrix resolution technique that factors an m by n matrix X into three matrices as follows.

The matrix S may be a square matrix containing the singular values of the matrix X, where square measures specifically r singular values, wherever r is that the rank of X. The rank of a matrix is that the range of linearly freelance rows or columns within the matrix. Recall that two-vector square measures linearly freelance if they will not be written because of the total or scalar multiple of the other vectors within the area. Observe that linear independence somehow captures the notion of a feature or clustered item that we tend to try to induce at. To come to our previous example, if each user who liked Star Wars additionally liked the Matrix, the two show vectors would be linearly dependent and would solely contribute one to the rank.

We can do a lot, however. We might love to check movies if most users who like one additionally just like the alternative. To accomplish, we will merely keep the primary k singular values in S, where k may provide US the simplest rank-k approximation to X, and so has effectively reduced the spatial property of our original area. So we have got: ratings Matrix[user][movie] = total (user Feature[f][user] * movie Feature[f][movie]).

5.3 Recommendations with the SVD

Given that the SVD somehow reduces the spatial property of our dataset and captures the “features” that we will use to check users, however, can we truly predict ratings? The primary step is to represent the info set as a matrix wherever the users square measure rows, movies square measure columns, and therefore the individual entries square measure-specific ratings. So as to supply a baseline, we tend to fill all told of the empty cells with the common rating for that shown and so figure the SVD. Once we tend to scale back the SVD to induce X_hat, we will predict a rating by merely wanting up the entry for the acceptable user/movie try within the matrix X_hat.

6 Conclusion

Sentiment analysis is a future of big data as per this paper purposed. AI techniques are well-organized technique for creating any kind of result. Whatever AI technique is imposed in this paper or on big data [14] for sentiment analysis presents basic concept for futuristic technique. Reduced matrix system is generated for user recommended data which is based on AI techniques. So this paper presents base for big data to find common sentiment using AI techniques for user recommended data.

References

1. R. Nicole, “Title of paper with only first word capitalized,” J. Name Stand. Abbrev., in press. E Fegienban, “Artificial Intelligence Research” *IEEE Trans. On Information Theory,* vol. IT-9, pp. 248–253, October 1963.
2. Sosinsky B, Cloud Computing Bible. 1st ed. Wiley; 2011.
3. G. Adomavicius and A. Tuzhilin, “Towards the Next Generation of Recommender Systems: A Survey of the State-of-the-Art and Possible Extensions”, *IEEE Transactions on Knowledge and Data Engineering* **17** (2005), 634–749.
4. http://www.eecs.wsu.edu/~cook/ai/hw/h1.
5. J. L. Herlocker, J. A. Konstan, A. Borchers and John Riedl, “An Algorithmic Framework for Performing Collaborative Filtering”, *Proc. 22nd ACM SIGIR Conference on Information Retrieval*, pp. 230 –237, 1999.
6. M. Saerens, F. Fouss, L. Yen, and P. Dupont, “The principal component analysis of a graph and its relationships to spectral clustering,” in: Proc. Eur. Conf. on Machine Learning, 2004, https://citeseer.ist.psu.edu/saerens04principal.html.
7. K. Goldberg, T. Roeder, D. Gupta and C. Perkins, “Eigentaste: A Constant Time Collaborative Filtering Algorithm”, *Information Retrieval* **4** (2001), 133–151.
8. R. Salakhutdinov, A. Mnih, and G. Hinton, “Restricte Boltzmann Machines for Collaborative Filtering”, *Proc.* 24th.
9. J. Konstan, B. Miller, D. Maltz, J. Herlocker, L. Gordon and J. Riedl, “GroupLens: Applying Collaborative Filtering to Usenet News”, *Communications of the ACM* **40** (1997),77–87, www.grouplens.org.

10. J. Nocedal and S. Wright, *Numerical Optimization*, Springer (1999).
11. Sarwar, B. M., Karypis, G., Konstan, J. A., and Riedl, J. (2000). Application of Dimensionality Reduction in Recommender System—A Case Study. In *ACM WebKDD'00 (Web-mining for Ecommerce (Workshop).*
12. Gilbert strang video lecture Lec 26 MIT 18.06 Linear Algebra, Spring 2005.
13. Wikipedia contributors. Singular value decomposition [Internet]. Wikipedia, The Free Encyclopedia; 2012 Apr 13, 07:46 UTC [cited 2012 Apr 19]. Available from: http://en.wikipedia.org/w/index.php?title=Singular_value_decomposition&oldid=487135458.
14. Dr. Arcot Rajasekar, "The Data Bridge: Sociometric Methods for Long-Tail Scientific Data." IEEE Trans (ASE/IEEE) International Conference on Big Data held in Washington, D.C. Sept. 8 – 14, 2013.
15. http://en.wikipedia.org/wiki/Cosine_similarity.
16. http://artificialintelligentsystems.wordpress.com/2010/09/04/types-of-ai-search-techniques/.

Negotiation and Monitoring of Service Level Agreements in Cloud Computing Services

S. Anithakumari and K. Chandrasekaran

Abstract SLAs are so significant in cloud computing because it establishes agreements between the cloud service providers and cloud consumers, about the quality of the providing service. SLA monitoring is the only available provision to check whether the agreed parties are following the agreement terms or not. A multistep SLA negotiation, which contains the selection of apt cloud service provider and the negotiation with the selected provider, is proposed here. An efficient SLA negotiation algorithm is also included in this negotiation method. Experimental evaluation shows that the proposed method is more efficient in resource allocation and it gives more revenue to the cloud providers.

Keywords Cloud computing · Service level agreement (SLA) · SLA negotiation · SLA monitoring · QoS

1 Introduction

Service Level Agreements (SLAs) plays a major role in maintaining the quality of service in cloud computing environment. An SLA contains a set of SLOs (Service Level Objectives) which define QoS properties for the agreed upon service. The usage of new improved mechanisms for negotiating and monitoring SLAs is highly essential because of the dynamic change in service requirements. Online monitoring of SLA is advantageous to both the involved parties because it detects possibility of violations in SLA and initiates some actions to correct or compensate it. SLA negotiation takes care of the conflicts in SLO values and tries to resolve them with some agreed upon values. The negotiation protocol controls the decision, desirability, and preferences of the parties to finalize an agreement.

S. Anithakumari (✉) · K. Chandrasekaran
NITK Surathkal, Surathkal, Karnataka, India
e-mail: lekshmi03@gmail.com

K. Chandrasekaran
e-mail: kchnitk@gmail.com

S.C. Satapathy et al. (eds.), *Proceedings of the International Conference on Data Engineering and Communication Technology*, Advances in Intelligent Systems and Computing 469, DOI 10.1007/978-981-10-1678-3_62

Dynamic SLA management includes a sequence of steps such as: (i) Negotiation and establishment of SLA with a suitable service provider, (ii) Monitoring the SLO parameters and assessing the performance of the delivered service and (iii) Renegotiation of SLO values on the detection of an SLA violation. In the literature, several techniques are discussed for SLA management, but none of these steps mentioned the multistep approach. In this paper, we introduce a multistep approach for dynamic SLA management where the selection of a suitable provider and negotiation of SLA are the different steps included. An efficient architecture and its working algorithm are included as part of this negotiation. The remaining part of the paper is organized as follows. Section 2 describes the related research work and Sect. 3 explains the significance of SLA negotiation and SLA monitoring. Section 4 gives a brief idea about dynamic SLA negotiation. Section 5 describes the proposed SLA negotiation architecture and the algorithm for implementing negotiation. Section 6 describes the experimental studies and finally Sect. 7 concludes the paper.

2 Related Work

Many research works are going on in the field of cloud computing regarding SLA negotiation and SLA management. In [1] the authors described a framework for defining and monitoring SLAs in interdomain fields. An approach for SLA-driven management suitable for distributed systems has introduced in [2] by A. Keller et. al. using CIM (Common Information Model). In [3], the authors tried to evaluate multiple architectures which perform SLA auditing by considering qualitative and quantitative aspects. An SLA management framework using agent systems has described in [4] in which the service provider (responder agent) markets its capabilities in service level and the service consumer (initiator agent) gets the marketed information to initiate the overall negotiation. The works explained in [5–8] describe runtime SLA renegotiation for managing violations in SLA. In [5], SLOs are modified and renegotiated at runtime and online services are adjusted to dynamically agreed SLOs. An almost similar approach is explained in [8], which allows the changes in SLO values by keeping the existing SLA. In [6], the authors explained a renegotiation protocol that permits the service provider or service consumer to begin with a renegotiation at the time of need.

3 SLA Negotiation and SLA Monitoring

SLAs are established among the involved parties through a set of iterative negotiations as shown in Fig. 1. With the agreement both the involved parties establish a consent on corresponding roles, rights, and obligations. The user specifies his requirements and the orchestrator of the resource starts the negotiation by using local

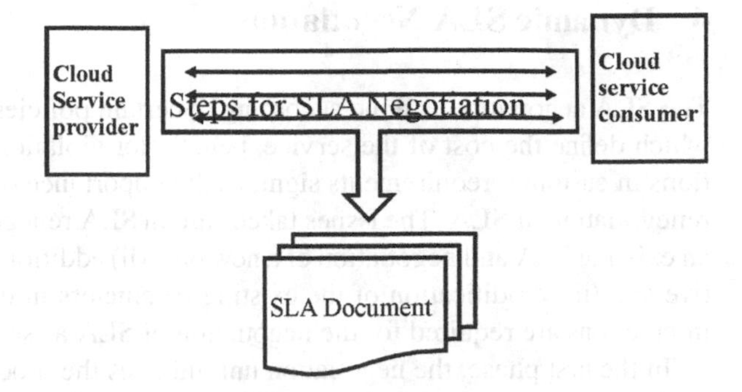

Fig. 1 Construction of SLA

resources and he tries to identify a common time slot where all the needed computing resources are available. This time slot identification means the orchestrator reserves the resources for assuring the agreement terms mentioned in the SLA.

The renegotiation of SLA is also possible in cases of lost, delayed, or reordered messages. The principle of renegotiation assumes an asymmetric nature of resource provisioning and consumption and gives more importance to the provider side. On initiation of renegotiation the agreement contract goes into a renegotiating state and after the completion of renegotiation, current state is viewed as superseded state as shown in Fig. 2.

SLA Monitoring is for determining whether the agreed parties are following the agreed terms and is done by checking the SLO values continuously. The different SLA monitoring schemes are online monitoring, proactive monitoring, and reactive monitoring. In online mode SLO values are monitored continuously and in proactive mode corrective actions are initiated before SLA violations. In proactive monitoring, the SLA negotiation is done immediately after service discovery for ensuring the availability of the cloud service. In reactive monitoring, the concept is different because on occurrence of SLA violation one of the involved parties complains to the monitor regarding violation. The reactive monitoring is better because the monitoring overhead is less and it gives an immediate response to SLA violations.

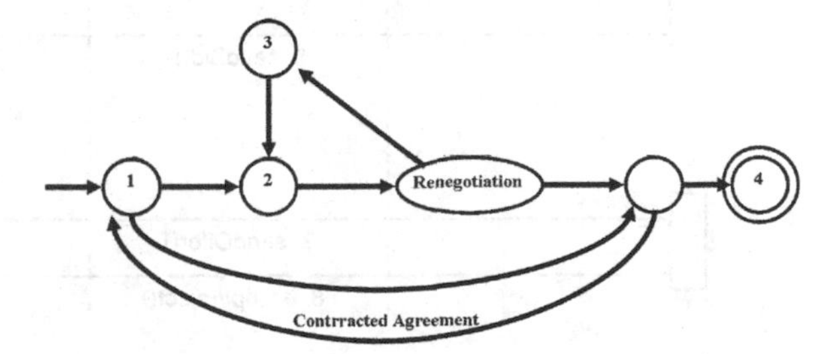

Fig. 2 Finite state machine for SLA negotiation

4 Dynamic SLA Negotiation

The SLA negotiation is done according to certain policies framed by the negotiator which define the cost of the service, penalty for violation, etc. The dynamic variations in customer requirements signifies the importance of dynamic negotiation and renegotiation of SLA. The issues taken care in SLA renegotiation are: (i) removal of an existing SLA and negotiation of a new one, (ii) addition/deletion of an SLA objective and (iii) modification of the existing parameters in the SLA. A set of message interactions are required for the negotiation of SLA as shown in Fig. 3.

In the first phase, the negotiation unit initiates the process by taking all available SLA templates from all providers. The initiator selects the most suitable template from this set as an initial point, which describes the background for all subsequent iterations. In the second phase, the initiator generates a new SLA template as per the selected template. The agreement initiator is free to modify the contents of the generated template and the new modified template is forwarded to agreement responder through a message for checking its validity. The service provider (sometimes the agreement provider) then checks whether the defined service can be provided or not and if possible the provider sends back the template to the client, to convey that the particular offer will be admitted. The provider responds with some counter offers if the services are not possible. The negotiation initiator checks the received counter offers in the third phase and stops the negotiation process if the requirements are not satisfying and repeats the whole steps from phase 1.

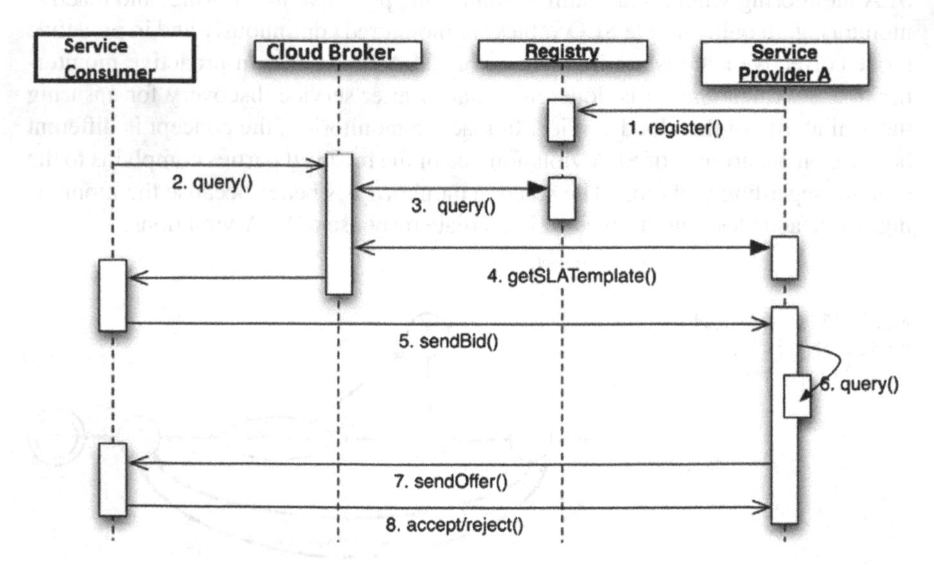

Fig. 3 Message interactions during SLA negotiation

5 Negotiation Architecture and Implementation

The SLA negotiation architecture (Fig. 4) contains the following main units such as: (i) service requisition unit, (ii) service listening unit, (iii) SLA negotiation unit and (iv) agreed SLA document. Service requisition unit is to identify the apt cloud service and service provider where the selection is based on QoS values, nature, and behavior of the service and service provider, provisioning interface, etc. Service listening unit checks external service registry periodically for the changes and updates of the service. SLA negotiation unit is responsible for finalizing the suitable interface for interacting with the service. The last unit, agreed SLA document, contains all details about the agreed upon SLA.

The process of SLA negotiation is done according to a set of rules defined using XML schema in the form of conditional statements. The format of a negotiation rule is:
if (condition) *then* **perform action** *else* **perform action**

Here the 'condition' deals with QoS values and the '**perform action**' decides the actions to be done on QoS. The operation of negotiation unit is controlled by this action part. Different rule actions are: (i) accept, (ii) reject, and (iii) set actions. Accept action is for accepting different SLO values in the SLA. Reject action takes care of rejecting values and set action takes care of proposing new values for QoS.

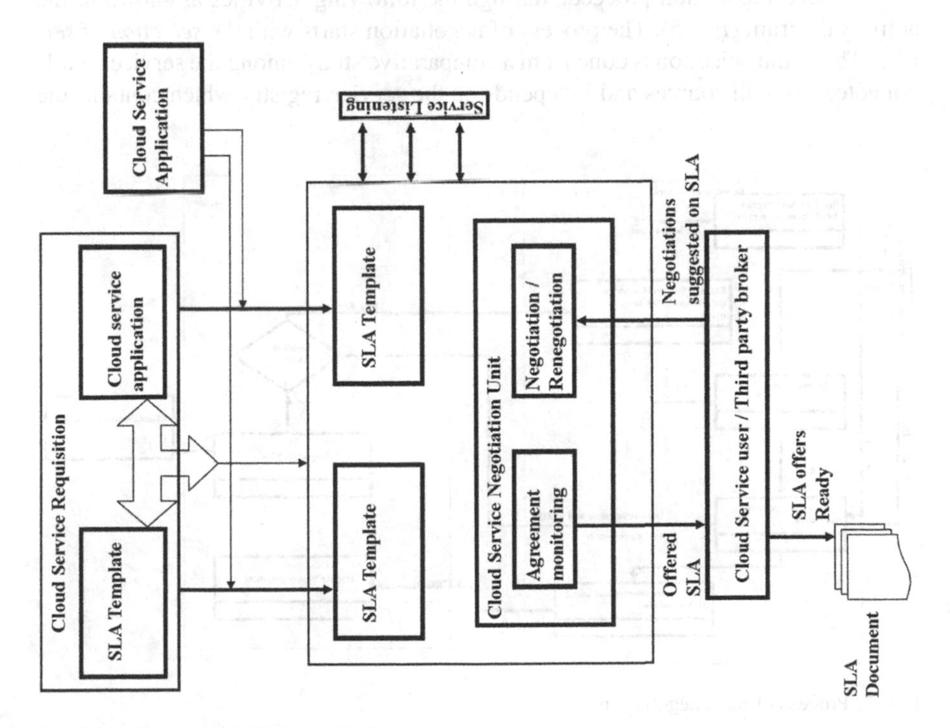

Fig. 4 Architecture of SLA negotiation

To establish an SLA negotiation, the system has to go through a sequence of steps. The proposed algorithm for SLA negotiation contains the following steps. The detailed algorithm is omitted from this paper because of space restrictions. In the first

Algorithm 1 SLA Negotiation

1. SLA negotiation at the user requirement level.
2. SLA negotiation at the service provider level.
3. SLA negotiation at the job execution level.
4. Confirmation of negotiated SLA.

step the user requirements in terms of service descriptions, particularly meta data of the service, has communicated to the cloud broker in the system. The meta data include: information about the computing service, estimated time for completing the service, estimated service cost, etc. The broker then checks for a cloud service provider, who can provide the service as per the user requirements. This is done in Step 2 and on completion of Step 2 the service broker come up with a particular service provider and in the third level, the actual SLA negotiation is executed. As part of the execution, both the involved parties institute a consent on the QoS parameters of the provided service and the agreement terms has finalized. This finalized SLA document has then communicated to the involved parties as a confirmation.

The SLA negotiation proceeds through the following activities as shown in the activity diagram (Fig. 5). The process of negotiation starts with the *selection of service*. The initial selection is done from a comparative study among the service details collected from all sources and it depends on the service registry which contains the

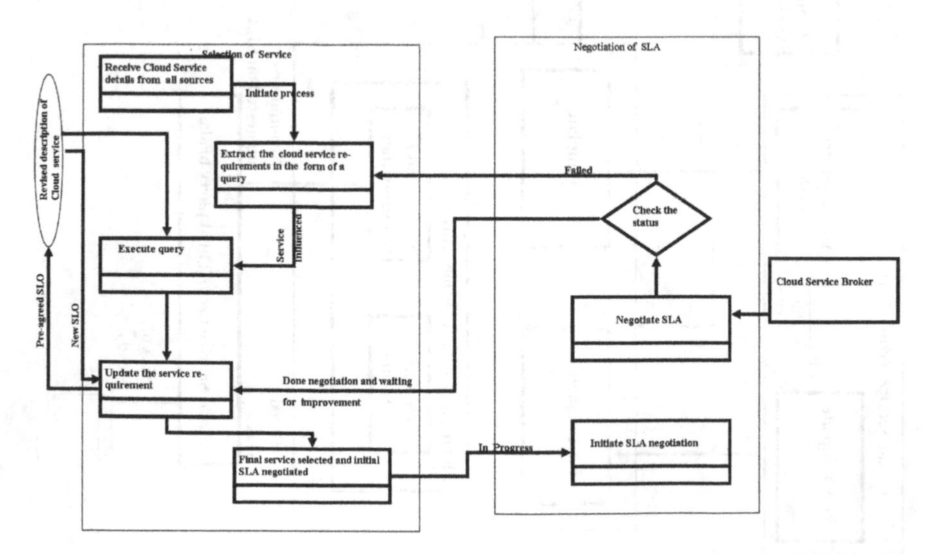

Fig. 5 Process of SLA negotiation

details about all existing services together with the description of new modified service. The QoS parameters of the selected service and its SLO values are negotiated in this phase to generate the most suitable SLA for both the parties. If the negotiation with the selected service is not going smoothly, then a next service is selected and same sequence of steps continued.

6 Experimental Evaluation

The SLA negotiation architecture has been developed and implemented in Java and is deployed as an application. The details of service registry are accessed using RMI and the service deployment has done through service applications, programmed for notifying queries about service discovery and SLA negotiation rules. The service registry details are stored in a database and all the parties involved in the negotiation communicate their objectives, requirements, context, preferences for negotiation and business rules before establishing negotiation. To evaluate the efficiency of the proposed framework, we have performed a sequence of experimental studies and measured the overhead of SLA negotiation.

6.1 Results and Discussion

We have considered a set of computing resources like RAM, number of CPU cores, memory, network bandwidth, processor availability, etc. kept as a stored database within the system. A service requesting environment where every service has its own executable code and a set of limitations on QoS measures is also assumed. For each service request, 10 resources were selected in a random fashion from the stored database, by assuming these resources as available resources for the selected request. The random process model we have used here generates 50 requests and chooses the resources in a random fashion from the stored database. SLA negotiation is iterated several times and the average values are taken for plotting the results. Figures 6 and 7 show the rate of sanctioning of the service requests with and without the newly devised negotiation method. From Figs. 6 and 7, it is clear that out of 15 and 25 arbitrarily generated requests, 22 and 34 % of the requests are serviced by the proposed method of SLA negotiation and the rate of sanctioning the resources (Fig. 8) is more when comparing with the resource allocation without negotiation. These results prove the efficiency of our proposed method and it is confirmed that the throughput of the scheduler gets improved with the usage of our method. This performance improvement shows that, with the use of our proposed negotiation method the available resources/services in the cloud environment can be allocated to the needy service consumers in an efficient way and it enhances the

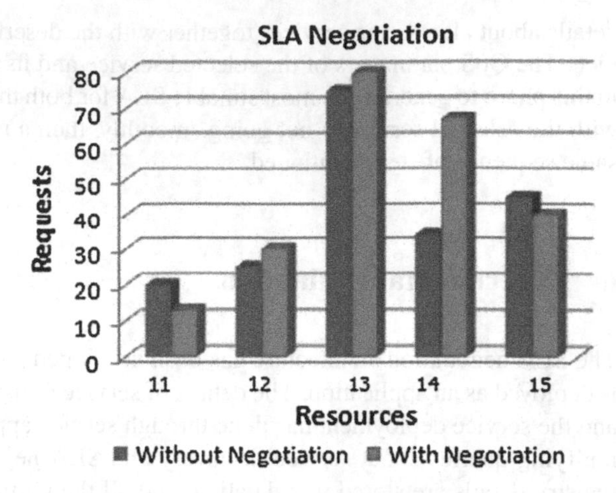

Fig. 6 Rate of serviced requests

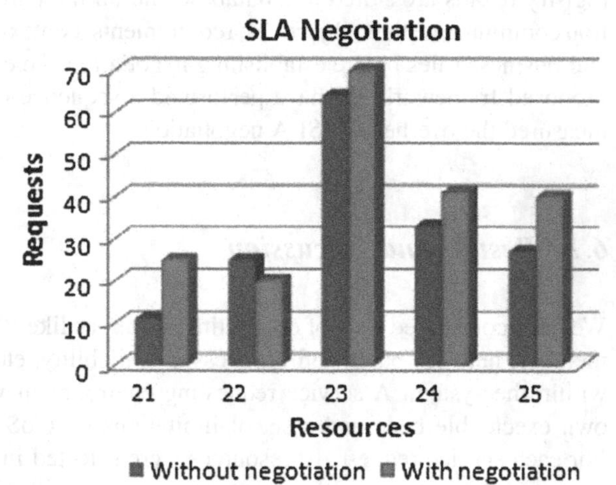

Fig. 7 Rate of serviced requests

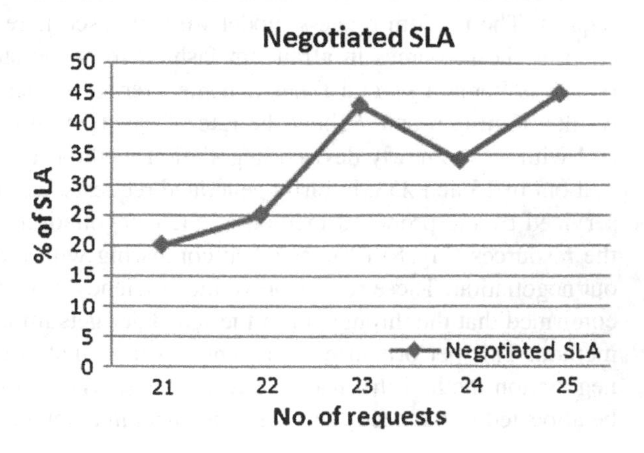

Fig. 8 Percentage of negotiated SLAs

utilization of resources which in turn increases the business volume of the cloud provider. Our newly proposed SLA negotiation method will increase the business revenue of the cloud provider and customer satisfaction of the cloud service user.

7 Conclusion

SLA negotiation has done over QoS values according to the alternative services determined by the service discovery mechanisms. The SLA negotiation ensures that a service could be utilized by an application and have a guaranteed set of provisions, according to the requirement, at the time of execution of the service. The negotiation approach introduced has analyzed with a valid set of experimental results which show that negotiation of SLA reduces the required time to do online service deployment. The performance of the designed framework might change dynamically according to the change in negotiation rules used by different designers. Our proposed SLA negotiation framework helps service providers and customers to finalize the agreements in a fast and efficient manner with decreased effort on both parties. The proposed method gives more efficiency to resource allocation and it improves the business revenue of cloud providers and satisfaction of cloud consumers.

References

1. A. Keller and H. Ludwig, "Defining and monitoring service-level agreements for dynamic e-business." in *Lisa*, vol. 2, 2002, pp. 189–204.
2. M. Debusmann and A. Keller, "Sla-driven management of distributed systems using the common information model," in *Integrated Network Management VIII*. Springer, 2003, pp. 563–576.
3. A. C. Barbosa, J. Sauvé, W. Cirne, and M. Carelli, "Evaluating architectures for independently auditing service level agreements," *Future Generation Computer Systems*, vol. 22, no. 7, pp. 721–731, 2006.
4. Q. He, J. Yan, R. Kowalczyk, H. Jin, and Y. Yang, "Lifetime service level agreement management with autonomous agents for services provision," *Information Sciences*, vol. 179, no. 15, pp. 2591–2605, 2009.
5. G. Di Modica, O. Tomarchio, and L. Vita, "A framework for the management of dynamic slas in composite service scenarios," in *Service-Oriented Computing-ICSOC 2007 Workshops*. Springer, 2009, pp. 139–150.
6. P. Hasselmeyer, B. Koller, M. Parkin, and P. Wieder, "An sla renegotiation protocol," in *Proceeding of the Second Non Functional Properties and Service Level Agreements in Service Oriented Computing Workshop*, 2008.
7. P. Wieder, J. Seidel, O. Wäldrich, W. Ziegler, and R. Yahyapour, "Using sla for resource management and scheduling-a survey," in *Grid Middleware and Services*. Springer, 2008, pp. 335–347.
8. R. B. Doorenbos, "Production matching for large learning systems," Ph.D. dissertation, University of Southern California, 1995.

Impact of Performance Management Process on Print Organizational Performance—In Indian Context

P. Iyswarya and S. Rajaram

Abstract Performance management is an important concept in Human Resource Management. It places a very big role in today's business world. In measuring organizational performance, performance of an individual takes place a very dominant role. This article conceptualizes and investigates the four dimensions of performance processes like performance planning, performance development, performance appraisal, reward, and recognition. These processes are linked to print organizational performance parameters like effectiveness, quality of news production, target achievements, and innovation on the part of news presentation. This article is an endeavor to provide the benefits and the drawbacks of performance management process and its relationship with print organizational performance. The results may be used as the stepping stone for further empirical research and it will help to formulate the strong and healthy organizational performance in print organization.

Keywords Impact · Performance management · Process · Print organization · Relationship

1 Introduction

In today's competitive world, each and every activity of a human being expected to gain profit for an individual or the benefit to the society as a whole. But in corporate scenario, these terms are totally different in standardized terms like employer and employee. To achieve the organizational goals as well as to increase the organizational performance, employer should have the continuous watch on the employee

P. Iyswarya (✉) · S. Rajaram
Department of Business Administration, Kalasalingam University, Anand Nagar, Krishnankoil, Virudhunagar 626126, Tamil Nadu, India
e-mail: saiaishwarya4@gmail.com

S. Rajaram
e-mail: pcsrajaram@yahoo.co.in

S.C. Satapathy et al. (eds.), *Proceedings of the International Conference on Data Engineering and Communication Technology*, Advances in Intelligent Systems and Computing 469, DOI 10.1007/978-981-10-1678-3_63

and should maintain the positive relationship with the employee. Performance management is a device to improve the performance of employees in all organizational levels. By using performance management as a base, this article attempts to the next level called performance management process.

2 Aim of the Study

1. To find out the relationship between performance management and print organizational performance.
2. To examine the benefits and the drawbacks of performance management process on print organizational performance.
3. To identify, which one of the performance process significantly takes a dominant position in print organizational performance.

3 Research Gap

Large number of studies has been based on the area of organizational performance. But none of the comprehensive study has been focused on the performance management process and print organizational performance. To overcome the gap of previous research, this article is an effort to identify the impact of performance management process on print organizational performance, it will systematically fulfill the gap of previous research and it gives a base for future research.

4 Relationship Between Performance Management and Print Organizational Performance

4.1 Print Organizational Performance

Organizational performance means an action taken by the employer and the employee to achieve the primary goals and objectives of an organization. In the book of Kirkman et al. [2], defined that organizational performance is the achievement of organizational goals in pursuits of business strategies that leads to sustainable competitive advantages. Print media coming under the subcategory of media. It includes magazines, newspapers, journals, books, and dailies. Performance of the print organization includes the work of publisher, editor, writer, reporter, photo journalist, designer, and marketing executives. All the activities and the performance of these people related to news production, presentation, and distribution to the end users, called print organizational performance.

4.2 Performance Management

Performance consists of both behavior and results. Behaviors are originated from the performer and it transforms the performance from the concepts to the action. Behavior is not an instrument for results but it is an instrument for physical and mental efforts applied to tasks and it can be judged apart from the results (In the book of Deb [3]). Performance management is a process of creating a knowledge about what is to be achieved and how is to be achieved. Deb [3] wrote that performance management embraces all the aspects of organization including organizational strategies, environmental responsiveness, business processes, innovation with employees and managers at the epicenter of the process.

4.3 Link between Performance Management and Print Organizational Performance

Print organization should have a relationship with performance management to achieve its organizational goals and objectives. Katou and Budhwar [5] Achieving the organizational goals depends only upon the extent to which the organizational performance is reached. Organization's sustainable competitive advantage is achieved with the source of highly committed workforce. A well designed and developed performance management system in organization can ensure the workers to do the actions related to organization's goals. Performance management helps the organization to achieve its organizational goals at the time of giving proper job analysis with effective recruitment policy, proper training and development programs, competitive pay structure, and maintaining good labor and employee relations. (In the book of Deb [3]).

5 Proposed Conceptual Framework

In the theoretical framework performance management processes (Independent variable) and print organizational performance parameters (Dependent variable) were analyzed by using the mediating variable called organizational performance. These two variables have been chosen to see the impact between these variables. The theoretical framework can also be seen from the Fig. 1.

6 Processes of Performance Management

The purpose of performance process is to transform the raw potential of human resource into performance by removing intermediate barriers as well as motivating and rejuvenating the human resource (Kandula [5]). Performance management is a

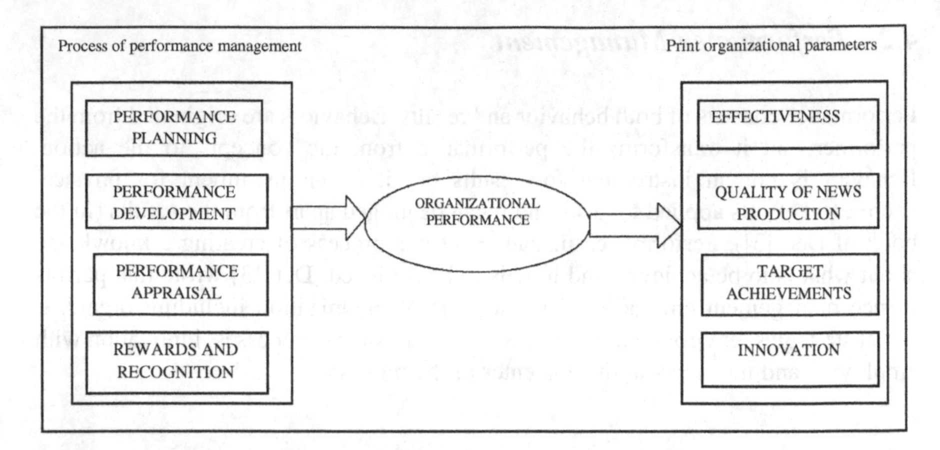

Fig. 1 Model to represent the impact of performance management process on print organizational performance

cyclical process in any organization. It is managed and maintained by the way of following performance processes.

6.1 *Performance Planning*

Performance planning is the first and foremost process in performance management. The aim of performance planning in print organization is to understand the meaning of organizational objectives, performance standards and competencies required to complete the task. In this stage, employer should describe about the job and expectations from the employee. In print organization, performance planning is very important for the workers to produce the news with quality and the news with expected standard. Armstrong [6] expressed that, performance planning is an agreement between manager and the individual to achieve organizational objectives, increasing production standards, improve performance, and develop the required competences for an organization. Performance planning in print organization is the basis for converting aims into actions.

6.2 *Performance Development*

Rao [7] explained that, development is a continuous process of developing, shaping, improving, as well as changing the skills, knowledge, creative ability, values, and commitment of employees based on the requirements of present and the future organizational commitments. In print organization employees should develop the skills, knowledge, qualifications and attitude. These kinds of developments are interconnected with the development of print organization and also the employee's development. Important thing to be considered in the performance development is, to increase the productivity of an employee by the prescribed standards and the norms of print organization. Productivity leads to the growth and expansion of an organization. Growth and expansion is the basic one for growing concern as well as going concern. By the growth of an employee, organizational growth is determined in print organization.

6.3 Performance Appraisal

Performance appraisal means, appraising the performance of an employee after the completion of prescribed job. In the book of Deb [3] defined that performance appraisal is the systematic evaluation of the individual with regard to his or her performance on the job and his potential for development. Performance appraisal in print organization based on the news presentation to the end users. While doing performance appraisal in print organization, organization should evaluate the work of editor, writer, reporter, photo journalists, designer, and efforts of marketing executives. Appraising the performance of an employee is based on the organizational process and procedures of performance appraisal. Performance appraisal is the formal measuring system of an employee. Employee's performance is measured after the completion of particular period especially 1 year. Employee's work completion is measured with the actual standard fixed by the print organization. Employees who complete the expected or fixed standard will be appreciated. But the employees those who are in below expected production will be under the review of appraising teams.

Rosemond and Asuinura [1] suggested the following steps for performance appraisal. Print organizations should follow these steps to validate the employee's performance. The first step in performance appraisal is scheduling: Scheduling means informing the employees regarding the review prior 10 days or 2 weeks. The instructions should be given to prepare for the annual performance appraisal. It includes the contents of job objectives and development goals of the employee for an individual and organization. The second step in performance appraisal is preparing for the review: In this step employees are asked to produce the documentation of work completed. It includes all the details of work patterns completed. The third step in performance appraisal is conducting the review: Employer reviews the performance of an employee. In this step, if employee's performance is below the expectations then employer gives the suggestion to improve the changes in performance.

6.4 Reward and Recognition

In print organization, there is a possibility of increasing the performance by proper appreciation to the employee. Proper rewards and recognition makes the employee to be more loyal toward the job, toward the employer, and toward the organizational goals. In print organization recognition by the way of intrinsic happiness or extrinsic happiness is very important for the employee to be loyal. In the book of Deb [3], rewards can be either intrinsic or extrinsic. Intrinsic happiness means making the employee to feel,“this job is highly suited for me and for my career.” Intrinsic happiness of employees automatically leads to job satisfaction. By achieving job satisfaction, employees surely moves to the next level, i.e., job involvement. If the work has been done with the involvement then organizational goals and the desired expectation from the employee’s job will be attained. Extrinsic happiness is very important for the employees to sustain in the current organization. Extrinsic happiness means providing reward in the form of monetary benefit and career advancement. Career advancement means, promotions from one job to another job or one job designation to higher job designation. Getting proper recognition by the way of monetary benefit makes the employees to realize the job requirements and expected standard of production from the job. Extrinsic happiness may be in the form of career advancement.

7 Positive Impact of Performance Management Process on Print Organizational Performance

7.1 Goal Setting and Development Related Plans

In print organization, under the performance planning, employer and the employee are in a position to set their goals to achieve the organizational objectives. Performance management process gives an idea for goal setting and development-oriented plans, because it provides the measurement processes to the organization to identify the efficient employees. Based on the efficiency level, employer can set the organizational goals. Development is very important for the individual and the organization. In print organization, employees get a chance to develop their innovative skills, technical knowledge for the news production and the news presentation.

7.2 *Effective Decisions Related to the Organizational Growth*

Employer of the print organization should take a valuable decision on the growth of an organization and benefit for the employee. Performance management process helps to identify the effective and the efficient employees. Based on the capacity of skilled employees, employer can determine objectives for an organization, production standard, and work quality. Performance management process gives an instruction as, production standards and objectives of an organization should be clear, concise, and achievable. It gives confidence among the employees to complete the job.

7.3 *Meaningful Measurements*

Under the criteria of measurements, all the efforts taken by the employees are considered. Right from the job roles to output given by the employee is considered in this criterion. Efforts taken by the employee is appreciated as well as absorbed. It gives a benefit as; employer can have the continuous watch on employee's activities. And the employees can correct their mistakes from the beginning itself. Performance management process makes the employees to be trained and well versed.

7.4 *Feedback and Coaching*

Managing committee of print organization may think that, all the initiatives and efforts taken by the management are correct. Performance management process helps to receive feedback from the employees regarding expectations from the management and gives feedback to the employees concerning the performance and suggestions to improve the performance to achieve the organizational targets. In print organization, receiving and giving feedback leads to smooth working conditions. Because in print organization, employees are running to capture the single second news as well as to achieve the work targets. By the review of performance management process, employer and employee can get a chance to interact with each other and low-performed employees can improve the performance by getting proper coaching and training from the management to do the work without errors and improve the performance.

7.5 *Documented History of Employee's Performance*

Performance management process provides the benefit as documented history of employee's performance. It gives the confidence among the employees who are in bottom level of operation. Documentation is very helpful at the time of promotion. From the management perspectives, employer can have a continuous watch on the employee's performance and the employer can understand the efforts of employees by documented evidence.

8 Negative Impact of Performance Management Process on Print Organizational and Employee's Performance

8.1 Discouragement and Lack of Credibility

Employees in print organization are under the control of continuous supervision. Performance and efforts of an employee are periodically reviewed and commenced by the immediate supervisors and the managers. But, employees in print organization expect freedom while news delivering and decision-making at the time of news production. Continuous performance management process gives a thought in the mind set of the employee as "whether I am producing the quality work to the society?" and "my performance acceptable and appreciable by the organization or not?" continuous performance management process gives discouragement and lack of credibility to an employee at the time of news production.

8.2 Implementation Failure

While using performance management process in print organization, all the activities of an employee are predefined and preplanned. Based on the performance planning, employees should work to achieve the desired organizational objectives. In print organization, the final outcome, i.e., newspaper includes the efforts given by publisher, editor, writer, reporter, photo journalists, designer, and efforts of marketing executives. If performance planning is not done in the proper manner then all the activities will get affected. Performance planning should be in the manner of acceptable as well as achievable. Otherwise implementation of performance management process leads to failure.

8.3 Bias

In print organization, all the production-related activities (i.e., from the news collection to news presentation) are carryover by the different designations of employees. Individual biases in print organization lead to the partial perspectives. By using performance management process, employees activities are assessed and reviewed by the management. At that time, bias places a dominant role. While assessing individual's performance, all the merits of an employee are only considered. Bias in print organization makes the employees to be discouraged and dissatisfied.

9 Discussions of the Research

From the study following statements have been identified

1. Performance management process has the direct and causal connection with print organizational performance. It has the impact on other core areas of HRM like talent management, employee services, organizational effectiveness, labor relations, and consulting.
2. At the same time, performance management process has an indirect influence on the operational performance of an organization.
3. Performance management process assists the gap between employee's performance and organizational expectations from the employees.
4. Rewards and recognition highly influence the print organizational performance. Based on the influence of rewards and recognition, all the processes of performance management moving in the correct direction.

10 Directions for the Future Study

This paper is a conceptual measure of evaluating the impact of performance management on print organizational performance. And this article depends only on the secondary data. Future studies may be conducted in the empirical view with larger sample size and particular sector wise longitudinal studies also be suggested.

11 Conclusion

To conclude that, this article shows the four processes of performance management and its impact on print organizational performance. It is mainly aimed to investigate the relationship between print organizational performance and performance management process. Findings of the article revealed that, performance management process has positive and meaningful impact on print organizational performance. And it highly influences the core areas of HRM like talent management, organizational effectiveness, employee services, and labor relations. This outlook suggests the employees to focus on organizational growth as well as individual growth. Three negative effects also highlighted by this article, i.e., discouragement, bias, and implementation failure. Employer and the employee should try to overcome the negatives proactively. As per the Indian print organizational culture all the activities of an organization are based on the human capital of that industry.

This article is a proof to reveal the power of human resources to the society. Overall there is a strong conclusion that, performance management process helps to increase the organizational performance. Considering the above-mentioned findings, to sustain in the current competitive world, employees of print organization should understand the values of performance management process.

12 Key Points

1. Organizational performance influences the employee to achieve organizational goals. So, the Print organization should focus on the organizational performance and performance of the employees.
2. Print organizational performance has positive relationship with performance management.
3. Performance management includes the processes of performance planning, performance development, performance appraisal finally rewards and recognition.
4. Performance management process gives the benefit as goal setting and development-related plans, effective decisions related to the organizational growth, meaningful measurements of employee's performance, individual's growth by following organizational growth, feedback and coaching, documented history of employee's performance.
5. Compared to other performance management process, rewards and recognition highly influence the employees to achieve job satisfaction and make the employees to be loyal to the organization.

References

1. Boohene Rosemond., Asuinura: The Effects of Human Resource Management Practices on Corporate Performance: A Study of Graphic Communications Group Limited. *International Business Research*, Vol. 4, no. 1, 1–7 (2011).
2. Bradley Lane Kirkman, Kevin B. Lowe, and Dianne P. Young: High performance Work Organizations: Definitions, Practices, and an annotated Bibliography. p. 7 (1999).
3. Deb Tapomoy. Performance and Reward Management. Ane Books Private Ltd. New Delhi. pp. 1–74, 76-80 (2009).
4. Kandula, Srinivas, R.Performance Management, New Delhi: Prentice Hall of India private limited. pp. 5–6 (2006).

5. Katou, Anastasia A and Budhwar, Pawan. S: The Effects of Human Resource Management Policies on Organizational Performance in Greek manufacturing firms. *Thunderbird International Business Review*, Vol. 49, no 1, pp. 1–35 (2007).
6. Michael Armstrong: A handbook of Human Resource Management practice. Kogan page India. (2006, 10th Ed. 495-519 and 2009, 11th Ed. pp. 507, 617-643).
7. V S P Rao Human Resource Management–Text and cases second edition. Excel books, New Delhi. pp-30. (2010).

Mobility Aware Path Discovery for Efficient Routing in Wireless Multimedia Sensor Network

Rachana Borawake-Satao and Rajesh Prasad

Abstract This paper proposes effective solution for routing information in wireless multimedia sensor network using multipath and multi-objective routing scheme. The ubiquitous nature of the future Internet demands multi-objective routing for serving the dynamic applications and new technologies. Multimedia data and scalar data should be treated differently while routing through WMSN. This separation of data requires multiple paths for multiple objectives. Objectives can be the speed of communication, the energy efficiency of the network, the lifetime of the network, reliability of communication, or load balancing in the network. This paper discusses the advantages of multipath routing and proposed an effective solution for finding multiple paths depending upon the demand of quality of service from the network. The path discovery methodology is evaluated using a mathematical model and the results are compared for the mobility of the network which is a demand of ubiquitous future Internet.

Keywords WSN: Wireless Sensor Network • WMSN: Wireless Multimedia Sensor Network • Ubiquitous future Internet

1 Introduction

Dynamic applications of future Internet are promoting use of wireless multimedia sensor network due to availability of high-quality multimedia services. Since low-cost multimedia devices are easily available the use of these devices is in

Rachana Borawake-Satao (✉)
Smt Kashibai Navale COE, (Savitribai Phule Pune University),
Vadgaon Bk, Pune 411041, Maharashtra, India
e-mail: rachana.borawake1@gmail.com

Rajesh Prasad
NBN Sinhgad School of Engineering, (Savitribai Phule Pune University),
Ambegaon Bk, Pune, Maharashtra, India
e-mail: rajesh.prasad@sinhgad.edu

S.C. Satapathy et al. (eds.), *Proceedings of the International Conference on Data Engineering and Communication Technology*, Advances in Intelligent Systems and Computing 469, DOI 10.1007/978-981-10-1678-3_64

demand. Application-specific QoS requirement, high bandwidth demand, multimedia source coding technique, power consumption, and multimedia in network processing are the factors which influence the design of routing algorithm [1]. Various algorithms are proposed and implemented for multipath routing in multimedia sensor network [2].

Wireless multimedia sensor network enhancing the capability of the wireless sensor network for modern applications such as smart home, smart city, advanced healthcare systems, and multimedia surveillance sensor networks [1].

1.1 Multipath Routing

Multipath routing provides better solution in terms of reliability, load balancing, high aggregate bandwidth, end-to-end delay, minimum energy consumption, and high throughput [2]. Performance evaluation of the various routing algorithms can be analyzed using parameters like routing load, average end-to-end delay, jitter, energy balancing, and average energy consumption. In [3] author compared multipath routing techniques based on energy efficiency, delay, fault tolerance, and data accuracy.

In [4] author proposes context aware routing which combines cluster formation algorithm with routing. During cluster formation algorithm the information required for routing is also preserved and later used for routing purpose. This definitely improves the lifetime of the network as remaining energy is considered for routing. This also resolves the energy hole problem in the network.

MEVI [5] is a multi-hop hierarchical routing protocol for efficient video communication (MEVI). This algorithm proposes cross-layer solution for the selection of the routes. Algorithm implemented two modes for video retrieval and transmission where it is event-based video transmission. The main addition of the paper is the cluster formation by sending a single beacon message, multi-hop communication between CHs and base station, and cross-layer scheme to acquire the network conditions for selecting the routes.

Multipath routing reduces delay in the network by processing delay estimation and finding alternate path if some paths or nodes are exhausted in the network. In wireless sensor network effective performance improvement is achieved through multipath routing [6, 7]. Similar approach can be applied in WMSN for dynamic protocols. Multimedia traffic can be classified into set of classes, and priorities can be assigned to each one for packet classification. Non-preemptive packet scheduling scheme gives improved results for routing multimedia data [8]. The DCM (Dynamic capacity multipath routing) algorithm uses anchor nodes for deciding duty cycle scheduling of the node in vicinity of the target and improves lifetime of the network [9].

Quality of service (QoS) is a key issue in WMSN, if packet classification will be based on QoS issues like delay, residual energy, and loss rate, targeting a particular application which will be effective. Cluster-based architecture is more suitable for

QoS-based routing [10]. Interference awareness, bandwidth awareness, congestion control schemes, and priority scheduling are some of the key aspects for design of routing protocols for application in future Internet [11–15].

The major communication challenges for QoS aware routing are energy consumption, application-specific requirements, resource constrains, variable link capacity and packet errors, dynamic network connectivity, and topologies [16]. Future Internet demands smart services in ubiquitous computing environment which introduces mobility awareness in basic architecture of wireless multimedia sensor network. In [17] author proposes mobile multimedia geographic routing (MGR) for QoS provisioning in MMSNs (Mobile multimedia sensor network). Mobility of nodes in the network and mobility of sink node in various applications are the challenges for researchers in MMSN.

2 Path Discovery Methodology

The important issue in communication is effective data dissemination and gathering. Various protocols are available for WSN and WMSN for routing the data efficiently. If we compare WSN with WMSN the multimedia data is a critical issue to address. The audio and video data transmissions require high data transmission rate and good quality of service. Hence, it is required that the design of routing protocol in WMSN must have effective methodology to handle multimedia data.

The path discovery process is achieved through multipath routing. There are many goals of multipath routing protocols to achieve such as reliability, load balancing, high aggregate bandwidth, minimum end-to-end delay, minimum energy consumption, and high throughput.

The proposed system uses multiple paths for routing. The captured data is divided into parts and forward through multiple paths. This will increase speed of data transmission as well as priorities can be assigned to the respective paths depending upon the parameter.

Following parameters are used for path discovery process:

(a) Link quality index (LQI)
(b) Remaining energy (Er)
(c) Hop count (HC)
(d) Speed of mobility (Mf)
(e) Location of node
(f) Movement direction of node

As shown in Fig. 1 the path discovery process starts after deployment of sensor network. After deployment the initiation process takes place which includes location awareness about the neighboring node, sink node, and the node itself. Once all locations are known the next hop is checked for whether it is a sink node or not. If next node is a sink node, the algorithm stops; otherwise path discovery takes place.

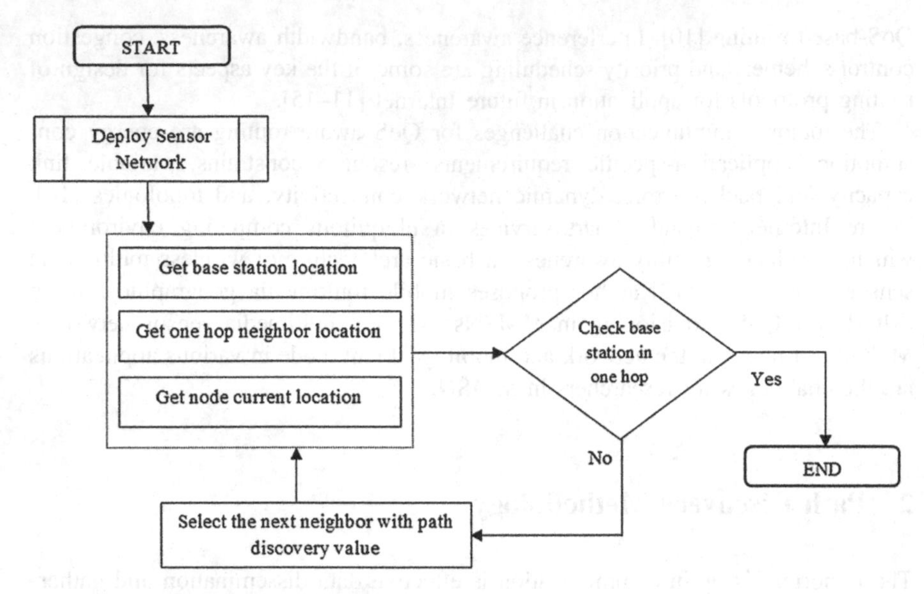

Fig. 1 Path discovery methodology

The path discovery process includes analysis of different parameters like remaining energy, hop count, LQI, and mobility value of neighboring node. Depending on these values the weight for multiple paths is calculated. One of the best solutions is used or multiple solutions are used for efficient routing.

The path discovery process makes use of formulae (I) in Sect. 3 for calculation of the $Node_{val}$. Various methods can be adapted to calculate the $Node_{val}$, but for this paper we are considering mobility as important factor as it is useful for ubiquitous computing in future network.

3 Mathematical Model for Performance Evaluation

As given in Eq. (1) the $Node_{val}$ is calculated using three terms: remaining energy (*Er*), current link quality (*LQ*), and current hop count (*HC*). The $Node_{val}$ ranges between 0 and 1. The multiplication factors α, β, and γ are used to assign priorities to the parameters respectively.

The equation *Er/Ei* gives value of remaining energy between 0 and 1 [5]. If application gives priority to energy saving, the value of the α will be more than β and γ. Possible values for α, β, and γ are {0.2, 0.3, 0.5}.

Similarly, the equation *curLQ/maxLQ* gives values for link quality ranges between 0 and 1 and equation *(totHC-curHC)/totHC* gives value for remaining traveling distance. As per the requirement of the application, three different paths

Table 1 Parameter description

Ei	Initial energy of node
Er	Residual energy of node
curLQ	Current link quality of node
maxLQ	Maximum link quality of node
totHC	Total hope count
curHC	Current hope count of node
M_f	Mobility factor
PT_r	Remaining pause time duration in milliseconds
totPT	Total pause time duration in milliseconds
$Node_{val}$	Node value for data transmission
PD_{val}	Path discovery value with mobility added to $Node_{val}$
Node with highest PD_{val} has better conditions to transmit packet	

are estimated and packet allocation is done according to the priorities assigned by the variables α, β, and γ:

$$Node_{val} = [(\alpha{*}Er/Ei) + (\beta{*}curLQ/maxLQ) + (\gamma{*}(totHC - curHC)/totHC))] \quad (1)$$

where $\alpha + \beta + \gamma = 1$.

As mentioned in Sect. 1 the mobility is a critical issue in design of routing path in case majority nodes in the network are mobile node. Here we introduce mobility factor (M_f) as ratio of remaining pause time (*PTr*) and total pause time of the node (*totPT*). This ratio gives prediction regarding mobility of the node. If the node is having possibility of changing its position from the current place within very short time, then possibility of the selection of that node is reduced using mobility factor (M_f) (Table 1).

$$\text{Mobility factor } (M_f) = (PTr/totPT) \quad (2)$$

$$\text{Path Discovery value } (PD_{val}) = Node_{val} * M_f \quad (3)$$

3.1 Experimental Setup

Multiply mobility factor (M_f) with $Node_{val}$ to get an exact value of every node with respect to the mobility prediction of the current node.

Figure 2 shows the experimental setup with total four nodes where node 1 is willing to transmit data, and various values related to all nodes in the scenario are describe in Table 2. Table 2 also describes the scenario if priority is assigned to a certain parameter which node is selected for transmission.

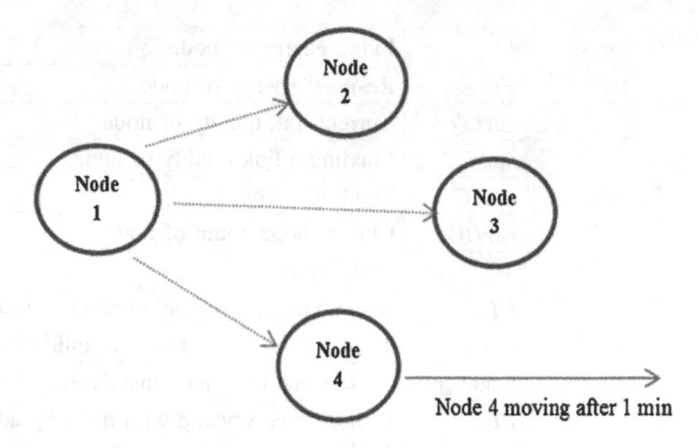

Fig. 2 Experimental setup for node mobility

3.1.1 Case 1: No Mobility Considered

As shown in Fig. 2 consider case 1 where the node 1 trying to select one node for transmission of data (either node 2 or node 3 or node 4). The decision depends on the values of remaining energy (*Er*), link quality (*LQ*), and hope count (*HC*) as shown in Table 2.

If energy saving is a priority, then value for α in Eq. 1 is selected as highest value and the node 4 is selected for transmission. Naturally, the node with highest remaining energy is selected which in this case is node 4.

If priority is high data accuracy (less packet loss) then Link Quality is important then value for β in Eq. 1 is selected as highest value and the result will be selection of node 3 for next data forward.

Similarly, if minimum delay is the requirement then value for γ in Eq. 1 is selected as highest value and result will be the selection of node 2 for next data forward.

Table 2 Values for no mobility in the network

Node Id	Remaining energy (*Er/Ei*)	Link quality (*LQ/maxLQ*)	(*maxHC* – *HC*)/*totHC* (distance to travel)
2	0.5	0.59	0.9
3	0.75	0.63	0.7
4	0.9	0.55	0.6
After applying Eq. 1			
If α is greater (priority = energy)		Node 4 is selected	
If β is greater (priority = LQI)		Node 3 is selected	
If γ is greater (priority = HC)		Node 2 is selected	

Table 3 Values for mobility in the network

Node Id	Remaining energy (Er/Ei)	Link quality (LQ/Mac LQ)	(Max_HC – HC)/tot_HC (distance to travel)	Mobility factor (PTr/tot_PT)
2	0.5	0.59	0.9	0.50 (moving after 5 min)
3	0.75	0.63	0.7	0.50 (moving after 5 min)
4	0.85	0.55	0.6	0.10 (moving after 1 min)
After applying Eq. 3				
If α is greater (priority = energy)			Node 3 is selected	
If β is greater (priority = LQI)			Node 3 is selected	
If γ is greater (priority = HC)			Node 2 is selected	

3.1.2 Case 2: Mobility Considered

As shown in Fig. 2, consider case 2 where the node 1 trying to select one node for transmission of data (either node 2 or node 3 or node 4). The decision depends on the values of remaining energy (*Er*), link quality (*LQ*), hope count (*HC*), and mobility factor (M_f) as shown in Table 3.

If energy saving is a priority then value for α in Eq. 1 is selected as highest value and the node 4 is selected for transmission. But if we are using Eq. 3 then as per the mobility factor of node 4, it will not be selected; instead node 3 will be selected as node 4 is moving early as compared to node 4. Hence, connection loss and data reliability will be increased.

Similarly, while giving priority to link quality (LQ) and hop count (HC) the mobility factor is considered and accordingly decision takes place and increases reliability of data transmission.

4 Conclusion

This paper has addressed issue of multipath and multi-objective routing in wireless multimedia sensor network (WMSN). In Sect. 1 we discussed various multipath routing schemes. Section 2 describes our proposal for path discovery mechanism which not only takes care of multiple objectives like remaining energy, link quality, and hop count for routing but also considers mobility factor of the node. Path discovery value makes this algorithm compatible for future Internet applications. Section 3 supports the algorithm by implementation of mathematical model. This section also describes in detail about the functionality of model in two different cases. We proposed effective mechanism for routing using path discovery algorithm.

Future studies in this research can be enhancement of the path discovery algorithm using packet classification and dynamic hole healing mechanism for expanding the application areas for the algorithm. Various duties scheduling mechanism can be applied to path discovery to improve the lifetime of the network.

References

1. Ian F. Akyildiz, Tommaso Melodia, Kaushik R. Chowdhury.: A survey on wireless multimedia sensor networks. Computer networks 51 (2007) 921–960. Journal, Elsevier publication.
2. Jayashree A, G. S. Biradar, V. D. Mytri.: Review of Multipath Routing Protocols in Wireless Multimedia Sensor Network –A Survey. International Journal of Scientific & Engineering Research Volume 3, Issue 9, September-2012, ISSN 2229-5518.
3. Prashant Chaudhari, HareshRathod, B. V. Budhhadev.: Comparative Study of Multipath-Based Routing Techniques for Wireless Sensor Network. Proceedings published by International Journal of Computer Applications® (IJCA) International Conference on Computer Communication and Networks CSI- COMNET-2011.
4. Longpeng Zheng, ZhipingJia, Ruihua Zhang, Hui Xia, Lei Ju, ChuanhaoQu.: Context-Aware Routing Algorithm for WSNs Based on Unequal Clustering. 12th IEEE International Conference on Trust, Security and Privacy in Computing and Communications 2013.
5. Denis do Ros´ario, Rodrigo Costa, HelderParaense, K´assio Machado, Eduardo Cerqueira and TorstenBrauny.: A smart multi-hop hierarchical routing protocol for efficient video communication over wireless multimedia sensor networks. 2nd IEEE international workshop on smart communication protocols and algorithms, 978-1-4577-2053-6/12 ©2012 IEEE.
6. S.Pratheema, K.G.Srinivasagan, J.Naskath.: Minimizing End-to-End Delay using Multipath Routing in Wireless Sensor Networks. International Journal of Computer Applications (0975 –8887) Volume 21– No.5, May 2011.
7. YacineChallala, AbdelraoufOuadjaoutb, NoureddineLaslab, Mouloud.: Secure and efficient disjoint multipath construction for fault tolerant routing in wireless sensor networks. Journal of Network and Computer Applications 34 (2011).
8. Qian Ye, Meng Wu, Yufei Wang.: Traffic Scheduling Scheme for Disjoint Multipath Routing Based Wireless Multimedia Sensor Networks. 2010 IEEE Asia-Pacific Services Computing Conference.
9. Nan Song, Xinyu Jin, Yu Zhang.: A Multi-path routing protocol for target tracking in WMSNs. 978-1-4244-3709-2/10/$25.00 ©2010 IEEE.
10. Jayashree Agrakhed, G. S. Biradar, V. D. Mytri.: Adaptive Multi Constraint Multipath Routing protocol in Wireless Multimedia Sensor Network. 2012 International Conference on Computing Sciences.
11. Ilnaz Nikseresht, Hamed Yousefi, Ali Movaghar, and Mohammad Khansari.: Interference-Aware Multipath Routing for Video Delivery in Wireless Multimedia Sensor Networks. 32nd International Conference on Distributed Computing Systems Workshops 2012.
12. Hongli Xu, Liusheng Huang.: Bandwidth-Power Aware Cooperative Multipath Routing for Wireless Multimedia Sensor Networks. IEEE Transactions on Wireless Communications, Vol. 11, No. 4, April 2012.
13. Guannan Sun, Jiandong Qi, Zhe Zang, Qiuhong Xu.: A Reliable Multipath Routing algorithm with related congestion control scheme in Wireless Multimedia Sensor Networks. 978-1-61284-840-2/11/$26.00 ©2011 IEEE.

14. Elham Karimi, Behzad Akbari.: Improving Video Delivery over Wireless Multimedia Sensor Networks Based on Queue Priority Scheduling. 978-1-4244-6252-0/11/$26.00 ©2011 IEEE.
15. Moufida Maimour, C. Pham, Julien Amelot.: Load Repartition for Congestion Control in Multimedia Wireless Sensor Networks with Multipath Routing. 978-1-4244-1653-0/08/$25.00 ©2008 IEEE.
16. Muhammet Macit a, V. Cagri Gungor a,b, Gurkan Tuna.: Comparison of QoS-aware single-path vs. multi-path routing protocols for image transmission in wireless multimedia sensor networks. http://dx.doi.org/10.1016/j.adhoc.2014.02.008, 1570-8705/, Ad Hoc Networks 19 (2014) 132–141 2014 Elsevier.
17. Min Chen, Chin-Feng Lai and Honggang Wang.: Mobile multimedia sensor networks: architecture and routing. Chen et al. EURASIP Journal on Wireless Communications and Networking 2011, 2011:159 http://jwcn.eurasipjournals.com/content/2011/1/159.

Emerging Internet of Things in Revolutionizing Healthcare

Poonam Bhagade, Shailaja Kanawade and Mangesh Nikose

Abstract In the era of science, technology plays the crucial role in healthcare for sensing devices, moreover in communication, storing, processing, and display devices. The advances in various streams such as sensing techniques, nanotechnologies, embedded systems, wireless communication networks, and miniaturization resulting to evolve intelligent systems to monitor and control various medical parameters in post-operational days. Hence, the IoT (Internet of Things) is emerged as a recent trend in healthcare communication systems. IoT serves as a catalyst for the healthcare and plays a major role in numerous healthcare applications. Wearable sensors recognize anomalous and unforeseen conditions by examining physiological parameters along with the symptoms and transfers the vital signs for medical evaluation. Hence, prompt provisional medication can be done immediately to avoid severe conditions. In the proposed system, a microcontroller is used as a gateway to communicate to the several sensors depending on parameters to be monitored such as pulse rate counter, temperature sensor, accelerometer, etc. The microcontroller acquires the sensor data, processes it, and transmits to the network through appropriate data transmission protocol, successively providing a real-time monitoring of the healthcare parameters for healthcare professionals. The doctor can access the data anytime by simply logging to the HTML (HyperText Markup Language) webpage or typing the corresponding unique IP (Internet protocol) address in Internet browser.

Keywords Internet of things · Microcontroller · Gateway · Wi-Fi module · Temperature sensor · Pulse rate counter · Accelerometer

---

Poonam Bhagade (✉) · Shailaja Kanawade · Mangesh Nikose
Electronics and Telecommunication Sandip Institute of Technology and Research Center, Nashik, India
e-mail: poonambhagade@gmail.com

Shailaja Kanawade
e-mail: kanawade.shailaja@sitrc.org

Mangesh Nikose
e-mail: mangesh.nikose@sitrc.org

S.C. Satapathy et al. (eds.), *Proceedings of the International Conference on Data Engineering and Communication Technology*, Advances in Intelligent Systems and Computing 469, DOI 10.1007/978-981-10-1678-3_65

1 Introduction

The Internet is the combinational result of a transformation of telecommunications and information made available by sustained, augmented development in the performance, moreover in cost of electronics devices. Considering, the Internet becomes very essential to each aspect of life like education, business, industries, finance, entertainment, social networking, e-commerce, shopping, etc. Thus, the recent new mega trend in an era of Internet is IoT [1]. Visualize a world where several objects or things can sense, analyze, communicate, and exchange data over a private Internet protocol (IP) or public networks. The interconnected objects acquire the information or data at specified intervals, analyze, and respond necessarily, providing a smart network for the purpose of analysis, planning, decision making, and responding. This summarizes the world of **Internet of Things (IoT)** [2]. The whole concept of IoT is based on sensors, gateway, and wireless networks, which permits user to communicate and access the information or application. Being that IoT offers more prominent assurance in the field of health awareness [3, 4].

Researches are still in progress to focus at upgrading quality of human life concerning to health by designing, developing, and fabricating invasive and non-invasive sensors [5]. Global population and noteworthy rise in aging population are the reasons behind the rapid progress in this field. As per the statistic given by the U.S. Department of Health, more than 20 % of world's population will go above 65 years of age by 2050 [6]. This results in an increasing demand for medical care. But this medical care is very expensive for durable surveillance furthermore time consuming for consultations with health professionals or doctors [7]. The cost of hospitalization is high and it increases in case of recovery after a serious illness or surgery. Nowadays, hospitals are looking for moving patients back to home imminently as possible after surgery to recoup. During this recovery time, numerous physiological parameters are needed to be continuously analyzed. Due to this, telemedicine and distant monitoring of patients from home are acquired added significance and necessity.

This paper is structured as follows. Section 2 defines problem statement. Section 3 offers a brief review of previous work done related to healthcare systems and sensor networks. In Sect. 4 visualize the complete system overview. It also gives an idea about the sensors that are used in this system. The conclusion is stated in Sect. 5 and paper ends with a brief discussion on future developments.

2 Problem Statement

Nowadays, the patient who stays in home during post-operational period, get checked by overseer or medical caretaker. But it failed to accomplish ceaseless analysis, on the grounds that anything can be changed in health parameter within seconds and during that period if guardian or caretaker is not in the premises cause harm or sometimes become a serious issue with patient's life and death. So, in this developing era, web administers of the entire world give a thought to provide keen health awareness framework in which time-to-time ceaseless checking of the patient is accomplished.

3 Related Work

3.1 Health Monitoring System

ZigBee-based transmission is a reliable IEEE protocol for wireless patient monitoring. It uses a ZigBee module for fall monitoring, which combines indoor positioning, fall recognition, ECG observation, etc. The time when the fall is recognized by a triaxial accelerometer of a device, the contemporary location of patient is transferred into the emergency center using a ZigBee network [6, 8].

In microcontroller-based health monitoring system which uses sensor network, when blood pressure reading, heart rate, or skin temperature goes out of the standard limit for a patient, a system is designed to indicate with the help of an alarming circuit.

In development of noninvasive continual blood pressure computing and monitoring system, it computes blood pressure using a volume oscillometric methodology and photoplethysmography (PPG) technique continuously. There is linear relationship between rate of change of blood volume in a body organ, e.g., finger, and blood pressure. The optical sensor network is not only utilized to examine the rate of change of blood volume but also for estimating the blood pressure.

In PPG-based methodologies for noninvasive and unceasing blood pressure measurement, the design and implementation issues in the body sensor networks, the PPG signal could be obtained easily from an optical sensor attached on epidermis [8]. It is used alone or combined along with the ECG signal to compute the blood pressure. Depending on this, sensor-based systems and new instruments could be developed as well as integrated along with computer-based health monitoring systems which aims to support continuous and distant surveillance of assisted livings.

3.2 Sensor Networks

Web services are broadly and victoriously used mechanisms in the field of Information technology (IT). They can be stated as process to establish practical and distributed applications imposing on web standards such as HTTP (Hypertext Transfer Protocol). Sensor networks can extremely benefit from their utilization, since web services permit an emergence of WSNs into any of the systems which is escalated on standard IT components like industry or home automation as well as in healthcare monitoring systems.

In the IoT-enabled distant healthcare monitoring systems, the data acquired from the several sensors should be accessible anytime and anywhere without any restrictions, which needs constant network connectivity. If the distant healthcare monitoring application transfers the data continuously, then the amount of data generated at database or central station will be huge. This contributes to a hyper-connectivity framework in which network will get connected. In remote healthcare monitoring application we are not able to make the use of total available bandwidth effectively; in case we use the traditional mode for transmission of data. This mode leads to loss of sufficient amount of data due to delay as well as buffer overloading, but it is not acceptable specifically in the healthcare applications [9].

The platform and architecture of sensor networks in healthcare system perform a prominent role for unceasing monitoring of physiological parameters primarily for chronic patients. The appropriate network has to be selected relying on the performance, cost, ease of access, ease of configuration, requirement of additional sensor nodes, energy consumption and range, etc. A comparison of IEEE protocols that are presently available is shown in Table 1 [10].

ZigBee is used to create PAN (Personal area networks) from tiny, low-power digital radios. It follows IEEE 802.15 standard. It allows short-range wireless data transfer at comparatively low rates. It allows data to reach over a longer distance by passing it through intermittent devices or nodes to go more distant ones. Due to lower data rate, i.e., 250 Kbps, it reduces power consumption. As a result, it provides longer battery life and secured networking applications. There are few constraints on ZigBee as it may be unable to carry vital signs especially emergency messages which may be critical and important for diagnosing the disease or clue to directing the required therapy [5, 8].

Bluetooth has packet-based protocol containing master–slave structure. The bluetooth devices communicate with one another on a securely using an unlicensed short-scale radio frequency. Similar to ZigBee, bluetooth also may not be suitable

Table 1 Comparison of different protocols

Standard	IEEE protocols			
	ZIGBEE (IEEE 802.15.4)	BLUETOOTH (IEEE 802.15.1)	WI-FI (IEEE 802.11)	WIMAX (IEEE 802.11)
Range	100 m	10 m	5 km	15 km
Data rate	250–500 kbps	1–3 Mbps	1–450 Mbps	75 Mbps
Bandwidth	2.4 GHZ	2.4 GHZ	2.4, 3.7, 5 GHz	2.3, 2.5, 3.5 GHz
Network topology	Star, Mesh, Cluster tress	Star	Star, Tree, P2P	Star, Tree, P2P
Application	Wireless Sensors (monitoring and control)	Wireless Sensors (monitoring and control)	PC-based data acquisition, mobile internet	Mobile internet

for sending vital signs, primarily for critical situations and emergency messages, as these messages are exceptionally important for diagnosing [5, 8].

Wi-Fi stands for wireless fidelity which is standardized by IEEE. It is used to transmit the data wirelessly. Wi-Fi has advantage of increased communication range so it can be used to transmit data up to the 5 km distance.

While WIMAX stands for worldwide interoperability for microwave access. It is a wireless MAN technology offering interoperable broadband connectivity to users. It is used to extend the transmission up to 15 km.

4 Proposed System

The basic architecture of human health monitoring system is as shown in Fig. 1 [10].

Sensors are selected based on the parameter to be monitored. Many sensors can be used to measure the physiological parameters such as heart rate, body temperature, accelerometer, etc. The processor collects the raw data from sensors, processes the data, and then displays on a display. A trans-receiver can be used to send the data to a central station, if the device having feature of wireless data transmission is used. The data may be completely processed or not at the sensing stage but it is stored in the memory and display either in numerical or graphical format.

The proposed system is shown in Fig. 2.

The proposed system comprises a temperature sensor to compute the body temperature, pulse rate counter to count the pulses, and accelerometers which recognize any fall that may occur. The microcontroller will gather the data from sensors to process and analyze. The processed data is then transmitted over internet through Wi-Fi protocol. Wi-Fi is used as a medium to connect with internet. The server is developed to store this data and update the data with respect to time. The website is developed to display the current data. Thus the doctor can access the data anytime and anywhere simply by accessing the corresponding website or typing the corresponding unique IP address in Internet browser.

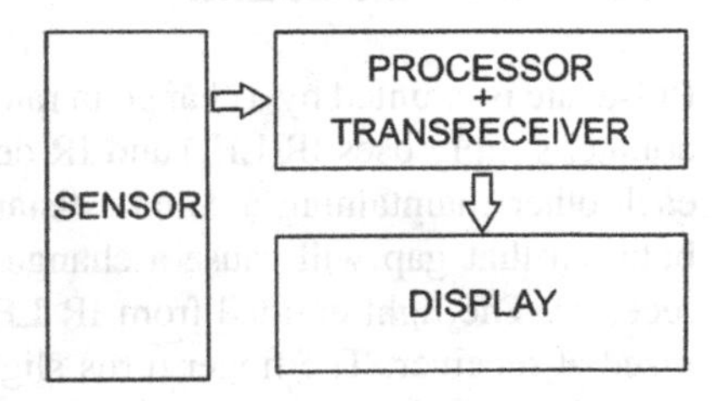

Fig. 1 The basic architecture of simple health monitoring system

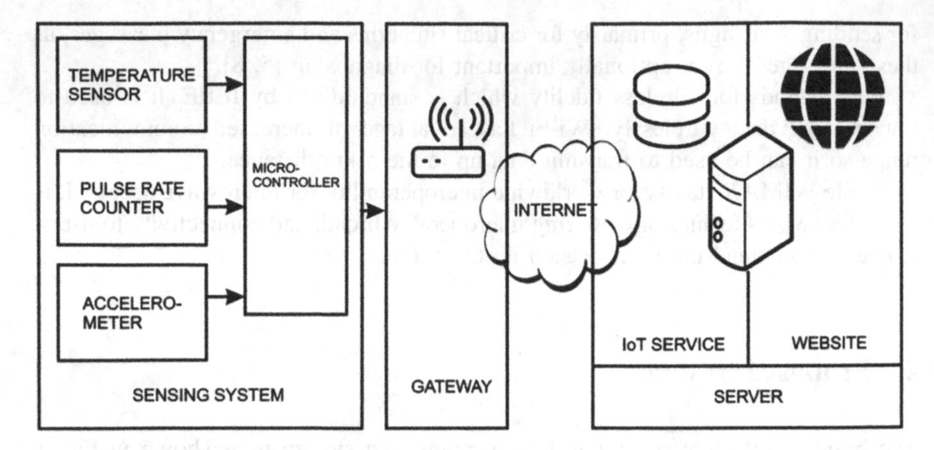

Fig. 2 The proposed system

4.1 Sensing System

Depending on the parameters to be measured, different sensors are used.

4.1.1 Body Temperature Sensor

It is most common physiological parameter measured by wearable sensors in health monitoring system. The variation in skin temperature can help to understand what is happening with patient's body and detect the various medical symptoms which may lead to heart attacks, shock, stroke, etc. Hence, it is exceptionally important and useful for estimating the physiological condition. The body temperature is commonly measured using different temperature sensors like DS600, DS18S20, and DS1621, which gives 9-bit temperature reading or simply LM35 sensor [6].

4.1.2 Pulse Rate Counter

Pulse rate is counted by a change in rate of blood flow in the blood vessels. Pulse rate counter simply uses IR LED and IR detector which are positioned to facing toward each other maintaining a certain distance in between them. By putting a finger in between that gap will cause a change in amount of IR light to be collected at IR receiver. The light emitted from IR LED must pass throughout a finger and recognized at receiver. The finger turns slightly opaque as the heart pumps the pulse of blood from the blood vessel. Due to this, the amount of light reaching to the detector is reduced. The detector signal varies with every heart pulse which is then converted into electrical pulse and amplified. Thus pulses are measured in healthcare systems. The signal of pulse rate counter is so weak or poor and noisy, and contains DC and

AC components, additionally systolic peak which measures the contraction of arteries, whereas diastolic peak for pressure is exerted on the wall of arteries [8].

4.1.3 Accelerometer

An accelerometer is a device that measures the vibration or acceleration of motion of a structure. These are generally used in healthcare systems to measure acceleration along with a sensitive axis. It is useful for body motion evaluation, detection of fall, and postural orientation. Various types of accelerometer are available depending on piezoresistive, piezoelectric, and variable capacitance methodology of transduction. Generally, all of them follow the same key idea of operation of a mass that reacts to acceleration by creating a spring or any identical component to compress or stretch corresponding to the measured acceleration. Accelerometers are attached to various parts of body to determine the energy expenditure during physical movement [11–14].

5 Conclusion and Future Work

The emergence of the IoT in healthcare is important for two reasons:

- Advancement of sensing and communication technology allows various devices to collect, maintain, and analyze data which was not accessible earlier. In healthcare, the patient's data is collected over the time which helps to provide preventive care and quick diagnosis of acute complications and promote understanding of patient's progress and estimating the success of treatment.
- The ability of devices to collect the data by their own removes the constraints of human-entered data—automatically gathering the data which doctors need at anytime and in the way they require it. This automation reduces the probability of error. This reduction results in increased efficiency, lower the cost, and better quality in any field.
- The future work of the project plays a crucial role in making the system more advanced and accurate. This system can be enhanced by embedding many more sensors such as sensors for breathing rate estimation, swallowing monitoring, hand gesture detection, and analysis as well as gait analysis to the internet which measures the corresponding medical parameters to make the system more beneficial and user friendly for human health monitoring. The restrictions over communication can be removed by establishing a Wi-Fi mesh network which in turn increases the communication range. The most important parameter in any transmission system is security, so further work can be done to allow the data to be transmitted securely over a long distance to the destination and provision for authorized accessing only which enables this system to be acceptable worldwide without any fear or anxiety.

References

1. Steven E Collier, "The emerging enernet: convergence of the smart grid with the internet of things," 2015 IEEE rural electric power conference.
2. Bhoomika. B.K, Dr. K N Muralidhara, "Secured smart healthcare monitoring system based on Iot," International Journal on Recent and Innovation Trends in Computing and Communication, Volume: 3 Issue: 7, 4958–4961.
3. Alexandre Santos, Joaquim Macedo, António Costa, M. João Nicolau (2014) Internet of Things and smart objects for m-health monitoring and control, Procedia Technology 16(0) p 1351–1360.
4. Feng Xia, Laurence T. Yang, Lizhe Wang and Alexey Vinel, "Internet of Things," International journal of communication systems, *Int. J. Commun. Syst.* 2012; 25:1101–1102.
5. Simone Corbellini, Franco Ferraris and Marco Parvis, "A system for monitoring worker's safety in an unhealthy environment by means of wearable sensors," IEEE International instrumentation and measurement technology conference victoria, vancouver island, canada, May 12–15, 2008.
6. Karandeep Malhi, Subhas Chandra Mukhopadhyay, Julia Schnepper, Mathias Haefke, and Hartmut Ewald, "A zigbee-based wearable physiological parameters monitoring system," IEEE sensors journal, vol. 12, no. 3, pp. 423–430, March 2012.
7. Lei clifton, David A. Clifton, Marco A. F. Pimentel, Peter J. Watkinson, and Lionel Tarassenko, "Predictive monitoring of mobile patients by combining clinical observations with data from wearable sensors," IEEE journal of biomedical and health informatics, vol. 18, no. 3, May 2014.
8. Sushama Pawar, P. W. Kulkarni, "Home based health monitoring system using android smartphone," international journal of electrical, electronics and data communication, vol. 2, issue. 2, Feb. 2014.
9. M. P. R. Sai Kiran, P. Rajalakshmi, Krishna Bharadwaj, Amit Acharyya, "Adaptive rule engine based iot enabled remote health care data acquisition and smart transmission system," 2014 IEEE World Forum on Internet of Things (WF-IoT).
10. Subhas Chandra Mukhopadhyay, "Wearable sensors for human activity monitoring: a review," IEEE sensors journal, vol. 15, no. 3, pp. 1321–1330, March 2015.
11. Sangil Choi, Richelle LeMay, and Jong-Hoon Youn, "On-board Processing of Acceleration Data for Real-time Activity Classification," IEEE 10th Consumer Communications and Networking Conference (CCNC) - Las Vegas, NV, USA, pp. 68–73, Jan 2013.
12. Tal Shany, Stephen J. Redmond, Michael R. Narayanan and Nigel H. Lovell, "sensors-based wearable systems for monitoring of human movement and falls," IEEE sensors journal, vol. 12, no. 3, pp. 658–670, march 2012.
13. Juha P¨arkk¨a, Miikka Ermes, Panu Korpip¨a¨a, Jani M¨antyj¨arvi, Johannes Peltola, and Ilkka Korhonen, "Activity classification using realistic data from wearable sensors," IEEE Transactions on information technology in biomedicine, vol. 10, no. 1, pp. 119–128, 2006.
14. K.C. Kavitha, A. Bazila Banu, "Wireless Health Care Monitoring," International Journal of Innovative Research in Science, Engineering and Technology, Volume 3, Special Issue 3, March 2014.

Swarm Intelligent WSN for Smart City

Shobha S. Nikam and Pradeep B. Mane

Abstract The smart city uses digital, computer, and communication technologies to enhance quality and performance of urban services by reducing costs and resource consumption. The smart city aims to drive economic growth and improve the living standards using state-of-the-art technology. Evolution of wireless sensor networks has been started from the idea that small wireless sensors can be used to gather information from the physical environment, in many situations ranging from monitoring the environmental, surveillance for security, health care, automated building control, traffic control, and object tracking. Sensors help each other to relay the information to the base station. The existing work includes simulation of routing protocols for WSNs in smart city and some hardware implementation. This paper presents a swarm intelligent routing algorithm for wireless sensor network to implement smart city. The particle swarm optimization (PSO) algorithm being most efficient is chosen for implementation (Yamille dl valle et al. in IEEE Trans Power Syst 12 (2):171–195, 2008 [1]). This paper has presented the implementation of prototype for smart homes and solid waste management system, and has also proposed the solution for street light intensity control and traffic congestion detection. For smart homes, fire detection, gas leakage detection, temperature control, and light intensity control are considered, while the designed solid waste control system is capable of detecting garbage level in dustbin and conveys it to the central monitoring system. The implemented prototypes for smart homes and waste management show that it will lead toward improvement of the living standard of the people living in smart city by way of energy conservation, health, and hygiene.

Keywords WSN · Swarm intelligence · PSO · Smart city

S.S. Nikam (✉) · P.B. Mane
Department of Electronics Engineering, AISSMS's Institute
of Information Technology, Pune, India
e-mail: shobha.nikam32@gmail.com

P.B. Mane
e-mail: pbmane6829@rediffmail.com

S.C. Satapathy et al. (eds.), *Proceedings of the International Conference on Data Engineering and Communication Technology*, Advances in Intelligent Systems and Computing 469, DOI 10.1007/978-981-10-1678-3_66

1 Introduction

The foundation elements in a smart city includes sufficient water supply, liquid and solid waste management, assured electricity supply, public transportation, economical housing, smart homes, powerful IT connectivity, digitalization, good governance, especially e-Governance and citizen assistance, security, particularly children and women and health and education.

A wireless sensor network (WSNs) is an upcoming technology, which has potential applications in surveillance (military/civil), disaster management, environment, structural monitoring, healthcare, and many more. In WSNs sensors are spatially distributed autonomous sensors. The sensors sense and acquire physical parameters, and transmit acquired information over wireless medium to monitor the environment. The data transmitted by the wireless sensors can be received by one or more base stations. The various challenges which need to be addressed in WSN are ad hoc deployment, dynamic topology, spatial distribution and constrains in computational resources, bandwidth, memory, and energy. In WSN the nodes are generally static and deployed over vast areas or they can also be mobile and capable of interacting with the other nodes in its vicinity. In the WSN when the nodes are mobile such networks are generally called as a robotic network or as a sensor actor network [2]. To obtain optimum results in WSN an individual sensor node must have moderate memory and computational resources.

It is observed that as compared to analytical methods, bio-inspired optimization methods are computationally efficient. Swarm intelligence is an approach to collective behavior of sensor nodes inspired from the self-organized behaviors of social animals like ants and birds [3]. The advantages of using swarm algorithms are that they are robust, scalable and flexible to model the collective behaviors for coordination of large number of sensor nodes. This paper presents an application of swarm intelligent sensor networks for smart cities.

This paper is presented in six sections. Section 2 gives a brief introduction to swarm routing algorithms. The computationally efficient PSO algorithm is described in Sect. 3. Section 4 provides smart city architecture. Section 5 gives prototype design of smart homes and solid waste management system using WSN. The conclusion of the implanted work is given in Sect. 6.

2 Nature-Inspired Routing Algorithms

1. **Ant Colony Optimization (ACO)**:

The ant colony optimization (ACO) methods are inspired by operating principles of ants, which are used to perform complex tasks like nest building and foraging [4].

2. **Ant Net Algorithm**:

The algorithm is applicable to asymmetric packet switched networks, and the primary objective of the algorithm is to maximize the performance of a network. The algorithm achieves load balancing by probabilistically distributing packets on multiple paths [4].

3. **Ant-Based Control (ABC)**:

ABC approach is used in circuit-switched telecommunication networks for routing and load balancing problems. As a symmetric network, a circuit-switched network forms a virtual circuit between a sender and a receiver by explicitly connecting them through crossbar switches [4].

4. **Particle Swarm Optimization (PSO)**:

Particle swarm optimization (PSO) is a simple and computationally efficient optimization algorithm. It has been applied to address WSN issues like optimal deployment, node localization, clustering, and data aggregation [3–5].

After comparing the various swarm routing algorithms, it is observed that PSO is uncomplicated and computationally capable algorithm for optimization. Hence, PSO is preferred for routing in WSN designed for smart city.

3 Particle Swarm Optimization

The mathematical model for the velocity of the particle in the particle swarm optimization algorithm is given by Eq. 3.1. In the number space, the vector $X_i \in R_n$ determines the position of every sensor particle and the vector $V_i \in R_n$ determines the movement of the particle by its velocity:

$$\overrightarrow{v_i}(t) = \overrightarrow{v_i}(t-1) + A_1.\text{rand}_1.(\overrightarrow{p_i} - \overrightarrow{x_i}(t-1) + A_2.\text{rand}_2.(\overrightarrow{p_g} - \overrightarrow{x_i}(t-1)) \quad (3.1)$$

where A_1 and A_2 represent positive numbers, and rand_1 and rand_2 are random numbers.

The equation is divided into three elements.

$\overrightarrow{v_i}(t-1)$: The first element shows the trend of the particle to stay in the same direction.

$A_1.\text{rand}_1.(\overrightarrow{p_i} - \overrightarrow{x_i}(t-1)$: The second element is a linear inclination headed for the best position denoted as BEST pat any time found by the given particle.

$A_2.\text{rand}_2.(\overrightarrow{p_g} - \overrightarrow{x_i}(t-1)$: The third element is a linear inclination toward the best position found by any particle considered to be its global best position denoted as BESTg.

Particle Swarm Topologies:

The different swarm topologies are shown in Fig. 1. The particles are in connection with two neighborhoods, i.e., local best (BESTp) and global best (BESTg).

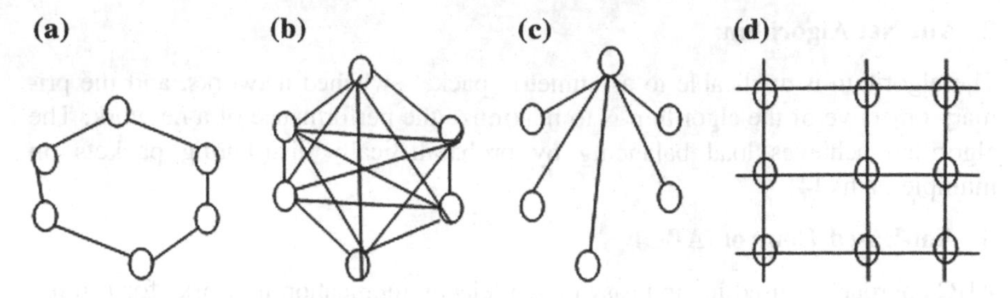

Fig. 1 Swarm topologies **a** Ring topology. **b** Global best. **c** Wheel topology. **d** Von Neumann topology

In the local best approach, each particle has access to the information of its immediate neighbors. The two most widely used local best topologies are ring topology and wheel topology. In the ring topology, each particle is connected with its immediate neighbors as shown in Fig. 1a. In the wheel topology the individual particles are connected to central particle and all the information is transferred to a focal/central individual [6] as shown in Fig. 1c.

The global best topology is a fully connected network. In this topology the particles are inclined toward the best position found by any particle of the swarm, where every particle is able to access the information of all other particles in the cluster as shown in Fig. 1b. The von Neumann topology shown in Fig. 1d can perform better than the other topologies for global best [1, 6]. The selection of topology depends on the application.

The prototype systems developed for smart home and solid waste management have taken into consideration the wheel topology.

4 Smart City Architecture

The clustered architecture of sensor network for the smart city is shown in Fig. 2. The city area can be divided into N trajectories depending on the need of monitoring data at the specific area, e.g., high traffic density of the roads, population of the area (smart homes and waste management). Clusters can be formed by grouping the sensor nodes. Every cluster consists of cluster head and sensor node, and each sensor network node has a microcontroller and a radio transceiver. Communication between cluster head and sensor node will be implemented using ZigBee module. Communication between clusters and base station can be implemented using GSM module.

The PSO algorithm for the clustered sensor network-based smart city can be implanted using the following logical steps:

i. Start the swarm by choosing a random position to clusters formed.
ii. For individual cluster, examine the fitness function.

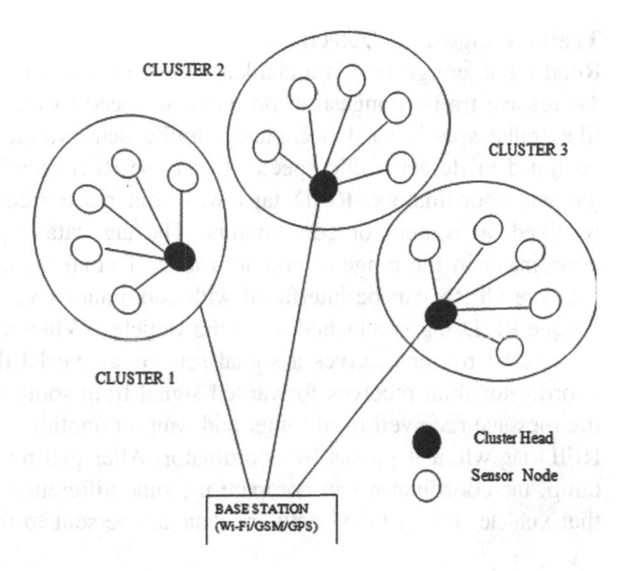

Fig. 2 Clustered architecture of sensor network

iii. Compare the fitness value of particle with its best position (BESTp) for every cluster. Whenever current value is better than the BESTp value, then fix the current value as a BESTp, and the particle's current position as, X_i, P_i.
iv. Find the cluster which has highest fitness value. This value is determined as BESTg and its position as P_g.
v. Revise the positions and velocities of all the cluster particles using i and ii.
vi. Iterate the steps from ii to v for getting the required result.

Sample smart city application areas are considered which includes street light intensity control, city traffic congestion, smart homes, and solid waste management.

Street Light Intensity Control:
The system enables the remote control of street lighting lamps, savings of maintenance costs, and electric power. The architecture uses integrated doppler sensors for vehicle detection and to complete the power efficiency objective. The system will have a gateway module installed at control center, and the control modules installed in the street lighting lamp system. For local communication through the WSN network, ZigBee is used and for communication between gateway module and control modules GSM will be used. The adjustment of the light intensity can be done by altering the PWM signal's duty cycle. This change will be induced remotely, from the control center. Thus, the command sent from the control center is transmitted through the GSM network to the WSN coordinating node. The coordinator sends the light intensity adjustment command to the first module and it will subsequently be sent throughout the network until the destination node is reached. When a vehicle is detected, the light intensity of the lamps is increased to a predefined level, not to affect road traffic safety, and reduces in the opposite case [7].

Traffic Congestion Detection:
Road traffic congestion is a challenge to the entire large and growing urban areas. To resolve traffic congestion problem, we need to consider the following factors like traffic speed, road occupancy, traffic density, etc. [8]. This system can be designed to detect traffic speed; it can use RFID devices like tags, routers, and gateway/coordinators. RFID tags will emit radio frequency signals that can be received by routers or coordinators. The tag data captured by routers relay to coordinator in the range or another router. To make it a dual-radio device external GSM or GPRS can be interfaced with coordinators via serial interface. Assuming unique RFID tag is attached to all the vehicles. When a vehicle crosses the router, the nearest router receives a signal sent by active RFID tag of that vehicle. The coordinator then receives forwarded signal from router. The coordinator can save the message received from router and wait for another message to come from same RFID tag when it passes by coordinator. After getting the signal, from its timestamp, the coordinator can compare the time difference and calculate the speed of that vehicle. Using GSM network data can be sent to the control room/station.

5 Prototype Design

In this paper an attempt has been made to design and implement smart homes and the waste management system which are important elements in a smart city.

5.1 Smart Home

In smart cities the smart homes will be a key element as it will lead to the quality enhancement of human life and will also be of importance with respect to the conservation of energy resources. In smart homes energy conservation and efficient utilization is very important. The system is designed for fire detection and energy saving by controlling temperature and light intensity.

The designed and implemented system works on master and slave principle as shown in Fig. 3. The master asks for information from the slaves by sending frames using PC; smart phones can also be used. The communication between the master and the slave devices will be formed through ZigBee device. This system is designed for one master and multiple slaves. Master sends the request to all the slaves. If the slave ID in request matches with device's own slave ID, then slave device accepts the frame and sends requested parameter to the master device. If ID does not match then the slave device discards the frame sent by master device. In slave 1 passive infrared (PIR) sensor, gas sensor (MQ6), and fire sensors (bimetallic strip) are used. In slave 2 IR sensor is used to detect human presence in the home. If count is zero then the relay1 is turned off which results in energy saving. For temperature sensing LM35 and for light detection LDR is used. Using the smart

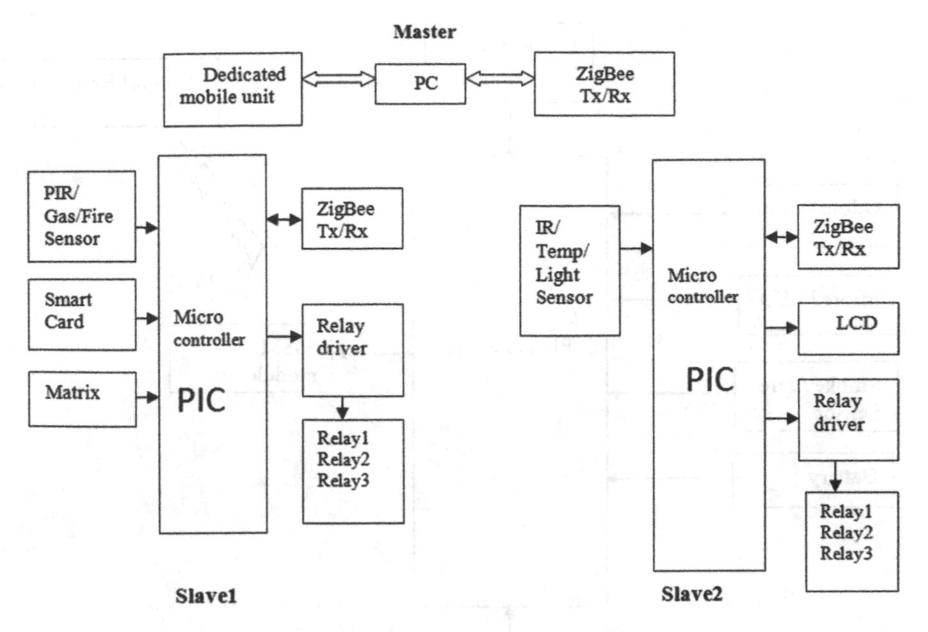

Fig. 3 Master–slave arrangement in smart home

card the threat of misusing the system can be minimized and only authorized person will have complete access to the system. Using the GSM module, measured parameters can be sent to the central database or authorized person via SMS.

5.2 *Solid Waste Management*

In smart cities waste management is of critical importance as it is directly related to the health and hygiene of the people staying in the city. Both a threat to environmental security, public health, and a strategic renewable resource, municipal solid waste is an inevitable byproduct of civilization and a target for clean technology innovation. The designed and implemented system shown in Fig. 4 consists of four pairs of infrared sensor to detect the level of garbage in the dustbin. As garbage level increases, the first infrared pair which is mounted at the bottom of dustbin at 10 % level will not receive a signal for 1 min and the system will sense the "bin is empty" and will start filling up. When garbage level is filled up to second infrared pair, it will not receive signal for 1 min indicating that garbage is "filled up to 80 %." Similarly, when it reaches up to third infrared pair, it detects that bin is "filled up to 90 %." It will display or inform the control room that "dustbin will fill soon."

For fourth infrared pair for garbage level detection, it will display "overflow of dustbin." The system cross checks filled dustbin using ultrasonic sensor and gives

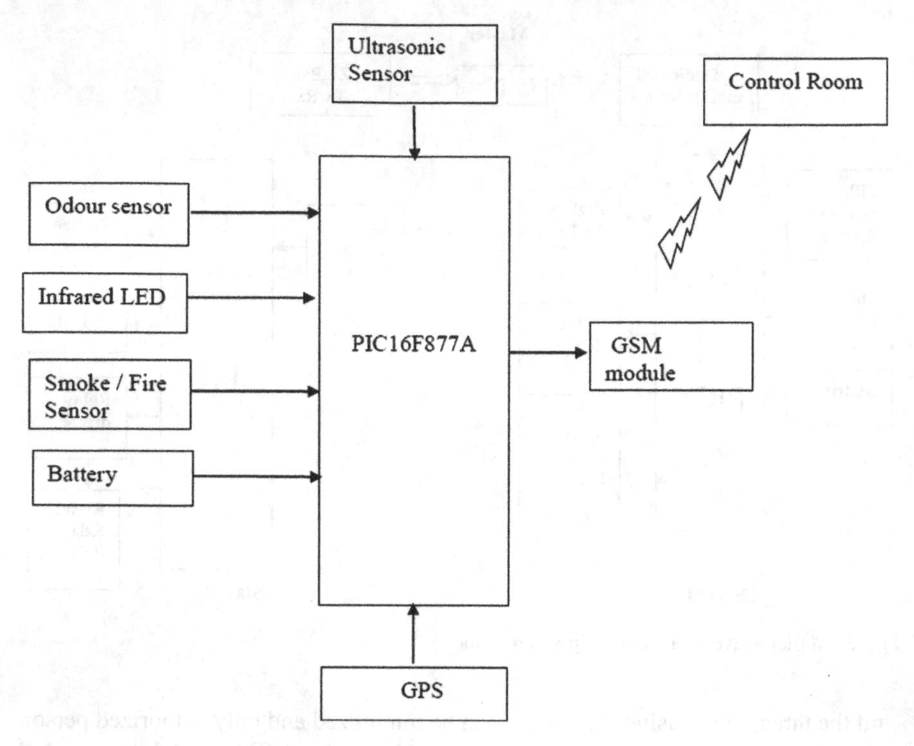

Fig. 4 Block diagram of solid waste management system

protocol "bin is full or overflow" through the GSM technology. With the help of GPS technology the garbage system is able to detect the location and send the protocol to the control room. If dustbin will not be clear by municipal sweepers within 1 h then another message will be sent to the controller that "unload dustbin within 1 h, garbage is already overflow." When first infrared pair receives signal, it indicates that bin is unloaded. It means "bin is unloaded." The designed system consists of two sensors namely smoke sensor (MQ2) and methane gas sensor (MQ6) continuously checking the level of gas in dustbin. When dustbin is not filled up for 2–3 days, decomposition of waste produces odor which contain methane which is detected using (MQ6) methane sensor. Similarly, in summer days garbage gradually burns to produce smoke/co2, which is detected by smoke sensor (MQ2). In such case designed system gives the alert by sending message to control room or fire-bridged. The supportive GPS technology is used to indicate location of dustbin, each and every message or protocol sent by the system through GSM which carries the GPS coordinates which will detect the location of dustbin. The designed system is working on battery source, so there is limited power consumption. The implemented sensor and technology make safe and real-time waste collection system (Fig. 5).

Fig. 5 Solid waste management system results

6 Conclusion

In this paper an attempt has been made to design prototypes for smart city which includes smart homes and solid waste management. Nature-inspired routing protocols can remove WSN design issues such as scalability, maintainability, battery life, adaptability, survivability, etc. Swarm intelligent routing protocols are compared and effective PSO algorithm is chosen and designed for smart city. The work has been done to design real-time automated homes and smart waste collection and management system which is cost effective, consumes low power, and improves human life.

Smart city should meet the demands of performance of urban services by reducing costs, resource consumption, and energy saving. From the results of the implemented work it is observed that the smart homes will lead to automation of the homes, hence leading toward energy conservation and the waste management system will lead toward improving public health and environmental security.

References

1. Yamille del Valle, Ganesh Kumar Venayagamoorthy, Salman Mohagheghi, "Particle Swarm Optimization: Basic Concepts, Variants and Applications in Power Systems", IEEE Trans. Power Syst., vol. 12, no. 2, pp. 171–195, April 2008.
2. Manuele Brambilla, Eliseo Ferrante, Mauro Birattari, Marco Dorigo, "Swarm robotics: a review from the swarm engineering perspective", Springer Science + Business Media New York 2013 Swarm Intell, doi:10.1007/s11721-012-0075-2.
3. Muhammad Saleem A, Gianni A. Di Caro b, Muddassar "Swarm intelligence based routing protocol for wireless sensor networks: Survey and future directions", Information Sciences 181 (2011) 4597–4624.
4. Gianluca Baldassarre, Vito Trianni, Michael Bonani, Francesco Mondada, "Self-Organized Coordinated Motion in Groups of Physically Connected Robots", ieee transactions on systems, man, and cybernetics—part B: cybernetics, vol. 37, no. 1, february 2007.
5. Shekh Md Mahmudul Islam, Mohammad Anisur Rahman Reza and Md Adnan Kiber, "Wireless Sensor Network using Particle Swarm Optimization", Proc. of Int. Conf. on Advances in Control System and Electricals Engineering 2013.

6. Raghavendra V. Kulkarni, Ganesh Kumar Venayagamoorthy, "Particle Swarm Optimization inWireless Sensor Networks: A Brief Survey", IEEE Trans on systems, man and cybernetics.
7. Alexandru Lavric, Valentin Popa, Stefan Sfichi, "Street Lighting Control System Based On Large-Scale WSN: A Step Towards A Smart City", 2014 International Conference and Exposition on Electrical and Power Engineering (EPE 2014), 16–18 October, Iasi, Romania.
8. Dang Hai Hoang, Thorsten Strufe, Thieu Nga Pham, Hong Ngoc Hoang, Chung Tien Nguyen, Van Tho Tran, Immanuel Schweizer, "A Smart Data Forwarding Method Based on AdaptiveLevels in Wireless Sensor Networks", Eight IEEE Workshop on Practical Issues in Building Sensor Network Applications 2013.

Representing Natural Language Sentences in RDF Graphs to Derive Knowledge Patterns

S. Murugesh and A. Jaya

Abstract English language has its own word order, which determines the way of ordering words in a sentence. This paper presents the manner by which sentences in English language can be represented as knowledge patterns by means of RDF graphs. Knowledge patterns are general patterns of knowledge that can be used in any knowledge base or ontology. Knowledge pattern is a logical design pattern which defines a formal expression that can be exemplified and morphed in order to solve a domain modeling problem. Knowledge patterns are proposed by ontology engineers. When using them, the general symbols in the pattern are renamed to special symbols from the modeled domain. The RDF graph language is used to represent knowledge pattern contained English sentences. The basic RDF model can be processed even in the absence of detailed information (schema) on the semantics. RDF model has the important property of being modular, i.e., the union of knowledge represented as directed graphs. The paper also provides examples of usage of knowledge patterns.

Keywords RDF graph · Ontology · Knowledge pattern

1 Introduction

Sentences are formed by sequence of words. Each word in a sentence belongs to a particular class of words, i.e., noun, pronoun, verb, adverb, adjective, preposition, etc. The ways in which sentences are to be constructed are laid down by the grammar of the particular language. Basically, there are two basic components of

S. Murugesh (✉) · A. Jaya
Department of Computer Applications, B.S. Abdur Rahman University, Chennai, Tamilnadu, India
e-mail: murugesh.here@gmail.com

A. Jaya
e-mail: jayavenkat2007@gmail.com

S.C. Satapathy et al. (eds.), *Proceedings of the International Conference on Data Engineering and Communication Technology*, Advances in Intelligent Systems and Computing 469, DOI 10.1007/978-981-10-1678-3_67

any sentence, i.e., subject and predicate. Apart from these there are extensions to the sentences like object, attribute, and sometimes adverb complements.

It is to be borne in mind that when a sentence is constructed, the order in which the words appear in a sentence is more important than the class to which the word belongs to. Word order indicates the order of members in a sentence. There are two kinds of word order, fixed word order where the order in which the members of a sentence appear are important, whereas in free word order, the order is not given importance. The members of a sentence are given a notational convention, i.e., subject is represented as S and predicate is marked by the letter V and O stands for object. The ways in which these three basic members of a sentence appear decide on the word order. There are six basic types of word order: they are subject verb object, subject object verb, verb subject object, verb object subject, object subject verb, and object verb subject. English, Roman, Bulgarian, and Chinese languages use the SVO word order. Even though there are many types of sentences, the basic focus of this paper is on declarative sentence since the subject of interest is ontology or knowledge base which contains positive or negative facts about a particular topic of interest. It is to be noted that questions or imperative sentences do not form a part of either ontology or knowledge base always. Negative sentences are identified by the presence of the word "Not" in the sentence.

Statements in RDF language contain facts and they are represented by RDF triples. A single RDF triple represents one statement and it consists of a subject, predicate, and an object. A directed graph is used to represent ontology or knowledge base in RDF graph language. Nodes in the directed graph represent subjects and objects of the statements and arcs or edges represent relations or predicate. Resource description framework (RDF) describes any concept, relation, or thing that exists in the universe. There are three things in RDF, subject (S), predicate (P), and object (O); subject is the thing you are describing, predicate is the attribute of the thing you are describing, and object is the thing we are referring to with a predicate. Consider the sentence, "ATM machine reads ATM cards"; ATM machine is the subject, predicate is reads, and object is ATM cards. There can be more triples associated with the subject, the more the triple, the more we know about the subject. Predicates point to the vocabulary. Vocabulary defines what the triples actually mean.

The English language makes use of fixed word order known as subject verb object. Here object refers to direct object. The basic word order may be little different since an auxiliary verb (do, have, be, will, can, etc.) or an indirect object may appear in the sentence. Sentences in each language may express a variety of things. It can state facts—declarative sentences (positive or negative). It can be asking about things' questions. It can be command or imperative sentences. The basic focus of this article is on knowledge patterns, which are closely related to knowledge bases or ontologies. We mainly look in for declarative sentence. Questions or imperative sentences do not form a part of any knowledge base or ontology. Declarative sentences usually states positive or negative facts. The structures of both positive and negative sentence are similar in English. A negative sentence consists of the word "not". The positive declarative sentence contains only

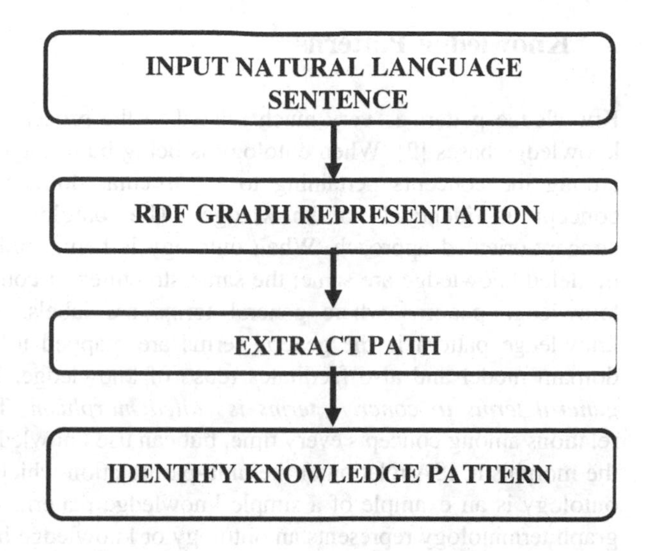

Fig. 1 Knowledge pattern discovery process

a subject and predicate. Another complex type of sentence that contains an object is of the form subject verb object, for instance "I see an ATM Machine," "ATM machine prints receipts," etc. A sentence may also contain an adverbial complement of place, manner, or time. A sentence may also have complements or combination of them. A sentence containing all three types of adverbial complements is labeled as subject verb object manner place time. "I withdraw money everyday." is an example of an adverbial complement sentence. Subjects and objects in sentences are extended by word features like "small" and "big." Negative declarative sentences are the same as a positive declarative sentence. When a verb in a sentence is changed to its opposite meaning, then it is a negative declarative statement [1]. Figure 1 depicts the knowledge pattern extraction/discovery process. The rest of the paper is organized as follows: Sect. 2 deals with literature survey, Sect. 3 about the knowledge patterns in general, and Sect. 4 is regarding the procedure to derive knowledge patterns for English sentences and Sect. 5 concludes the paper.

2 State of Art

In research conducted by Rostislav Miarka et al., they suggested use of RDF graph for representing knowledge patterns. Clark, Thompson et al. recommended that knowledge patterns can be used in any knowledge base or ontology. Antonin Delpeuch proposed that RDF is widely accepted as a standard for linked representation of information, i.e., words in a sentence can be represented using RDF graph. Peter Exner proposed an end-to-end system that extracts RDF triples for describing entity relations and properties from unstructured text. Wikipedia and DBpedia had created voluminous amounts of linked data which can be used as model training sets for information extraction tasks.

3 Knowledge Patterns

Knowledge pattern is very much related to the process of creating ontologies or knowledge bases [2]. When ontology is being built, the concepts and the relations among the concepts pertaining to a particular domain are modeled. The term concept is identical to knowledge. The ontology building is based on concept-oriented approach. When ontology is being built, some structures of the modeled knowledge are same; the same structures of concepts can be identified as knowledge patterns where general terms are labels. The advantage of using knowledge pattern is the general terms are mapped to concrete terms from the domain model and also facilitates reuse of knowledge. *The process of renaming general terms to concrete terms is called morphism*. Thus we need not create relations among concepts every time, but can use knowledge pattern and make only the morphism. ISA relation or the unit–part relation which is found in almost every ontology is an example of a simple knowledge pattern. A directed graph in RDF graph terminology represents an ontology or knowledge base. Nodes correspond to subject and an object of statements, and arc denotes relations or predicates. RDF contains only binary relations; complex relations should be broken down to a set of binary relations. An URI, Uniform resource identifier, identifies nodes and edges in an RDF graph. A standard RDF graph uses only solid lines. To differentiate between knowledge patterns from the conventional RDF statements, dashed lines are used. The resource description framework (RDF) is an artificial language that takes the form of knowledge graphs. A triple links two objects, which consists of one entity, i.e., the subject of the predicate, the type of the link represented by a string refers to the property and a second entity is the object of the predicate. In a graph a triple is represented as a directed property from subject to object labeled by the predicate. Example of a typical RDF triple and a triple which represents knowledge pattern is shown in Fig. 2.

Figure 2 shows the classical RDF triple (above) and RDF triple representing knowledge pattern (below).

Morphism is a technique using which the general terms from the pattern are mapped to the special terms from the problem domain. Knowledge patterns can be represented using an RDF Graph [3, 4]. Mapping of one term will be denoted by one RDF triple identified with solid/thick line. The subject of the triple is the special

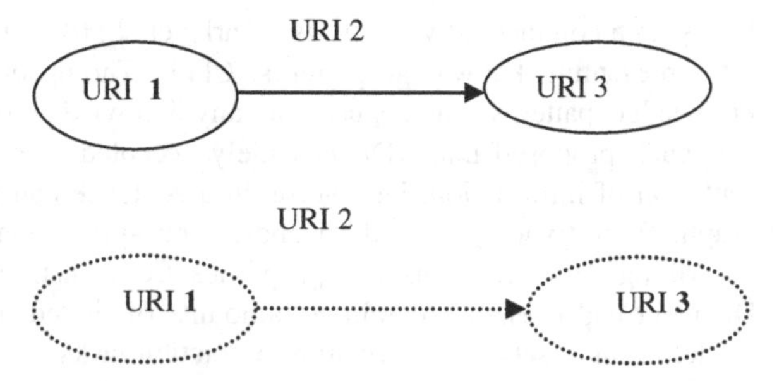

Fig. 2 Classical RDF triple

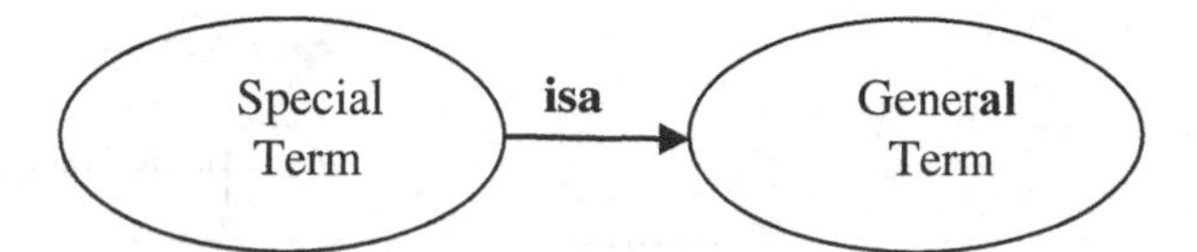

Fig. 3 Morphism

term from the domain, predicate will be "isa" relation, and object is the general term from the pattern. An example of morphism of one term is shown in Fig. 3.

4 Knowledge Patterns for English Sentences

Since our focus is on building an ontology which contains statements of fact, the main concern is on positive declarative sentence. A sentence contains three members, subject, predicate, and object, which facilitate us to represent the sentence as RDF graph as shown in Fig. 4.

The subject from sentence is denoted by the node "subject," the object is symbolized by the "object" node, and the predicate by the arc "predicate."

An example of the sentence "ATM machine reads ATM card" is represented in RDF graph in Fig. 5 and as RDF/OWL representation in Fig. 6.

In some cases a negative declarative sentence also states some negative fact—the fact that some statement does not hold good, i.e., not true [5]. RDF graph notation provides a symbol known as falsum. The standard form of representing a negative declarative sentence in RDF graph is shown in Fig. 7

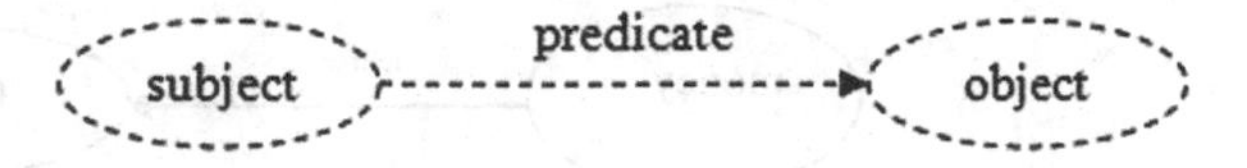

Fig. 4 Knowledge pattern for a sentence in basic form

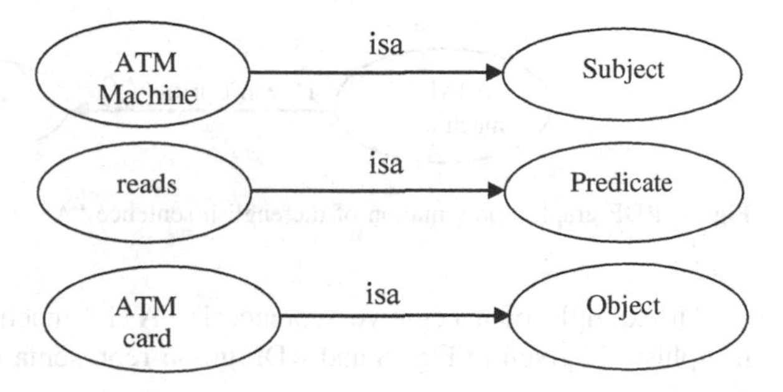

Fig. 5 RDF graph for the sentence "ATM machine reads ATM card"

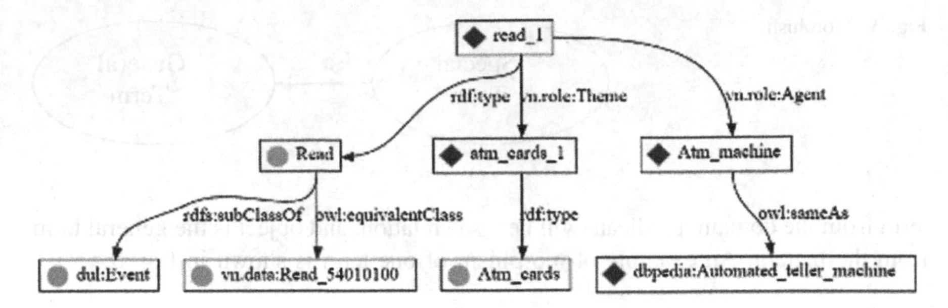

Fig. 6 RDF/OWL ontology and linked data from natural language sentences

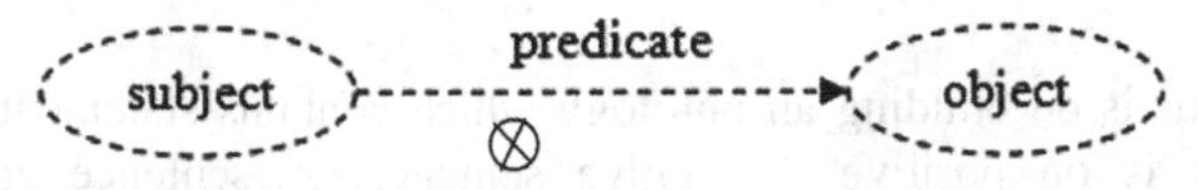

Fig. 7 Basic form of a negative sentence

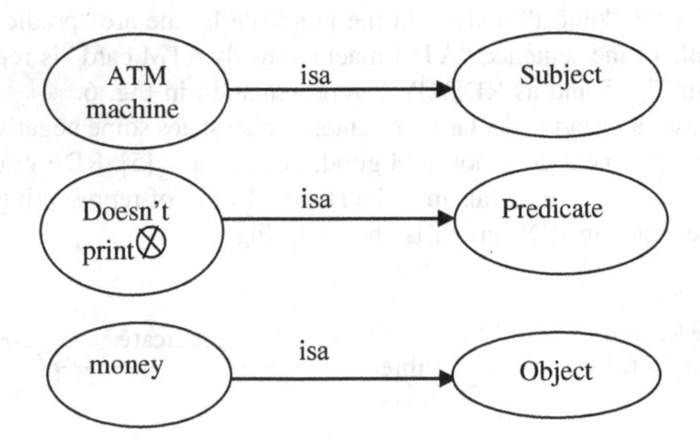

Fig. 8 Morphism

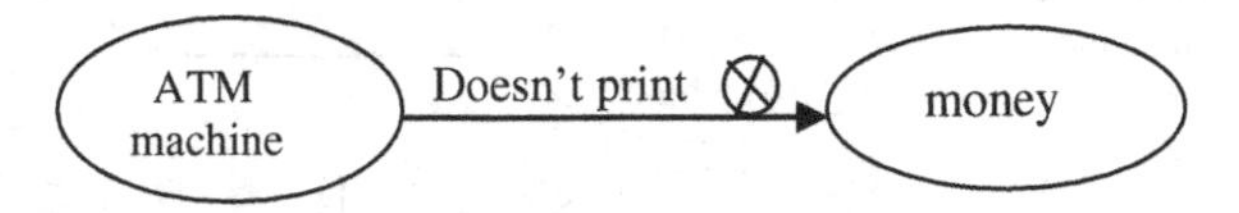

Fig. 9 RDF graph representation of the english sentence "ATM machine doesn't print money"

An example of a negative sentence is "ATM machine doesn't print money," morphism is given in Fig. 8 and RDF graph representation of the same is given in Fig. 9.

5 Conclusion

English has its own vocabulary and grammar. Grammar plays a key role in a language, and it determines the particular order of a word in a sentence. This paper has introduced the concept of representing natural language sentences with the help of RDF graph. Thus knowledge patterns can be extracted from plain text and can be represented in RDF/OWL.

References

1. Rostislav et al, "Knowledge patterns in RDF graph language for English Sentences", Proceedings of the federated conference on Computer Science and Information systems pp 109–115, IEEE (2012).
2. P. Clark, J. Thompson and B. Porter, "Knowledge patterns," in Handbook on Ontologies, S. Staab and R. Studer, Eds. Berlin: Springer-Verlag, ISBN 3-540-40834-7, pp. 191–207. (2004).
3. W3C: Resource Description Framework (RDF): Concepts and Abstract Syntax, http://www.w3.org/TR/2004/REC-rdf-concepts-20040210/.
4. W3C: RDF Primer, http://www.w3.org/TR/2004/REC-rdf-primer20040210/.
5. Peter Exner and Pierre Nugues, Department of computer science, Lund University, Entity Extraction: from unstructured text to DBpedia RDF triples.

A Framework to Enhance Security for OTP SMS in E-Banking Environment Using Cryptography and Text Steganography

Ananthi Sheshasaayee and D. Sumathy

Abstract Authentication of user in online banking is a major issue in recent days where transactions are carried out using insecure Internet channel. The modern communication medium is very much exposed to various threats. One time password (OTP) is used to prove one's identity over the wireless channel. The OTP sent to user's registered mobile number as SMS is most commonly used technique for user authentication. OTP SMS sent normally as plain text is vulnerable to various attacks along the communication channel. To solve this problem, this research aims to provide a technique to transform the OTP using a lightweight cryptography and hide the cipher text using text steganography and send the stego text as SMS to user mobile. Personal Identification Number (PIN) supplied by the bank during registration is used for ciphering. The user needs to know the PIN to read the OTP. The user can proceed with the business transaction, only after this authentication. This process provides end-to-end-encryption of the OTP SMS.

Keywords OTP SMS • Cryptography • Text steganography • User-authentication • Security principles

1 Introduction

Internet is a boon to the modern society. It provides a platform to obtain variety of services including information gathering, online chats, emails and financial services like online shopping, bill payments, e-banking and many more. It is widely used as it provides all services at ease and comfort. The people using internet for their financial transactions and to manage their accounts online is growing rapidly

Ananthi Sheshasaayee · D. Sumathy (✉)
Department of Computer Science, Quaid-E-Millath Government College for Women, Chennai 600002, India
e-mail: sumathy.research@gmail.com

Ananthi Sheshasaayee
e-mail: ananthi.research@gmail.com

S.C. Satapathy et al. (eds.), *Proceedings of the International Conference on Data Engineering and Communication Technology*, Advances in Intelligent Systems and Computing 469, DOI 10.1007/978-981-10-1678-3_68

because of the numerous advantages it provides. The security concerns of Internet, poses serious threats to the financial transactions carried out. Online banking system provides with various mechanisms for authentication in an anonymous network [1]. The user needs to prove his identity before gaining access to his confidential data. The basic level of authentication is static password that needs to be changed only when needed. Static passwords are easy to remember and used every time for login. These passwords are prone to various attacks like shoulder surfing, brute force attack, birthday attack and are easy to guess by the intruder. To reduce the risks with static passwords, One-Time passwords (OTP) were introduced [2].

OTPs provide another level of authentication that is to be provided in order to proceed with the online business transaction. When the user requests for online banking facility, it required to register his personal information along with his personal IMSI (International Mobile Subscriber Identity) number with the bank. Transactions carried out to the bank account are intimated through SMS to the IMSI. When online business transaction need to performed, it is required to login with the static password. An OTP SMS is sent to the registered mobile number which is valid only for a specific time units and session. OTP is used to provide second level of authentication and transaction proceeds if only validated [3].

2 Cryptography and Text Steganography

Cryptography and Steganography are the two widely used techniques for information security. Cryptography is secret writing that scrambles the secret text into meaningless cipher text. Steganography is covered writing where the secret text is hidden under a cover which hides the very existence of the secret message. The cover media used to hide decides the type of steganography namely image, audio, video or text [3]. Text Steganography is the simplest form of steganography as it involves less overhead but difficult to hide, due to lack of redundant bits unlike image, audio or video. Text Steganography uses text to hide the secret text. The cover text can be an email, spam text or an innocent looking SMS. The meaningless information of the cipher text leads to suspicion and hence prone to attacks for deciphering. The advantage of steganography is that the attacker may overlook the existence of the secret text itself.

The idea of combining both techniques for the advantages each possess would be a highly secured system for securing OTP SMS which is vulnerable to various threats over the insecure communication medium.

3 State of Art

In this section, this paper presents a brief overview of the related works in the area of OTP SMS security. A multichannel system for proving one's authenticity using RC4 encryption of OTP and QR code to hide the cipher text is proposed in [1]. To counter the attacks on SMS, an innovative scheme in [2] of encrypting the OTP using AES and hiding it under LSB steganography that is sent to user email is suggested. A visual cryptography combined with biometrics with PIN and for hiding OTP with image for better security is presented in [3]. A system for authenticating clients using digital watermarking is presented in [4]. A trust worthy m-banking system to verify using OTP with biometrics is proposed in [5]. It checks OTP with webpage id code along with biometrics for proving user identity. To safeguard SMS OTP from attacks two methods were proposed in [6]. End-to-end-encryption and an exclusive channel on the handset are proposed for transferring confidential data. Combining of text steganography with cryptography in visual form for the online shopping payments is proposed that provides secrecy of one's data is suggested in [7]. In [8] a technique to encrypt OTP using symmetric cipher with time synch module that provides efficient end-to-end-encryption is suggested. A simple method to secure OTP with secret key and trans-code to counter MITM attacks was suggested in [9]. It also provides efficient user authentication.

4 Motivation

During online banking transactions, the OTP sent from banking server to customer registered mobile number for user authentication. The secret OTP travels as plain text along the communication channel has to pass through many intermediate points like SMSC (Short Message Service Centre), SMS GMSC (SMS gateway MSC), MSC (Mobile Switching Centre), BSS (Base Station System) (Fig. 1 illustrates the diagrammatic representation) [10]. There are many different threats to OTP at these points namely wireless interception-eavesdropping of communication between mobile phones and base stations due to protocol inefficiencies, mobile malwares-intercept SMS messages containing OTP, 'mSpy'—a mobileapp that can be installed on the device to be hacked [6] and many more. As a part of MITM attack, a mobile malware is installed in the user's mobile which allows the attacker to capture the user's mobile and traces, call logs, inbox, contacts, and all the event happenings. The user is forced to reveal his credentials either using social engineering or by shoulder surfing. With the credentials known and mobile activities captured, OTP SMS is not spared.

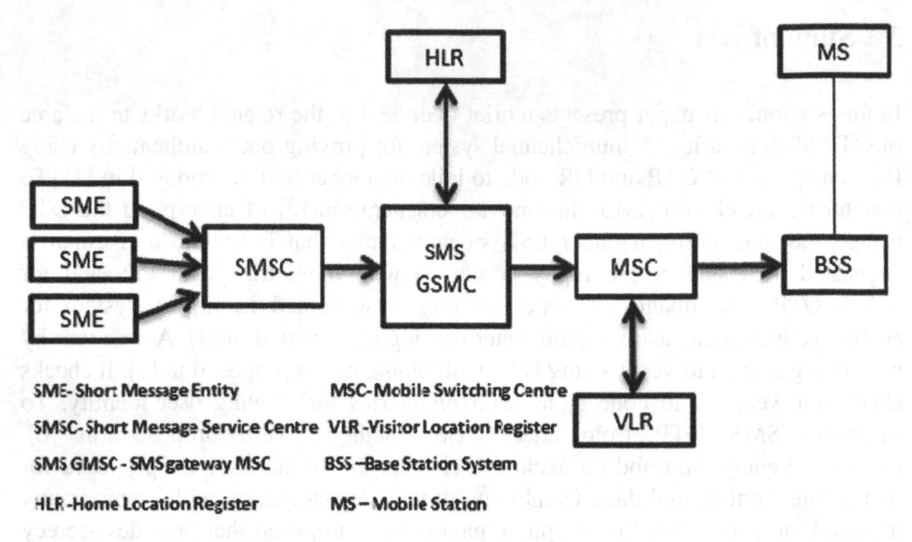

Fig. 1 Network nodes in GSM network for SMS

5 Methodology for Secure Internet Banking

Cryptography and Steganography can be combined to secure OTP SMS for the advantages each technique has. Ciphering scrambles the secret text into meaningless code and text steganography hides the very presence of secret text under an innocuous text carrier.

The step-by-step transaction procedure (Fig. 2 depicts the flow) of using mobile phones for internet banking is given as below:

- Step 1: The registered customer uses the bank portal to enter into his personal homepage by using his login and password.
- Step 2: The bank server verifies the user credentials and validates the user if found correct and allows the user to access the web page or generate queries. If not access is denied.
- Step 3: When a customer initiates an online bank transaction, the request reaches the bank server, and OTP is generated.
- Step 4: The key for encryption is generated by combining Personal Identification Number (PIN) and date-of-birth (DOB) in the specified format.
- Step 5: The bank server then encrypts the OTP using light weight cipher using the keys generated in step 4.
- Step 6: The cipher text generated in step 5 is further hided using text steganography using stego-key under an innocent looking SMS.
- Step 7: This stego text with OTP hidden is sent as SMS to the user's registered mobile through Wireless Application Protocol (WAP).
- Step 8: The user receives the SMS and the cipher text is extracted using the same stego-key.

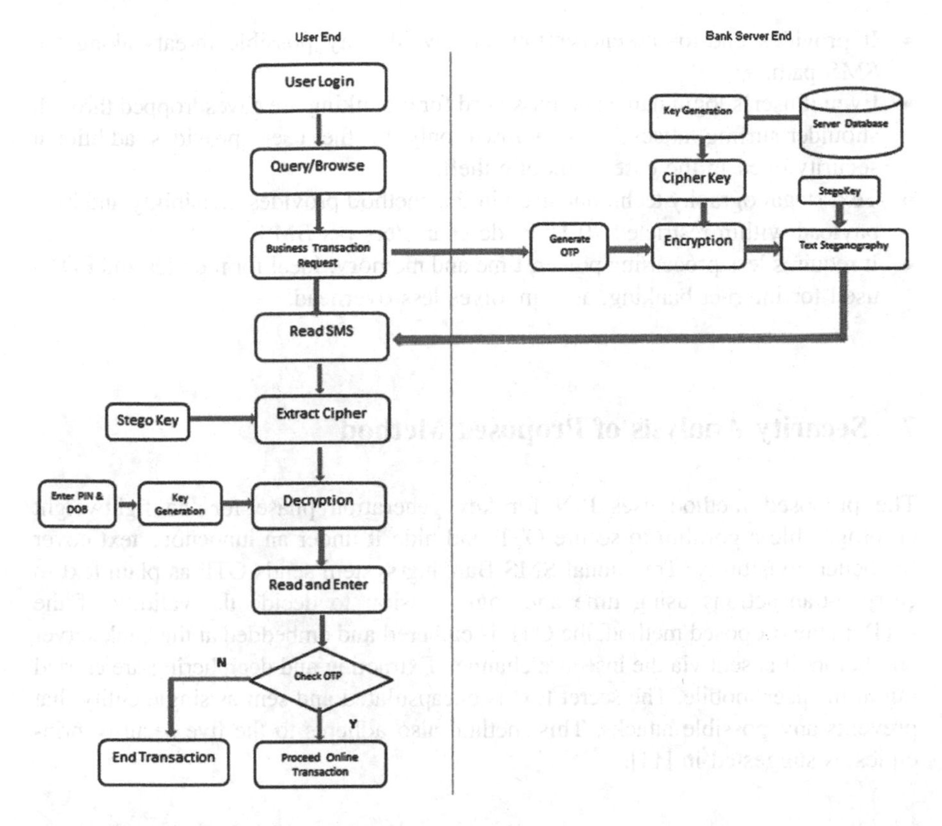

Fig. 2 Proposed Method for securing OTP

Step 9: The user is asked to enter the PIN and DOB. This is used for deciphering the cipher text.

Step 10: If PIN and DOB is correct, then OTP is deciphered and visible to the user. He proceeds with the online business transaction.

Step 11: If the PIN typed is wrong, cipher text C cannot be deciphered. Error is prompted and the transaction process comes to a halt.

These steps are carried out at the bank server and at user mobile, every time the particular customer initiates an online transaction.

6 Advantages of Proposed Work

- The 4-digit PIN and ciphering with steganography provides three levels of security. PIN—assumed to be kept as secret, assures flawless user identity and ciphering with stego under a simple text maintains invisibility.

- It provides end-to-end-encryption and avoids any possible threats along the SMS path.
- Even if user's login name and password for e-banking are eavesdropped through shoulder-surfing attack, PIN, known only to the user, provides additional security even in the case of mobile theft.
- Text steganography technique used in this method provides invisibility and high payload within restricted 70 Unicode characters per SMS.
- It requires less processing power, time and memory. Ideal for mobiles and PDAs used for internet banking, as it involves less overhead.

7 Security Analysis of Proposed Method

The proposed method uses PIN for key generation phase for the lightweight cryptographic algorithm to secure OTP and hide it under an innocuous text cover for better invisibility. Traditional SMS Banking system sends OTP as plain text to carry e-transactions using time and login session to decide the validity of the OTP. In the proposed method, the OTP is ciphered and embedded at the bank server end before it is sent via the insecure channel. Extracting and deciphering are carried out at the user mobile. The secret text is encapsulated and sent as single entity that prevents any possible attacks. This method also adheres to the five security principles as suggested in [11].

7.1 Integrity

The OTP travels as stego text under an innocent SMS from the bank server to the user mobile. If the content is altered midway the extraction of the OTP will not be possible even for the genuine user and hence attack is understood. Request for new OTP or terminating the session can proceed.

7.2 Confidentiality

In the proposed method, the OTP is encrypted with the secret parameters that are assumed to be a secret between the user and the server. The level of confidence depends on how secret these parameters are kept. The storing of PINs at the bank server should be fully protected and the user should not reveal the PIN to anyone. If any one of these happens then confidentiality is compromised.

7.3 Authentication

For authentication, the proposed method uses PIN supplied by the bank at the time of registering his personal identity, As PIN is not stored in mobile application the attacker cannot use this to decrypt OTP even if he steals the user's mobile.

7.4 Availability

The availability of OTP SMS depends on the factors listed below.

Mobile device: A smart phone with a mobileapp installed for extracting and decrypting is required. The algorithms are lightweight, thus not much processing power is required.

Network Service Provider: The signal strength, proper network elements and traffic handling power of the service provider decides the time within which the SMS delivered.

Capacity of the bank server: If the OTP validation fails, the proposed method discards the process and thereby reduces the load on the server. The bank server must be equipped with high speed processing elements to handle multiple requests generated at any point of time.

7.5 Nonr-Repudiation

The user need to register in person for the e-banking services by giving his mobile number, IMSI (International Mobile Subscriber Identity). The bank also provides the secret PIN during registration. Since the OTP SMS is sent to this unique number and the secret PIN must be used for reading the OTP, the user cannot deny the business activity.

8 Discussion

Table 1 clearly explains that the proposed algorithm has comparatively better secrecy, security, and authentication, invisibility for the minimal overhead in terms of time, processing power, and no extra hardware and software costs.

Table 1 Comparison of existing algorithms with the proposed method

Item	Traditional OTP	Technique 1 [2]	Technique 2 [3]	Technique 3 [9]	Proposed method
End-to-End encryption	No	Yes	Yes	Yes	Yes
OTP transfer via Internet	Plain text	Plain text	OTP under image	cipher text	cipher-stego text
Cryptography	No	Yes. Visual cryptography	Yes. AES encryption	Yes. RSA algorithm	Yes. Light weight feistal cipher
Steganography	No	Yes. Image steganography. Finger prints as cover image	Yes. Image steganography using LSB technique.	No	Yes. Text Steganography
H/W circuitry	Basic h/w	h/w to capture biometrics	No extra h/w	No extra h/w	No extra h/w
S/W requirements	Basic s/w	s/w for preprocessing of image	s/w for OTP retrieval from image	mobileapp to generate passcode	mobileapp to extract and decipher OTP
Sending media	SMS	SMS	Email	SMS	SMS
Levels of authentication	Two	Three	Two	Three	Three

9 Conclusion

Even though Internet is used extensively for banking and shopping transactions, research proves it to be very insecure. It is exposed to various threats like phishing, spoofing, hacking, and many more. As the enhanced technology improves the way banking and financial transactions are carried out, the attackers also use sophisticated techniques to access confidential data. SMS-based OTP was introduced to mitigate the attacks. Recently, research proves that OTP SMS is also under threat. An approach is proposed to secure OTP SMS through cryptography and text steganography with an additional authentication using a PIN. The method is very simple to use and can be used in all mobile phones with less time complexity and does not require extra overhead for its implementation but counters many of the attacks. It also addresses the five principles of secure services, and hence proves to be reliable over other protocols.

References

1. Ashraf Aboshosha et. at.: Multi-Channel User Authentication Protocol based on Encrypted Hidden OTP. International Journal of Computer Science and Information Securitty Vol 13. No. 6, 14–19 (2015).

2. Ms. Ankita R Karia et.al.: SMS-Based One Time Password Vulnerabilities and Safeguarding OTP Over Network. International Journal of Engineering Research & Technology (IJERT), Vol. 3 Issue 5,1339–1343 (2014).
3. A. Vinodhini, M. Premanand, M. Natarajan: Visual Cryptography Using Two Factor Biometric System for Trust worthy Authentication, International Journal of Scientific and Research Publications, Volume 2, Issue 3, 1–50 (2012).
4. Md. Nazmus Sakib et. al.: Security Enhancement Protocol in SMS-Banking using Digital Watermarking Technique, UKSim Fourth European Modelling Symposium on Computer Modelling and Simulation, IEEE 170–173 (2010).
5. Chang-Lung Tsai, Chun-Jung Chen: Trusted M-banking Verification Scheme based on a combination of OTP and Biometrics, Journal of Convergence. Vol 3, No. 3, 23–29 (2012).
6. Collin Mulliner et. al.: SMS-based One-Time Passwords: Attacks and Defense, Technical report (2014).
7. Souvik Roy, P. Venkateswaran: 2014 IEEE Students' Confernce on Electrical, Electronics and Computer Science (2014).
8. Farzad Tavakkoli et. al.,: A Novel Model for Secure Mobile SMS Banking, Journal of Information Assurance and Security, Vol 6, 369–378 (2011).
9. Safa Hamdare et al.: Securing SMS Based One Time Password Technique from Man in the Middle Attack, International Journal of Engineering Trends and Technology Vol 11, No 3, 154–158 (2014).
10. D Sumathy et al: A Framework of Security Issues and Standards for Efficient SMS, Int. J. Computer Technology & Applications, Vol 5, No 2, 469–478 (2014).
11. W. Stallings: Network Security Essentials-Applications and Standards, 4th edn. Pearson (2011).

Study on the Use of Geographic Information Systems (GIS) for Effective Transport Planning for Transport for London (TfL)

M.S. Mokashi, Perdita Okeke and Uma Mohan

Abstract This paper is intended to illustrate the impacts of geographic information systems in the planning of transport services, having the Transport for London as the precise case study. The study entailed an identification of GIS features, which support the planning of transport services. These features were studied in line with the current challenges faced by TfL, and were practically recommended as a means of tackling the problems. As a means of recognising bottlenecks, likely to be encountered in the future, an integration of GIS into travel demand analysis should be done (Alterkawi in GBER J 1(2):38–46 [2]). Both qualitative and quantitative methodologies are used as a means of collecting data useful to the research area. It was found that TfL is already in use of GIS, but to a minimal extent, the organisation is open to recommendations in this regard.

Keywords Geographic information systems · Transport networks in Europe · Transportation services · Web mapping · Data capture · Transport for London (TfL)

Track: *GIS for effective transport planning.*

M.S. Mokashi (✉)
Independent Consultant PI, Bangalore, India
e-mail: mohan_mokashi@yahoo.co.in

Perdita Okeke · Uma Mohan
London School of Commerce, London, UK
e-mail: okekeperdita@yahoo.com

Uma Mohan
e-mail: uma.mohan@lsclondon.co.uk

S.C. Satapathy et al. (eds.), *Proceedings of the International Conference on Data Engineering and Communication Technology*, Advances in Intelligent Systems and Computing 469, DOI 10.1007/978-981-10-1678-3_69

1 Introduction

Over the years, there has been an extensive development in the planning of transportation services in London. This change is reflected in the regularity of transport services, the ability of individuals to track public transport (buses and trains), and also get live updates on departure times, arrival times, routes, diversions, and changes to services.

This paper analyses means through which the use of geographic information systems could aid the Transport for London, in planning transport services more effectively.

1.1 Introduction to the Research Field

The transport network in London is extensively developed, and it includes private and public services. 41 % of London journeys are made using private transport systems, while 25 % are made using public services [23].

In London public transport network is the principal pivot for the UK in land and air transports. These services are rendered by TfL. Most transport networks, such as buses, trains and river services, are controlled by Transport for London [4].

Other rail services are either wholly operated by National Rail or contracted to external transport firms by the Department for Transport. Furthermore, there are numerous self-regulating airports functioning in London, for instance Heathrow, which happens to be the busiest in UK [8].

As a result of the 1986 privatisation, bus services were twirled off to a discrete operation grounded on viable proffering. Included in the establishment of the Greater London Authority, in 2000, TfL took over the management of London transport. It remains the publicly owned transport corporation for the city of London till date [21].

1.2 Introduction to the Research Focus

Through the use of Information Technologies, it has become possible to track most public transport (e.g., buses and coaches), get live departure times online, and live feeds. The transport service in London, alongside that of Paris, is one of the greatest in Europe [4].

At peak time, rush hours, and weekdays, certain stations tend to be busier and overcrowded; Geographic information systems could help TfL to decide the busiest areas, and provide better services in such areas [14].

Also, in cases of emergency, or scenarios where people will need to be in a certain place in the earliest time possible, better services could be provided. For example, better 'city to airport' shuttle services.

TFL already provides an express service to Heathrow Airport, which guarantees a 15-min journey to Heathrow [23]. However, this service is only guaranteed from Paddington. Through the use of geographic information system, the express service could be better planned to serve more areas across London.

Research Aim
This research aims to investigate means through which the efficient use of geographic information systems (GIS) could yield better planning of transport system, and resolve the current problems with the transport system in London.

2 Literature Review

2.1 Geographic Information Systems

Geographic information systems (GIS) are systems modelled to capture, operate, store and scrutinise various categories of topographical and spatial data [17]. GIS is a science of basic perceptions, application and structures.

GIS applications are tools, which enable organisations to initialise collaborative query, edit maps, scrutinise spatial data and render more competitive services [1, 9, 25, 26].

Maliene et al. [15] describe geographic information systems (GIS) as the basis of several services, reliant on location, which depend on visualisation and analysis, stating that such applications are often related to transport planning, management, telecommunication and other businesses.

2.1.1 Uses and Examples

According to Alsaheel [1], the execution of GIS is regularly prompted by application-specific requirements. Thus, most GIS applications are custom designed to meet the specific needs of an organisation. In essence, GIS developed for a transport organisation might not really be comparable with that, which has been developed for a telecommunications company. However, Wolfgang [28] suggested that GIS is always suitable and compatible with all sectors of an organisation. Nonetheless, GIS could help in the creation of explicit decision support systems (DSS) at an enterprise level.

2.2 Transport for London (TfL)

Transport for London (TfL) is the organisation responsible for managing the transport system in London. Basically, the organisation is constantly developing new strategies to manage transport services in London, and the headquarter is located in Windsor House, in the city of Westminster [23].

In 2000, Transport for London was created by the Greater London Authority Act 1999. London Regional Transport was the predecessor, and from it, TfL gained most of the foundational structures and functions. Bob Kiley, Ken Livingstone and Dave Wetzel were the first commissioner, chairperson and vice-chairperson of Transport for London. Until 2008, when Boris Johnson was elected as the Mayor of London, Wetzel and Livingstone remained in office [10].

Although TfL was in charge of transport networks since 2000, it was not responsible for the London Underground until after the agreement of the public--private partnership (PPP) contract for maintenance, which was very confidential, in 2003. Previously, the Metropolitan Police was responsible for managing the public carriage office [22].

2.3 GIS as an Information System for Transport for London (TfL)

Alterkawi [2], in agreement with Alsaheel [1] stated that the purpose of integrating GIS to travel demand analysis process is for the forecasting of future congestion areas, shortest path and travel time allocation for popular centres.

Not only does GIS include features, which improve access to services, they present information accurately, according to positioning and location, which can be used to enhance planning [27].

2.3.1 Key Features of GIS Used in Transportation

GIS is capable of portraying graphics, while connecting features with attribute table. This specific feature is valuable for the maintenance and update of roadway network. GIS further recognises errors in associations and trips loading from traffic analysis zone [2]. Antenucci et al. [3] highlighted road design, traffic volumes, highway mapping and analysis of accident data as some transportation applications of GIS.

3 Research Methodology and Data Collection

Deductive or Inductive Approach?
While deductive approach involves the analysis of already existing hypothesis, inductive approach starts with specific research questions, upon which data is collected and researcher either analyses data from a new perspective or builds on (enhances) an already existing foundation [5].

A deductive approach therefore begins with a general level of focus, while an inductive approach kicks off with a specific focus level.

Deductive Approach
This research was not based on existing hypothesis, neither is it concerned with testing such hypothesis for a point to be proved. Deductive approach is therefore not the suitable design model for this research.

Theory	⟶	Analyse Data	⟶	Support Hypothesis or not
General Focus Level		*Analysis*		*Specific focus level*

Inductive Approach
This research commenced with a main research question, from which sub-questions were formed, and data collected for the purpose of creating or developing a theory. The inductive approach is being used for this research.

Data Collection	⟶	Pattern Discovery	⟶	Develop Theory
SpecificFocus Level		*Analysis*		*General focus level*

3.1 Data Collection Method

3.1.1 Survey

Both Gerring [11] and Remenyi et al. [19] agree that surveys are a set of predetermined questions. As a quantitative means of obtaining data, a **survey** was created to analyse the experience of customers as they used public transport services within London.

3.1.2 Interviews

Boyce and Neale [6] and Kahn and Cannell [12] agree that an interview is a focused dialogue involving multiple parties. Saunders et al. [20] categorise interviews as structured, unstructured and semi-structured. The latter is used for this research.

3.1.3 Choice of Sampling

Sampling types	Explanation	Advantages	Disadvantages
Random sampling	The chances of each member of the population to be chosen is equal E.g. Pulling names out of a cap	For larger samples, it renders the best chance of unbiased representative sample	It could be difficult to achieve being that it is time consuming
Stratified sampling	Splitting the population into subcategories, such that members are selected in the proportion that they occur in the population E.g. 2.5 % of population are blacks, so 2.5 % of chosen sample should be blacks	An intent effort is made to make the sample a fair representation of the population	It takes time to identify subcategories and calculate proportions
Volunteer sampling	This comprises individuals who have volunteered to be involved in a study. E.g. Responses from volunteering advert	It is convenient	This could lead to a participant being biased, and is therefore considered to be unrepresentative
Opportunity sampling	A selection of people who are available at a given time E.g. Walking into an organisation and interviewing available individuals	This is economical, convenient and quick	This could lead to biasness, as the researcher might partially select only those that are helpful

Opportunity sampling was used for this research because it is a very quick method of sampling, which does not require ambiguous procedures, and will not put an economical constraint on the researchers.

3.1.4 Sampling Size

According to Proctor [18], there is no ideal sample size, as it depends on the nature of problem investigated. Thus, having considered the research objectives of this study, 100 survey responses and 10 interviews are regarded as ideal, within the given time.

A total of 784 survey invitation emails have been sent out, and a survey web link is made available online.

3.2 Data Analysis Tools

Survey monkey is embedded with great analytical tools, with which the data received from the survey filled in by respondents were analysed, classified and simplified. Through the use of ***excel spread sheet***, a summary of the survey was portrayed in bar charts, showing the number of collections received on each day covered by the survey period.

In line with the inductive approach, using these tools, a pattern is discovered after the collection of data, and a theory established.

3.3 Ethical Aspect

Collecting data regarding TfL, being a well-known established organisation, had to be done with extra care and precaution, to ensure that their terms and conditions are not breached. Confidentiality was a major area of concern, specifically considering the fact that as part of the research, a survey was sent to emails of various individuals, and it was necessary to ensure that their emails are not used for any other purposes beyond this research, and are not transferred to third parties, in line with the Data Protection Act.

4 Discussion of Main Findings

4.1 Discussion of the Main Findings in the Context of the Literature Reviewed

Bulc [7], Turro [24] and several other authors of publications gave their views on Transportation Networks in Europe; their views did not particularly contrast, but were complementary.

Alterkawi [2] particularly discussed the suitability of GIS for transport planning and elaborated on the specific features of GIS, which support the planning of transport services. This tallied with some of the findings of the research, as indeed, GIS contains web mapping, travel demand forecasting, shortest route algorithm, spatial data analysis, data capture, amongst other features, which are useful for the planning of transport services.

In agreement, Alsaheel [1], Maliene et al. [15], Volk [25] and Warite [26] further discussed the benefits of GIS and their suitability for planning transport services. Indeed, through the efficient use of GIS, forecasting future travel activities will be made possible, thus making the planning of transport services more effective.

However, Wolfgang [28] insinuated that GIS was fit for the purpose of every challenge an organisation may have. This was found to be a limited statement as TfL had some challenges, which GIS was not the solution to. For example, increasing the capacity of tube and buses is not so much a GIS-solvable challenge as managing underground traffic.

Although Koontz [13] and O'Shaughnessy [16] did not entirely agree with Wolfgang [28], they admitted that GIS is suitable for various other purposes aside transportation. However, they mentioned that the implementation of GIS is very challenging.

Nevertheless, during the course of the research, it was found that there is no specific application, for which GIS is more suitable than others. In essence, GIS is suitable for various sectors, and is designed according to the requirements of the sector. Therefore, a geographic information system designed for an engineering firm might not perfectly be suitable for use in a transport network. This finding was very much in alignment with the argument of Alsaheel [1].

Alterkawi [2], Volk [25] and several other authors perceived GIS to be a technology, without which transport services cannot be operated. This was right, but limited, as TfL is actually making use of GIS (to a minimal extent), but does not rely on it. Therefore, the use of GIS is not the foundation of transport planning, but enhances and improves it.

4.2 Conclusion

This project entailed investigating means through which an efficient use of geographic information system could help the Transport for London to plan transport services more efficiently and effectively.

It was discovered that TfL is already making use of GIS, but to a limited extent; thus recommendations were made for the organisation to optimise its use of GIS for more effective results. Recommendations were also made on how GIS could be used to tackle some of the specific challenges faced by the organisation.

There is scope for further research in this area.

References

1. Alsaheel, M. (2014). GIS implementation in Organizations: Strategies and Issues at Departmental Level. *Geographic System "GIS"*. 14 (5), 10.
2. Alterkawi, M (2001) Application of GIS in Transportation Planning: The Case of Riyadh, the Kingdom of Saudi Arabia. *GBER Journal.* 1(2). pp. 38–46.
3. Antenucci, J., Brown K., Croswell, P. Kevany, M. & Archer, H. (2012) Geographic Information System: A guide to the technology. New York. Springer US.
4. Balaban, D. (2010) Transport for London to Discard MIFARE Classic. *NFC Times Paris.* 22 (14). p. 18.
5. Blackstone, A. (2015) Principles of Sociological Inquiry: Qualitative and Quantitative Methods. New York. Flat World Education Inc.
6. Boyce, C & Neale, P, (2006) Conducting in-depth Interviews: A Guide for Designing and Conducting In-Depth Interviews, USA, Pathfinder International Tool Series.
7. Bulc, V. (2015). Trans-European Transport Network. *Mobility and Transport.* 10 (2), p. 33–55.
8. Civil Aviation Authority (2014) Terminal and Transit Passengers 2012 comparison with previous year.
9. Clarke, K.C. (2012) Advances in Geographic Information Systems, Computers, Environment and Urban Systems. Vol. 12, pp. 175–184.
10. Curtis, S. (2014) Transport for London goes Contactless. The Telegraph. 61(4). p. 9.
11. Gerring, J. (2007) *Case study research: Principles and practices,* Cambridge, Cambridge University Press.
12. Kahn, R.L. & Cannell, F. (1957) The Dynamics of Interviewing: Theory, Technique and Cases. New York. Wiley.
13. Koontz, L.D. (2003). Geographic Information Systems: Challenges to Effective Data Sharing. *United States General Accounting Office*. 8 (3), 9.
14. Maguire, D.J. & Kouyoumjian, V & Smith, R. (2008) The Business Benefits of GIS: An ROI Approach. New York, USA. PA Consulting Inc.
15. Maliene, V., Grigonis, V., Palevicious, V. & Griffiths, S. (2011) Geographic Information System: Old Principles with new Capabilities. Urban Design International 16 (1). pp. 1–6. doi:10.1057/udi.2010.25.
16. O'Shaughnessy, B. (2003). Challenges of implementing GIS in a gas utility: Realising the Benefit. *the agi conference at GeoSolutions 2003*. 2 (3), 11.
17. PolicyMap (2015) Fully Web-Based Geographic Information System (GIS) and Mapping. http://www.policymap.com/ [Retrieved 26th March, 2015].
18. Proctor, T, 2003, Essentials of Marketing Research, 3rd edition, New Jersey, Prentice Hall.
19. Remenyi, D., Williams, B., Money, A. & Swartz, E. (2003) *Doing research in business and management: An introduction to process and method,* London, SAGE Publications.
20. Saunders, M., Philip, L. & Adrian, T. (2009) Research Methods for business students. 5th ed. Harlow, Pearson Education.
21. Sinha, S.P., Faguni, R., Prasad, M. (1993). Instant guide of geography. 23. Transportation geography, 1st Ed. New Delhi: Mittal publ.
22. TfL Investment Programme (2010) *London Underground*. Available at: http://www.ascenda-mcl.com/IMG/pdf/TfL-investment-programme-london-underground.pdf [Retrieved on: 4th May, 2015].
23. TfL (2014) TfL introduces Oyster 'One More Journey' on London's buses. Available at https://www.TfL.gov.uk/info-for/media/press-releases/2014/June/TfL-introduces-oyster-one-more-journey-on-london-s-buses [Retrieved on 13th May, 2015].
24. Turro, M. (1999) Going Trans-European: Planning and Financing Transport Networks for Europe. Oxford. Pergamon Press.
25. Volk, K. (2014). Benefits of using GIS mapping to manage city infrastructure. *Moore Engineering Inc.*. 31 (23), p 54–76.

26. Waite, R. (2011). The Benefits of GIS. *Enhance your IT Strategy*. 20 (11), p 99–100.
27. Winther, R.G. (2014) "Mapping Kinds in GIS and Cartography" in Natural Kinds and Classification in Scientific Practice, edited by C. Kendig.
28. Wolfgang K. (2010) The Mathematics of Geographic Information Systems (GIS) Vienna Austria, University of Vienna.

An Integration of Big Data and Cloud Computing

Chintureena Thingom and Guydeuk Yeon

Abstract In this era, Big data and Cloud computing are the most important topics for organizations across the globe amongst the plethora of software's. Big data is the most rapidly expanding research tool in understanding and solving complex problems in different interdisciplinary fields such as engineering, management health care, e-commerce, social network marketing finance and others. Cloud computing is a virtual service which is used for computation, data storage, data mining by creating flexibility and at minimum cost. It is pay & use model which is the next generation platform to analyse the various data which comes along with different services and applications without physically acquiring them. In this paper, we try to understand and work on the integration model of both Cloud Computing and Big Data to achieve efficiency and faster outcome. It is a qualitative paper to determine the synergy.

Keywords Big data • Cloud computing • Software • Research • Integration

1 Introduction

Many organizations have grown and made the data associated with them also grew exponentially. Almost all the MNC's have data for multiple applications and in variety of formats. Sometimes, the corporates face many hurdles to keep all data securely. In 2011, research work by Mr. Gartner showed how there is a steep rise in data generation which implies that more and more data is emerging from all the spheres and it goes unchecked without analysis. This induced a need to develop a new technique to handle the enormous data available so that maximum utilization

Chintureena Thingom (✉) · Guydeuk Yeon
Centre for Digital Innovation, Christ University, Bangalore, India
e-mail: chintureena.thingom@christuniversity.in

Guydeuk Yeon
e-mail: yeon@christuniversity.in

S.C. Satapathy et al. (eds.), *Proceedings of the International Conference on Data Engineering and Communication Technology*, Advances in Intelligent Systems and Computing 469, DOI 10.1007/978-981-10-1678-3_70

can be made as such it yielded to new development named Big Data. It is the most advanced data mining technique to analyse to the best as compared to all other research methods [1]. Enormous data available to the corporates which need to be used for the development is about to turn imperfect, complex, often unstructured. For this, they used advanced computational tools which have developed to reveal and identify trends and pattern within and across large database that would often otherwise remain undiscovered. It is difficult to execute the work by using the existing database management systems and other applications and desktop statistics and visualization packages, which takes lots of simultaneously run software in different servers and consumes a very large storage size. "Cloud computing is a model for on demand network which provide access to a plethora of different configuration computing resources which can be provided with least management effort" [2]. Before cloud computing, orthodox business applications have always been not so user friendly and complicated. The huge amount of hardware and software required to run them and it is difficult to manage all the components to gather together. The system requires a large team of experts to configure or install and then to process it. With cloud computing, the user can ease the process as the requirement or the component needed to run the system will be provided by the experienced vendor. In this digital age, the aspect of work activities and personal life are heading towards the availability of everything in the virtual online world. Many organizations and web based companies like Google, Yahoo, and other IT firms comes with the most powerful computing system "Cloud Computing" which means that there can be sharing of web infrastructure like IaaS, Paas and SaaS for data storage, computations and its scalability. The shared infrastructure means it works like a utility. In cloud computing, the model works on pay-per-use. Also it has automatic upgrades and scaling up or down is easy. Cloud-based apps can be made and execute in very less time with minimal cost [3]. All the process like anything from the basics to integration into the multimedia processing can be accomplished more effectively using cloud computing than using one's personal computer.

As the cloud computing is an internet-based services, it offers various companies an effective method of conducting their business by promoting their product or peoples without having a physical building or infrastructure of their own and it builds the economies of scale while enabling to achieve greater efficiency and by reducing costs. Now-a-days most of the world's largest companies are simply moving their non-cloud products and services to cloud computing and running all kinds of apps in the cloud, like Human Resource planning, Accounting, Material distribution and much more.

2 Big Data Technologies

Big data is a vital system which requires advanced technologies to efficiently process the enormous quantities of data.

2.1 Column-Oriented Databases

Traditional, row-oriented databases are excellent for OLTP (Online Transaction Processing) with high update speeds, but less on query performance as when the volume of data to be analyse grows, the data leads to an unstructured system and gives slower execution. In this method of Column oriented databases, it will store the data on the columns, instead of rows, allowing for huge data compression and the outcomes come out very fast. The disadvantage of these databases is that they will generally takes the updates on batch wise which gives slow update time than traditional models.

Map Reduce

This is a programming model that allows for massive data processing, scalability against thousands of servers or clusters of servers. Map Reduce implementation consists of two tasks:

- "Map" task: The input dataset will be converted into a different set of key/value pairs, or tuples;
- "Reduce" task: The results of the "Map" task will be combined to form a reduced set of tuples (hence the name).

2.2 Hadoop

Hadoop is by far the most popular implementation of MapReduce. This is developed by Google but Apache in developed a generalized software framework today called as Hadoop, being an entirely open source platform for handling Big Data. It has several different uses but it is the most efficient applications for large volumes of dynamic data which changes in real time basis, such as location-based data, weather related data or the real data of web-based or social media data, or machine-to-machine transactional data.

2.3 Hive

Hive is a data warehouse infrastructure system to provide data summarization and supports analysis of large datasets stored in the systems. It consist of Hive QL which is provided by Hive system and it also integrate map/reduce factor. To accelerate queries, it provides index. It is initially developed by Facebook.

2.4 Pig

SQL is one of the data flow language which is not suitable for big data so PIG was developed. It consists of a "Perl-like" language that allows faster activity execution over the data analysis on a Hadoop cluster which also can be written in Java and other languages also. PIG was developed by Yahoo!

3 Cloud Computing Providers Offer Their Services According to Several Fundamental Model

3.1 Software as a Service (SaaS)

It provides user with facility to run software on a cloud infrastructure without physically acquiring it. The applications are accessible anytime anywhere. It just need to have internet connectivity. Clients need not required to install any software on their device. They just need a web browser and network connection [4]. SaaS applications are designed and delivered to the user over the internet. The client has to pay only for services which he used as it is pay-per-use model. For example, with Microsoft Office 365 you can use services of Microsoft word, power point or excel without actually installing it on your computer. You just need to pay the fee which will be charged only for the services you had used. It might be a monthly fee. You can use this software anytime from anywhere.

3.2 Platform as a Service (PaaS)

PaaS is one of the best service which is provided through cloud where the set of tools and services designed to make coding and deploying those applications quickly and efficiently. Paas through cloud provides a platform for researcher or a developers to execute or create their applications on provider's platform over the internet [5]. PaaS provides physical server for software over the web. The user can control the deployed application but does not have control over cloud infrastructure. PaaS is mostly used by software development teams. Ex. Google Apps.

3.3 Infrastructure as a Service (Iaas)

Customer can rent a data center environment without worrying about maintenance i.e. it provides users with virtual servers. IaaS generally can help to achieve higher efficiency as it can includes multiple users on a single hardware [2]. IaaS can be implemented on the individual servers, computing resources, networks or messaging systems. All application are provided to the clients after charging a minimal fee. User has to pay only for services that he/she uses. i.e. pay for what you use model is applied. An organization can build a complete infrastructure using IaaS.

The early driver of widespread adoption of cloud computing was the SaaS delivery model. It offered the user scalability and customization based on their needs and goals. Since then, a combination of technologies has emerged which have further increase the demand for cloud computing.

- Server virtualization: Through this, multiple users can access a single server with their various hardware and software as they are pooled together. It gives the same feel and the nature of the server is maintained like an individual dedicated server, but without incurring any additional cost.
- Service-oriented architecture (SOA): Organizes software code so that one set of data, and the code written to process it, can be reused by other applications in the organization [6].
- Open source software: The software is made available to all the cloud users with no restriction or limitation of copyright duplication.
- Web development: The platform for web development is provided through the cloud have reduced the organization cost and made available to use it with more efficient manner.

4 Big Data Characteristics

Big Data has the following characteristics i.e. Variety, Velocity and Volume

4.1 Variety

Big Data can analyse any form of Data. It can be stored in multiple formats in databases, or excel access or in a simple text file. Sometimes, the data is not stored in the traditional format as we assume, it can be in any other form also. It can be a short video clips, a text message or in pdf or jpeg format. The organization need to gather these data and form a structured or meaningful data for the effective processing. It will be easy if we have data in the same format, but it may not be in all the time. This kind of challenges we can overcome with the Big Data. These varieties of the data represent Big Data.

4.2 Velocity

In simple words, velocity means "The speed of data when it comes in and out". The enormous growth of data in social media explosion have changed how we look at the data. We feel that the yesterday data is recent. The matter of the fact newspapers is still following that logic [7]. On social media sometimes a few seconds old messages is not something interests users of social media. The users ignores old messages and pay attention to recent updates. The movement of data is now dynamic and almost real time. This high velocity data represent Big Data.

4.3 Volume

We are currently seeing that there is very fast growth in data storage not only text data but data in different formats like videos, images and music formats. It is very common to have Terabytes and Petabytes of the storage system for enterprises. The applications and architecture built to support the data needs to be revaluated quite often as the data volume increases. The big volume indeed represents Big Data.

5 Cloud Computing Characteristics

5.1 On Demand Self-services

We would like to mention that the computing environment such as email services, applications development platform, network or server service can be provided to the user even without requiring user interaction with the service provider. On demand self-services include Amazon Web Services (AWS), Google. Cloud services are available and shared over the internet or virtual network and can be accessed through any standard equipment such as mobile phones, laptops and PDAs. The resource provider's computing devices are gathered together and then extend its services to multiple consumers from different physical locations and the services are provided through the virtual resources dynamically after assigning according to consumer demand [8]. The resources shared over the internet includes storage, processing, memory, network bandwidth, virtual machines and email services. The pooling together of the resource and sharing the services with different user on pay-per-use model builds the efficiency and increases economies of scale (Gartner).

5.2 *Rapid Elasticity*

Cloud services can be rapidly and elastically unrestricted, in some cases automatically, to suit the economies of scale. The scalability of the cloud computing services can be made to the capacity required by the users. As the cloud services requirement comes to the user, its services can be extended to any number of user. The Cloud capabilities which are available in the virtual platform can be bought anytime anywhere by anyone.

5.3 *Measured Service*

Cloud computing resource is a pay-per-use model and it thus provides an effective and transparency for the service provider and also the consumer of the utilized service. The cloud usage can be measured, controlled, and reported to meet the user's requirement and in its comfort level. Cloud computing services use a metering device which enables to control and optimize resource use. This implies that just like electricity or municipality water services are charged per usage metrics-pay per use. The more you utilize the higher the bill [9].

5.4 *Multi Tenacity*

It is a new addition in the features of cloud computing. It governs the use and security of data shared over the internet in a cloud system. It refers to the policy and the process for the enforcement, segmentation, isolation, billing models for all the different users.

6 Combined Working of Big Data and Cloud Computing—Benefits

Integration of Big Data and Cloud Computing is the two most vital technique in the world of Information Technology. These two IT initiatives can change the entire business world and also influence the way we can analyse a data. Most cloud vendors are already offering hosted Hadoop clusters that can be scaled on demand according to their user's needs [10]. As cloud computing continuous to mature, a growing number of enterprises are building efficient and agile cloud environments i.e. cloud computing has a structure to support their big data projects. Large to

Medium sized companies are getting more value from their data than ever before with the Cloud computing, by enabling intense fast analytics at a fraction of previous costs [11]. Thus, Big Data & Cloud technologies using as a pair organizations can make big data analytics in clouds a reasonable option, since, data is becoming more valuable. This, in turn drives companies to acquire and store even more data, creating more need for processing and retrieving [12]. Thus, Big Data and cloud computing go hand-in-hand.

7 Conclusion

In my view this paper presents an outlines of cloud computing and big data. Cloud computing has become major discussion thread in the IT world. It is reduces the cost of purchasing physical infrastructures like email servers and software. Big data is best for fast query performance, massive data processing and scalability. Hence, it is not suitable for small business i.e. where fewer amounts of data are capture, manage and process.

References

1. White, Tom (10 May 2012). Hadoop: The Definitive Guide. O'Reilly Media. p. 3. ISBN 978-1-4493-3877-0.
2. PS Ryan, S Falvey, Sarah and R Merchant (2013-12-19)."When the Cloud Goes Local: The Global Problem with Data Localization". IEEE Computer. Retrieved 2013-02-17.
3. M. Sharma, H. Bansal, AK Sharma, Cloud Computing: Different Approach & Security Challenge, IJSCE, 2012.
4. T. B. Winans, J. S. Brown, Cloud Computing, A collection of working papers.
5. Prathmesh Arnikar, Siddharth Sonawane, Ankita Fale, Sagar Aghav, Shikha Pachouly "Load Balancing In Cloud Computing" in Journal Of Information, Knowledge And Research In Computer Engineering.
6. Meenakshi Sharma, Anitha Y and Pankaj Sharma "An Optimistic Approach for Load Balancing in Cloud Computing" in International Journal of Computer Science and Engineering, Volume-2, Issue-3 March 2014.
7. Sumit Khurana and Khyati Marwah "Performance Evaluation Of Virtual Machine (Vm) Scheduling Policies In Cloud Computing (Spaceshared & Timeshared)" in IEEE July 2013.
8. Enda Barrett, EndaHowley, Jim Duggan "A Learning Architecture for Scheduling Workflow Applications in the Cloud" 2011 Ninth IEEE European Conference on Web Services978-0-7695-4536-3/11 $26.00 © 2011 IEEE.
9. BoonyarithSaovapakhiran, George Michailidis†, Michael Devetsikiotis "Aggregated-DAG Scheduling for Job Flow Maximization in Heterogeneous Cloud Computing" 2011 IEEE 978-1-4244-9268-8/11.
10. Hsu Mon Kyi, Thinn Thu Naing "An efficient approach for virtual machines scheduling on a private cloud environment" 2011 IEEE 978-1-61284-159-5/11.

11. A nist national definition of cloud computing, http://www.csre.nist.gov/groups/SNS/cloudcomputing.
12. Tejinder Sharma, Vijay Kumar Banga "Efficient and Enhanced Algorithm in Cloud Computing" in International Journal of Soft Computing and Engineering (IJSCE) ISSN: 2231-2307, Volume-3, Issue-1, March 2013.

A New Approach for Rapid Dispatch to Remote Cooperative Groups with a Novel Key Archetype Using Voice Authentication

T.A. Amith, J. Prathima Mabel, Rekha Jayaram and S.M. Bindu Bhargavi

Abstract Any newly emerging network must deal with problems of effectively and securely broadcasting data to remote cooperative groups. Limited or no network connectivity, absence of valid encryption/decryption protocols, and the individual metrics concerning the sender are some of the problems that plague newly established networks [1]. Existing techniques to overcome these problems do not address future concerns such as scalability, security, etc. The generation of a valid key that does not need to be changed or managed regularly is essential to a newly established network as it offloads the computational overhead and reduces the connection cost to a fraction of what it was before. This paper explores the possibility of generating such a key making use of acoustic model training and voice-tag creation. It also provides an overview for simple, efficient, and easy addition or deletion of members to cooperative groups, while keeping the rekeying mechanism as flexible as possible so as to increase the network's robustness. The implementation is designed in such a way that the group size does not factor into the computational overhead or the communication cost. The strong encryption protocol detailed in this paper, along with a simple and easy way to manage the implementation, makes this a viable solution to the problems that are associated with new networks.

Keywords Ad hoc networks · Broadcast · Cooperative computing · Access control · Information security · Key management

T.A. Amith (✉) · J. Prathima Mabel · Rekha Jayaram · S.M. Bindu Bhargavi
Dayanand Sagar College of Engineering, Bengaluru, Karnataka, India
e-mail: amith.tallanki@gmail.com

J. Prathima Mabel
e-mail: prathimamabel@gmail.com

Rekha Jayaram
e-mail: rekhajayaram20@gmail.com

S.M. Bindu Bhargavi
e-mail: bindu.sm@gmail.com

S.C. Satapathy et al. (eds.), *Proceedings of the International Conference on Data Engineering and Communication Technology*, Advances in Intelligent Systems and Computing 469, DOI 10.1007/978-981-10-1678-3_71

1 Introduction

The current uptick in the adoption of wireless mesh networks (WMN's) has exposed how deeply the problem of security and privacy is rooted in any newly established network. A WMN is by its very nature an open and distributed system. A WMN is multihop layered network which provides high-speed Internet access points with dedicated stationary mesh routers [2]. The final band consists of a significant quantity of portable device users. As such, it is vital to administer network access control to deal with the problem of malicious attackers and/or freeloaders who inadvertently introduce congestion into WMN's. A WMN also contains enormous amounts of information that is user-specific and user-sensitive. Thus, it is imperative that one enforces adequate security protocols to protect against attacks on the network and privacy control mechanisms that ensure that the user-specific data is not revealed to others [3]. The attacks that can be carried out on a WMN vary from run of the mill passive eavesdropping techniques to highly specialized and targeted phishing to message interception and alteration. The success in the adoption rate of WMN's depends on successfully rooting out these problems along with mounting a solid line of defense against all attacks. The most popular way of distributing keys is the Diffie–Hellman protocol [4]. The algorithm enables two users to compute a common key using both secret and publicly exchanged information. However, this method does not scale well. When the number of users is extended even by as less as two additional members, the Diffie–Hellman protocol fails as there is no provision for selecting groups or subgroups of members to assign the keys to. Our project aims to extend the Diffie–Hellman algorithm in a simple yet flexible way by making use of an acoustical model to train an heuristic algorithm.

The new hybrid archetype that is prospective in this paper is a combination of the established broadcast encryption and GKA (Group Key Agreement). In the prospective system each end user in the group maintains a key pair composed of a secret key and a public key. The remote sender chooses his group in an ad hoc way and then determines the public key of each of his recipients. This is a substantiated way to secure transmissions. The transmitted messages cannot have any useful information extracted from them even if some or all of the end users other than the intended recipients are in collusion to intercept messages. This paper presents a comprehensive implementation of the above prospective of hybrid archetype. The structure of the paper is as follows: Sect. 2 gives details about the shortcomings that persist in the various existing systems and Sect. 3 deals with the prospective system. Sections 4, 5, and 6 elaborate on the implementation of the prospective system. Finally, we conclude the paper in Sect. 7.

2 Existing Systems

The existing systems currently use a group key agreement technique that provides an adequate solution that secures communication amongst members of the same group. However, for a remote sender, it is required him or her to stay online for multiple rounds along with the rest of the users who are members of the group to negotiate the key and all members of the group are required for decryption, and thus the messages which have undergone encryption. This removes the reliance on a server-generated key system and ensures that proper coping mechanisms are implemented for the interactions between the recipients and the sender.

The new key management system is of a hybrid form that has been derived as a result of extensive experiments that have been performed in context to mobile ad hoc networks [5–7].

The prospective protocol has been enumerated in Sect. 3 (Fig. 1).

3 Prospective System

During the first trial, the public group encryption key is extracted, and a constant complexity is enforced between the sender and the receiver. This applies even if the members of the group/subgroup are changing constantly or the system updates for rekeying. The prospective system scores highly on that security front as, even with collusion from non-intended recipients, an attacker performing an active attack will be unable to extract any valid information from the messages transmitted with the aid of the public/private keys. A variation of the Diffie–Hillman algorithm serves as

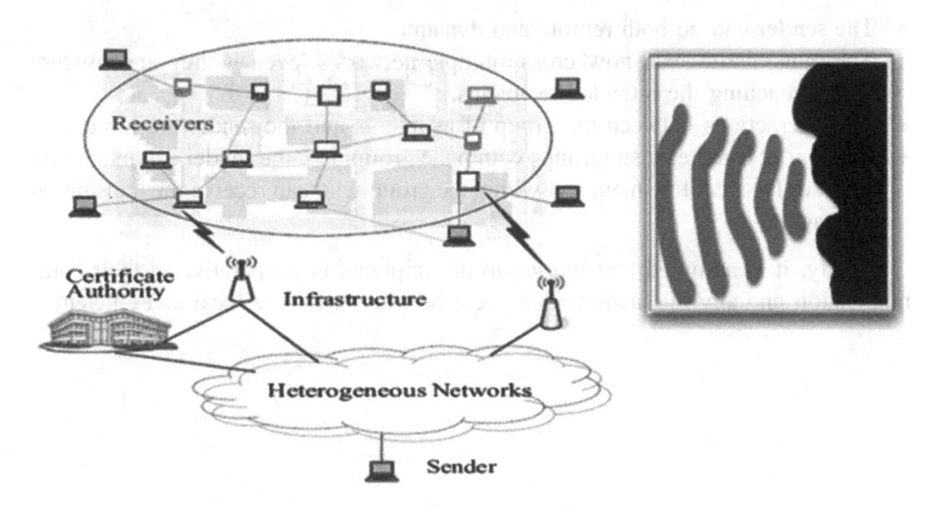

Fig. 1 Prospective system architecture

the proof in this instance. This paper proposes two different conversion algorithms that make use of voice-to-phoneme conversion.

Abstractions from voice tags which are independent of the speaker applications in embedded platforms are extracted. This in turn lessens CPU loads and ensures faster memory retrievals. A phonetic decoder which is independent of its speaker undergoes a batch conversion of multiple phonetic hypotheses. Repeated utterances of the same voice tag will result in a general hierarchy of the phonetic consensus that is preserved by the conversion algorithm. The hypotheses which are phonetically admissible are then automatically chosen to represent the voice tag. As a control to the above-described procedures, an expert phonetician was employed to produce similar transcriptions of the voice-tag references. A detailed comparison was then performed for the voice-tag representations obtained using the above-mentioned algorithms against the manual transcription obtained from the expert. With only a marginal degree of error, both the algorithms performed exceptionally well against the manual transcription method. This leads to the favourable conclusion that both algorithms are effective in their targeted ranges and for their specific purposes. Our contribution includes three aspects. A formalization statement for secure dispatch to secluded cooperative groups, given certain constraints, and the implementation of a one-to-many efficient and secure channel is established. This is done so that we can implement a voice-based security system which consumes less computer resources. Second, a new key management archetype is prospective that allows transmissions to be delivered without compromising either security or efficiency [8]. This is done by effectively circumventing constraints and taking advantage of some of the mitigating features. A hybrid approach that is characterized by a GKA system and a public key broadcast encryption technique lies at the crux of both the algorithms [9]. This is highly effective, if in case either member changes or system updates need to be performed. This new approach yields the following advantages:

- The sender can be both remote and dynamic.
- The transmission can now cross-multiple networks, even if they are insecure before reaching the intended recipients.
- The interactions between the group of recipients and the sender are limited.
- The sender can create subgroups within the group, i.e., the sender can make sure that it only select few members within the group who can receive the transmitted message.

Finally, it is assumed that the group of recipients is cooperative in their communication and any interactions that occur between them are local and efficient.

4 Implementation

Implementation can be defined as the actualization of a theoretical concept. It is the execution of a plan, idea, or model in such a way that the results obtained can be considered successful. Many implementations may exist for a given standard or a specification. Implementation is one of the most important steps in the software development life cycle (SDLC). This phase of the system is conducted with the idea that whatever is conceived, designed, and developed must be implemented, keeping in mind that it fulfills scope, objective, and user requirements of the initial problem. The implementation phase produces a complete solution to the initially defined problem.

The implementation in this paper is carried out along the following lines. In the first step, multiple nodes are created which act as access points to the multiple portable devices such as PDA's, smart phones, tablets, wearables, etc. Then, a connection is established amongst these devices. Once the connection is established, a private/public key pair is generated. This is further clarified through key management to make sure that every recipient of the intended group is assigned a key. Then the group is divided into various subgroups in an ad hoc way. A rekeying process takes place to make sure none of the non-intended members are assigned keys. A message authenticator is used to verify the validity of each user to give accurate results.

5 Key Management

The most important aspect of group-Oriented communications when it comes to providing or restricting access is concerned with key management. An ideal key management archetype allows no compromise on security or efficiency during transmissions to groups even when they are remote. This is done through various methods, by either effectively exploiting mitigating factors or circumventing constraints [10]. This system securely distributes session keys to the intended recipients, the system in of itself acts as the session key and simultaneously encrypts any messages under the session key, which can be decrypted only by the intended recipients. In the prospective scheme, an authentication key is a pair of public/private keys and a certificate signed by the base station which is distributed before hand in each cluster head. Member sensor node identities are verified using the authentication key. The base station and all the cluster heads are made aware of the authentication key beforehand. Pair-wise keys are established amongst the cluster heads using the public/private key. Each sensor node is also made aware of an authentication key and the public key of the base station. Public key is used to verify the certificates of the cluster heads (Fig. 2).

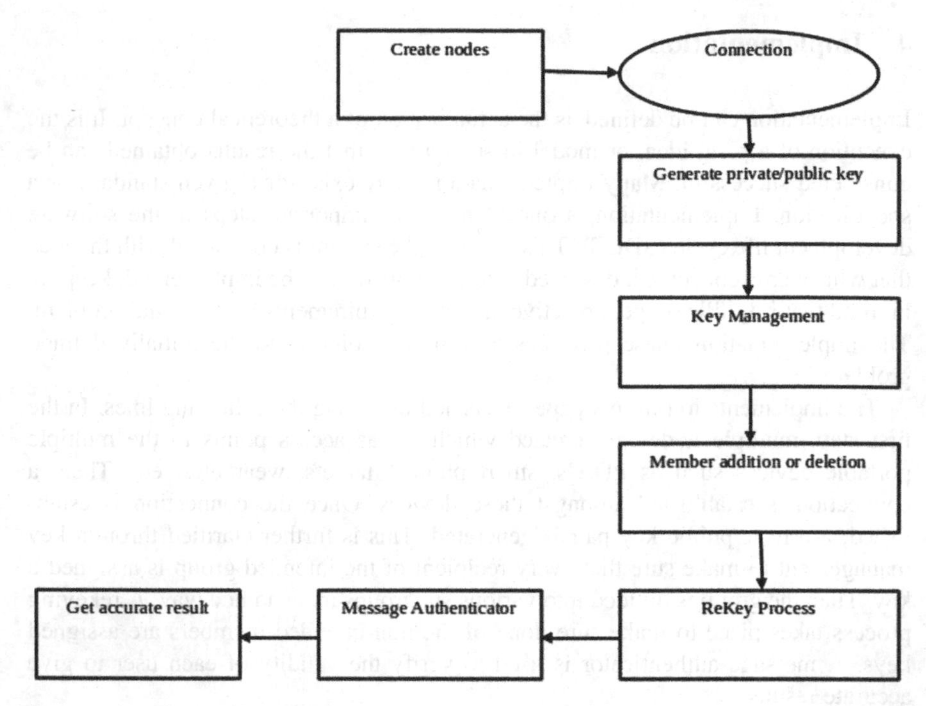

Fig. 2 Design flow of the graph

6 Algorithms Used

The following is an explanation of the batch mode conversion employed by the voice-to-phoneme algorithm: Let us consider that there are M example utterances, X(m[1, M])m ε available to a voice tag (batch mode). M X is a set of distinguishable vectors corresponding to an example utterance. From here on, for the purposes of this paper we consider a set of distinguishable vectors and an utterance to be one and the same. The objective here is to find N phonetic strings, P(n[1, N]) n, following an optimization criterion. In previous works enumerating a batch mode voice-to-phoneme conversion method, the authors modify a tree-trellis search algorithm for a single utterance which is used to modify multiple phonetic strings [11]. The newly generated tree-trellis search algorithm results in a tree containing partial hypotheses that are phonetic in nature. Each of the hypotheses is directly related to the example utterances and makes use of the time-synchronized Viterbi decoding with a phoneme loop grammar in the forward direction. The M trees of phonetic hypotheses of the example utterances are used jointly to estimate the admissible partial likelihoods from each node in the grammar to the start node of the grammar. Then, utilizing the admissible likelihood of partial hypotheses, an A* search in the backward direction is used to retrieve the N-best phonetic hypotheses, which maximize the likelihood. This modified tree-trellis algorithm falls firmly into the category of probability combination algorithms. Because this algorithm requires storing M trees of partial hypotheses simultaneously, with each tree being rather large in storage, it is very expensive in terms of memory consumption. Furthermore, the complexity of the A* search increases significantly as the number of example utterances increases. The above-proposed algorithm is disregarded because of its many downsides namely the space and time complexity of the DTW implementation. Almost every embedded platform has its own dedicated library to manage its specified hardware. Given two utterances, iX and j are upper-right corners of the trellis. A new utterance ijX can be formed along the best path of the trellis, *ijij* X = X ⊕ X, where ⊕ is denoted as the DTW operator. The length of the new utterance is the length of the best path choices of both the principal ideas of a sequential voice-tag creation strategy via voice-to-phoneme conversion which is described seed phonetic of a voice-tag abstraction. The high-frequency phoneme n-grams correspond to voice tag via a phonetic decoder; the best phonetic string is used to create the current (or seed) phoneme n-gram histogram for the voice tag.

1. An example by the end user gives us positive confirmation of the speech recognition result and a new voice-tag utterance. This new voice tag for example in turn gives rise to N new phonetic strings which get created via the phonetic decoder; and the best amongst those N is used to update the current histogram of the utterances. In reality though, many utterances are of a low frequency and require a garbage collection process to be disposed of.
2. Every phonetic string for N current and N new voice-tag phonetic strings gets a histogram estimated.
3. Each phonetic string histogram is compared with the current histogram by making use of a distance metric, usually the divergence measure. Multiple metrics such as distance metric and divergence measures are utilized to compare the current histogram of every phonetic string with the current histogram of the voice tag. N phonetic strings whose histograms are the closest match to the voice-tag histogram are selected.
4. The steps from 1 up to 3 are iterated whenever a new utterance is made available.

The RSA algorithm has long been in use to both encrypt and decrypt messages. It is a very popular asymmetric cryptographic algorithm. An asymmetric algorithm contains two different keys, namely a public key and a private key. The public key is made available to everyone, while the private key as the name suggests is kept private. The strength of the RSA algorithm lies in the fact that anyone with a public key can encrypt a message; only someone with the knowledge of the previously selected prime factors can decode the message, provided that the initially selected prime factors are large enough. The RSA algorithm can roughly summarized in the following steps:

1. p and q are chosen randomly with the condition that both p and q are prime and are relatively large in value.
2. The product of p and q is assigned to the variable n. $n = pq$.

- The variable n acts as the modulus for generating both the sets of private and public keys.

3. The totient is calculated such that $\varphi(n) = (p - 1)(q - 1)$
4. An integer e is chosen such that $1 < e < \varphi(n)$, e also has to be co-prime to $\varphi(n)$
   - e is the public key component
5. The variable d is computed so as to satisfy the congruence relation which is $de \equiv 1(\mathrm{mod}\varphi(n))$
   - d is kept as private component.

Encrypting messages: Haley gives her public key n & e to Andy and does not share her private key with anyone. Andy wants to send message M to Haley. Andy makes use of a reversible protocol such as the padding scheme to turn the message M into a number smaller than n. Andy makes use of the following formula:

$c = m\ e \bmod n$ to compute the cipher text c.

Exponentiation by squaring is the quickest method of solving the above formula.

Andy then sends c to Haley. Decrypting messages: Haley can recover M from c using her private key in the following procedure: $m = c\ d \bmod n$

By making use of m, she can recover the original message M.

7 Conclusion and Future Enhancements

The research in this paper has proposed a new type of key management archetype to enable dispatch broadcasts to secluded cooperative groups along with a sufficient security authority. Future enhancements that can be recommended to our project are as follows:

- The admin login can be provided with better security than the traditional password-based system.
- The system can be extended to send all kinds of files (E.g., GIF, JPEG's, EXE file, MpeG, and Mp4's.).
- The system can be implemented with more efficient routing algorithms.
- This will result in a higher standard of security being implemented in group transmission, allowing less chance for data to be altered or leaked by hackers.
- Making use of the above algorithms will significantly reduce the overhead required during communication and result in successful and secure transmissions even with lower bandwidths.

References

1. Qianhong Wu, Member, IEEE, Bo Qin, Lei Zhang, Josep Domingo-Ferrer, Fellow, IEEE, and Jess A. Manjn Fast Transmission to Remote Cooperative Groups: A New Key Management archetype- IEEE/ACM TRANSACTIONS ON NETWORKING, VOL. 21, NO. 2, APRIL 2013.
2. Y. Zhang and Y. Fang, ARSA: An Attack-Resilient Security Architecture for MultiHop Wireless Mesh Networks, IEEE J. Sel. Areas Commun., vol. 24, no. 10, pp. 1916, Oct. 2006.
3. K. Ren, S. Yu, W. Lou and Y. Zhang, PEACE: A Novel Privacy-Enhanced yet Accountable Security Framework for Metropolitan Wireless Mesh Networks, IEEE Trans. Parallel Distrib. Syst., vol. 21, no. 2, pp. 203–215, Feb. 2010.
4. B. Rong, H.-H. Chen, Y. Qian, K. Lu, R. Q. Hu and S. Guizani, A Pyramidal Security Model for Large-Scale Group-Oriented Computing in Mobile Ad Hoc Networks: The Key Management Study, IEEE Trans. Veh. Technol., vol. 58, no. 1, pp. 398–408, Jan. 2009.
5. Y-M. Huang, C.-H. Yeh, T.-I. Wang and H.-C. Chao, Constructing Secure Group Communication over Wireless Ad Hoc Networks Based on a Virtual Subnet Model, IEEE Wireless Comm., vol. 14, no. 5, pp. 71–75, Oct. 2007.
6. Q. Wu, J. Domingo-Ferrer and U. Gonzalez-Nicolas, Balanced Trustworthing Safety and Privacy in Vehicle-to-vehicle Communications, IEEE Trans. Veh. Technol., vol. 59, no. 2, pp. 559–573, Feb. 2010.
7. M. Burmester and Y. Desmedt, A Secure and Efficient Conference Key Distribution System, in Advances in CryptologyEUROCRYPT94, LNCS, vol. 950, pp. 275–286, 1995.
8. M. Waldvogel, G. Caronni, D. Sun, N. Weiler and B. Plattner, The VersaKey Framework: Versatile Group Key Management, IEEE J. Sel. Areas Commun., vol. 17, no. 9, pp. 1614–1631, Sept. 1999.
9. M. Steiner, G. Tsudik and M. Waidner, Key Agreement in Dynamic Peer Groups, IEEE Trans. Parallel Distrib. Syst., vol. 11, no. 8, pp. 769–780, Aug. 2000.
10. A. Sherman and D. McGrew, Key Establishment in Large Dynamic Groups Using One-way Function Trees, IEEE Trans. Software Eng., vol. 29, no. 5, pp. 444–458, May 2003.
11. Whitefield Diffic and Martin E Hellman New Directions in Cryptography, Invited Paper, IEEE.

ICT Enabled Proposed Solutions for Soil Fertility Management in Indian Agriculture

B.G. Premasudha and H.U. Leena

Abstract An intensive implementation of Information and Communication Technologies is highly required for greater beneficiary in agriculture. To achieve this in a sustainable way, implementation of ICT-based technology in precision farming at the earliest is required. More specifically, Indian farming requires efficient usage of fertilizer recommendations for each crop type at time intervals during the harvesting period. India needs to go mature in taking agriculture technology a bit further by using technologies like Multi-model decision support system, GPS (Global Positioning System), GIS (Geographic Information System) Cloud, Spatial data analytics, and various other technological wonders. This paper proposes a best available edge cutting and benefiting ICT initiatives by using cloud-based web GIS, Mobile, and Kiosk in fertilizer recommendations for precision farming and primarily aiming at the soil health in Indian agriculture.

Keywords ICT · GPS · Web GIS · Cloud · DSS · Fertilizer recommendations

1 Introduction

Soil Fertility management plays an important role in precision farming. The soil testing is used for estimating the fertility status of the soil and soil health. Presently, the available Indian soil testing laboratories are facilitating the nutrient values based on the soil samples collected from the farmers cropped land. This soil testing report is used to recommend the fertilizers required for the harvesting. Improper usage of fertilizers in farming is leading to environmental pollution and large yield gap that

B.G. Premasudha · H.U. Leena (✉)
Department of Master of Computer Application, Siddaganga Institute of Technology, Tumkur 572 103, Karnataka, India
e-mail: leenalatesh@gmail.com

B.G. Premasudha
e-mail: bgpremasudha@gmail.com

S.C. Satapathy et al. (eds.), *Proceedings of the International Conference on Data Engineering and Communication Technology*, Advances in Intelligent Systems and Computing 469, DOI 10.1007/978-981-10-1678-3_72

still exists across India for major crops [1]. To overcome this, a complete and effective decision support system is highly required for a balanced fertilizer recommendation system. At present, Government of India has taken several measures to overcome consequences in agricultural industry such as setting up STL (Soil Test Laboratories), fertilizer recommendations, issue of Soil Health Card in many states.

Multiple measures and inventions are undergone in ICT (Information and Communication Technologies) in Indian agriculture as such introduction of NAT (New Agricultural Technology), largely recognized as HYV (High Yielding Varieties)—fertilizers [2].

A cloud-based web GIS (Geographical Information System) for fertilizer recommendations using DSS (Decision Support System) DSS for fertilizer recommendations are predominantly essential for India because an enormous amount of overseas trade is depleted each year on fertilizers [3]. DSS assists in providing information on fertility status, expected crop yield, soil health of a particular village, Taluk, District, State or country. Use of ICT helps to provide recommendations to farmers on soil fertilizers.

This paper showcases the need of ICT in soil fertilizer recommendations for precision farming with the literature survey in Sect. 2, Sect. 3 with ICT enabled proposed multi-model DSS and Sect. 4 with conclusion.

2 Literature Survey

India ranks third largest fertilizer user and average rate of nutrient application is only 85 kg/ha [4]. The major nutrients available in Indian soil are classified into two types of macro (nitrogen, phosphorous, potash) and micro nutrients (sulfur, zinc, boron, iron). Consumption of these nutrients is concentrated in certain areas and varies from state to state. To manage a balanced fertilizer usage in Indian agriculture, the Government of India has taken several measures to promote Integrated Nutrient Management (INM) involving soil test to improve soil fertility, National Project on Management of Soil Health & Fertility (NPMS&F) was launched during the year 2008–09 to develop soil test and soil fertility (State of Indian Agriculture, 2012–13, Government of India).

Several studies have shown methodical soil testing followed by proper utilization of nutrients (N, P, K, S, Zn, and B) can increase the efficiency level by 2–3 times in most of the states of India [5]. At present, many SHC (Soil health Centers) has been setup at state levels of India to the needs of state farmers in respect of soil testing. The fertilizer recommendations are generally made in conjunction with soil test reports provided by SHC without using ICTs.

The applicability of ICT in expansion services has been researched by a number of researchers [2]. However agricultural innovations and agricultural technologies are mainly produced from ARIs (Agricultural Research Institutes) but not much is documented on use of ICT tools related to crop variety, land use, soil health, soil nutrients requirement, and irrigation.

Few initiatives are taken by Indian Government for providing online and mobile messaging services to farmers related to agricultural queries, agro vendor's information to farmers [4], it provides static data related to soil quality at each region.

At present, several fertilizer recommendation support systems are available across countries and each has several limitations. Many developed countries based on ESRI website are already using web-based online systems integrated with spatial database for mapping [3], GPS, GIS, etc. in the betterment of agriculture.

In [5, 6] authors have described the Fertilizer recommendations System development by Bhopal–India. This system is based on STCR (Soil Test Crop Response) for targeted yield approach of fertilizer recommendations for different crops. This software is limited only for few states of India. The system lacks to provide geospatial data and nutrients of soil shown based on region for each district are static which are based on the data collected by agricultural departments of the respective district.

In [6] authors have described sustaining soil fertility using ICT in Indian agricultural fields which proposes a tool for fertilizer recommendations which support multi-language but fails to provide spatial data. In [7], the paper describes about use of cloud computing for agriculture sector in developing Multidisciplinary Model for Smart agriculture using Internet-of-Things (IoT), sensors, cloud-computing, mobile-computing, and big-ata analysis for storing details of agriculture information.

Currently in India, different DSS Models are incorporated in various areas of agricultural management [3] such as land and water management, soil and water resource management, crop production, and management.

Although researchers have proposed many models in agriculture sector using one or more of ICT technologies mentioned; the dynamic multi-model is needed that provides an integrated approach to fertilizer recommendations which help to maintain soil fertility, increasing crop yield, and helps in distributing Soil Health Card.

2.1 Importance of DSS in Fertilizer Recommendations

- DSS intelligent system helps in optimizing and balancing soil nutrients.
- Automates the estimation of fertilizer requirement based on parameters (land, soil, crop, and season).
- It enables complete multi-model web GIS and smart database across cloud.
- Provides complete visualization and spatial decision support system which helps farmers and e-governance.
- Inscribing DSS into smart cloud-based web GIS application, smart phone application, and kiosk system.

2.2 *Challenges for ICT in Indian Farming*

Though technology is part of the agriculture, the efficiency of the technology is not up to the mark it is expected since the systems that are in place are not such user effective, customizable, open to new trends, and customer friendly. The legacy solutions induce and empower the static means of help to agriculture that is insufficient for the current and future trends of farming methods.

- Lack of centralized technology system in agriculture.
- Limited accessibility to ICT tools to farmers.
- Lack of appropriate ICT-based service offers targeting rural sectors.
- Scarcity of sophisticated technical centers and poor economic condition of farmers.
- Low awareness and usage of technology among farmers.

3 Proposed Multi-model Decision Support System for Fertilizer Recommendations

With the basic background discussed earlier, it is observed that the increased ICT demands in agriculture will enable farmers to bring more sophistication in modern farming. Using recommendations from the proposed solution, farmers are handed with best recommendations, like fertility usage, soil test data, expected yield, insects, and weeds. Two types of fertilizer recommendations can be done, one by using generic recommendations using SHC reports which may not be accurate and the other one which is more efficient way using STCR targeted yield equations to provide precise amount of fertilizers required based on crop type, variety, soil type and season. Farmers get awareness in the proper usage of fertilizers required for each crop. Using this technology, farmers will have better control over yields sooner than they start farming.

3.1 *Cloud-Based Web GIS Multi-model Decision Support System for Fertilizer Recommendations*

Today, inscribing intelligent system using GIS cloud aided, large-scale database is an important characteristics of individual farms, e.g., soil, air, weather condition, and water quality. The cloud database should be spatially hierarchical (that should be able to provide data for geographical in zoom controls regions) and dynamic (data should be updated regularly). This database will serve as a repository and store spatial agricultural datasets.

Geospatial computing is an effective approach to be implemented in agriculture sector. GIS applications are best integral on cloud due to the use of wide-ranging and chronological agricultural datasets, processing capabilities, and an intelligent optimized data storage [6] method that helps to get fertilizer recommendations based on available agricultural datasets. Many companies in IT, such as Amazon, HP, Google, Microsoft, IBM, etc., have realized potential need in the field of GIS-based cloud computing, and are being used in various areas [2].

A complete GIS cloud-based multi-model decision support system will help farmers in farming and sustaining agriculture as depicted in Fig. 1. This system allows farmers to access, update, retrieve information related to their farm, get fertilizer recommendations, and can visualize spatial data. The cloud database would be initialized with what is already available with the various organizations, such as Soil Testing Labs, State, and central agriculture departments, ICAR institutes, and agriculture universities (field trials, soil sample data). The contents of the database are continuously updated by the results generated by the Soil Testing Departments, individual farmers, NGOs across India.

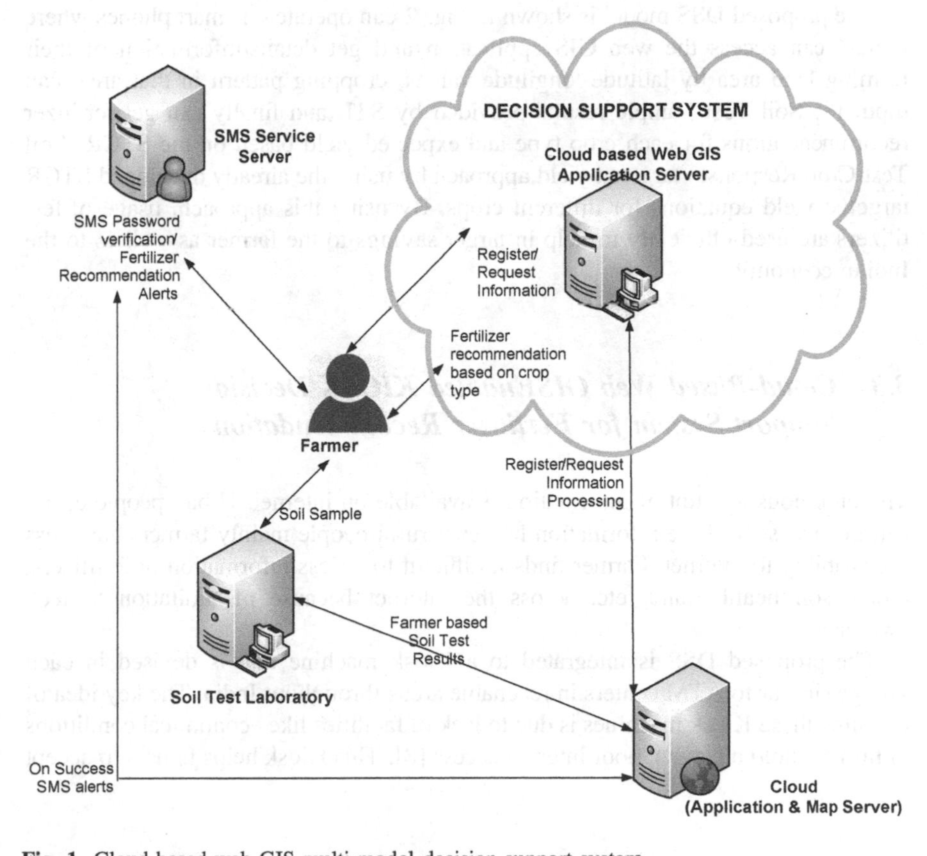

Fig. 1 Cloud based web GIS multi-model decision support system

3.2 Cloud-Based Web GIS Enabled Mobile Decision Support System for Fertilizer Recommendations

As per current situation in India, there have been significant gains in usage of ICT; especially mobile users are increasing day by day. In India, the number of smart phone users is nearly 50 % in rural area and it may increase in near future. With this aspect, smart phone plays a vital role to farmers in providing better agricultural information to help economic growth as well to get better yield to farmers. This DSS enabled on mobile computing will help farmers to access the system, view their land using spatial maps, get details on fertility status of their land, update soil test data, and retrieve fertilizer recommendations.

Development of android application to get fertilizer recommendations based on Geospatial Cloud gives more flexibility and reliability to the proposed DSS. Integration of message convey systems (SMS) that assists farmers in time of alerts and useful short message in farmer understandable and communicable regional language helps in betterment of getting best fertilizer recommendations and expected yield.

The proposed DSS model is shown in Fig. 2 can operate on smart phones where famers can access the web GIS application and get details/information of their farming land area by latitude-longitude values, cropping pattern in that area, can input the Soil Test Sample results provided by STL and finally can get fertilizer recommendations for each crop type and expected yield based on the STCR (Soil Test Crop Response) targeted yield approach by using the already developed STCR targeted yield equations for different crops. By using this approach, usage of fertilizers are used efficiently to help in larger savings to the farmer as well as to the Indian economy.

3.3 Cloud-Based Web GISEnabled KIOSK Decision Support System for Fertilizer Recommendations

The enormous amount of information is available on internet. Urban people easily can get access to these information however rural people mainly farmers have less accessibility to internet. Farmer finds it difficult to access information of fertilizers, crops, soil health, land, etc. across the internet because of limitation to tech awareness.

The proposed DSS is integrated to a Kiosk machine that is devised in each village similar to ATM centers in reachable areas throughout India. The key idea of enabling these Kiosk machines is due to lack of facilities like economical conditions of farmers hold a gadget, poor internet access [8]. This Kiosk helps famers to accept

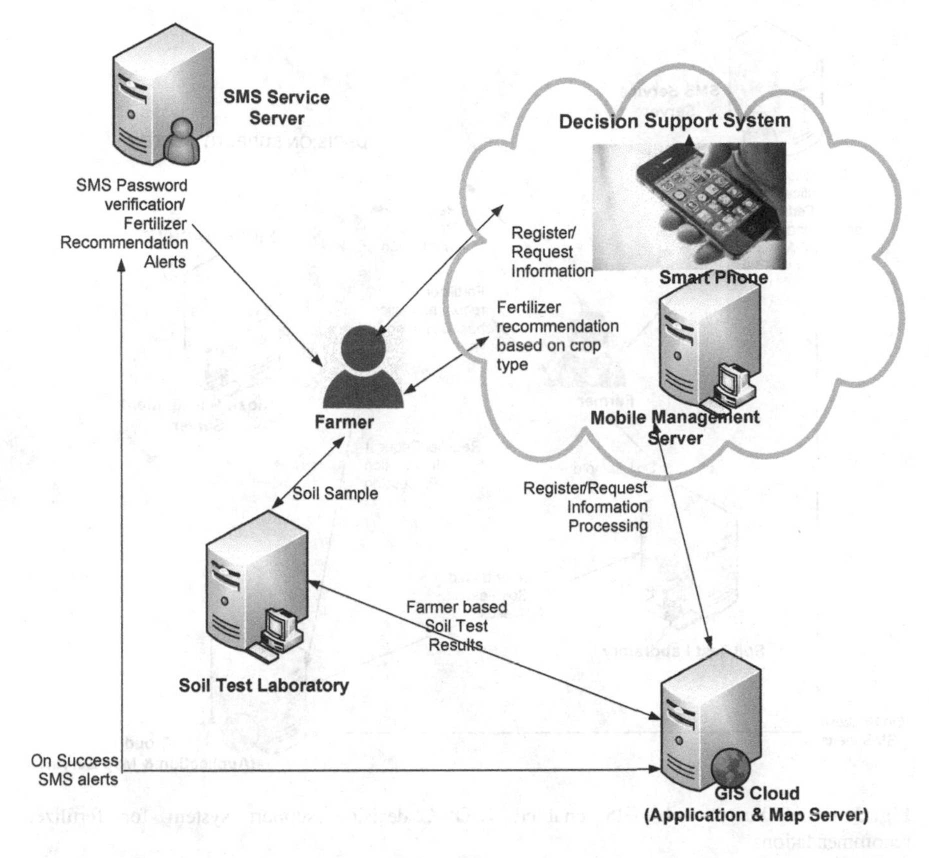

Fig. 2 Cloud based web GIS enabled mobile decision support system for fertilizer recommendations

and implement the DSS to get fertilizer recommendations in their local languages as Indian farmers are less communicable in English. Live demos can be viewed by the farmers on fertilizers utilization which helps farmers to get clear picture about their farming. Vending machine option can be induced to the Kiosk which helps farmers to get fertilizers through their Kiosk centers itself which also saves time in buying fertilizers outside.

The proposed Kiosk enabled DSS model shown in Fig. 3 is a touch screen, voice-based system supporting regional language which helps famers access the web GIS application and get details of their farming land area by latitude-longitude values and get balanced fertilizer recommendations. By placing this Kiosk in each village, farmers can use this like touch screen systems (kiosks) similar like existing touch screen systems in the railway stations, coffee vending machines, and ATM centers.

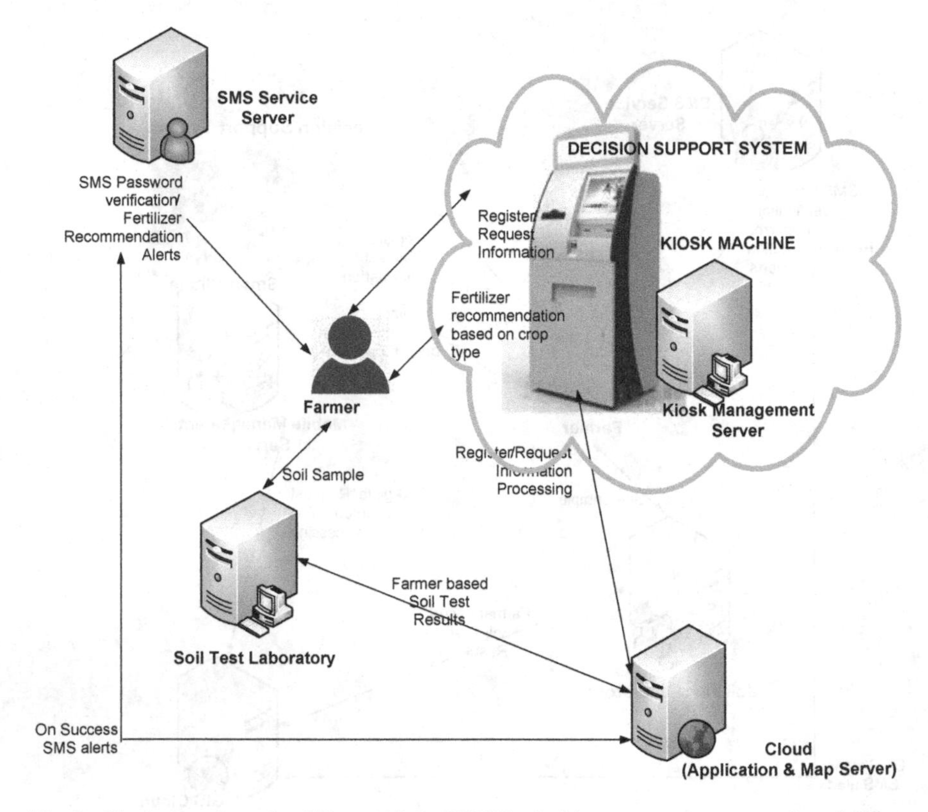

Fig. 3 Cloud based web GIS enabled KIOSK decision support system for fertilizer recommendations

4 Conclusion

This article aims toward the proposition of Decision Support System for optimizing usage of fertilizers in Indian agriculture using various ICTs: Cloud-based web GIS Server, Mobile application, and Kiosk System. Using this interactive system, farmers are facilitated with recommendations about the proper use of fertilizers for a particular crop using real-time soil sampling data and enabling to get maximum crop yield, cost efficient, and increase farmer economic growth.

All these proposed methods are highly depending on huge agricultural data sets and active technology will keep the services up and available online, further proposed methods must be engaged to global active to update farmer and agricultural database is highly required all time in Indian agriculture. This interactive system will be helpful to e-governance to make decisions in distributing required fertilizers for each state by using analysis report provided by the interactive system.

References

1. P. Mala, "Fertilizer Scenario In India" (2013).
2. Gorakhnath U. Waghmode, Avinash D. Harale, "A survey on: Cloud Enabled Agricultural Management", International Journal of Research (IJR), e-ISSN: 2348-6848, p- ISSN: 2348-795X, Vol. 2, Issue 09 (2015).
3. Vidya Kumbhar, T.P. Singh, "A Comprehensive Study of Application of Decision Support System in Agriculture in Indian Context", International Journal of Computer Applications, Vol. 63(2013).
4. Wev-Yaw Chung, Pei-Shan Yu, Chao-Jen Huang "Cloud computing system based on wireless sensor network", IEEE, Vol. 2, (2013).
5. PK Basavaraja, RC Gowda, DC Hanumanthappa, H. Yogeeshappa, G.V. Gangamrutha, Pradip Dey, A. Subba Rao, "GPS-GIS based soil fertility maps of Tumkur district for precised fertilizer recommendation ", University of agriculture sciences, GKVK, Bangalore (2013).
6. Deepak Rao B, Dr. Nagesh H R, Dr. H G Joshi, "Sustaining Soil Fertility Using ICT in Indian Agricultural Fields", IJER, Vol. 3, pp. 18–20 (2014).
7. Hemlata Channe, Sukhesh Kothari, Dipali Kadam, "Multidisciplinary Model for Smart Agriculture using Internet-of-Things (IoT), Sensors, Cloud-Computing, Mobile-Computing & Big-Data Analysis", International Journal of Computer Technology and Applications (IJCTA), ISSN:2229-6093, Vol. 6, pp. 374–382 (2015).
8. T.V. Subrahmanyam, K. Satish, Y.K. Viswanadham, "K-RIAD Kiosk for Rural India Agricultural Development", International Journal of Engineering Research & Technology (IJERT), ISSN: 2278-0181, Vol. 1, Issue 6 (2012).

Software Maintenance: From the Perspective of Effort and Cost Requirement

Sharon Christa, V. Madhusudhan, V. Suma and Jawahar J. Rao

Abstract Software and software deliverables have high impact on all fields. Once software is deployed, it has to be maintained continuously, till it becomes obsolete. Various activities that come under maintenance include adaptive, corrective, and predictive maintenance. Even though software maintenance is not tagged as a core field in software engineering compared to other software-related activities, almost 70 % of time and resources are allotted for maintenance activities. According to the related work, very little research is going on in the field of maintenance and this article highlights the scope of maintenance-related research. Identifying the factors that directly and indirectly affect the maintenance activity will in turn make the estimation activities easy. Implementation of an effective software maintenance model will have a very high impact in the quality of software and thereby with customer satisfaction. This article aims to project an effective maintenance model which reduces cost of rework and improves customer satisfaction index.

Keywords Software maintenance · Adaptive maintenance · Maintenance cost estimation · Effort estimation

---

Sharon Christa (✉) · V. Madhusudhan · V. Suma
Department of Information Science and Engineering,
Dayananda Sagar College of Engineering, Bengaluru, India
e-mail: sharonchrista@gmail.com

V. Madhusudhan
e-mail: vmadhusudhan91@gmail.com

V. Suma
e-mail: sumavdsce@gmail.com

J.J. Rao
Department of Industrial Engineering and Management,
Dayananda Sagar College of Engineering, Bengaluru, India
e-mail: jawahar_rao@yahoo.com

S.C. Satapathy et al. (eds.), *Proceedings of the International Conference on Data Engineering and Communication Technology*, Advances in Intelligent Systems and Computing 469, DOI 10.1007/978-981-10-1678-3_73

1 Introduction

Maintenance, as the word says, is the process of preserving a condition or situation or the state of being preserved. Maintenance has to be performed for the perfect working of any system. Like any material or object in use, software and software deliverables also require maintenance. From the time software is released to the time it is taken out of use the whole software system evolves, thus making it in shape for use. There are not many domains that still have to make use of the possibilities that software can provide. From robotic surgery to defense to embedded systems to astronomy, software systems are in use some or the other way.

Customer satisfaction is the main objective of any service provider and same is the case of a software developer and service provider. This has to be taken care even after the software is deployed. The features that add to customer satisfaction varies for type of software, domain, etc. Good process and products will always produce better and more reliable results that in turn make customers "happy". In order to keep the customers happy, one should be able to answer the following questions:

- What makes a customer happy?
- How a software product is able to bestow reliable results?
- What makes software sustain for the continuous changes?

The only answer to the above questions can be obtained by the activities carried out as post deployment activities in software and software deliverables. These activities may be treated as software evolution or software maintenance, which makes a system fit for use in any environment as well as any domain. It may be recalled here that evolution demands transformation of a software or process or product from one technology to another technology while maintenance may demand modifications to the existing systems or fixing of defects or even to adapt to a new environment [1]. Software needs to be enhanced for its functionality not just in preventing the errors that may occur in future but for making it compatible with different environment and also to fix the errors observed in software that is in use.

Even though maintenance of software enables software to sustain, it is not tagged as a core software engineering activity. Act of software maintenance is challenging due to lack of proper documentation, unstable team, unskilled staff, etc. Almost 70 % of time and resources are allotted for maintenance activities. After all these support, maintenance activities are facing real-time problems due to quality of original programming, programmer availability, hardware/software reliability, documentation quality, etc.

Efficacious software stands the test of time irrespective of the hardware or the environment it was designed for. In 1969, Lehman addressed the issues related to software maintenance and till date the core issues in maintenance remain the same. As the software ages with a complex structure, it becomes difficult to understand the system and thereby maintain the system. The characteristics of high-quality software are not just the development of the software product but also to maintain it with changing requirements which the customer dictates from time to time.

2 Why Software Maintenance?

Software maintenance is a postdelivery activity and its main purpose is to preserve the value of software over time [2]. All the issues related to software, from character enhancement to defect fixing are handled as a maintenance activity once the software is delivered to the customer. It is a fact that world is never static and perfect. Hence, requirements and enhancements always keep evolving during post-release period or the product is prone to failure due to creep of latent bugs. Requirements on which product was initially defined, built and released, hence will undergo modifications.

An efficient maintenance team should perform the functional and performance enhancements raised by the customer. They should make changes according to the environment and should be able to fix the postproduction defects. If maintenance is not performed reliably in the specified time, it will result in business down time and it directly or indirectly result in customer dissatisfaction. It further raises questions on product quality that directly affects the organization.

Taking the example of scenario that happened in UK on 12 December 2014, when they had to shut down five International Airports because of a software glitch. The air traffic control system of UK dated back to 1960s with its source code written in redundant JOVIAL language. The supercomputer that runs the software crashed from 15:30 until 16:30 just for an hour. The incident happened due to one line error in the software source code. This 1 hr window of software failure resulted in loss of business and almost 10,000 customers were directly affected.

The above-stated example solely reveals the importance of effective and efficient software maintenance.

3 Related Works

In general, more than half of the development time of a software engineer is spent for understanding, modifying, and retesting existing code, which is maintenance activity. Therefore, it is very important to identify the factors influencing maintenance process and also appropriately calculating the effort, time, and cost associated with it [3]. Analysis of maintenance work performed on several products helped the authors of [3] to put forth the lacuna in the maintenance phase. Software maintenance activities can be viewed in different perspectives. Authors of [4] has very clearly stated the problems in external and internal perspectives of software maintenance includes high maintenance cost, delayed maintenance service, prioritizing the change request and poorly designed and coded software. Further, there is lack of proper documentation.

Authors in [5] have classified various problems encountered by software maintainers which include perceived organization alignment problems, process problems as well as technical problems. The authors of [6] have pointed out that very less research work is carried out in software maintenance compared to software development. Even the academic works does not highlight the maintenance process in depth.

Software maintainers provide services on daily basis based on various contexts and interfaces. Even though International standards body has well-defined specifications for maintenance-related activities, the SEEBOK initiative identified a large number of software maintenance specific activities not covered under it [4].

Authors in [7] clearly state that it is an unmanageable task to estimate the size and thereby the effort related to it with a degree of accuracy. Along with that, accuracy of estimation in maintenance increases by various other factors such as complexity and functionality of the system. The authors of [7] also stated that maintenance is an evolutionary activity that is entirely different from development process and also has different inherent characteristics and requires more attention in context of estimation models.

Authors in [8] have stated the lacuna in software cost estimation model and its importance. They have put forth the categories that come under cost estimation, which includes size, effort, and project duration. Authors have also mentioned the factors which rise as issues to maintenance team. Some such issues include selection of software cost estimation model, software size measurement, accurate estimation, and so on.

Authors of [8] have also mentioned that real-time data from software maintenance projects are not available because of organizational constraints. They also state that very less research work is happening in the field of maintenance compared to estimations carried out in software development.

Gerardo Canfora and Aniello [9], in their article "Software Maintenance" have mentioned maintenance task as an iceberg to specify and highlight the herculean problems and costs that relates to it.

Thomas M. Pigoski [10] has pointed out very clearly the breakdown of the maintenance phase. The key issues, according to him include measurement, cost, estimation, technical, and management. The author mentioned the various process models as well as activities related to it. The author mentioned the lacuna in the existing activities related to maintenance. The author mentioned the unique activities as well as the supporting activities related to maintenance. He has mentioned the lack planning in the maintenance phase and has related it to the missing attributes associated with maintenance. He further mentions in detail the key issues related to technical level, managerial level as well as organizational level.

Author of [10] has cited the key issues that come under technical level in the maintenance phase as limited understanding, maintainability, testing, and impact analysis. The author states that the technical staff will have very less or zero knowledge about the software under maintenance which will intensify the problem.

From the organizational aspect, cost, and cost estimation are major factors. Since major share of life cycle cost is consumed by maintenance phase, and all organizations completely depend on the project turn over, accuracy of proper cost estimation is a major factor that concerns them.

Depending on the above-stated aspect of the software, post deployment is outsourced or a maintenance team is assigned. In the managerial perspective, there are more complex challenges that include process, staffing, training, experience of staff, etc.

Measurement and monitoring of maintenance process is the one area where there is a major deficit in research. According to the authors in [11], the current maintenance practitioners are not able to keep up with the requirements in the maintenance field. Because of the lack of documents as well as design details, it will become further complex.

Even though software engineering is a well-defined area, evaluation of software maintenance activity is not well defined. Authors of [11] have proposed a quantitative-based approach. But it lacks in analyzing the status of maintenance activities. Outlier behavior again is not considered in this study.

Thus, above works indicate the scope for research in software maintenance and the drastic need for accurate estimation of maintenance in order to reduce the rework cost, time, and effort.

4 Maintenance Process

In the current scenario, maintenance activity is either performed by software developing organization or can be outsourced to a third party. Whoever does the maintenance, the whole set of activities is same. The whole set of maintenance activity aims at maintaining the reliability and quality with minimum effort, cost, and time. In real world, even organizing the maintenance activity and finding the right person is a difficult task [3]. In case of a third party doing software maintenance, all the resources related to the software to be maintained is handed over to them.

Once customer raises any issues, it is registered formally as a modification request and is first analyzed and cross verified to check if the issue is relevant. A change request form is generated based on that. It is passed on to the maintenance team after verifying its relevance, where it is classified into major enhancement, minor enhancement, or defect fixes. Defect fixes are further classified depending on their severity or priority. These fixes will be carried out as per the Service Level Agreement defined by the customer. As for minor enhancement, the effort will be estimated and the customer will be informed about its development and deployment in the required state or version of the software. Major enhancement will be estimated and taken to the change control board for approval and actions by the development team. The resources required for the corresponding request as well as its impact on the system are analyzed. Required resources are allocated. For each defect identified a report is generated.

Software maintenance is a continuous effort, and the whole set of maintenance activities costs five times more than the cost of whole development process [4]. Hence, it is important that maintenance has to be performed with care, which demands a clear, consistent, and complete knowledge of the requirements. The main necessity for an efficient maintenance work is proper communication of the requirement from customer to the maintenance team. The complete set of documents related to the product development must be available to the maintenance team. The team members should have appropriate skill set to handle the product under maintenance. In most of the cases, people who do maintenance may not be a part of the product development, thereby demanding enough awareness to be given to the team about the tools and techniques used [12]. In case of a feature enhancement, the maintenance team should be well aware of the modules that directly or indirectly get affected by it. Failure of which may lead to the condition called software regression.

However, software maintenance was never considered a rewarding job for the reasons like less creativity involved and also the work load. This results in the likelihood of staff leaving the job or changing the domain, thus directly affects the maintenance activity. For the realization of high-quality software that is dependable, understandable, and efficient and satisfying to the customer, it is required to have accurate estimation of effort, time, and cost involved in the maintenance activity, as well as the quality of people involved in the maintenance. This directly results in good quality of software maintenance.

5 Types of Maintenance Activities

Based on the survey by Lientz and Swanson in the late 1970s the whole set of maintenance activities are classified into four broad categories [2]. Corrective maintenance is actually defect fixing postdelivery. The whole process involves reproducing the failure reported and finds out the cause of the failure. If the defect is in code, verify and validate its corresponding documentation. Fix the defect without altering anything, since defects are injected causing regression failure. Update the documentation accordingly. Once the maintenance is done, it is tested again to make sure the fix works and no regression faults have been introduced.

Adaptive maintenance is performed to make a computer program usable in a changed environment which includes hardware upgrade, software platform changes, or policy changes. The changes should preserve existing functionality and performance, otherwise it is required to perform adaptive maintenance. Perfective software maintenance is performed to improve the performance, maintainability, or other attributes of a computer program that includes functional or nonfunctional

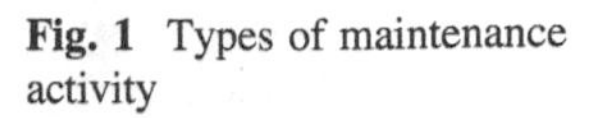

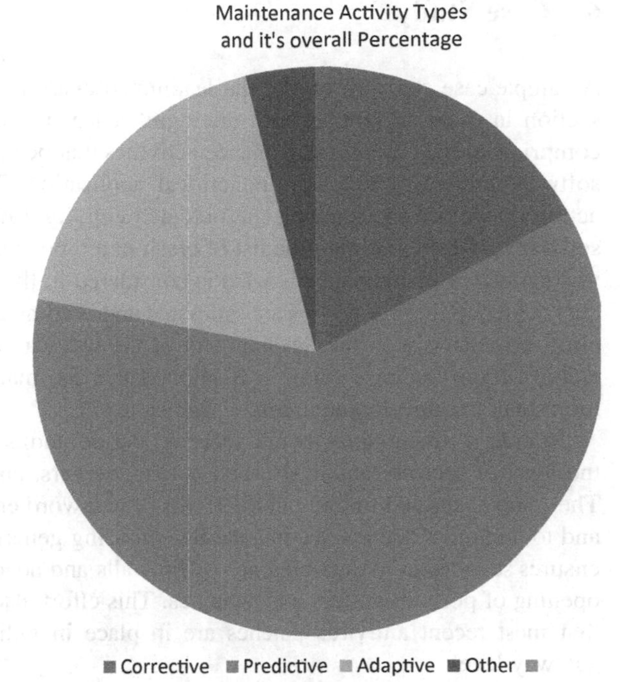

Fig. 1 Types of maintenance activity

requirements. Preventative maintenance is performed for the purpose of preventing any maintenance issues before they occur. It involves changing a software system in such a way that it does not alter the external behavior of the code yet it improves its internal structure [2]. Figure 1 depicts the maintenance effort required for various types of maintenance activities as explained above.

Figure 1 infers that 75 % of the maintenance effort is spent on adaptive and perfective maintenance whereas 17 % of effort is spent on error correction. It is an indisputable fact that estimating software metrics within software development and maintenance project is important. Since, software effort has a direct relation to the overall cost of the project, it is important to predict apt and effective software effort metrics. Effective prediction of effort can help allocate proper resources required and also helps in realistic cost estimates. Since, effort estimation is not exactly based on the actual statistics, but is computed based on domain knowledge, the estimated effort will not reflect the complexity or skill set required to perform a particular maintenance task. However, the estimation model that exists in the industry has lacuna since they use multivariate linear regression techniques, and this technique itself has drawbacks demanding the need for an accurate software estimate [13].

6 Case Study

A sample case study of one of the leading software industry is considered in this section in order to comprehend the significance of software effort estimation. It comprises of the type of maintenance activities that occur in network security-based software project, which is a noncritical application. The range of maintenance activity varies from changing the outdated copyright date to fixing the alarm for software crash to retaining the list of crash in a repository. Corrective maintenance and perfective maintenance is what is considered in this case study. The effort and cost related to these activities are estimated and is compared with the actual cost and effort associated with the development of the software and related software deliverables. From the case study, it is implicit that the maintenance cost and effort is more than the development cost of the same.

In order to maintain network security, period audits are conducted to ascertain the level of security amongst users, system, servers, and across different systems. The setup is audited to find out the levels of password encryption, unique password and to find out if there is a complete ban on using generic password. The audit also ensures security in maintaining across firewalls and adequate protection is given in opening of ports in routers and switches. This effort also investigates to make sure that most recent antivirus patches are in place in individual systems and at the gateway level.

Table 1 Maintenance activities that arises in the network security application along with it's severity

Domain	Issue	Severity
Network security	Enhance the application to retain alarms once the application is restarted	Major
Network security	Copyright dates in login web page are outdated	Minor
Network security	Password complexity modification enhancement	Medium
Network security	Update install/uninstall mechanism to be able to stop and restart application during migration	Major
Network security	Software monitarization mechanism	Medium
Network security	Remove unwanted messages during installation and uninstallation	Medium
Network security	Upgradation of the supporting Apache software to support latest version of the application	Major
Network security	Upgradation of the supporting SANE software to support latest version of the application of the application	Major
Network security	Upgradation of the supporting openssl software to support of the application latest version	Major
Network security	Adapt application to work in a client without Java	Major

This case study classifies network security issue into major, minor, and medium severity and thereby analysis its effect. The results of this analysis as depicted in Table 1 thus indicates the amount of maintenance effort that one would have to put in as per the issues raised and also indicates the degree of accuracy that one should have while estimating such efforts.

This paper limits to bring in awareness of the maintenance effort and further papers will bring out predictive model for better estimation of maintenance effort. It may be noted that the impact of maintenance effort varies from application to application and hence it further leads towards scope for research pertaining to various domains of applications.

7 Conclusion

Software is one of the highly beneficial introductions of human thoughts to the society. In fact generation of software and application of software in all the domains of livelihood has turned out to be a panacea. Hence, it is the rudimentary responsibility of every software developer to develop software projects which is going to be the best fit for purpose. Hence, every software industry strives towards all those strategies which lead toward the development of good acceptable software. These strategies include both preproduction and postproduction actions one has to follow for every software development.

Nevertheless, such quality gates are emphasized, yet there is always a proneness to overlook postproduction action points. This is because of the investment on time and effort involved in enhancing productivity in the company rather than looking at rework under maintenance activity. However, it is proven that the cost, time, and effort required for maintenance is very high and endorses the reputation of the company too.

This paper therefore brings in awareness toward formulating those strategies which aims at reducing the rework expense due to maintenance and uphold the flag of the company in the industrial quality market.

References

1. Ian Sommerville, "Software Engineering", Pearson Education, Eighth Edition.
2. sce.uhcl.edu/helm/swebok_ieee/data/swebok_chapter_06.pdf on 27-08-2015.
3. Gerardo Canfora and Aniello Cimitile, "Software Maintenance", Software Maintenance: Research and Practice Journal, November, 2000.
4. Alain April, Jane Huffman Hayes, Alain Abran, and Reiner Dumke, "Software Maintenance Maturity Model (SMmm): The software maintenance process model", 2004.
5. Bennett, K.H. Software Maintenance: A Tutorial. In Software Engineering, edited by Dorfman and Thayer. IEEE Computer Society Press: Los Alamitos, CA, 2000; 289–303 pp.

6. Pigoski TM. Practical software maintenance: Best practice for managing your software investment. John Wiley & Sons: New York, NY, 1997; 384 pp.
7. Pankaj Bhatt, Gautam Shroff, Arun K. Misra, "Dynamics of Software Maintenance", ACM SIGSOFT Software Engineering Notes Page 1 September 2004 Volume 29 Number 5.
8. Ruchi Shukla Arun Kumar Misra, "Estimating Software Maintenance Effort -A Neural Network Approach", ISEC'08, February 19–22, 2008, Hyderabad, India.
9. Gerardo Canfora and Aniello Cimitile, "Software Maintenance", Article, 2010.
10. Rajiv D. Banker, Srikant M. Datar, Chris F. Kemerer, "A Model to Evaluate Variables Impacting the Productivity of Software Maintenance Projects", A Journal on Management Science, Vol 37, No. 1, January 1991.
11. Suma. V, Pushpavathi T.P, and Ramaswamy. V, "An Approach to Predict Software Project Success by Data Mining Clustering", International Conference on Data Mining and Computer Engineering (ICDMCE'2012), Bangkok (Thailand), December 21–22, 2012.
12. Henk van der Schuur, Slinger Jansen, Sjaak Brinkkemper, "Sending Out a Software Operation Summary: Leveraging Software Operation Knowledge for Prioritization of Maintenance Tasks", Joint Conference of the 21st International Workshop on Software Measurement and the 6th International Conference on Software Process and Product Measurement, 2011.
13. Márcio P. Basgalupp, Rodrigo C. Barros, Duncan D. Ruiz, "Predicting Software Maintenance Effort through Evolutionary-based Decision Trees", SAC'12, Riva del Garda, Italy, March 25–29, 2012.

FPGA Implementation of Low Power Pipelined 32-Bit RISC Processor Using Clock Gating

R. Shashidar, R. Santhosh Kumar, A.M. MahalingaSwamy and M. Roopa

Abstract Here we developed the RISC 32-bit processor architecture using Clock gating technique to perform logical memory and branching instruction. The different blocks are using to fetch, decode, execute, and memory read/write to execute four stage pipelining. The Harvard architecture used which contains memory space for data and program. To reduce the power of RISC core, clock gating technique is used in the architectural level as an effective low power method. The further enhancement of pipeline architecture can be done using Verilog and simulation is carried out using Model sim tool and implemented on FPGA board.

Keywords Architectural power reduction • Clock gating

1 Introduction

The Reduced Instruction Set Computer (RISC) is a microprocessor CPU design philosophy that favors a smaller and simpler set of instructions that take the exactly similar amount of time to achieve. ARM, DEC Alpha, SPARC, MIPS, and IBM's PowerPC the most common RISC microprocessors.

Nowaday's electronics industry appears with low power. Necessity for low power has activated a major typical example where power dissipation has become as important consideration, such as area and performance. As RISC architecture

R. Shashidar (✉) · R.S. Kumar · A.M. MahalingaSwamy · M. Roopa
Department of ECE, Dayananda Sagar College of Engineering,
Bengaluru 560078, India
e-mail: Shashidhar.rin@gmail.com

R.S. Kumar
e-mail: rsanthosh.kumar665@gmail.com

A.M. MahalingaSwamy
e-mail: smasuma@gmail.com

M. Roopa
e-mail: surajroopa@gmail.com

S.C. Satapathy et al. (eds.), *Proceedings of the International Conference on Data Engineering and Communication Technology*, Advances in Intelligent Systems and Computing 469, DOI 10.1007/978-981-10-1678-3_74

supports predefined set of instruction it follows processor. Also the lengths of the instructions are same. The features considered in traditional CPU architecture were inspired by researchers. These features were used to build the coding were being ignored by the programs. Several processor cycles were performed for the more complex features. As a result gap between the main memory and processor was increased. This lead to streamline processing within the CPU, in parallel time taken to minimize the total number of memory accesses [1].

People started searching other alternative methods because of completed design of controller in CISC and reduced performance. By keeping the very simple instruction set on CPI, we can overcome the speed reduction when a processor talks to the memory. This is the reason why very few instructions are typically RISC architecture, here processor gets data from memory not from store and load [2]. Hence such addressing modes are kept limited. By using operands and opcode, complexity of controller design is reduced with bits determined in instruction register, Pipelining with additional register added a new dimension in the speed. The concept called super scaling is used to enhance the pipeline architecture, by allotting more than one execution unit [3].

The word Computer Architecture was first defined by Amdahl, Blaauw and on April7, 1964 Brooks of International Business Machines (IBM) Corporation introduced IBM System-360 computer family. The Computer Architecture which includes Operation codes, Instruction format, Instruction set, Addressing modes with register memory and memory locations.

The RISC architecture has been developed as a result of many projects started in 1975 at the IBM T.J. Watson Research Center and was completed by early 1980s. The term RISC (Reduced Instruction Set Architecture) architecture was widely known after the use of RISC the Berkeley research project and well recognized today.

The reduced instruction set computer (RISC) is a microprocessor CPU design that takes identical amount of time for execution with simpler and smaller set of instructions. PA-RISC, ARM, MIPS, DEC Alpha, SPARC, and IBM's PowerPC are the most common RISC microprocessors. The classic RISC pipeline consists of Instruction decode, Instruction fetch, and register fetch, Execute, Memory access, and Register write back [4].

2 Methodology

Simultaneous execution of instructions of parts or stages more effectively and efficiently is carried with the help of technique called pipelining that uses RISC processor. During the first instruction exaction the next instruction will be decoded

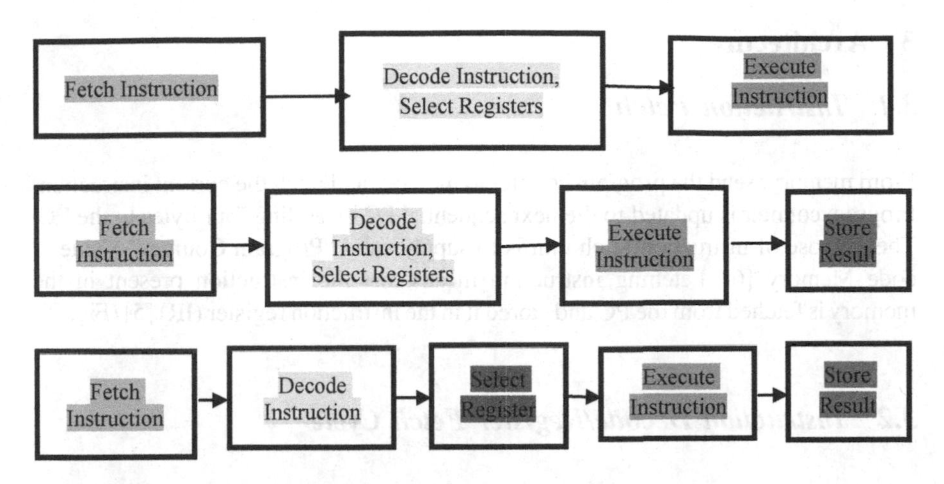

Fig. 1 Block diagram of pipelined architecture

and in parallel operands are loaded and following instruction is being fetched [4]. Figure 1 shows the three, four, and five stage RISC Pipeline, respectively.

Branch Prediction

a. This is a digital circuit which tries to guess in which way the branch as to go.
b. Increase flow in instruction pipeline.
c. High performance is achieved effectively.
d. Two-way branch predictions are implemented with conditional jump.
e. Without branch prediction, after the conditional jump the processor has to execute cycle, making sure that the next instructions have not entered the fetch stage in instruction Pipeline.
f. Branch predictor identifies that conditional jump is most likely to be taken to avoid the waste of time and identified branch to be fetched and speculatively executed.
g. Pipeline start over the correct branch by increasing the delay, this happens when the detected guess is wrong. In this case partially executed instructions are desirable [5].
h. Miss prediction of branch causes waste of time which is equal to the number of stages in the pipeline from the fetch stage to the execution stage.
i. More the pipeline greater the need for good branch predictor.
j. Conditional jump is identified by the Branch predictor.
k. Target of conditional or unconditional jump prediction is attempted by branch target predictor prior to the decoding and execution of the instruction itself.
l. Often in same circuit Branch prediction and branch target prediction are combined.

3 Architecure

3.1 Instruction Fetch

From memory, send the program counter to memory and fetch the current instruction. Program counter is updated to the next sequential PC by adding four bytes to the PC. The purpose of instruction fetch unit is to supply valid “Program Counter” Value to code Memory [6]. Fetching instruction means that the instruction present in the memory is fetched from the PC and stored it in the instruction register (IR) [5] (Fig. 2).

3.2 Instruction Decode/Register Fetch Cycle

Read the registers and decode the instruction comparable with register source specified from the register file. The purpose of instruction Decode and register fetch unit is to get operand value from CPU registers. The ID cycle decodes the instruction. During this cycle the instruction stored inside the Instruction register (IR) gets decoded [7].

3.3 Execution/Effective Addresses Cycle

Prior cycle contains operands which are operated by ALU for performing one of three functions depending on the instruction type. The purpose of instruction Execute is to perform required operation based on the opcode and store the result in immediate register.

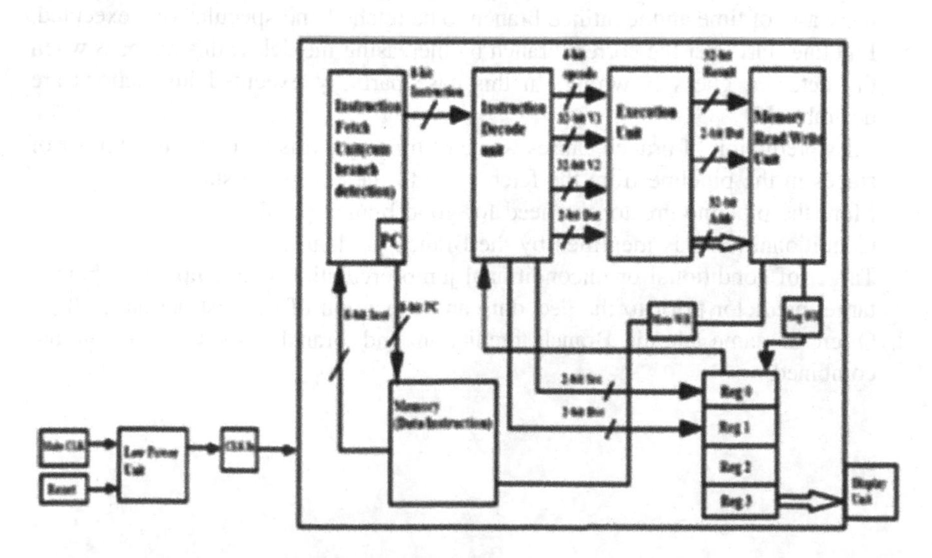

Fig. 2 Architecture of pipelined RISC processor

4 Clock Gating Methodology

4.1 Clock Gating

The clock gating applies to synchronous load enable registers; the HDL variable decides synchronous clock and control signals are used in a flip flop. Synchronous load enable, synchronous reset, synchronous set, and synchronous toggle come under the synchronous control signals. Register banks disabled during some clock cycles. Typical implementation uses multiplexers. Clock gating cell replaces multiplexers [8].

Design compiler by use of feedback loops to implement the register. Unnecessary power can avoid by maintaining the same register logic level through multiple cycles. Clock gating eliminates the unnecessary activity associated with reloading register banks to save power (Fig. 3).

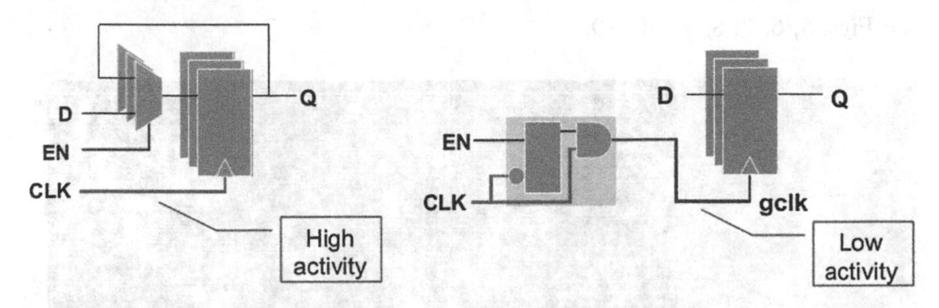

Fig. 3 Clock gating

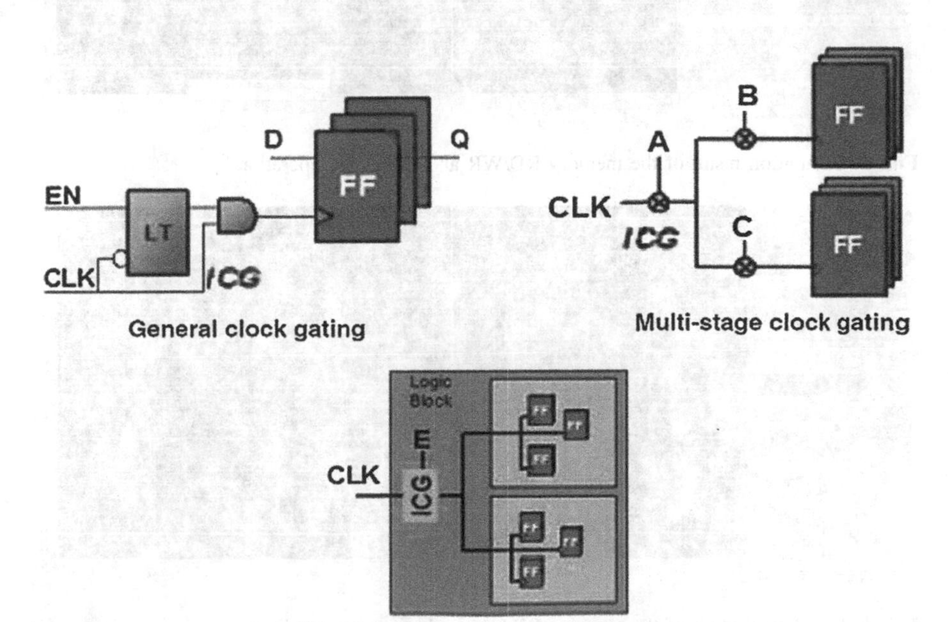

Fig. 4 Clock gating styles

4.2 Clock Gating Styles

From the standard style, one clock gating cell is applied per register bank, More dynamic power saving can be gained by implementing clock gating in different modes. Figure 4 shows three different clock gating styles such as general, multi-stage, and hierarchical.

In multistage clock gating, further gating is done for the clock gating cells which have common enable signals. As a result, more dynamic power can be saved.

In the hierarchical clock gating style, clock gating is done at the logic hierarchy level by gating hierarchical logic modules that share a common enable and the same clock group. This reduces the number of clock gating cells in the design [8].

5 Simulation Results of RISC Processor

See Figs. 5, 6, 7, 8, 9, and 10.

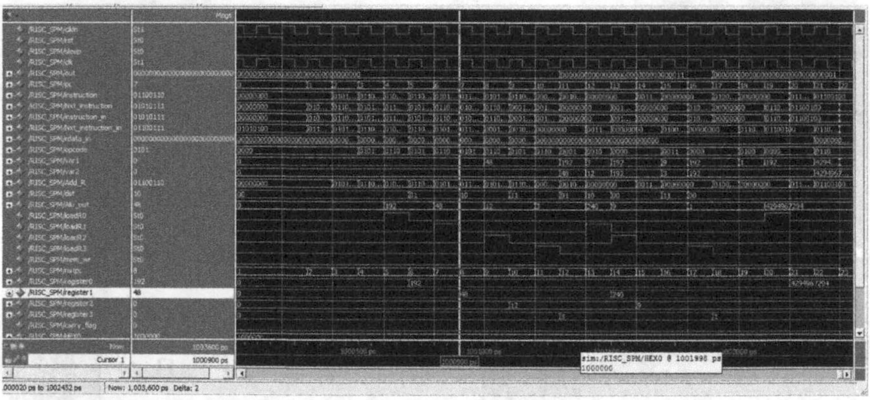

Fig. 5 Simulation result of the memory RD/WR and arithmetic operation

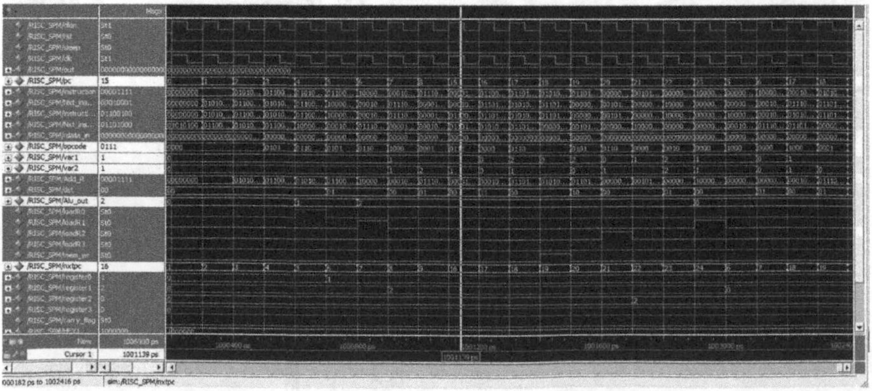

Fig. 6 Simulation result of the memory RD/WR and operation condition

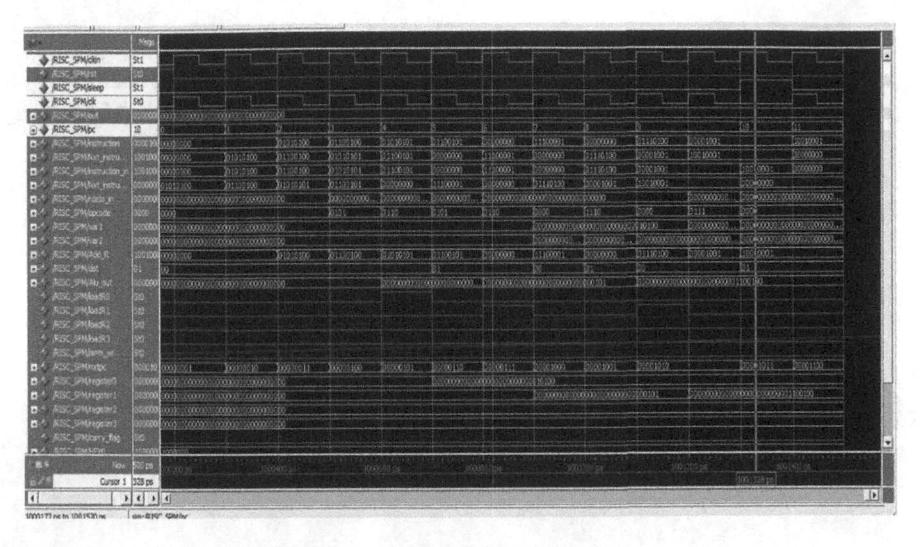

Fig. 7 Simulation result of the how power unit using clock gating

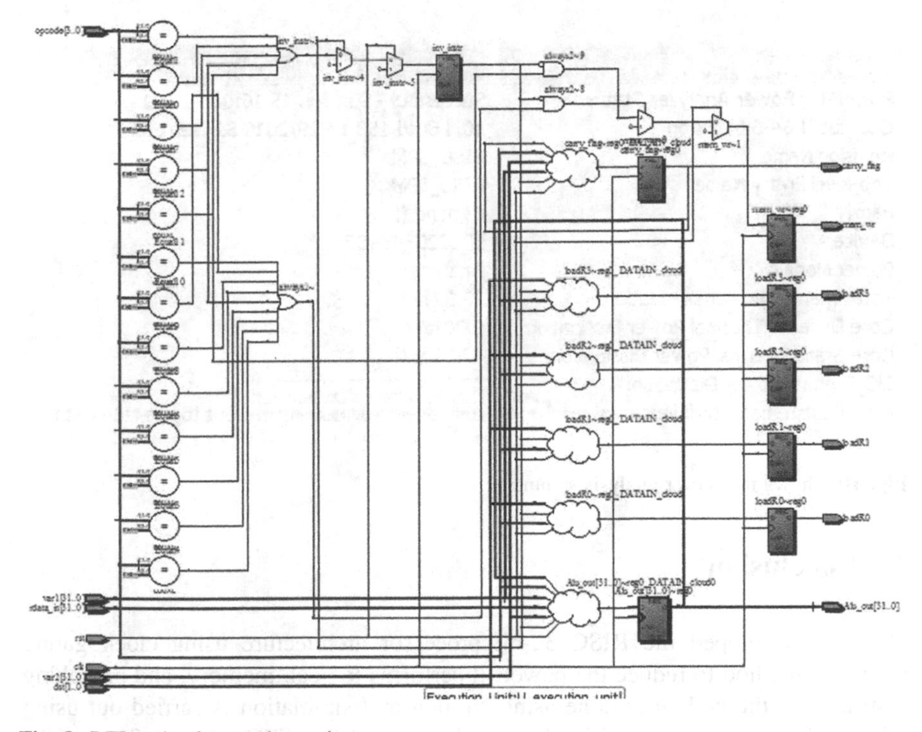

Fig. 8 RTL unit of execution unit

Analysis & Synthesis Summary	
Analysis & Synthesis Status	Successful - Sun Jun 15 15:41:12 2014
Quartus II 64-Bit Version	10.1 Build 153 11/29/2010 SJ Web Edition
Revision Name	RISC_SPM
Top-level Entity Name	RISC_SPM
Family	Cyclone II
Total logic elements	16,161
Total combinational functions	7,873
Dedicated logic registers	8,493
Total registers	8493
Total pins	31
Total virtual pins	0
Total memory bits	0
Embedded Multiplier 9-bit elements	2
Total PLLs	0

Fig. 9 Shows the device utility summary

PowerPlay Power Analyzer Summary	
PowerPlay Power Analyzer Status	Successful - Sun Jun 15 16:06:42 2014
Quartus II 64-Bit Version	10.1 Build 153 11/29/2010 SJ Web Edition
Revision Name	RISC_SPM
Top-level Entity Name	RISC_SPM
Family	Cyclone II
Device	EP2C20F484C7
Power Models	Final
Total Thermal Power Dissipation	70.37 mW
Core Dynamic Thermal Power Dissipation	0.00 mW
Core Static Thermal Power Dissipation	47.36 mW
I/O Thermal Power Dissipatio[illegible]	[illegible]
Power Estimation Confidence	[illegible]nt toggle rate data

Core Dynamic Thermal Power Dissipation

Fig. 10 Shows the power analysis summary

6 Conclusion

Here we developed the RISC 32-bit processor architecture using clock gating pipelined method to reduce the power. It performs logical, memory, and branching instructions, the coding is done using Verilog and simulation is carried out using Model SIM Se6.4e tool and implemented on ALTERA FPGA board. The results can be obtained from the synthesis report for the proposed architecture speed which is approximately 111.286 MHz.

In future, one can extend this project to 64-bit RISC processor with high performance and lower power consumption using clock gating techniques or other lower power techniques which can be implemented in both back end and front end with 45 nm technology.

References

1. M. Kishore Kumar, "Fpga Based Implementation of 32 Bit RISC Processor." in 2012.
2. PreetamBhosle, Hari Krishna Moorthy "FPGA Implementation of Low Power Pipelined 32-Bit Risc Processor" in 2012.
3. Neenu Joseph, Sabarinath. S "FPGA Based Implementation Of High Performance Architectural Level Low Power 32-Bit Risc Core" in 2009.
4. Indu. M, ArunKumar. M "Design of Low Power Pipelined Risc Processor" in 2013.
5. Sivarama P Dandmundi Guide to RISC processor for programmers and engineer.
6. Kui YI and Yue-Hua Ding "32-Bit Risc Cpu Based On Mips Instruction Fetch Module Design" in 2009.
7. Rupali S. Balpande and Rashmi S. Keote. "Design of Fpga Based Instruction Fetch & Decode Module Of 32-Bit Risc (Mips) Processor" in 2011.
8. MahendraPratap Dev, Deepak Baghel, Bishwajeet Pandey, Manisha Pattanaik, Anupam Shukla "Clock Gated Low Power Sequential Circuit Design" in 2013.

Dynamic Software Aging Detection-Based Fault Tolerant Software Rejuvenation Model for Virtualized Environment

I.M. Umesh and G.N. Srinivasan

Abstract Cloud computing has emerged as one of the most inevitable technologies that encompasses huge software components and functional entities, which during operation accumulates errors or garbage, thus causing software aging. Software aging can cause system failure and hazardous consequences, and therefore to deal with it, a technique called software rejuvenation is proposed that reboots or re-initiates the software to avoid fault or failure. Conventional approaches still suffer from higher downtime resulting in decreased performance and unavailability. In this paper, a controller based dynamic software aging detection and fault tolerant rejuvenation model has been proposed. This model has been implemented in a live virtual migration environment to ensure resource security during virtualization. In addition, the proposed checkpoint and log detail based migration has eliminated the probability of downtime which is highly significant for realistic applications.

Keywords Software aging · Rejuvenation · Virtualization · Fault tolerant · Live migration

1 Introduction

Cloud computing has emerged as an efficient and large scale computing system that provides multi-level services through virtualization while ensuring fair resource (software applications, data, bandwidth and storage) allocation to the users. A cloud-based software application, in general is developed as a distributed system that encompasses multiple cloud service segments where it communicates with each other through certain interfaces or connections. On the other hand, the comprising

I.M. Umesh (✉)
Bharathiar University, Coimbatore, Tamil Nadu, India
e-mail: umesh.mphil@rvce.edu.in

G.N. Srinivasan
Rashtreeya Vidyalaya College of Engineering, Bengaluru, Karnataka, India
e-mail: srinivasangn@rvce.edu.in

S.C. Satapathy et al. (eds.), *Proceedings of the International Conference on Data Engineering and Communication Technology*, Advances in Intelligent Systems and Computing 469, DOI 10.1007/978-981-10-1678-3_75

software components (SCs) in their long-run operation accumulate numerous internal errors and garbage, thus resulting into software aging and probable failure or performance degradation [1]. Numerous large scale software applications and associated systems do suffer from performance degradation or even failure because of premature resource exhaustion. The huge resource consumption, fragmentation and error accumulation causes software aging.

The above issues demand certain optimal and effective fault tolerant technique to avoid software failure [2] at runtime. To deal with the situations of aging, a number of fault tolerant approaches have been proposed such as redundancy oriented software failure and its recovery, rollback scheme based on check points. Recently, a novel preventive and proactive approach called 'software rejuvenation' has been proposed to deal with these aging issues. Software rejuvenation enables freeing up the resources, storage and deletion of garbage, system reconfiguration, etc., that significantly reduces the probability of premature system failure and performance degradation caused due to software aging.

In this paper, three predominant issues have been considered.

1. Prediction of probable aging in online software components on cloud platform.
2. Performing software rejuvenation using certain efficient virtualization paradigm.
3. Optimal resource utilization and its security during live migration.

The proposed rejuvenation model examines dynamic behaviour and recent resource utilization of a host VM and predicts aging. On the other hand, the rejuvenation controller schedules migration of the services on aged VMs to newly created virtual machines using a live migration paradigm. The novelty of the proposed system is its dynamic fault prediction, secure migration and zero downtime.

Section 2 presents related work, Sect. 3 represents the contribution and proposed system, which is followed by Sect. 4 that discusses results and the conclusions are given in Sect. 5. The relevant references used in this manuscript are given at the end.

2 Related Work

To deal with software aging, software rejuvenation has emerged as a key technique that significantly reduces performance degradation, resource depletion and premature system failure [1]. A number of approaches have been developed for enhancing software rejuvenation by means of appropriate scheduling to reduce downtime and availability maximization. These approaches are classified into two major classes. The first is periodical rejuvenation method that considers time and workload to perform scheduling and the second one incorporates adaptive and proactive rejuvenation where the rate of resource depletion and performance degradation is examined dynamically. Some researchers have suggested rejuvenation based on the time estimation [3–8], and the predictive measurement approach [9, 10]. The time-based scheme intends to estimate various dynamic parameters such as workload, mean time to repair (MTTR), and failure distribution over certain defined period. A number of tools have been developed for scheduling, such as continuous-time Markov chain (CTMC) [3], a Markov regenerative process (MRGP) [5], stochastic reward nets (SRNs) [4, 6, 7] and others [8, 11, 12]. On the other hand, predictive paradigm continuously monitors the operational behaviour and performs triggering for rejuvenation in case of any fault occurrence resulting in degradation or downtime.

The virtualization-based rejuvenation has emerged as a significant approach to achieve higher resource availability and minimal downtime [12–15]. The earlier studies [14] proposed a CTMC model to monitor behaviour of the virtualized system, which was later implemented for rejuvenation scheduling. The system availability on the basis of time-based rejuvenation was employed and consolidation of virtual machines (VMs) was performed. An approach, VMSR was developed for software rejuvenation for application server systems [16]. CTMC approach to consolidate multiple VMs on a single host server was employed in some studies. The work done in [13, 14] employed only the time-based rejuvenation policy without taking into consideration of the VMM failure and its rejuvenation problem and since VMM being a very critical point of faults often plays significant role has not been addressed.

3 The Proposed Model

This section discusses the proposed rejuvenation model.

3.1 System Model

In this paper, a software rejuvenation model has been developed for applications running on virtualized environment. The proposed model encompasses multiple interconnected components or services with numerous functional software components. A software module was developed to monitor virtual machine (VM) functions for predictive fault analysis and for rejuvenation scheduling. Figure 1 represents the overall architecture of the proposed software rejuvenation model. As a significant contribution of this research a fault tolerance technique was developed that continuously operates for aging detection to avoid premature failure, fault or any probable hazardous consequences. Here, fault and aging has been estimated using aging detector that periodically estimates CPU and memory utilization along with a message signal arrival time. The aging detector retrieves key performance indices about VMs, namely, CPU utilization, other resource consumption, and associated context of probable failure, and uses these metrics for labelling the VMs for its fault status. The proposed dynamic fault estimation unit examines the degree of fault at a certain defined interval, iteratively in the course of operation and the respective degree of fault is automatically updated in a queue in the decreasing order of criticality. Thus, the fault label reflecting highest probability of failure and service disruption has been scheduled for prioritized rejuvenation at the first instance. The overall proposed scheme of the software rejuvenation is given in Fig. 1.

The overall rejuvenation monitoring-based execution and scheduling is controlled and performed by Rejuvenation Controller. Once any software component is detected with highest fault probability or failure prone, its fault status label is updated to the controller which then initiates rejuvenation process. The proposed mechanism has the ability to send the tasks and service requests of aged VM to the new VM in the same host and even to the other host VMs as per resource availability and scheduling. It significantly reduces the computational cost and probability of downtime. The services of aged VM are migrated to the newly created VM so as to continue normal software running operation. To ensure resource security and availability, a live migration technique is developed to avoid copying of resources or significant information. Once the VM migration has been done, controller schedules for rejuvenation where, the cache, garbage data and fault causing errors are deleted by means of rebooting the VM that finally destroys the aging

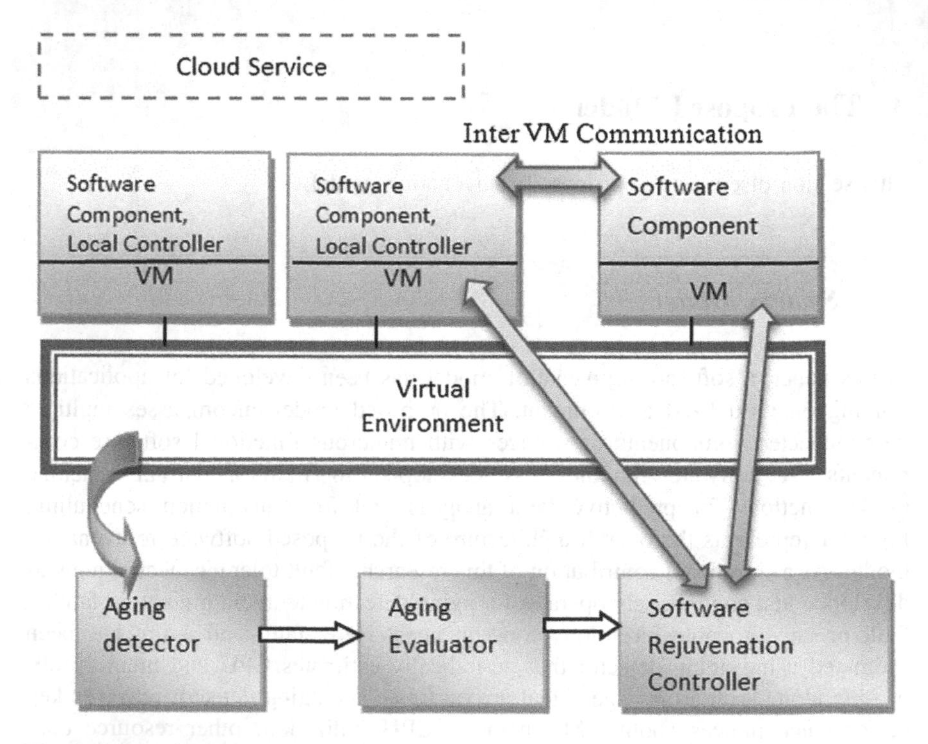

Fig. 1 Proposed rejuvenation model

effects. In our proposed scheme, it has been ensured that during rebooting of the aged VM, the software components keep running on the newly created virtual machine and thus a replica is formed on the new VM. To ensure non-disrupted migration, a replica of the communication trace back and logs have been generated that helps VMs to operate at the newly created virtual machine. Once, the rebooting of the initial VM is done, the replica of the running software components are migrated to the original host and newly created trace backs, log files, check points, etc., are deleted to make it available for next rejuvenation. The detailed discussion of the proposed research methodology and implementation strategies is given below.

3.2 Dynamic Aging and Fault Prediction

As illustrated in Fig. 1, the functional software components execute on individual VMs. Considering realistic scenarios; it is probable that some of the VMs might be active, some may be underutilized and some VMs are overloaded. It is noted that overloaded nodes tend towards higher probability of aging. The dynamic functional behaviour of cloud systems and large scale online software demand certain predictive model for aging detection. To achieve this, a dual threshold scheme was developed for static aging threshold and dynamic aging threshold estimation. For dynamic aging detection, the CPU utilization is considered as the indicator for aging. Rising CPU utilization infers higher aging. A robust local regression (LRR) algorithm is used for CPU utilization threshold-based aging detection. Robust local regression provides a simple but powerful approach for estimating the aging prediction.

The static threshold scheme estimates a message receiving time called transmission delay at rejuvenation controller. It was also taken into consideration the transmission delay using a message packet between host VM on which the software is running and the fault aging detector. Local controller on host VM collects runtime metrics such as CPU utilization and memory occupancy of the software components. In case any VM is suffering from aging, it is reflected in higher CPU and memory consumption. Furthermore, the inability of a host VM to send message to the aging detector might state aging condition and thus demanding rejuvenation.

Expected message arrival time (EMAT) represents the expected time for reaching the next packet at the aging detector. Here, it is assumed that in case aging detector doesn't get message at certain predefined time interval from a host VM, it represents the probability of aging or failure. The actual message arrival time of the recent packets was considered as the historical statistics in a sliding window. In other words, the recent n messages $m_1, m_2, \ldots, m_n$ and $T_1, T_2, \ldots, T_n$ are their respective actual arrival time at the aging detector. Here the value of $\mathrm{EMAT}_{k+1} (k > n)$ is the expected arrival time of the next message. Mathematically, it can be expressed as

$$\mathrm{EMAT}_{k+1} = \frac{1}{n}\left(\sum_{i=k-m+1}^{k} (T_i - \Delta t_{k+1} \times i)) + (k+1)\Delta t_{k+1}\right) \quad (1)$$

3.3 Evaluation of Software Aging

Retrieving the significant metrics such as CPU Utilization, memory availability, expected message arrival time (EMAT), etc., the software aging evaluation has been done and the respective aging label is assigned based on metrics and behaviour. The proposed model defines three aging levels for scheduling rejuvenation. Level 1 refers to the critical aging scenario reflecting maximum failure probability and thus demanding instant rejuvenation. Level 2 refers to the serious situation that may lead to serious outcome, such as higher performance degradation and higher transmission delay or high performance loss. Level 2 states the requirement of rejuvenation as early as possible. Level 3 represents the normal functional environment and thus rejuvenation is not needed in this case. In addition to it, rejuvenation is also scheduled based on dynamic thresholding done on the basis of predictive analysis. At certain defined time interval, the proposed aging detector estimates the aging level for each VM and updates the event queue, where it is ordered automatically in ascending order of the aging level.

3.4 Live Migration Based Software Rejuvenation

A robust live migration scheme has been developed to reduce the problem of re-execution and shutting down of the running application or VMs thus reducing the downtime significantly. To achieve this, a checkpoint-based mapping scheme has been implemented that stores the functional records and supports live migration during rejuvenation. The functional rejuvenation events for the proposed system are given in the Fig. 2.

Time	Original VM Status	Transitory VM
T1	Normal functioning	
T2	Aging	Created
T3	Generate checkpoint	Trace-log mapping between 2 VMs
T4		Live Migration from Original VM
T5	Rejuvenation done	
T6		Transmit trace-log to Original VM
T7		Shutdown
T8	Normal functioning	

Fig. 2 Rejuvenation events

The proposed system incorporates mapping of the functional metrics (EMAT, CPU utilization, resource availability, etc.), which is used by the rejuvenation controller to schedule live migration followed by rejuvenation. Once any fault probability (Level 1 and Level 2) is detected with software components on VM, the rejuvenation controller triggers the rejuvenation scheduler to start migrating on-going software components to the newly created virtual machine (Figs. 1 and 2). The newly created virtual machine establishes a connection with the software components of the original VM to generate the mapping of the on-going process, and ultimately the image of the mapped checkpoint is transmitted to the newly created VM. Once these activities are over, rejuvenation of original VM is done. Finally, once the original VM having software components continues functioning, the newly created virtual machine is refreshed by deleting log files, check points and other images before shutdown. In case of multiple hosts and aged software components, the use of genetic algorithm (GA) is being proposed to implement an evolutionary computing based migration scheduling scheme for VM allocation during migration. Due to the space constraints, the discussion of GA based resource allocation in multiple host scenarios will be discussed in separate paper.

4 Discussion

The proposed fault tolerant paradigm successfully incorporates multi parameter (CPU Utilization, memory and expected message arrival time (EMAT)-based aging detection, which has been followed by a live migration scheme. Thus, the proposed system can accomplish realistic and better aging prediction, dynamic scheduling and secure live migration. The controller based rejuvenation scheme proposed in this paper, where the local controller which is connected to each functional software components, collects dynamic behavioural statistics and transmits it to the aging detector which is a part of rejuvenation controller. The local controller has been deployed not only to retrieve the real time functional behaviour of software components but it also maintains log files and checklist details which is further used for continuing communication while performing live migration.

Unlike traditional replication oriented approaches which require huge replication host node for the individual software component, this scheme does not require any such nodes and hence, it is highly cost effective. Now, focusing on the downtime during rejuvenation, the proposed system employs log files and checkpoint based synchronization that enables undisrupted function of software components. Interestingly, the proposed system performs better than generic rebooting based rejuvenation as rebooting based scheme introduces higher downtime and re-initialization of software can be time consuming and costly. On the contrary, the approach can significantly deal with multiple rejuvenations of independent software components or VMs. In addition, the implementation of multiple parameters for aging detection enables the proposed system to be more realistic and responsive. Hence this model can be a novel alternative for large scale software applications or cloud application utilities.

5 Conclusion

The overall proposed system can be a potential candidate for fault tolerant software aging detection and software rejuvenation for cloud infrastructures as well as online/offline software applications. A future proposal has been planned to enhance the fault detection accuracy, dynamic scheduling and resource allocation for live migration. Furthermore, it is intended to employ certain evolutionary computing scheme for live migration and task placement to enhance rejuvenation and down-time reduction.

References

1. Cotroneo, D., Natella, R., Pietrantuono, R., Russo, S.: Software Aging and Rejuvenation: Where we are and where we are going. in Proc. Of the 3rd International Workshop on Software Aging and Rejuvenation (2011) 1–6
2. Jhawar, R., Piuri, V.: Fault Tolerance and Resilience in Cloud Computing Environments. in Cyber Security and IT Infrastructure Protection, J. R. Vacca, Eds. Elesvier: USA, (2014) 1–28
3. Huang, Y., Kintala, C., Kolettis, N., Fulton, N.: Software Rejuvenation: Analysis, Module and Applications. Proc. Int'l Symp. Fault-Tolerant Computing, (1995) 381–391
4. Vaidyanathan, K., Harper, R.E., Hunter, S.W., Trivedi, K.S.: Analysis and implementation of software rejuvenation in cluster systems. ACM SIGMETRICS Performance Evaluation Review, Joint International Conference on Measurement and Modeling of Computing Systems, (2001) 62–71
5. Wang, D.Z., Xie, W., Trivedi, K.S.: Performability analysis of clustered systems with rejuvenation under varying workload. Performance Evaluation, vol. 64, (2007) 247–265
6. Liu, Y., Trivedi, K., Ma, Y., Han, J., Levendel, H.: Modeling and analysis of software rejuvenation in cable modem termination systems. The 13th International Symposium on Software Reliability Engineering (ISSRE), (2002) 159–172
7. Liu, Y., Trivedi, K., Ma, Y., Han, J., Levendel, H.: A proactive approach towards always-on availability in broadband cable networks. Computer Communications, vol. 28. (2005) 51–64
8. Salfner, F., Wolter, K.: Analysis of service availability for time-triggered rejuvenation policies. Journal of Systems and Software, vol. 83. (2010) 1579–1590
9. Li, L., Vaidyanathan, K., Trivedi, K.S.: An Approach to Estimation of Software Aging in a Web Server. Proc. Int'l Symp. Empirical Software Eng. (ISESE), (2002) 45–52
10. Kourai, K.: Fast and Correct Performance Recovery of Operating Systems Using a Virtual Machine Monitor. Proc. of the 7th ACM SIGPLAN/SIGOPS international conference on Virtual execution environments, vol. 46. 7 (2011) 99–110
11. Okamura, H., Dohi, T.: Comprehensive evaluation of a periodic checkpointing and rejuvenation schemes in operational software system. Journal of Systems and Software, vol. 83. (2010) 1591–1604
12. Vaidyanathan, K., Trivedi, K.S.: A Comprehensive model for software Rejuvenation. IEEE Transaction on Dependable and Secure Computing, vol. 2. 2(2005) 124–137

13. Machida, F., Kim, D., Trivedi, K.: Modeling and Analysis of Software Rejuvenation in a Server Virtualized System. Proc. of 2nd Workshop on Software Aging and Rejuvenation (2010) 1–6
14. Thein T, Park J.S.: Availability Analysis of Application Servers Using Software Rejuvenation and Virtualization. Journal of Computer Science and Technology, vol. 24. 2 (2009) 339–346
15. Thein, T., Chi S., Park, J.: Availability Modeling and Analysis on Virtualized Clustering with Rejuvenation. Int'l Journal of Computer Science and Network Security, vol. 8. 9 (2008) 72–80
16. Kourai, K.: Fast and Correct Performance Recovery of Operating Systems Using a Virtual Machine Monitor. Proc. of the 7th ACM SIGPLAN/SIGOPS international conference on Virtual execution environments, vol. 46. 7(2011) 99–110

Analysis of Group Performance by Swarm Agents in SACA Architecture

K. Ashwini and M.V. Vijayakumar

Abstract This paper focuses on to develop a cognitive architecture which demonstrates a broad range of cognitive issues such as decision-making, learning, problem solving, planning, using the concepts of swarm intelligence. To investigate the behavior of swarm intelligent agents in ambient environment with respect to motivation, coordination and performance. To check how performance of agents in a group varies with proper motivation and coordination. Since group of agents are performing together the intelligence performance exhibiting from the group is always greater than individual intelligence performance. The question is, what exactly is happening in these groups that their performance is superior to the individuals? With better group members, performance of the group will be better. What makes the individual member better? This research work provides the solution to the stated problem by developing the cognitive architecture and found that the actual dependents are type of individuals in a group and type of the task performed by the agents.

Keywords Cognitive architecture • Motivation • Agent • BDI • Swarm intelligence • Perception

1 Introduction

According to the research of social psychology of groups, social loafing is the concept where people exert less effort when they work in a group than when they work individually. But we know that group performance is always greater than the

---

K. Ashwini (✉)
Jain University, Bengaluru, India
e-mail: ashwini.kodipalli@gmail.com

M.V. Vijayakumar (✉)
Dr. Ambedkar Institute of Technology, Bengaluru, India
e-mail: dr.vijay.research@gmail.com

S.C. Satapathy et al. (eds.), *Proceedings of the International Conference on Data Engineering and Communication Technology*, Advances in Intelligent Systems and Computing 469, DOI 10.1007/978-981-10-1678-3_76

individual performance. In 1913, Ringelmann says people put less efforts when they work in group than they work individually by considering the rope pulling experiment [1]. But he did not say the reason why people put less effort. In 1974, Alan Ingham proved that performance decrease depends not only on the communication but also on the motivational losses by considering the same rope pulling experiment [2]. Hence motivation is a force acting on or within a person that causes the voluntary efforts toward the direction of the goal. Karau and William found motivation was highest when individual believed that the goal was easily attainable and very valuable [3]. According to Richard M Ryan and Edward L Deci, motivations are of two types—Intrinsic and Extrinsic [4]. Intrinsic motivation is an internal force acting within a person to perform the task without expecting the reward. The person performs the task due to his or her own interest. Extrinsic motivation is an external force acting on the person to perform the task so that reward is achieved or to avoid punishment. A motivational theory explains why and how human behavior is activated. There is no single theory that is universally accepted. Motivational theories are categorized into two-content and Process theories. Content theories say what motivates the people to perform the task. Some of the examples are Maslow, Alderfer, Herzberg, and Mclelland. Process theories deals with how motivation occurs. Examples are Expectancy Theory, Goal Setting Theory, and Equity Theory. This research has concentrated on the goal setting theory which is proposed by Edwin Locker in 1960. According to Fred C. Lunenburg, the goal setting plays a significant role in the performance. He says performance is always higher for specific and challenging task with the appropriate feedback [5]. According to Mark R. Lepper, David Greene, and Richard Nisbet when an expected external reward decreases, a person's intrinsic motivation to perform a task decreases. Reward has the impact on the performance [6]. According to C. Bram Cadsby, Fei Song, Francis Tapan individuals were more motivated to perform at the high level when they were paid more based on their performance by experimenting the Pay-For-Performance verses fixed pay, and showed that Pay-For-Performance achieved higher productivity significantly through sorting and incentive effects. Productive employees selected PFP [7]. According to Barrie L Cooper, Pamela Clasen, Dora. E Silva-Jalonen, Mark Butler, when an individual were told that the feedback will be given, then they were more motivated and performed the task in a better way. Performance is high [8]. According to Jurgen Wegge, Alexander Haslam, all groups with the specific and difficult group goals performed better than a group asked to do the best. Group goal setting counteracts motivational losses like social loafing, group goal setting increased team identification, compensate for other weak team members, value of group success and failure. Group goal setting is a robust strategy for motivation and performance [9].

2 Proposed System

The design of Swarm Ambient Cognitive Architecture (SACA) draws heavily on natural swarms and our immediate surroundings which are termed as ambient environment. Hence the name SACA. The SACA is developed from the Society of Mind Cognitive Architecture developed by Vijayakumar [10]. The proposed cognitive architecture has collections of motivated and non-motivated intelligent agents working together for the predefined goal. The implemented SACA is a four-layered architecture with reflexive, reactive, deliberative, swarm, and meta-learning agents. The reflexive, reactive, and deliberative agents based on given situation in the environment exhibit cognitive skills like planning, decision-making, reasoning, problem solving. At the higher level in the proposed cognitive model, the swarm and meta-learning agents exhibit skills like learning, communication, and capability to move. The behavior of the agents in the environment is analyzed based on how well the agents are self or externally motivated to perform a given task which is defined as competition, survival of the agents in the environment. In the cognitive architecture, the test beds are used for simulating, comparing the outcomes of the architecture.

In the developed simulated test bed using prolog graphics the agents are created. The swarm agents are represented in circle shapes. Different parameters considered for agents are food defined as energy resource and are represented in the small green square shape and diamonds defined as the goal parameter and are represented by small while circle shape as shown in the Fig. 1.

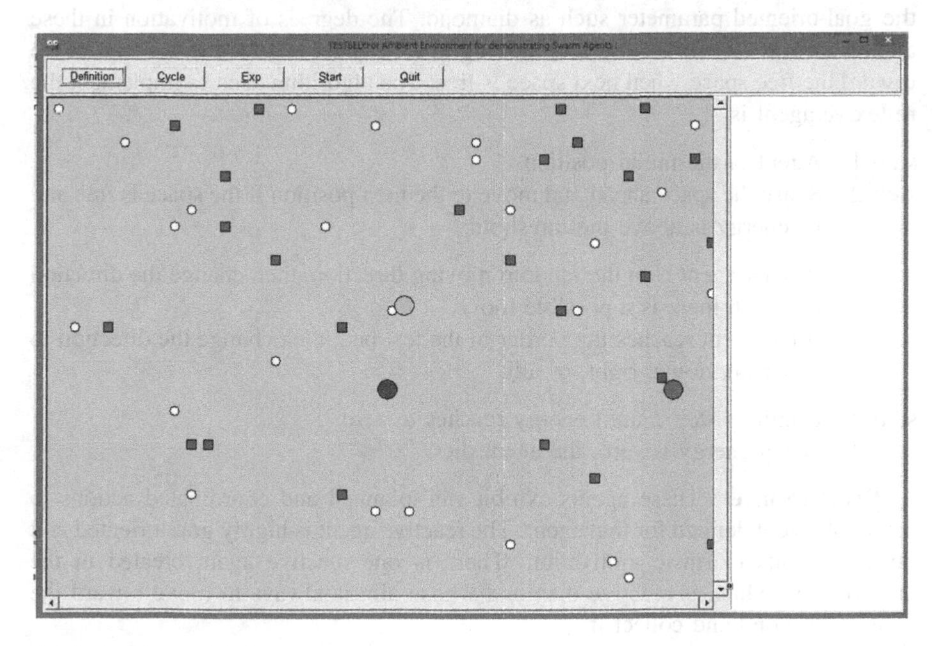

Fig. 1 Ambient test bed

In the test bed, the agents have to collect as many diamonds as possible. Since group of agents are performing together, the performance of the group will be high. Different metrics are used to analyze the behavior of the swarm agents. The major metrics are performance with respect to competition, level of motivation, and social interaction with each other and with the environment. Performance of the swarm agents depends on the number of cycle the agents has survived in the environment. The performance of the agents is evaluated based on the diamond collection. The performance of the group depends on the level of motivation of individuals and type of motivation of the individuals. Resource collection and life expectancy are two metrics used in the SACA test bed experimental results. The life expectancy is defined as agents' survival in a test bed for fixed energy. Resource collection is defined as the number of resources collected in a given time cycle.

SACA architecture consists of reflexive, reactive, deliberative, swarm, and meta-learning agents.

Reflexive layer: These agents are in lower level and exhibit simple reflexive behavior. These agents exhibit low level of motivation or same times zero level of motivation toward the collection of goal parameters. The reflexive agents can sense the edges of the environment and center of the environment. These agents can move in all the four directions. If the next space is free, it moves to that position without any deliberative intensions. The agents exhibit simple reflex behaviors, the actions which happen before thinking. These behaviors are implemented using Finite State Machine. Based on the input, the output of FSM is directly mapped on to the agents' actions.

This agent does not understand the energy bound parameters such as food and the goal-oriented parameter such as diamond. The degrees of motivation in these agents are zero. Zero because it is moving toward edges where it is idle and move toward the free space when next space is free. The algorithm used to implement the reflexive agent is

step 1: Agent in the initial position

step 2: Sense the space ahead and move to the next position if the space is free and its energy is above the threshold.

If other agent is in the random moving direction, then change the direction in which there is a possible move.
If the agent reaches the border of the test bed, then change the direction to either up, down, right, or left.

step 3: Continue step 2 until energy reaches to zero.

step 4: When energy is zero, the agent dies.

Reactive layer: These agents exhibit well-planned and coordinated actions to satisfy the goal defined for that agent. The reactive agent is highly goal-oriented and hence exhibits extrinsic motivation. There is one reactive agent created in the environment which is reactive diamond whose aim is always to move toward the nearest diamond and collect it.

These agents understand the affecting parameters of their behaviors. Through the perceptual range, the agents calculate the shortest distance toward the diamond and it always moves in the direction in which the nearest diamond is available. Within the perceptual range, if there are no resources available, then the agent move toward the edges of the environment like reflexive agent.

These agents lack in intrinsic motivation such as understanding that its energy level has gone down below the threshold and it requires food to survive longer in the environment. The agent at this level shows the extrinsic motivation. The algorithm used to implement the reactive agent is

step 1: Agent in the initial position.
step 2: Sense the goal parameter in the environment and move in the direction.

If there are many goal parameters in the perceptual range, calculate the shortest distance and move toward the nearest goal.
If wall is found in its move, then change the direction and move toward the free space.
If agent 2 is found in the direction, then move with the agent.

step 3: When energy is zero the agent dies.

Deliberative layer: The second layer of architecture is the deliberative layer which has a BDI agent who is capable of reasoning about their own internal task and plan. There is one BDI agent, BDI-diamond. BDI-diamond selects and controls reactive diamond.

The BDI agent first maps its internal state on to the belief set. Based on what is its internal stale, it creates an appropriate desire set from the belief set. From the desire set it sets an intension to the BDI agent.

For example, If an agent is intended to collect the diamond and if its energy is less than the threshold value, it changes its desire to collect the nearest food. The agent at this layer is highly (Intrinsic) motivated comparing to the other two layers below.

Due to highly intrinsic motivation, the agents collect the parameters available in the environment to achieve their goals according to their intensions. Agents are highly motivated as they understand their energy levels. They intelligently change their desire to collect food where they are hungry. The algorithm used to implement the deliberative agent is

step 1: Agent in the initial position.
step 2: Sense the goal parameter in the environment and move in the direction.

If there are many goal parameters in the perceptual range, calculate the shortest distance and move toward the nearest goal. (Planning)

step 3: If energy is less than the threshold (less than 40) then agent moves toward the shortest food. On consumption of food the agent energy increases. (decision-making)

step 4: If energy is greater than the threshold (more than 40) then agent moves toward the shortest goal.
step 5: Continue until all foods are collected or till the agent dies.

Swarm layer: This is the layer where agents will coordinate and perform the assigned task. Suppose if the task assigned to the agents is to find and consume the treasure food, then the working of the agents is as follows:

step 1: Agents at random initial position.
step 2: The agent which can sense the treasure food first will update its current position to the memory and all agents will follow the position of the agent which has been sensed. As first sensed agent moves near its position, it will be keep updating and all other agents will follow it.
step 3: Continue till all agents reach the treasure. Energy level can be considered as described in the deliberative layer.

Meta-learning layer: This is the fourth layer of architecture. Consider the scenario where in the test bed there is a decayed food. On consuming the decayed food the agent loses its current energy by 10 units. So, immediately the agent has to move toward the food. Consume the food and increase the energy by 45 units. Next time if an agent comes across same decayed food it avoids consuming it as it learnt from its previous experience. This is reinforcement learning. Reinforcement learning algorithm can be used in this layer. It is a learning through interacting with the environment and selects it actions based on the past experiences and new exploitation. The agent receives the reward which helps in learning for future. The algorithm of each layer is implemented using prolog programming language.

Bar graphs are used to design the graphs because bar graph displays the data in separate columns and are used to compare two data sets.

3 Results

The results will demonstrate that group performance depends not only on the communication among the agents but even the individual's level of motivation adds to the performance. SACA architecture is designed to check how individual agents will behave in a group, how agents behavior will have an impact on the group performance. The graphs are generated by experimenting with the following data: Initial energy 100 units, Cycle 500 units, Food 25 units, Diamonds 25 units. 3 agents are considered: Reflexive, Reactive, and Deliberative. The experiment is conducted and the output text file is generated with the data. This data is plotted to the excel sheet and the graphs are generated. Figure 2a gives the comparative study to find out the highly motivated agent among the group of agents. Figure 2b shows the group performance by considering the motivated agents. Figure 2c shows the group performance by considering the non-motivated agents.

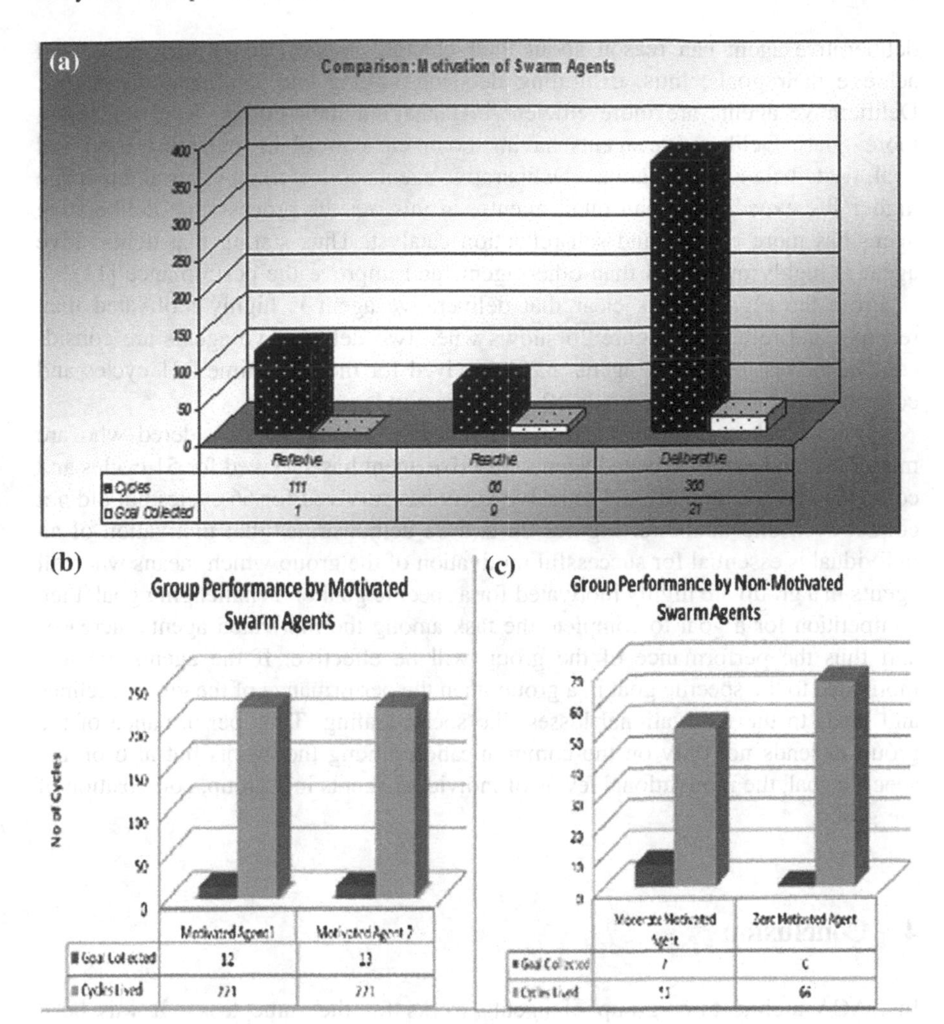

Fig. 2 **a** Comparison of swarm agents. **b** Group performance by motivated swarm agents. **c** Group performance by non-motivated swarm agents

When the experiment is conducted with the same input, the Fig. 2a shows that deliberative agent is highly motivated agent than reactive and reflexive agents. This graphs shows that intrinsic motivation is better than extrinsic motivation. After running the experiment, the deliberative agent has survived for 366 cycles out of 500 cycles and has collected 21 diamonds out of 25 diamonds. The reactive agent has survived for 66 cycles and collected nine diamonds. The reflexive agent has performed for 111 cycles and collected one diamond. Since the deliberative agent is intrinsically motivated, the agent has collected more number of diamonds and survived for longer cycle than reactive and reflexive; hence the deliberative agent is highly motivated agent than reactive and reflexive. This graph also shows deliberative agent can reason about their change of aims, sense their state, and achieve their goals, thus, exhibiting decision-making and intelligent behaviors. Deliberative agents are more efficient in managing their energy level collecting more goals. Deliberative agents having complete control in managing food and goal, try to balance motivations. Deliberative agents collect more goals and manage higher life expediency than other agents. In this, results process that deliberative agent has more control and self-reflection catalyst. Thus stating that deliberative agent is highly motivated than other agents and improve the performance [11].

From the Fig. 2a, it is clear that deliberative agent is highly motivated than reactive and reflexive. Figure 2b shows when two deliberative agents are considered in the test bed, both agents have survived for the same time 221 cycles and collected almost the same goals 12 to 13 diamonds.

Figure 2c shows when reactive and reflexive agents are considered who are moderated and zero motivated agents, reactive agent has survived for 51 cycles and collected seven diamonds and reflexive agent has survived for 66 cycles but did not collect even one diamond. Figures 2b and 2a demonstrates that motivation of an individual is essential for successful motivation of the group which means when all agents in a group are highly motivated for a specific goal or a challenging goal, then competition for a goal to complete the task among the motivated agents increases and thus the performance of the group will be effective. If the agents are less motivated to the specific goal in a group, then the performance of the group declines and leads to the motivational losses like social loafing. Thus performance of the group depends not only on the communication among the agents but also on the specific goal, the motivational levels of individual agents in a group, composition of the group.

4 Conclusion

In SACA architecture, group of agents works for the same tasks. It was been demonstrated in SACA that the group performance was high when they had a specific goal and their performance was high. The group's goal setting also increases the individual performance in SACA. It motivates agents to achieve the goal. Performance of swarm agents in SACA is high, because individuals are more responsible of their task when they work in a group. Therefore, the performance is more when they have a well-desired goal setting with highly motivated agents. SACA model clearly demonstrates that communication alone among the agents is not sufficient for the performance to increase but even on the type of the task (specific goal), the motivational levels of individual agents in a group and composition of the group(type of individuals in the group). The main contribution of the work has made several contributions to the field of motivational theories and cognitive architectures. It is found that the performance of the group will be high when the individuals are highly motivated to the goal and the goal defined should be specific.

References

1. Ringelmann, M.: Recherches sur les moteurs animés: Travail de l'homme, 2nd series, vol. 12, pp. 1–40, Annales de l'Institut National Agronomique (1913)
2. Ingham, A.G.: The Ringelmann effect: studies of group size and group performance. Vol. 10, pp. 371–384, Journal of Experimental Social Psychology, (1974)
3. Karau, S. J., Williams, K. D. : The effects of group cohesiveness on social loafing and social compensation., vol. 1, pp. 156–168, Group Dynamics: Theory, Research, and Practice (1997).
4. Richard, R., Edward, D.: Self-determination theory and facilitation of intrinsic motivation, social development, and well-being, vol. 55, pp. 68–78. American Psychologist, (2000)
5. Fred, L.C.: Goal-Setting Theory of motivation, vol. 15, pp. 1–6. International Journal of Management, Business, and Administration, (2011)
6. Mark R, L., David, G., Richard, Nisbet.: Undermining Children's Intrinsic Interest with Extrinsic Reward: A test of over justification Hypothesis, vol. 28, pp. 129–137, Journal of Personality and Social Psychology (1973).
7. Bram, C.C., Fei, S., Francis, T.: Sorting and incentive effects of pay for performance: An experimental investigation, vol. 50, pp. 387–405, Academy of Management Journal, (2007).
8. Cooper, Barrie L., Pamela, C., Silva-Jalonen, Dora. E., & Butler, Mark.: Creative performance on an in-basket exercise: Effects of inoculation against extrinsic reward, vol. 14, pp. 39–57, Journal of Managerial Psychology, (1999)
9. Jurgen W., Alexander H.: Improving work motivation and performance in brainstorming groups: The effects of three group goal-setting strategies, vol. 14, pp. 400–430, European Journal Of Work and Organizational Psychology, (2005)
10. Vijayakumar, M.V.: Society of Mind cognitive architecture, PhD Thesis. (2008), http://www2.dcs.hull.ac.uk/NEAT/dnd/agents/agents.html
11. Ashwini, K., Dr Vijayakumar, M.V.: Communication Among Agents Using SACA Architecture, Vol. 6, pp. 3571–3576, International Journal of Computer Science and Information Technologies, (2015)

Background Modeling and Foreground Object Detection for Indoor Video Sequence

N. Satish Kumar and G. Shobha

Abstract This paper proposed efficient and reliable algorithm for background subtraction for indoor video sequences. The paper mainly focused on compensating the illumination variation in the frame and then applying improved Gaussian mixture model to build the background model and then detect the moving foreign objects in indoor video sequences by using the Euclidian distance as metric. To compensate the illumination variation, homomorphic filtering algorithm for color image in HSV color space was proposed in the paper. The paper also reported the performance evaluation of the proposed method and existing and found that the proposed algorithm achieved 80 % of improvement in detecting actual foreground objects in indoor video sequences having illumination variations.

Keywords Background subtraction · Homomorphic filter · Euclidian distance

1 Introduction

Background subtraction and foreground detecting are very preliminary and essential steps in many computer vision applications. The frames of the given video with and without the intruding objects show some different behaviors that can be well defined and described by a statistical model. The statistical model has been built for the first initial frame of the given video, and then new foreign foreground objects can be detected by comparing them pixel by pixel from the next frame of the same video which does not match with the model. This process of extracting foreground objects from the background is usually called as background subtraction. This process of segmenting foreground and background is often one of the first and foremost task in computer vision applications such as activity recognition, video

---

N.S. Kumar (✉) · G. Shobha
CSE Department, R V College of Engineering, Bangalore, India
e-mail: satish.rmgm@gmail.com

G. Shobha
e-mail: shobhatilak@rediffmail.com

S.C. Satapathy et al. (eds.), *Proceedings of the International Conference on Data Engineering and Communication Technology*, Advances in Intelligent Systems and Computing 469, DOI 10.1007/978-981-10-1678-3_77

summarization, key frame extraction, etc. [1] The background subtraction method's performance relies upon the type of the background model applied. The background model should be capable of handling video types with objects overlapping, cluttered areas in the visual field and shadows [2]. The extraction of moving foreign objects may result in extraction of relevant information and can be used for many computer vision applications such as recognition, classification, and activity analysis. There are many existing basic models which can be applied for initial frame of the video. The basic model is taking the average of the images over time [3, 4]. The initial frame to be selected for background model is the set of initial frames where there is no motion of objects. Even though there are many existing background subtraction algorithms which can handle all the issues and challenges, still many algorithms are more sensitivity to illumination changes for indoor video sequence [5, 6].

This paper proposed efficient and reliable method of background modeling and foreground object detection which can solve the problem of gradual illumination variation for indoor video sequence. This paper also reported performance evaluation of the proposed method in comparison with existing methods.

2 Related Work

Mixture of Gaussian (MoG), which is very popular model, is to build initial frame of a given video with an only one Gaussian [7], which is a basic technique that calculates mean and standard deviation of each pixel; this is called background model. Incoming frame is subtracted with the background model and gets the threshold of the result. Algorithm also updates the model parameters recursively that can be adapted to the scene change in the video; the model does yield accurate results when the video contains dynamic content. In order to avoid this situation multi-Gaussian distribution per pixel (MoG) was used to build the model, which can handle such situations [8]. MoG to build background model is proved successful for outdoor video sequence but not feasible for indoor video sequences because of more illumination variations. MoG method results in lot of noise or false alarms. To overcome many false alarms, texture features have been used to model background statistics [9].

Even if color is significant information to extract features from images, there are some constraints where color information is not enough. Combining with color information texture information and analysis plays a vital role for many computer vision applications. Texture measures can also be used to build background model which can compensate variation in illumination. Texture can become useful tools for interpretation of high-level context for natural image. The approach employed local binary pattern (LBP) operator, which has shown excellent performance when applied for background subtraction and has several properties that favor its usage in background modeling. But this method also fails to capture gradual illumination variation which is normal in indoor video sequences.

Zhang et al. [10] proposed fuzzy logic approach which combines the gradient and color information for background subtraction. Raja et al. [10] make use of Hue Saturation Intensity (HSI) color space which results in limited level of intensity invariance and achieved moderate results for indoor video sequences. Ivanov presented a method where fast background subtraction is carried out by incorporating disparity. This method is invariant to changes in illumination, but cannot achieve accurate results for indoor environment. Lim and Mittal analyzed the method proposed by Ivanov [11] and suggested a better sensory placement strategy which can help in reducing errors and results are enhanced. Hanzi and Suter [12] evaluated the performance proposed by (GMM method) and suggested certain modifications in order to get better performance of the algorithm. [13] done extensive error analysis of statistical background modeling and pixel labeling and provides both theoretical analysis and experimental validation. All the above-said background methods can perform well for the outdoor video sequences but cannot be applied for indoor environment [14]. The next section describes about the proposed method which is the modification of existing improved Mog method with illumination compensation technique.

3 Proposed Method

All background subtraction algorithms track objects by applying statistical background model. There are many background subtraction methods but all background subtraction methods failed for indoor video sequences because of illumination variation. To overcome the illumination problem, homomorphic filter [14] on HSI color space has been proposed, which extracts the intensity, lightness, and brightness of each scene. Then, Adaptive Gaussian mixture model was applied on the enhanced illumination compensated frame. The proposed methodology is given in the Fig. 1.

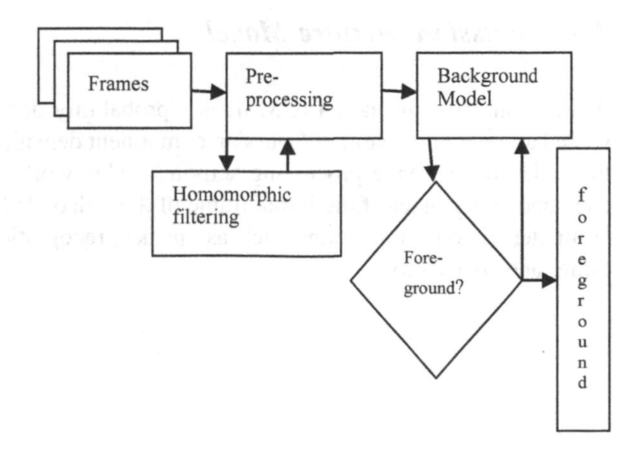

Fig. 1 Proposed methodology

3.1 Preprocessing

This is the preliminary step in the proposed method, all frames have been preprocessed to eliminate illumination component, which can be a factor under different illumination factors. The main aim of preprocessing step is improvement of the input data by suppressing undesired illumination variation.

3.2 Homomorphic Filter for Color Images

This paper considered and analyzed many preprocessing image enhancement algorithms and found that homomorphic algorithm is best to remove illumination variation component from an image. This filter normalized the brightness and increases the contrast of an input frame. Homomorphic filter has to be directly applied on color images without converting color image to gray scale and convert back to color image.

Following is the developed homomorphic algorithm for color images:

Step 1: Get background frame from video
Step 2: Separate color and intensity of image

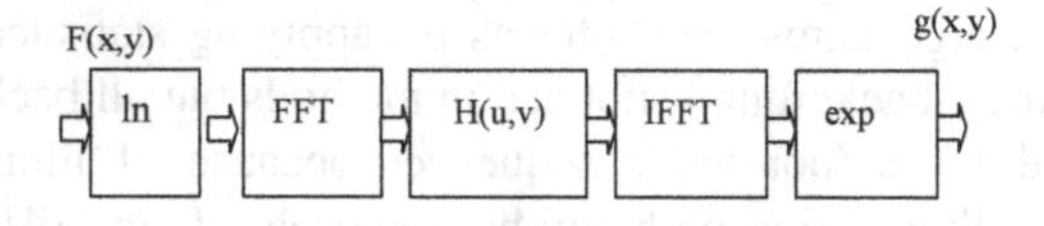

Step 3: Apply homomorphic filter on intensity part of the image
Step 4: Integrate color and intensity component of image.

3.3 Gaussian Mixture Model

A Gaussian mixture model (GMM) is a probability density function that is represented as a weighted sum of Gaussian component densities. GMMs are usually used to model many image processing activities. This work employed GMM to model each and every pixel of the initial frame of the video [15]. GMMs are used in many computer vision applications such as speaker recognition system, behavior modeling, and many more.

Principle:
This paper describes the principle to model each pixel of image using GMM. The probability of getting the current pixel whether background or foreground is given by Eq. (1):

$$P(X_t) = \sum_{i=1}^{K} \omega_{i,t} n\left(X_t, \mu_{i,t}, sd_{i,t}\right) \tag{1}$$

where K is number of Gaussians to be used to model, $\omega_{i,t}$ is weight associated with ith Gaussian with mean $\mu_{i,t}$ and standard deviation, n is a Gaussian pdf

$$n(X_t, \mu, sd) = \frac{1}{(2\pi)^{n/2}|sd|^{1/2}} e^{-1/2(X_t-\mu)sd^{-1}(X_t-\mu)} \tag{2}$$

The covariance matrix is given by Eq. (3)

$$\sum_{i,t} \sigma_{i,t}^2 I \tag{3}$$

The first B Gaussian distributions which go beyond certain threshold T are termed as background distribution

$$B = \text{argmin}_b \left(\sum_{i=1}^{b} \omega_{i,j} > T \right) \tag{4}$$

Where T is threshold value.

Case 1: If match is found with one of the K Gaussians of background model, then for the matched component updating has to be done.

$$\omega_{i,t+1} = (1-\alpha)\omega_{i,t} + \alpha \tag{5}$$

Learning rate is given by α.

$$\mu_{i,t+1} = (1-\rho)\mu_{i,t} + \rho X_{t+1} \tag{6}$$

$$\sigma_{i,t+1}^2 = (1-\rho)\sigma_{i,t}^2 + \rho\left(X_{t+1} - \mu_{i,t+1}\right).)(X_{t+1} - \mu_{i,t+1})^T \tag{7}$$

3.4 Foreground Detection

In this method, foreign moving objects are located by comparing the behavior of the next incoming pixel with the background model. If it behaves similarly, that means, if incoming pixel matches with the background model, then it classifies as background otherwise new foreign pixel has arrived into the scene. Euclidian distance metric is used to find out the match with the model and identifies the foreground moving objects from the frame. Euclidian distance is given by the following Eq. (8).

$$d = \sqrt{\sum_{i=1}^{n} (X_i - Y_j)^2} \tag{8}$$

4 Result and Performance Evaluation

A lot of experimentations have been carried out with the proposed method and performance was evaluated with improved GMM and GMM using different types of videos [16]. This paper also reported the experimental results and showed that, the proposed methods can outperform existing methods for videos which contain illumination variations. Figure 2 shows the comparison of the various BS algorithms.

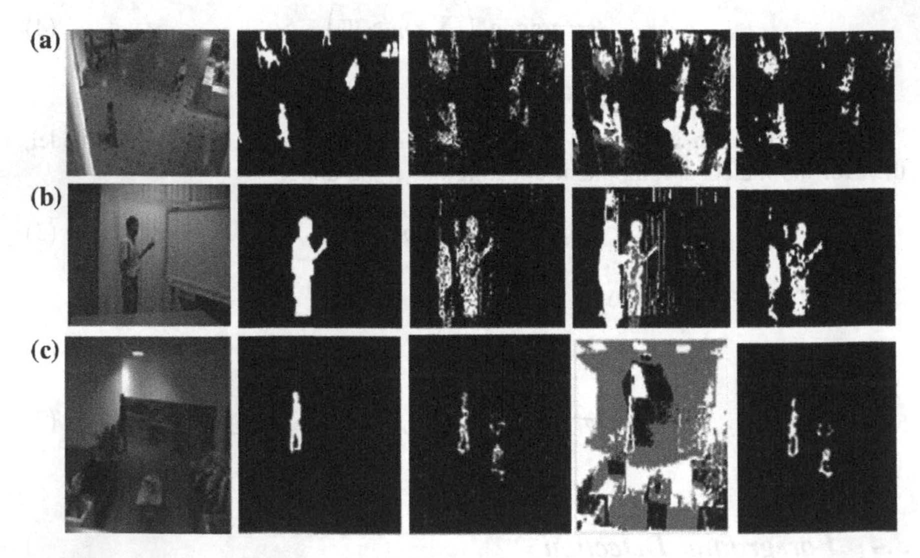

Fig. 2 Background subtraction results from *left* to *right*: **i** reference frame, **ii** ground truth, **iii** GMM, **iv** improved GMM, **v** proposed method

Table 1 Values for different BS algorithms

FPR		
GMM	IGMM	Proposed
1913	1628	1496
580	755	468
117	127	100
611	233	202
1783	928	590

4.1 Performance Evaluation

Following are the standard metrics which can be used for performance evaluation of the background subtraction algorithms. The following table reported the false positive ratio values of three methods and found that proposed method outperforms all the existing methods.

- False Positive Ratio (FPR): which are not actually foreground pixels, is given by **FPR = FP/(FP + TP)**.
- False Negative Ratio (FNR): which are not actually background pixels.
- True Positive Ratio (TPR): which are actually foreground pixels. (Table 1)

$$\text{Recall} = \frac{tp}{tp + fn} \tag{9}$$

$$\text{Precision} = \frac{tp}{tp + fp} \tag{10}$$

where *tp* is true positives, *fp* is false positives, and *fn* is false negatives.

The following Fig. 2 shows the results of our proposed method and Table 2 gives the comparison with GMM and IGMM methods, and found that proposed method outperforms the other two methods in terms of precision and recall values, the same is reported in Table 2.

The Fig. 3 demonstrated recall and precision values for all the video sequences and found that for the proposed algorithm, recall and precision values are better than others. The above graph shows the values of recall and precision for Dataset#1, Dataset#2, and Dataset#3. The graph shows that for all the datasets both precision and recall values are better for the proposed method.

Table 2 Precision and recall values for different dataset and BS algorithms

	Dataset#1		Dataset#2		Dataset#3	
	Pre	Rec	Pre	Rec	Pre	Rec
Proposed	0.94	0.91	0.93	0.89	0.91	0.88
IGMM	0.85	0.86	0.83	0.88	0.89	0.89
GMM	0.76	0.80	0.78	0.80	0.75	0.79

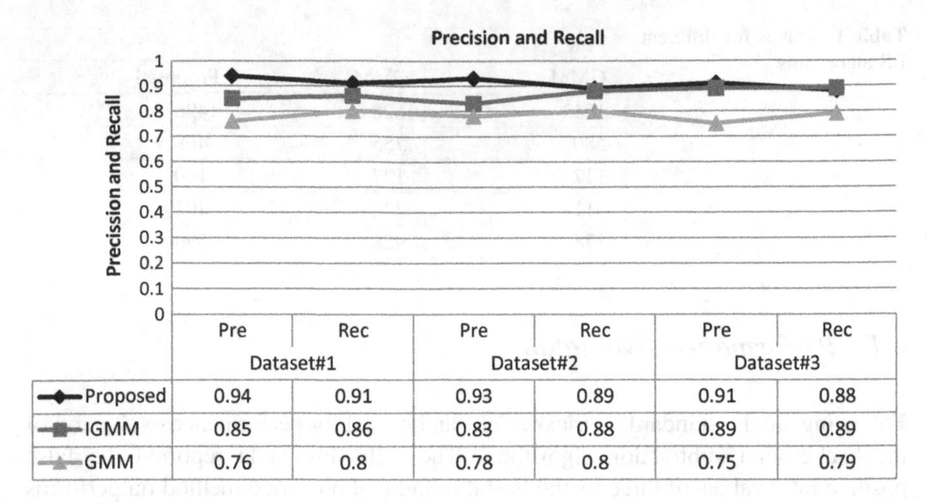

Fig. 3 Graph for precision and recall

5 Conclusion

This section concludes that very less research has been done specifically for indoor video sequence where there is lot of illumination variation. The proposed approach of detection moving objects using background subtraction is a contribution to the background subtraction for indoor environments. To compensate illumination variation, homomorphic filter has been applied directly on color frame. This paper also proposed homomorphic filter as a preprocessing step before background subtraction. The major strength of the proposed method is its invariant for illumination changes for indoor sequence.

References

1. Kumar, N.S.; Shobha, G.; Balaji, S., "Key frame extraction algorithm for video abstraction applications in underwater videos," in Underwater Technology (UT), 2015 IEEE, vol., no., pp. 1–5, 23–25 Feb. 2015
2. M. Piccardi, "Background subtraction techniques: A review," in Proc. IEEE Int. Conf. Syst., Man Cybern., Oct. 2004, pp. 3099–3104.
3. R. J. Radke, S. Andra, O. Al-Kofahi, and B. Roysam, "Image change detectionalgorithms: A systematic survey," IEEE Trans. Image Process., vol. 14, no. 3, pp. 294–307, Mar. 2005.
4. A.M. Elgammal, D. Harwood, and L.S. Davis, "Non-ParametricModel for Background Subtraction". ECCV, 2000, pp. 751–767.
5. Y. Benezeth, P. M. Jodoin, B. Emile, H. Laurent, and C. Rosenberger, "Review and evaluation of commonly-implemented background subtraction algorithms," in Proc. IEEE Int. Conf. Patt. Recog., Dec. 2008, pp. 1–4.
6. B. Rinner and W. Wolf, "An introduction to distributed smart cameras," Proc. IEEE, vol. 96, no. 10, pp. 1565–1575, Springer, 2009.

7. Stauffer and W. E. L. Grimson. Adaptive background mixture models for real-time tracking. IEEE Computer Society Conference on Computer Vision and Pattern Recognition, 1999, 2:252, 1999. URL: http://doi.ieeecomputersociety.org/10.1109/CVPR.1999.784637.
8. S.E. Umbaugh, "Computer Imaging: Digital Image Analysis and Processing", CRC Press, Florida, 2005
9. Hong-xun Zhang and De Xu. Fusing color and gradient features for background model. In 8th International Conference on Signal Processing, 2006, 2006.
10. Y. Raja, Stephen J. Mckenna, and S. Gong. Segmentation and tracking using colour mixture models. In In Asian Conference on Computer Vision, pages 607–614, 1998.
11. Y. Ivanov, A. Bobick, and J. Liu. Fast lighting independent background subtraction. In IEEE Workshop on Visual Surveillance, pages 49–55, 1998.
12. Hanzi Wang and D. Suter. A re-evaluation of mixture of gaussian background modeling [video signal processing applications]. In IEEE International Conference on Acoustics, Speech, and Signal Processing(ICASSP '05), 2005, volume 2, pages 1017–1020, 2005.
13. Ser-nam Lim, Anurag Mittal, Larry S. Davis, and Nikos Paragios. "Fast illumination-invariant background subtraction using two views: Error analysis, sensor placement and applications". IEEE Conference on Computer Vision and Pattern Recognition (CVPR'05), 1: 1071–1078, 2005.
14. R. Hummel, "Image enhancement by histogram transformation," Comp. Graph. Image Process., vol. 6, pp. 184–195, 1977.
15. P. Noriega, and O. Bernier, "Real Time Illumination Invariant Background Subtraction Using Local Kernel Histograms". BMVC, 2006, volume 3, pp. 979–988.
16. Shobha. G, N. Satish Kumar "Recent Trends in Machine Learning for Background Modeling and Detecting Moving Objects" International Journal of Advanced Engineering and Technology (IJAET), vol. 3, Issue 2, pp. 309–313.

Agri-Guide: An Integrated Approach for Plant Disease Precaution, Detection, and Treatment

Anjali Chandavale, Suraj Patil and Ashok Sapkal

Abstract Agriculture growth is the key component in socioeconomic growth of our country due to liberalization and globalization. Gradually with significant increase in technology, advanced telecommunication services assist plant disease treatment at remote locations. Earlier systems were designed for either monocot or dicot plant family disease detection. The paper proposes an integrated approach for monocot and dicot plant disease detection and treatment along with precautionary measurement through smartphone and image processing techniques. The paper mainly focuses on plant disease detection technique based on integrated approach of K-means segmentation algorithm and SVM Classifier. The proposed and developed approach gives 83 % accuracy for plant diseases recognition.

Keywords Classification · Dicot plant disease · Feature extraction · Monocot plant disease · Preprocessing · Segmentation

1 Introduction

In India, our farmers suffer due to lack of knowledge in context with crop diseases along with proper cultivation process, which in turn results in reduction in crops at higher investment [1].

Plant disease is one of the important factors which causes significant reduction in the quality and quantity of plant production. Classification and recognition of plant

Anjali Chandavale (✉) · Suraj Patil
Department of Information Technology,
MIT College of Engineering, Pune, India
e-mail: c.anjali38@gmail.com

Suraj Patil
e-mail: surajpatil024@gmail.com

Ashok Sapkal
Department of E&TC, COEP, Pune, India
e-mail: ams.extc@coep.ac.in

S.C. Satapathy et al. (eds.), *Proceedings of the International Conference on Data Engineering and Communication Technology*, Advances in Intelligent Systems and Computing 469, DOI 10.1007/978-981-10-1678-3_78

diseases are important tasks to increase plant productivity and economic growth. Detection and classification of plant diseases are one of the challenging topics and much more interesting topics in Engineering and IT fields.

Due to Digital India Initiative, each and every person irrespective of his age and literacy is using finger tip access through mobile or smartphones. Therefore the paper presents framework titled as 'Agri-Guide v1.0' for farmers based on smartphone with the help of image processing techniques. Agri-Guide v1.0 provides precautionary measures to avoid plant diseases for monocot and dicot family along with disease detection and the treatment.

The related work is briefed in Sect. 2. Section 3 presents a proposed and developed work for precaution, recognition of plant diseases and treatment. Section 4 discusses the result and analysis and Sect. 5 concludes the paper.

2 Related Work

Ferzli et al. [2], contributed in the development of agriculture with the integration of image processing, visualization, and mobile cloud computing.

Dubey and Jalal [3] investigated the approach of detection and classification of apple fruit diseases which cannot be extended to other fruit diseases. Revathi and Hemalatha [4] explored advance computing technology to assist the farmer in the plant development process which is suitable for dicot family plant. The approach based on enhanced particle swarm optimization (PSO) feature selection method with skew divergence method is applied to identify the visual symptoms of dicot family: cotton plant diseases [5] whereas the approach [6] based on hybrid algorithms of template matching and support vector machine is suitable for sugar beet. Yan et al. [7] proposed a method to select features of cotton disease leaf image by introducing a fuzzy selection, fuzzy surfaces, and fuzzy curves. The extracted features were used for diagnosing and identifying diseases. The features were extracted from the fuzzy selection approach. This approach removes the dependent features of an image for reducing the number of features in classification process.

In [8], they investigated an approach for automatic identification of chilies plant diseases. For that, color features were extracted with the help of CIELAB color transformation model and comparing the color features for identification of chili plant disease. The accuracy can be increased with the enhancement in extracted features.

Next paper [9] proposed an approach to diagnose the grape plant leaf disease using image processing and artificial intelligence techniques. Remote sensing technology is used to identify plant diseases at the local level [10]. In current scenario, the available systems for plants are Spray Guide and CropInfo. Spray Guide needs user intervention where as CropInfo provides information about production aspects, post-harvest technology, processing possibilities, and market information of plants.

3 Proposed Work

The proposed approach named as ‘Agri-Guide v1.0’ shown in Fig. 1 has considered the plants such as onion, wheat, and potato under monocot and dicot family plants, respectively. The farmer sends image of affected plant leaf to the server through mobile camera. At server side the features from image are extracted and the disease is detected with the help of SVM Classifier. The server retrieves an appropriate treatment based on disease detection and sends to farmer. The processing at the server side is explained in later section.

3.1 Server Side Processing

The processing at server side has three phases Disease Precaution, Disease Detection, and Treatment as shown in Fig. 2.

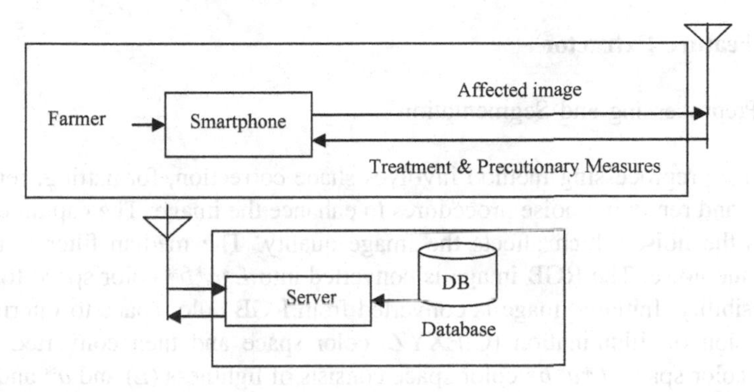

Fig. 1 Architecture

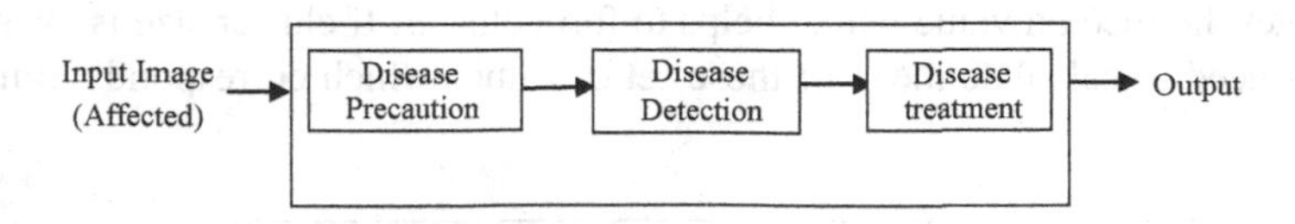

Fig. 2 Processing at server side

3.1.1 Disease Precaution

The precautionary measures provide the usage and proportion of pesticides, insecticides, and water as per the monocot/dicot plants. The history of crops corresponding to usage and proportion of pesticides, insecticides, and water is stored. The storage of history is beneficial for disease detection in the future. For each farmer separate records are maintained.

3.1.2 Disease Detection

The automatic detection of plant disease is an integrated algorithm based on existing K-means segmentation algorithm and support vector machine (SVM) classifier as shown in Fig. 3. In the initial phase, the features such as color, texture, and shape feature values are extracted with the help of L^*a^*b color space and gray level co-occurrence matrix (GLCM). The extracted features are sent to SVM classifier to detect particular plant disease.

3.1.3 Feature Extractor

Image Preprocessing and Segmentation

The image preprocessing method involves shade correction, formatting, removing artifacts, and removing noise procedures to enhance the image. The captured image contains the noise which affects the image quality. The median filter is used to remove the noise. The RGB image is converted into $L^*a^*b^*$ color space for better color visibility. Initially image is converted from RGB color space to International Commission on Illumination (CIEXYZ) color space and then converted into to $L^*a^*b^*$ color space. $L^*a^*b^*$ color space consists of lightness (L) and a^* and b^* for the color opponent dimensions. After converting into $L^*a^*b^*$ color space, the image is segmented into sub-images of same color using K-means clustering algorithm. The odd cluster size $k = 3$ is used for clustering. The odd cluster size calculates the median value which helps to form cluster. If cluster size is even, then there is need to calculate mean of these set of values which corresponds to median

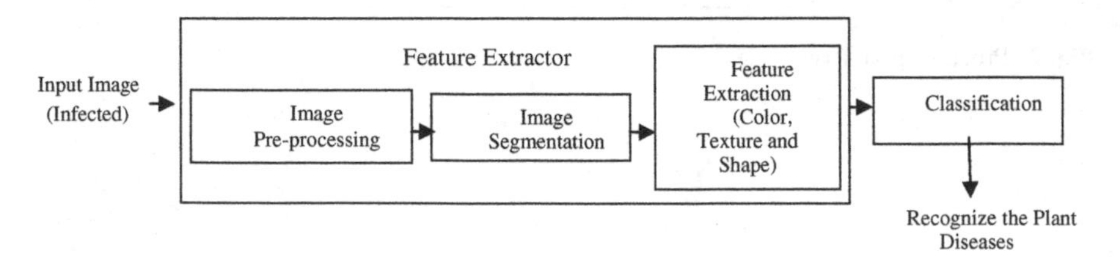

Fig. 3 Algorithm for disease detection

value to form cluster. After segmentation, the infected regions of plant leaf are clearly visualized. Clustered regions are those areas in the image that represents visual symptoms of plant disease.

Feature Extraction

The plant leaf disease symptoms are classified based on variations in color, texture, and shape features. These are dominant in classifying plant leaf disease symptoms. Color features are extracted with help of $L^*a^*b^*$ color space. First convert the image into 8 bit gray scale image and then use the gray level co-occurrence matrices (GLCM) to extract texture features. To extract shape features, four feature vectors namely area, perimeter, skewness, and kurtosis [6] are considered. The following feature vectors are used to extract the texture feature from GLCM matrix,

1. Contrast: Contrast is the difference in luminance or color that makes an object distinguishable which is calculated using Eq. 1.

$$\text{Contrast} = \sum_{i,j} |i-j|^2 p(i,j) \quad (1)$$

2. Angular Second Moment (ASM): ASM is calculated using Eq. 2.

$$\text{ASM} = \sum_i \sum_j p(i,j)^2 \quad (2)$$

3. Energy: The square root of angular second moment is known as energy and calculated using Eq. 3.

$$\text{Energy} = \sqrt{\left[\sum_i \sum_j p(i,j)^2\right]} \quad (3)$$

   Where, μ_x, μ_y, σ_x, and σ_y are the means and standard deviations of p_x and p_y coordinates.
4. Homogeneity: Homogeneity measures the nearest distributed elements in the GLCM to the GLCM diagonal which is calculated using Eq. 4.

$$\text{Homogeneity} = \sum_{i,j} \frac{1}{1-(i-j)^2} p(i,j) \quad (4)$$

5. Entropy: Entropy measures the loss of information and also measures the image information and calculated using Eq. 5.

$$\text{Entropy} = -\sum_i \sum_j p(i,j) \log(\mathrm{p}(i,j)) \quad (5)$$

   Where, element $[i, j]$ is generated by GLCM matrix. 'i' is x coordinate and 'j' is y coordinates. 'p' is the probability value of element $[i, j]$.

6. Correlation: Correlation measures the linear dependency of gray levels of adjacent pixels and calculated using Eq. 6.

$$\text{Correlation} = \left[\sum_i \sum_j (ij)p(i-j) - \mu_x\mu_y\right] / \sigma_x\sigma_y \tag{6}$$

7. Inverse Difference Moment (IDM): IDM is also known as local homogeneity and calculated using Eq. 7.

$$\text{IDM} = \sum_i \sum_j \frac{1}{1+(i-j)^2} p(i,j) \tag{7}$$

IDM is high when inverse GLCM is high and local gray level is uniformed.

Classification

Support vector machine (SVM) is a set of related supervised learning method used for classification and regression. SVM is used to classify the infected plant leaf diseases. In this process, color, texture, and shape features extracted from segmented image are considered. The SVM classifier is trained for fifty patterns of each plant leaf disease. It compares the extracted features of an input image with the stored features in database. If the input image features are matched with stored features in database, then it gives the result of corresponding plant leaf disease.

3.1.4 Disease Treatment

Based on the detection of disease, the treatment is provided. The treatment corresponding to plant leaf diseases for each stage is stored based on the symptoms and features extracted. Once the disease is detected, the corresponding treatment is retrieved from database and sent to the farmer.

4 Result and Analysis

The experiments and their validations are carried out using Android Software Development Kit (SDK), Eclipse Integrated Development Environment (Helios IDE), Android Honeycomb (3.0) to Lollipop (5.0) Operating System (O.S.) Platform. The hardware computing platform used for experiments is Intel Core-i5, 2.5 GHz with 4 GB RAM. The variations in the plant leaf image diseases are taken into account from time to time along with guidance and consultation from more than 4 Agricultural Domain experts. The input image to proposed and developed system, 'Agri-Guide v1.0,' is of either of Joint Photographic Expert Group (JPEG), Graphic Interchange

Format (GIF), and Portable Network Graphic (PNG) format. Thus the input database has more than 1000 leaf images of wheat, onion, and potato plants. The input image is an RGB image which varies from 60 × 20 pixels of size ~1.5 Kb up to 300 × 200 of size ~44 Kb. The output of proposed and developed system is precautionary measures for disease prevention and treatment in case of disease detection.

The wheat, onion, and potato plants leaf image dataset was collected from National Agricultural Research Center, Pune. The samples collected are of diseases namely purple blotch, Stemphylium Blight, Basal Rots, Black Mould disease of onion plant, stripe rust, powdery mildew, aphids, yellow rust disease of wheat plant and bacterial wilt (brown rots), black scurf, dry rot disease of potato plant.

The proposed and developed system is trained for 50 patterns of wheat, onion, and potato plants leaves diseases. The accuracy increases with training of more number of patterns. It is observed that accuracy of known patterns is 83 % whereas accuracy unknown patterns is 73 % as shown in Fig. 4. The response time of Agri-Guide v1.0 is less than 1 ms. Figure 5 shows graphical user interface

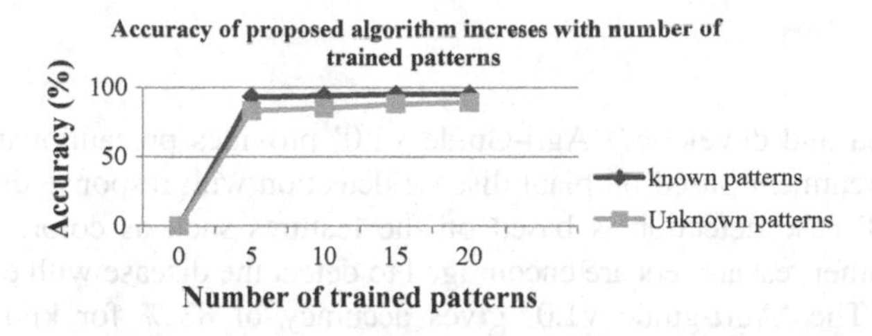

Fig. 4 Accuracy of proposed and developed approach

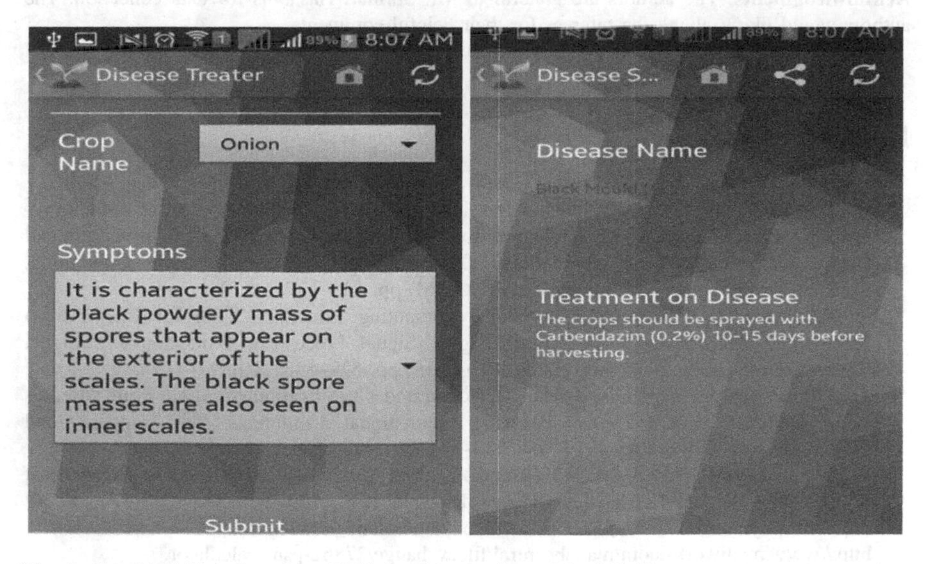

Fig. 5 Graphical User Interface (GUI) implementation of Agri-Guide

Table 1 Comparative study of existing system and Agri-Guide

	Spray Guide	CropInfo	Agri-Guide
Purpose	Provides information about usage of pesticides and insecticides and water	Provides information about post harvesting technologies and production aspects	Provides precautionary measures and treatment for plant disease
Accuracy	–	–	83 % Known pattern 73 % Unknown patterns
Family	Monocot	Monocot and Dicot	Monocot and Dicot

(GUI) implementation of proposed and developed Agri-Guide. The comparative study of Agri-Guide v1.0 with existing systems is as shown in Table 1.

5 Conclusion

The proposed and developed 'Agri-Guide v1.0' provides precautionary measures along with treatment based on plant disease detection with response time less than 1 ms. The disease detection is based on the features such as color, texture, and shape. The other researchers are encouraged to detect the disease with enhancement in features. The 'Agri-guide v1.0' gives accuracy of 83 % for known patterns whereas 73 % for unknown patterns of monocot and dicot plant diseases.

Acknowledgments The authors are grateful to Mr. Shrihari Hasabnis for data collection. The authors would like to thank the referees for their helpful comments.

References

1. Biswajit Saha, Kowsar Ali, Premankur Basak, Amit Chaudhuri: Development of m-Sahayak-the Innovative Android based Application for Real-time assistance in Indian Agriculture and Health Sectors, The Sixth International Conference on Mobile Ubiquitous Computing, Systems, Services and Technologies (UBICOMM) pp. 133–137, (2012).
2. R. Ferzli and I. Khalife: Mobile Cloud Computing Education Tool for Image/Video Processing Algorithms, Proceeding of Digital Signal Processing Workshop and Signal Processing Education Workshop (DSP/SPE) IEEE, pp. 529–533, (2011).
3. Shiv Ram Dubey, Anand Singh Jalal: Detection and Classification of Apple Fruit Diseases using Complete Local Binary Patterns, International Conference on Computer and Communication Technology IEEE, pp. 346–351, (2012).
4. P. Revathi and M. Hemalatha: Advance Computing Enrichment Evaluation of Cotton Leaf Spot Disease Detection Using Image Edge detection, IEEE, (2012).
5. http://murphylab.web.cmu.edu/publications/boland/boland_node4.html.
6. http://www.mathworks.com/matlabcentral/fileexchange/27862-psnr-calculator.

7. Yan-Cheng Zhang, Han-Ping Mao, Bo Hu, Ming-Xi Li: Features Selection Of Cotton Disease Leaves Image Based On Fuzzy Feature Selection Techniques, International Conference on Wavelet Analysis and Pattern Recognition IEEE, pp. 124–129, (2007).
8. Rong Zhou, Shunichi Kaneko, Fumio Tanaka, Miyuki Kayamori, Motoshige Shimizu: Early Detection and Continuous Quantization of Plant Disease Using Template Matching and Support Vector Machine Algorithms, International Symposium on Computing and Networking IEEE, pp. 300–304, (2013).
9. Sanjeev S. Sannakki, Vijay S Rajpurohit, V. B. Nargund and Pallavi Kulkarni: Diagnosis and Classification of Grape Leaf Diseases using Neural Networks, International Conference on Computing, Communications and Networking Technologies IEEE, (2013).
10. Jingcheng Zhang, Jinling Zhao, Dong Liang, Linsheng Huang, and Dongyan Zhang: New Optimized Spectral Indices for Identifying and Monitoring Winter Wheat Disease Applied Earth Observation and Remote Sensing IEEE, (6, JUNE 2014).

Determination of Area Change in Water Bodies and Vegetation for Geological Applications by Using Temporal Satellite Images of IRS 1C/1D

Mansi Ekbote, Ketan Raut and Yogesh Dandawate

Abstract Watershed management is a topic of great concern in the areas coming under shadow zone of rain. Watershed management needs the information about the area covered by water bodies and the relative change in the water body area over the years. This information can be obtained from field survey or by remote sensing. In remote sensing the satellite image is classified into a range of types of land covers by land use and land cover segmentation. The area of segmented water bodies is calculated and validated with the ground truth obtained by field survey. Pondhe village in Pune district is taken into consideration in this particular study. The temporal satellite images of this area are classified into various land covers, and area calculation is carried out. The reduction in water bodies and in the vegetation area is observed over the years by 2.94 % and 27.10 % respectively. The same technique can be used for various regions in future to provide change in areas and to establish a link between geology and image processing.

Keywords Land use land cover (LULC) classification · Image segmentation · Satellite images · Watershed management

1 Introduction

Every continent has been affected by water scarcity. About 1.2 billion people that are around one-fifth out of total world's population live in physically water scare areas. Around 500 million people are soon to face the same situation. About 1.6 billion persons that are nearly equal to a 25 % of world's overall population suffer

Mansi Ekbote (✉) · Ketan Raut · Yogesh Dandawate
Vishwakarma Institute of Information Technology, Pune, India
e-mail: ekbotemansi@yahoo.com

Ketan Raut
e-mail: ketan.raut@viit.ac.in

Yogesh Dandawate
e-mail: yogesh.dandawate@viit.ac.in

S.C. Satapathy et al. (eds.), *Proceedings of the International Conference on Data Engineering and Communication Technology*, Advances in Intelligent Systems and Computing 469, DOI 10.1007/978-981-10-1678-3_79

from economic water shortage. Economic water storage occurs when countries fail to build the infrastructure which is necessary to take water from rivers and aquifers.

In urban India scarceness of water is becoming number one woe at a very fast rate. It has been revealed that out of 32 major cities, 22 have to face water shortage on daily basis. The condition is even worse in rural India. 73.2 % of rural population has undersupply of water. Therefore, watershed management becomes India's crying need. Planning and development of watersheds needs proper understanding of movement and occurrence of water in surface and subsurface systems and also earth and nutrient fatalities in watershed.

The proposed work aims to categorize the land use and land cover. Land cover indicates the physical land type, like open water, forest, or bare land. Land cover can be determined by studying the satellite and aerial imagery. Land cover maps present data to assist managers best comprehend the current landscape. To get an idea about change over time, land cover maps of several different years are required. With this data, managers can weigh up past management decisions as well as gain insight into probable effects of their present decisions before they are implemented. Land use shows how people use or utilize the landscape—whether for conservation or development or mixed uses. Land use cannot be determined from satellite imagery. Land use and land cover segmentation is done based on different intensities of various regions in satellite image. The images are taken from IRS 1C/1D satellite launched by ISRO (Indian Space Research Organization). The satellite images taken here for study are in GeoTIFF format. Each satellite image is divided into three areas, viz., land, vegetation, and water bodies and each area is represented by on separate plane. The rate of degradation of vegetation and water bodies is calculated over the period of 15 years. The study area for this project is Pondhe village in Purandar taluka, Pune district. This area has extremely fertile land but its utilization is limited due to scarcity of water. Thus this area seeks for some measures to conserve the water resources already available and also to find some ways to enhance the water resources.

This paper proposes an efficient way of classifying satellite image in different land use and land cover which is used to calculate the areas covered by various land covers. The calculated area in number of pixels is then linked to the actual area in square meters obtained by actual field survey. The change in area covered by water bodies and the area under vegetation is calculated and the rate of change is deliberated. This offers an important tool for planning and management of watershed. The flow of this method is divided into five main stages which are preprocessing, segmentation, classification, area computation, and accuracy validation. In the first stage, preprocessing, original satellite image is applied with wiener filter. This removes blur from input image if any. Then the image is resized to a lower size. Satellite images are low contrast images by default. So in order to make the image suitable for segmentation, satellite images are subjected to contrast stretching and histogram equalization. For segmentation, using intensity-based classification on these three planes the land use and land cover classification is performed. After segmentation the area of various land covers is computed in number of pixels. Finally, this area is linked to the physical area in square meters and a geological relation between land use and land cover is derived. To avail this study in geological point of view and to provide an aid to water conservation, rate of change of various land use is computed. The land use and land cover classification to be useful is associated with the geological observations. The relation between geographical and geological data is derived by this segmentation process. The changes in various areas over the years are observed and a relation is established to see how various geological aspects are changing over time.

The remaining paper is organized as follows: literature survey is presented in Sect. 2. The proposed scheme is described in Sect. 3. Section 4 illustrates the segmentation process. Area calculations are included in Sect. 5. Section 6 presents a conclusion of the method proposed and study carried out.

2 Literature Survey

Various techniques are evolving day by day to classify satellite images into various land use and land covers. The goal is to reduce computational complexity of the method while increasing the accuracy. Various traditional as well as new methods of classification and segmentation are studied and implemented to check their suitability and accuracy of segmentation for this work proposed. Few of the important literatures studied are included here.

Pradhan et al. [1] proposed a technique to categorize Google earth images to calculate roof area in aim of estimate rainfall and to help in rainwater harvesting. They made use of GLCM, k-means clustering algorithm for this work. In this method the parts of the satellite images are segmented and classified by the use of k-means algorithm. The residential areas and built up areas are then separated. Then the roof top area is calculated. Using this calculated area of this roof tops, the water runoff is computed based on the rainfall. Rainwater harvesting schemes can be designed by this estimation, for water conservation, recycle, and percolation.

Prasad et al. [2] proposed a method for classifying land use and land cover in multispectral satellite images using SVM classifiers. They used SVM, Hierarchical Clustering Algorithm, Fuzzy C-Means Clustering. The accuracy of this method was 83.2 % and this method was not suitable for low contrast images.

Shiraishi et al. [3] proposed a scheme for fully supervised SVM Classifiers, ensembles of S one-class SVDD Classifiers. The specially aimed area is classified into various land cover classes by supervised classifiers.

Longbotham et al. [4], proposed a dexterous system for land use image classification. They made use of image various methods for processing the images and Support Vector Machines (SVM). The way calls for Training of the SVM and Testing of SVM. The image data is subjected to nonlinear isotropic diffusion and unsharp filtering in the stage of training. For training the SVM input is the segmented image pixels which match the land use parts. To make the testing automated, the regions segmented were mined out by using the active contour model.

The SVM which has been trained is then correctly classified into the regions of land cover on the basis of values of pixels of the area which is mined out.

Many researchers have presented various methods for land use and land cover classification [5–8]. Out of number of techniques studied, the methods useful for satellite images are shortlisted and tried on the set of satellite images of desired area. In our case, due to the eastward slope of land, active contouring method failed to present accurate results. Also due to very low change in intensities, k-means clustering gave misclassified clusters. The methods though suggested good means of classification, did not present any way of area computation and making it significant for geologists, water managers and various committees of water conservations. Thus there is a need of tool which can establish a link between image processing and geology. Once, this method works out well for area under study, the same method can be used for different areas and a geological survey can be made easier. The proposed work focuses on this application.

3 Proposed Scheme

A general flow of method to obtain the desirable results from the input is explained in this section. The complete method is organized in four main steps: preprocessing input satellite images, segmentation of preprocessed image into different land covers computation of area of each area and finally validates the accuracy of results obtained by comparing with ground truth data. The block diagram of technique being proposed is represented in Fig. 1.

In the presented approach, the input is satellite image of area under study. The satellite images are in GeoTIF format. The satellite image is of very high resolution. These images are of same area taken in different years. The time series of images covers a period of 15 years.

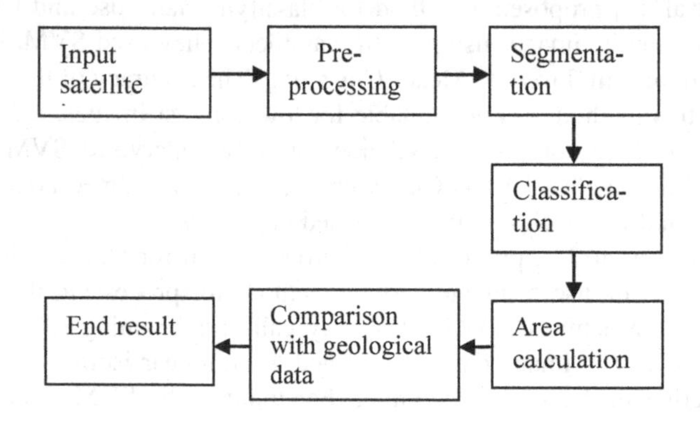

Fig. 1 General block diagram of system

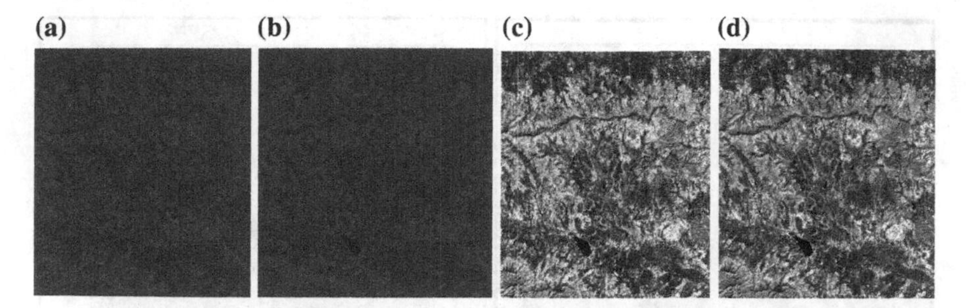

Fig. 2 Preprocessing results. **a** Original satellite image. **b** Grayscale image. **c** Contrast stretched image. **d** Histogram equalized image

The input image is then undergone to preprocessing. The input images readily are not well suited for further processes, thus there arises a need to preprocess the input satellite image and make it ready for next stage. The blur, low contrast, darker intensities, and high dimensionality of satellite images make the segmentation a great challenge. Thus following preprocessing steps are conducted. In preprocessing step, the image undergoes resizing, contrast stretching, and deblurring. In preprocessing a good care has been taken to maintain the quality of the image. The results are shown in Fig. 2.

The preprocessed image is then given to segmentation stage where it is segmented into various land use and land covers. The segmentation is done based on intensity value of each pixel. The satellite image is classified into three land covers, viz., bare land, water bodies, and vegetation. The area of interest, that is, vegetation and water bodies are observed over a period of time to determine what change has been taken place and at what rate and also to estimate the possible condition and suggest protective measures for avoiding problems.

Once the image is classified into desired classes, the next stage is area computation. The area of each land cover is calculated in terms of number of pixels. The already available area in square meters obtained via field survey is linked with this calculated area and a relation between image area and physical area is established.

The relation established between image area and physical area in the previous stage is checked in the last step of accuracy validation. In this stage the hypothesized relation is tested for rest of the images in the data set and the correctness is verified before presenting the end result of classified satellite image.

3.1 Study Area

In Pune district of state Maharashtra, a classic village, named Pondhe is situated in eastern parts. The watershed of Pondhe is positioned about 60.5 km starting Pune

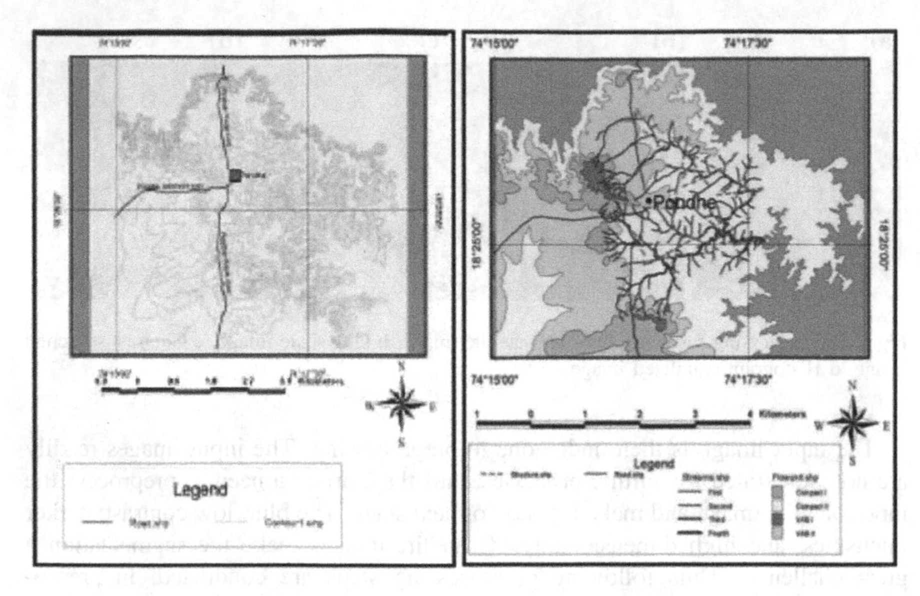

Fig. 3 Base map and geological map of Pondhe village (*Source* ACWADAM, p. 5, 10)

town. Figure 3 is a base map and geological map of the watershed which has been arranged with the reference of Survey of India Toposheet numbered 47/J/7/NW [9]. At the bottom of the wide crest which characters the western partition of the watershed, Pondhe village is situated. The area is encircled by the latitudes 18° 25′N and 18° 27′30″N and by the longitudes 74° 15′E and 74° 17′30″E [9].

4 Segmentation

Segmentation is the heart of the proposed work. The accuracy of the whole technique depends on the accuracy of segmentation. There are numerous segmentation techniques for land use and land cover. But our goal is to provide a technique which is computationally simple and accurate. Thus after a comparative evaluation of various methods following method is devised and proposed.

First, the input image is divided into three color planes red, green, and blue. Then on each plane, segmentation is carried out based on the value of intensity separately. An average intensity value of the entire plane is calculated first and then the threshold for segmentation is decided. Once the threshold has been set, the regions of the image plane having intensities above the selected threshold are set to 255, i.e., the maximum intensity and rest are set to zero, i.e., minimum intensity.

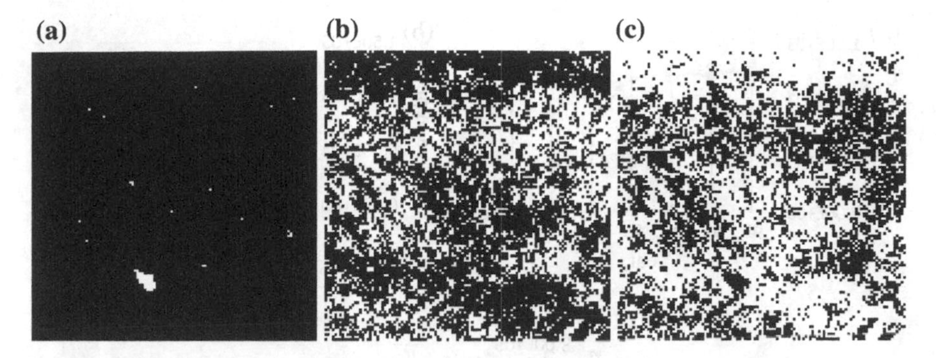

Fig. 4 Segmentation results. **a** Water bodies. **b** Vegetation. **c** Bare ground

Mathematically representing the above relation for segmenting water bodies from the satellite image,

$$i'(s,j) = \begin{cases} 255, & r(s,j) < b(s,j) \text{ and } g(s,j) < r(s,j) \\ 0, & \text{otherwise} \end{cases} \quad (1)$$

where,
$i'(s, j)$ is the new intensity value for the pixel (s, j) and
$r(s, j)$, $g(s, j)$, $b(s, j)$ are the red, green, and blue components of the pixel (s, j).

Similar relations are derived for segmenting other areas too.

Thus as an output of segmentation we get, three planes each representing one area or one land cover. For segmenting water bodies, blue plane of image is used. For segmenting vegetation part, green plane is used. Whereas, these two areas when subtracted from original image we obtain the remaining area which is nothing but the bare land. The results of segmentation are shown in Fig. 4.

5 Area Computation

To make segmentation results beneficial for end user, we need to represent various areas statistically. Here, the area of each region is calculated in terms of pixels. The area in pixels is mapped with area in square meters. This mapping establishes a link between the area computed by the proposed method and the physical area of that particular region. Once the mapping is done, the relation is tested for several images for accuracy validation. The results are then compared with the ground truth, which is nothing but area calculated by actual field survey. The variation of various areas over the course of 7 years is represented in the graph below in Fig. 5.

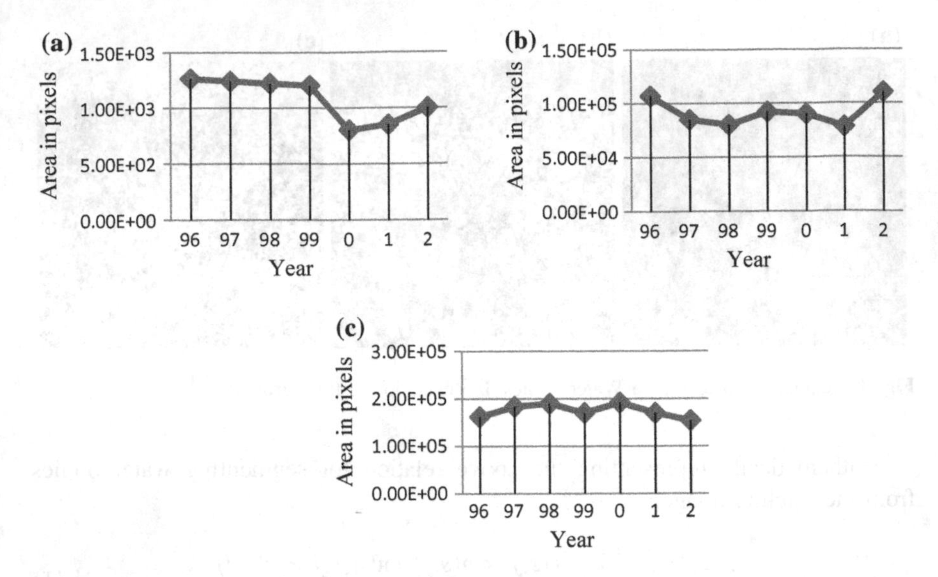

Fig. 5 Variation of land covers. **a** Water. **b** Vegetation. **c** Bare land, over 1996–2002

6 Conclusion

The approach described, a classification system, based on processing of the images, for land use and land cover and computing areas of land, water bodies, and vegetation in Pondhe village located in Purandar taluka in Pune district, is presented. The algorithm uses satellite images of Pondhe village obtained from NRSA database. The main aim of the presented work is to help geologists, water resource engineers, builders, and architects for taking crucial decision such as water content in particular area, for study of water flow, soil quality, and texture analysis respectively. The degradation of water bodies and vegetation is observed over the period of 7 years by 2.94 % and 27.10 % respectively. Thus, a tool is presented to aid water management and with the verified accuracy the technique can be used for different areas too.

Acknowledgments We would like to express our gratitude to Dr. Himanshu Kukarni and ACWADAM for their guidance and contribution.

References

1. Sadashiv Pradhan, Dhiraj Patil, A. V. Chitre, Y. H. Dandawate: Rainfall Water Runoff Determination using Land Cover Classification of Satellite Images for Rain Water Harvesting Application. International Journal of Scientific Engineering and Technology. vol. 2, no. 5, pp. 404–410. (2013)

2. S.V.S. Prasad, T. Satya Savitri, I.V. Murali Krishna: Classification of Multispectral Satellite Images using Clustering with SVM Classifier. International Journal of Computer Applications. vol. 35, no. 5, pp. 32–44. (2011)
3. Tomohiro Shiraishi, Takeshi Motohka, Rajesh Bahadur Thapa, Manabu Watanabe, Masanobu Shimada: Comparative Assessment of Supervised Classifiers for Land Use–Land Cover Classification in a Tropical Region Using Time-Series PALSAR Mosaic Data. IEEE Journal of selected topics in applied earth observations and remote sensing. vol. 7, no. 4, pp. 1186–1199. (2014)
4. Nathan Longbotham, William Emery, Fabio Pacifici: Multispectral land-use/land-cover model portability in multi-temporal multi-angle very high resolution imagery. In: IEEE International Geoscience and Remote Sensing Symposium, pp. 1115–1118. (2013)
5. Bharathi S., Manju M., Vasavi Manasa C. L., Mallika H. M., Maruti M. K., Deepa S. P., Venugopal K. R., Patnaik L. M.: Automatic Land Use/Land Cover Classification using Texture and Data Mining Classifier. In: IEEE Region 10 Conference, pp. 1–4. (2013)
6. Mattia Marconcini, Diego Fernández-Prieto, Tim Buchholz: Targeted Land-Cover Classification. IEEE Transactions on Geoscience and Remote Sensing. vol. 52, no. 7, pp. 4173–4193. (2014)
7. Saroj K. Meher, B. Uma Shankar, Ashish Ghosh: Wavelet-Feature-Based Classifiers for Multispectral Remote-Sensing Images. IEEE Transactions on Geoscience and Remote Sensing, vol. 45, no. 6, pp. 1298–1307. (2007)
8. Medha Aher, Sadashiv Pradhan, Yogesh Dandawate: Rainfall Estimation Over Roof-top Using Land-Cover Classification of Google Earth Images. In: International Conference on Electronic Systems, Signal Processing and Computing Technologies, pp. 111–116. (2014)
9. Vinit Phadnis, Himanshu Kulkarni, and Uma Badarayani: Study of Pondhe watershed, Purandar taluka, Pune district, Maharashtra. Technical Report, Advanced Center for Water Resources Development and Management (ACWADAM), http://www.acwadam.org/publication.asp

Significance of Independent Component Analysis (ICA) for Epileptic Seizure Detection Using EEG Signals

Varsha K. Harpale and Vinayak K. Bairagi

Abstract Electroencephalography (EEG) is a measurement tool to measure electrical activity generated by translating chemical variation in brain into voltage. EEG signals are measured with multielectrode placed at properly localized part of the brain with either intracranial or extracranial (Scalp EEG) method. EEG analysis has become very important to detect various human diseases. Usually, EEG signals are recorded with the multichannel acquisition module. It is very data intense, time and resource consuming because it should handle a heavy workload of computations. As EEG signal composed of various random signals, independent component analysis (ICA) is considered to be very important method. This paper specifically studies significance of ICA in Epileptic seizure detection using EEG signals. Mostly ICA is used for EEG artifact removal from raw EEG signals. Thus ICA is useful in removing artifacts and improving epileptic seizure detection accuracy.

Keywords Electroencephalogram (EEG) · Seizures · Independent component analysis (ICA)

1 Introduction

Various real-time biomedical devices use the EEG information for diagnosis of brain disorders, thus demands real-time digital signal processing of EEG signals. Intracranial EEG is preferable during surgeries and scalp EEG signals are most commonly used for diagnosis and monitoring central nervous system activities [1]. The electroencephalogram (EEG) is a well-proven measurement tool to measure

---

V.K. Harpale (✉)
E&TC Department, PCCOE, Pune-44, India
e-mail: varshaks3@gmail.com

V.K. Bairagi
E&TC Department, AISSMS, IOIT, Pune-01, India
e-mail: vbairagi@yahoo.co.in

S.C. Satapathy et al. (eds.), *Proceedings of the International Conference on Data Engineering and Communication Technology*, Advances in Intelligent Systems and Computing 469, DOI 10.1007/978-981-10-1678-3_80

electrical activities in brain. These measured signals containing various artifacts such as muscle movements, eye blinking, and instrumental electrical noise [2]. ICA is a mathematical tool for extracting individual and independent components from mixed EEG signal. Most of the random signal analysis uses ICA for extraction of independent sources from high-dimensional recorded data as offline signal analysis [3]. Most of the research carried in this area uses only two channels, whereas for better spatial resolution and improved accuracy of detection or prediction of diseases more numbers of channels are preferable [4].

Development of epileptic seizure initiates ictal activity, this is analyzed to detect epileptic zone of each channel [5]. There are various high performance tools such as Electroencephalography (EEG), functional Magnetic Resonance Imaging (fMRI), Near Infrared Spectroscopy (NIRS), and Magnetoencephalography (MEG) for brain signal analysis. Recent brain research demands the design and implementation of real-time system. While implementing such real-time system computational complexity, hardware optimization, and power dissipation are the challenges. Designing real-time system with asynchronous circuit's benefits in power optimization as clock is measure component of power consumption.

Epilepsy is considered to be major neurological disorders among all brain diseases [6]. Currently, numerous antiepileptic drugs are available for seizure control, but 30 % of epileptic patients remain either drug-resistant or develop limiting adverse effects [7]. The occurrence of epileptic seizure is detected and suppressed using an electrical stimulator [8]. High performance and accuracy algorithms or devices are required in order to use electrical stimulator. Earlier research demonstrates various real-time seizure detector methods [9] but still further improvement is expected.

Since artifacts because of eye blinking and muscle movements affect the accuracy of epileptic seizure detection, it is required to separate from EEG signals of interest. Principle of artifact removal using independent component analysis (ICA) [10] is based on decomposition of mixed signals recorded from different sources. ICA processor implementation with closed-loop system can be designed to enhance the performance of epileptic seizure detection [11].

This paper includes Sect. 2 which gives idea about selection of features for EEG signals, Sect. 3 explains about concept of independent component analysis; Sect. 4 covers significance of ICA, Sect. 5 gives recent trends in EEG analysis architectures on the basis of various performance parameters and Sect. 6 concludes this work.

2 Selection of Features for EEG Analysis

Electroencephalogram (EEG) is a tool which records a multifrequency nonstationary brain signal from various localized channels. Feature extraction is very important aspect of the analysis and classification of EEG signals. Repeatability and regularity (periodicity) of the signal are well analyzed by autocorrelation function for differentiating brain signal into seizure, non-seizure, and pre-seizure EEG. The autocorrelation for nonstationary signal is not robust because of highest changes during seizure, thus during seizure regularity of the signal increases and EEG signal behave more oscillatory in nature [12].

EEG is a nonstationary signal, comprises of events at different frequencies. Most commonly used frequency domain feature is Power Spectral Density (PSD). PSD has the limitation that it cannot isolate the time at which frequency of interest occurs. Spectrogram, sequential computation of PSD over small window is calculated to extract frequency and temporal information for detection of seizures in EEG signal.

EEG related time frequency features such as event-related shifts in the power spectrum or event-related spectral perturbation (ERSP) and inter-trial coherence (ITC) (event-related phase-locking) can be used as shown in Fig. 1.

Various methods are adopted to extract time–frequency information such as Gabor Transform, Hilbert Houng Transform, Wigner Ville Distribution, and wavelet Transform.

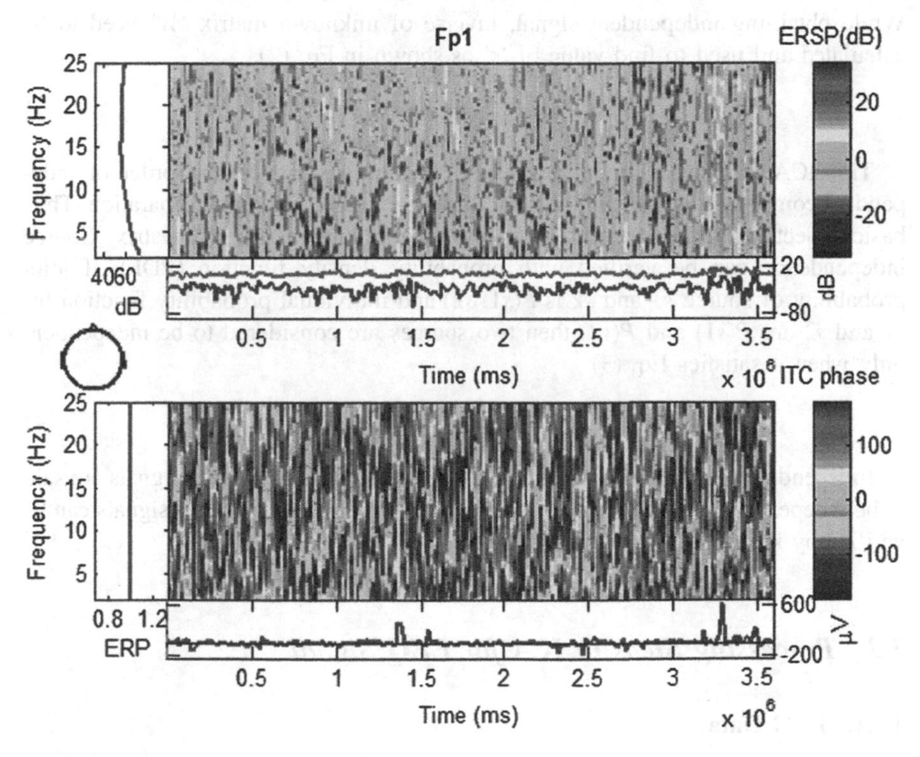

Fig. 1 Time frequency analysis of ERSP and ITC

3 Independent Component Analysis (ICA)

Independent component analysis (ICA) is a statistical tool, normally used for blind source separation [13]. ICA also has significant application in EEG artifact removal and feature extraction.

3.1 Concept of ICA

ICA model assumed that source signals are statistically independent and has non-Gaussian distribution. It is composed of independent source signals s_i, unknown mixing matrix A_i and mixed signal x, where $i = 1, 2, 3 \ldots n$. This mix signal component can be represented as per Eq. (1).

$$x = \sum_{i=1}^{n} A_i s_i \tag{1}$$

Here A is assumed to be square matrix, x and s are considered to be vectors. While obtaining independent signal, inverse of unknown matrix 'W' need to be calculated and used to find value of 's' as shown in Eq. (2).

$$s = A_i^{-1} x \, or \, s = Wx. \tag{2}$$

This ICA has some limitations like determining variance and order of independent component is difficult, but still this is preferable for signal separation. Thus basic objective of ICA model is to estimate unknown mixing matrix. Source independence can be verified with probability density function (PDF). If joint probability of source $s1$ and $s2$ is $P(s1, s2)$ and individual probability function for $s1$ and $s2$ are $P(s1)$ and $P(s2)$ then two sources are considered to be independent only when it satisfies Eq. (3)

$$P(s1, s2) = P(s1)P(s3) \tag{3}$$

Independence also can be checked with covariance of $s1$ and $s2$. Signals are said to be independent if its covariance is zero. Non-Gaussian property of signals can be verified by kurtosis function.

3.2 Processing Steps in ICA for EEG Signal

3.2.1 EEG Data

Standard database [14] of epileptic seizure collected at the Children's Hospital Boston is used to understand ICA concept applied to EEG signal. EEG recording of 22 patients with sampling rate of 256 samples per second and 23 channels is used for testing. The International 10–20 system for EEG electrode location was adopted to record the data. Here experimentation is performed with a single subject and data of 1 h recording.

3.2.2 Kurtosis for Non-Gaussianity

An arbitrary distribution of integrated independent signals tends to Gaussian distribution. If Independent signals are mixed then they can be separated by linearly transforming signals as non-Gaussian. Thus non-Gaussianity is very important in calculation of ICA components. Normalized signals thus tested for non-Gaussianity using kurtosis. If signal ‘*x*’ has unit variance its kurtosis is the fourth moment of that signal as defined by formula given in Eq. (4)

$$\text{kurt}(x) = E\left(x^4\right) - 3\left(E\{x^4\}\right)^2 \tag{4}$$

Figure 2 shows statistic of one patient recorded with 23 channels and thus clears non-Gaussianity of individual signals. This statistics is calculated by EEGLAB Toolbox, which shows kurtosis is nonzero and thus non-Gaussian behavior of signals.

3.2.3 Independent Components

Independent Component Analysis (ICA) is a signal processing tool that separates statistically N independent inputs which have been mixed linearly in N output channels, without knowing their dynamics. The EEG signal frequency range is the mixing of brain fields at the scalp electrodes and their distribution is linear. ICA gives the scalp topographies of the independent components thus support source localization. If we apply ICA to the single continuous recording of patients, it defines power distribution, frequency distribution, and topographic presentation of independent components. Figure 3 shows spectrograph for power density function for 6, 10, and 22 Hz frequency. Similarly, all independent components and their spatial characteristics can be observed with Fig. 4.

Mean:	0.155	0.025-quantile: -42.1
Trimmed mean:	0.049	0.5 -quantile: 0.417 (median)
		0.975-quantile: 38.3
Standard dev.:	22.43	Skewness: 1.38 (near 0 if Gaussian)
Trimmed st.d.:	13.84	Distribution is right-skewed
Variance:	503.1	Excess kurtosis: 26.7 (near 0 if Gaussian)
Range:	714.4	Distribution is super-Gaussian
Data points:	921600	Kolmogorov-Smirnov test: not Gaussian

Fig. 2 Parameters for testing non-Gaussian distribution

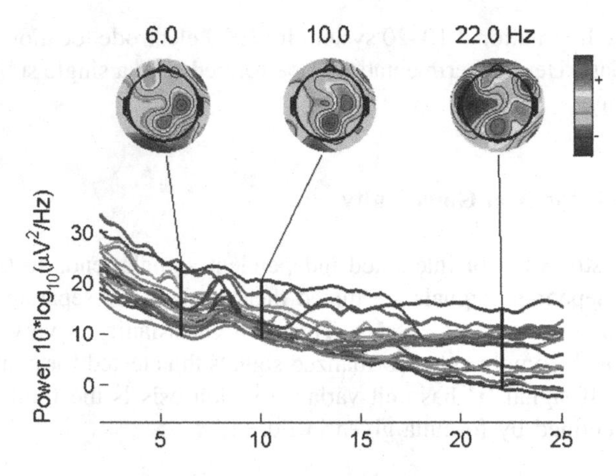

Fig. 3 Specto map of EEG signal

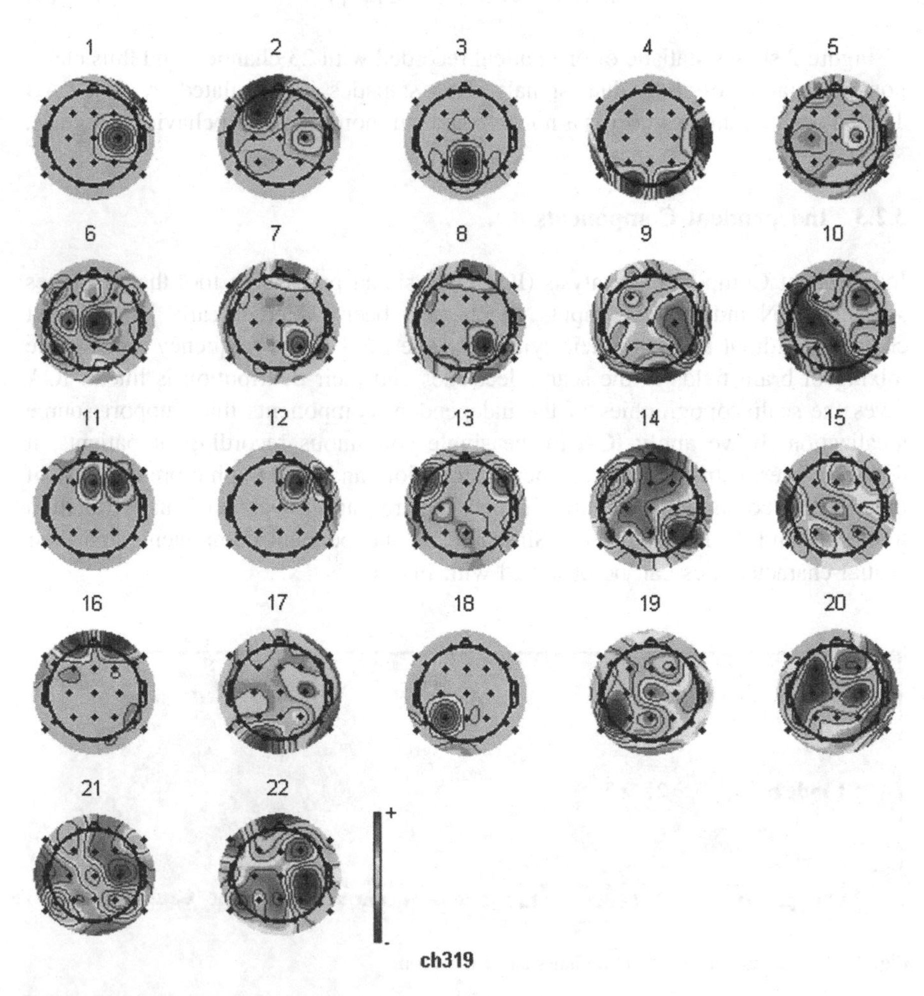

Fig. 4 Independent components and their spatial mapping

4 Significance of ICA in EEG Analysis

Thus ICA plays very important role in various mixed signal analysis. In EEG analysis ICA model is mostly being used for feature extraction, seizure detection, and artifacts removal. Mammone and Foresta et al. [15], proposes hybridization of wavelet and ICA for significant artifacts rejection from EEG signals recorded by multichannel scalp EEG electrode. This automatic wavelet independent component analysis (AWICA) provides full automated solution for artifact removal from scalp EEG with improved performance.

Arunkumar and Balaji et al. [16] uses principle component analysis (PCA), independent component analysis (ICA), hurst exponent (H) for epileptic seizure detection. The system could achieve 73.8 % seizure detection rate by ICA, hurst component with thresholding method and 87.9 % with wavelet transform.

Soomro and Badruddin et al. [17] proposes eye blink artifact removal method using Empirical Mode Decomposition (EMD) and ICA. The system could achieve signal to artifact ratio (SAR) as −19.1673 dB for noncorrected EEG and 2.71 dB for corrected EEG. Thus most of the research in the field of ICA, FastICA and EEG carries artifact removal with good rejection capability as shown in Table 1.

Table 1 Application of ICA for artifact removal from EEG signal

Authors	Methodology	Application and performance parameters
Liao and Shih et al. [18]	System on chip architectural design with 90 nm CMOS tech. for online recursive ICA(ORICA)	Power consumption 8.56mW and latency 0.2532 s
Soomro and Badruddin et al. [19]	Hybrid model of empirical mode decomposition (EMD) and advanced canonical correlation analysis (CCA)	2.2–6.0 dB improvement in signal to artifact ratio (SAR)
Jirayucharoensak and Israsena et al. [20]	Neurofeedback system is designed for artifact removal using ICA and wavelet transform	Attention analysis with good artifact removal
Liao and Fang [21]	Online recursive ICA(ORICA) for artifact removal	BCI application
Bedoya and Estrada et al. [22]	Fuzzy classification based on (Learning Algorithm for Multivariate Data Analysis) along with ICA	BCI application
Zachariah, and Jai et al. [23]	Artifact removal with hybrid model of wavelet transform and ICA	Increase in accuracy and reduction in processing time achieved.
Mammone and Foresta et al. [15]	Hybrid model of ICA and Hurst Exponent	Epileptic Seizure Detection with 87.9 % detection rate
Heute and Guzmán et al. [24]	Combination of ICA and Adaptive Weiner Filter is used	Improved artifact rejection ratio
Mahajan and Morshed [25]	Combination of ICA, modified Multi-scale entropy (mMSE) and wavelet denoising	average sensitivity: 90 %, average specificity: 98 %, average execution time: 0.06 s

Table 2 Comparison of various EEG analyzer architectures

Parameters	Huang 2008 [2]	Van 2011 [26]	Yang 2015 [11]
Implementation approach	FPGA	ASIC	ASIC
No. of channels	4	8	8
Operating freq (MHz)	68	100	11
Power dissipation (mW)	–	16.35 W	0.0816
Gate count	315 K	272 K	69.2 K
Core area (m^2)	–	1.4872	0.3869
Max. latency(ms)	–	290	84.2
Algorithm	Info Max	Fast ICA	Fast ICA
Application	EEG	EEG	ECoG

5 ICA Architectures of EEG Analyzers

The EEG signal analysis provides a proper detection of the mental disorders of human being. Recent studies include combination of analysis of EEG and heart rate variability to improve diagnosis and treatment methods.

Chen et al. [3] presents SoC architecture designed using with ICA and loseless data compression. This architecture is complexity-efficient with an average correlation of 0.9044 and power consumption optimized by 41.6 %. Jenihhin et al. [4] was based on an original patented algorithm for SASI calculation. For the proof of concept it was, first, implemented as a MATLAB program running on a computer connected to a commercial EEG signal capturing equipment.

Chen et al. [9] presented a solution for epilepsy seizure detection. The proposed SoC is a combination of signal processing closed loop system with an 8-channel Intracranial EEG (iEEG) and 10-bit ADC. It also includes an electronic stimulator based on adaptive high voltage and a wireless transmission link. To improve the performance of epileptic seizure detection, independent component analysis (ICA) is applied by Yang et al. [11] to multichannel signals.

Table 2 shows the comparison of different SoC architecture designed for EEG signal analysis by various authors and concludes Yang et al. [11] shows better performance, with limitation of low operating frequency.

6 Conclusion

The electroencephalographic signals are random signals collected from different locations of electrodes placed on the scalp. Various time and frequency domain features can be calculated and selected as per requirements. These also comprise of various artifacts such as eye blinking and muscle movements. Independent Component Analysis (ICA) is one of the well-proven artifact removal methods. Combination of ICA and other methods such as ESD, entropy, and kurtosis improves signal to artifact ratio and assure corrected signal without artifacts. ICA can be also used for epileptic seizure detection, but the accuracy improvement still is a challenge in this method. There is one more scope of system on chip (SoC) and FPGA implementation of ICA method for seizure detection with greatest accuracy.

References

1. K. Shyu and M. Li: FPGA Implementation of FastICA based on Floating-Point Arithmetic Design for Real-Time Blind Source Separation, Int. Joint Conf. on Neural Networks, Canada (2006).
2. W. Huang, S. Hung, J. Chung, M. Chang, L. Van, C. Lin: FPGA implementation of 4-channel ICA for on-line EEG signal separation, Biomedical Circuits and Systems Conf, (2008).
3. C. Chen, E. Chua, S. Tseng, C. Fu and W. Fang: Implementation of A Hardware-efficient Eeg Processor For Brain Monitoring Systems, IEEE Biomedical Conference, (2012).
4. M. Jenihhin, M. Gorev, V. Pesonen,: EEG Analyzer Prototype Based on FPGA, 7th International Symposium on Image and Signal Processing and Analysis (2010).
5. M. Sun, and R. Sclabassi: Precise determination of starting time of epileptic seizures using subdural EEG and wavelet transforms, IEEE-SP International Symposium on Time Frequency and Time Scale Analysis, pp. 257–260, (1998).
6. W. Chen, H. Chiueh, T. Chen, C. Ho, Chi Jeng, M. Ker, C. Lin, Y. Huang, C. Chou, T. Fan, M. Cheng, Y. Hsin, S. Liang, Y. Wang, F. Shaw, Y. Huang, C. Yang, and C. Wu: A Fully Integrated 8-Channel Closed-Loop Neural-Prosthetic CMOS SoC for Real-Time Epileptic Seizure Control, IEEE Journal Of Solid-State Circuits, Vol. 49(1), (2014).
7. P. Kwan, S. C. Schachter, and M. J. Brodie: Current concepts: Drug resistant epilepsy, N. Engl. J. Med., vol. 365(10), pp. 919–926, (2011).
8. C.Y. Lin, W.L. Chen, and M.D. Ker: Implantable stimulator for epileptic seizure suppression with loading impedance adaptability, IEEE Trans. on Biomed. Circuits Syst., vol. 7(2), pp. 196–203, (2013).
9. W.M. Chen: A fully integrated 8-Channel closed-loop neural prosthetic SoC for real-time epileptic seizure control, Int. Solid State Circuits Conf., pp. 286–287, (2013).
10. A. Hyvärinen, J. Karhunen, and E. Oja: Independent Component Analysis, Wiley, (2001).
11. C.H. Yang, Y.H. Shih, and H. Chiueh: An 81.6 microW FastICA Processor for Epileptic Seizure Detection, IEEE Tran. On Biomedical Circuits And Systems, Vol. 9(1) (2015).
12. Andrea Varsavsky Iven Mareels Mark Cook: Epileptic seizures and the EEG measurement, models, detection and prediction, CRC Press (2011).
13. Hyvärinen A, Oja E.: Independent component analysis: algorithms and Applications". Pubmed journal of Neural Network. Vol 13(4–5), pp. 411–30 (2000).
14. Database: Goldberger AL, Amaral LAN, Glass L, Hausdorff JM, Ivanov PCh, Mark RG, Mietus JE, Moody GB, Peng CK, Stanley HE. PhysioBank, PhysioToolkit, and PhysioNet: Components of a New Research Resource for Complex Physiologic Signals.
15. Mammone, N.; La Foresta, F.; Morabito, F.C.: Automatic Artifact Rejection From Multichannel Scalp EEG by Wavelet ICA, IEEE Journal of Sensors, pp. 533–542, (2012).
16. Arunkumar, N.; Balaji, V.S.; Ramesh, S.; Natarajan, S.; Likhita, V.R.; Sundari, S.: Automatic detection of epileptic seizures using Independent Component Analysis Algorithm, in International Conference on Advances in Engineering, Science and Management (ICAESM), pp. 542–544, (2012).
17. Soomro, M.H.; Badruddin, N.; Yusoff, M.Z.; Malik, A.S.: A method for automatic removal of eye blink artifacts from EEG based on EMD-ICA, IEEE 9th International Colloquium on Signal Processing and its Applications (CSPA), pp. 129–134, (2013).

18. J-C. Liao; W-Y. Shih; K-J Huang; W-C. Fang,: An online recursive ICA based real-time multichannel EEG system on chip design with automatic eye blink artifact rejection, International Symposium on VLSI Design, Automation, and Test, pp. 1–4, (2013).
19. Soomro, M.H.; Badruddin, N.; Yusoff, M.Z.; Jatoi, M.A.: Automatic eye-blink artifact removal method based on EMD-CCA, International Conference on Complex Medical Engineering (CME), pp. 186–190, (2013).
20. Jirayucharoensak, S.; Israsena, P.; Pan-ngum, S.; Hemrungrojn, S: Online EEG artifact suppression for neurofeedback training systems, 6th International Conference (BMEiCON) on Biomedical Engineering, pp. 1–5, (2013).
21. Jui-Chieh Liao; Wai-Chi Fang: An ICA-based automatic eye blink artifact eliminator for real-time multi-channel EEG applications, IEEE International Conference on Consumer Electronics (ICCE), pp. 532–535, (2013).
22. Bedoya, C.; Estrada, D.; Trujillo, S.; Trujillo, N.; Pineda, D.; Lopez, J.D.: Automatic component rejection based on fuzzy clustering for noise reduction in electroencephalographic signals, XVIII Symposium of Image, Signal Processing, and Artificial Vision (STSIVA), pp. 1–5, (2013).
23. Zachariah, A.; Jinu Jai; Titus, G.: Automatic EEG artifact removal by independent component analysis using critical EEG rhythms, in International Conference on Control Communication and Computing (ICCC), pp. 364–367, (2013).
24. Heute, U.; Guzman, A.S.,: Removing "cleaned" eye-blinking artifacts from EEG measurements, International Conference on Signal Processing and Integrated Networks (SPIN), pp. 576–580, (2014).
25. Mahajan, R.; Morshed, B.I.: Unsupervised Eye Blink Artifact De-noising of EEG Data with Modified Multi-scale Sample Entropy, Kurtosis, and Wavelet-ICA, IEEE Journal of Biomedical and Health Informatics, vol.19(1), pp. 158–165, (2015).
26. W.C. Fang, C.K.Chen, E. Chua, C.C.Fu, S.Y. Tseng and S. Kang, "A Low Power Biomedical Signal Processing System-on-Chip Design for Portable Brain-Heart Monitoring Systems", IEEE Biomedical Conference, Jun 2010.

Author Index

S.C. Satapathy et al. (eds.), *Proceedings of the International Conference on Data Engineering and Communication Technology*, Advances in Intelligent Systems and Computing 469, DOI 10.1007/978-981-10-1678-3

Printed in the United States
By Bookmasters